VETERINARY
APPLIED PHARMACOLOGY
AND THERAPEUTICS

THE AUTHORS

G. C. Brander is a member of the scientific staff of the Research Division of Beecham Pharmaceuticals, specializing in the problems of animal diseases, and special Professor of Animal Health of the University of Nottingham. After graduating from the Royal (Dick) Veterinary College and Edinburgh University in 1938, he worked at the Central Veterinary Laboratory, Weybridge, and in the Sudan Veterinary Service as a Research Officer. After the war he joined the Cooper Technical Bureau as a veterinary adviser and from 1955 to 1958 was Technical Director of William Cooper & Nephews, Australia. His interests have been in the field of epidemiology, particularly with reference to the control of animal parasites, and latterly the control of the bacterial infections of farm animals. His present work is related to the use of modern therapeutic drugs in the prevention of animal disease. He has published work on the control of external parasites in animals, and on the use of antibiotics in the control of mastitis. He has collaborated with the National Institute for Research in Dairying in a series of studies on the control of mastitis by the use of antibiotics and systems of hygiene. He was appointed a member of the Medicines Commission in 1976.

D. M. Pugh graduated from the Veterinary School of Bristol University in 1957 and, after a short period in general practice, returned to the Department of Pharmacology at Bristol as Junior Fellow in Veterinary Pharmacology. There, in addition to teaching, he undertook a study of adrenal cortex function in sheep. After a further period in general practice he joined the Veterinary Science Division of Boots' Pure Drug Co. Ltd, Nottingham, and for two years was engaged in the evaluation of several therapeutic agents in domesticated species. Since 1964 he has been responsible for the course in Veterinary Pharmacology at Trinity College, Dublin. Following a study leave in 1967, he was granted a Doctorate in Veterinary Medicine by the University of Zurich for his thesis on the abortifacient action of warfarin. Current research interests include the study of factors involved in the implantation of ova with special reference to those affecting the blastocyst surface coat and the role of prostaglandins in animal reproduction. He has published the results of all aspects of his research, is a member of his School's Syllabus Sub-Committee, is keenly interested in the future of veterinary education, and is an examiner in veterinary pharmacology.

VETERINARY APPLIED PHARMACOLOGY AND THERAPEUTICS

G. C. BRANDER
B.Sc. (Edinburgh), M.R.C.V.S.
Controller of Veterinary Research and Development,
Beecham Pharmaceuticals,
Nutritional Research Centre, Walton Oaks, Tadworth, Surrey
Special Professor of Animal Health, University of Nottingham

AND

D. M. PUGH
M.A. (Dublin), M.Sc. (Bristol), B.V.Sc. (Bristol),
Dr.med.Vet. (Zurich), M.R.C.V.S.
Senior Lecturer in Pharmacology, School of Veterinary Medicine,
Trinity College, Dublin

THIRD EDITION

BAILLIÈRE TINDALL · LONDON

Printed by
PHILIPPINE Graphic Arts, INC.
151-163 Tandang Sora, Caloocan City
Tels. 35-45-91 to 95

A BAILLIÈRE TINDALL book published by
Cassell & Collier Macmillan Publishers Ltd
35 Red Lion Square, London WC1R 4SG
and at Sydney, Auckland, Toronto, Johannesburg
an affiliate of
Macmillan Publishing Co. Inc.
New York

First published 1960 (by P. W. Daykin)
Second edition 1971
Third edition 1977

Spanish edition (CECSA, Mexico) 1965
ELBS edition 1977

ISBN 0 7020 0657 2

Published in the United States of America by
Lea and Febiger, Philadelphia

Printed in Great Britain
at the University Press, Aberdeen

**Reprinted under Authority of
Presidential Decree 285 as
amended by P. D. 400 as
further amended by P. D. No. 1203.
by**

994

NATIONAL BOOK STORE, INC.
701 Rizal Ave., Cor. Soler St., Manila
Tel. 49-43-06 to 09

CONTENTS

PART III
THE CONTROL OF INFECTIOUS DISEASE: CHEMOTHERAPY

PART IV
THE ESSENTIALS OF PHARMACY IN VETERINARY PRACTICE

PART V
APPENDICES

PREFACE

Today's pharmacology has evolved from the older, more general, drug science known as *materia medica*. This evolution has been characterized by a reduction in concern with the physical characteristics, properties and preparations of drugs and with drug indications and dosages. Instead, emphasis has been placed on the modes of action, the effects, the side effects and the potential or actual toxicities of the compounds used. This desirable change has been possible because the evolution of any science is a moving away from simple morphological descriptions of materials or events towards a position where the factors underlying an observed system become elucidated and finally begin to yield to quantitative expression. Pharmacology is now in this satisfying position and at this stage of development it is preferable to deal with the principles and structure of the science rather than with a mountain of factual minutiae. Part I of this book describes those contributions which are unique to pharmacology and describes the origins of the science and its relation and indebtedness to cognate disciplines.

The fully revised text adheres to the pattern, established in the previous editions, of grouping drugs on the basis of their actions and of the body systems in which they find their major application in therapeutics, rather than on the basis of clinical relationships. For this new edition the text has been both condensed, in the case of obsolescent material, and considerably expanded by the inclusion of numerous new preparations now available for veterinary use, as well as by discussion of recently acquired knowledge affecting our understanding of drugs and their actions. Particularly significant additions or substantial revisions concern bioavailability, pharmacokinetics, the mode of action of vitamin D, the role of prostaglandin in reproduction and inflammation, the hypothalamic releaser hormones and modern developments in anthelmintics, antibiotics and antibacterials.

The book is divided into five sections. The first of these deals with general pharmacology, i.e. the background of the subject, how a drug acts, what happens to a drug in the body and the use of drugs as therapeutic agents. Part II deals in detail with the actions and uses of named drugs in those systems of the body to which they are normally applied as medication. The drugs described are of natural and synthetic origin and include the vitamins and hormones, but exclude drugs used to combat infectious disease as these are discussed separately in Part III. The frequency with which new remedies are introduced into the agro-veterinary area together with variation in the useful life-span of these preparations combine to make selection of examples suitable for inclusion in a book of this type rather difficult. In general, well established remedies are discussed in detail while newer and, perhaps, superior agents are described more briefly, usually in comparison with the older, standard preparation. The relevance of drugs, particularly antibiotics, to disease control systems is discussed, especially the use of mastitis preparations in a disease control programme. It will be noted that a major change has taken place in the field of anthelmintic therapy with the development of a range of new drugs which, in addition to being relatively non-toxic, have wide anthelmintic activity. The essentials of pharmacy in veterinary practice, that is prescription writing, drug dispensing and the legal aspects of the use of drugs in animals, are briefly considered in Part IV. The relevant tables of prescription-writing in Latin and metric weights and measures, together with some useful notes on the treatment of poisoning are included as appendices in Part V. With the needs of the clinician in mind, and in response to several requests, we have added to this edition a therapeutic index for increased ease of reference.

The content of the book has not, however, been planned solely to ensure compatibility with the present scope of pharmacology teaching in the veterinary schools, because the authors believe that this would have been an undesirable truncation. Just as histology may be seen as a preparation for histopathology, and anatomy as a preparation for surgery, so is pharmacology the basis for therapeutics and it is hoped that this book will be of assistance to those now involved in the reintroduction of therapeutics into the curricula. Because it attempts to describe the state of the drug sciences as they apply to veterinary medicine it is hoped that this book will be of value to all those concerned with the administration of drugs to animals, be they students, teachers, manufacturers, investigators or legislators.

During the sixties, the vital importance of a full and accurate understanding of a drug's potential was tragically emphasized by the thalidomide disaster. In addition, the concern felt over the occurrence of tissue residues following the long-term administration of drugs as aids to animal production led to further emphasis being placed on the essential role of studies designed to examine the long-term and delayed effects of drugs. A further salutary experience was obtained when the extreme persistence of the organochlorine pesticides in body fat was discovered. Fears as to the extent to which we are damaging ourselves, our animals and our whole environment are now widely discussed not as academic issues but as matters of public concern. The Medicines Commission, which has taken over the valuable interim roles of the Committee on the Safety of Drugs (The Dunlop Committee), the Veterinary Products Safety Precautions Scheme and the Pesticides Safety Precautions Scheme, is the body now charged with ensuring the safety of medical, veterinary and agricultural products in the U.K. A similar function is carried out by the F.D.A. in the United States and in Europe the European Commission is responsible for establishing standards for the examination of drugs for use in man and animals. In the exhaustive investigations required by these authorities much of the work of the evaluation of the safety of medicinal products has fallen to pharmacologists. The consequent rapid development of the subject has done much to ensure that the veterinary graduate of the future will be fully trained not only to utilize to the full the powerful drugs now available, but also to recognize clearly the risks consequent upon their use and abuse.

April 1977

G. C. Brander
D. M. Pugh

PART I

INTRODUCTION

1

General Introduction

PHARMACOPOEIAS

Medicines and salves are as old as history and many of the drugs mentioned in the earliest records continue to render good service. Numbers of loathsome remedies, mainly introduced by the charlatans of the Dark Ages, were also included in medical textbooks as recently as the middle of the nineteenth century and it was not until 1864, by virtue of the *Medical Act* of 1858 and of the *Medical Council Act* of 1862, that a reference book was published detailing all the officially accepted drugs: this book was entitled the *British Pharmacopoeia* (B.P.). Superseding the three major pharmacopoeias then in use, i.e. those of London, Edinburgh and Dublin, it gave detailed descriptions, stated the composition, established standards of purity, and provided human dosage rates for all the drugs considered by the Pharmacopoeia Committee of the General Medical Council to be adequately tested and of known therapeutic value. At varying intervals since 1864 revised editions have appeared and it is now published by the Medicines Commission.

The *British Pharmacopoeia* was, and is, of little direct value to the practitioner and in 1883 William Martindale published his *Extra Pharmacopoeia* with the purpose of supplying practitioners and pharmacists 'with practical and up-to-date information on drugs and galenical preparations to supplement that contained in the *British Pharmacopoeia* . . .': so states the preface to the twenty-third edition published in 1952. That aim is continued with each succeeding edition, but the onus of production has been borne by the British Pharmaceutical Society since 1934 when the Society purchased the copyright after the death of William Martindale's son. In 1948 the Pharmaceutical Society also purchased the copyright of *Squire's Companion*, a similar work and a strong competitor of the *Extra Pharmacopoeia* which had appeared first in 1864, some 19 years before Martindale's first edition, but which lapsed on the death of Squire's second son shortly after publication of the nineteenth edition in 1916. Both are now produced as a combined edition in one volume.

Neither Martindale's *Extra Pharmacopoeia* nor *Squire's Companion* were accorded official recognition: the need for an official reference book for the use of those involved in 'the prescribing and dispensing of medicines' was recognized in 1907 when the Pharmaceutical Society published the first edition of the *British Pharmaceutical Codex* (B.P.C.). This book, besides giving details of manufacturing, physical and chemical characteristics, and standards of purity in much the same manner as the B.P. also contains monographs on actions and uses as well as details of toxicity and antidotal measures. Information is contained 'on all drugs and medicinal preparations in common use throughout the British Commonwealth and on all the principal substances which were official in the pharmacopoeias of France, Germany and the United States of America'. It is specifically stated that the inclusion of a drug 'is now regarded as an indication that it is of value or is an established constituent of a useful preparation'. These quotations are taken from the preface of the Codex of 1954. A number of the drugs detailed in the B.P.C. are not included in the B.P., either having been deleted or having been considered not sufficiently proved to warrant inclusion: the Codex has included them because of their wide usage and general clinical acceptance.

Both the B.P. and the B.P.C. are now republished once every five years to keep pace with the rapid advances in knowledge, but even so discoveries are numerous and five years between editions is too long. More frequent publication is uneconomical and therefore addenda to the B.P. and supplements to the B.P.C. are published as interim measures between the quinquennial revised editions.

These official and semi-official publications are the British milestones on the road of progress in the treatment of human diseases: the publication of the first edition of the *British Veterinary Codex* in 1953, modelled on the *British Pharmaceutical Codex* and published by the same authority, was a similar milestone marking the advance of veterinary therapy. The first British Pharmacopoeia (Veterinary) will appear in 1977.

These books of reference are widely accepted throughout the world and especially in the British Commonwealth. Among the national pharmacopoeias, the *United States Pharmacopoeia* (U.S.P.) remains of great importance. 1969 saw the first edition of a cooperative *European Pharmacopoeia*. In the U.S.A., the *National Formulary* (N.F.), which is analogous to the B.P.C., establishes legal standards and formulations of accepted useful drugs: other American publications of importance are *AMA Drug Evaluations* which provide authoritative information on newly discovered drugs, and the *United States Dispensatory* (U.S.D.) which is a huge volume incorporating information on all the drugs officially recognized in the U.S.P., B.P., I.P. and N.F. as well as many other new, not officially recognized drugs. It provides detailed commentaries on therapeutic theories and methods, with numerous references, in far greater detail than any similar semi-official publication and includes a sizeable section on veterinary applications.

RATIONAL THERAPY

These books, then, establish the value of the majority of the drugs now in use, but it should be remembered that the emergency treatment of diseased animals is but one veterinary function. The aim of all physicians, human or veterinary, is increasingly directed towards the prevention of disease. The application of hygiene to husbandry and the calculated use of mass vaccination, pest control and eradication by slaughter, have achieved the complete elimination of some diseases.

However, preventive measures are not always successful, and once disease has overtaken an animal curative measures must be applied. These may be surgical or medical, the latter being the concern of this book.

Curative methods can be *rational* or *empirical.* By rational medicine is meant the use of therapeutic methods which are founded on a knowledge of the normal physiological processes, of abnormal physiological processes (pathophysiology), and on a knowledge of the pharmacological actions of drugs. Empirical methods, still used in practice because of our incomplete knowledge of many of the vital body processes, are not founded on a comprehensive understanding of the condition under treatment and the drugs in use, but on experience and tradition. Unfortunately, the knowledge of disease processes and pharmacology is frequently incomplete and non-specific or symptomatic treatment must be applied: this treatment of symptoms alone verges on empirical therapy. Also, in veterinary medicine more than in human medicine, treatment is conditioned by economic factors, and although specific therapy may be possible, symptomatic treatment may be necessitated because of the high cost of the specific therapeutic agent.

It is important not to forget that drug therapy alone is rarely able to achieve an absolute cure without adequate and skilled nursing. In many cases, especially with small animals, meticulous and sustained nursing has achieved a cure when drugs have been of no value. On the other hand, unskilled and misguided nursing can equally reduce the value that rational therapy achieves. It is therefore important to impress upon owners the necessity of providing care and attention for their animals and also to make sure that they understand what nursing is required and how it should be carried out.

When rational methods are applied in curative medicine there are several stages of examination, observation and deduction which must be passed through before the point of drug administration is reached. Each of these stages represents a whole field of study and, together they constitute the processes which result in a diagnosis.

To trace the causes of the symptoms of a disease the veterinarian must have an understanding of *tissue pathology*, and also of *functional* or *metabolic pathology*. This knowledge must itself be founded on a study of normal tissues—*macroscopic and microscopic anatomy*—and on a study of a normal metabolism which is *physiology*.

When a diagnosis has been made the next decision will concern the choice of treatment. This will often incorporate measures which fall within the spheres of *hygiene* and *animal husbandry;* the former discipline is concerned with the housing and control of the living conditions, whereas the latter relates to the feeding, breeding and rearing of animals. It will be seen that advice on nursing is closely concerned with these two subjects. Finally a decision must be made on the drug to be used, and the method of its administration. These decisions are respectively the concern of the related subjects of *pharmacology* and *therapeutics.*

THE DRUG SCIENCES

Pharmacology is defined as the study of the effect of drugs on the structure and metabolism of normal tissues. A drug can be defined as any substance which is capable of modifying a biological activity. Within the scope of this book however the term is most frequently applied in the narrower context of substances used in the treatment of disease. The study of the actions of drugs under laboratory conditions has been termed *experimental pharmacology*, whilst those that relate to the effects of drugs on normal or diseased humans and domestic animals has been termed *applied pharmacology* or *pharmacotherapeutics.*

Therapeutics is concerned with the application of drugs in the treatment of disease. Such details as the choice of drug, the route of administration, the form in which the drug is applied, and the frequency of administration, all come within the scope of therapeutics.

Posology is that discipline which is concerned with dosage of drugs, whilst *metrology* is the study of the scales of weights and measures.

Both posology and metrology are integral parts of the wider subject of *pharmacy* which is concerned with the compounding, standardization and dispensing of drugs. Associated with pharmacy, and an essential part of it, is *pharmacognosy* which, literally, is the study of the recognition of drugs and is related to a knowledge of the chemical and physical characters and sources of drugs.

The old term *materia medica* was widely applied to the study of all factors relating to drugs, including their actions, and was usually extended to include therapeutics.

Pharmacognosy is of diminishing significance to the practitioner, because more and more pharmacy is performed by the manufacturing drug houses and less and less by the veterinary surgeon. A detailed knowledge of pharmacology and therapeutics is however essential if the opportunity provided by an accurate diagnosis is not to be wasted. The continuing efforts of the drugs industry to improve upon existing remedies and to discover drugs which possess novel action have tended to increase the degree of specialization within pharmacology. *Pharmacodynamics* is that branch which is concerned with the characterization of the action and effect of a drug, while the investigation of its fate within the body is *pharmacokinetics. Pharmacometrics* is the study of techniques used in the measurement of drug effects. *Pharmacogenetics* is the study of the effects of genetic constitution on drug action. *Immunopharmacology* relates to the use of drugs in the investigation and control of immune responses. *Neuropsychopharmacology* is concerned with the action of drugs upon the nervous system and behaviour.

Toxicology was originally a science whose scope was essentially that of pharmacology, differing only in that the subject matter was the study of poisons. Nowadays the definition also includes the study of the toxic effects of drugs and other chemicals in the body and the environment.

MODERN THERAPEUTICS

The development of rational therapeutics commenced towards the end of the nineteenth century because by that time a number of subsidiary sciences had advanced sufficiently to provide the basic information which was previously lacking. The study of bacteriology and pathology provided details on the cause and course of diseases, and thus facilitated the rapid advances which could be made in preventive medicine and aseptic surgery. Pharmacology started to examine critically the mode of action of drugs on body systems by the use of experimental procedures and thereby added to physiological knowledge. The advances in pharmacology were greatly aided in this by the concurrent growth of organic chemistry which not only supplied the compounds which extended the pharmacopoeias, but also made possible a rational association of a drug's molecular structure with its biological activity. The final contributors to the treatment or prevention of disease with drugs were the biochemists, who were largely responsible for the elucidation of the function and the isolation of the endocrine hormones and the vitamins. The interfaces of biochemistry, pharmacology and therapeutics are major contributors to experimental medicine today.

DRUG SOURCES

Plant Sources

Despite the modernization of the pharmacopoeias which resulted in the removal of the 'eye of newt and leg of toad' class of remedy, many old-established drugs remain 'official'. In the early days of medicine, the only useful drugs were those derived from *plant sources*, e.g. the purgative castor oil is mentioned in the Ebers papyrus while the term galenical dates back to the second century. Active principles can occur in all plant structures, but concentrations are often highest in one particular part. Such parts are preferred for therapeutic purposes and are exemplified below:

Root (*radix*)*:* e.g. Gentian, B.Vet.C.
Corms: e.g. Colchicine, B.P.
Bulbs: e.g. Squill, B.Vet.C.
Rhizomes: e.g. Ginger, B.Vet.C.
Bark (*cortex*)*:* e.g. Cascara, B.Vet.C.
Wood (*lignum*)*:* e.g. Tar, B.Vet.C. (Stockholm Tar).
Leaves (*folia*)*:* e.g. Cocaine, B.Vet.C.
Flowers(*flos*)*:* e.g. Pyrethrum Flower, B.Vet.C.
Fruit (*fructus*): e.g. Anise, B.Vet.C. (Aniseed).
Seeds (*semen*)*:* e.g. Physostigmine Salicylate (B.Vet.C.).

The therapeutic usefulness of plants is due to the fact that they contain a variety of active principles, of which the alkaloids and glycosides are the most potent.

Alkaloids

Members of this group of compounds, widely distributed throughout the plant kingdom, consist chemically of nitrogenous organic bases which react with acids to form salts. The structure of these alkaloids can be rather complex and includes many organic ring structures of which the following are examples:

Pyridine (nicotine)

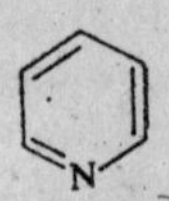

Piperidine (pethidine-synthetic)

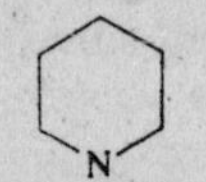

Pyrrolidine (cocaine, atropine and nicotine)

Quinoline (quinine)

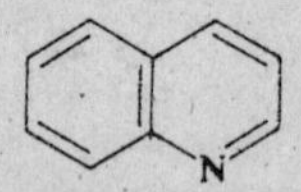

Isoquinoline (narcotine and papaverine)

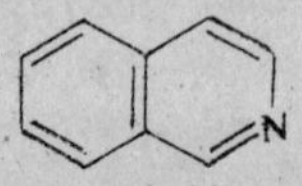

Phenanthrene (morphine)

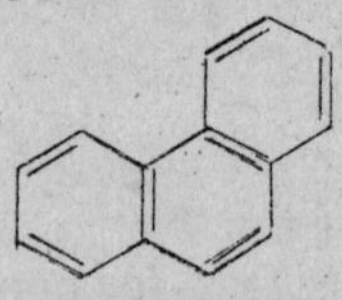

Glycosides

The members of this group of compounds, also sometimes known as glucosides, vary considerably in their chemical composition but have in common the production of a sugar and another substance, known as a *genin* or *aglycone*, on enzyme or acid hydrolysis. The genin may be an alkaloid, a tannin or compounds of unknown structure, collectively called *neutral* or *bitter principles*. Some glycosides, following enzyme hydrolysis, produce hydrocyanic acid, and are therefore known as *cyanogenetic glycosides*. Those which have a specific action on the heart are known as *cardiac glycosides*.

The ester-like linkage of the sugar and genin may be illustrated by reference to the simplest glycoside:

Phenanthrene (morphine)

$$CH_2OH\,.\,\overbrace{CH(CHOH)_3CH}^{O}\,.\,OH + CH_3OH$$

$$= CH_2OH\,.\,\overbrace{CH(CH\,.\,OH)_3CH}^{O}\,.\,OCH_3 + H_2O$$

In this way methyl glucoside is formed, i.e. glucose plus methanol. If the sugar were galactose or mannose, the glycosides formed would be known as methyl galactoside or methyl mannoside, respectively. Isomeric forms exist. By substituting other substances which contain hydroxyl groups, or will react to produce the hydroxyl groupings, other glycosides varying from simple alkyl compounds to complex aromatic and heterocyclic compounds can be formed.

Saponins

These are now classed as a subgroup of the glycosides, the group characteristic being the ability to cause foaming when shaken with water. They are sometimes used as emulsifying agents, e.g. Quillaia, B.P. This material contains a large proportion of saponins and has been used frequently as an emulsifying agent for volatile oils. Saponins frequently occur along with other glycosides, as in digitalis, and increase their absorption rate.

Tannins

These are a heterogenous group of compounds, some of which are glycosides. They have certain chemical and pharmacological characteristics in common, such as astringency, producing inks with iron salts, and high water solubility. The tannins are widely distributed throughout the plant kingdom and are usually named after their source, e.g. *catechutannin* from catechu.

Carbohydrates

Cellulose, starch and sugars are well-known examples. Gums are complex carbohydrates, often glycosides, which form viscous adhesives with water. Pectins are similar carbohydrates derived from fruit.

Fats and Fixed Oils

These are, respectively, solid and liquid esters of fatty acids and glycerin. They are found in the most concentrated form in the seeds of plants but may occur throughout the structure, e.g. Olive Oil, B.Vet.C.; Castor Oil, B.Vet.C.; Linseed Oil, B.Vet.C.; Theobroma Oil, B.Vet.C.

Waxes

Waxes are similar in consistency and character to fats but differ chemically in being esters of fatty acids and alcohols other than glycerin.

Volatile Oils

These are also referred to as Essential, Aromatic or Ethereal Oils. They are obtained usually by steam distillation. Chemically they are mixtures of aldehydes, esters, ketones, alcohols and terpenes. They are volatile, have a characteristic odour specific to each oil, are colourless, highly mobile liquids which leave no permanent greasy mark on paper and are non-lubricant. They are only slightly soluble in water, but miscible with fixed oils and organic solvents. On cooling they usually separate to give a supernatant liquid or *elaeoptene*, and a solid or *stearoptene*, e.g. Anise Oil, B.Vet.C.; Chenopodium Oil, B.Vet.C.; Peppermint Oil, B.Vet.C.; Turpentine Oil, B.Vet.C.

Resins

These compounds are brittle, amorphous compounds mainly derived from the oxidation of the terpene components of volatile oils. Resins fracture giving smooth, shiny surfaces; they are insoluble in water but soluble in most organic solvents and oils, e.g. Podophyllum Resin, B.Vet.C.

Oleoresins

Oleoresins are mixtures of volatile oils and resins. Turpentine Oil, B.Vet.C., is the best example in veterinary use.

Animal Sources

Animal sources contributed to the variety of useless remedies which packed the old pharmacopoeias with a selection of animal excreta, tissues or even whole creatures from the lower orders. A few of these have persisted into relatively recent times, e.g. the use of raw liver in the treatment of pernicious anaemia. The major products of animal origin are nowadays exemplified by the hormones, vitamins and antisera.

Hormones, e.g.
- Pituitary (Posterior Lobe) Injection, B.Vet.C.
- Insulin Injection, B.Vet.C.
- Thyroid, B.Vet.C.
- Chorionic Gonadotrophin, B.Vet.C. (LH)
- Serum Gonadotrophin, B.Vet.C. (FSH)

Vitamins, e.g.
- Veterinary Cod-Liver Oil, B.Vet.C. (vitamins A and D)
- Halibut Liver Oil, B.Vet.C. (vitamins A and D)

Antisera, e.g.
- Canine Distemper Antiserum, B.Vet.C.
- Lamb Dysentery Antiserum, B.Vet.C.

These products are all obtained by extraction from the relevant endocrine gland or biological fluid, and by subsequent purification.

Other biological sources of significance are the artificially cultured viruses, bacteria, fungi and yeasts. The viruses and bacteria which cause infectious disease can often be modified to yield products capable of stimulating immunity; the fungi have yielded antibiotics; and brewers yeast is a valuable source of vitamins.

Virus	Canine Contagious Hepatitis Vaccine, B.Vet.C.
Bacteria	Tetanus Vaccine, B.Vet.C. Mallein, B.Vet.C.
Fungi	Benzylpenicillin, B.Vet.C. Streptomycin Sulphate, B.Vet.C.
Yeasts	Dried Yeast, B.Vet.C. (B group vitamins)

Mineral Sources

Mineral sources have contributed a number of simple compounds which remain useful in many fields, e.g.

Antacid	Light Magnesium Oxide, B.Vet.C.
Purgative	Magnesium Sulphate, B.Vet.C.
Disinfectant	Sodium Hydroxide, B.Vet.C.
Electrolyte	Sodium Chloride, B.Vet.C.
Expectorant	Potassium Iodide, B.Vet.C.
Diuretic	Potassium Nitrate, B.Vet.C.
Haematinic	(Veterinary) Ferrous Sulphate, B.Vet.C.
Styptic	Silver Nitrate, B.Vet.C.

The contribution of the chemists does not end with these inorganic examples, but could be considerably extended were one to include the the long list of compounds which have flowed from organic chemistry, usually as a result of the planned research and purposeful searches carried out by the chemical manufacturer. It would be invidious to list individual examples from this group as compounds are to be found on nearly every page of the B.Vet.C.! The first synthetic organic drugs introduced into medicine were the *volatile anaesthetics* and these were followed by the *phenolic antiseptics*. The synthesis of the *chemotherapeutic agents*, which began with the organic arsenical, salvarsan and continues today with the long-acting sulphonamides, constitutes the biggest single advance in the treatment of infectious disease since the advent of antisera. Less dramatic, but of considerable importance in human medicine has been the synthesis of a wide range of substances capable of controlling or altering body functions, e.g. the *diuretics*, *tranquillizers*, *analgesics*, *anaesthetics* and the many compounds capable of modifying autonomic nervous system function. Recently, a valuable series of *antibiotics* has been created from fermentation-produced 6-aminopenicillanic acid through a series of synthetic modifications. The attention of pharmacologists is nowadays directed more and more to the activity of drugs at the molecular rather than the organic level and this is well exemplified by drugs which have been shown to be able to interfere with amino acid sequence in the nucleic acids.

DRUG NOMENCLATURE

This is an area in which considerable confusion

exists because any one chemical compound may be known as a drug under a variety of names. The blame for this surfeit can be laid at the doors of the drug manufacturers, who, for protection and standardization purposes, find it advantageous to give 'proprietary names' to each product. Thus one chemical compound can be produced as a drug under a series of unrelated names by a series of pharmaceutical manufacturers. Even more confusing is the situation where the same drug may be used as a component in a number of 'shot-gun' preparations (products containing more than one active principle) produced by the same company but each preparation possessing a distinctive proprietary name. Proprietary names are usually registered as trademarks.

In an attempt to rationalize this situation, the pharmacopoeia commissions give each chemical compound an 'approved name' (also known as an official name or a trivial name) and the compound is entered under this name in the pharmacopoeia. This does not occur immediately after the introduction of a new drug, but takes place when the drug is considered to have established itself. Such a rationalization is very useful at the national level, but the international situation has also to be considered, in particular in the context of the free movement of veterinarians between the European Economic Community countries. At present, there are several national pharmacopoeias and each commission can coin its own approved name. The World Health Organization at Geneva periodically publishes lists of approved names meant for international use in an attempt to limit further international diversification.

The approved name is usually derived from the full chemical name of the compound by a process of contraction. One might think that the best means of positive identification of drugs would be to use chemical names only, but these are usually long and difficult to remember. Additionally, there are several naming conventions used by chemists and this can result in the existence of several widely differing, correct chemical names for one compound. Therefore an internationally acceptable approved name would appear to be the best solution available. But manufacturers prefer to use brand names as they consider that proper formulation of a product is often as important as its active ingredient. The outcome is that drugs are advertised under proprietary names and referred to by these names by most clinicians (sometimes in ignorance of a product's active constituents) while the approved names are used among research workers and non-clinical teachers. Whenever an approved name is available for a drug, it is used in this book. A drugs index is provided and this gives some of the approved name/proprietary name equivalents.

Bibliography

AMA Drug Evaluations (1975) Acton, Mass.: American Medical Association Publishing Sciences Group.

British Pharmacopoeia (1973) London: HMSO.

Extra Pharmacopoeia (1972) 27th ed. London: The Pharmaceutical Press.

The British Pharmaceutical Codex (1973) London: The Pharmaceutical Press.

The British Veterinary Codex (1965) London: The Pharmaceutical Press.

United States Dispensatory (1973) Philadelphia: Lippincott.

2

The Essentials of Drug Action

DRUG CLASSIFICATION

Unlike the situation applying to nomenclature, there is no international attempt to provide a unified system for the classification of drugs. To some extent the classification tends to relate to the area of interest of the user, e.g. the physician may describe magnesium sulphate as a laxative when given by mouth; the pharmacist may describe it as a saline or inorganic purge; the physicist as an osmotic purge; the pharmacologist as a bulk purge. The physician categorizes on the basis of the *drug's effect*, e.g. a bacteriostat, a diuretic, a stimulant, or on the symptomatic basis of the *indication for the use* of a drug, e.g. an analgesic, an antacid, an antispasmodic. The chemist's interest lies in a *drug's structure* rather than in its action. This is evident from the use of chemical group names, e.g. the steroids, the barbiturates, the glycosides. *Drug action* often forms the basis for different approaches to classification, e.g. to describe a drug as a surfactant, an osmotic diuretic, or an emollient is to refer to it in *physical terms*. The description of a drug as a parasympathomimetic, an adrenergic or a neuromuscular blocker are all functional *physiological* terms; the older *therapeutics* terminology is exemplified in describing liquid paraffin as a lubricant purge, or digitalis glycosides as heart tonics. Another early classification was based on the *source of the drug* or preparation, e.g. identification by naming the plant source and plant structure—gentian root, juniper berries, pyrethrum flowers. The development of families of drugs with similar actions has lead to the establishment of *type-compounds* against which the newer additions are compared. This practice has given rise to such terms as histamine-like, atropine-like, chlorpromazine-like.

It is often the case that descriptive terms derived from different systems are applied to the one drug, e.g. a bulk, saline, osmotic purge; a sympathomimetic amine; an atropine-like antispasmodic. Carbachol can be described in *whole body* terms as a purgative, in *organ* terms as a gut stimulant; in *functional* terms as a parasympathomimetic; in *histological* terms as a smooth muscle stimulant; in *physical* terms, as a membrane depolarizer; in *comparative* terms, as possessing muscarinic activity; and in *chemical* terms, as a choline derivative.

Two examples of progress in pharmacology are seen in the increasing precision with which the sites and modes of action of drugs are defined, and in the incorporation of this knowledge into the descriptive terms applied to drugs. When, *digitalis leaves* were used as such in the treatment of heart failure, they were described as a *heart tonic*, and later as a *myocardial stimulant*. The active principles were eventually isolated and they became known as the *digitalis* or *cardiac glycosides*. Investigation into the mode of action of the glycosides has yielded several explanations one of which has lead to their description as *membrane stabilizers*. This action refers to the effect digitalis has on sodium and potassium concentration differentials across the cell membrane, and it is suggested that the precise action is the inhibition of an enzyme (transport ATPase) which normally functions in the extrusion of intra-cellular sodium.

THE MODE OF ACTION OF DRUGS

The previous section described the profusion of descriptive terms applied to drugs and related this in part to the increasing specificity of our knowledge both of the sites and mechanisms of drug action. The term *drug action* is used to describe the method by which the drug influences a cell. The *drug effects* are a sequel to this action.

The effects are the detectable changes which follow an alteration in cellular function and which were initiated by the combination of the drug with the cellular components or contents. Our knowledge of drug actions is still inadequate, but it is customary to call the site of combination with the drug in or on the cell the *receptor* area and to speak of the resulting union as a *drug–receptor complex*. This provides a chemical or physico-chemical basis for explaining drug action which removes medicine from the realms of witchcraft, to the objectivity expected in the science laboratory. Drug action can be shown to be affected by the same factors as other chemical reactions, e.g. temperature (usually controlled by the body), concentration (shown to be dependent upon dosage and fate of the drug) and catalysts (enzymes).

The receptor concept has been developed as a result of a study in depth of the mechanisms of drug action relative to the autonomic nervous system function. This has given rise to a model for drug action which has been applied to other classes of drugs. However, this is not applicable to all drugs as certain effects are explainable without the need to invoke specific receptors. These so called non-specific drug actions will be described before discussing receptor theory.

Structural Non-Specificity

An early proposal to explain the mode of action of drugs involved the volatile anaesthetics, the anaesthetic potencies of which were found to be related to their *oil–water partition coefficients*. The more oil soluble members were the most potent, and potency was believed to be dependent on the ease of entry of these drugs into the lipoid material of nerve cells, accumulation in the cell, and eventual interference with normal metabolic processes. Structurally non-specific drugs share the ability to accumulate, in certain cells not because of specific receptor attachments, but because of a shared physico-chemical property. Ferguson's principle of structural non-specificity relates a drug's action to its *thermodynamic activity* in the biophase (i.e. the phase in which a drug reaches its target) rather than its concentration. This implies that when a number of drugs can produce the same effect via a physico-chemical mechanism, then although their concentrations in the biophase may differ for the same degree of effect, their thermodynamic activities would be the same. The thermodynamic activity may be regarded as a measure of the number of drug groups which are free to react.

More recent theories used to explain the mode of action of general anaesthetics also have a physical basis. It has been suggested that the molecular volume of the drug is important in that, if suitable, it is able to occupy a membrane pore, block the movement of ions, and so interfere with cell function. The clathrate theory suggests that the build-up of structured water within the cell is the factor which disrupts cellular activity. A clathrate is formed when water molecules become arranged in a cage-like manner enclosing a second substance, in this case, the anaesthetic. These water micro-crystals would increase the electrical impedance of nervous tissue. However, no clathrate has yet been shown to be stable under physiological conditions.

Cell Function

A small number of drugs are said to act by direct *irritation* of the cell. This action may be mild, e.g. the purgatives, or severe, e.g. iodide blisters for horses. It may even result in cell death, e.g. the keratolytics dissolve keratin, the styptics arrest bleeding by causing denaturation of protein, while phenolic disinfectants are strongly caustic.

Many drugs, however, can act extra-cellularly on the cell wall, or intra-cellularly, and the effect is seen as a decrease or increase in some functional activity of the cell. This has given rise to the description of many drugs as *stimulants* or *depressants*. Investigations into the mode of action of drugs continue to yield considerable evidence of the ability of drugs to modify enzyme-catalysed reactions. While it is often impossible to correlate the significance of an in vitro enzyme inhibition, with the in vivo action of a drug, there is a growing body of evidence which supports the assertion that drug action is always dependent upon the modification of an enzymic function. This may occur as a result of a direct action on the enzyme, coenzyme or co-factor. An *indirect drug action* in which the drug does not act directly on host cells, but allows host cells to return to normality by the removal of pathogenic organisms or material is exemplified in the treatment of infectious diseases, e.g. the use of drugs which kill parasitic viruses, bacteria, fungi, protozoa, metazoa, helminths, nematodes and arthropods. *Replacement therapy* involves supplying the animal with the electrolytes, hormones, vitamins or nutrients which it lacks and thus allows a return to cellular normality.

THE FOCUS OF DRUG ACTION

Methodological advances continue to allow greater precision in the localization of the sites of action of drugs. Formerly drugs were simply related to *target organs*, e.g. a gastric sedative, then to *tissues*, e.g. a myometrial stimulant, or *elements* of *tissues*, e.g. a depressor of fibroblastic activity, and more recently to *sub-cellular sites* e.g. a microsomal inhibitor, or even a to precise location in a *metabolic* pathway, e.g. inhibitors of cholinesterase, aminoxidase and carbonic anhydrase. Many drugs act directly on the tissues in which the characteristic action of the drug is seen, e.g. a bronchodilator. There are, however, other drugs which act at sites remote from the one in which the effect is seen, e.g. vomiting centre stimulants. A third type of action occurs when a drug does not directly elicit its effect, but causes the effect by allowing another drug or an endogenous agent to increase its action, e.g. an anticholinesterase would inactivate body cholinesterase and thus allow endogenous acetylcholine to accumulate and act for an increased duration at an elevated concentration, or heparin acts in alimentary lipaemia because it increases the free concentration of lipoprotein lipase by displacing it from its binding site.

Structure–Activity Relationship

One of the most useful concepts in the evolution of pharmacology was the recognition that drugs with a similar structure often exert the same action. Organic chemists were able to produce compounds in a homologous series, and it was often possible to point out that activity increased with increasing substitution up to a maximum and then declined, e.g. a 6-carbon chain separating quaternary ammonium heads produces optimal ganglionic blockade. Activity could also be shown to be seated in one particular grouping or fragment of a molecule, the masking of which would lead to inactivation, e.g. the amino group of the sulphonamides. This finding explained why compounds of differing chemical structure could exert the same drug action. When a drug possesses more than one action, it may be possible to localize these in different molecular areas. It is possible for instance to subdivide the muscarinic and nicotinic areas of molecules related to acetylcholine, and by structural modifications to enhance or diminish either activity. It is apparent that molecular manipulation can affect pharmacological activity either qualitatively or quantitatively. This feature is well illustrated by the steroid hormones. The specificity of drug action leads to the key and lock analogy, where the drug is the key and the lock its site of action. The specificity of a key for the lock is demonstrated by the difference in activity which often exists between two optical isomers, e.g. *laevo*-rotatory adrenaline is ten times more active than the *dextro*-rotatory enantiomorph. Valuable though structural studies are for identifying a molecule already shown to possess interesting biological activity, our general knowledge in this field is still so slight that random trial-and-error screening tests are found to be more valuable by drug companies in their searches for new drugs or new activities, than is the construction of novel molecules.

Receptor Theory

In the lock and key analogy, the drug has been described as the key while the lock itself is the ultimate site of action, known as the *receptor*. The receptor may exist in a histologically recognizable specialization on a cell, e.g. the motor end plate of voluntary muscle cells, but is often a functional rather than anatomical locus. The β-adrenergic receptor for example is now believed to be the enzyme adenyl cyclase. The groups which combine with the drug together with adjacent groups which may facilitate or impede the process of combination are known as the ***receptor groups*** or ***receptor site.*** The union of drug and receptor yield the *drug–receptor complex* which is responsible for the initiation of an action. The fact that the specificity of key for lock is not absolute is shown by the ability of drugs from a homologous series to produce similar effects. There are, however, a large number of distinct receptors which are conveniently labelled with the name of their respective 'keys' e.g. histamine receptors, atropine receptors.

A drug which complexes with a receptor and elicits a response is called an *agonist*. Such a drug has both *affinity* for the receptor, in that it can complex, and *efficacy*, in that the resulting complex can initiate a biological response. It is possible for a drug to combine with a receptor but to elicit no response. Such a drug has affinity but lacks efficacy and may behave as a *competitive antagonist*. Curare, for example, competes antagonistically with acetylcholine at the motor end plate, in that it can occupy receptors but can elicit no muscle contraction. A *partial agonist* is a drug which can complex with a receptor but which elicits only a weak response, i.e. it has affinity but only limited efficacy. A partial agonist can compete with an agonist for a receptor and may thereby reverse the action of the agonist, e.g. the morphine antidote nalorphine acts in this way.

The Drug–Receptor Complex

The current view of the combination of drug with receptor is of a union or interaction of forces. These forces are contributed by the effector cell, in the receptor, and by the drug, because of its molecular configuration. When the spatial relationship of these forces is optimal, complexing takes place and a tissue response is noted. The forces which bind drug to receptor are weak reversible bonds and are exemplified by van der Waal's, ionic, dipole and hydrogen bonds. More rarely, covalent bonds are formed. These high-energy bonds are stable and are usually formed by agents with a long duration of action.

Quantitative Considerations

For a given concentration of drug in the biophase (i.e. that portion of a drug dose which reaches its receptors in an active state) a certain tissue response may be measured. Increasing the concentration will usually increase the response. At each level of response there is an equilibrium state between the drug in the biophase and that bound to receptors. The complexing of an agonist particle with a receptor may be regarded as an all-or-none event in that the drug–receptor complex initiates a unit of response characteristic of that particular agonist. On this basis, the increase of response associated with an increased agonist concentration can be visualized as being due to the occupation of a greater number of receptors—the *occupation theory* of drug action. An alternate theory states that the response depends upon the rate of formation of drug–receptor complexes—the *rate theory* of drug action. If one assumes the activation of all receptors when a drug is eliciting its maximal response, then the difference between the maximal responses to different drugs must reflect the differing efficacies of the drug–receptor complexes. The situation is, however, complicated by the observation that on occasion a maximal response is attained with a less than maximal receptor occupancy. This anomaly is known as the spare receptor concept and can be taken to mean that the responding tissue has already reached the limit of its capacity to respond.

Drug Action Antagonized

The concept of *competitive antagonism* was introduced in a previous section and explained in terms of the occupation theory. In terms of the rate theory, an agonist complexes with, activates and dissociates from a receptor, thus permitting further series of activations, but the antagonist dissociates slowly. When an agonist and its competitive antagonist are present in the biophase, the tissue response will depend upon the relative affinities and concentrations of the two drugs. The reduced effect of the agonist can always be improved by increasing agonist concentration, i.e. the antagonism is reversible at a rate which now depends upon the rate of antagonist dissociation from the receptor. *Non-competitive antagonism* is also reversible, but at a rate which depends only on the rate of removal of the antagonist from its receptors.

Such reversal is usually slower than that of a competitive antagonist. The usual example of non-competitive antagonism to acetylcholine is that produced at the neuromuscular junction by the depolarizing drug, succinylcholine. Raising of agonist concentration does not overcome non-competitive antagonism as it does in the case of competitive antagonism.

Competitive or non-competitive reductions of response to an agonist both exemplify types of *pharmacological antagonism*. When a drug reduces another's effect by eliciting an opposing response due to the activation of a second species of receptors, the reduction exemplifies physiological antagonism. The antihistamine drugs block the action of histamine pharmacologically while the effects of histamine action can be countered physiologically by adrenaline.

The use of a second drug whose effects are opposite to those produced by the first is an example of *physiological antagonism*. The characteristic effects of histamine-release in an experimental animal can be suppressed by the prior administration of an antihistaminic drug; this is *competitive antagonism*. The effects of histamine-release in anaphylaxis can be countered physiologically by the administration of adrenaline.

In the treatment of poisoning, the further absorption of material from the gastro-intestinal tract is sometimes prevented by converting the poison into an insoluble form. The administration of a second drug for the purpose of changing the nature of the first is an instance of *chemical antagonism*. The increasing application of chelates in toxicology provide an interesting method of removing potentially damaging agents from the body, e.g. sodium calcium ethylenediamine-tetracetate removes lead and penicillamine removes copper.

DRUG QUANTITATION

In discussing drug action and effect, it was stated that the giving of increasing quantities of drug usually resulted in an increasing effect. This factor necessitates for every case the choosing of a dosage regimen, the suitability of which may be investigated by estimating the drug concentration in body fluids. The methods used for measuring drug concentration or in determining drug activity find application both during the experimental examination of a drug and in its later production.

Chemical Analysis

This is the method most frequently used for establishing drug levels. Drugs of sufficient purity for chemical analysis can be dispensed according to weight or volume. Chemical methods are usually highly specific, and the activity in use of a chemically assayed drug is therefore extremely reliable.

Biological Assay

Some preparations contain active principles which are not amenable to chemical assay, and this has led to the development of techniques which measure biological activity rather than ingredient concentration. The techniques rely upon the measurement of an effect produced in a biological system, sometimes an isolated tissue preparation, sometimes entire laboratory animals and they are referred to as *bioassays*. When chemical estimation is impossible or unsuitable, as in the standardization of vaccines and antisera, of drugs of unknown composition, of some hormones, vitamins and antibiotics, a suitable bioassay is devised. Many new remedies have been introduced on the basis of a bioassay, which was later replaced by chemical quantitation when the active principle of the remedy had been characterized.

Although the error inherent in a bioassay is much greater than that found in a chemical technique, reliable results can be obtained by adequate replication (repetition of the analysis) and the statistical analysis of the results. Bioassay techniques are based on a comparison of the activity of the substance under test with that obtained using a standard preparation of the same material under closely controlled experimental conditions. The standard preparation of a drug is often defined by international agreement within the World Health Organization, and is maintained by a reference laboratory. Its potency is expressed in units of activity per unit weight or volume.

Bioassay can also be applied to the problem of identification of a drug whose effect is characteristic, or for which a known antagonist is available. For such a comparative assay, it is usual to use more than one type of test. Analysis of the mode of action of a new drug may be based on comparative bioassay, and so was the screening procedure used during the initial investigations which led to the detection of activity. A similar technique will be used to assess the drug as a therapeutic agent.

Variation in Response to Drugs

There are many factors which can affect the response of individual animals to a standard dose of a drug. Such factors (detailed in Chapter 4) contribute to the well-known *individual variation* in response and are responsible for the need for cautious treatment with those drugs in which variation is marked. This is one reason why there is not always one correct dose, but a range of doses from which the physician has to choose. Variations can exist in the response of one animal to a standard drug dose, perhaps because of the time of day of dosing, or diet, or temperature. The variation in biological experiments is large in comparison to that encountered in the more precise sciences, so it is customary to measure and describe not only an effect but also the variability of the effect and even to attempt to distinguish between the error inherent in the technique and that contributed by the experimenter—the so-called personal error. The more common statistical procedures used in quantitative pharmacology (or pharmacometrics) are described briefly in the following section.

Statistical Methods

Normal distribution. Biological variants frequently fit what is known as a *normal distribution curve*, whose bell-shaped symmetry is illustrated (Fig. 1). Data which yield a normal curve owe their variability to the operation of random factors. A normal curve is one example of what is known as a *frequency distribution.*

Such a curve illustrates the scatter of the results which would be obtained if one plotted the frequency of occurrence on the vertical axis against a measurable variable on the horizontal. From the data it is simple to compute an *average* or *mean result*. This mean would, however, give no measure of the spread or scatter of the results. One solution to this is to quote the *range* of the measured effect, but even this gives no measure of the way in which the individual results are dispersed throughout the range and, if normally distributed, the majority would in fact be clustered about the mean.

Standard deviation. Scatter is usually expressed in standard deviation (SD) units. The standard deviation is the square root of the mean of the squares of the deviations of the observations from their arithmetic mean. When data are normally distributed, 95 per cent of the values will lie within ± 2 SD of the mean. This could also be expressed as there being a 5 per cent (or 1 in 20) chance that any observation will fall outside the ± 2 SD limits. It is also true for normally distributed data that the range of distribution closely approximates to 6 SD units.

Variance. This is the mean of the squares of the deviations of the observations from their arithmetic mean.

The standard error. This value is obtained by dividing the standard deviation by the total number of observations less one. As the experimental mean is obtained by the examination of but a sample from the population, then it is possible that the experimental mean will differ from the true mean. The standard error of the mean quantitates this possibility.

Significance. The concept of statistical significance is based on estimates of the probability of a particular occurrence being due to the

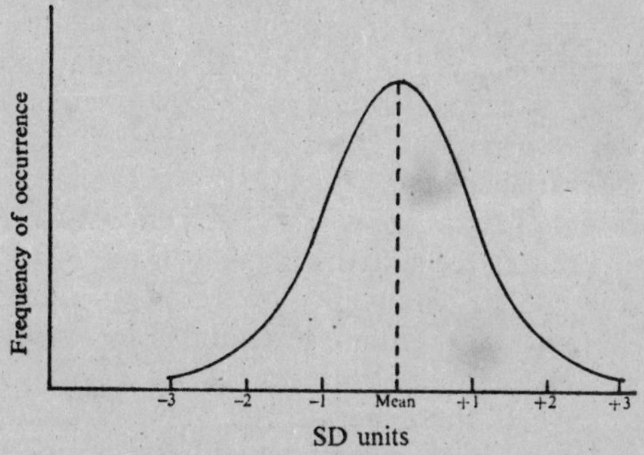

Fig. 1. A normal distribution curve. The range of distribution is expressed in standard deviation (SD) units.

operation of chance factors. The smaller the probability, the greater is the likelihood that the occurrence is due to the operation of non-random factors, i.e. the occurrence may be described as significant. In general, the minimal acceptable odds for significance in biological experimentation are 1 in 20, i.e. the probability of the deviation being due to the operation of random factors is 1 in 20, which may also be written as 5 per cent, or $P = 0{\cdot}05$. This means that there are nineteen chances out of twenty that the deviation occurred for non-random reasons, i.e. the assumption being that the deviation is due to factors introduced intentionally by the experiment.

The *t*-test. A common problem in experimentation is to decide whether the difference between two means is significant. When dealing with fairly small groups, this is accomplished by calculating the numerical value of '*t*', which is the ratio of the difference between the means to the standard error of the difference. The experimentally obtained '*t*' value is compared with the appropriate value of '*t*' to be found in statistical tables. If the experimental value is greater than the table value, then the difference between the means is statistically significant at the stated levels of probability, i.e. the means were not obtained from two samples from the same population of results, but from two populations with different means.

The chi-squared test. This procedure is used to test the significance of the difference of two groups of results when the results are expressed numerically, e.g. as percentages. Such data are often obtained as the end-point of a therapeutic trial, e.g. the different frequencies of a cure under two different treatments. The procedure is similar to a '*t*' test in that an experimental value for 'chi' is calculated and this is then compared to values given in a chi-squared table.

The variance ratio. The ratio of the variances of two sets of data yields a statistic known as '*F*'. This value is compared with that in an '*F*' table of the chosen probability level and at a cross-reference appropriate to the number of observations in each set (= number of degrees of freedom). If the experimental value is greater, then there is a significant difference in the variability of the two sets of data, i.e. they are drawn from two populations with different variances. An '*F*' test is often used to investigate the suitability of data for the subsequent application of a '*t*' test.

THE DOSE–RESPONSE RELATIONSHIP

When the size of an effect can be varied with increase or decrease of drug dosage, it is possible to examine the dose–response relationship. This parameter is the starting point of the characterization of a drug's activity and is utilized in the computation of further characteristics.

In drug testing, a positive result can be the simple presence of a response while an increase in dosage will cause a greater percentage of the population to respond, i.e. the response is *quantal.* This type of response is obtained in establishing the lethal dose of a drug, i.e. the animal either lives or dies. Such an all-or-none response is in contrast with the situation where all units may respond but with an intensity which varies with dosage. The diminishing increase in response with arithmetic increase in dosage obtainable from a piece of isolated guinea-pig ileum when exposed to histamine exemplifies a *graded* or *gradual* response. When this data is plotted as a graph, a *hyperbolic* curve is obtained (see Fig. 2). When the data is converted so that response is plotted as a percentage of maximum response instead of response in absolute units, and dose is represented not arithmetically but logarithmically, then the curve becomes sigmoid and is known as a *log dose–response curve.* The central portions of such curves are usually linear and the curves produced by different concentrations of the same drug (as in a bioassay) are parallel. Both the dose–response and the log dose–response curves of a quantal response are S-shaped. The information can, however, be transformed mathematically to yield a straight-line relationship by applying a probit-transformation to the response data. To demonstrate the value of a dose–response curve as a source of information in therapeutics, we will consider such a curve obtained from the evaluation of an anaesthetic (curve A, Fig. 3).

The *potency* of the drug, a property determined by its pharmacokinetic behaviour as well as its ability to occupy and then activate receptors, is

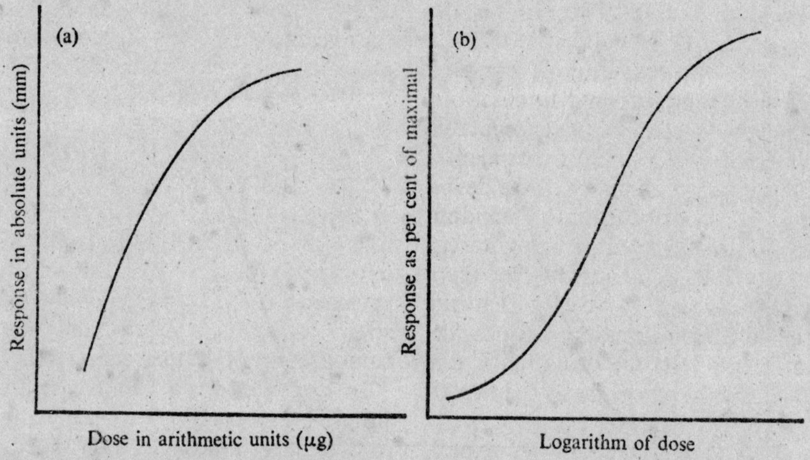

FIG. 2. When a measurable drug response is plotted in absolute units against dosage, a hyperbolic curve (a) is obtained. When the response data is plotted as the percentage of the maximal response and dosage is entered logarithmically instead of arithmetically, the curve becomes sigmoid (b).

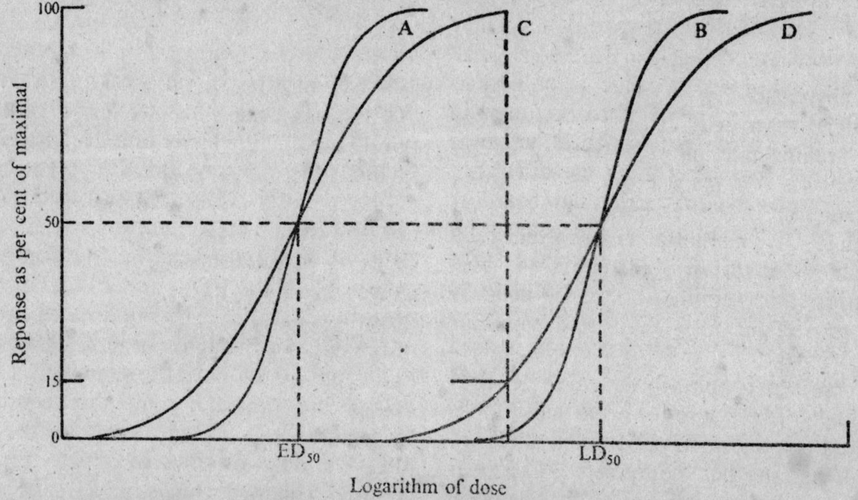

FIG. 3. The dose–effect curves (A) and (C) and dose–lethality curves (B) and (D) of two anaesthetics which have the same ED_{50} and LD_{50} values, but whose curves differ importantly in slope. These drugs would share the same therapeutic index, but the superiority of drug A would be revealed by its better therapeutic ratio.

represented by the distance between the origin and the foot of the curve. Potency is of importance only when considered in relation to the separation between the *dose–effect curve* and curve B, the *dose–lethality curve* (a curve in which the response recorded is death) of the same drug.

The *steepness* or *slope* of the linear part of the curve indicates the extent by which dosage must be increased to secure an increase in response. The vertical dropped from the maximum effect of curve A falls far short of the origin of curve B, showing that anaesthesia can safely be induced without the risk of mortality. Not so with anaesthetic C whose vertical cuts its corresponding lethality curve (curve D) at a point which indicates that 15 per cent of animals in which anaesthesia was induced would die!

The *maximum effect* is that point at which increasing dosage will not secure any further

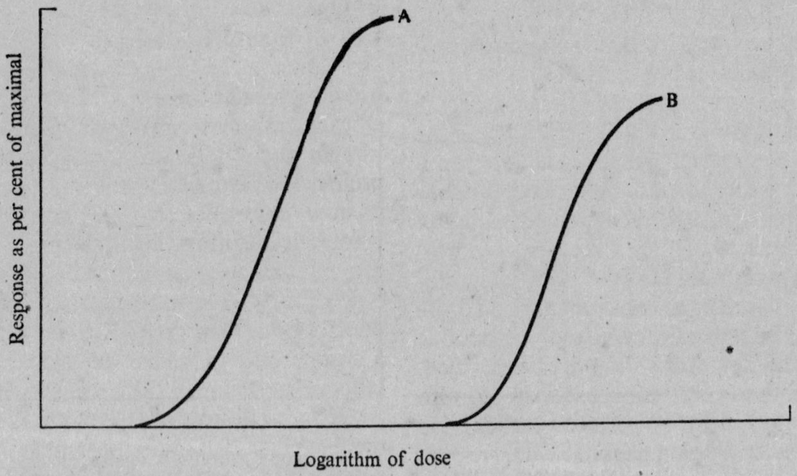

FIG. 4. Dose–effect curves of two cough depressants with different potencies and maximal effects. If abolition of the cough reflex is taken as the maximum response, then B, although less effective than A, could be the safer anti-tussive in that despite possible overdosage, the protective function of the cough reflex would not be completely suppressed.

increase in response. Fig. 4 shows that drug A is more potent than drug B and is capable of exerting a greater maximum or ceiling effect. The fact that the curves are parallel suggests that both drugs act in the same manner, but that assuming that no differential is introduced by factors such as absorption, distribution, biotransformation and elimination, then drug A has both greater affinity than and superior efficacy to drug B.

The *variability* of a point in the curve depends upon the range of response for a particular dose or the range of dose which produces a standard response. It is another composite parameter as it includes individual variation between animals, variation inherent in the drug and variation introduced into the experiment by the experimenter, e.g. the difficulty of determining a repeatedly reliable end-point. It is nonetheless an indicator of a drug's reliability in use.

In Fig. 3, the dose which anaesthetized 50 per cent of the sample of animals is an example of an ED_{50}, i.e. a dose of a drug which will produce a response which is 50 per cent of the maximal possible. The ED_{50}s of several drugs which produce the same effect can be used to compare their relative potencies.

Similarly, the LD_{50} is the dose of a drug which kills half of a test population and a comparison of LD_{50}s is a measure of relative toxicities.

The *therapeutic index* of a drug is obtained by dividing the LD_{50} by the ED_{50}. The higher the value, the safer is the drug and so this is a measure of a drug's safety. However, the therapeutic index is a valid basis for the comparison of drugs only if the ED_{50} and LD_{50} curves for the drugs are parallel. The importance of the slope of the curve is brought out in Fig. 3 which shows two drugs which have the same ED_{50} and LD_{50} values and thus the same therapeutic indices. In general, the flatter the curve, the safer the drug, i.e. dosage is not so critical. However, as already explained, this is only valid if the top of the dose–effect range does not overlap the bottom of the dose–mortality range!

The *therapeutic ratio* is a safety measure which includes a consideration of curve steepness and so is preferable to the therapeutic index. It is obtained by dividing the LD_{25} by the ED_{75}. Computation of therapeutic ratio values would reveal that drug A (Fig. 3) is a much safer compound than drug C although C has the flatter curve. Ideally, the division of LD_1 by ED_{99} would give an even better index of safety, but the difficulty of determining such values with accuracy excludes this possibility.

EVALUATION OF NEW REMEDIES

The continual growth of the pharmaceutical industry has resulted in the frequent appearance of new drugs, and on each occasion the clinician is faced with the problem of deciding whether the new product is superior to the old. Such decisions are best based on adequate comparisons between the new drug and the type compound of its class. The results of these drug trials are of most value when expressed in a factual manner and reinforced by statistical examination. To be really meaningful, the efficacy, toxicity and mode of action of the new drug should be described. The details of the formulation used should also be given, together with dosage recommendations for all the domesticated species.

Drug Development

The initial discovery of the possession of a biological activity by a compound is usually made in a routine *screening procedure*. This is often a qualitative bioassay designed to demonstrate the presence of one particular activity. The active compound is then further examined both for efficacy and toxicity and these properties are expressed in the terms described in the previous section. At this stage it is customary to begin examining the mode of action of the compound, the methods of its formulation and the stability of the formulated material. If the therapeutic index of the compound is acceptable and the efficacy is confirmed, the next step is a limited trial in domesticated animals. If the compound still seems promising, its usefulness is further examined in a *field trial* which is conducted under normal clinical, as opposed to research, conditions. The final value of a compound is usually established only after it has been in use as a product over several years.

Control of Drug Development

The severity of side-effects associated with the clinical use of some drugs coupled with misgivings concerning the adequacy of the evaluation of others has lead to the introduction of legislation in the U.K. *The Medicines Act* (1968), provides for the control of the safety, quality and efficacy of medicinal products intended for human and veterinary use by means of a comprehensive licensing system. The sale of drugs is now controlled by product licences, while wholesale dealing also requires a wholesale dealers licence. A manufacturers licence is required for the manufacture of medicinal products. Should the manufacturer wish to develop a new product, he has to submit evidence of laboratory examination of the drug to the licencing authority who, in turn, if satisfied with the evidence, will grant an animal test certificate which will allow the manufacturer to carry out a field trial. The practising veterinary surgeon is exempted from many of the restrictions imposed by the act, but only in relation to the sale or manufacture of drugs within his own practice. The act, by virtue of the evidence required by the licencing authority for the granting of a product licence, and by virtue of the requirements concerning premises, equipment, staff and records for the granting of manufacturers' and wholesale dealers' licences, does much to regulate the safety, efficacy, quality and description of medicinal products.

Bibliography

BACQ, Z. M. (1971) *Fundamentals of Biochemical Pharmacology*. Oxford: Pergamon.

BURGER, A. (ed.) *Selected Pharmacological Testing Methods*. Vol. 3. Medicinal Research Series. London: Edward Arnold.

GOLDSTEIN, A., ARONOW, K. & KALMAN, S. M. (1974) *Principles of Drug Action*, 2nd ed. New York: John Wiley.

RANG, H. P. (ed.) (1973) *Drug Receptors*. London: Macmillan.

STEWART, G. A. (1963) Statistics as applied to pharmacological and toxicological screening. *Prog. medicinal Chem.*, **3**, 187–235.

TURNER, R. A. & HEBBORN, P. (ed.) (1971) *Screening Methods in Pharmacology*, Vol. 2. London: Academic.

UVAROV, O. (1972) The organisation and conduct of trials for assessing new therapeutics. *Vet. Ann.*, **13**, 316–20.

3

The Fate of Drugs in the Body

INTRODUCTION

When a drug is used in therapeutics, it is important to know that a certain period—the so-called *latent period*—will elapse before the drug will be seen to elicit its effect. Many factors contribute to this delay, among which are the rate of absorption from the site of administration, the ease with which the drug can move through the various body-water compartments, the rate at which the drug is inactivated in the body and the rate at which it is excreted from the body. These factors contribute to what is known as the fate of the drug within the body. From a knowledge of the fate of the drug it is possible for the physician, by considering dose size, dose frequency and route of administration, to maintain an effective concentration at the site of action for as long as is desirable. To this end he is assisted by the pharmacist who formulates the drug into an appropriate preparation and by the chemist who synthesizes new compounds whose molecular configuration will make them more suitable than existing drugs, e.g. in having an increased duration of action. In this chapter it is proposed to discuss those aspects of body function and those physico-chemical properties of the drug which determine the fate of the drug within the body.

ABSORPTION OF DRUGS

Drugs and Membranes

The rate of absorption depends upon the route by which the drug is administered, the way in which the drug is formulated and both the physico-chemical and pharmacological properties of the drug. When a local action is required, absorption from the site of application could be a disadvantage. However, when a systemic effect is required, absorption is essential. A major determinant of systemic usefulness is the ability of the drug to cross the various barriers which could prevent its diffusing throughout the entire body water. One important consequence of the choice of route of administration is that this determines the number and nature of the barriers which the drug has to cross and thereby determines the latent period of drug action. As the mechanisms by which drugs cross membrane barriers are in part dependent upon the physical properties of the barrier, it is proposed firstly to discuss the barrier, then the mechanisms of penetration and, lastly, the influence of the route of administration on absorption.

The Membrane Barrier

The barrier to drug penetration is usually regarded as being a cell wall. In the case of absorption from the gastro-intestinal tract the drug has to penetrate the cell membrane of the lining epithelium, cross the cell, pass through the basal membrane, enter the fluid of the lamina propria and finally enter blood or lymph capillaries via the vessel wall. We often describe the cell membrane as a double layer of lipoid molecules containing and also overlaid by proteinaceous material. The membrane is punctured by pores whose normal function is to allow the passage of hydrophilic, i.e. water-soluble material such as urea and potassium ions. These pores are generally about 0·4nm in diameter, but those of the glomerulus and capillary wall are much wider and thus allow larger molecules to pass. Drugs and normal cellular metabolites cross membranes either passively or actively.

Lipid Diffusion

Most drugs cross the cell membrane by simple diffusion through the lipoid material. The non-ionized form of the drug dissolves in lipid and diffuses down a concentration gradient into the aqueous phase beyond the barrier. Many drugs are either weak acids or weak bases and the non-ionized fraction is preferentially soluble in lipid, i.e. it has a high lipid–water partition coefficient. One practical consequence of these facts is that drugs which are weak acids, e.g. salicylates, are present almost entirely in the non-ionized state in the strongly acid pH of gastric juice and are thus absorbed very rapidly. Once across the gastric mucosa the drug encounters slightly alkaline tissue fluid and ionizes freely. As the rate of diffusion depends only on the differential concentration of the non-ionized forms on either side of the membrane, further diffusion is not impeded by the material which has already crossed because this material has, to a large extent, ionized.

Filtration

Lipid insoluble drugs and ions enter cells through the water-filled pores at rates which depend on molecular size and the concentration gradient. Both lipid diffusion and filtration are passive processes.

Active Transport

When molecules move across a membrane against a concentration or electrical gradient, a mechanism which actively transports the material is postulated. Such mechanisms occur for cell nutrients and these transport chemically related drugs. Active transport differs from diffusion in that it requires the expenditure of energy, is selective, is saturable and may be blocked by metabolic inhibitors, e.g. the sodium pump of nerve and kidney. In active transport systems, a specific carrier material is believed to unite with the substance to be transported, carry it across the membrane and release it at the membrane boundary.

Facilitated Diffusion

This differs from active transport only in that the carried material does not move against a concentration gradient. The rates of passage of the material are greater than is explainable by the laws of diffusion and the mechanism shares with active transport the properties of selectivity, saturability and susceptibility to blockade.

Pinocytosis

It has been shown that some cells can engulf macromolecules and transport them in vesicles. Pinocytosis can apply to large drug molecules.

ROUTES OF ADMINISTRATION

When the condition which requires therapy is on the body surface or is in an organ or cavity communicating with the surface, the drug can be introduced directly to the desired site of action and treatment is described as *local* or *topical*. When it is desired to produce an effect throughout the body, the treatment is called *systemic*.

Systemic effects are much dependent upon factors such as absorption and consequent distribution of the drug within the body, and they can be produced when the drug is given by the oral route or by injection. Parenteral administration means the use of any route other than the oral one, but in general usage it usually refers to injection. The most common routes of injection are intradermal, subcutaneous, intramuscular, intravenous, intraperitoneal, intra-articular and epidural.

The choice of the route of administration can be influenced by such practical factors as the nature of the formulation, the patient's temperament, economic factors and convenience both to the owner and veterinary surgeon. Additionally there are pharmacokinetic factors which are considered in the following discussion of the more common routes.

Oral

There are a large number of pharmaceutical preparations available for oral administration, e.g. mixture, drench, electuary, bolus, pill, tablet, powder or capsule.

In man certain drugs are absorbed from the *mouth* via the sublingual mucosa, e.g. nitroglycerine, but this route is impracticable in veterinary medicine. Its advantage is that the absorbed material does not pass through the liver before entering the arterial side of the circulation. This is of value for drugs which are very rapidly detoxified by the liver.

The *stomach* can delay the absorption of drugs in relation to the times of feeding, e.g. as the pylorus may be closed for some time following feeding, the absorption of drugs whose principal site for absorption is the small intestine would be delayed. Additional hazards are that irritant drugs may cause vomiting and that the low pH of stomach content, or bacterial action may destroy certain drugs, e.g. benzyl penicillin. Delays in absorption frequently occur in animals with gastro-intestinal disease or fever. Other hazards to oral dosing common both to stomach and *intestine* are that the drug may either bind to or dissolve in the contents, and thus fail to be absorbed, e.g. fat soluble vitamins, or may form an insoluble derivative, e.g. calcium phytate.

To be absorbed a drug has to be soluble either in the minute fat droplets or the aqueous phase of gut content. Truly insoluble compounds are not absorbed, e.g. the X-ray contrast medium barium sulphate. On the other hand, limited absorption can be utilized to maintain certain drugs in high concentration in the intestine, e.g. streptomycin is highly ionized and lipid insoluble.

As the pH of gastric content is usually in the range 1 to 3 and that of the intestinal contents is 7 to 8, it is to be expected that the rates of absorption of the same drug from both sites will differ widely. This difference depends on the pK_a of the drug. The pK_a is a factor which expresses the degree of ionization of a compound, being, for acidic compounds, the negative log of the ionization constant, K_a, and for basic compounds it is 14—pK_b. When the pH of a medium is so adjusted that the numbers of ionized and non-ionized molecules of a dissolved drug are equal, the pK_a is numerically equal to the pH. Thus, knowing the pK_a of a compound, it is possible to predict its pattern of absorption from the various compartments of the gastro-intestinal tract. When pH is higher than pK_a, an acidic drug will become more ionized while a basic drug becomes less ionized. The alkaloids are basic drugs and have pK_a values in the range 6 to 9. This results in their being highly ionized in and, consequently, poorly absorbed from the stomach. In the alkaline small intestine the ionization of weak bases is suppressed and, therefore, absorption of such compounds is rapid.

Investigations have shown that prediction of absorption rates must take into consideration the extent to which the non-ionized moiety is preferentially fat soluble, e.g. barbiturates have similar pK_a values and molecular weights, yet, differ in their rates of absorption. The difference is related to fat solubility, i.e. the more fat soluble members are the most rapidly absorbed.

On occasion, it has been shown that the expected rate of absorption of a drug is greater than that achieved on administration. An explanation for this discrepancy is that the rate of blood flow is inadequate to remove drug which has already crossed a membrane barrier and so diffusion ceases when concentrations of non-ionized material are equal across the membrane.

Parenteral Injection

Preparations designed for injection are always sterile and are frequently adjusted to body tonicity and pH. If inadequate systemic concentrations are achieved following oral administration, perhaps due to incomplete absorption or destruction in the gut, then the drug must be injected.

Parenteral administration offers the additional advantages of a more rapid and more reliable response, greater accuracy of dosage and is essential in a patient already vomiting or unconscious.

The rates of absorption of different drugs differ widely for the reasons already described, but in general, are slowest following subcutaneous injection, more rapid following intramuscular injection while onset of effect is usually immediate following intravenous injection.

Intravenous injection is used for emergency treatment and where a rapid effect is required when the drug is available in aqueous solution, e.g. calcium borogluconate for milk fever; when a drug is too irritant for administration by any other route; for very precise dosage control, e.g. sodium pentobarbitone anaesthesia; for long-term administration of drugs with transient action by an intravenous drip, e.g. heparin. The injection is usually made into the jugular vein in the horse, cow, sheep and goat. The use of the milk vein is not to be recommended owing to the increased chances of haematoma formation and increased exposure to contamination. In the pig the ear veins are accessible while in the dog and cat, the cephalic and recurrent tarsal veins are most suitable. All parenteral routes eliminate the requirement of a drug having to cross a mucosa as a first step in absorption. The use of the intravenous route eliminates the necessity of the drug having to cross the endothelial barrier.

An *intramuscular injection* is used for relatively irritant drugs, for drugs whose rate of absorption from subcutaneous injection would be inadequate, or for the administration of depot preparations, e.g. iron–dextran for piglet anaemia. In all species the gluteals and upper thigh muscles are used, while the neck muscle may be also used in the pig, horse and cow.

The *subcutaneous injection* route is chosen when a slow and continuous absorption of a drug is required, provided that the formulation approaches body pH and tonicity is not otherwise irritant or vasoconstricting. The injected material disperses through the loose connective tissue, and the drug must dissolve in tissue fluid before it can enter either capillaries or lymphatics. Lipid soluble drugs enter by diffusion while hydrophilic material can enter the relatively large aqueous pores of the capillary endothelium. The rate of absorption can be increased by enhancing drug dispersion either by massage of the injection site or by the inclusion of the enzyme *hyaluronidase* in the formulation. This enzyme hydrolyses the hyaluronic acid polymers which comprise the intercellular cement and thus facilitates diffusion through the tissues. Subcutaneous injections can be made at any point of the body surface where there is adequate loose skin and preferably in an area where body movement will aid dispersion, e.g. the axillary fold. Injection into fat depots is to be avoided because of poor blood supply. Following injection, the blood supply to the area can be increased and absorption enhanced by the application of warmth or vigorous massage. Such precautions are valuable following the subcutaneous infusion of large volumes of mineral solutions, e.g. magnesium sulphate.

The *intraperitoneal injection* route is of increasing importance in large animal practice for the administration of large volumes because of its great absorbing surface and because the absorption rate is rapid. The injection is made via the sub-lumbar fossa, care being taken to avoid delivering the solution into an abdominal organ.

Intrathoracic and intracardiac injections are occasionally made when performing euthanasia in small animals.

The *epidural injection* route is used in cattle at parturition when it is desired to eliminate straining. A local anaesthetic is introduced between the first and second coccygeal vertebrae. Anterior epidural anaesthesia is produced when the drug is given via the lumbosacral space.

Intra-articular injections are given when an anti-inflammatory preparation is delivered directly into a joint capsule.

Intrathecal injection involves penetration of the membranes which enclose the central nervous system. This route is occasionally used for radiographic purposes.

The rectal, vaginal and intramammary routes are all used when therapy is required in these regions. Special pharmaceutical presentations are used and it is important to remember that absorption into the systemic circulation can occur, e.g. penicillin can be detected in the blood following intramammary mastitis therapy.

Controlled Absorption of Injected Drugs

It is often desirable to delay the rate of absorption of a drug for a variety of reasons. The incorporation of a vascoconstrictor in local anaesthetic solutions both prolongs action and reduces toxicity. When absorption is slowed, it is possible to administer larger amounts of a drug, maintenance doses have to be given less frequently, and blood concentrations remain steadier. This can be achieved by administering the drug in the form of a less soluble salt, e.g. procaine penicillin rather than the crystalline salt, or the drug can be formulated in a water-repellent vehicle, e.g. the steroid hormone esters are often presented as solutions in oils. The ultimate in delayed absorption is the implantation of sterile tablets subcutaneously. Such preparations provide slow release of drug over several months and are best exemplified by the sex hormone esters. A newer alternative is the intramuscular injection of a poorly soluble micro-crystalline suspension.

External Application

In addition to producing a superficial effect, absorption can occur following application of certain drugs to the skin. The sebaceous secretions and keratinized epithelium of skin limit penetration to the lipid soluble drugs. The penetration of such drugs to the dermis is aided by formulation in fats or fat solvents. The absorption of some drugs is so great as to constitute a toxic hazard, e.g. organophosphate insecticides in cattle, and organochlorine insecticides in cats and so treatment of the whole body in one application may have to be avoided. Furthermore, it should be realized that drugs applied to broken skin could be absorbed much more rapidly, and that animals can lick off sufficient drug to constitute a toxic hazard. Drugs intended for superficial activity are usually applied as dusting powders or solutions while, when penetration is required, they are presented as ointments.

Inhalation

When a drug is a gas or vapour or can be presented as a dispersion of minute droplets in air, it can be inhaled and absorption is very rapid because of the enormous surface area offered by the respiratory surfaces. Volatile and gaseous anaesthetics and bronchodilators in man are routinely administered by this route.

Summary

Irrespective of the route of administration, if a drug is to be absorbed, i.e. to leave its site of administration and enter the blood stream, then it must be soluble in either oil or water. The rate of absorption depends upon the pH at the absorbing surface, the pK_a of the drug, its oil-water partition coefficient, its concentration, its rate of dissolution, the blood supply behind the absorbing surface and the area of the absorbing surface.

THE DISTRIBUTION OF DRUGS IN THE BODY

Once a drug has entered the bloodstream certain factors which tend to decrease its active concentration come into operation. These will be discussed under the headings of 'Protein-Binding', 'Binding at Other Sites' and 'Drug Dilution'. Via the circulation, the drug is able to gain access with varying degrees of success to all parts of the body. This dispersion which ends with penetration into the site of action is known as drug distribution.

Protein-Binding

This term refers to the variable and often significant portion of absorbed drug which becomes physically bound to plasma proteins. The active concentration of the drug refers to that part which is not bound because it is only this fraction which is free to leave the plasma and enter the site of action. There is a dynamic equilibrium between the bound and free drug fractions. When free drug leaves the circulation, bound drug is released to restore the balance. In this way the bound drug can be regarded as a storage depot, e.g. a single dose of the anti-trypanosomiasis drug, suramin, could give protection for up to 3 months. The relationship of free to bound drug is dependent upon total concentration, i.e. as the sites at which drug may be bound become increasingly occupied, the proportion of free drug rises rapidly when more drug is absorbed. A practical consequence of binding to plasma protein is that the toxicity of drugs which are normally highly bound will be greatly increased in hypoproteinaemia assuming that dosage is not reduced. Similarly, it is possible to increase the toxicity or efficacy of a highly bound drug by administering a second drug which has a greater affinity for the same binding sites.

Drug binding in the circulation is usually to plasma albumin but binding to blood cells does also occur. The bond is usually freely reversible and is of the ionic, hydrogen or van der Waals' type. Binding can take place in excess of 99 per cent, e.g .with dicoumarol, or can be virtually nil as with amino-antipyrine.

Binding at Other Sites

Certain drugs tend to accumulate at sub-cellular structures, e.g. the acridine dyes have recently been shown to enter *nucleoproteins* while many drugs exert their action at the *mitochondria*. Other drugs tend to be concentrated at their site of action, e.g. the mercurial diuretics in the *kidney*, iodine in the *thyroid*. On occasion the body is unable to distinguish between normal and undesirable agents, e.g. the sequestration of lead in *bones*. The antimalaria drug, mepacrine, leaves the blood in a matter of minutes and becomes fixed to various tissue cells, e.g. the *liver* concentration can reach 20 000 times that of the plasma. In recent years much attention has been focused on the persistence of organochlorine pesticides in body *fat depots*. Periods of starvation or inappetence could result in the release of such agents, with consequent toxicity problems. Dissolution in fat is a common occurrence with highly lipid soluble compounds, e.g. the volatile anaesthetics. The fact that griseofulvin is deposited in *keratinizing epithelium* is the basis for its value in the treatment of ringworm infection.

Drug Dilution

By no means all drugs diffuse throughout the total body water. The *volume of distribution* of a particular drug refers to that portion of total body water which it is capable of entering. Some drugs are unable to leave the plasma, e.g. the plasma expander dextran sulphate; others are limited to the extracellular fluid, e.g. sodium iodide, while others distribute so rapidly throughout the total body water that they can be used to estimate its volume, e.g. amino-antipyrine, an antipyretic analgesic. When a drug leaves the plasma to enter other body fluid compartments, its blood concentration will tend to fall. When absorption ceases, this tendency can no longer be averted.

Physico-Chemical Factors and Drug Distribution

The movement of absorbed drug out of the circulation into the extracellular fluid and thence into the site of action is dependent in terms of rate and extent upon the same factors as those already described as determining drug absorption. The fact that most drugs tend not to be distributed in equal concentration throughought the total body water is an indication of the operation of these factors, the more important of which are the presence of an active transport mechanism, the pH differential between the aqueous phases, the pK_a and lipid solubility of the drug and the presence of binding sites.

Fig. 5 shows how a drug moves from its site of administration to its site of action. The lipid-soluble, non-ionized fraction is shown to be in equilibrium between all the compartments of distribution. This should not confuse the fact that although there is a free to bound equilibrium within each compartment, the total concentration of drug can differ widely between the compartments. There can be significant pH differences between compartments and these would allow widely differing ionized to non-ionized equilibria. As there is an equal concentration of non-ionized drug in each compartment, then the pH effect is seen as widely differing concentrations of ionized drug from one compartment to the next. In this way, the total

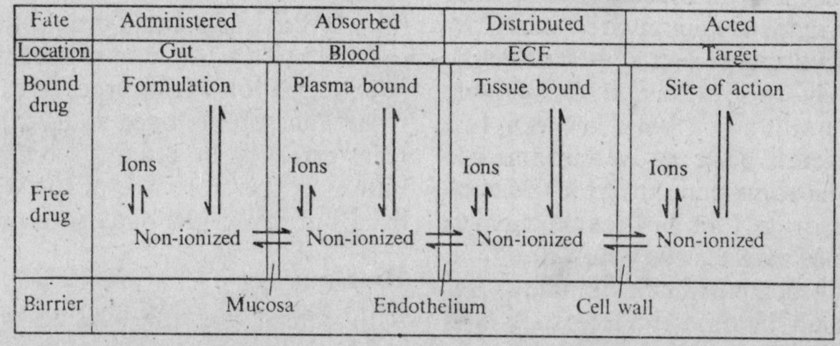

FIG. 5. A summary of the drug equilibria which exist both within and between the various compartments of body water. In this example the drug has been given by mouth and its fate is followed up to its entry into its site of action.

amount of free, i.e. active, drug (ionized + non-ionized) can vary between the compartments.

Such an effect is well exemplified when the pH difference between two phases is considerable, e.g. between stomach (pH 2) and extracellular fluid (pH 7). A weak acid, pK_a 4, would exist almost entirely in the non-ionized state in the stomach, and would be highly ionized in the ECF, see below:

Stomach: $[HA]/[H^+][A^-] = 1 : 0{\cdot}01$
Concentration = 1·01

ECF: $[HA]/[H^+][A^-] = 1 : 1000$
Concentration = 1001

This prediction is based on the Henderson–Hasselbalch equation which, for an acid, can be expressed as:

$$\log(\text{Non-ionized}/\text{Ionized}) = pK_a - pH$$

The wide pH differential in the above example gives a 1000-fold difference in concentration across a membrane. Although other situations provide smaller pH differentials, it is generally the case that acidic drugs tend to accumulate in phases of high pH while basic drugs tend to localize in phases of low pH.

The binding sites in the compartments of distribution also exert an effect on the total amount of drug to be found in each compartment when equilibrium exists between the compartments. By this means, it is possible for drug concentrations to vary between two compartments even if both are at the same pH and thus have the same concentrations of non-ionized and free drug. This is because there is usually a difference between the binding abilities of the individual, compartmental binding sites which would result in different free to bound ratios from one compartment to the next.

The remaining factor which can cause an unequal distribution of a drug between two compartments is the presence of a suitable active transport system in the dividing membrane.

Specialized Barriers

Beyond the normal factors which interact to govern distribution are certain functional barriers which possess the ability to exclude certain substances. The blood–brain barrier is one example, but further barriers occur in the placenta and in the secretory epithelia.

The blood–brain barrier. This barrier exists between the plasma and the extracellular space of the brain, while the *blood–cerebrospinal fluid barrier* occurs in the choroid plexus. These barriers can be regarded as being protective in function, as nutrients usually penetrate with ease. In general, lipid-insoluble and highly ionized drugs penetrate the brain very slowly, while lipid soluble agents enter very rapidly, e.g. the volatile anaesthetics. Certain drugs have great difficulty in passing from the plasma into the brain, e.g. penicillin and iodides.

The placenta. This barrier is also of clinical importance, e.g. in choosing an anaesthetic for caesarean section. Lipid soluble drugs diffuse across the placenta with ease and most anaesthetics will therefore cause some respiratory depression in the newborn.

Re-Distribution

This phenomenon is well illustrated by a consideration of the pharmacokinetics of thiopentone. When this agent is given intravenously, it diffuses rapidly into the highly-vascular, lipid-rich CNS and brings about immediate anaesthesia. The blood–brain equilibrium becomes disturbed when the drug also enters body fat. At this stage some drug diffuses back from the CNS to the blood to restore the equilibrium. A continuation of this process results in recovery from anaesthesia once sufficient anaesthetic is transferred from brain to body fat. Redistribution is therefore one of the ways by which a drug's action in the body is terminated.

Drug Storage

In considering the binding of drugs, it was shown that drug concentration can vary considerably in different sites. This accumulation may occur at the site of action, e.g. chlorpromazine in the brain, or the drug may simply become sequestered far from its site of action, e.g. phenylbutazone is an anti-inflammatory agent but is very highly protein bound. These two examples show that the blood concentration can give misleading impressions both of the concentration of a drug at its site of action and of its duration of action. Storage of drug away from the site of action has to be considered in the calculation of dosage. It is important to provide enough drug to saturate body stores and still have sufficient free drug to produce the desired effect. When this process is achieved by repeated dosage, as in the induction of anaesthesia with pentobarbitone, it is essential to recognize the approach of saturation because at this point, the concentration of free drug will rise rapidly with a consequent dramatic increase in effect.

Summary

The ability of a drug to leave the bloodstream and enter other body compartments depends on the drug's lipid solubility and pK_a, the extent to which it becomes protein bound or the existence of a suitable active transport system. The concentration which a drug can achieve in the entered compartment depends on the difference in pH across the separating membrane, the different binding affinities across the membrane, the existence of a suitable transport system or the presence of specific barrier in the separating membrane.

THE METABOLISM OF DRUGS IN THE BODY

The various enzymatically mediated chemical changes which a drug undergoes in the body before final elimination from the body are collectively known as its *metabolism* or *biotransformation.* These changes usually result in a diminution of activity or a reduction in toxicity of the administered compound and are then called *detoxication.* Occasionally, biotransformation can yield an intermediate of increased activity or toxicity. Some drugs are inactive or of a low order of activity in the form administered and are metabolized to the active form in the body. It is generally true that the rate at which a drug is metabolized is concentration dependent.

The enzymes which carry out drug metabolism are found throughout the body, but are abundant in the liver. The frequency of liver damage as a manifestation of toxicity indicates that the liver often removes from the blood agents which it cannot detoxify. The enzymes involved in drug metabolism can be those which normally handle the products of body metabolism or may be specific drug metabolizing enzymes. Drug detoxifying activity can be found in kidney and lung, blood and gut content as well as in the liver, e.g. plasma pseudocholinesterase inactivates procaine, the local anaesthetic.

Liver Microsomes

These are sections of hepatocyte endoplasmic reticulum and are obtained by ultracentrifugation of liver homogenate. The endoplasmic reticulum is known to be continuous both with the nuclear and cell membranes and has been shown to be the seat of drug-metabolizing enzymes. The membraneous nature of the reticulum means that lipid-solubility is a requisite for a drug if it is to gain access. Electron microscopy has revealed that the appearance and number of the microsomes are influenced by factors such as drugs and hormones, age and sex, stress and temperature, nutritional status and pathological state. Microsomes have a need for NADPH and oxygen.

Many of the differences which exist between *species* in respect of drug toxicity can be traced back to enzymic differences either qualitative or quantitative in character, e.g. cats lack glucuronide-forming enzymes.

Repeated administration of a barbiturate results in a diminishing response, believed to be due to an increase in barbiturate metabolizing enzymes. This is *enzyme induction* and is an explanation for the existence of drug tolerance. Several drugs have been shown to be able to increase microsomal activity, others to decrease it.

Sex can be shown to influence the rate of drug metabolism in the rat where the female metabolizes drugs more slowly than the male—a depressant action of oestrogens.

The effect of *age* is well exemplified by the increased toxicity of chloramphenicol in the young pig because of its low glucuronyl transferase activity. Many newborn or foetal animals have low drug-metabolizing ability.

The reduction of the lethal dose of amphetamine in cooled mice is an example of the effect of *environmental temperature* on enzyme activity.

The importance of *nutrition* is exemplified by the increased incidence of carbon tetrachloride-induced liver damage in under-fed sheep. Organ *pathology* is also important as any degree of dysfunction will decrease the organ's drug detoxifying ability, e.g. the increased duration of barbiturate anaesthesia in dogs with liver and kidney damage.

Biotransformations

The types of reaction found are oxidation, reduction, hydrolysis and conjugation. Conjugation, which is the coupling of the foreign substance with an endogenous substrate, is usually a last step in detoxication and almost always results in loss of biological activity. Conjugation may, however, be preceded by one or more stages involving oxidation, reduction or hydrolytic cleavage which may result in activation, change of activity or inactivation of a drug.

These reaction types each contain several mechanisms by which various classes of compounds become modified, e.g. oxidation may be achieved by hydroxylation of aromatic compounds or alkyl groups, oxidation of alkyl groups, sulphoxidation, the exchange of sulphur for oxygen, *O*- and *N*-dealkylation, oxidative deamination or dehydrogenation depending on the molecular configuration of the substrate. In general, however, detoxication yields a metabolite more polar than its predecessor. Although most drugs cross the glomerular membrane freely, their lipid-solubility allows them to leave the tubule and re-enter the peri-tubular vasculature, but their more highly-ionized, hydrophilic metabolites are less easily reabsorbed and are therefore more easily excreted. This alteration of the drug's physico-chemical characteristics also reduces its ability to penetrate other membranes and decreases its tendency to become protein bound or dissolved in body fat. Drugs which are highly ionized, e.g. decamethonium, or which possess biochemically unreactive groups, e.g. phenobarbitone, often fail to undergo chemical change within the body. Very few drugs undergo complete degradation within the body and so for most drugs some part of the dose is excreted unchanged. Forensic analyses usually rely on demonstrating the presence of unchanged drug.

The examples given in Table I by no means exhaust the known drug biotransformations and although all except one (the reduction of chloral hydrate is an activating biotransformation) are one-stage inactivations, it should be noted that many drugs suffer a multiple-stage biotransformation. Additionally, the one substrate may be subject to more than one degradation pathway in which case a variety of metabolites may be found in the urine, e.g. imipramine yields fourteen metabolites in the rat.

Summary

The factors which determine a drug's distribution determine also the enzymes to which it will be exposed and, to some extent, its degradation pathway. Liver microsomes are the major site for biotransformations of lipid-soluble drugs. As soon as free drug reaches a detoxication site, the body total of active drug starts to fall. Detoxication is a major mechanism for lowering the concentration of active drug in the body.

Table I. Examples of drug biotransformation reactions

Reaction	Substrate	Metabolite
Oxidation (hydroxylation)	Pentobarbitone	Hydroxypentobarbitone
Oxidation (deamination)	Adrenaline	Dihydroxymandelic acid + methylamine
Oxidation (sulphoxidation)	Chlorpromazine	Chlorpromazine sulphoxide
Reduction (aldehyde to alcohol)	Chloral hydrate	Trichloroethanol
Reduction (cleavage and reduction of azo group)	Prontosil	Sulphanilamide
Hydrolysis	Procaine	*para*-Aminobenzoic acid + diethylaminoethanol
Hydrolysis	Acetylcholine	Acetic acid + choline
Conjugation (*N*-acetylation)	Sulphanilamide	Acetylsulphanilamide
Conjugation (glucuronic acid)	Chloramphenicol	Chloramphenicol glucoronide
Conjugation (*O*-methylation)	Adrenaline	Metanephrine

THE ELIMINATION OF DRUGS FROM THE BODY

Drug clearance from the body follows two major pathways, viz. excretion of unaltered drug or preliminary metabolism followed by excretion. The main organ of excretion is the kidney, which eliminates water soluble material, followed by the liver, which eliminates metabolized material via the bile, while volatile agents are excreted through the lungs. Small quantities of drug can be detected in secretions such as saliva, milk and sweat, and these may be regarded as further routes of excretion. A small number of substances are neither metabolized nor excreted but become fixed by tissues with the consequent risks of tissue damage and cumulative poisoning. All the foregoing mechanisms contribute to the lowering of plasma free drug concentration, the effectiveness of which is called *clearance*.

Factors Affecting Excretion in Urine

Glomerular Filtration

If a substance is in solution and has a molecular weight below about 68 000, then it will appear in glomerular ultrafiltrate in direct relation to the plasma concentration of its free form. Drug bound to plasma protein is not filtered and this is another mechanism by which binding prolongs a drug's stay in the body.

Tubular Reabsorption

When the concentration of lipid-soluble, non-ionized material in the tubule rises, as a consequence of water reabsorption, the drug will tend to diffuse back through the tubular wall, into the renal interstitium. In alkaline urine, acidic drugs ionize and are therefore less well absorbed than would be the case if the urine were acid and the drug non-ionized. The opposite situation applies to basic drugs. Intentional manipulations of urinary pH can be used to favour the excretion or retention of drugs and poisons.

Tubular Secretion

In the proximal convoluted tubule are active transport systems which carry ionized acids and bases from plasma into ultrafiltrate. These systems operate both for endogenous substrates, e.g. uric acid and choline, and drugs, e.g. glucuronic acid conjugates and quaternary ammonium compounds. Penicillin is carried by the acidic transport system and is very rapidly lost into the urine. When penicillin production was limited, it was customary to extend its duration of action by the concurrent administration of caronamide, a drug which competed with penicillin for the same tubular transport system.

Renal Blood Flow

Just as efficient haemodynamics are essential for normal kidney function, so they influence the excretion rates of drugs.

Solubility in Ultrafiltrate

Hydrophilic drugs are often excreted in the urine unchanged, whereas lipid-soluble drugs may undergo metabolism to a more water-soluble form before they can escape in the urine. Occasionally, as is the case with acetylsulphanilamide, a metabolite is less soluble than was the parent drug in the concentrated, acidic ultrafiltrate of the proximal convoluted tubule. There is then a risk of precipitation of drug in the tubule with consequent impairment of renal function. This problem is avoided by alkalinization of the urine, unrestricted water intake, and the use of sulphonamide mixtures. These allow the same therapeutic effect at a reduced dose of each sulphonamide in the mixture and hence a reduction in concentration of the individual sulphonamide in the ultrafiltrate. The possibility of precipitation depends on the concentration of each member, not on the total sulphonamide concentration—the principle of independent solubilities.

Further Excretory Pathways

Excretion of Drugs into the Gut

The quantity of a drug which, following oral administration, can be recovered from the faeces is comprised of unabsorbed drug and drug which has re-entered the gut. Drugs gain access to the intestinal tract via its wall, its secretions and the bile, either by diffusion or active transport. Active transport is largely confined to drug or metabolite entering via the bile, e.g. penicillin, bilirubin, glycocholic and taurocholic acids. Diffusion can take place through the whole gut surface dependent on the usual considerations, i.e. lipid-solubility and pK_a, pH and concentration gradients across the mucosa. Certain metals are excreted into the large gut, e.g. calcium.

Entero-Hepatic Circulation

Some drugs are excreted via the bile as glucuronide conjugates. The conjugate is usually reabsorbed from the gut and excreted in the urine as is the case with the steroid hormones. Sometimes the conjugate can be broken in the gut by the enzyme β-glucuronidase and the liberated drug can be reabsorbed to act again. In this way, the action of a single dose of a drug can be extended, e.g. the purgative phenolphthalein acts in man for several days. As some of the drug is lost at each circulation the decline in effect is gradual.

Excretion into Milk

The fact that antibiotics can appear in cows milk following both parenteral injection and local instillation has lead to anxiety over the possibility of health hazards to man, e.g. induction of sensitivity to penicillin. In the United Kingdom there is now a permitted maximum antibiotic activity in milk and suppliers who exceed this limit are liable to penalty. To avoid these problems, milk from antibiotic treated cows should not be included in town supplies for a period which will depend upon both the drug used and its formulation. Another hazard is the excretion of radioactive strontium as a consequence of nuclear fallout having contaminated pastures. Cows milk is usually slightly acid relative to plasma and therefore tends to concentrate basic, fat soluble drugs.

Excretion Via the Lungs

This route is of major importance for the volatile anaesthetics. The large surface area presented by the alveoli together with the pulmonary vasculature allow very rapid equilibration of volatile agents between blood and alveolar air. This is well exemplified in the rapid recovery from an anaesthesia induced by gaseous or volatile agents. Some drugs are at least in part oxidized to carbon dioxide and this fraction is then excreted via the lungs.

Summary

Excretion is most rapid when a drug or its metabolite is present in the blood in an ionized, highly polar form as it is then only poorly reabsorbed from tubular ultrafiltrate. The excretion of lipid-soluble material is influenced by pH and concentration differentials across the kidney tubule wall. The excretion of some drugs is greatly accelerated by active transport systems. Drug elimination depends on the above factors which operate at the excretory surface, the maintenance of a good blood supply to the kidney and the extent to which the drug is bound.

Termination of drug action is brought about by metabolic inactivation, storage in the body other than at the site of action and simple excretion. The process begins as soon as absorbed drug enters the circulation.

Bibliography

di Carlo, F. (ed.) (1974) *Drug Metabolism Reviews*, vol. 2. New York: Marcel Dekker.

LaDu, B. N., Mandel, H. G. & Way, E. C. (1971) *Fundamentals of Drug Metabolism and Drug Disposition*. Baltimore: Williams and Wilkins.

Notari, R. E. (1975) *Biopharmaceutics and Pharmacokinetics, An Introduction*, 2nd ed. New York: Marcel Dekker.

Rabinowitz, J. L. & Myerson, R. M. (ed.) (1971) *Absorption Phenomena. Topics in Medicinal Chemistry*, vol. 4. London: Wiley Interscience.

4

Principles of Pharmacotherapeutics

PLASMA DRUG LEVELS

The various facets of a drug's life in the body which were discussed in the preceding chapter have now to be related to the use of the drug as a therapeutic agent. Selection of the route of administration can now be seen as a potent factor in determining the rate of absorption of the drug, the latent period before drug effect is seen, the maximal attainable blood concentration and the rate of clearance and, consequently, the maximal obtainable effect for the dose selected and its duration of action.

Latency

One can usually relate the onset of drug action to a plasma concentration peculiar to that drug. The likely overall patterns of plasma drug concentrations following the administration of the same dose by intravenous, intramuscular or subcutaneous injection are illustrated in Fig. 6. The diagram also demonstrates the effect of the route of administration on the length of the latent period, the maximal attainable concentration and the duration of action. The dissimilarities between the three curves are due to the combined effects of the different rates of absorption and clearance of the drug referable to the different routes of administration. The latent period is seen as the time interval between completion of administration and the point at which drug concentration at the site of action is high enough for the drug to be able to exert its characteristic effect. Latency is minimized by using the intravenous route.

Duration of Action

The effect which a drug can produce directly

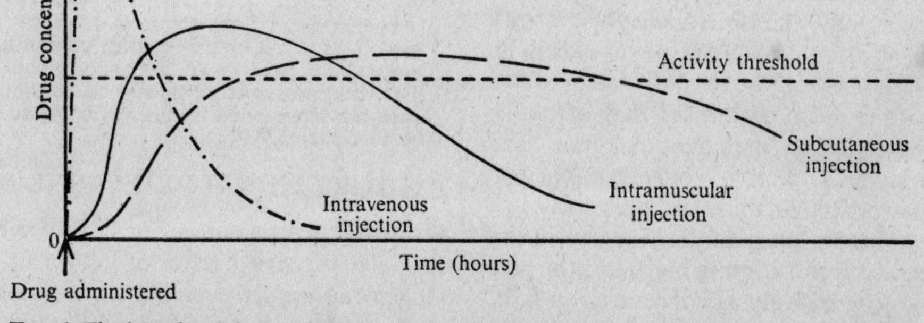

FIG. 6. The interdependence of the route of administration of a standard dose of drug and the latent period, the maximal attainable concentration and the duration of action. The latent period is the interval between drug administration and the crossing of the activity threshold. The period of drug action ends when concentration falls below the threshold.

is limited to those tissues which it is able to enter in its distribution phase. At this point, storage or binding will protect the drug from biotransformation or excretion and so prolong its duration of activity. On the debit side, binding or storage will lower the concentration of free drug and so reduce the intensity of effect and, following re-distribution, may terminate drug action.

The duration of drug action is clearly shown to be dependent on route of administration in Fig. 6. Following intravenous administration, the whole of the dose is present in the bloodstream and is immediately available for excretion. The blood concentration at any point in time will reflect the rate of clearance. When the drug is given in the same dose either subcutaneously or intramuscularly, then absorption is continuous and the blood level will reflect a balance between the rate of absorption and rate of clearance. When absorption is gradual, then rate of excretion is reduced because the concentration of drug in the blood does not reach the same high level which follows intravenous administration. Together, gradual absorption and reduced clearance result in the increased duration of action which follows both intramuscular and subcutaneous injection.

Drug Clearance

When deciding on the frequency with which further doses of a drug are to be given, it is clearly important to know the extent to which residual quantities of previous doses remain in the body. Drug concentration in blood can usually be estimated and, borrowing from isotope terminology, the time taken for a drug's initial concentration to be halved is called its *half-life*. In most cases, the rate of loss of active drug follows the same pattern as that of radioisotope decay, i.e. an exponential curve. The rate of loss depends on blood concentration, as a constant proportion of the drug present is lost in unit time. There is a rapid fall in concentration when plasma concentration is high, but the rate of fall slows progressively as concentration falls. Clearance is a composite parameter which depends upon the rate of enzymic inactivation of the drug, the extent to which the drug becomes protein-bound or otherwise stored in the body and the rate at which the drug is excreted from the body.

The concentration of drug in plasma is frequently measured as an indicator of the duration of a drug's useful life in the body and it should now be apparent that forces which tend to alter this level are constantly in operation. These forces can be subdivided into those which tend to increase drug concentration—principally absorption, activating biotransformations and release of bound drug, and those which tend to decrease the concentration—principally inactivating biotransformations, storage in the tissues and excretion. The interplay of those forces which determine plasma drug concentration is schematically represented in Fig. 7.

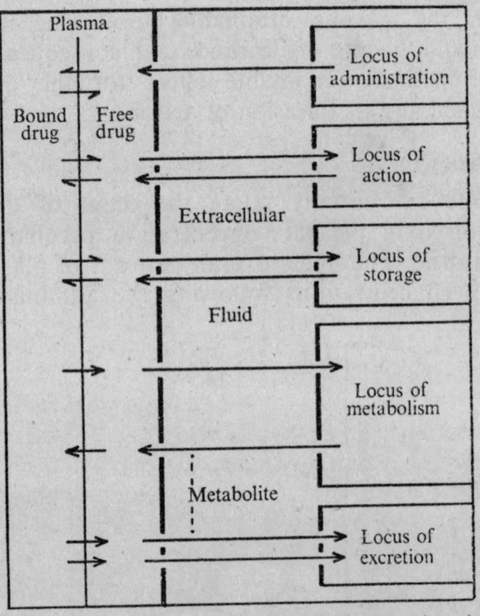

FIG. 7. The plasma free drug concentration is always the product of a dynamic equilibrium. The drug movements which tend to modify this state are represented in this figure by the arrows.

FACTORS WHICH DETERMINE DOSAGE

A satisfactory response can be expected only if the drug reaches its site of action in adequate concentration. In deciding on the quantity of drug to administer the factors already discussed in the previous chapter have to be considered along with the other influences now to be described.

Body Size

Dosage is usually related to body weight as this is a convenient indicator of volume of distribution. Within a species, this is a practical way of achieving parity of tissue concentrations between individuals. However, allowance has to be made for fatness of the individual. When the drug does not penetrate fat, then no dose increase is necessary, but when the drug is fat-soluble, then increased dosage is necessary to allow for that portion of the dose which will become sequestered in the body fat. This problem is exemplified by the susceptibility of the fat-free greyhound to thiobarbiturate overdosage.

Genetic Factors

Certain families or races within a species may be unduly sensitive to drug action. This can sometimes be explained by the absence of a particular enzyme, e.g. the absence of glucose-6-phosphate dehydrogenase in Negroes increases the toxicity of the antimalarial drug primaquine. Such anomalies have lead to the emergence of the new science of pharmacogenetics. When the response to drug administration is quantitatively or qualitatively abnormal, *idiosyncrasy* is said to have occurred, e.g. as occasionally happens in sheep with the fasciolicide carbon tetrachloride and which necessitates test-dosing of a small group from the flock. Idiosyncrasy is sometimes explainable on a genetic basis. *Susceptibility* usually describes a quantitatively abnormal response and would be demonstrated by a *hyper-reactor*. A *hyporeactor* is unusually resistant to a drug.

Species Variation

Within the domesticated animals are several examples of extreme resistance or sensitivity to drugs, e.g. cats are very sensitive to phenolic disinfectants, organochlorine pesticides and are excessively stimulated rather than depressed by morphine. Such variation militates against the direct extrapolation of dosage from species to species on simple body weight-derived factors. Drugs for veterinary use should be properly evaluated in any species in which they are to be used.

Alimentary Anatomy

This is a special example of a species-based consideration and is directed at the ruminant, whose rate of passage of digesta is very slow and whose volume of gut content is large relative to overall weight. Another problem is that an oral dose can enter different compartments depending on the operation of the oesophageal groove.

Age

Very old and very young animals usually require a reduced dose because of organ dysfunction. In the old this is usually degenerative change in liver or kidneys and in the young it is because full function has not developed, e.g. the toxicity of chloramphenicol in baby pigs.

Sex

Pregnancy is a contra-indication to the use of certain drugs, e.g. the purgatives may induce abortion. The thalidomide tragedy has added a new dimension to this consideration and examination for teratogenicity is now a routine part of the evaluation of a new drug. The excretion of some drugs into the milk is another example of a sex-based toxicity hazard. The rates at which certain drugs are broken down differ between the sexes, e.g. the sex hormones.

Timing

Orally administered drug is absorbed more rapidly and completely when the upper digestive tract is empty, but is often more irritant then. Hence the common directions of 'give the medicine 1 hour before (or after) a meal'.

Pathology

The sick animal usually has an impaired drug detoxicating ability. Any condition directly affecting the liver and/or kidneys is especially important, e.g. liver damage would prolong the action of pentobarbitone.

Tolerance

When a drug is administered regularly over a period of time, it is often necessary to increase the dose in order to maintain a steady level of response. The importance of the liver microsomes in tolerance to barbiturates has already been described. Tolerance to a drug usually disappears on withdrawal of the drug. Dogs can become tolerant to the sleep-inducing effect of barbiturates, but tolerance is rarely a problem in veterinary medicine as prolonged periods of drug administration are rarely undertaken. Resistance to the action of certain drugs can occur for other reasons, e.g. when a drug is antigenic, specific antibodies can be produced and these will inactivate the drug, e.g. insulin. Resistance of bacteria to the antibiotics is an important problem which is fully discussed in Section III.

Temperature

The temperature dependance of the lethal dose of amphetamine in mice has already been mentioned, but another example is available in those drugs which have to volatilize in order to act, e.g. the volatile anaesthetics. The rate of volatilization of chloroform is an important variable in the rate of induction of anaesthesia in the open air.

Indication

This consideration is pure therapeutics and includes dose adjustment with reference to the nature of the disease and in particular with reference to the causal agent, e.g. the treatment of acute fascioliasis requires higher dosage rates than the chronic condition. Very often the use of elevated dosage rates means an assessment of the increased toxic hazard against the increased benefit. Certain antibiotics are so toxic that their systemic administration is undertaken only in emergency situations, e.g. polymyxin. Other situations require a careful assessment of drug dosage in relation to desired effect, e.g. barbiturates are capable of inducing levels of CNS depression ranging from sedation to euthanasia.

Bioavailability

As knowledge has advanced so it has been applied within the drug sciences, with the result that by a combination of initiative, ethics and the force of law, the quality and reliability of remedies has been continuously improved. This process began with the chemical identification and quantitative determination of active principles and was continued with the establishment of standards of purity. More recently many countries have adopted rigorous testing procedures for establishing the safety and clinical effectiveness of medicines. For much of this time it has been believed that the equivalent products of differing manufacturers would be equally effective clinically if administered so as to supply the same dose of active principle. Unfortunately this is not always the case and the more extensive use of pharmacokinetic studies in recent years has revealed the reason. It is that between apparently equivalent formulations there may be differences both in terms of the percentage of the active principle which is released and in terms of the rates at which the active principle is released. These variations are reflected in differing maximal plasma concentrations of drug and differing durations of drug presence in plasma. Obviously such variations will result in differing intensities and durations of drug effect and hence in clinical usefulness. There is said to be a difference between the products in respect of *bioavailable dose*, i.e. the dose obtained by the subject as opposed to the notional dose contained in that quantity of product delivered as a dose. *Biopharmaceutics* is a specialization which is concerned with the study of factors which influence bioavailability and with controlling the tendency of such factors to complicate drug therapy. It is much applied to comparing the *bioequivalence* of competing products.

The drug manufacturer can only manipulate those factors which influence the bioavailability of active principle from his product and it is to be anticipated that current drugs regulations will ensure bioequivalence. It remains with the clinician to use his knowledge of the ways in which other factors may influence bioavailability, e.g. inter-actions with other drugs or dietary components. Despite the existence of books devoted to bioavailability it remains for the clinician just one more factor which can influence his choice of product, dose size and dose frequency. It is for him to make those ultimate decisions and his monitoring of effect is his most useful guide to the adequacy of his choice.

MULTIPLE DRUG THERAPY

The use of a mixture of drugs is liable to introduce further variables into dosage computation because of the possibility of a variety of interactions between the components and the patient. Before the potent and specific agents now in use became available, it was common practice to use remedies which contained a large number of components. Nowadays, the use of a number of drugs is justified only if this achieves greater efficacy or safety. The use of so-called 'shot-gun therapy' is a poor substitute for skilful and accurate diagnosis. The following sections will deal with some of the hazards and concepts peculiar to multiple drug therapy. The discussion under 'Receptor Theory' in Chapter 2 (p. 12) provides the theoretical background.

Increased Response

In order to reduce the incidence of drug toxicity, two or more drugs which can elicit the same type of response may be administered. If the final response is quantitatively equal to the sum of the expected responses had the drugs been used singly, then *summation* is said to have taken place. Should the actual response be greater than is explainable on the basis of simple summation, then *synergism* has occurred. Summation is exemplified by the use of streptomycin and dihydrostreptomycin mixtures and implies that the component drugs share the same mode of action. Synergism is exemplified by the use of penicillin and streptomycin mixtures. One explanation for potentiation of action is that two drugs may act sequentially to produce the same effect, e.g. pyrimethamine and a sulphonamide sequentially inhibit the synthesis of nucleic acids in coccidia.

Decreased Response

In multiple drug therapy it is also possible that the actual response is less than the sum of the component responses, in which case there has been *antagonism* between the drugs used. Antagonism can sometimes be explained on the basis of one drug's interfering with or even reversing the action of the other, i.e. it is often dependent on a mechanism involving pharmacological or physiological incompatibility. The use of a mixture which contains a bacteriostatic sulphonamide and a bactericidal antibiotic exemplifies this possibility in that the greatest antibacterial effect from penicillin is achieved when the organism is multiplying rapidly. A sulphonamide would arrest cell division and so could reduce the usefulness of the antibiotic.

Incompatibility

This is basically a pharmaceutical problem. but can be included conveniently at this point. The components of a mixture may be physically incompatible, e.g. oils and water, or chemically incompatible when they would react together, e.g. acids and alkalis or oxidizing and reducing agents. Of greater relevance is the question of physiological or pharmacological incompatibility of drugs given sequentially.

Increased Toxicity

This could arise on several counts. Two drugs both of which use the same excretory pathway would be metabolized at a reduced rate and, therefore, would be potentially more toxic. Pharmacological incompatibility is exemplified by the use of adrenaline as cardiac stimulant in an animal anaesthetized with a drug which sensitizes the heart to the action of adrenaline, e.g. cyclopropane.

Decreased Toxicity

Human patients suffering from atrial fibrillation often receive treatment with digitalis glycosides before receiving the heart muscle depressant quinidine sulphate as this reduces the incidence of acute congestive heart failure. A more commonplace example is the use of a premedicant tranquillizer before inducing anaesthesia. This simplifies the process of induction and reduces the dose of barbiturate necessary, a valuable gain in the poor-risk subject. The possibility of sulphonamide metabolites coming out of solution in the kidney tubules is much reduced by using sulphonamide mixtures rather than one of the components at equivalent dosage.

FACTORS WHICH DETERMINE DOSE FREQUENCY

In the treatment of disease, the initial problem is to achieve an adequate response and this has been shown to depend on an adequate concentration of drug reaching its site of action. An adequate therapeutic effect often demands that the drug act over an extended time course. A satisfactory duration of action can sometimes be achieved by techniques already discussed, such as formulating the drug in such a way as to ensure a slow release of active principle, or the use of an implant. However, it is more commonly necessary to extend the duration of action by administering the drug repeatedly.

Drug Clearance

The factors which operate so as to necessitate repeat dosing have already been discussed. The more efficiently a drug is cleared from the body, the shorter will be the interval between doses. An initial loading dose followed by smaller, daily maintenance doses is a regimen commonly used in therapy with sulphonamides and is illustrated in Fig. 8(a). With this method, the first dose is aimed at fulfilling therapeutic requirements while the subsequent doses merely replace that proportion of the loading dose which is lost through binding, metabolism and excretion. In this way the plasma concentration is maintained at a level compatible with a continuous bacteriostatic effect.

When clearance is also aided by the storage of drug in the body, including binding to plasma proteins, then eventual saturation of the storage sites will lead to a rapid increase in blood concentration unless dose size or frequency is carefully adjusted. If, in addition, either the degradation or the excretion of the drug is slow, the situation becomes even more critical. Binding to plasma protein would of itself depress the rates of inactivation and excretion. If these factors operate in such a way that one dose of a drug is not eliminated before the next is administered, then the concentration of drug in the body will increase with each successive dose. This process is known as *cumulation* and is represented schematically in Fig. 8(b). The significance of cumulation is that the concentration of the drug might rise to toxic levels. Cumulative toxicity is a particular problem with compounds which have a long half-life, e.g. thyroxine, digitalis glycosides, lead, copper and the organochlorine pesticides.

Stability of Tissue Concentration

In addition to the factors previously discussed, frequency of dosing can also exert a considerable effect not only on the duration of drug action but also on the character of drug effect. The action of the sulphonamides, for example, depends on their being present at their site of action not only for an adequate period but also at an adequate concentration throughout that period. This is in distinction to a drug like penicillin which needs to be present at an adequate concentration only intermittently throughout its period of use. In general, greater stability is achieved when the total daily dose is given as several divided doses throughout the 24 hr rather than as a single daily dose. However the onset of action is slower when divided doses are used as a longer interval elapses before a suitable drug concentration is achieved at the site of action. This delay may be avoided by increasing the size of the first dose.

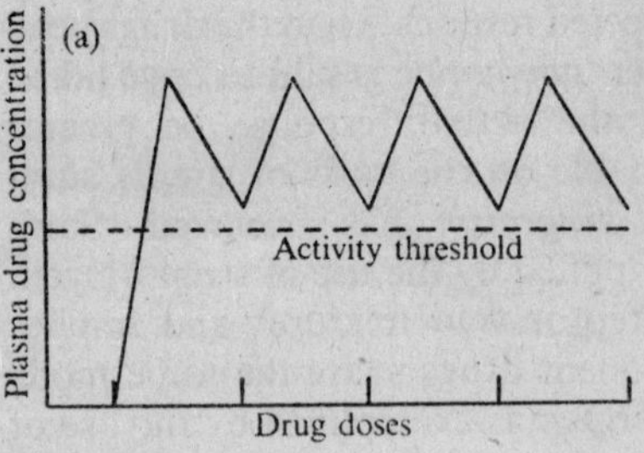

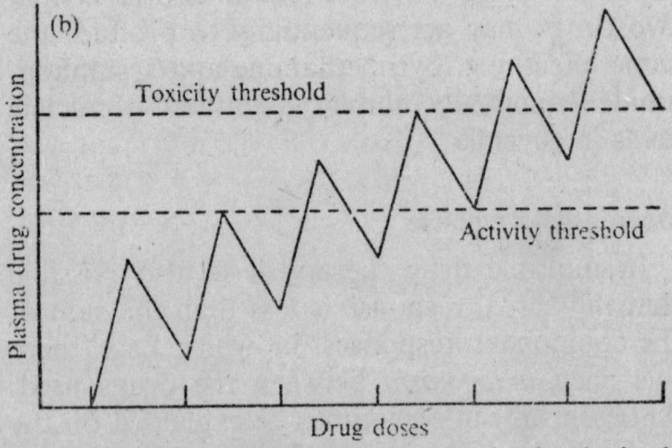

FIG. 8. (a) The probable plasma drug levels obtained following the administration of an initial loading dose, which achieves the desired plasma concentration, followed by smaller maintenance doses which maintain this concentration. (b) When the increase in plasma drug concentration caused by one dose of a drug is not reversed before the next dose is given, the concentration can increase with each successive dose. This can produce cumulative toxicity.

PHARMACOKINETIC PREDICTIONS

From studies of the plasma concentrations of drugs versus time, it has become apparent that in some cases the rate at which drug is lost from plasma due to the effects of metabolism and excretion is constant in terms of *weight per unit time* and *zero-order* kinetics are said to be in operation. Such a situation is seen when the processes of elimination become saturated, i.e. the rate of elimination can no longer increase in response to an increase in plasma drug concentration and may be encountered when renal failure is present or when elimination is heavily dependent on a saturable active transport system.

More frequently the decline in plasma drug concentration follows *first order*, i.e. exponential, kinetics in which a constant *fraction* of the drug concentration is lost in unit time. An exponential process can be defined by its *rate constant*, which expresses the fractional change per unit time or by its *half-time* ($T\frac{1}{2}$) which is the time for the process to be half-completed and *both* are unrelated to drug concentration.

The body eliminates all drugs even though at widely varying rates and is therefore described as an *open system*. For the purposes of kinetic studies the body is regarded as being composed of *compartments* which may or may not equate with anatomical sub-divisions of the body. The drug enters the compartment at administration and leaves by elimination and it is possible to apply mathematical expressions to this simplified model of the body without having to identify either the compartment in anatomical terms or the relative contributions of those processes, the total effect of which is expressed as elimination. The *central compartment* is the blood plus all those tissues whose drug concentrations rapidly equilibrate with plasma levels, i.e. distribution is presumed to be uniform and to occur so rapidly that drug loss by metabolism or excretion during distribution is insignificant. A third basic assumption is that only a small proportion of any drug dose is in combination with receptors, that this proportion is in equilibrium with the bulk of drug and that its fate, and hence the time course of drug effect, will depend entirely on the fate of this unattached portion of the dose. Accepting these assumptions it is possible to use drug plasma concentration as a valid indicator of what is happening to drug at the receptor level.

If the log of drug concentration in plasma is plotted against time for a drug which fits all the foregoing assumptions, then the points would fall on a straight line, i.e. *a one-compartment open model* would be operating.

In fact, although this ideal might be achieved for a drug administered by rapid intravenous injection and which was not able to escape from the circulatory system, by far the most common curve is *bi-exponential*. In such a plot (Fig. 9) there is an obvious elbow in the more-or-less straight line. The initial rapid fall in concentration represents the *distribution phase*, while the drug is equilibrating throughout the central compartment. The elbow is taken to represent the completion of this process and the onset of the *elimination phase*. This latter phase is a composite of drug loss by metabolism, excretion and distribution into the so-called *tissue compartment* which represents the less vascular body components which equilibrate with plasma much more slowly than the lean, well-perfused tissues. In this case the body is represented as a *two-compartment open model*. Certain drugs will reveal a three-phase curve and for them a third compartment has to be postulated and usually indicates extensive penetration by the drug into a rather non-vascular location.

Fig. 9 shows how the experimental data will yield the drug's half-time. It is also possible to derive from the graph the drug's volume of distribution, i.e. that volume in which the dose of drug would have to be dissolved in order to yield the zero-time plasma drug concentration which is obtained by extending the elimination phase straight line back to its point of intersection with the concentration axis of the graph. This value is also called the apparent volume of distribution because it can sometimes exceed total body water volume. This anomaly is due to the effect on the calculation of that amount of drug which is not actually in solution but has been lost elsewhere, e.g. by binding to proteins.

The parameters defined in pharmacokinetic studies are applied to therapeutics in the construction of dosage regimens. If the desired plasma concentration of a drug is known, then it is possible to compute a dose which will yield this concentration and it is possible to predict the rate at which this concentration will decline. From this knowledge it is further possible to predict the interval between the first and second doses so that the fall from the desired plasma concentration will not be such as to be therapeutically disadvantageous. Provided that there is an acceptable margin between effective and

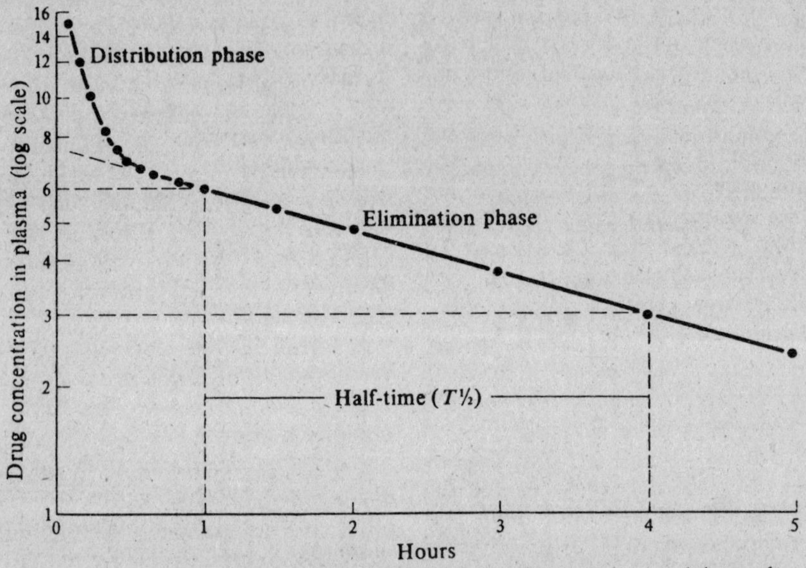

FIG. 9. Idealized plot of plasma concentrations obtained at timed intervals following the intravenous injection of a drug whose behaviour fits a two-compartmental open model.

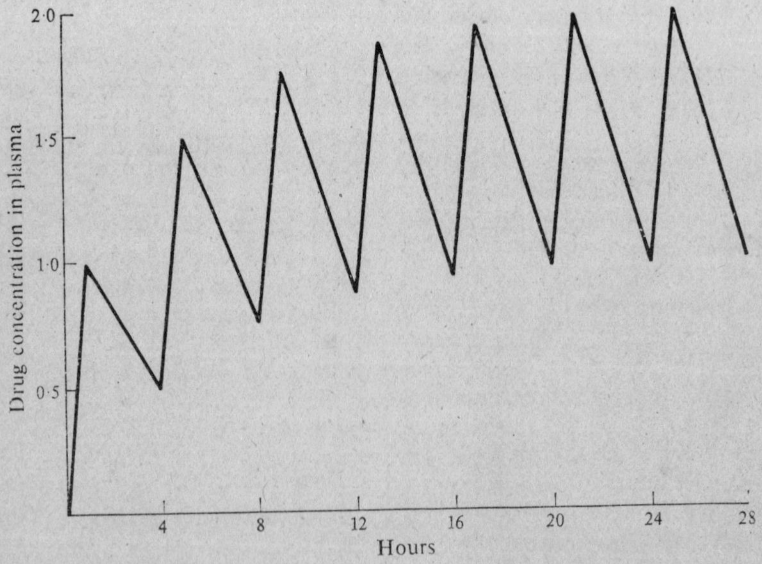

FIG. 10. Diagrammatic representation of predicted fluctuations in plasma concentration of a drug with a $T\frac{1}{2}$ of four hours given four-hourly at a constant dose rate. The mean concentration is almost constant after four half-times.

toxic drug concentrations in plasma, dose interval for most drugs is equal to or less than the half-time if the intention is to maintain relatively constant plasma drug concentrations.

Clearly, if a drug is administered at a constant dose rate at an interval equal to its half-time, the amount of drug present in the body will increase with each dose. The increase in plasma concentration does not in most cases rise continuously but tends to level off—the so-called plateau effect. This happens because as the amount of drug in the body increases so the amount of drug eliminated during one $T\frac{1}{2}$ increases. When the amount eliminated in one $T\frac{1}{2}$ approximates to the dose then the plasma concentration will fluctuate between reasonably constant maximum and minimum levels. If drug is administered at a constant dose level at intervals equal to its half-time then it will have achieved about 90% of the plateau concentration after three to four doses (Fig. 10). If absorption is complete (or the drug is administered intravenously), then the amount present in the body mid-way between two doses will stabilize at 1·44 times the dose in the plateau state. If the dose interval is reduced to half of the half-time, then the drug will accumulate to 2·88 times the dose. In repeat dosing, the plasma concentration will be determined by dose size, dose frequency, the fractional absorption of the drug and its half-time. If the apparent volume of distribution, the half-time and the desired plasma concentration are known, it is possible to predict the dosage regimen which would achieve this mean plasma concentration. The ready availability of computers has greatly simplified the solution of pharmacokinetic predictions infinitely more complex than the foregoing example.

DRUG TOXICITY

All drug actions can be considered as manifestations of toxicity in that, with the exception of replacement therapy, the drug acts because it is able to modify some aspect of cellular function. Such actions are, however, desirable and are responsible for the characteristic effect of the drug, the so-called *main* or *major effect*. Drugs which destroy the causal organisms of disease and yet do not harm the host are said to exhibit *selective toxicity*. However, this ideal state has yet to be achieved as the vast majority of drugs possess toxic actions. Such action may be the drug's main action in excess when the drug has a narrow therapeutic ratio, or may follow accidental overdose or cumulation. When a drug produces undesirable effects as a part of its main action, these are described as *side-effects*, e.g. motor incoordination following phenobarbitone sedation in the dog. When the undesirable effect is a consequence of the main action, the term *secondary effect* is sometimes applied, e.g. super-infection by insensitive organisms following treatment with broad-spectrum antibiotics. These various toxic effects may appear early in a course of treatment, or only after chronic exposure or, occasionally, following the withdrawal of a drug.

Toxicity Testing

The toxicity of a new drug is carefully examined and quantitated before it is released as a product. This examination determines not only the acute toxicity, expressed as LD_{50} values, but also the chronic toxicity. When the drug is administered regularly in sub-lethal doses it is possible to detect toxicity in terms of depression of growth rate of young animals, loss of weight in adults, inappetence, condition of coat, appearance of faeces and urine, abnormality or depression of movement or reproduction, or changes in behaviour and temperament. At the pre-determined end of the trial, the dosed animals are killed and compared with untreated controls. The comparison is made on the basis of gross appearance at post mortem, organ weights and histopathological examination of all tissues. The routine criteria used in these laboratory tests are selected for their relation to toxic phenomena previously identified in practice. The more common manifestations of drug toxicity are described in the next section.

Common Drug Toxicities

Hypersensitivity reactions. The induction of hypersensitivity to penicillin in man following the consumption of contaminated milk has lead to attempts to provide antibiotics for intramammary use in formulations designed to reduce their persistence in milk. Very few cases of drug allergy in domesticated animals have been described in the literature. In hypersensitization, the antigen is a complex of drug and an endogenous protein.

Drug dependence. The ability of certain drugs to induce a state in which the continued presence of the drug is essential for the taker's well-being has caused the sale of such drugs to be stringently controlled. In veterinary medicine this problem is not encountered because the long-term treatment of severe pain is rarely undertaken. However, the prescription, storage and dispensing of the drugs of addiction by veterinarians is regulated by specific legislation. The *Misuse of Drugs Act* defines the restrictions imposed on opium alkaloids, their derivatives and analogues, and cocaine. The free sale of other drugs, including some habit forming drugs is also controlled, e.g. by orders made under the *Pharmacy and Poisons Acts*.

Organ toxicity. When a drug is concentrated in any site, the chances of a localized toxicity are increased. This can occur in the kidney where the excreted drug is rapidly concentrated by the resorption of water from the proximal tubule. Degenerative change in the tubule has been recorded after many drugs, e.g. the mercurial diuretics and several antibiotics. The liver is a second very common site of damage due its receiving drugs in high concentration via the portal vein and due to its function in removing toxic substances from the circulation as a preliminary to detoxication, e.g. carbon tetrachloride causes centrilobular fatty change in the sheep even after a normal dose.

Blood abnormalities. Occasionally, the use of a drug is followed by disturbances of the formed elements of blood, most commonly a depression of white cell count as a consequence of bone marrow depression. In man, long term exposure to chloramphenicol has produced this response. Bracken-poisoning of cattle is a clinical example of severe bone-marrow depression.

Central nervous system. Drugs can induce changes in the normal pattern of activity. These are frequently CNS depressants to which occasional animals respond abnormally, e.g. motor-incoordination as a consequence of tranquillization. However, other classes of drugs can also affect the CNS, e.g. many of the sulphonamides exert a depressant effect, while convulsions can be induced by streptomycin or procaine. Measurement of the effect of the drug on voluntary motor activity is a test procedure commonly used to quantify this toxic manifestation.

Teratogenicity. The necessity for examining the possible damaging effect of a drug on the foetus was tragically realized after the thalidomide affair. The discovery that many drugs can on occasion induce foetal abnormalities has lead to a considerable reduction in the prescription of antemetics for pregnant women. In veterinary medicine the foetus is more likely to be affected by the induction of abortion by purgatives administered in late pregnancy.

Carcinogenicity. The ability to induce neoplastic change has been associated with certain compounds. Although the latent period between contact with drug and the appearance of neoplastic change may be very long, a search for such change forms part of the routine post mortem and histopathalogical examination of test animals.

Bibliography

GILLETTE, J. R., MITCHELL, J. R. & BRODIE, B. B. (1974) Biochemical mechanisms of drug toxicity. *A. Rev. Pharmac.*, **14**, 271.

MORSELLI, R. E., COHEN, S. A. & GARATTINI, N. S. (ed.) (1974) *Drug Interactions*. New York: Raven.

NOTARI, R. E. (1975) *Biopharmaceutics and Pharmacokinetics, An Introduction*, 2nd ed. New York: Marcel Dekker.

SWARBRICK, J. (ed.) (1973) *Current Concepts in the Pharmaceutical Sciences: Dosage Form, Design and Bioavailability*. Philadelphia: Lea & Febiger.

WADE, O. L. (1970) *Adverse Reactions to Drugs*. London: Heineman.

PART II

DRUGS USED IN THE MODIFICATION OF CELL, TISSUE, ORGAN AND SYSTEM FUNCTION

5

The Alimentary Canal

I. The Mouth and the Stomach

To assist in classification the alimentary canal has been divided into four sections: (a) mouth and pharynx, (b) oesophagus, (c) stomach; and (d) intestines. Obviously there will be some drugs which will have an effect on more than one section but, as in other chapters, each drug will be placed in that section where its most important action is exerted. The importance of the action will be determined from a therapeutic view-point.

THE MOUTH AND PHARYNX

Sialagogues or Sialics

These are drugs which are used to increase the volume and fluidity of saliva. This definition excludes any drug which effects an increase in only one of these characteristics; it particularly excludes sympathetic nerve stimulants which, though they cause a slight increase in volume, tend to decrease the fluidity.

The aim of all sialagogues is to improve a patient's digestion and appetite by increasing saliva production. In addition to causing salivation these drugs usually increase gastric secretion and for this reason some may also be classed as stomachics or gastric stimulants.

Sialagogues may be divided into two groups according to the means by which salivation is induced.

1. *Reflex sialagogues* are those which, by stimulating the taste buds on the tongue, initiate a reflex which results in salivation. Gentian, calumba, quassia and nux vomica preparations are commonly used examples of this group.

2. *Direct sialagogues* are those which by-pass the sensory half of the salivary reflex and cause salivation by direct stimulation of the parasympathetic nervous system. Carbamoylcholine, arecoline, pilocarpine and physostigmine are examples, but not one of these drugs is used solely for this purpose (see 'The neuromuscular purgatives', p. 60).

Reflex Sialagogues

These are most commonly referred to as 'the bitters'. The 'bitter principles' of these drugs vary considerably and only those which contain an alkaloid (an organic nitrogenous base which reacts with acids to form salts) can be grouped according to the chemical nature of the drug which gives them their bitter taste. The others called the 'simple bitters' are divided according to whether or not the preparations made from the plant possess a characteristic odour. There are thus *aromatic simple bitters* such as orange or lemon peel, and *non-aromatic simple bitters* such as gentian, calumba and quassia.

Because they act locally on the taste buds of the tongue all bitters must be administered orally in the form of a powder, drench, or electuary. Tablets, pills, capsules, boluses or administration by stomach tube are useless, for the drugs in these forms do not contact the taste buds or mucosa of the mouth and pharynx.

In order to achieve their full effect bitters should be administered either 5 to 15 min before feeding or actually mixed in with the food.

The value of bitters as therapeutic agents has been questioned in recent years, but, though their use in the smaller species has justifiably lapsed, cattle drenches and horse powders containing gentian and nux vomica are still extensively employed. They are relied upon by experienced practitioners to stimulate and improve the digestion of convalescent cattle and horses. The opponents of therapy with bitters claim that these two species dislike these drugs, but horses, well known for their selective feeding habits, will take them mixed in bran mashes and 'chop'. So readily are they taken that the best of the simple bitters, gentian powder, in spite of its relatively high cost, is frequently used as a bulking and disguising agent for less acceptable drugs.

Dogs and cats object to the taste of bitters, and meat or meat extracts are far more useful salivary and gastric stimulants for carnivorous species.

The simple bitters. The preparations of *gentian* are representative of those of the whole group of simple bitters. These are: a powder, Powdered Gentian, B.Vet.C.; a compound tincture, Compound Tincture of Gentian, B.Vet.C.; a soft extract, Extract of Gentian, B.P.C.; a compound strong, or concentrated, infusion, Concentrated Compound Infusion of Gentian, B.Vet.C.; and a weak, or fresh, compound infusion.

The greenish brown powder and the green tinged brown compound tincture are the two preparations commonly used. Their toxicity is non-existent and the dosage-level, which is the same for all the simple bitters, is given below.

The powder:
- Horse and Cow: 15–60 g (½–2 oz.)
- Sheep and Goat: 4–8 g (60–120 gr.)
- Pig: 3–7 g (40–100 gr.)
- Dog and Cat: not used

The tincture:
- Horse and Cow: 16–60 ml (½–2 fl. oz.)
- Sheep and Goat: 8–45 ml (¼–1½ fl. oz.)

The alkaloidal bitters. This group is represented by the preparations of Nux Vomica, B.Vet.C., which are the most frequently used examples of reflex alkaloidal sialagogues. The tincture and powder are the preparations most commonly used; their dosage scales are:

The powder:
- Horse and Cow: 2–8 g (30–120 gr.)
- Sheep and Goat: 0·6–1·3 g (10–20 gr.)
- Pig: 0·6–1 g (10–15 gr.)

The tincture:
- Horse and Cow: 16–32 ml (½–1 fl. oz.)
- Sheep and Goat: 1–4 ml (15–60 m.)
- Pig: 0·3–1·3 ml (5–20 m.)

The popularity of nux vomica preparations is not entirely due to their bitter action, for the two alkaloids, which are the bitter principles, *strychnine* and *brucine*, have a powerful action on the spinal cord and peripheral nerves once they are absorbed. Brucine is much less active than strychnine but similar in effect. The effect of strychnine is to speed up the progress of nerve impulses, particularly at nerve junctions, so that the general tone of the whole body is improved. In pharmacological terms the alkaloid causes facilitation of nerve impulses across the synapses.

The combined bitter and 'nerve tonic' actions of nux vomica are widely used in the treatment of convalescents by both human and veterinary physicians.

Appetite Suppressants

Certain sympathomimetic amines reduce the desire for food and have been used for this purpose in obese dogs, e.g. dexamphetamine and chlorphentermine. Such compounds are discussed in Chapters 8 and 24.

Antisialagogues or Antisialics

These are drugs which reduce the flow and fluidity of saliva. There is often a reduction at the same time of all respiratory and alimentary gland secretions.

The therapeutic aim of antisialagogues is to reduce salivation, particularly during a period of excessive salivation such as may occur during inhalation anaesthesia with ether or chloroform, and also during morphine narcosis. Although there are several agents which will either mechanically, or by their astringent properties, reduce salivation—kaolin and alum are respective examples, only atropine and related alkaloids such as hyoscine and hyoscyamine have much practical value.

Atropine

Derived primarily from *Atropa belladonna*, deadly nightshade, and in lesser quantity from related solanaceous plants such as henbane and thorn apple, atropine is a typical alkaloid; it is therefore a white crystalline solid which is odourless when pure, insoluble in water but soluble in alcohol, and reacts with acids to form salts, e.g. atropine sulphate and hydrochloride, which are more frequently used then the parent alkaloid. The incompatibilities of atropine are also those of a typical alkaloid: atropine will be precipitated from alcoholic solutions such as tinctures, by the addition of water; also salts of heavy metals, particularly silver, lead, mercury and zinc, tannic acid and related acids, will precipitate atropine and its salts.

In addition to the pure atropine salts used in the form of injections, the alkaloid is commonly used as the relatively crude belladonna-plant preparations. The more popular preparations are the powdered leaf, the tincture and a soft extract. The green, soft extract of belladonna, for example, is frequently used to prepare electuaries for the alleviation of symptoms in 'broken-winded' horses.

Actions and Uses (Internal)

Following absorption the effects of atropine can be divided into two groups;

1. *The action on the central nervous system.* In normal therapeutic doses no cerebal effect is apparent. Intravenously in small over-doses atropine has a stimulant and sometimes convulsant action on the cerebral cortex. Larger doses will excessively stimulate the medulla and still larger doses will affect the spinal cord. This sequence is typical of convulsants or analeptics of which picrotoxin, and lepatzol are further examples.

Excessive overdosage produces cortical and medullary depression, coma and death.

2. *The action on the peripheral nervous system.* This is one of paralysis of the muscarinic part of the parasympathetic division of the autonomic nervous system. The muscarinic effects of acetylcholine are seen when the postganglionic parasympathetic receptors are activated, e.g. by the alkaloids muscarine or pilocarpine. The paralysis is not complete when single therapeutic doses are used, but even partial blocking of the parasympathetic impulses allows greater action of the sympathetic system. The major effects are:

(i) Removal of vagal tone, which results in: (a) an increase in heart rate equivalent to the amount of vagal depression previously exerted; (b) relaxation of intersphincter intestinal muscles and constriction of sphincters; (c) shutting down of the splanchnic circulation, which turns large quantities of blood into the circulation; (d) shutting down of the alimentary secretory glands, including the pancreas, following the lack of blood supply and digestive activity; and (e) salivary secretion is proportionately controlled.

(ii) Cutaneous circulation is usually increased, even by the therapeutic doses, but sweating is suppressed so that the temperature may rise in those species which rely on the production of sweat for some heat loss. The dog is one species in which the lack of normal sweat glands is responsible for no temperature variation following atropine administration.

(iii) The respiratory rate, and sometimes the depth, is increased. This is associated with a relaxation of the bronchi and suppression of bronchial secretions.

Uses

The main internal uses of atropine are: (1) The alleviation of pulmonary emphysema (broken wind in horses) in the form of belladonna electuaries, e.g.:

Green Extract of Belladonna, B.P.C.	0·6 g (10 gr.)
Prepared Nux Vomica, B.Vet.C.	2·0 g (30 gr.)
Powdered Gentian, B.Vet.C.	8·0 g (120 gr.)
Excipient	q.s. (q.s.)

(2) The restriction of salivation and bronchial secretions during anaesthesia, especially with ether and chloroform. This is of particular value in protracted anaesthesia of horses, when copious amounts of saliva and bronchial exudates may be drawn into the lungs and produce pneumonia. Atropine sulphate injections are given subcutaneously 15 to 20 min before induction of anaesthesia using the following doses:

Horse and Cow: 15–30 mg ($\frac{1}{4}$–$\frac{1}{2}$ gr.)
Sheep: 4–8 mg ($\frac{1}{16}$–$\frac{1}{8}$ gr.)
Pig: 2–4 mg ($\frac{1}{30}$–$\frac{1}{16}$ gr.)
Dog: 0·3–0·6 mg ($\frac{1}{200}$ gr. up to $\frac{1}{100}$ gr.)
Cat: 0·3 mg ($\frac{1}{200}$ gr.)

These doses can be used in cases of pytalism—excessive salivation—which may occur in some types of poisoning such as lead and mercury poisoning, in diseases of the salivary glands such as strangles in horses, and in severe affections of the oral mucosa and tongue particularly if swallowing is affected, as in 'wooden tongue' of cattle.

(3) The use as an *antispasmodic* where excessive parasympathetic stimulation is responsible for 'tonus rings' or circular muscle spasm. This applies to all smooth muscle of hollow organs, of the uterus, ureter, bronchioles, but particularly of the intestines. Whenever parasympathetic stimulants such as carbamoylcholine are administered as purgatives to horses, atropine should always be available to counteract excessive stimulation. The quantity of atropine necessary is not proportional to the level of overstimulation but is a fairly constant proportion of the animal's weight, 2 to 3 mg per cwt, that is ½ to 1 gr. per animal.

Actions and Uses (External)

External actions are confined to the effect of atropine on the eye, with the exception of some slight dulling of peripheral sensory receptors. This ocular effect is well known, for atropine has long been used by oculists to dilate the pupil, enabling the retina and other deep structures of the eye to be easily identified and examined. This dilatation of the pupil or *mydriatic* action is due to paralysis of the oculomotor nerve and hence of the circular muscles of the iris. This allows the radial, sympathetic innervated muscles to respond to the impulses which are normally present, and which are responsible for maintaining tonus. (The iris gives a perfect illustration of the balanced control which is exerted over all involuntary muscle by the two divisions of the autonomic nervous system. In many cases the two act on the same muscle fibres but here that is not the case.)

As a result of this dilatation of the pupil, the muscles of the iris become localized at the periphery of the chambers and must consequently bulge at right angles to the radius. This fills in the filtration angle of the eye and constricts the spaces of Fontana which drain into the circular canal of Schlemm. In the normal eye this is of no significance for the effect usually passes off within 2 to 3 days, although exceptionally it may persist for up to a week, and no more than a temporary slowing of aqueous humour circulation occurs, with no ill effects. In eyes where there is any tendency to *glaucoma* (high intra-ocular pressure) the extra pressure resulting from the iris collapse may be sufficient to cause closure of the spaces. This closure becomes permanent owing to the building up of a pressure which is too great to allow the iris to return to its normal position following dispersal of the atropine effect.

Besides affecting the iris, the ciliary muscles which control the accommodation, or focus, of the lens by pulling on the suspensory ligaments, are also affected; the lens surfaces become maximally convex and the eye then has a fixed long distance focus. This ciliary paralysis also materially affects filtration.

Eserine, or physostigmine, has been used to reverse the atropine effect where paralysis of the iris for 24 to 48 hr was contra-indicated. Eye Ointment of Physostigmine, B.Vet.C. (0·125 per cent) and Eyedrops of Physostigmine, B.Vet.C. (0·5 per cent) were used for this purpose, but the effect was not completely satisfactory and a synthetic agent was devised similar in effect to atropine. This is *homatropine*. Used in the form of 1 to 2 per cent eye drops or eye ointments, homatropine effects which, like atropine effects, are maximal in about 30 min, last for, at the most, 24 hr and any hyperactivity can be adequately controlled and reversed (cancelled) by physostigmine. Indeed, where there is a danger of iris adhesion, alternate dilation and constriction of the pupil can be achieved by the alternate use of homatropine and physostigmine. The iris is thus in constant movement and adhesions are less likely.

Demulcents

These are agents which are used to lubricate, soothe and protect an alimentary mucous membrane, but the term is usually applied only to those affecting the mouth, pharyngeal, oesophageal and gastric mucosae.

Demulcents serve several purposes in that they may be used to protect mucosae prior to, and during, administration of irritant substances; to allay irritation and inflammation already caused either by chemical or bacterial action; to act as innocuous bulking agents for more active drugs; and to act as sweetening, flavouring and disguising agents: some are also easily digested foods, *vide* the carbohydrates and the animal proteins.

They consist of several groups of compounds which differ very widely in their chemical structure.

1. The sugars: this group includes sucrose, lactose and related carbohydrates such as honey, treacle (molasses), malt extracts, starch, flour, and linseed mucilages.

2. The gums, such as acacia and tragacanth.

3. Animal proteins such as gelatin and egg albumen.

4. Vegetable oils and glycerol.

5. Miscellaneous compounds such as liquorice powder, the hydrocarbon liquid and soft paraffins, and colloid forming compounds such as bentonite (a hydrated aluminium silicate *q.v.*) and carboxymethylcellulose.

Groups 1, 2, and 3 and liquorice powder can be conveniently mentioned at this point, but the other groups will be mentioned elsewhere—see Index.

The Sugars

All the sugars, in solution, and their relatives, starch, flour and linseed, in gruel form, have good demulcent properties. For this reason *sucrose* in the form of Syrupus, B.Vet.C., a 66⅔ per cent sucrose solution in water, is used as a vehicle and bulking agent for many official and unofficial syrups, the active ingredients of which are often quite irritant. Syrup of Squill, B.P.C., is used as an irritant expectorant in both medical and veterinary cough mixtures. Its sweet disguising taste is very useful in this type of preparation, and also in those containing alkaloids, such as Syrup of Codeine Phosphate, B.Vet.C. Honey, treacle, and malt have similar characteristics but their semi-solid nature makes them especially useful as electuary bases for both both small and large animals.

As a food, *malt* has advantages over other demulcent carbohydrates for, besides various sugars, it contains diastase, an enzyme which is capable of converting large quantities of starch into easily assimilated sugar. It also contains appreciable proportions of the B complex of vitamins. As a convalescent treatment *doses* of up to 30 g (1 oz.) may be given to dogs, calves, and foals, but large doses tend to have an aperient effect and it is this which controls dosage in any particular individual.

Treacle is frequently used as an electuary base, for it is a readily available household food and the owner is able, in an emergency, to make his own electuary from powder left by the veterinary surgeon. The sugar content of treacle and molasses used to be used as a treatment of acetonaemia in cattle and pregnancy toxaemia in ewes but it has, in recent years, been established that these remedies aggravate the conditions. This is because the main fermentation product of treacle, or even glucose, in the rumen is β-hydroxybutyric acid which, when absorbed, is added to that already present in the blood. Moreover, continued dosing with molasses or treacle in the quantities recommended in the past soon produces scouring.

Dextrose, B.Vet.C.

Formerly known as glucose, is not so useful as a sweetening agent, for neither lactose nor dextrose are markedly sweet. This is often an advantage, for dextrose is more acceptable to small animals, enabling larger quantities to be given as a food and as a *gastric sedative* or *antemetic*. Dosage is limited only by the amount accepted by the individual. Parenterally, dextrose, either in a 20 per cent solution subcutaneously or 40 per cent intravenously, supplies carbohydrate for the treatment of debility or acetonaemia. It is not advisable to use the 40 per cent concentration subcutaneously, for the osmotic pressure is high and abscess formation may occur. Dextrose is also used as a food and partly as a blood volume maintaining substance in the form of 'glucose-gum-saline', which contains 5 per cent dextrose and 6 per cent gum acacia in normal saline. (This is no longer considered a safe blood-volume expander.) Glucose saline is a similar preparation. These are available for the treatment of shock and debility in all species.

The Gums

Acacia, B.Vet.C., commonly known as 'Gum Arabic', and Tragacanth, B.Vet.C., or 'Gum Dragon', which are shrub exudates, are of less importance than formerly. They are useful in the making of emulsions *for internal use only* in animals (they matt the hair too much for external use); they combine emulsifying and demulcent properties. Mucilage of Acacia, B.Vet.C., is useful as a suspending agent in drenches containing heavy powders such as heavy magnesium oxide. Tragacanth is used in 'barium meal' as a suspending agent for barium sulphate; the meal is used as a radio-opaque diagnostic drench.

Liquorice

Powdered Liquorice, B.Vet.C., an aromatic brown powder, is used as a demulcent, flavouring and bulking agent for powders to be administered in the food of farm animals. It contains besides the sweet glycerin, a cortisone-like, substance, glycyrrhetinic acid, which is used as a cortisone substitute. As a demulcent and diluent the dosage varies from about 2·6 g (40 gr.) for the dog, to 85 g (3 oz.) for the horse and cow.

Aniseed

Powdered Anise, B.Vet.C., is sometimes used instead of liquorice, for its stronger flavour is more readily accepted by animals, and its lower dosage of 30 g (1 oz.) maximum for horses and cattle is useful where large powders are not required. It has no demulcent properties, but is carminative and antispasmodic in the stomach and intestine. Anise Oil, B.Vet.C., has a strong odour which has suggested its use as a fly repellant, although it is not very efficient for this purpose. Some slight expectorant action occurs during excretion.

Gelatin, B.Vet.C., and egg albumen

Both these are extremely good demulcents but the latter is the only one used to any marked degree in veterinary practice. If egg albumen is whipped up with a little glucose a very good demulcent and nutritional supplement is made. A teaspoonful administered every hour or so will alleviate emesis of local, but not central, origin in dogs and cats.

THE OESOPHAGUS

In ruminants, the closure of the oesophageal groove is an important factor in therapy, particularly against helminth parasites, for this closure directs fluids straight into the abomasum, the home of some of the most important ruminant worms. The groove in young calves and lambs is closed by stimulation with milk or water, but adult sheep require 1 per cent copper sulphate or merely a crystal rubbed on the oral mucosa. Adult cattle show no oesophageal reaction to copper, but it has been shown that 60 ml (2 fl. oz.) of a 10 per cent sodium bicarbonate solution will give rapid closure. There are indications that the closure reflex can be conditioned.

Parasympathomimetic drugs such as arecoline or pilocarpine have been used in 30-mg; (½ gr.) doses to stimulate salivation and increase oesophageal peristalsis for the relief of choke in the cow and horse; choke may be caused by foreign bodies such as potatoes, pieces of turnip, sugar beet pulp, etc., in the oesophagus. Carbamoylcholine is used in small doses repeated at 6 to 12 hourly intervals (1 to 2 ml of 1 : 1000 solution, i.e. 1 to 2 mg). The phenothiazine derivative tranquillizers will often relieve choke by causing the oesophagus to relax.

THE STOMACH

The drugs which act on the stomach are classified into carminatives, antispasmodics and antizymotics, emetics and antemetics (or gastric sedatives), amongst which are included the antacids.

Carminatives

These are drugs which, when administered orally, cause the expulsion of gases from the stomach, politely known as eructation. They have also been used in flatulent colic to assist evacuation of gases from the large intestine.

They are mainly volatile compounds. There are thus compounds which contain volatile oils such as powdered aniseed and ginger, pure volatile oils such as oil of turpentine and oil of peppermint, and substances derived from volatile oils by cooling, the stearoptenes camphor and menthol. Certain volatile anaesthetics and related compounds also have a carminative action: chloroform, ether and alcohol are examples.

These compounds in the correct dosage cause

mild irritation of the gastrointestinal mucosa which results in vasodilatation. This is probably responsible for the well known 'warm' feeling which follows the swallowing of these compounds. Other actions which result are the relaxation of gastrointestinal musculature, but particularly the cardiac sphincter of the stomach, for a period of up to 30 min, which probably plays a large part in releasing gases from the stomach.

The antispasmodic action of some of these drugs, particularly ginger, is due to the same factors which give the carminative effect. The rhythm and tone of peristalsis may be initially increased, but this is often followed by a decrease of movement which will help in relaxing colic spasms. Atropine and homatropine which exert the effect by a parasympatholytic action have already been mentioned as antispasmodics, see p. 46. Hyoscine and hyoscyamus have a similar effect. Pethidine also has a marked antispasmodic action.

The carminative actions of these drugs are not constant and their use for this specific purpose is decreasing, particularly in small animals. Many of these drugs are retained for other purposes. For example, ginger is used in horse and cattle practice as a flavouring and bulking agent, whilst turpentine is used as an antizymotic and antispasmodic for horses and cows as well as for its external counter-irritant action.

Ginger

Zingiber, familiar to all as root, stem, and powdered ginger, is still widely used in horse and cattle practice. The yellow-brown powder is used in doses of 15 to 60 g (½ to 2 oz.) for horses and cattle, incorporated in powders for oral use. Where cattle are drenched, the ginger can be included in the form of either the strong tincture or, better than this, the weak tincture, the dose of which for horses and cattle is 15 to 60 ml (½ to 2 fl. oz.) For small animals the doses are 0·3 to 4 ml (5 to 60 m.) of either the weak tincture or the syrup prepared from the strong tincture; these are not frequently used.

The use of ginger is restricted to internal administration as a flavouring and bulking agent, and as a carminative and antispasmodic.

Antizymotics

Strictly defined, these agents are drugs which prevent or decrease bacterial or enzyme fermentation. Their use is aimed at the prevention of further gas production in tympany and bloat in ruminants, and flatulent or tympanitic colic in horses. These conditions occur only occasionally in other animals and treatment follows the same principles as in the larger species.

Bloat and Tympany in Ruminants

Historically these have been treated, irrespective of aetiology, with volatile oils, especially turpentine, with formalin, phenol, and with phenolic compounds such as salol, cresol and lysol. Chloroform, chloral hydrate and ethyl alcohol have also been used (the last of these often in the form of gin, whisky or ale!).

More recent research has indicated the need for accurate diagnosis of both the exciting cause of ruminal distension and also the immediate cause.

The exciting cause is usually ruminal stasis, which may be due to one of several circumstances: oesophageal obstruction, too little or too much bulk in the rumen, too high or too low a pH, abnormal diet, and foreign bodies in the reticulum are some examples.

The immediate causes can be divided into three groups.

1. Free gas in the rumen (tympany).

2. Gas locked in the bubbles of a frothy mass (bloat).

3. Mechanical distension by abnormal food eaten in quantity following accidental access.

The treatment of groups 1 and 3 are similar and follow the historical methods, but frothy bloat must be treated as specifically as possible, and for this there are newer agents which are listed as 'surface tension increasing/decreasing' or 'antifrothing agents'.

Free gas may be removed by stomach tube or trocar and cannula. If a cannula inserted through the left flank is used, it is usual to leave it in place for 12 to 24 hr: a longer period predisposes to aeration and a change of ruminal flora. Although this period of change may be short-lived it can aggravate the existing condition.

Following removal of the gas, and therefore

any danger of asphyxia by diaphragmatic pressure, the next procedure will depend on the initial or 'exciting' cause. Where the cause requires surgical interference, such as oesophageal obstruction and foreign bodies in the reticulum, the treatment is obvious. Where no pathological changes in structure occur such as those caused by actinomycosis of the rumen and oesophagus or enlargement of mediastinal lymph glands, some attempt at diagnosing the physiological and pathological changes must be made before rational treatment can be applied. The history of the case will often indicate dietary causes, and a clinical examination will help to assess the volume of ruminal contents and diagnose the milk fever and febrile conditions. By a process of elimination a hard core of undiagnosable cases will be left. This nucleus of cases is treated empirically.

Where dietary causes are involved the remedy is obvious, but with more abstruse physiological causes, pH changes, etc., the diagnosis and treatment is more difficult. In the United States it is common practice with all these cases to remove a regurgitated bolus from the mouth of a normal ruminating cow and administer it immediately to a non-ruminating animal: modifications of this treatment are the oral administration of samples of ruminal fluid from slaughtered normal animals or exciccated ruminal contents. Whole ruminal content has been the most effective material used in this form of treatment. Improvement follows the re-establishment of a normal ruminal micro-organism population with consequent normalization of pH and fermentation products. In certain cases, surgical replacement of ruminal contents will be necessary.

Undue reliance for the reinstatement of ruminal movement in cattle is placed on such drugs as powdered nux vomica in 1 to 3 g doses (30 to 90 gr.) by mouth, liquor strychninae hydrochloridi (1 per cent solution of strychnine hydrochloride) in 4 to 6 ml doses subcutaneously, and barium chloride in 1·5 to 3 g (15 to 30 gr.) doses by mouth. (This last treatment is not recommended.) Perhaps the most popular treatment, however, is the subcutaneous injection of 1 to 3 ml of carbamoylcholine, Carbachol, B.Vet.C., 1:1000 injection. This injection, repeated at 1 or 2 hourly intervals until rumination is re-established, is probably the best available for restarting ruminal movement. Two or three injections are sometimes required.

If the fermentation processes still continue, it is usual to administer an antizymotic; 15 to 30 ml of turpentine in 300 to 600 ml of linseed oil (1 fl. oz. in ½ to 1 pt.) is the basic formula of most cattle drenches used. These may be given by stomach tube, as a drench or, if a trocar and cannula are used, they may be injected directly into the rumen through the cannula. In the last case care must be taken to ensure that the cannula has not been displaced by abdominal movements lest the fluid is introduced into the peritoneal cavity. A 3 in. wide-bore needle is very often used. In sheep the doses will be smaller: 4 to 8 ml in 60 to 300 ml (60 to 120 m. in 2 to 10 fl. oz.) Formalin in doses of 4 ml (60 m.) three times daily for cattle in 300 ml (½ pt.) of water, or 0·6 to 1 ml (10 to 15 m.) in a few ounces of water for sheep have also given good results. The use of phenol, lysol and similar agents is not recommended nowadays.

The treatment of conditions in group 3 is complicated, for several factors must be considered: the distension of the rumen by the swelling mass of material, which leads to pressure paralysis of the muscle; the possibility of these substances—sugar beet pulp, potatoes, wheat and barley, fermenting and producing gas; and the danger of asphyxia, particularly as the animal is usually recumbent.

The distension and asphyxia can be relieved only by rumenotomy, and the practitioner must decide when a dangerous degree of distension has been reached. The fermentation and further swelling of grains can be controlled within limits by formalin solutions and by restricting fluid intake, respectively.

Frothy Bloat

Frothy bloat is typified by the inability to relieve the ruminal distension by stomach tube, probang, or trocar and cannula. All that exudes from any of these instruments is a small amount of fluid and bubbles. The condition often appears and matures very rapidly.

The bubbles are thought to result, in many cases, from a high intake of saponins and similar compounds which only slightly reduce or even increase the surface tension of the fluid within the rumen and produce a high viscosity foam. It has been observed for centuries that bloat occurs most frequently on lush clover pastures. All clovers, and alfalfa particularly, have a high saponin content. Other exciting causes such as lack of roughage in the diet and toxic substances causing ruminal stasis are also implicated.

The rational treatment for this high viscosity foam is to reverse the process by diffusing throughout the rumen agents which cause a reduction in foam stability. Turpentine and paraffin (kerosene) have been in use for many years but more recently organic silicones, particularly methylsilicone in a fine state of division, have been added to these. Fixed oil–detergent mixtures have been sprayed on pastures and administered orally, with fairly good results. Polyethylene glycol surfactants, other emulsifying agents, and penicillin in 50 to 200 mg doses are also used successfully.

Oil of Turpentine

This is the most important volatile oil in veterinary medicine, and its characteristics are often quoted as typifying the group characters. It has a typical odour, is a colourless mobile liquid which yellows with age, is insoluble in water but miscible with organic solvents including alcohol, benzol, chloroform and fixed oils. The characteristic odour is capable of tainting flesh and milk, and turpentine therefore should neither be administered within 4 days of slaughter, nor given to animals where the milk is to be sold. Its irritant properties make a demulcent vehicle essential for oral administration.

Dosage

The horse and cow will need from 15 to 60 ml (½ to 2 fl. oz.) and 15 to 90 ml (½ to 3 fl. oz.), respectively, the sheep up to 12 ml (180 m.) and the pig up to 4 ml (60 m.).

Turpentine acts before absorption as a reflex sialagogue, a carminative, and an antizymotic, when it is administered mixed in linseed oil. A representative formula for a horse colic treatment administered by stomach tube would be:

℞	Chloral hydrate	15 g (½ oz.)
	Oil of turpentine	30 ml (1 fl. oz.)
	Linseed oil to	600 ml (1 pt.)

Further actions of less importance are: (1) An anthelmintic effect has been claimed against lung-worms when intratracheal injection is made but the whole principle of these injections is not recommended.

(2) A reflex cardiac and respiratory stimulant from gastric irritation.

Absorption from the intestine is quite rapid, as is the case for all readily fat-soluble non-electrolytes. No clinically discernible effects are seen at normal dosage levels.

Excretion is typical of all volatile oils and commences quite soon after administration. The general effect is that of stimulating all secretions, due to irritation of the various glands and organs which provide excretory routes—the salivary glands, the lungs, the milk, urine and sweat glands. Thus an expectorant effect occurs in the bronchi, and the breath has an odour of turpentine by which undisclosed therapy can be confirmed. Diuresis results from dilation of the renal blood vessels due to irritation, and also resorption by the tubules is decreased. An odour, classically described as that of violets, is given to the urine. During excretion by the skin irritation of the sweat glands occurs, along with dilation of the blood vessels and sweating (diaphoresis)—hence turpentine can be classed as a mild *diaphoretic*.

Overdosage will give symptoms which one would expect from a powerful irritant—severe gastroenteritis, liver damage, and nephritis, with an increased flow of dark, and possibly bloodstained, urine initially, followed by a decrease, and possibly a cessation, of urine excretion. Nervous excitement symptoms are also reported.

The Silicones

These are organic compounds of silicon with a basic structure of alternating silicon and oxygen atoms. The silicon atoms have attached to them various organic groups such as methyl and phenyl groups. It is those silicones which possess methyl groupings which are of importance in veterinary work because of their important property of increasing surface tension of fluids when added in very small amounts. Their action in bloat is probably one of decreasing foam stability.

Polymerized methyl silicone is a viscous translucent liquid which is non-volatile and chemically inert, tasteless and non-toxic. It is used for the treatment of bloat in concentrations of 2 to 5 per cent in fluid hydrocarbons such as purified kerosene. Dosage varies from 100 ml of the lower concentrations to 50 ml of the higher ones. Administration is by stomach tube or drench, diluted with water, or more rapidly by injection directly into the rumen using a cardiac needle or fine cannula.

Other silicones are used for coating glassware from which complete fluid evacuation is required as with antibiotic and similarly expensive and active drugs. When syringe plungers are coated certain silicones will act as almost permanent lubricants. The lubricant, non-reactive and coating properties are also utilized when coating needles and cannulae which are intended to remain inserted within blood vessels for long periods. This coating smooths the needle surfaces so well that fibrin will not deposit on the surface of the metal and clotting will not therefore take place in contact with, or inside the needle.

Acid and Enzymes

In certain cases, production of gastric acid is inadequate and attempts have been made to correct this situation by the oral administration of dilute hydrochloric acid. In man, many enzyme containing remedies are available for the relief of digestive dysfunction. The lack of pancreatic enzymes is most important and this is dealt with in Chapter 14.

Emetics

True emesis is not possible in the horse, ruminants and rats. Regurgitation of stomach contents sometimes occurs in all these species particularly in young ruminants, but where it occurs in adults the animal seldom lives long after the event. Any regurgitation in the horse occurs via the nose. True emesis is, of course, described as a reflex in which the pylorus and the pyloric area of the stomach contract, the fundus relaxes and the cardia opens; abdominal contraction allied with thoracic and diaphragmatic relaxation drives the stomach contents into the oesophagus; the oesophagus and thorax then contract, forcing some of the material back into the stomach and the rest into the mouth.

Emesis may be caused reflexly by irritation of the stomach mucosa, or centrally by the stimulation of the vomiting centre situated in the medulla just below the third ventricle.

Reflex Emetics

The irritant substances used are mainly inorganic salts, sodium chloride, sodium carbonate (washing soda), copper and zinc sulphates; mustard is also used. Many other substances such as powdered digitalis, aspirin, ipecacuanha, and antimony and potassium tartrate can also induce vomiting; the last two examples were once used as emetics, but they are unreliable and dangerous, particularly the last, 'tartar emetic', and are no longer in use for this purpose.

The sodium salts and mustard are usually available in the home, and are therefore always at hand when an emergency treatment is necessary.

Sodium chloride may be used in crystalline form as half a teaspoonful thrown onto the back of the tongue or as a solution, when 1 to 2 teaspoonsful may be dissolved in half a cupful of tepid water and given as a drench. The carbonate is used in the form of one sizeable crystal pushed over the back of the tongue. Copper sulphate is used in the form of a 1 per cent aqueous solution of which up to half a cupful may be required. Zinc sulphate is similarly administered but mustard requires ½ to 1 tablespoonful in ½ to 1 cup of tepid water.

Vomiting usually occurs within 15 min of administration, but most of these drugs are unreliable as emetics. For this reason the sodium salts are preferred as no deleterious effect results from non-emesis, whereas purgatives or gastric lavage are needed if vomiting does not follow the administration of copper or zinc salts, which are all toxic following absorption.

Central Emetics

Drugs in this group stimulate an area in the medulla known as the chemo-sensitive trigger zone which, in turn, stimulates the vomiting centre. Destruction of the trigger zone eliminates the effect of central emetics.

Apomorphine hydrochloride. This is an alkaloid

derived from morphine by reaction with hydrochloric acid. It has all the properties of alkaloids, and though derived from opium it is, for obvious reasons, not subject to the full Misuse of Drugs Regulations.

Actions and Uses

The actions of apomorphine are peculiar in that the initial stimulatory properties of morphine are intensified, whereas the secondary depressant effects are minimal. The stimulatory effect is almost entirely specific to the vomiting centre, and, following administration, vomiting occurs within 2 or 3 min preceded by nausea and salivation; emesis may continue for 5 to 15 min depending upon the size of dose and susceptibility of the individual. Exhaustion is often quite marked and is intensified by the depressant after-effects of the apomorphine.

The use of this drug is indicated where poisons have been recently ingested or where smooth foreign bodies have been swallowed. It should not be used in advanced poisoning cases for the additional shock may kill the patient.

Administration should not be repeated if the first dose fails in its object, for excess first causes excitement and subsequently marked depression, coma and death.

Dosage

For the dog the dose varies from 1·5 to 7·5 mg ($\frac{1}{40}$ to $\frac{1}{2}$ gr.) subcutaneously, with an average of 3 mg ($\frac{1}{20}$ gr.) Oral dosages are approximately twice these quantities, but less reliable effects are achieved by this route.

Antemetics

Drugs which assist in suppressing emesis can be classified under several headings.

1. Demulcents which coat, protect, lubricate and soothe the gastric mucosa.
2. Local gastric sedatives, which include the groups of antacids and alkaline stomachics and act either by acid neutralization, by coating the gastric mucosa, or by local nerve sedation.
3. Central gastric sedatives, such as hyoscine and certain tranquillizers which exert their effects via the central nervous system.

The aim of antemetic administration is to suppress the vomiting reflex, so conserving the animal's strength and the chloride content of the stomach, blood and tissues. Continued vomiting is extremely exhausting and causes severe shock. Once foreign material has been removed, the reflex should be suppressed as soon as possible. Where emesis is initiated centrally the cause must be diagnosed and treated, for in such conditions as nephritis with uraemia no amount of local gastric sedation, and only rarely central sedation, will prevent vomiting.

Demulcents

These have already been mentioned and amongst them dextrose, and possibly glycerin, are the most effective antemetics in this group.

Local Gastric Sedatives

These may be divided into antacids, inert coating agents and local nerve sedatives.

Antacids. These are widely used in both human and veterinary medicine for the neutralization of the hydrochloric acid of the stomach. Although gastric hyperacidity does occasionally occur in small animals, it is doubtful whether wide use of antacids is justified, for it is usually only a symptom of far greater disturbances. In ruminants the effect of pH on rumination is only just becoming known, but the long established use of bicarbonates and carbonates—both ammonium and sodium salts—can often be justified where excessive carbohydrates have been fed or where such conditions as ketosis are present.

Rapid acting compounds are those normally soluble in water, such as sodium bicarbonate and ammonium bicarbonate. (Ammonium carbonate consists of a mixture of bicarbonate and carbamate and constantly evolves ammonia. The bicarbonate is stable and almost odourless yet retains the activity of the other compound.) The slower antacids are the less soluble compounds such as the carbonates of calcium and magnesium, the hydroxides and oxides of these, and the even less soluble silicates of aluminium and magnesium. Aluminium hydroxide gel and, more recently, the amino acid coordination compounds of bismuth and aluminium—bismuth and aluminium glycinates for example—are more modern additions to the range.

The antacid effect of carbonates is only

temporary, as the carbon dioxide released by the chemical reaction collects in the cardia, distends the stomach before it is released by sphincter relaxation, and in doing so causes a reflex secretion of acid and digestive enzymes. This is the so-called 'acid re-bound' action.

It is usual, where true neutralization is required, to combine the rapidly acting sodium bicarbonate with the slower acting oxides, carbonates or trisilicate of magnesium, and kaolin (a hydrated aluminium silicate). The choice of slow acting drugs depends on whether a laxative or antidiarrhoeic action is required in the intestines, where the basic ion is of more importance than the acid radical. Calcium, bismuth and aluminium tend to have an antidiarrhoeic action, whereas magnesium is laxative. (Although the trisilicate dissociates so poorly that the ions are relatively inactive and mechanical coating effects superimpose themselves to give an antidiarrhoeic action. This effect is aided by the adsorptive forces which attract and fix toxins present in the intestines. This toxin adsorption property is also possessed by kaolin, but to a lesser degree).

For the treatment of most gastritis conditions the glycinates and trisilicates are superior to the simple alkalis.

Local mucosal nerve sedatives. Where irritation of the gastric mucosa is excessive, local nerve sedatives may be used to assist the demulcent and mechanical coating agents. These consist mainly of less soluble, topically applied, local anaesthetics such as benzocaine and amethocaine, used for small animals in doses of about 0·3 g (5 gr.) three times daily. Chlorbutol is also used in similar dosage. In general this method of dealing with the problem is rarely satisfactory.

Central Gastric Sedatives

Many central nerve sedatives will suppress the vomiting centre in the medulla along with other centres and brain areas. Chlorbutol (Chloretone) has already been mentioned for its local effect on the gastric mucosa, but, following absorption, this drug also has a central sedative effect, including depression of the vomiting centre. The compound has thus been widely, and often effectively, used in human and veterinary medicine as a travel sickness remedy, which, as with most central antemetics, must be administered 15 to 30 min before starting the journey. Further investigation into central antemetics has shown that hyoscine, in doses up to 0·6 mg ($\frac{1}{100}$ gr.) and its near relatives, and, more recently still, chlorpromazine, have definite antemetic properties superior to the older compounds. The many phenothiazine derivatives now marketed for their antiemetic effects are of value in the prevention of motion sickness and in the treatment of persistent vomiting of whatever cause. Antiemetic potency can be assayed by the suppression of apomorphine-induced vomiting in dogs.

Bibliography

Bass, P. (1974) Gastric antisecretory and antiulcer agents. *Adv. Drug Res.*, **8,** 205–329.

Brand, J. J. & Perry, W. L. M. (1966) Drugs used in motion sickness. *Pharmac. Rev.*, **18,** 895.

Dunlop, R. H. (1972) Pathogenesis of ruminant lactic acidosis. *Adv. vet. Sci. comp. Med.*, **16,** 259–93.

Inch, T. D. & Brimblecombe, R. W. (1974) Anti-acetylcholine drugs: chemistry, stereochemistry and pharmacology. *Int. Rev. Neurobiol.*, **16,** 67–139.

Sellers, N. F. & Stevens, C. E. (1966) Motor functions of the ruminant forestomach. *Physiol. Rev.*, **46,** 634.

Spiro, H. M. (1970) *Clinical Gastroenterology*, London: Macmillan.

Wang, S. C. (1965) Emetic and antiemetic drugs. In: *Physiological Pharmacology* (ed. W. S. Root & F. G. Hofmann), vol. 2, pp. 256–316. New York: Academic.

6

The Alimentary Canal

II. The Intestines

The drugs used expressly for action on or in the intestines are divided into a number of groups: (1) purgatives and laxatives, (2) astringents and antidiarrhoeics, and (3) antiseptics. The anthelmintics and antibiotics are dealt with in Chapters 27 to 32, and Chapter 37.

PURGATIVES AND LAXATIVES

A *purgative* is a drug which causes marked intensification of intestinal activity and results in the expulsion of intestinal contents from the colon and rectum. The terms *aperient* and *cathartic* are also used to describe this group of compounds.

A *laxative* causes a similar action but the effect is milder and rarely results in more than a slight intensification of intestinal activity. Frequently a laxative action is achieved by small doses of drugs which in larger doses will cause purgation, but there are some agents which will only produce a laxative effect no matter how large the dose. Liquid paraffin is one of these.

Superpurgation may follow overdosage of, or idiosyncrasy to, purgatives. It consists of extreme and continued intestinal activity indicated by frequent and painful evacuation of fluid and possibly dysenteric (bloodstained) faeces. This causes shock and dehydration which may result in collapse and death. It is a difficult condition to treat and mortality is fairly high. Horses are more susceptible than other species and less responsive to treatment.

It follows that purgatives must be used intelligently and only when necessary. At one period of medical and veterinary history purging was, apart from bleeding, the only remedy for the majority of illnesses, but this position happily no longer exists and the frequent use of these drugs is not justified.

Before prescribing or administering purgatives, particularly in small animals, the veterinary surgeon must be certain that purgation is required, for owners frequently are unable to distinguish between the straining due to constipation and that which follows diarrhoea (frequent evacuation of fluid or semi-fluid, faeces).

There are a considerable number of purgatives, and classification is essential for a clear understanding. The four main groups of intestinal stimulants are:

1. Lubricant laxatives
   Liquid paraffin, white and yellow soft paraffin
2. Bulk purgatives
   (a) Simple: Agar-agar, pysllium, sodium carboxymethylcellulose, wheat bran
   (b) Saline: Sodium chloride, magnesium oxide, carbonate and sulphate, potassium and sodium tartrate
3. Irritant purgatives
   (a) Direct: Mercury compounds, sulphur, phenolphthalein, vegetable oils
   (b) Indirect: Anthracene compounds—aloes senna, cascara, rhubarb, synthetic anthraquinones
   (c) Drastic: Jalap, colocynth, scammony, podophyllin, croton oil, barium chloride
4. Neuromuscular purgatives
   Parasympathomimetic drugs—carbaminoylcholine, arecoline, physostigmine, pilocarpine

Lubricant Laxatives

The drugs which comprise this group have a laxative rather than purgative action, for their effect is entirely mechanical. The chemical and physical characteristics of these drugs governs their activity. They are long chain hydrocarbons which are extremely resistant to chemical and physiological reactions, and particularly digestion. This ensures that once administered they remain almost entirely unchanged and unabsorbed throughout the length of the alimentary canal; their oily and lubricant character is therefore used to the maximum throughout. This oily property enables the drugs to coat the mucosa and intestinal contents, and so smooth the passage of faecal material.

These compounds have a number of disadvantages: (1) As the intestinal contents are essentially aqueous, there can be little mixing of inert hydrocarbon and faeces. Chronic constipation or impactions will not benefit or may even be aggravated by the oil, for the coating of the mucosa decreases the effect of any irritant action the intestinal contents may have. This decrease in irritability may result in the further slowing of intestinal movements.

(2) Coating of the mucosa prevents contact with food in the intestine so that absorption may be hindered. Not only normal protein, carbohydrate and fat absorption, but that of essential supplements, particularly the fat soluble vitamins, may be radically affected. Consequently continued use, especially in young animals, is contra-indicated.

(3) Undue reliance is sometimes placed on the degree of activity of these compounds when mild purgation is required. This may result in dangerous and severe affects. A good example is the use of liquid paraffin with oil of chenopodium for worming young puppies. The paraffin is insufficient to expel either the excess of the worming mixture or the worms, but more important still, the excess chenopodium is absorbed and causes poisoning. Some of the 'patent remedy' worming mixtures have in the past been guilty of this error.

(4) In small animals, liquid paraffin may actively encourage the continued existence of faecoliths, for the oil aids the passage of soft material past the solid mass. The owner—and sometimes the veterinary surgeon—can be misled by the fact that the animal continues to defaecate.

Liquid and the soft paraffins, white and yellow, are the main examples of this group. Of these, liquid paraffin is the most widely used.

Liquid Paraffin, B.Vet.C.

This is a clear, 'water white' fluid residue from petroleum distillation, of varying density, with no odour, but a bland oleaginous 'taste'. There are available two grades, heavy and light. The fluid is often referred to as mineral oil.

Internally liquid paraffin has a demulcent action in the mouth, oesophagus and stomach. In the intestines it is a typical lubricant laxative. It has the disadvantages of the group but, additionally, large doses or continuous administration will result in anal seepage and the soiling of bedding, furniture and carpets when administered to small animals. Some slight absorption is reported to produce a benign tumour of the intestine when administration is continued over a long period.

Dosage

The pig with a dose of 60 to 300 ml (2 to 10 fl. oz.) and the dog 4 to 30 ml (60 m. to 1 fl. oz.), are the animals in which liquid paraffin is most frequently used.

No matter by how much the dose is increased, the laxative action cannot be increased to one of purgation.

Externally the drug has very wide *emollient* uses. (An emollient is a fat or an oil used on the skin for its softening and protecting effect. The oil film acts by preventing the evaporation of water from the skin.) A wide variety of pharmaceutical preparations use liquid paraffin as a base: creams, emulsions, pastes and suspensions.

For the removal of foreign bodies from the eye, flooding with liquid paraffin is quite efficient and usually available as an emergency home remedy.

Finally, liquid paraffin is a useful lubricant for diagnostic instruments—probangs, stomach tubes, catheters, urethral sounds, etc.

Petroleum jelly and white soft paraffin have occasionally been used in pill or bolus form as laxatives.

Bulk Purgatives

The members of this group depend for their action on an increase in volume of intestinal contents. This increase causes distension of the intestines, which induces a reflex countering contraction of musculature and an increase in power and speed of peristalsis.

The methods by which this increase in volume is achieved divide bulk purgatives into two groups. Both groups utilize water, but the *simple bulk purgatives* adsorb water and themselves increase in bulk. *Saline bulk purgatives*, by virtue of the high osmotic pressures which they exert as hypertonic solutions, not only prevent the absorption of water which normally takes place but also withdraw water from the tissues into the intestines. It is therefore the bulk of the water which is the stimulating factor.

The simple bulk purgatives. Indigestible cellulose is responsible for the water-holding properties of the traditional bran mash and the newer alternatives agar, plantain seeds or psyllium, and sodium carboxymethylcellulose.

As a group these compounds are mainly of use in small animals. They have only mild stimulant effects which are maintained throughout the whole length of the intestines, and are particularly useful when sharp foreign bodies have been swallowed—needles, sharp bones and stones. Doses of up to 10 g (150 gr.) of Agar, B.Vet.C., are recommended for dogs.

Saline bulk purgatives. All members of this group owe their laxative and purgative properties to their ability to draw water from the tissues into the lumen, or to retain there fluid already present. This ability is shown only when quantities large enough to create hypertonic concentrations in the intestines are given. It follows that no saline purgative should be used—nor would one be effective—in dehydrated animals where intestinal and tissue fluids are depleted.

A minor part of the purgation is in some cases due to a slight irritant action of the drugs, either directly or because of their hypertonicity.

It is usual to administer saline purgatives in hypertonic concentrations, for it would be impracticable to administer the quantity of water necessary to achieve isotonicity relative to the doses of the salts. This delays the action of the compounds, for time is needed for the osmotic pressure to exert its full effect and withdraw into the lumen sufficient water to distend the intestinal wall and institute, or increase, peristalsis.

Saline purgatives will mainly affect the small intestine. It is obvious that volumes capable of great distension of the small intestine could be easily accommodated in the large intestines of most animals. However, the hypertonicity of the solution prevents the normal dehydration of faeces which occurs in the large intestine, and the lower colon and rectum will therefore be stimulated to some degree by the volume of fluid retained. The action will be noticed normally after 3 to 12 hr in simple stomached animals, but nearer to 18 hr after dosing ruminants. Saline purgatives are less reliable in horses.

As the osmotic pressure of saline solutions is related to the number of ions present in those solutions, the strongest and most persistent purgative effect will be exerted by those compounds which are not rapidly absorbed into the tissues and yet are freely soluble and ionized. The best effect therefore is exerted by magnesium sulphate (Epsom's salts) which is freely soluble, strongly ionized and produces two ions both of which are only slowly and sparingly absorbed. Sodium sulphate (Glauber's salts) is both a saline purge and a diuretic.

Sodium chloride, the salts of tartaric acid, and the carbonate and oxide of magnesium, are all less efficient and tend to be laxative rather than purgative.

Dosage

Dosage of Magnesium Sulphate, B.Vet.C., is representative of this group of saline purges.

As a purgative:
- Cow: 0·24–0·48 kg (½–1 lb.)
- Sheep: 60–120 g (2–4 oz.)

As a laxative:
- Horse: 30–60 g (1–2 oz.)
- Cow: 60–120 g (2–4 oz.)
- Sheep: 7·5–15 g (¼–½ oz.)
- Pig: 15–30 g (½–1 oz.)

Irritant Purgatives

This group of purgatives can be further subdivided to distinguish between those compounds which: (1) do not require absorption, i.e. *direct irritant*; and (2) those which do require absorption, i.e. *indirect irritant purgatives*. A further division includes a number of miscellaneous *drastic purgatives* which are no longer in general use, and which are, in the main, complex vegetable compounds: the group includes such drugs as *podophyllin, scammony, colocynth,* and *croton oil.* Barium chloride, which was once used as a smooth muscle stimulant and intestinal irritant, is an exception to this vegetable origin; 0·5 g (7½ gr.) doses were injected intramuscularly in horses, and 2 g (30 gr.) are sometimes still used as a ruminal and intestinal stimulant in cattle.

Direct irritant purgatives. This group includes mercury compounds, sulphur, vegetable oils, and phenolphthalein, all of which affect the small intestine. Some are therefore used as adjuvants for the indirect irritants as they commence their action as soon as they arrive in the small intestine.

Mercury compounds which can be used are those which are largely insoluble and do not ionize to any extent. Thus, metallic mercury and calomel are the most commonly used examples. The metal itself has no purgative or laxative action unless the globules are broken up and the dispersed particles prevented from recoalescing. Chalk and liquorice are used as dispersing agents in Grey Powder and Mercury Pill, B.P.C., respectively. The increase in surface area allows the gastric juices more opportunity to attack the metal or salt, and to dissolve and ionize minute quantities which are sufficiently irritant to cause an increase in peristalsis.

Calomel, or mercurous chloride, is slightly more soluble than metallic mercury and is used in 2·3 g (40 gr.) doses as an adjuvant to aloes for the horse, in 0·3 to 2·3 g (5 to 40 gr.) doses for the pig, 15 to 120 mg (¼ to 2 gr.) for the dog, and 10 to 65 mg (⅙ to 1 gr.) for the cat. Action can usually be expected within 6 hr in the smaller species, for irritation of the small intestine commences immediately. The whole of the alimentary tract will be affected.

The possibility that a portion of the dose may be changed to the soluble, toxic mercuric salt has caused the mercury purgatives to fall into disuse.

Sulphur has never been extensively used as a laxative in veterinary medicine. As with mercury, pure sulphur has little or no action but, when converted into sulphides and polysulphides by intestinal secretions, these sulphides exert an irritant and therefore stimulant effect on the whole length of the intestines.

Phenolphthalein is a white powder, soluble in dilute alkali and mainly used as a laboratory indicator. It possesses useful small and large intestine irritant properties.

The effect on different species of animals is not constant, and the use of phenolphthalein is mainly confined to the pig, dog and cat, for which the doses are 1 to 2 g (15 to 30 gr.), 0·03 to 0·2 g (½ to 3 gr.) and 15 to 65 mg (¼ to 1 gr.), respectively.

The main advantage of phenolphthalein is the prolonged period of its action. A proportion of any dose is absorbed, and of this a considerable amount is re-excreted via the bile into the small intestine in conjugated form. The conjugate can be split by bacterial β-glucuronidase and so some of the dose will be re-absorbed. With each cycle, however, some drug is lost. As faecal and urinary excretion deplete the dose, the colonic concentration drops below an effective level. This cyclical action may continue for several days after an initial action within 4 to 6 hr and the compound is therefore of most use in the treatment of chronic constipation.

Vegetable oils are probably the most efficient and safe of the direct irritant purgatives. As a group they act by combining in a saponification process with the alkaline bile salts of the small intestine. This process produces monobasic, dibasic and tribasic soaps. Sodium and potassium soaps are the most important ones as they are soluble and capable of exerting an irritant action on the intestinal mucosa. Glycerol is a by-product of this saponification and is useful in lubricating and breaking up the faecal masses; it also assists defaecation by exerting a strong osmotic pressure to retain fluid within, and possibly attract fluid into, the intestinal lumen.

The soaps produced by each oil differ in their irritant properties so that we have olive oil producing the mildly irritant salts of oleic acid, and at the other extreme castor oil giving ricinoleates and linseed oil producing linoleates and linolenates, all of which are relatively strong irritants. Small amounts of stearates and palmitates are also produced.

Oleaginous purgatives can be used in all species, but are mainly of value in non-ruminants. The oils in common use are linseed oil for the horse, 0·5 to 1 litres (1 to 2 pt.) and castor oil 30 to 180 ml (1 to 6 fl. oz.) for the foal, calf and pig, 4 to 30 ml (60 m. to 1 fl. oz.) for the dog.

The effect is seen within 4 to 8 hr in the smaller species, and in 12 to 18 hr in the horse. There is an almost complete clearance of the intestines when these oils are used, and a temporary suspension of defaecation follows until sufficient bulk reaches the small colon to restimulate the defaecation reflex. For this reason oleaginous purgatives are not indicated in chronic constipation, where the condition would be aggravated after a temporary relief. They should be followed by moist bulky food.

Both linseed and castor beans figure in plant toxicology. Linseed contains a cyanogenetic glycoside which gives prussic acid when acted upon by enzymes also present. The enzymes are activated by steeping the seed in luke warm water, and sufficient acid can be produced to kill adult cattle. Linseed tea for nursing convalescent animals should therefore always be prepared with boiling water to inactivate the enzymes.

Castor bean husks contain a toxalbumen *ricin* which is one of the most toxic substances produced by any plant. It occasionally causes trouble when cottonseed cake is contaminated with castor beans.

Indirect irritant purgatives. These are agents which require absorption and metabolization before substances are produced which will, on excretion into the intestine, exert an irritant action and stimulate intestinal movement. The main members of the group are aloes, cascara sagrada, senna and rhubarb—all plant derivatives. Synthetic replicas of the substances into which the active ingredients of the plant compounds are metabolized—anthraquinones—are now used more extensively. From these substances the common title of the group is derived—the *anthracene purgatives*.

Aloes, B.Vet.C.

This is the type compound and was the most widely used anthracene purgative until the introduction of synthetic anthraquinones to veterinary medicine. When the aloes reaches the small intestine the soluble active principle is dissolved and absorbed. At this stage the compound is inactive and relatively non-irritant. Within the body the principle circulates and is metabolized, probably in the liver, releasing into the system the active agents, or emodins, which in fact are anthraquinones. These are excreted into the large intestine which is irritated and stimulated.

Because of this excretion into the large intestine, these anthracene purgatives are of most value in those animals which suffer from impactions of the large intestine. The horse, with its numerous colon flexures and its decrease in diameter at the pelvic flexure, benefits particularly from this type of purgative. All non-ruminants respond well to anthracene purgatives.

Because of this complex cycle, purgation of the horse is delayed for at least 18 hr. In other animals the action of aloes is more rapid, although its use in these species is now infrequent.

The dose of aloes for the horse varies from 8 to 30 g (120 gr. to 1 oz.).

Cascara sagrada, B.Vet.C. (Sacred Bark) of Spanish South-American origin has been used in liquid extract and pill form for dogs and cats. The dose is 2 to 4 ml (30 to 60 m.) of the extract for the dog, and 0·1 to 0·5 g ($1\frac{1}{2}$ to 8 gr.) of the dry extract. **Senna,** used in the form of a fresh infusion of the seed pods, is not often used in veterinary medicine.

The final example **Rheum** (Rhubarb, B.Vet.C.), the rhizome of a Far Eastern rhubarb, contains various emodins including chrysophanic acid, a tetrahydroxy-anthraquinone. This is a dye, and may stain alkaline faeces and urine a reddish brown. It is also peculiar in that the powder and other derivatives contain a high proportion of *tannic acid.* This is a powerful intestinal astringent and will exert a constipatory action following the initial purgative effect of the anthraquinone.

These anthracene purgatives of plant origin are now superseded by those of synthetic origin, e.g. **1 : 8-dihydroxy-anthraquinone**. It is used mainly in the form of a dispersible powder, when a wetting agent such as the detergent sodium lauryl sulphate is included.

The advantages of synthetic compounds apply to the anthraquinones. They are of standard unvarying composition; solution and absorption rates are approximately constant under the same conditions of administration. It follows that, with the exception of idiosyncratic individuals, a forecast of the effect for a given dose can be given with assurance. Small doses will be laxative and large doses purgative. The time elapsing between dosing and action will vary according to the fullness of the intestine and the dilution of the drug. If dissolved in large quantities of water, action in the horse will follow 10 to 12 hr after dosing. If given in powder form to an animal with a full stomach, 24 to 36 hr may elapse. Young animals and race-horses are susceptible, and doses should be split and given at intervals of 8 to 12 hr. This is routine practice for the small species.

Dosages in the horse are approximately the same as for aloes; those for pigs are almost equal to the horse doses.

Horse and Cow: 10–30 g (150–480 gr.)
Foals and Calves: 3–5 g (45–75 gr.)
Sheep and Goats: 4–6 g (60–90 gr.)
Pigs (up to 8 weeks): 3–5 g (45–75 gr.)
  Porkers: 5–15 g (15–25 gr.)
  Adults: 10–20 g (150–300 gr.)
Dogs: 0·15–3 g (2½–45 gr.)
Cats, Rabbits, etc.: 0·15–0·3 g (2½–5 gr.)

The Neuromuscular Purgatives

This group of drugs is composed essentially of parasympathomimetic agents and the reversible cholinesterase inhibitors. The parasympathomimetics or cholimimetics are drugs which are able to reproduce the effects of stimulation of the parasympathetic nervous system and are usually structurally related to acetylcholine. These are described first.

Used as a purgative in the horse the first effect noticed is an increase in discomfort. The animal may kick at the flank, stamp, lie down and get up, and show general signs of uneasiness. It is at this time that walking the animal up and down distracts his attention and prevents him attempting to go down and roll. This would be dangerous because of the possibility of 'twist', that is, torsion of the intestines. Salivation commences usually within 5 min of injection and is quite profuse, clear, thin and watery after the initial frothing. Patchy sweating occurs after about 10 min. Within 20 min of injection soft and sloppy faeces are passed, quite often mixed with fluid; the faeces usually consist of undigested foods and are light green in colour. The effect begins to wear off in half an hour and within the hour, if there is no impaction, the animal is back to normal.

Where intestinal impaction is present most practitioners consider it unwise to use neuromuscular purgatives, owing to the danger of excessive muscular contractions resulting in rupture of the intestine. The intensification of the symptoms immediately after injection is also rather alarming to the owner, and sometimes to the veterinarian! Where death follows the veterinary surgeon may be held responsible, and it should be remembered that death is not an infrequent sequel to an attack of colic in the horse.

Acetylcholine and Pilocarpine. These drugs are too transient in effect and too toxic, respectively, to be of use as purgatives.

Arecoline. Betel nut has been known for many years as a digestive stimulant and is still chewed by the natives of the East Indies and associated areas. Besides its intestinal stimulant effect areca nut is also well known for its anthelminthic properties, particularly in respect of tapeworms. The smooth muscle stimulant effect of arecoline is similar to pilocarpine, although some actions, such as the miotic effect, are not so marked.

In horses arecoline can be used in doses of up to 0·14 g (2 gr.) as a neuromuscular purgative. In cattle its use in doses of 4 to 33 mg ($\frac{1}{16}$ to ½ gr.) as an oesophageal and ruminal stimulant was at one time widespread but, as in other species, carbamoylcholine has almost entirely replaced it.

Carbachol, B.Vet.C. Also known as carbamoylcholine. A synthetic drug with parasympathomimetic actions, carbachol is now the most commonly used drug of the neuromuscular purgatives. The usual form for large animals is a subcutaneous injection containing 1 mg per ml, sometimes expressed as a 1 : 1000 solution.

Administered subcutaneously the drug has an action similar to, but more prolonged, than other drugs of this group. The side actions on the bronchi, heart and circulatory system and the secretory system are said to be less than those of arecoline; there is relatively no action on the eye. The drug is also active by mouth.

In cattle some stimulant effect is exerted on the rumen, and it is particularly used to recommence rumination when stasis has accompanied milk fever.

In dogs the 1 : 1000 solution has the effect of causing vomition, whereas a 1 : 10 000 solution has the ordinary purgative effect.

Dosages for subcutaneous administration are:

Horse and Cow: 2–5 mg
Foal: 0·5–1 mg
Pig: 0·5–2 mg
Sheep: 0·25–0·5 mg

and as an emetic in the dog 0·25 to 1 ml of a 1 : 1000 solution, or as a purgative 0·25 to 1 ml of a 1 : 10 000 solution.

Cholinesterase inhibitors are represented by Physostigmine Salicylate, B.Vet.C., obtained from the Calabar, or Ordeal bean (so called because of its use by native witch doctors for trial by ordeal), and two synthetic compounds with similar actions, Neostigmine Bromide, U.S.P., and Neostigmine Methylsulphate, B.Vet.C.

These exert their action by blocking the enzyme cholinesterase, which normally breaks down the acetylcholine released at nerve endings. The inhibitors have a great affinity for the enzyme and combine with it, so preventing the breakdown of acetylcholine; the action of normally produced acetylcholine is therefore prolonged.

Physostigmine Salicylate, B.Vet.C.

This is alternatively named *eserine salicylate*. It has all the actions of a parasympathetic system stimulant, but the action on smooth muscle is greater than its secretory action so that salivation is not so profuse as with pilocarpine. Despite this fact, the parenteral injection of physostigmine is rare and the miotic action achieved by the use of the 0·125 per cent eye ointment, or the lamellae containing 0·06 mg, is the only one worth mentioning. **Neostigmine** is sometimes preferred for this miotic action as its antagonistic action is greater. The milder stimulant effect on smooth muscle is utilized in the treatment of atony and obstruction of the urinary tract rather than for a purgative action, although some practitioners, particularly in the United States, prefer neostigmine for the treatment of ruminal stasis and impactions. In the dog and cat atony of the bladder and sometimes urethral calculi obstructions can be successfully treated with this synthetic compound. Obtainable in ampoules and multi-dose vials neostigmine is used in doses of 4 to 25 mg in the horse and cow, and 0·25 to 2·5 mg in the dog and cat.

Atropine and curare overdosage are best counteracted with neostigmine injections.

Enemata

On occasion, e.g. for the removal of impacted rectal and colonic contents in the dog or in the treatment of colic, it is not advisable to rely on a purgative. In such cases, it is preferable to infuse large volumes of warm fluid per rectum. The distension and in some cases slight irritation caused by the fluid stimulates contractions in the colon and rectum while the fluid also softens the faecal mass and helps to float out such material as undigested bone fragments. Warm, soapy water or glycerin are frequently used for this purpose. Softening of rectal contents may also be achieved by the administration of a surfactant, e.g. sodium dioctylsulphosuccinate.

Cholagogues

This old term refers to drugs which stimulate bile flow. It is nowadays preferable to distinguish between drugs which merely cause contraction of the gall bladder, the cholekinetics, and those which actually stimulate the secretion of normal bile, the choleretics.

Cholecystokinin is the physiological cholekinetic but certain drugs are able to cause the discharge of stored bile, e.g. arecoline and oxytocin.

The bile salts have been used in human medicine for many years as choleretics. Compounds traditionally used as cholagogues include magnesium sulphate, calomel and rhubarb.

The compound 4(4-methoxynaphthalene-(1))-4-oxobutyric acid (Genebile) has been promoted as a true choleretic as distinct from a hydrocholeretic, i.e. a drug which stimulates the secretion of dilute bile. Experiments on conscious ruminants with exteriorized bile ducts have not supported this claim.

These agents are used in man in the treatment of gall bladder disease and to promote the absorption of dietary fats and vitamins.

INTESTINAL ASTRINGENTS

An astringent is a compound which precipitates protein, whether intracellular or extracellular. If intracellular precipitation occurs the process should be reversible and the cell should not be destroyed. Should the cell be destroyed, then the substance is not an astringent but a caustic. The protein precipitate is usually soluble, either in excess of protein or excess of astringent. If excess of astringent is present the re-solution may involve destruction of the cell and irritation and damage to the tissue or organ.

The precipitate of protein covering the surface of a cell or tissue is relatively impermeable to the passage of fluids in either direction. It also acts as a barrier between tissue and irritant, and the underlying tissue is therefore soothed and protected from further damage, so that healing can take place. The protecting and soothing effect is enhanced by an astringent action on any exposed nerve endings so that pain is lessened; the toxins and histamine-like substances released in damaged tissues are not absorbed. Local and generalized shock are both reduced whilst the recovery rate is speeded up.

Intestinal astringents are subdivided again into *vegetable* and *mineral astringents*. Tannic acid and catechu are commonly used examples of the first group, whilst the sulphates of iron, copper, zinc and aluminium, with one or two other compounds such as ferric chloride, are examples of metallic astringents in common use.

Vegetable Astringents

Tannic acid is found in most plants in small quantities, but in oak gall (oak apples) the concentration is very high as it is also in catechu. The tannic acid content of rhubarb has already been mentioned, and the bitter astringency of stewed tea is known to all. Gallic acid is usually present along with tannic acid, of which it is a breakdown product; its astringent properties are poor and it is not used on its own.

Tannic Acid, B.Vet.C.

Tannic acid is a light amorphous powder which is freely soluble in water and alcohol. It is used internally in doses up to 24 g (360 gr.) in cattle and horses, 2·6 g (40 gr.) for sheep and goats, up to 1·3 g (20 gr.) for the pig and 0·6 g (10 gr.) for the dog. The solubility is such that the action is rapid in onset but conversion into gallic and pyrogallic acids, amongst others, is also rapid and, as these compounds are not astringent, the action quickly passes off. Continuous action, prolonged into the large intestine, is not usually achieved by tannic acid, and catechu, krameria, kino, and synthetic products of tannic acid are preferred, their lesser solubility and slow continuous release of tannic acid being more desirable. In the stomach, large tannic acid doses tend to cause irritation, nausea and sometimes emesis in the vomiting species. Precipitation of food protein which may occur in the stomach is followed by re-solution in the small intestine with the release of tannic acid. In the intestines precipitation of protein, particularly on the mucosal surface, produces a protective layer almost impervious to the passage of fluid in either direction. Stimulation of the inflamed mucosa is decreased, with a proportional decrease in peristalsis.

Precipitation of alkaloids and heavy metals by tannic acid is utilized fully in the treatment of poisoning. Insoluble tannates are usually formed and when precipitated can be removed by enemas, gastric lavage or mild purgation.

Catechu, or gambir is preferable to tannic acid for internal administration. It contains concentrations of catechutannic acid varying from 20 to 50 per cent. In veterinary practice catechu nigrum, a dark brown gritty powder, is more frequently used than the pale catechu of the B.Vet.C.

An ingredient of many astringent cattle drenches, catechu is usually combined in doses of 8 to 16 g (120 to 240 gr.) with an equal quantity of ginger, and up to 90 or 120 g (3 or 4 oz.) of a chalk and kaolin mixture. If in fluid form either chlorodyne or tincture of opium is also used, and the catechu is usually present in the form of the tincture, of which the dose for cattle is approximately 30 ml (1 fl. oz.). Catechu is preferable to tannic acid because the slow release of the active principle in the gut reduces its toxicity. The slow release of catechutannic acid is responsible for the prolonged effect of catechu, and inflammation of the large intestine can be materially affected by the solution of catechutannic acid from the catechu which has reached the lower part of the tract in an unchanged state.

Metallic Astringents

Aluminium Hydroxide Gel, B.Vet.C., is an antacid and mild intestinal astringent whose popularity as an antidyspepsia agent in human medicine rests also in part on the suggestion that it adsorbs pepsin.

The gel is a thixotropic and viscous, aqueous suspension of the hydrated oxide and carbonate of aluminium. (Thixotropism is the phenomenon shown where changes in character, related to changes in molecular shape and size, result when certain liquids or semi-liquids are agitated. Thus liquids become solid, and vice versa, when shaken. Many preparations of aluminium exhibit the phenomenon.)

ANTIDIARRHOEICS

Literally, these are substances which are useful in decreasing diarrhoea, but the term is more narrowly applied to those drugs which are almost inert and therefore coat and protect the intestinal mucosa. The qualification 'almost' is applied because some of the drugs have additional properties. For example, aluminium silicate in its various forms, and magnesium trisilicate are capable of adsorbing toxins; this property is extremely useful in bacterial enteritis.

The drugs involved are the inert salts of the alkaline earths and metals, calcium, magnesium, aluminium and bismuth. There are thus the carbonates of calcium and bismuth, the subnitrate and salicylate of bismuth, and the silicates of magnesium and aluminium: aluminium hydroxide, and combined bismuth and aluminium salts are also mild antidiarrhoeics as well as antacids.

Bismuth Salts

Only three salts of bismuth are nowadays used in veterinary practice; the carbonate, the salicylate and the subnitrate. These are tasteless and odourless, white or off-white powders which are insoluble in water, and tend to darken on exposure to light.

Carbonate. This can be used in small animals for its mild, slow antacid action in conjunction with other compounds. It also has some coating antidiarrhoeic properties. Another important feature, which should be remembered when suspecting intestinal foreign bodies, is that the compound is radio-opaque and, should X-ray photographs be taken, interpretation of these must take this fact into consideration. (Owners may administer gastric sedatives containing bismuth to allay the vomiting which is a cardinal feature of acute intestinal obstruction.) Where barium sulphate is not available, bismuth carbonate is a useful radio-opaque substitute.

Salicylate. Salicylate splits in the small intestine to give salicylic acid salts and bismuth carbonate. The salicylic acid salt is reputed to be mildly antiseptic, its irritant action being masked by the sedative carbonate. The compound is combined in scour mixtures with other intestinal sedatives and antiseptics, the commonest being compound mixture of bismuth salicylate with brilliant green. This mixture has been widely used for the treatment of white scours in calves, but has been largely superseded by the antibiotics. The salicylate is decomposed by bicarbonates such as those of potassium and sodium, so that mixing of these together is not advisable.

Subnitrate. Also known as the oxynitrate, is the most active antidiarrhoeic of the bismuth salts, but even so is not strongly active and its use is almost entirely confined to small animals. It is usually combined with chalk or kaolin.

Dosages of the three compounds are identical:

Horse and Cow: 15–30 g (½–1 oz.)
Calf and Foal, Sheep and Pig: 2–4 g (30–60 gr.)
Dog: 0·3–2 g (5–30 gr.)
Cat: 0·1–0·3 g (1½–5 gr.)

Calcium Carbonate, B.Vet.C. and Chalk, B.Vet.C.

These are white or off-white powders, the former being microcrystalline and the latter, which is a native form of the carbonate, amorphous. Both are almost insoluble in water but in other respects act in the same way as any carbonate. Both may be used as cheap skin sedatives and diluents for more expensive and more active powders for external application. Internally, the antacid effects of all carbonates are exerted, including the reflex stimulation of more acid. They have a general mechanical gastrointestinal sedative action by coating the mucous membrane with an inert, protective layer: this also has a mild antidiarrhoeic effect. Chalk is also used as a calcium supplement in feeding, but steam bone flour (sterilized bone meal) is preferred because of its phosphorus content and the correct physiological ratio of the two.

The dosages for cattle and horses are 30 to 120 g (1 to 4 oz.), for sheep and pigs, 8 to 15 g (120 to 240 gr.), and for dogs and cats up to 4·0 g (60 gr.). The doses of these compounds are determined more by their bulk than their activity.

Chalk is incorporated in Aromatic Powder of Chalk, B.Vet.C., and Aromatic Powder of Chalk and Opium, B.P. The latter contains the equivalent of 0·25 per cent anhydrous morphine and is therefore controlled by the Misuse of Drugs Regulations. Both drugs are used in dogs at the same dosage rate as the plain chalk and for the same gastrointestinal sedative action.

Magnesium Carbonate, B.Vet.C.

This exists in two forms, a light, fluffy, very bulky powder, and a heavy finely granular amorphous powder which is about 3½ times less bulky than the light form. From these two carbonates are prepared, by heating (calcination), the light and heavy magnesium oxides respectively.

All four may be *used* externally as skin sedatives, their action being purely mechanical. Internally, they have antacid properties: for true antacid action the oxides are preferred for they cause no reflex hydrochloric acid secretion as do the carbonates. Any magnesium chloride formed acts as a saline laxative in the intestines.

Magnesium Trisilicate, B.Vet.C.

Less soluble than either the oxide or carbonate. It has good antacid properties, though the action is a slow one. The salt reacts with gastric acid to yield the mildly laxative magnesium chloride and a mildly adherent gel of silicon dioxide. The gel is adsorbent (pepsin included) and protective. Magnesium trisilicate has relatively strong toxin adsorptive properties for which it is preferred before kaolin (aluminium silicate). Magnesium trisilicate is about twice as expensive as the carbonate or oxide.

Dosages of these five compounds are more or less identical: for foals and calves up to 16 g (240 gr.), pigs up to 8 g (120 gr.), lambs and dogs up to 2 g (30 gr.), and cats up to 0·6 g (10 gr.). Doses for the trisilicate are usually slightly less than those for the other four compounds.

Kaolin, B.Vet.C., is a natural aluminium silicate. In two forms, the heavy, mainly for external, and light, for internal use, kaolin is a smooth, odourless and tasteless, off-white powder. The light form is useful as an antidiarrhoeic in all species and ages. Like most silicates, kaolin possesses the property of adsorbing other compounds, including bacterial toxins and vitamins. Its adsorptive properties are not so powerful as those of magnesium trisilicate. Kaolin is, however, relatively cheap when compared with the magnesium compound; it is consequently very widely used in cattle and calf scour powders and mixtures, and in most gastrointestinal mixtures for the smaller farm animals and household pets. Chalk, kaolin, catechu and chlorodyne or morphine sulphate, makes a very useful gastrointestinal sedative combination.

The dosage of kaolin is determined more by its bulk and effect than by toxicity considerations, for kaolin is one of the least toxic of all drugs:

Horse and Cow: 50–250 g (1¾–8 oz.)
Calf and Foal, Sheep and Goat: 15–60 g (½–2 oz.)
Pig: 15–30 g (½–1 oz.)
Dog up to: 8 g (120 gr.)

Kaolin shares with all other adsorbents the problem of non-selectivity in its adsorbent action and decreases the absorption of tetracyclines and anticholinergics. Bentonite, fuller's earth, attapulgite and activated charcoal are further examples of adsorbents.

Constipants

Certain drugs possess the ability to decrease peristalsis. When intestinal passage is delayed, a greater amount of fluid is absorbed from the contents and this results in a marked anti-diarrhoeic effect. The alkaloids of opium delay the passage of intestinal contents by increasing the tone of intestinal smooth muscle and so impeding peristalsis. Compounds in this group are discussed in Chapter 21. It is also possible so to reduce intestinal mobility and secretory activity with atropine-like substances, that faecal consistency will return to normal.

Antispasmodics

The ability of atropine to interfere with the peripheral actions of acetylcholine, its so-called parasympatholytic activity, is also utilized to diminish intestinal motility and secretions. Spasmolytics diminish gut contractions and thereby reduce the pain of colic and, by reducing the rate of passage of intestinal content, exert a constipant action. Unfortunately, the unwanted blockade of muscarinic sites elsewhere in the body gives many side effects. *Methantheline*, *propantheline* and *oxyphenonium* exemplify the many atropine-like compounds for which a degree of gut-selectivity is claimed. Many such analogues have direct smooth muscle relaxing activity unrelated to muscarinic receptor blockade. *Benzetimide* is a very recent analogue for which a useful degree of gut-specificity in ruminants is claimed. Other alkaloids with atropine-like activity are hyoscine (scopolamine) and *l*-hyoscyamine.

Opium and certain of its alkaloids have contributed valuably to the non-specific control of scouring for many years (see Chapter 21). *Diphenoxylate* is structurally related to the synthetic analgesic pethidine. It has powerful constipant activity at doses substantially lower than those necessary for CNS effects. It is also noteworthy that the very limited solubility of this analogue means that it is unlikely to be a candidate for drug abuse.

Intestinal Antiseptics

This group of drugs was believed to have an antiseptic action against infections of the intestines. The historical drugs associated with intestinal antisepsis are salol (phenyl salicylate) and guaiacol, but although salol releases phenol when split in the intestines this is produced so slowly that little antibacterial action results. The feeble action of the intestinal antiseptics has resulted in their being replaced by more potent antibacterial agents such as the antibiotics and sulphonamides.

Bibliography

FAIRBAIRN, J. W. (ed.) (1976) The anthraquinone laxatives. *Pharmacology*, **14**, Suppl. 1.

HOLTON, P. (1973) *Pharmacology of Gastrointestinal Motility and Secretion*, vols. 1 and 2. Section 39a in *International Encyclopaedia of Pharmacology and Therapeutics*. Oxford: Pergamon.

ROOT, W. S. & HOFMANN, F. G. (ed.) (1967) Cholinergic drugs. In: *Physiological Pharmacology*, vol. 3, pp. 1–263. New York: Academic.

TRUELOVE, S. C. (1966) Movements of the large intestine. *Physiol. Rev.*, **46**, 457.

WARE, M. (ed.) (1971) Gastro-intestinal agents. *Today's Drugs*, **2**, 143–4.

7

The Respiratory System

The circumstances in which drugs may be used to influence the respiratory system are firstly, where there is inflammation of the lungs and the airways leading to the lungs, and secondly, where there is interference with the mechanics of respiration. Inflammatory conditions include pneumonia, bronchitis, tracheitis and laryngitis, caused by bacteria, viruses and worm parasites. Interference with the respiratory rate and depth is usually related to depression or stimulation of the medullary centres, the former occurring most commonly during general anaesthesia whilst stimulation may occur physiologically in fevers and diseases which interfere with oxygenation of the blood. The mechanics of respiration are seriously impaired by conditions which lead to broncho-constriction, e.g. hypersensitivity reactions.

The early stages of inflammation of the lungs and air passages are usually accompanied by an increase in viscidity, and a decrease in volume of mucus. This results in a dry, inflamed, and painful mucosa which is irritated at each successive inhalation and results in a harsh, dry, painful and entirely useless cough—frequently referred to as an 'unproductive' cough. In the treatment of this condition the aim is to cause an increase in mucous secretion to protect the mucosa and at the same time reduce the frequency of the damaging cough, partly by preventing stimulation of the mucosa by coating it with mucus and partly by centrally depressing the cough reflex. This can be achieved by the combined use of expectorants such as ipecacuanha, ammonium chloride and potassium iodide, with either codeine or opium. In the recovery stages of respiratory tract inflammations there is usually an excess of mucus and tissue débris in the airways, which prevents free passage of air and stimulates coughing with each respiratory movement. The aim of therapy in these circumstances is to increase the fluidity of the mucus, stimulate the activity of the cilia of the mucosa so that all the débris may be expelled by coughing and ciliary movement. Ipecacuanha, squill, ammonium chloride, and potassium iodide, may be given orally for this purpose. At this stage coughing should be reduced only if it is continuous and preventing the animal and its owner from resting and sleeping.

Expectorants will increase the fluidity and volume of respiratory secretions, and in the early stages of inflammations will relieve pain and tend to reduce slightly the incidence of coughing. Chronic inflammatory conditions will also benefit.

Cough sedatives reduce the incidence of coughing usually by central depression of the cough centre. They should not be used when respiratory secretions are copious unless coughing is excessive and causing exhaustion—this state is uncommon in animals.

Bronchodilators will cause the smooth muscle of the airways to relax and can therefore contribute significantly to the relief of dyspnoea.

By destroying the infecting agents in the respiratory system, **chemotherapeutics** which are effective against bacterial and helminth parasites will assist in respiratory sedation but are not considered as cough sedatives in normal classifications. The agents which are used against helminth parasites and the antibacterial agents are considered in Part III.

In respiratory depression during anaesthesia **analeptics,** which cause convulsions if administered to normal animals, will stimulate the respiratory centre in the medulla, increasing rate and depth of respiratory excursions. Analeptics are detailed along with other stimulants in Chapter 24. Various other compounds such as ammonia and carbon dioxide may be used as respiratory stimulants during anaesthetic depression.

EXPECTORANTS

The expectorants may be classified according to their route of administration and their site of action.

1. *Local expectorants:* these act directly on the mucosa of the respiratory airways and may be applied: (a) by inhalation, e.g. steam, terebene, eucalyptus, benzoin, iodine vapour, etc., or (b) by excretion via the mucosal cells: volatile oils, potassium iodide.

2. *Reflex expectorants* which act by irritation of the oesophagus and stomach causing irritation, nausea and a concurrent increase in the rate of production and the volume of bronchial secretions. Ipecacuanha, squill, antimony and potassium tartrate are widely used examples of the so-called emetic expectorants, whilst ammonia compounds and many volatile oils also cause reflex effects.

Inhalants

This group of drugs is not frequently used in veterinary medicine, mainly because animals resent the strong odours of inhalants, but also because the steam is usually generated from water in a bowl or bucket and the animal, at any rate initially, has to be restrained in order to keep its head in the vapour. Steam kettles are rarely used because of the difficulty of restricting the area and maintaining the animals living quarters at a high enough temperature.

By far the best method of vaporizing inhalants is to fill a bucket (for large animals) or a bowl (for small animals) with steaming hot water and pour a few drops of the inhalant on to the water. The heat vaporizes the inhalant with the steam, and when the animal's head is held in the steam both this and the vapour must be inhaled. In large animals the bucket may be put in a sack and the animals head introduced into the sack. This is frequently resented, and is an unsafe method because of the danger of upsetting the steaming water. An alternative is to heat a building brick, put it in a bucket in a sack; the inhalant is then placed on the brick and the animal's head held in the sack.

In small animals the steam method is quite practicable and safe. Dogs in particular will soon learn that relief can be gained by holding their noses over the steaming bowl and will cooperate actively.

Inhalants are of most value for clearing nasal and laryngeal exudates but tracheal and bronchial exudates are also affected. In large animals, equine influenza is probably the condition for which inhalants are most indicated, but strangles cases and similar conditions would also benefit. In small animals the distemper complex of dogs and feline influenza benefit most.

Benzoin, B.Vet.C.

This balsamic resin is obtained by incising the bark of a tree native to the East Indies and Siam. The crude substance occurs as resinous masses of a greyish or reddish brown colour with whitish 'tears' embedded in it. Benzoin is used in the form of Compound Tincture of Benzoin, B.Vet.C. (Friars' Balsam), or Inhalation of Benzoin, B.Vet.C., which are very similar, the first containing balsam of tolu and aloes, as well as the benzoin, storax and alcohol of the inhalation.

Actions and Uses

Externally the compound tincture has been used as an antiseptic dressing for small cuts, but this is now obsolete. When given by mouth benzoin is reputed to cause a reflex expectorant action but this effect is uncertain. Following absorption, excretion occurs mainly via the kidneys, which are slightly stimulated by irritation: at the same time a slight urinary antiseptic effect is exerted. When administered by inhalation, loosening of mucus and débris occurs mainly in the nasopharyngeal and laryngeal regions, but some action occurs lower in the respiratory tract. The steam mixed with the vapour is possibly the major therapeutic agent. Two to four teaspoonfuls of either inhalant per litre of water is sufficient.

Eucalyptus Oil, B.Vet.C.

In the fresh, pure state, oil of eucalyptus is colourless but, as with all volatile oils, it turns yellow with age and oxidation. It has a characteristic pungent, penetrating odour and taste. The solubilities are typical of volatile oils—very slightly soluble in water but soluble and miscible with organic solvents.

Actions and Uses

Eucalyptus oil is frequently incorporated in ointments and oils for the treatment of catarrh in humans by application around the nostrils. This can be used in animals but is usually resented; occasionally applications to the muzzle will cause blistering and this mode of administration should not be encouraged amongst owners. In small animals, if this type of application must be used, rubbing on the fore legs is preferable to muzzle applications: volatilization and inhalation is usually just as good. Steam inhalation is superior in effect though more troublesome in application Eucalyptus has also been included as a fly repellant in skin dressings and sprays: it is also included for antiseptic and deodorant properties in skin and wound dressings. These uses are obsolescent.

Terebene, B.Vet.C.

This compound is a mixture of hydrocarbons obtained by distillation of turpentine which has been treated with sulphuric acid. It is a colourless or pale yellow liquid, with an odour and taste resembling those of turpentine. It is almost insoluble in water but soluble in organic solvents.

Actions and Uses

Terebene can be used in place of turpentine (see Chapter 5), but its main indication is as an irritant expectorant, administered by inhalation or by mouth. If given by mouth it has the carminative and antispasmodic action of turpentine before absorption; after absorption, excretion is via the kidneys, when it has the irritant diuretic and urinary antiseptic action of volatile oils, and also via the lungs, when it has an irritant expectorant action. It has been included in intratracheal injections for 'husk' in cattle.

Dosage

Horse: 30–60 ml (1–2 fl. oz.)
Cow: 30–120 ml (1–4 fl. oz.)
Sheep and Goat: 4–12 ml (60–180 m.)
Pig: 2–8 ml (30–120 m.)
Dog: 0·25–1 ml (4–15 m.)

Non-Inhalant Local Expectorants

These expectorants are administered orally and excreted in small amounts by the lungs. The group includes potassium and sodium iodides, potassium citrate, and creosote.

Potassium Iodide, B.Vet.C.

This is the most important member of the group. After oral administration the iodide ion can be detected in the bronchial secretions within 30 min. The increase of secretions, which has been reported, is in dispute, but the liquefaction which occurs is most beneficial in chronic bronchitis. Irritation makes it contra-indicated in acute inflammatory conditions.

Reflex Expectorants

The reflex expectorants are a group of compounds which promote an increase in volume and fluidity of respiratory secretions and an increase in ciliate activity by stimulation of nerve endings in the pharyngeal, oesophageal and gastric mucosa. This stimulation initiates reflexes which result in stimulation of the respiratory mucosal cells.

Ammonia

Ammonia in the form of carbonate or chloride is frequently used as a reflex expectorant in both the human and veterinary fields. It has been shown that in rabbits large doses (0·5 g per kg of the carbonate and 0·4 g per kg of the chloride) will cause profuse respiratory secretion by stimulation of the vagal nerve endings in the stomach. The viscosity and pH are reduced, whilst the volume of secretion and activity of the cilia of the mucosa are increased. While these doses are far greater than those used clinically, the long and widespread use of ammonium salts indicate some clinical benefit. Camphor when given by mouth is also capable of acting as a reflex expectorant.

These compounds are indicated in both acute and chronic bronchitis, especially where the respiratory mucosa is dry and secretions sparse.

Emetic Expectorants

Ipecacuanha, squill, balsam of tolu, and antimony and potassium tartrate are all emetic expectorants. Their action is one of irritation of the oral, pharyngeal and gastric mucosa to the extent of causing nausea, one of the physiological results of which is an increase in all body secretions, including sweat, saliva and respiratory secretions. They are useful as expectorants for all types of respiratory system inflammations, usually at sub-emetic dosage.

Ipecacuanha

In small doses ipecacuanha is a reflex expectorant with an action extending over a period of several hours. Some diaphoretic or sweating action is also exerted in man, but this has not been confirmed in animals. Small doses are also reputed to stimulate gastrointestinal activity but this action is of little therapeutic significance.

In large doses the drug causes emesis in the vomiting species, but the drug is rarely used for this purpose: this emetic action is mainly a local one although some central stimulation also will occur.

Tincture of Squill or Syrup of Squill, B.P.C.

These are common constituents of cough mixtures and are included for their emetic expectorant action.

Balsam of Tolu

In the form of syrup (Syrup of Tolu, B.P.) is also a frequent ingredient of cough mixtures for dogs because of its slight reflex expectorant action.

Mucolytics

Acetylcysteine

This drug exemplifies a modern approach to management of the sticky mucus component in cough. Inhalation of 20 per cent solution as a spray gives liquefaction of mucopurulent material in 1 min.

RESPIRATORY STIMULANTS

When respiration is depressed, particularly in anaesthetic depression, or when animals are suffering from anoxaemia as in coal gas poisoning, respiratory stimulants may be useful. These stimulants may be local irritants causing reflex stimulation (ammonia), they may be analeptics given intravenously (picrotoxin, etc.), or they may be 'natural' stimulants in the form of oxygen plus 5 per cent carbon dioxide administered by inhalation.

Local Irritants

These are applied by inhalation. Ammonia gas evolved from the stopper of a bottle of Strong Solution of Ammonia, B.Vet.C. or from a sample of carbonate of ammonia, is the irritant usually given. This causes reflex stimulation of the respiratory centre as well as the vasomotor centre: both respiratory and circulatory improvement follows. Once breathing has stopped ammonia is not very effective clinically, but whilst breathing and cardiac action continues improvement in respiratory rate and depth can be achieved with ammonia.

The Analeptics

These medullary stimulant and convulsant drugs are detailed in Chapter 24. They are administered intravenously or intramuscularly and cause a stimulation of the respiratory and vasomotor centres in the medulla. They are far more effective than is the inhalation of ammonia.

The 'Physiological' Stimulants

These consist of mixtures of Oxygen, B.Vet.C. and Carbon Dioxide, B.Vet.C.

Oxygen–Carbon Dioxide

Oxygen is normally stored compressed in cylinders which are painted black over the lower two-thirds and white over the upper third. It is an odourless and tasteless gas.

Actions and Uses

Oxygen is the first essential of all life and no animal can survive for longer than a few minutes with its source of oxygen cut off. During anaesthetic respiratory depression, in coal gas poisoning, in drowning, during severe pneumonia or cardiac decompensation with venous congestion and stagnation, amongst other conditions, oxygenation of the blood and tissues is decreased.

Unless this anoxia is counteracted rapidly there is great danger, firstly of irreversible damage to brain tissue, and secondly, if anoxia is continued longer, of death from asphyxia.

Anoxia of tissues and anoxaemia may be counteracted in one of two ways. The first consists of stimulating the medullary centres with drugs; this causes an increase in the respiratory intake of air, and speeds the circulation of blood through the lungs and into the body after oxygenation. The analeptic drugs and ammonia will do this. The second way, which is becoming increasingly favoured in both the medical and veterinary fields, is to increase the oxygen content of the inhaled air, using artificial respiration where necessary. Considerable benefit may be derived by severe cases of parasitic bronchitis in calves, following the administration of oxygen directly into the trachea from an oxygen cylinder via tubing and a needle inserted between two tracheal rings.

Hyperbaric oxygen therapy is the term used to describe the use of oxygen at pressures greater than atmospheric. This therapy has been used with benefit in man to treat ischaemic ulcers, to preserve tissues rendered ischaemic through trauma, in the treatment of gas gangrene, to increase the radiosensitivity of malignant cells and in the relief of loss of mental function in senility. The value of hyperbaric therapy is presently a matter for controversy owing to the problem of oxygen toxicity.

The administration of oxygen at 2 atmospheres pressure makes an animal independent of haemoglobin due to the oxygen dissolving in the plasma. It follows from this that in **carbon monoxide** (coal gas) poisoning, where the haemoglobin carrying capacity is reduced by the formation of carboxyhaemoglobin, the administration of oxygen at a pressure greater than 1 atmosphere would be of considerable benefit. In practice the administration of oxygen with 5 per cent carbon dioxide for approximately 1 hr using a mask and artificial respiration is the best that can be done, unless blood transfusion is possible. Neither oxygen administration nor transfusions should be continued for beyond 1 hr, for respiration even in fresh air for 1 hr is sufficient to reduce blood concentrations of carbon monoxide to below the danger level, and 3 hr sees the elimination of almost all the carbon monoxide. Animals should be kept warm and allowed to rest for 24 hr, after which further trouble is unlikely.

Asphyxia by drowning is more complicated, for water must first be expelled from the lungs before oxygen can be administered satisfactorily. Injection of oxygen intravenously or subcutaneously at the rate of 500 ml per hr has given immediate relief from cyanosis and dyspnoea in man but care is required to guard against gas emboli.

The use of high concentrations of oxygen alone in threatened respiratory collapse, pulmonary oedema, bronchopneumonia, etc., is not advisable if administration is prolonged, but for the short periods for which oxygen may be used in animals, no serious side actions should be anticipated. It is, however, advisable in all cases to include **carbon dioxide** at a concentration of 5 per cent. This causes greater pulmonary ventilation by its physiological stimulation of the respiratory centre: a secondary result is that pulmonary secretions are more easily expelled because of the increase in volume, and decrease in viscosity, of these secretions which follows the hyperaemia induced by the carbon dioxide: the increased pulmonary ventilation also provides increased propulsion of these secretions through the respiratory airways.

Dosage and Administration

A concentration of 40 to 60 per cent oxygen (normal air concentration 20 per cent) is the optimum for practical purposes if administration is by an intratracheal tube or oxygen tent, but concentrations higher than this can be achieved by use of a mask. A flow of 5 to 8 litres per min will supply the necessary concentrations when the tube or tent are used. All oxygen administered should be warmed and humidified by passing through warm water.

Where carbon dioxide is used it is preferably administered from a combined cylinder which contains oxygens 95 per cent, carbon dioxide 5 per cent. The cylinder is painted black in the lower two-thirds, and the upper third is in alternate quadrants of grey and white.

Pure carbon dioxide gas is contained in cylinders painted completely grey. It may be used to supply the carbon dioxide in mixed gases, but there is a danger of over- or under-dosing which is avoided by using the gases already mixed. Besides its use as a respiratory stimulant, carbon dioxide may also act as caustic for destroying warts: it is not advisable to use it as a local anaesthetic spray although it has been occasionally used for this purpose.

COUGH DEPRESSANTS

Cough sedatives are used to reduce the frequency of coughing when it is causing pain and exhaustion. This depression may be achieved by: (1) *Local action* using demulcents, such as glycerin, honey, etc., expectorants such as ammonium chloride which produces profuse and thin respiratory secretions to coat and soothe the membranes when dry and inflamed. Atropine, hyoscine and belladonna can be used to reduce profuse secretions but they are rarely indicated in respiratory inflammations: they are useful mainly as bronchial relaxants in pulmonary emphysema or 'broken-wind' in horses, and occasionally in cattle. (2) *Central action* using the opium derivatives such as codeine, diacetylmorphine (heroin), dihydrocodeinone and Pholcodine, B.P.C.

Codeine

This alkaloid of opium is not controlled under the Misuse of Drugs Acts, for though addiction can develop, it does so only very slowly: withdrawal is not accompanied by the degree of disturbance that occurs with stronger addictive drugs, nor are the clinical results of addiction so disastrous.

The alkaloid is most commonly used as Codeine Phosphate, B.Vet.C., which is a white crystalline powder with a bitter taste, no odour and a solubility of 1 : 4 in water: it is far less soluble in organic solvents. Aqueous solutions are acid to litmus.

Actions and Uses

Codeine exerts its effects centrally having a sedative and, in large doses, a hypnotic effect. It has relatively little depressant effect on the respiratory centre and on respiratory secretions, but the cough centre is selectively depressed and the analgesic effect is appreciable. It has far less constipatory effect than the other alkaloids of opium.

Codeine phosphate is used most commonly in the dog, either as the syrup alone, in cough mixtures, or in the form of tablets in combination with antipyretic analgesics. The dosage required for cough suppression or antitussive effect is about one-quarter of that required for analgesia.

Dihydrocodeine Bitartrate, N.F.

The problem of addiction led to the search for derivatives with antitussive effect but with reduced liability to dependence. Dihydrocodeine is more active than codeine when equal weights of base are compared. Its withdrawal effects are intermediate in severity between morphine and codeine.

Diamorphine Hydrochloride

Many of the alkaloids of opium are cough centre depressants but unfortunately their addictive properties limit their usefulness to those situations where their even more powerful effect is required. Diamorphine, also known as heroin, is in this respect three to four times more powerful than morphine and thirty to forty times more powerful than codeine. Because of its severe addictive effects, world organizations are endeavouring to stop all manufacture and usage. In veterinary medicine heroin has no advantages over other cough sedatives and is now rarely used.

Pholcodine, B.P.C.

A morpholinoethyl ether of morphine, pholcodine is used as a cough sedative at a dosage about half that of codeine. It is an effective cough supressant and lacks analgesic and addictive properties.

Dextromethorphan

This is the *d*-isomer of the codeine analogue of the potent synthetic analgesic levorphanol; while retaining powerful antitussive action, it is non-analgesic and non-addictive.

BRONCHODILATORS

The autonomic nervous system is a mechanism for the regulation of function in many of the divisions of the body. The therapeutic benefits deriving from the ability to increase or decrease parasympathetic function in the alimentary system have already been described. In relation to respiratory function, parasympatholytic activity is of importance for the reduction of bronchial secretions during prolonged anaesthesia, while stimulation of the sympathetic division results in bronchodilation. Although the bronchodilators have little indication in veterinary medicine, they are widely used in the symptomatic relief of asthma in man.

The transmitter liberated at the post-ganglionic endings of the sympathetic division of the autonomic nervous system is noradrenaline while the adrenal medulla liberates in addition adrenaline. Sympathetic stimulation is followed by a spectrum of changes both motor and inhibitory in character. The discovery that it was possible to stimulate either the motor or inhibitory effects lead to the postulate that there are two types of adrenergic receptor, viz. α and β receptors. The first drug shown to block α receptors was ergotoxine. The subsequent administration of adrenaline results in the activation of the β receptors and the appearance of effects characteristic of their stimulation. The major effects characteristic of α and β site activation are listed in Table II. In general the α effects are excitatory while the β effects are depressant. The cardiac and intestinal receptors are exceptions to this generalization. Many tissues possess both types of receptors but usually one species is in preponderance.

Adrenaline is able to activate both α and β receptors and so is able to bring about bronchodilation. Noradrenaline can be regarded as an α-site activator and is therefore not an effective bronchodilator. The synthetic amine Isopropylnoradrenaline (isoprenaline) is, principally, a β-stimulator and has seen considerable use in the relief of bronchospasm. Unfortunately, bronchodilation can be accompanied by an unpleasant and potentially dangerous tachycardia elicited by the actions of isoprenaline on the cardiac β-sites. This has lead to the development of structurally-related drugs in which the cardiac actions are reduced relative to those on bronchial smooth muscle, e.g. orciprenaline and salbutamol. This pharmacological segregation of adrenergic β-sites has led to their being subdivided into β_1 (heart and small intestine) and β_2 (lungs, uterus, blood vessels).

Table II. Major effects of α and β site activation

α	β
Contracts dilator papillae	Relaxes ciliary muscle
Contracts splenic capsule	Relaxes bronchioles
Vasoconstriction (skin, gut)	Vasodilatation (muscle)
Contracts uterus	Relaxes uterus
Pilomotor contraction	Increased force and rate of heart contractions
Gut sphincter contraction	Gut wall relaxation
Gut wall relaxation	Insulin secretion
Suppressed insulin secretion	Hyperkalaemia
Pancreatic juice reduced	Glycogenolysis
Sweating (horse)	Lipolysis

Another approach to bronchodilation is the use of spasmolytics to reduce parasympathetic broncho-constricting activity. Atropine has been used for many years in the treatment of pulmonary emphysema in horses, usually in the form of Belladonna Leaf given by mouth. Although atropine is not such a powerful bronchodilator as the sympathomimetics, it has given considerable symptomatic relief in cases of 'the heaves'.

In that bronchospasm is to a degree a contributory feature of dyspnoea in coughing, these agents are indicated in the management of severe cough. While spasmolytics are always administered systemically, β-site activators can in addition be administered in a spray or aerosol.

Bibliography

AUSTEN, K. F. & LICHTENSTEIN, L. M. (ed.) (1973) *Asthma: Physiology, Immunopharmacology and Treatment*. New York: Academic.

BURN, J. H. (1968) Catecholamines: (a) pharmacology. In: *Recent Advances in Pharmacology* (ed. J. M. Robson & R. S. Stacey), 4th ed., pp. 155–77. London: Academic.

LEDINGHAM, I. McA. (1968) Hyperbaric oxygen. In: *Recent Advances in Pharmacology* (ed. J. M. Robson & R. S. Stacey), 4th ed., pp. 479–502. London: Academic.

SALEM, H. & AVIADO, D. M. (ed.) (1970) *Antitussive Agents*, vols 1–3. Section 27 in *International Encyclopaedia of Pharmacology and Therapeutics*. Oxford: Pergamon.

TRIGGLE, D. J. (1972) Adrenergic receptors. *A. Rev. Pharmac.*, **12**, 185.

8

The Heart and Circulatory System

Although this system has been arbitrarily divided into three sections: (1) heart, (2) peripheral vascular system, and (3) blood, it is rarely possible to influence one part of the cardiovascular system without affecting the remainder in some way. For example, if the potassium content of the blood is abnormally increased, as could happen with a potassium supplemented gluco-corticoid treatment regime for ketosis there is an action on the heart resembling that of vagal stimulation in which the contractions decrease in power, lose their rhythm and eventually cease altogether. Likewise, if the peripheral vessels are dilated as in shock, the heart rate will increase in an endeavour to maintain the blood pressure and to keep the blood moving through the vessels; at the same time the dilated vessels allow the passage of fluid from the blood into the tissue spaces, and a decrease in circulatory volume occurs with an increase in the proportion of cells (haemo-concentration). This integration of the whole system cannot be overstressed, for too frequently practitioners are surprised when an animal dies of heart failure during or shortly after an intravenous injection—often of calcium or magnesium preparations. (Calcium ions have the opposite effect on the heart to potassium and cause an increase in the power and speed of contraction, but in excess produce a permanent contraction with arrest of the heart in systole. Magnesium is similar in effect to potassium and causes incoordination, sudden dilation, and then arrest of the heart.) In recognizing this integration it should be noted that many of the drugs which affect the heart also act on the vascular system directly.

Drugs Affecting the Heart

The agents which affect the heart have been subdivided into: (1) *heart stimulants*, (2) *heart tonics*, and (3) *heart depressants*. Members of the first group exert mainly rapid but transient stimulant effects on the heart whereas those of the second group have an augmenting action which is slow in onset but prolonged in duration. The last group is used in attempts to control certain cardiac arrhythmias.

HEART STIMULANTS

A heart stimulant has been defined as an agent which increases the heart rate and the contractile power of heart muscle.

(1) Sympathomimetic amines which include adrenaline, isoprenaline, mephentermine and metaraminol. (2) Parasympatholytics such as atropine, hyoscine and synthetic analogues with similar actions. (3) Cardiac muscle stimulants of the purine or xanthine series such as caffeine and theophylline. (4) Reflex stimulants which act by local irritation. In this group are classed the so-called 'diffusible stimulants' such as ammoniacal and alcoholic preparations, and also the irritant injections of camphor and ether, although these injections are of doubtful value.

Analeptics such as picrotoxin, nikethamide or leptazol have a slight stimulant action on the heart although it is an indirect effect, partly attributable to respiratory effects and partly to the vasomotor stimulant action possessed by some of the group (e.g. leptazol).

The Sympathomimetic Amines

Of the physiological sympathomimetics adrenaline is used traditionally as a heart stimulant, for administration of noradrenaline does not produce an increase in cardiac output but, on the contrary, results in a steady rise in intracardiac blood pressure and a slowing of the heart rate. The other amines are of more significance as peripheral vasoconstrictors.

Adrenaline is mainly of use in acute heart failure associated with anaesthesia although its use in light chloroform and cyclopropane anaesthesia is contra-indicated. Administration may be by the intravenous route or by direct intracardiac injection: doses for the dog: 0·1 to 0·3 ml, and for the horse and cow: 2 to 4 ml of a 1 : 1000 solution of adrenaline (Injection of Adrenaline, B.Vet.C.). The injection should be combined with cardiac massage through the chest wall and artificial respiration.

Adrenaline is of little use as a cardiac stimulant in other circumstances for its action is very short lived. Other uses are mentioned in the section on the sympathomimetic amine effects on the peripheral circulation.

The cardiac stimulant activity of those sympathomimetic amines which activate β-receptors has led to their use in maintaining heart rate during heart block. Increased cardiac output coupled with peripheral vasodilation is a more modern approach to the management of shock. The compounds isoprenaline and metaraminol have been used for this purpose in man.

The Parasympatholytics

Atropine and hyoscine produce an increase in heart rate equivalent to the degree of vagal depression of the heart. However, the side actions are so numerous and the action so prolonged that the parasympatholytics find no place in veterinary medicine as heart stimulants.

The Xanthine Derivatives

The xanthine derivatives caffeine, theophylline and theobromine have a direct action on the heart muscle as well as some central stimulation of the vasomotor and vagal centres so that the total effect is unpredictable. In general, the increased muscular power which results from the direct action on the heart muscle causes an increase in ventricular output even though the heart rate may not increase appreciably. There is also some 'heart tonic' action, for the coronary artery is dilated, allowing an increased blood supply to the heart itself.

Theophylline is the most active and caffeine the least active of the group when cardiovascular effects are compared. Theophylline is now rarely used in human medicine, for it has been replaced by aminophylline which has a greater solubility but similar effects. Aminophylline has been recommended for pulmonary emphysema (broken wind) in horses and also for pulmonary congestion (cardiac asthma) in the dog when this is due to cardiac insufficiency: doses used for the horse have been 2 to 5 g (30 to 75 gr.) in food or by capsule daily for 2 or 3 weeks: in the dog 50 to 100 mg ($\frac{3}{4}$ to $1\frac{1}{2}$ gr.) by mouth three or four times daily have been suggested.

Although these drugs may be injected, the peripheral vasodilatation which occurs makes them unsatisfactory for use in heart failure and in other cases where such vasodilatation is detrimental.

Reflex Stimulants

Stimulants such as camphor and ether injections were frequently used in the management of various diseases. They were reputed to have a cardiac stimulant action by virtue of their local irritant effect which, it was suggested, caused reflex cardiac and respiratory stimulation.

Also classified in this group are the 'diffusible stimulants' which, when they reach the stomach, 'diffuse' through the wall causing irritation of nerve endings and initiating impulses which cause reflex stimulation of the heart. These gastric reflexes augment those initiated by irritation of the mouth and nasopharynx during administration. It is probable that those reflexes originating in the nose, mouth and pharynx are more powerful than those originating in the stomach.

The diffusible stimulants consist mainly of volatile, mildly irritant fluids and gases: ammonia, acetic acid, and alcohol are perhaps the most frequently used compounds. Administration may be by inhalation of ammonia—or by the oral administration of a compound such as Aromatic Spirit of Ammonia, B.Vet.C., the dose of which is: Horse and Cow: 30 to 120 ml (1 to 4 fl. oz.), Sheep and Pig: 4 to 16 ml (60 to 240 m.), Dog: 0·6 to 4 ml (10 to 60 m.). Dogs and cats usually object to the volatile irritants.

These reflex stimulants should now be regarded as obsolete.

Cardiac arrest should, logically, be treated by the administration of heart stimulants, but it will be seen from the preceding comments on the various groups that there is no stimulant which is suitable for this purpose. The value of adrenaline is debated because of the danger of causing ventricular fibrillation, but calcium chloride or calcium borogluconate in doses of 2 to 4 ml of a 10 per cent solution preceded by a few millilitres of 4 per cent potassium citrate solution has been suggested as a possible means of restarting a heart, even when all other methods fail. Ventricular fibrillation carries a grave prognosis, for electric shock therapy or procaine (several millilitres of a 2 per cent solution) injected into the left ventricle are only rarely effective. In veterinary surgery, positive pressure apparatus is increasingly available, but in the absence of such apparatus artificial respiration and cardiac massage must be limited to external pressure: where mechanical means of maintaining respiration are available direct manual massage of the heart through an incision in the diaphragm or chest wall can be tried and distinction between arrest and fibrillation effected.

CARDIAC GLYCOSIDES

The term 'heart-tonic' has been applied to those drugs which improve the functional capacity of the cardiac muscle. The cardiac tonics are slow in taking effect but the duration of action is prolonged, an attribute which is conducive to cumulation. The only heart tonics of veterinary significance are the *cardiac glycosides* derived from three plants—species of digitalis, strophanthus and squill. The digitalis glycosides are the most active and important.

Digitalis and its glycosides. Two species of the plant genus *Digitalis* provide a series of glycosides, the action of which cannot be paralleled by any synthetic compound: their mode of action is still unclear despite a lengthy clinical use and intensive study.

The main sources of the digitalis glycosides are the purple foxglove, *Digitalis purpurea*, and *Digitalis lanata*, the woolly foxglove. The leaves and seeds are normally used for processing.

From the seeds of *D. purpurea* is obtained a mixture of glycosides which when standardized is known pharmaceutically as Digitalin, B.Vet.C. The glycosides contained are *Digitalinum verum* and a number of water soluble glycosides of unknown constitution.

The leaves of *D. purpurea* contain a number of glycosides in variable proportion, three examples of which are, digitoxin, gitoxin and gitaloxin.

The seeds and leaves of *D. purpurea* also contain several saponins including digitonin, gitonin and tigonin which, though they have no heart tonic action themselves, probably increase the activity of the glycosides by affecting solubilities, absorption rates and similar factors.

The leaves of *D. lanata* produce glycosides, known as lanatosides A, B and C, from which, by hydrolysis, are obtained digitoxin, gitoxin and digoxin, respectively. Lanatosides A and B are acetyl derivatives of the natural purpurea glycosides A and B of *D. purpurea* which also release digitoxin and gitoxin respectively during extraction.

The complex molecule of a glycoside, as already defined in Chapter 1, consists essentially of a sugar and another compound which exerts the activity characteristic of the glycoside and is known as a *genin* or *aglycone*. Although the aglycone has this characteristic action and the sugar has no specific action their combination as a glycoside is far more active than the aglycone, indicating the synergistic action of the two. It is probable that the sugar increases the solubility of the aglycone and its ability to enter cells; the absorption and utilization of the aglycone are therefore improved. The sugars and genins produced by hydrolysis of the digitalis glycosides are set out in the Table III along with the saponins and their products of hydrolysis.

Structurally, the digitalis glycosides have a steroid form with a lactone ring attached at carbon 17 (see sex steroids and corticosteroids p. 119 *et seq.* and p. 140 *et seq.*) and the sugars attached through an oxygen atom at carbon 3. The formulae of digitoxin (i) and digoxin (ii) illustrate these points.

These glycosides are white odourless, crystalline preparations which have an intensely bitter taste. Digitalin is very soluble in water and alcohol but only very slightly soluble in other solvents. Digitoxin and digoxin are almost insoluble in water but both are soluble in alcohol.

Table III. The sugars and genins produced by hydrolysis of digitalis glycosides

Glycoside or saponin	Genin	Sugar(s)
Digitoxin	Digitoxigenin	3 molecules digitoxose, $C_6H_{12}O_4$
Gitoxin	Gitoxigenin	3 molecules digitoxose
Gitalin	Gitalligenin	2 molecules digitoxose
Digoxin	Digoxigenin	3 molecules digitoxose
Digitalin	Gitoxigenin or digitaligenin	1 molecule each digitalose $C_{17}H_{14}O_5$ and glucose $C_6H_{12}O_6$
Digitonin	Digitogenin	4 molecules galactose and 1 molecule xylose
Gitonin	Gitogenin	3 molecules galactose and 1 molecule pentose
Tigonin	Tigogenin	2 molecules each glucose and galactose with some rhamnose

They are available as preparations for oral and parenteral administration.

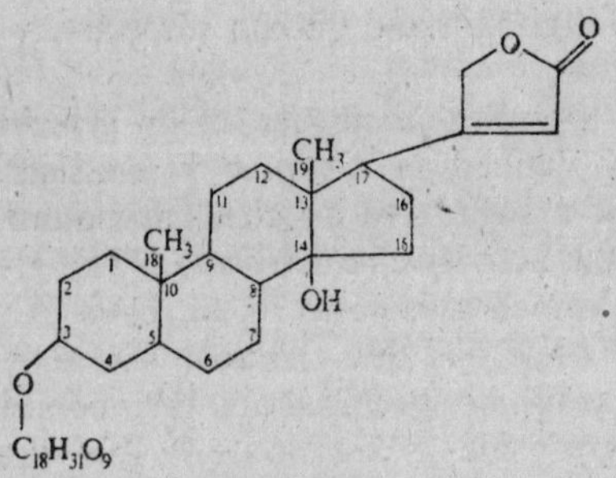

(i) Digitoxin

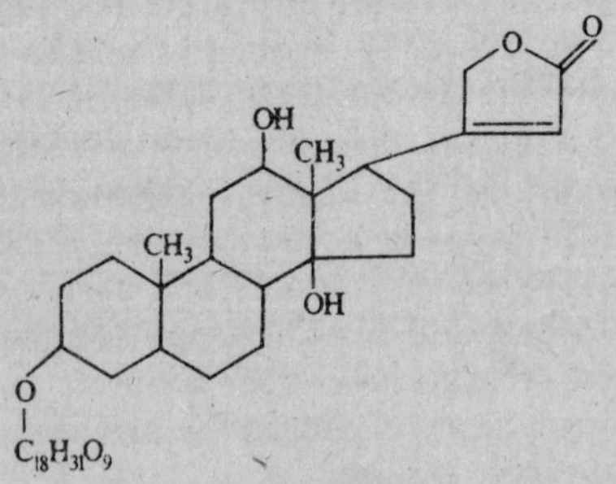

(ii) Digoxin

Prepared Digitalis, B.Vet.C.

This glycoside which has a standard digitalis activity, is a green powder with a slight odour, a bitter taste and a markedly irritant effect on all mucous membranes. The presence of saponins is indicated by frothing when shaken with water.

Assay. The glycosides digitoxin and gitoxin are available in pure form and estimation by weight is therefore possible, but formerly the potencies of all digitalis preparations and derivatives were estimated biologically.

Actions

The actions of all the digitalis glycosides are qualitatively the same but individual variations in potency exist. Although the potency ratios vary according to the type of assay (cat, frog, pigeon, etc.), it is generally accepted that digitoxin is the most powerful of the cardiac glycosides but is closely followed by digoxin: digitalin is far less powerful than either. Digitoxin provides full effects within 6 to 8 hr of either oral or parenteral dosing. It has a half-life of around 7 days, largely because it is extensively protein bound, and is therefore a strongly cumulative drug. Digoxin exerts some cardiac action within 1 hr of an oral dose and maximum effect follows within 6 hr: intravenous administration gives some immediate effect and reaches a maximum in about 2 hr: the majority is excreted within 2 days and the drug is therefore far less cumulative than digitoxin.

Digitalin, B.Vet.C.

This is readily soluble in water, is not so predictable in its effect and is rapidly excreted so that its cumulative effect is relatively slight.

Digitoxin is less inclined than the other glycosides to cause nausea and vomiting; it is very efficient whether administered orally or by injection (dose ratio 1 : 1) and is therefore normally given by mouth. Digoxin causes vomiting far more frequently than digitoxin and has a less efficient route dose ratio (1·5 oral to 1·0 intravenous) indicating a lower alimentary absorption: digoxin is therefore of most use for achieving digitalization (maximum digitalis action) rapidly by the intravenous route in cases requiring immediate relief from the effects of cardiac insufficiency.

The pharmacological actions of the digitalis group affect the cardiovascular system: the effects result partly from central stimulation of the vagal centre of the medulla and partly by a direct action on the heart muscle and nervous control mechanisms.

The effect of digitalis glycosides is primarily exerted on the muscle fibres of the heart, for heart muscle has a great affinity for cardiac glycosides which are probably assisted in entering the muscle cells by the sugar component and then bind to the contractile protein. The glycosides have been shown to increase the concentration of intracellular calcium ions, to increase the contractility of heart muscle proteins in vitro and to impede the function of the sodium pump in myocardial membrane with consequent increase in intracellular sodium and decrease in intracellular potassium ion concentrations. Despite this, their mode of action in vivo remains undecided. The effect on the muscle fibres is indicated by an increased tonus during diastole due to a slight shortening of each individual fibre, and a strong and sustained systole due to a prolonged and greater shortening of the fibres during systole. This increased effort is followed by an increased refractory period with an increased conduction time in the bundle of His, but a more powerful impulse. During the prolonged diastole which provides a long resting time, there is an increased coronary circulation and therefore an improved nutrition of the cardiac muscle. All these factors favour an improved heart action and a greater throughput of blood with an increase in blood pressure.

These effects are thought to promote a reflex vagal response (in addition to some direct stimulant action on the medullary centres) in consequence of which increased vagal tone results in slowing of the sinus rate, impairment of A-V conduction and shortening of the atrial refractory period. Vagal induced slowing of the heart rate, although it can be demonstrated experimentally with glycosides plus atropine, is not important in their therapeutic use. Slowing of A-V conduction is important and can progress to heart block. Similarly an undesirable sequel to the reduction of atrial refractory period is the conversion of an atrial flutter to a fibrillation. The central and local actions therefore combine to give a less frequent heart contraction, a stronger systolic phase, and a prolonged diastole with an increased ventricular filling time which in turn brings into play Starling's law of the heart whereby increased filling promotes increased muscular reaction in systole: venous pressure is reduced, arterial blood pressure is increased and cardiac throughput is also increased.

A subsidiary effect is exerted on the peripheral vessels which are differentially constricted, the arteriolar constriction being greatest in the intestines and spleen, weakest in the skin and brain, and intermediate in the kidneys. This action serves to augment the blood pressure.

An effect which is secondary to and consequent upon the cardiac effect, is a diuretic action in animals suffering from oedema due to cardiac insufficiency. There is no such action in normal animals. This diuretic effect is due to the resorption of transudates following the decrease in venous pressure; the increased blood volume allied with the rise in arterial blood pressure causes an increased renal throughput of blood, an increased filtration, and therefore greater urine production. The diuretic effect is therefore incidental and not, as was originally thought, a primary action of digitalis.

Uses

Diseases and malfunctions of the heart in domestic animals have received little attention mainly because, with the development of cardiac lesions and of venous congestion, the animal becomes a liability and is usually destroyed. It sometimes happens, however, that aged horses, especially at stud, are treated and usually some attempt is made to maintain the health of dogs affected with valvular discrepancies. The main value of digitalis is in the treatment of cardiac oedema and venous congestion consequent on cardiac decompensation: the so-called 'cardiac asthma' in old dogs is due mainly to congestion of the lungs and is not a true asthma. Digitalis is rarely used in cats because of the danger of inducing emesis by central stimulation of the vomiting centre, although digoxin by injection is less likely to cause emesis and has been used occasionally.

Dosage

Prepared Digitalis, B.Vet.C. will cause local irritation and vomiting in those species capable of emesis—pig, dog: in these species the glycosides should be used. In ruminants the glycosides are largely destroyed in the rumen and parenteral administration is obligatory. The intravenous route is preferred for all parenteral dosing.

Prepared Digitalis, B.Vet.C.:
- Horse: 1–4 g (15–60 gr.) orally
- Dog: 30–250 mg ($\frac{1}{2}$–4 gr.)

Digitalin, B.Vet.C.:
- Horse: 15–60 mg ($\frac{1}{4}$–1 gr.)
- Dog: 1–10 mg ($\frac{1}{60}$–$\frac{1}{6}$ gr.)

The officially recommended routes of administration are the subcutaneous and intramuscular injection routes, but this compound is most frequently administered orally in the dog in tablets containing 0·5 to 1 mg (approximately $\frac{1}{100}$ to $\frac{1}{50}$ gr.), the dose being repeated three times daily.

Digitoxin, B.Vet.C.
- Dog: 0·1–1 mg ($\frac{1}{600}$–$\frac{1}{60}$ gr.) orally or intramuscularly

Digoxin, B.Vet.C.
- Dog: 0·25–0·6 mg ($\frac{1}{240}$–$\frac{1}{100}$ gr.) orally or intravenously

The official doses recommended for oral or parenteral administration are:

Initial dose:
- Dog: 0·05–0·1 mg/kg
- Cat: 0·05 mg/kg

Maintenance dose:
- Dog: 0·025 mg/kg
- Cat: 0·01 mg/kg

The methods of administration adopted in the United States are based on a dose calculated to achieve maximum safe effect (digitalization), a working average for Digitoxin, U.S.P., (contains not less than 90 per cent pure glycoside $C_{41}H_{64}O_3$) being taken as 0·14 to 0·28 mg per 10 lb. body weight (U.S.D. 25th edn, 1955, p. 1974). In the 'intensive' method half the calculated dose is given immediately, followed by a quarter of the dose 6 hr later and by one-eighth every 4 to 6 hr thereafter until signs of toxicity occur (vomiting, diarrhoea, depression, and cardiac irregularities). A second method, known as the 'rapid' method, consists of one-seventh of the calculated digitalization dose given every 6 hr, while the 'slow' method provides for one-tenth of the calculated dose administered every 8 hr until toxic symptoms are noticed. Once the animal is saturated, the daily dose is assessed according to individual response and with due regard for the cumulative action of the glycosides used. The dose which will maintain a level just short of toxicity should be aimed at: it usually falls between the limits of one-quarter and one-tenth of the quantity required for digitalization.

Toxicity. Digitalis toxicity is mainly the result of cumulation and is associated most commonly with digitoxin. The symptoms consist of anorexia vomiting, diarrhoea, mental confusion and depression, a slowing of the heart rate (bradycardia), and prolonged systole with a short diastole. If the toxicity is not recognized, fibrillation and heart failure in the systolic phase occur. Treatment consists of stopping further dosing, but in the presence of ventricular arrhythmias it is necessary to administer potassium-rich solutions, a β-blocker such as propranolol or a myocardial depressant such as procainanide.

Strophanthus

The seeds of a tropical East African climbing plant *Strophanthus kombé* are the source of a standardized mixture of glycosides formerly recognized as **Strophanthin K.**

Strophanthus gratus yields the glycoside **ouabain,** now the most commonly used member of the group. The main advantage of ouabain lies in its very rapid onset of action following intravenous injection, making it the glycoside of choice for emergency digitalization. Ouabain is the least protein-bound of the cardiac glycosides and, consequently, has the shortest duration of action.

Squill

The bulb of white squill or *Urginea maritima*, a native of the Mediterranean area, contains two cardiac glycosides with relatively weak heart tonic actions. The powder and tincture are therefore rarely used for their effect on the heart but are frequently administered as emetic–expectorants in cough mixtures: the emetic–expectorant action is due to local irritation of the stomach which stimulates reflexly a bronchiolar secretion. The dose of the tincture is 8 to 30 ml (120 m. to 1 fl. oz.) in horses and cattle, and 0·3 to 1 ml (5 to 15 m.) in dogs.

A red variety of squill was used in the manufacture of certain rat poisons because of a toxic glycoside known as scilliroside contained in the bulb. Red squill rat poison was reputed to be non-lethal to domestic animals because its irritant action precipitates vomiting before any harm can be caused—rats are unable to vomit. In practice the theory was not invariably accurate and poisoning of household pets did occur.

HEART DEPRESSANTS

A problem more recently recognized in clinical practice than heart failure is that of cardiac arrhythmias. It is of interest that it has been possible to eliminate atrial fibrillation in the horse and to protect against ventricular fibrillation consequent upon digitalis overdosage or anaesthesia. The latter two indications have been mentioned in connection with the β-adrenergic blocking drug propranolol and this section will be devoted to those drugs which act because of their ability to depress myocardial irritability directly.

Quinidine Sulphate B.Vet.C.

This drug is used to correct atrial fibrillation. It acts by virtue of its ability to prolong the refractory period of heart muscle and by depressing its excitability and that of the pacemaker. This depression is due to a reduction in the rate of repolarization of cardiac membranes as quinidine makes membranes less permeable to sodium ions. The final effect is to reinstate effective atrial contractions or to slow atrial rate. Prolongation of the P–R interval of the ECG is evidence of slowing of atrio-ventricular conduction while wider QRS complexes indicate slower conduction in the ventricles.

Administration is by the oral or parenteral routes. Usually treatment is discontinued when fibrillation is arrested, but can be reinstated if fibrillation should recur. The use of this drug is sometimes accompanied by hypersensitivity reactions, e.g. urticaria or laminitis.

Procainamide

The antifibrillatory properties of the local anaesthetic procaine lead to the introduction of the less rapidly inactivated procainamide. The actions of this compound are very similar to those of quinidine, but its use is mainly in the control of ventricular arrhythmias, e.g. ventricular tachycardia during surgery.

The local anaesthetic **lignocaine,** an amide, has also been used to control arrhythmias. Local anaesthetics act by stabilizing nerve membrane, i.e. they interfere with permeability to sodium ions.

Propranolol

Propranolol, the β-adrenergic blocking drug, has recently been used to control sinus tachycardia, atrial flutter and atrial fibrillation as well as certain ventricular arrhythmias. It is interesting to note that this compound, in addition to its β-blocking properties, possesses considerable local-anaesthetic and quinidine-like activity.

DRUGS AFFECTING THE PERIPHERAL CIRCULATION

The drugs of clinical importance are those which will produce vasoconstriction and an increase in blood pressure. In human medicine and surgery, *hypotensive* agents are of great importance in the treatment of permanent high blood pressure (hypertension), and in the creation of hypotension as a means of keeping an operation site clear of blood in areas where tourniquets cannot be used. So far, neither of these uses of hypotensive drugs has been applied in practice but the creation of hypotension has been investigated as an anaesthetic adjunct in the veterinary schools.

There are several methods of increasing blood pressure.

1. The first is purely mechanical and consists of intravenous injections of plasma extenders (expanders) such as saline–dextran solutions, polyvinyl pyrrolidone, and blood products including whole blood, plasma and serum, and various plasma and serum fractions. These will be referred to in Chapter 9.

2. The second group are the vasomotor stimulants normally referred to as analeptics. Of these, nikethamide probably has the greatest effect on the peripheral vascular system. The analeptics are referred to in Chapter 24.

3. The third group comprises the vasoconstrictors including the alkaloids of ergot and vasopressin (Chapter 12) and angiotensin (Chapter 15) but mainly the sympathomimetic amines.

4. The fourth group consists of antihistaminics which could antagonize the hypotensive effect of histamine in anaphylaxis.

The sympathomimetic amines are dealt with in this section.

SYMPATHOMIMETIC AMINES

This is the term applied to a group of compounds with basic attributes which, when administered to an animal, produce effects similar to those which follow the stimulation of the sympathetic nervous system.

Chemically, the group is described as consisting of three sections, one of which contains alkyl amines (iii) with from four to eight carbon atoms:

$$CH_3{\cdot}(CH_2)_{1-5}\text{—}\underset{\displaystyle NH_2}{\underset{|}{CH}}\text{—}CH_3$$

(iii) Alkyl amines

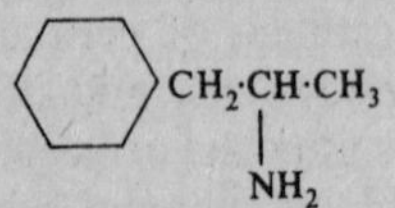

(iv) Cyclo-alkyl amines

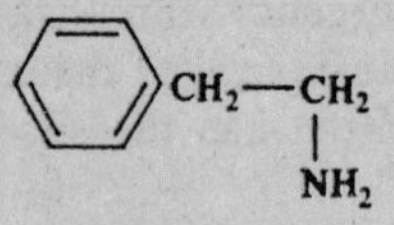

(v) Aryl alkyl amines

a second which contains cyclo-alkyl amines represented by the general formula as in (iv), and the third section which consists of aryl alkyl amines represented by (v).

It is the third group (v) with which the clinician is mainly concerned, for most of the drugs used clinically occur in this group, including adrenaline, noradrenaline, ephedrine and the amphetamines. The group has been classified firstly by the character of the aromatic nucleus, e.g. phenyl, monohydroxyphenyl or dihydroxyphenyl, alkyl amines: variations in the alkyl side chain provide a further means of subdividing these subgroups and there are thus ethylamines and isopropyl amines—primary and secondary amines are the most important.

The sympathomimetic amines of veterinary significance fall into two groups, the phenyl- and dihydroxyphenyl- groups. In the phenyl group are amphetamine and ephedrine, which in general have low pressor activity ($\frac{1}{100}$th of adrenaline) but are absorbed unharmed from the gastrointestinal tract and have a relatively prolonged action. They are β-phenyl-α-methyl-ethylamines (or β-phenyl-isopropylamines). In the dihydroxyphenyl group occur adrenaline and noradrenaline which are the most potent but the least stable of these amines.

Adrenaline, B.Vet.C.

This compound, known in the United States as epinephrine, is one of the active principles of the adrenal gland from which it may be extracted, though the adrenaline used commercially may also be prepared synthetically. It is a white or creamy white powder which is only slightly soluble in water and insoluble in organic solvents. With acids, adrenaline forms salts which are water soluble: the bitartrate or acid tartrate is the one now most frequently used, for it is more stable than the hydrochloride formerly used.

Chemically, adrenaline is (—),1,(3 : 4 dihydroxyphenyl),2,methylaminoethanol, represented graphically as in (vi). It is synthesized in the adrenal medulla and elsewhere starting with the amino acid tyrosine. This is oxidized to dopa (dihydroxyphenylalanine) which is decarboxylated to dopamine. Oxidation of dopamine yields noradrenaline which, in turn, is methylated to yield adrenaline.

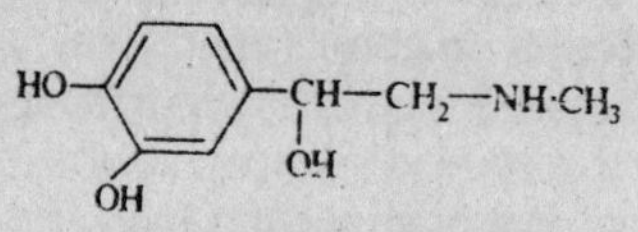

(vi) Adrenaline

Actions

Adrenaline can mimic all the effects produced by stimulation of the sympathetic nervous system. It activates α and β adrenergic receptors (p. 72) and is either destroyed by enzyme oxidation (amine-oxidase) or excreted in conjugated form (catechol O-methyl transferase).

Quantitatively more important in terminating the action of endogenous adrenaline is its reuptake by the adrenergic neurone into storage granules. Adrenaline has a mild local haemostatic effect. When applied to the eye in concentrations of 2 per cent, mydriasis (dilatation of the pupil) occurs, but this is of little veterinary significance.

Internally, the actions depend on the route of administration. Adrenaline is destroyed in the gastrointestinal tract and oral dosing is therefore of little value. Subcutaneous and intramuscular injections cause local vasoconstriction which slows absorption, and marked systemic effects should not be expected. The full systemic effects follow intravenous injection, but persist for only a few minutes due to rapid removal and oxidation of the adrenaline unless continuous infusion is carried out, when the effects will persist for as long as the infusion continues.

The systemic effects of adrenaline have been described as those involved in 'fear, fright and flight', so that all mechanisms involved in exertion are improved whereas other unnecessary functions are reduced to a minimum. Thus, following the intravenous injection or the natural release of adrenaline by fear, the following reactions occur:

(1) Passive dilatation of the vessels of the brain, lungs, and also of the coronary system, occurs as the increased blood pressure overcomes the resistance of what few constrictor fibres exist in these vessels. This is accompanied by an active dilatation of the vessels to the skeletal muscles.

(2) A reduced blood supply to the skin, gastrointestinal canal and kidneys follows powerful vasoconstriction of arteries, veins and arterioles.

(3) An increased heart rate is followed by a slower rate but a more powerful contraction as the aortic and carotid reflexes impose vagal slowing on the heart, initially accelerated by adrenaline's ability directly to stimulate the myocardium.

(4) The spleen contracts to increase the number of circulating blood cells.

(5) The bronchi relax making aeration of the lungs easier and, in the horse, hyperpnoea accompanies the rise in blood pressure, whilst, in other species apnoea may occur.

(6) The glycogen of the liver is depleted with the formation of glucose, necessary for the supply of energy required during the increased exertion. Muscle glycogen is converted into lactic acid and lipolysis also occurs.

(7) The activity of the stomach and intestines is reduced; relaxation of the main musculature with constriction of the sphincters is the usual response.

Other actions are variable or of slight clinical significance.

Uses

The actions of adrenaline are utilized clinically as follows:

(1) Haemostasis may be achieved in capillary haemorrhage by local application. It should be remembered that a temporary overdilation occurs as the effects of the adrenaline wear off and secondary bleeding may result unless coagulation is achieved while the affects of adrenaline last. In epistaxis—nose bleeding—which is not uncommon in high spirited thoroughbreds before races and may occur as a result of accidents in all species, adrenaline can be instilled or sprayed into the nostrils in concentrations

of 1 : 5000 (one part Solution of Adrenaline, B.Vet.C., plus four parts sterile water). Concentrations of 1 : 1000 may be used in severe haemorrhages, but the intense ischaemia caused may lead to local necrosis and retard the healing of damaged tissue. If adrenaline is used to arrest capillary haemorrhage in operation fields, dilutions of 1 : 30 000 to 1 : 100 000 are suitable. Adrenaline solutions may be applied by swabbing or spraying.

To minimize the incidence of secondary haemorrhage, about 1 per cent of ephedrine hydrochloride may be incorporated with the adrenaline solution: the effect may be prolonged for several hours and no secondary dilatation follows recovery.

(2) When incorporated in local anaesthetic solutions, adrenaline hydrochloride causes local vasoconstriction and slows absorption of the anaesthetic, thereby prolonging its action. Strengths of 1 : 10 000 to 1 : 50 000 are normal. Although the vasoconstriction also keeps the operation field free of blood, secondary dilatation may result in post-operative capillary bleeding.

(3) The use of adrenaline in threatened and actual cardiac collapse has been mentioned earlier in the chapter: intravenous or direct intracardiac injection of 1 : 1000 solutions in doses of 0·25 ml in the dog, and up to 4 ml in the horse, have been used but are definitely contra-indicated during collapse associated with chloroform, cyclopropane and chloral hydrate anaesthesia because of the probability of causing irreversible cardiac fibrillation. This danger is always present if adrenaline is used in anaesthetic collapse and the drug is therefore best ignored in favour of artificial respiration with cardiac massage.

(4) Adrenaline has been used in the treatment of allergic conditions in all species. Serum allergy in all species, blue nose, purpura haemorrhagica and laminitis in horses have all been treated with adrenaline: subcutaneous injection of up to 4 ml of 1 : 1000 solutions three or four times daily have been used in blue nose and purpura; in laminitis 2 ml of 1 : 2000 solutions injected subcutaneously on either side of the fetlock and repeated 24 hr later have been suggested.

Dosage

Injection of Adrenaline, B.Vet.C.:
- Intravenously:
  - Horse and Cow: 2–4 ml
  - Dog: 0·1–0·3 ml
- Subcutaneously:
  - Horse and Cow: 2–8 ml
  - Dog: 0·1–0 5 ml

This injection contains sufficient Adrenaline Acid Tartrate, B.Vet.C., to provide a concentration of 1 : 1000 (0·1 per cent) adrenaline. Solution of Adrenaline, B.Vet.C., also contains 0·1 per cent adrenaline and may be used for external application in place of the official injection.

Toxicity. Adrenaline is non-irritant but if injected subcutaneously in high concentrations an ischaemia may result which is sufficiently prolonged to cause necrosis or at least an area in which the coat colour is permanently bleached. Massage may help to avoid these effects.

The fibrillation which may follow the use of adrenaline in cardiac collapse during anaesthesia has been mentioned previously. Adrenaline administration in hypotension associated with the phenothiazine preanaesthetic sedatives is ineffective if not actually contra-indicated and *l*-noradrenaline should be used.

Noradrenaline

The adrenal medulla contains noradrenaline in addition to adrenaline: it is also released at the nerve endings of post-ganglionic sympathetic nerve fibres. Chemically, the hormone differs from adrenaline in that the terminal amino group is not methylated (vii). Noradrenaline, unlike adrenaline, acts mainly at α-receptors.

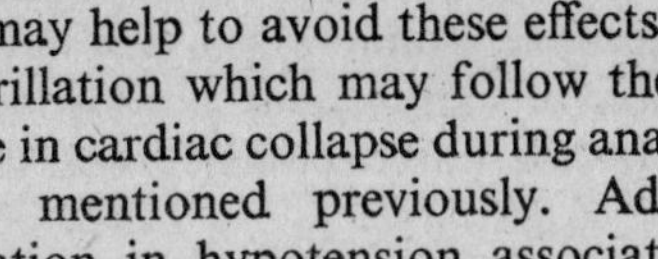

(vii) Noradrenaline

Racemic noradrenaline consists of equal parts of the relatively inactive *dextro*- and the active *laevo*-isomers, and is therefore only half as active as *l*-noradrenaline. The acid tartrate is the usual salt from which injections are prepared: this salt is a colourless or cream crystalline substance which is odourless and has a solubility in water of about 1 : 5 (U.S.P. 1 : 2·5). Solutions are unstable and darken on exposure to light and air: sodium metabisulphite assists in maintaining acidity and in preventing oxidation.

Actions

The action of noradrenaline is mainly one of peripheral vasoconstriction which differs from that caused by adrenaline in that the vasoconstriction is general and occurs in the vessels supplying skeletal muscle. As a result of this, there is a progressive rise in arterial and venous blood pressure which causes a reflex slowing of the heart: there is little increase in cardiac output. Noradrenaline is less active than adrenaline in causing bronchodilatation and gastrointestinal stasis: it has little effect on the blood supply to the brain but reduces that to the kidney and thus decreases filtration.

Uses

Noradrenaline is used as an agent for countering hypotension in cases of surgical and non-surgical shock. In surgical and traumatic shock, up to 10 μg per kg (5 μg per lb.) can be given intramuscularly. Where intravenous drips are in use, surgical shock may be treated by adding noradrenaline to the infusion to a final concentration of 8 μg per ml. The solution may be infused at a rate of 2 ml per min.

Ephedrine

One of the β phenylalkylamines, ephedrine is derived from various species of *Ephedra* shrubs. The alkaloid is now widely prepared by synthetic processes.

Chemically, the compound is the hemihydrate ($\frac{1}{2}H_2O$) of (—)-2-methylamino-1-phenylpropanol and may be graphically represented as in (viii). Reduction of ephedrine results in the replacement of the hydroxyl group by a hydrogen atom and produces desoxyephedrine or methylamphetamine (ix).

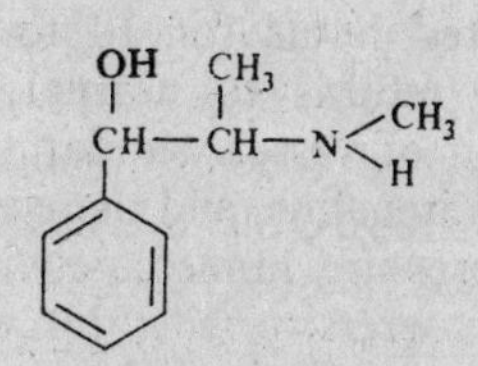

(viii) Ephedrine

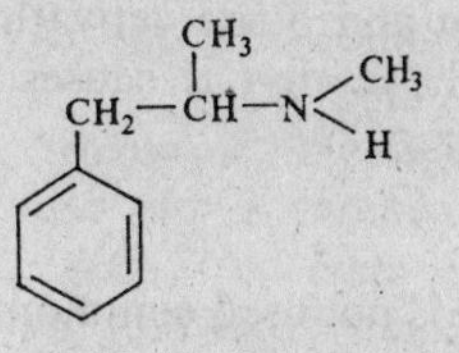

(ix) Methylamphetamine

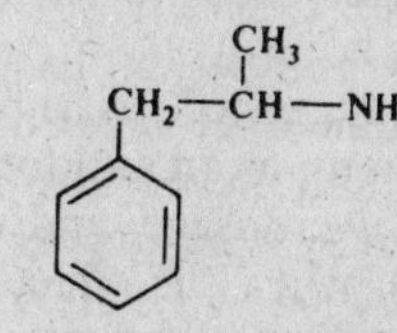

(x) Amphetamine

Actions

Ephedrine is active whatever the route, but effects are delayed for up to 1 hr after oral administration: although ephedrine is less powerful in its effects than adrenaline the action is prolonged and effects may be present as long as 4 or 5 hr after oral administration. Part of the action of ephedrine is related to its inhibition of the amine-oxidase destruction of adrenaline. This potentiation of the effects of adrenaline is not the whole effect, for adrenaline has very little central nervous system effects whereas ephedrine has: this action is directly related to the absence of hydroxyl groups from the benzene ring nucleus. The compound is excreted by the kidneys unchanged: this lack of metabolization no doubt has a strong bearing on the prolonged action of the compound.

Ephedrine has little effect on the heart in therapeutic doses beyond perhaps a slight increase in rate and a dilatation of the coronary vessels, but its effect on the blood pressure is different from that of adrenaline in that the rise in systolic and diastolic pressure following oral administration is sustained for as long as 5 hr. Following intravenous injection of small doses, the pressure rises quite strongly, is maintained at the elevated levels for 10 to 15 min and subsequently returns to normal: there is no overdilatation of vessels similar to that which occurs with adrenaline, and there is therefore no subnormal pressure immediately following the increase. A progressive reduction of response to successive doses—tachyphylaxis—occurs with ephedrine. This is because ephedrine, in addition to its direct α and β site activating ability, also acts indirectly in that it causes the release of stored noradrenaline.

Uses

Ephedrine is not used commonly in veterinary medicine. In man, its central stimulant action is an undesirable side-effect to its major use as a broncho-dilator.

It has been suggested as a treatment for urticarial conditions, as an antidote to morphine and barbiturate overdosage, and as a mydriatic (pupil dilatator), when the effects are likely to develop in about 30 min and persist for as long as 12 hr following instillation of a few drops of solutions containing up to 5 per cent of the hydrochloride or base.

Dosage

Oral or parenteral: 15 to 30 mg (¼ to ½ gr.). If the drug is to be injected, the hydrochloride should be used and any subcutaneous injections should be carefully dispersed by massage. Oral dosing is preferable and effects may be maintained for up to 3 days by repeating the dose every 4 to 8 hr.

Amphetamine, B.P.

Amphetamine (x) is racemic but the *dextro*-rotatory form, which is approximately twice as active as the racemic form though qualitatively identical is known as Dexamphetamine, B.P.C.

Actions

The actions of amphetamine are similar to those of adrenaline and ephedrine in type but there are marked quantitative differences. Amphetamine is active when administered by mouth, the base is locally active when the vapour is inhaled, and the salts may be given by intravenous or intramuscular injection. Following oral administration, effects commence within about 5 min and reach a maximum in about 30 min: cardiorespiratory effects have lasted for as long as 7 hr. About 50 per cent is excreted within 48 hr of dosing, whilst the remainder is slowly metabolized. The actions of amphetamine are significant in four directions: (1) cardiorespiratory stimulation (2) central nervous system stimulation (3) antispasmodic and (4) metabolic stimulation with a decrease in appetite.

The peripheral vasoconstriction caused by amphetamine results in a slow but considerable rise in blood pressure which may persist for several hours. It also has a respiratory stimulant effect due to central stimulation.

The cerebral and medullary stimulant effect of amphetamine and methylamphetamine is greater than that of all other sympathomimetic amines and superior to ephedrine, nikethamide, leptazol and caffeine in antagonizing the respiratory depressant effect of morphine.

The antispasmodic action is not so marked as are the previous two actions but the delayed emptying of the stomach no doubt has a bearing on the appetite reducing effect, which is exerted on dogs as well as humans. The metabolic rate is increased at the same time and the combined effect assists in reducing body weight. The *dextro*-isomer is more effective for this purpose.

Because of its misuse by people desiring CNS stimulation supplies of amphetamine and methylamphetamine are no longer generally available.

Other Sympathomimetics

Although there are very many sympathomimetic agents in existence, the 1965 *Veterinary Codex* describes only adrenaline, noradrenaline and ephedrine in this category. This is in part due to the lack of applications in veterinary medicine comparable to such human needs as nasal decongestion, asthma, vasodilation in peripheral vascular disease, the treatment of depressive states and the control of obesity.

The mode of action of the sympathomimetics is worthy of comment. Some act solely at the adrenergic receptor and are known as 'direct-acting', while others act because they are able to displace noradrenaline from its storage sites within the body and are said to be 'indirect-acting'. Repeated doses of an indirect-acting amine elicit decreasing responses (tachyphylaxis) as the stored amine becomes used up, e.g. ephedrine.

Isoprenaline

Isoprenaline (xi) is an amine which directly activates β-receptors (see p. 72). Its uses as a bronchodilator have been described. β-Stimulant action is now being utilized in the treatment of shock. Stimulation of the heart and vasodilation are thought to be more valuable in improving peripheral circulation and prevention of stagnation than the vasoconstriction and reduced heart rate which usually accompany the infusion of noradrenaline and other α-site activators, e.g. methoxamine.

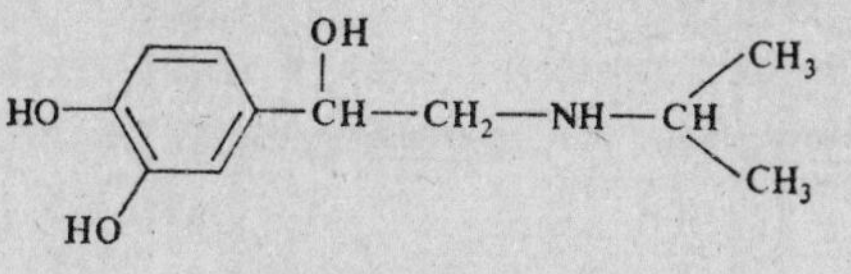

(xi) Isoprenaline

Metaraminol

Metaraminol (xii) possesses powerful direct and indirect actions. It is used to treat hypotensive states and as a nasal decongestant in man and has both α- and β-stimulant properties. It shares with all other α-methylated amines

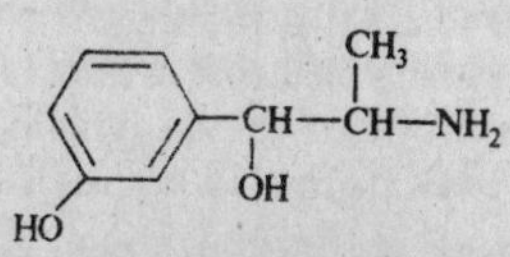

(xii) Metaraminol

the ability to resist monoamine oxidase and therefore has an increased duration of activity.

ADRENERGIC BLOCKADE

The early drugs used to reduce or eliminate sympathetic tone, did so by competitively occupying the adrenergic α-receptors. Although the compounds used possessed undesirable side-effects or were rather toxic they were used in man in an attempt to control hypertension or persistent peripheral vasoconstriction. The α-blockers are exemplified by **ergotamine, piperoxane, phentolamine, phenoxybenzamine,** but only the last two still find application, principally in the preoperative control of symptoms of tumour of adrenaline producing tissue (phaeochromocytoma).

Dichloroisoprenaline can block adrenergic β-sites but also acts as an agonist. Propranolol is more suitable for clinical use as it is without agonist activity. β-Blockers are used in man to control angina pectoris by reducing exercise-induced increases in heart rate, to control tachycardia and arrhythmias, e.g. due to digitalis overdose or during anaesthesia. Practolol exemplifies β-site blockers for which relative specificity to cardiac β-sites is claimed.

Adrenergic Neurone-Blocking Drugs

These drugs are treated separately as they exert their sympatholytic activity by mechanisms other than an action at the adrenergic receptor.

In the normal animal, noradrenaline is absorbed by and stored in adrenergic neurones. Re-absorption is an important mechanism in the termination of the action of the noradrenaline released by an activated sympathetic fibre. Certain drugs, e.g. **cocaine, imipramine** and **chlorpromazine** block the access of noradrenaline to the sites of binding and storage.

Before noradrenaline is released from a sympathetic nerve there is an initial release of acetylcholine and this triggers the subsequent noradrenaline discharge. The drug **hemicholinium** prevents noradrenaline release via a mechanism which involves the prevention of re-synthesis of intra-neuronal acetylcholine by blocking the mechanism which transports choline back into the cell. The mode of action of **bretylium** is believed to be an interference with the action of releaser acetylcholine in the neurone.

It is possible to cause a gradual leaching away of stored amines with the alkaloidal drug **reserpine**. This compound causes a progressive depletion of both central and peripheral noradrenaline stores and has been used in man as a tranquillizer and anti-hypertensive agent. It has seen veterinary application in the prevention of aortic rupture in turkeys. As the release of amines is slow, no sympathomimetic effect is seen, probably because the rate of release does not exceed the capacity of endogenous inactivating pathways. Reserpine can be used to distinguish between direct and indirect-acting sympathomimetics.

The ability to deplete body noradrenaline stores is also possessed by **guanethidine** and the more transient **bethanidine.** These anti-hypertensives deplete only peripheral stores and, unlike reserpine, have no effect on central nervous system (CNS) function. Additionally, they possess a bretylium-like action in diminishing the neuronal release of noradrenaline.

Noradrenaline is synthesized in the body along a pathway which begins with the amino acid phenylalanine (xiii).

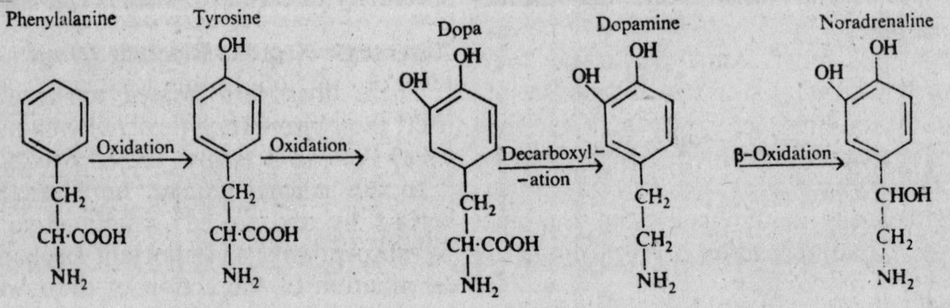

(xiii) The metabolic conversion of tyrosine to noradrenaline

The metabolic conversion of tyrosine to noradrenaline. **α-Methyldopa** is a drug which competes with dopa for dopa decarboxylase and so reduces dopamine and noradrenaline production. Additionally, the α-methyldopamine is oxidized to α-methylnoradrenaline. Although this is released along with noradrenaline when the nerve is stimulated, the end result is a reduction in sympathetic activity as α-methylnoradrenaline possesses only about one-tenth the transmitter activity of noradrenaline. Because the so-called *false transmitter* α-methylnoradrenaline is not destroyed by monoamine oxidase, it is more persistent than noradrenaline and so tends gradually to replace noradrenaline in the storage granule. Although its onset of effect is slow, this drug is much used in the control of hypertension in man.

VASODILATORS

This term is applied to an assortment of drugs which bring about vasodilation usually by means of a direct action upon smooth muscle in blood vessel walls, but can also include a group of drugs which diminish peripheral sympathetic tone by interfering with the transmission of impulses in the autonomic ganglia.

The Ganglion-Blocking Drugs

The autonomic nervous system differs from the somatic nervous system in having certain of its neurones aggregated in clusters called ganglia outside the CNS. The myelinated autonomic efferent leaving the spinal cord synapses in a ganglion with a neurone whose non-myelinated fibre (the so-called post-ganglionic fibre) innervates the effector organ. The parasympatho- and sympathomimetic agents, and the parasympatholytics and the α and β adrenergic blockers all act at the post-ganglionic fibre effector cell junction. The ganglion-blocking drugs reduce both sympathetic and parasympathetic effects by means of a competitive block of the ganglionic transmitter substance, acetylcholine. Although it is possible to bring about a depolarizing block of this nicotinic action of acetylcholine, e.g. with nicotine itself, this approach is not used clinically.

Although ganglion-blockers were used to control hypertension in man, the variety of side-effects consequent upon disruption of autonomic function made them far from ideal for this purpose and they are now superseded by the adrenergic neurone-blocking drugs. The first ganglion-blocking drugs were the symmetrical *bis*-quaternary ammonium compounds **pentamethonium** and **hexamethonium.** These differ only in the number of carbon atoms separating the quaternary nitrogens, five in the case of pentamethonium (C_5) and six in hexamethonium (C_6) (xiv). **Pentolinium** is a more potent member of this family, while **chlorisondamine** is an even more potent but asymmetric compound.

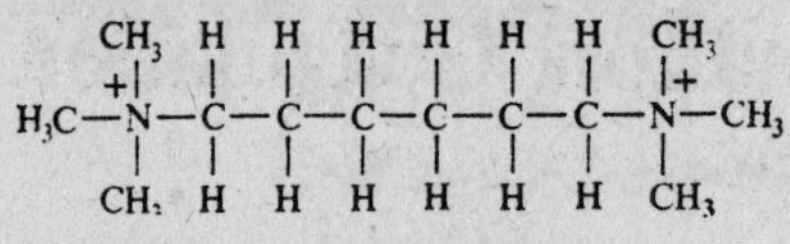

(xiv) Hexamethonium

These compounds have found no use in veterinary medicine beyond their application in experimental situations.

Smooth Muscle Relaxing Drugs

Angina pectoris is a common problem in man and various nitrites and organic nitrates have been used for its relief. **Amyl nitrite** was dispensed for inhalation. It causes relaxation of smooth muscle in blood-vessels, bronchi and the gut and a dramatic fall in blood pressure which can cause fainting.

Glyceryl trinitrate is absorbed from a tablet placed under the tongue so as to avoid immediate inactivation by the liver.

Erythrityl tetranitrate is used prophylactically as it has a delayed onset but prolonged duration of action.

Although all these drugs can cause coronary dilation in a healthy heart, and would therefore improve myocardial blood flow and so reverse any tendency to anoxia, it is now believed that the peripheral vasodilation which they cause also significantly reduces the work load of the heart by reducing blood pressure. The nitrites all tend to cause methaemoglobinaemia and as this can form a stable complex with cyanide, nitrites have a theoretical place as cyanide antidotes.

Papaverine. This opium alkaloid is a powerful smooth muscle relaxant. Apart from its ability to cause coronary vasodilation, it is useful in reducing a previously high tone in bile duct, ureters and intestine and has been used to treat colic. Although the action of papaverine is a direct one, its precise mechanism is not known.

Theophylline and aminophylline have already been mentioned in relation to their cardiac stimulant action. They can cause, in addition, both coronary and peripheral vasodilation.

Although papaverine and theophylline preparations could be used in the relief of peripheral vasoconstriction or cramps, it is likely that β-adrenergic stimulants would be more effective.

The vasodilator and cardiac stimulant actions of the xanthines are related to the β-site actions of adrenaline which are now believed to be mediated via the initial activation of the membrane-bound enzyme adenyl cyclase by adrenaline. This enzyme accelerates the intracellular conversion of ATP to cyclic AMP (cyclic adenosine 3′,5′-monophosphate) which then acts as a 'second messenger' inside the cell by triggering a reaction which modifies the cell's level of function. In this context the effect would be vasodilation but it, like all others, is terminated by another enzyme, phosphodiesterase. The xanthine derivatives are inhibitors of phosphodiesterase and by this means can amplify β adrenergic effects.

Bibliography

BARRETT, A. M. (1975) Catecholamines and the cardiovascular system. In: *Modern Trends in Cardiology*, vol. 3, pp. 44–67. London: Butterworths.

HIGGINS, C. B., VATNER, S. F. & BRAUNWALD, E. (1973) Parasympathetic control of the heart. *Pharmac. Rev.*, **25**, 119.

JEWITT, D. E. (1975) Antiarrhythmic drugs and their mechanisms of action. In: *Modern Trends in Cardiology*, vol. 3, pp. 333–76. London: Butterworths.

KHARKEVICH, D. A. (1967) *Ganglion-blocking and Ganglion-stimulating Agents*. Oxford: Pergamon.

MARLEY, E. (1964) Adrenergic system and sympathomimetic drugs. *Adv. Pharmac.*, 3, 168–252.

MYERSON, R. M. (1967) Digitalis. In: *Topics in Medicinal Chemistry*, vol. 1, pp. 303–39. New York: Interscience.

ROOT, W. S. & HOFMANN, F. G. (ed.) (1967) The adrenergic blocking drugs. In: *Physiological Pharmacology*, Vol. 4, Part III. New York: Academic.

SMITH, A. D. (1972) Subcellular localization of noradrenalin in adrenergic neurons. *Pharmac. Rev.*, **24**, 435.

TAVERNOR, W. D. (1962) Management of cardiac arrest in the dog. *Adv. small Anim. Pract.*, **3**, 158.

WARE, M. (ed.) (1971) Cardiovascular agents. *Today's Drugs*, **2**, 85–125.

9

The Control of Blood Formation, Coagulation and Volume

The blood is probably the most important 'tissue' in the animal body. Its functions are wide and varied and any change in the blood is soon reflected in tissue function just as changes in tissue metabolism are rapidly followed by changes in the blood. The diversity of the responsibilities of the blood are indicated by the more important functions listed below:

1. The transport of oxygen from the lungs to the tissues is the most vital function performed by any body tissue.
2. Nutritional substances are transported from the intestines to the tissues and nutritional products synthesized in the body are distributed from their points of manufacture to their sites of utilization.
3. Metabolites produced by the various tissues are carried to the excretory organs.
4. Changes in the fluid balance between cells, extracellular fluid and the blood are neutralized by the blood in conjunction with the organs of excretion.
5. The pH of the tissues and tissue fluids is maintained constant by means of the electrolyte and buffer mechanisms of the blood.
6. Body heat is regulated by the take up and release of heat by the blood.
7. The hormones which are essential for the control and coordination of the body functions are all distributed by the blood.
8. Antibody fractions and macrophages in the blood are important factors in the defence of the body against the attack of pathogenic organisms.

The variety and importance of these functions and the ubiquitous distribution of blood and the tissue fluids with which it is in constant equilibrium make the blood an ideal means of circulating soluble therapeutic agents to all cells of the body, but it must be remembered that any damage to the blood will be immediately projected into the tissues. The rapidity with which cyanides, non-volatile anaesthetics or strychnine cause generalized effects bears striking witness to this statement.

In carrying out intravenous injections, therefore, although the blood is capable of accommodating prodigious amounts of abnormal substances without any material change in its physical characters, this accommodation should be imposed upon as little as possible. In order to aid accommodation assistance may take the form of making solutions isotonic, neutral in reaction, equal in temperature to that of the body, or failing these, of giving the injections so slowly that the blood can dilute and accommodate the fluid as it is injected.

HAEMATINICS

The blood of domestic animals carries between 15 and 20 ml of oxygen per 100 ml of arterial blood. Of this, only a small proportion is in physical solution in the plasma, the remainder being combined with haemoglobin within the red blood cells, 1 g haemoglobin combining with approximately 1·36 ml of oxygen. It is therefore most important that the haemoglobin content of the blood should be maintained at its optimum for each species (see Table IV).

Haemoglobin consists of an iron-containing pigment and a protein, the pigment forming about 5 per cent of the haemoglobin molecule.

From these few facts it will be seen that the oxygen carrying capacity of the blood is proportional firstly to the number of circulating erythrocytes and secondly to the haemoglobin content of each individual erythrocyte. The number of erythrocytes is governed by the rate of formation and of destruction of red blood cells, whilst the haemoglobin content is determined both by the amount of available iron and also of the specific protein. A deficiency of iron or inability to utilize it because of the absence of other factors, such as copper or cobalt, and a deficiency of dietary protein, will both cause a hypochromic anaemia. Disorganization of erythropoiesis or extensive haemolysis or haemorrhage will result in anaemia due to a reduction in the numbers of circulating erythrocytes.

Table IV. Haemoglobin content of blood for different species

Species	Approximate haemoglobin (g/100 ml)	
Horse	10·0	±1·5*
Cow	12·03	
Sheep	12·4	±1·4
Goat	10·9	
Pig	12·0	
Dog	13·0	
Cat	10·5	
Cockerel	13·5	
Hen	10·0	

* Thoroughbreds are somewhat higher.

Perhaps the commonest form of anaemia in Great Britain is **piglet anaemia** which is usually a microcytic hypochromic condition due to lack of available iron but is sometimes associated with other factors, including deficiencies of the vitamin B complex and of dietary protein. The condition is a definite penalty of intensive indoor pig rearing, though it is ironic that the cleaner and more hygienic the conditions, the higher is the incidence of anaemia. Where free access to soil is available from birth it is uncommon to find anaemia as a rearing problem.

During the first week to 10 days after birth the normal haemoglobin level of between 10 and 11 g per 100 ml of blood decreases to between 5 and 6 g per 100 ml. This is due in part to a redeployment of the available haemoglobin but also to a reduction in iron intake associated with rapid growth and increased demand for haemoglobin. If extrinsic iron is available to supplement the daily oral intake of about 1 mg. from the sows milk the haemoglobin level will increase; should there be none then the level will continue to fall and precipitate subclinical and later clinical anaemia. Symptoms may be expected when the level falls below 5·0 g per 100 ml and consist most frequently of scours, in which the semi-fluid faeces are grey or yellowish, dyspnoea on exertion, a disinclination to play, a pallid skin and pale mucous membranes. These symptoms are most marked at about 3 weeks when the piglet may look fat and well fed: such piglets may die suddenly or become progressively thinner assuming a stunted runt-like appearance. It is the piglets which grow rapidly in their early days which are most severely affected—their requirements are proportionately greater.

Iron Compounds

The following methods of treatment have been used.

1. A combination of **ferrous sulphate** and **copper sulphate,** usually in proportions of about 5 to 1, have been administered in various ways. Most popular has been a mixture of 105 g (3¾ oz.) iron sulphate with 21 g (¾ oz.) copper sulphate in 1114 ml (2 pt.) of treacle, some of which is smeared on the sows teats once or twice daily. The B.V.A. Publication 14, 3rd edn, p. 103, suggests one teaspoonful (4 ml) daily to each piglet of a mixture each pint of which contains 180 gr. of ferrous sulphate and 60 gr. of copper sulphate (21 and 7 g per litre respectively). In Scandinavia and elsewhere it is a common practice to spray grass sods and soil with a mixture of ferrous and copper sulphates. This medicated soil is thrown into the pens daily. Dressing of the sow's teats is a very easy method but tends to make the skin sore and dirty, for the sticky syrup picks up straw and dirt very rapidly.

2. Oral dosing with **reduced iron** or **iron pyrophosphate** has also been widely used. The B.V.A. Publication 14 (*loc. cit.*) suggests reduced iron 0·5 to 1 g daily or every second day for a week to 10 days. This method is expensive in labour and has given place to the almost universal use of intramuscular iron injections.

3. Intramuscular and intravenous injections of **iron carbohydrate complexes** have been used in the field of human iron deficiency anaemias since about 1948, but it was not until the introduction in 1955 of an **iron-dextran** injection for

piglet anaemia that such preparations had been used in the veterinary field.

These complexes must be administered intramuscularly, for subcutaneous injection or leakage may cause tissue staining which is unsightly until it is completely absorbed some 6 weeks or so later. Absorption of the dextran and dextrin complexes is sufficiently rapid to prevent any fall of the haemoglobin content below the critical level and to restore acutely pathological levels to normal. The majority of the injected iron is used within about 1 week of intramuscular administration. The normal doses is 100 mg of ferric iron usually contained in 2 ml of the injection; this dose is often given within 3 days of birth and in severe cases repeated after about 10 days. However, muscular degeneration and death have occasionally followed administration of some organic iron compounds to very young pigs. In view of the rapidity of the therapeutic action of such compounds and the fact that blood iron rarely becomes depleted to danger level before about 10 days of age there seems little point in injecting piglets before they are about 5 days old.

Other iron preparations, all of which are astringent both externally and internally, include **Ferrous Sulphate, B.Vet.C.** which may be either a water soluble green crystal or powder with an astringent metallic taste, or a heavy greyish-white anhydrous powder with similar properties although it dissolves very slowly when compared with the hydrated salt. Ferrous sulphate was commonly used as an intestinal astringent in cattle and horses and also as a haematinic (a haemoglobin forming and erythropoietic substance). The astringent nature of iron salts is responsible for their undesirable side-effects on the digestive tract, i.e. gastric pain, vomiting, constipation or diarrhoea. Ferrous iron salts are better absorbed than ferric and are frequently prepared with a reducing agent such as ascorbic acid to prevent oxidation.

The dose of the exciccated salt as a haematinic is 4 g (60 gr.) given twice daily to a horse, usually in food with gentian and nux vomica powder.

Ferrous carbonate (Blaud) is used in the form of a pill or tablet, often in conjunction with arsenic, as a haematinic in small animals, and **Compound Syrup of Ferrous Phosphate, B.P.C.,** is also used in doses of 2 to 4 ml (30 to 60 m.): most of these preparations are being displaced by organic acid salts such as **ferrous gluconate** and **succinate** claimed as less irritant, less constipating and more effectively absorbed, especially if combined with therapeutic amounts of the vitamin B complex. Doses of 0·3 g (5 gr.) may be given twice or three times daily to small animals.

Cobalt

Cobalt deficiencies associated with a marked anaemia, loss of appetite, progressive wasting, decreases in blood volume and a drop of oxygen carrying capacity to 30 per cent of normal or lower are reported in sheep. The condition is widely known as 'pine' in Great Britain and elsewhere, but also as 'bush sickness' in New Zealand, 'nakuruitis' in Kenya and 'grand traverse disease' in the United States.

Symptoms in cattle are less severe than in sheep but are similar in type. Milk yields are reduced and calves are weak when born—if they survive they are unthrifty and anaemic.

As the condition is essentially related to deficiencies in the soil and herbage, preventive measures must include pasture dressing with **cobalt sulphate** or **cobalt chloride,** the quantity and frequency of dressings will vary with the area and degree of deficiency. Spraying solutions from the air has proved an efficient and rapid method of applying the salt. For animals not at grass, oral dosing twice weekly or mixing cobalt salts in the food is the best means of administration, for it has been shown that pining animals do not all use salt licks. Parenteral injection is not effective in alleviating the effects of cobalt deficiency, which is not surprising for the eventual deficiency is one of vitamin B_{12} which is manufactured from cobalt in the rumen by the bacterial flora and then absorbed.

Dosage

For prophylaxis:

Cattle: 25 mg ($\frac{2}{5}$ gr.)
Calves: 10 mg ($\frac{1}{6}$ gr.)
Sheep: 5 mg ($\frac{1}{12}$ gr.)
Lambs: 2·5 mg ($\frac{1}{24}$ gr.)

These doses must be given orally each day either in food or as a drench.

For oral treatment:

Cattle: 0·5 g (7½ gr.)
Calves: 0·2 g (3 gr.)
Sheep: 0·1 g (1½ gr.)
Lambs: 0·05 g (¾ gr.)

Repository 'bullets' have been developed which are administered orally and, theoretically, because of their weight, remain in the reticulum releasing cobalt slowly over a long period. Their main disadvantage applies to all oral tablets in that some individuals will return the bullet some time after dosing and it is not possible to detect the offenders until clinical signs of deficiency appear.

Copper

Copper is also necessary for the normal utilization of iron in haemoglobin formation in mammals and birds. Copper and iron are together associated in various enzyme activities, whilst copper is an essential component of other enzymes including ascorbic acid oxidase, polyphenoloxidase, lactase and copper protein compounds of unknown function. Through various enzymic and catalytic processes, copper is closely associated with normal wool growth, coat and skin colouring and normal bone metabolism. Derangement of all these processes are noticeable in clinical cases of copper deficiency.

Symptoms of copper deficiency in sheep vary according to whether the deficiency is complicated with cobalt deficiency or molybdenum excess, or other unidentified complications (cf. swayback in which no symptoms occur in adults but lambs are affected in utero by demyelination of the brain tissue). In uncomplicated deficiencies the wool changes its appearance and becomes straight with a steely sheen; this is followed by severe hypochromic and microcytic anaemia; lambs may be affected with demyelination of the brain and spinal cord.

In cattle, the picture of copper deficiency in the United Kingdom is frequently associated with excess molybdenum (teart). The symptoms in cattle are those of unthriftiness, anaemia, temporary sterility and, in some areas (Western Australia) sudden death ('falling disease'). Straightness of leg joints and stiffness are recorded, and also fading of hair colour, particularly around the eyes which is responsible for the popular name 'spectacle disease'. Scours are also a significant symptom in many areas.

Treatment has consisted of mineral licks containing 0·3 per cent or more of **copper sulphate;** intravenous injections of 20 mg copper (80 mg Copper Sulphate, B.Vet.C.) for sheep, and 50 mg (200 mg Copper Sulphate, B.Vet.C.) for cattle, are effective for about 4 months. Very effective is the use of **copper glycinate** intramuscularly in cattle and sheep, the doses being 120 and 45 mg of copper, respectively, equivalent to 400 and 150 mg of the glycinate. Copper sulphate may also be given as a food supplement at the rate of 2 g daily for adult cattle and 1 g daily for young stock. Top dressing of pastures is widely practised to offset soil deficiencies: in New Zealand, 5 lb. copper sulphate per acre annually (4 kg per ha) is found to be adequate, whilst in Australia as much as 20 lb. per acre (18 kg per ha) and as little as 7½ lb. per acre (6 kg per ha) have been reported adequate for a 2-year period.

Vitamin-Deficiency Anaemias

In contrast to the microcytic, hypochromic iron-deficiency anaemias, those due to a deficiency of **cyanocobalamin** (vitamin B_{12}) or **Folic acid** are macrocytic and hyperchromic. In addition, the total red blood cell (RBC) count is often very low. Folic acid (as Folinic acid) and cyanocobalamin act sequentially in the pathway which leads to the synthesis of nucleoproteins in cell division. Erythropoiesis is arrested at the intermediate normoblast stage. These cells are later released into the circulation as macrocytes and have a short life-span. The vitamins are more fully described in Chapters 17 and 18.

The problem of pernicious anaemia has not been recognized in domesticated animals, but vitamin B_{12} is the resultant deficiency of a dietary deficiency of cobalt. Additionally, vitamin B_{12} has been used to increase the stamina of race-horses and as a tonic after debilitating disease, e.g. calf scours.

THE COAGULATION OF BLOOD

The maintenance of an adequate haemoglobin

concentration in the blood will be able to maintain satisfactory cellular oxygenation only if the blood volume is not allowed to fall below a critical level. Blood loss is arrested by the process of coagulation; loss of volume is made good by fluid movements and, later, by protein synthesis and cell regeneration. In this section are described those drugs which can hinder or accelerate blood coagulation, and the various means by which plasma volume and osmotic pressure can be restored.

Before describing the mechanism of action of these drugs, it is necessary to describe the stages of the normal process of blood coagulation. For coagulation to proceed normally there are three pre-requisites.

1. The vessel wall has to be damaged so that collagen is exposed. This initiates platelet aggregation. Vascular endothelium is negatively charged and when this charge is lost in injury, the endothelium loses its ability to repel platelets. The haemorrhage in scurvy is associated with vessel wall defects.

2. Platelet count must be adequate. The platelets adhere to collagen, then to each other and finally undergo lysis liberating ADP, 5-hydroxytryptamine (5-HT) and histamine. Platelet insufficiency always results in haemorrhagic manifestations.

3. Plasma coagulation factors must be present in adequate concentration. Absence of one or more factors leads to impairment of coagulation, e.g. in the heritable coagulation defects.

Over the years, the number of factors believed to be involved in the clotting mechanism has steadily increased. As nomenclature is rather confused, the international convention relating names to Roman numerals is given in Table V.

Table V. Synonyms and numerals for coagulation factors

Factor no.	Names
I	Fibrinogen
II	Prothrombin
III	Thromboplastin
IV	Calcium ions
V	Labile factor or proaccelerin
VI	Existence questionable
VII	Stable factor, proconvertin or autoprothrombin I
VIII	Antihaemophilic globulin (AHG or AHF) or platelet cofactor I
IX	Plasma thromboplastin component (PTC) Christmas factor, platelet cofactor II, or autoprothrombin II
X	Stuart factor or Koller's factor
XI	Plasma prothrombin antecedent (PTA)
XII	Hageman factor or glass activation factor
XIII	Fibrin stabilizing factor

The sequence of events in the coagulation process is as follows. Thromboplastin activity can arise through two routes a slow (4 to 8 min) or intrinsic route and a rapid (10 to 20 sec) or extrinsic route, the starting point of which is the activation of proteolytic enzyme Factor XII by a wettable surface. Thromboplastin, in the presence of calcium ions, catalyses the conversion of inactive prothrombin to the enzyme thrombin. Due to the proteolytic action of thrombin, the soluble precursor fibrinogen is converted to fibrin monomers which, under the influence of activated Factor XIII, transform to polymerized, insoluble fibrin. The fibrin threads undergo shortening and so increase the strength and density of the clot.

The complexity of this sequential activation of precursors to yield catalysts provides a variety of possible sites for defects of coagulation. Many heritable defects have been identified in man and the majority of these have their veterinary counterparts, e.g. deficiencies of Factors VII, VIII, IX, XI, XII have been reported and the newborn of many species are deficient in Factors II, V, VII, IX and X.

It is essential that a haemostatic plug or thrombus be removed once it has fulfilled its function and the removal mechanism can be regarded as the final stage of the coagulation process. Fibrin is broken down by the enzyme plasmin which is derived from an inactive, circulating precursor, plasminogen. Plasminogen activators are found in vessel walls and are recoverable from many tissues, secretions and urine. These convert plasminogen to plasmin.

The activity of plasmin is normally limited by a circulating antiplasmin. It is interesting to note that this pattern of an enzyme, derived from an inactive precursor under the influence of an activating system, and held in check by a circulating antagonist is common to several stages of the coagulation process (see Fig. 11)

THE COAGULANTS

It would be only logical to expect that preparations of the substances involved in natural clotting processes would be available for arresting traumatic and surgical haemorrhage. Such preparations are available but should only be used as local applications: they are of little value for internal haemorrhage and should not be injected intravenously because of the danger of causing widespread intravascular clotting and death.

pack with a suitable diluent. It is of value in arresting capillary haemorrhage and may be used in conjunction with fibrinogen or fibrin foam for this purpose, for filling in 'dead space' during surgery, for grafting, in fact, for any surgical operation where it is necessary to prevent haemorrhage or produce rapid and strong adhesion of two surfaces.

Fibrinogen from human sources is available as

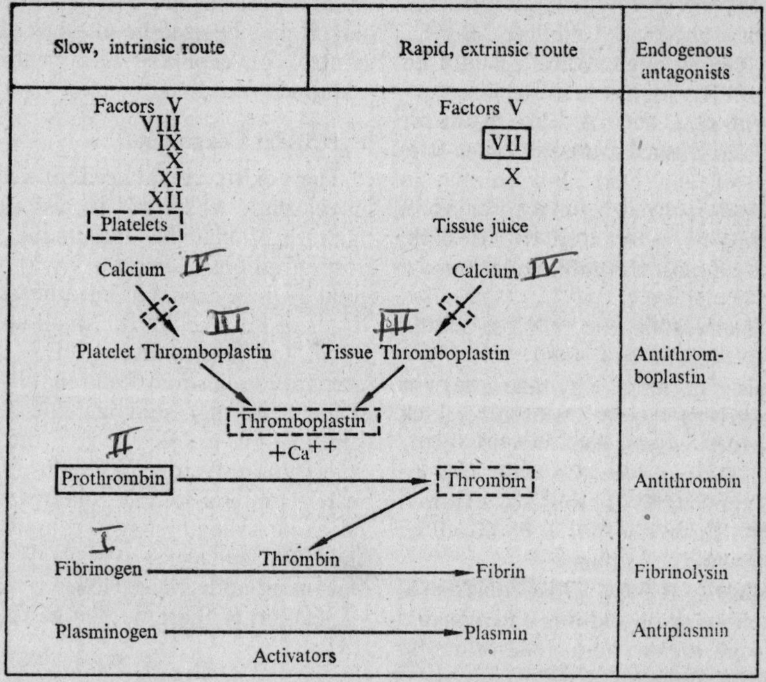

FIG. 11. The stages by which fibrin is formed and removed. The sites of action of coumarins are indicated by continuous line boxes and those of heparin by dotted line boxes.

The Physiological Coagulants

Thromboplastin, also known as thrombokinase, thromboplastic or zymoplastic substance, is produced naturally by platelets and damaged tissue, but commercial thromboplastin is most frequently an extract of cattle brain in normal saline. It is applied locally as a haemostatic, being of most use in capillary oozing: it is also used for treating epistaxis and in human surgery of the nose and throat, of glandular tissue, and of bone.

Thrombin is the enzyme which converts fibrinogen into fibrin. Bovine thrombin is available commercially as a powder supplied in a dual

a white powder or friable solid prepared from human plasma by precipitation with organic solvents and freeze drying. It is readily soluble in normal saline; solutions should be freshly prepared before use. Its main use is in assisting the adhesion of grafts of skin and mucous membranes, for which 2 per cent solutions are used: the amount of thrombin and the temperature of the solution determine the rate of clotting, and the amount of fibrinogen the strength of the clot.

Fibrin is available as Fibrin Foam (human) B.P. It is an insoluble material in the forms of strips of a fine white sponge. Fibrin foam strips are useful for filling in dead space, particularly where capillary oozing occurs; the strips are soaked in a thrombin solution and placed over the site of haemorrhage or in the cavity: coagulation will occur immediately and the strip can then be buried by suturing. These methods have been used extensively in human lung and brain surgery.

Other Surgical Haemostatics

These comprise Oxidized Cellulose, B.P.C., Absorbable Gelatin Sponge, B.P.C., and Calcium Alginate, B.P.C. All are heat sterilized.

Oxidized Cellulose, B.P.C. A white gauze or cotton with a faint odour and acid taste. It is insoluble in water and acids but soluble in dilute alkalis. The oxidized cellulose is acid in reaction and must be neutralized with sodium bicarbonate if used with thrombin solutions; it also inactivates penicillin.

If used as a haemostatic, the cellulose should be applied dry, when it will swell, forming a brown gelatinous mass which may take 7 days or more to be absorbed. It may be used to pack cavities and dead space, the wound being closed over it: surface bleeding will also be arrested by an application. The use of oxidized cellulose in bone surgery is not encouraged as it delays the formation of callus.

Calcium Alginate, B.P.C. This consists of calcium, or a mixture of sodium and calcium alginates—salts of alginic acid. The white to buff powder or fibrous 'wool' is almost odourless and tasteless; it is insoluble in water but soluble in sodium citrate solution. Calcium alginate is an absorbable haemostatic which may be applied in two ways—either sodium alginate is applied as a 4 per cent solution over the area which is then sprayed with a 2 per cent solution of calcium chloride, or it may be applied as a fibrous calcium alginate dressing soaked in sodium alginate solution. The dressings do not inactivate penicillin and may be buried in the same manner as other absorbable dressings. Removal of alginate dressings (aided by sodium citrate solutions) does not interfere with granulation tissue which incorporates any of the adherent dressing. Powdered alginates may be used to dust surgical gloves.

Absorbable Gelatin Sponge, B.P.C. Prepared from gelatin treated with formaldehyde, heated, whisked into a foam and freeze dried. The sponge is insoluble in water but can be wetted: it is a finely porous, white sponge-like material which will absorb about thirty times its own weight of water. The sponge may be applied dry or soaked in thrombin solution containing 100 units per ml, when clot formation occurs rapidly. It may be buried in dead spaces, etc., but takes 6 weeks or more to be absorbed. Nevertheless, it may be used in all types of surgery where control of capillary and venous bleeding is required.

Parenteral Coagulants

There exists a great need for a safe and efficient haemostatic which can be given parenterally. A mixture of **malonic** and **oxalic acids** in equal concentrations (5 per cent each) has been widely used by intravenous or intramuscular injection in doses of 0·5 to 3 ml for small animals and 5 to 20 ml for large animals. The efficacy of this preparation has not been reliably confirmed, but it has been suggested that the acids may cause rupture of platelets with the liberation of coagulation factors. **Vitamin K** (*q.v.*) and its water- and oil-soluble analogues are useful in checking haemorrhage caused by hypoprothrombinaemia such as may occur with Warfarin poisoning and 'sweet clover poisoning'. Details of vitamin K therapy may be found in Chapter 18.

Local Haemostatics

All astringent metals and plant substances will act as haemostatics but the main one used clinically is **ferric chloride** (*q.v.*) which may be used as Strong Solution of Ferric Chloride, B.Vet.C. **Alum** has also been used but is somewhat irritant, and **tannic acid** may be applied. With all these astringent compounds the action is one of precipitating sufficient protein in the open end of the vessel, both to block it, and to encourage normal clotting mechanisms. They should only be applied to clean surfaces, haemorrhage being temporarily prevented by pressure. Too much blood prevents direct application to the vessels involved and covers the area with an ineffective mass of precipitated protein under which bleeding continues. Minor arterial bleeding such as may occur during the docking of puppies can be arrested with ferric chloride solution.

Sclerosing Agents

A number of irritants when injected into dilated blood vessels cause their obliteration by fibrosis, e.g. benzyl alcohol, quinine and urea hydrochloride and sodium tetradecyl sulphate. Side effects include pain, sloughing and anaphylaxis.

THE ANTICOAGULANTS

Compounds used to prevent blood from clotting were developed initially so that blood samples might be maintained in a fluid state for subsequent laboratory examination. At a later stage, the introduction of blood transfusion dictated the need for a non-toxic anticoagulant. Finally, drugs became available for reducing the coagulability of blood in vivo. Such compounds are used in attempts to limit the spread of a pre-existing thrombus, e.g. in embolism or thrombosis, or in surgical procedures involving the vascular system.

The In-Vitro Anticoagulants

It is convenient first to consider those agents which are, for the most part, unsuitable for administration to a living animal. All the compounds share a common mode of action in that they render ionic calcium (Factor IV) non-available. **Sodium fluoride** or **oxalate** cause calcium to precipitate. **Sodium edetate (EDTA)** or **citrate** bring about the chelation of calcium, i.e. the calcium is still soluble but is no longer a free ion. Fluoride and oxalate are suitable for the preservation of samples for diagnostic examination. Edetate is used for this purpose but has also been used experimentally to induce hypocalcaemia. Citrate is used to prevent coagulation in blood collected for transfusion purposes. The citrate radical enters the Krebs cycle.

The Systemic Anticoagulants

Heparin is an anticoagulant which was originally isolated from liver during a search for agents which would promote blood clotting. It occurs in many tissues of the body and is stored in mast cells, the granules of basophils and in the intima of blood vessels. The physiological role of heparin remains in debate, as, apart from its role in maintaining the fluidity of blood, heparin can reduce blood fat levels by activating the enzyme lipoprotein lipase.

Heparin is a mucopolysaccharide closely related to chondroitin sulphate of cartilage, and is composed of units of glucosamine, glucuronic acid and sulphuric acid. These groups give the molecule its strongly acid (negatively charged) character. The anticoagulant action of heparin is terminated by contact with basic substances and such compounds are used as heparin antidotes, e.g. toluidine blue, protamine or hexadimethrine. It is suggested that heparin's anticoagulant action is due to its combination with the basic groups of enzymes involved in the coagulation process. In the body, heparin complexes with a plasma α-globulin. The resulting complex is antithrombin in action, but this is not the only point of action of heparin (see Fig. 11) which also interferes with the formation of thromboplastic activity and the action of thromboplastin on prothrombin. Additionally, heparin prevents the clumping of platelets.

The drug is usually administered by intravenous infusion as the sodium or lithium salt and the onset of action is almost immediate. The dosage is expressed in international units as the potency of heparin is determined by bioassay. Owing to its rapid inactivation by desulphation and depolymerization in the liver, the duration of action of heparin is short. As an alternative to intravenous drip administration, a depot formulation is available for intramuscular injection. Dosage ranges from a design concentration of 1 unit per ml for intravenous injection to 10 units per ml for maintaining blood samples in a fluid state for laboratory examination.

The coumarins are synthetic anticoagulants based on the molecule **4-hydroxycoumarin** (i). The series was developed from dicoumarol which was identified as the causal agent of a haemorrhagic disease of cattle in the U.S.A. Coumarin is the substance responsible for the smell of new-mown hay and it is envisaged that when sweet clover hay spoils, coumarin is converted to ***bis*-hydroxycoumarin** or **dicoumarol** (ii).

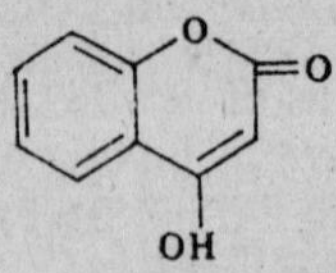

(i) 4-Hydroxycoumarin

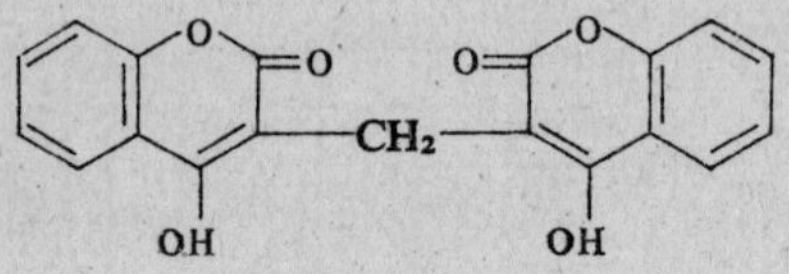

(ii) Dicoumarol

Dicoumarol was the first of this series to be used as an anticoagulant, but the mode of action of the compound had been pinpointed by the veterinarian who first described the disease as a prothrombin deficiency. It has since been shown that the coumarins cause a depression not only of plasma prothrombin levels but also of other coagulation factors, notably Factors VII, IX and X. It is now known that the coumarins do not increase the rate of prothrombin breakdown nor do they prevent the release of prothrombin from the liver, but they interfere with the synthesis of prothrombin in the liver. It is for these reasons that the effects of coumarin administration may not be seen for 48 hr, i.e. until plasma prothrombin levels have fallen sufficiently to reduce clotting efficiency. Many coumarin derivatives have been examined in attempts to find compounds with a more rapid onset of action, which cause a more regular and predictable coagulation defect and which have a lesser tendency to cumulate than dicoumarol. These compounds are exemplified by ethyl bis-coumacetate, cyclocoumarol, acenocoumarol and phenprocoumarin.

Warfarin sodium (iii) is an anticoagulant which, because of its potency and freedom from smell and taste, was originally developed as a rodenticide and, as such, has been incriminated in fatal poisonings and as a causal agent of foetal death and abortion is the domesticated species. It is now used in human therapeutics as, unlike dicoumarol, it can be administered by all routes, has a shorter latency and is less persistent in the body.

The toxicity of the coumarins is characterized by haemorrhage which initially may be a diathesis (blood is often first detectable in urine) but can become massive, e.g. haemorrhage into brain or body cavity would cause sudden death.

It has long been the practice to treat coumarin overdosage or toxicity with **vitamin K.** The structural similarity between coumarin and vitamin K molecules (iv) lead to the suggestion that coumarins act as antivitamins by competing with vitamin K for a liver proenzyme involved in prothrombin synthesis.

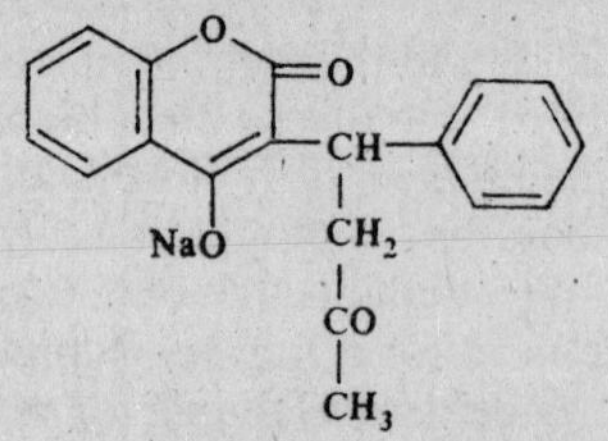

(iii) Warfarin sodium

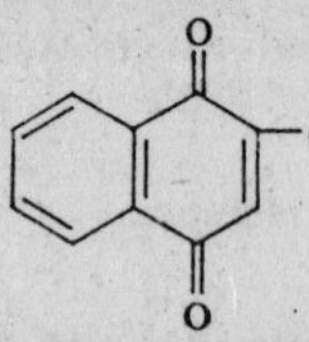

(iv) Menadione (vitamin K analogue)

The natural, oil-soluble vitamin K (phytomenadione) has been shown to be more effective in treating poisoned dogs than its synthetic water soluble analogues. The role of vitamin K is discussed more fully in Chapter 18.

The **indanediones,** e.g. phenindione are structurally related to the coumarins and produce anticoagulation of the same type. These compounds have been reported to have caused kidney damage, sensitivity reactions and leukopenia.

Anticoagulants are contra-indicated in liver and kidney disease, pregnancy or where a coagulation defect may exist. They are not much used in veterinary medicine and this may in part be due to the need for regular laboratory determinations of prothrombin time (Quick test) as a check on the suitability of the treatment regimen.

Fibrinolysis

In situations where haemorrhagic episodes occur because of overactivity of the fibrinolytic system, it is of value to be able to control plasmin production. On the other hand, there are also situations where the removal of thrombus, pus, or necrotic tissue is clinically desirable. **ε-Aminocaproic acid** can inhibit both the formation and action of plasmin. As plasmin and thromboplastin are formed at the same time, aminocaproic acid can be used to augment haemostasis in normal subjects.

Urokinase is a plasminogen activator recoverable from urine and is neither antigenic nor pyrogenic.

Streptokinase and **staphylokinase** are plasminogen activators isolated from bacteria. Streptokinase, which occurs with **streptodornase,** has been both pyrogenic and antigenic. Streptodornase reduces the viscosity of pus by depolymerizing nucleic acids.

Increased plasminogen activation can occur in stress, anoxia, fever and in lung or prostrate surgery owing to activator release. Chloroform and niacin are also plasminogen activators. The pig fibrinolytic mechanism is very poorly developed and this may well account for the frequency of occurence of fibrinous complications in pathological states in this species.

Defibrinogenation

Under conditions of severe tissue damage or extensive demands on the coagulation process, it is possible so to deplete fibrinogen levels that clotting can no longer occur. Naturally-occurring coagulation defects due to a lack of fibrinogen or to the presence of abnormal fibrinogen have been reported. As the formation of haemostatic platelet plugs takes place normally in such subjects, spontaneous haemorrhage occurs less frequently than in haemophiliacs.

Arvin is the trade name of a fraction isolated from the venom of the Malayan Pit Viper. This material has thrombin-like activity in that it causes precipitation of fibrin in circulating blood. The drug is given intravenously, is very rapid in action and is claimed to be free of side effects. The absence of fibrinogen renders the blood incoagulable and this is of advantage in surgical situations where post-operative thrombus generation significantly affects the chances of survival, e.g. in organ transplantation.

BLOOD VOLUME EXPANDERS

Variously referred to as plasma extenders, plasma expanders and plasma substitutes, blood volume expanders may either be fluids obtained from blood by various processes or synthetic colloidal solutions which have osmotic characteristics approximating to those of plasma: they are capable of creating and maintaining an increase in the circulating blood volume and are therefore intended for the treatment of the decreases in both circulatory volume and blood pressure which are associated with shock. They cannot increase the oxygen carrying capacity of the blood and cannot replace any deficiency of blood cells, though they may be used as vehicles in which to suspend erythrocytes for intravenous infusion.

The efficiency of a substance when used as a plasma expander depends to a great extent on the degree and persistence of the osmotic pressure exerted by its macromolecules within the vessels. The persistence is determined by the rate of loss from the circulation by one of the following means: (1) across the capillary wall into the interstitial fluids, (2) by excretion from the body, (3) by selective storage in certain tissues, or (4) by metabolic degradation. The value of any expander is therefore determined by the shape, structure, weight and electric charge of the macromolecule or particle, for these are the factors which govern the passage of those particles across a membrane.

The ideal plasma expander would have a molecular weight of about 70000 (that of plasma albumin), exert an osomotic pressure equal to that of normal plasma proteins, persist in the circulation long enough to allow mobilization into the circulation of normal plasma protein, and yet be eliminated at a rate identical with that of the plasma protein mobilization so that the osmotic pressure remains constant; in addition, the solution must have a viscosity which allows free administration through the finest needle with a minimum pressure.

As regeneration of plasma protein will take place at varying rates according to the individual and the occasion, it is unlikely that a single type of synthetic macromolecule will be able to achieve this ideal, and only a natural blood fraction which can take part to the full in normal blood metabolism could comply with all the requirements.

Indications

The use of blood volume expanders is primarily indicated in the decrease of circulatory volume associated with 'shock'. Shock is an extremely complex physiological reaction to acute physical stress which may take the form of severe and sudden pain, usually associated with trauma, tissue damage, again associated with trauma or extremes of temperature, and haemorrhage. Although plasma expanders may be of temporary value in histamine shock the specific antagonists and vaso-active drugs are preferentially indicated: shock associated with bacterial toxins cannot be reversed by plasma expanders alone but must be additionally countered by antitoxic remedies.

The response to a shock stimulus follows in each instance a similar but not identical course: there is a fall in blood pressure, a progressive haemoconcentration with a decrease in plasma volume and a fall in body temperature: other symptoms which occur less constantly are a decrease in plasma protein, a decrease in venous blood oxygenation, an increase in pulse rate, a darkening of the venous blood, a marked increase in coagulation time, haemolysis, and a decrease in the rate and depth of respiration. These conditions may persist for several hours in a reversible state, but as tissue anoxia, and especially brain and circulatory system anoxia, supervenes the state becomes irreversible and the animal dies. While shock is reversible, treatment will save the animal's life, but it should be borne in mind that the animal at this stage is very susceptible to further stimuli which may cause a rapid progression to the irreversible state.

In traumatic shock there may be extensive haemorrhage into the traumatized area or considerable extravasation of fluid and frequently plasma protein. (Extensive burns are extremely dangerous because of marked loss of plasma into the burnt area.) In all types of shock, but particularly in shock associated with nerve damage, there is a vasodilatation and an increased permeability of capillaries, and, as capillary anoxia develops, there may be protein loss also. The tissue anoxia develops owing to the decrease in circulatory volume which is followed by a fall in blood pressure, a decreased venous return, and a lower cardiac output; this in turn leads to a further decrease in blood pressure, decreased rate of circulation and therefore tissue anoxia.

The clinical symptoms of shock are primarily a collapsed animal which has cold extremities, a low general body temperature, pale or, later, cyanosed dry mucous membranes, a weak, frequent and rapid pulse, associated with shallow breathing and acute depression.

The Treatment of Shock

The immediate aim is first to remove the cause of shock, if possible, arresting any haemorrhage and administering analgesics (not the tranquillizing drugs, for they will prevent the normal vasoconstricting physiological reaction) where pain is severe—these should be given intravenously, for subcutaneous injections may not be absorbed until the shock is reversed and there may be a great temptation to give a second dose because of the lack of response; where the shock is reversed both doses would then be absorbed, with dangerous effects. One of the historic methods of treating shock has been to apply warmth: this can be dangerous and even fatal, for it may induce vasodilatation of the skin vessels and therefore decrease the volume of blood available for circulation in the vital organs. Optimum temperature would be a normal comfortable working temperature.

Efforts to increase the circulatory volume should begin as soon as possible and should continue until the volume administered is approximately equal to that lost, if haemorrhage has been the cause of shock. With other forms of shock, and where there is no means of estimating blood pressure beyond digital estimation of the pulse, it is difficult to estimate the amount of fluid which can be given by intravenous injection: should the plasma protein be depleted, as in haemorrhagic shock, or should protein have been lost into the interstitial tissue, subcutaneous administration is likewise of doubtful value for there will be insufficient osmotic pressure exerted by the blood to withdraw sufficient fluid from the tissue spaces. In these circumstances one can only estimate by digital pressure on the pulse the extent of pressure and volume increase, the infusion end point being determined in conjunction with the clinical response.

The dangers of infusions are those associated with over dilution of the blood and over increase of the volume. The former results in anaemia and loss in oxygen carrying capacity—only whole blood transfusions or time can rectify this state of affairs: the latter is more serious, for, if excess of an expander is given, the volume of circulating blood increases hourly as the plasma protein is regenerated, withdrawing large amounts of fluid from the tissues, and the animal literally drowns: the capillary bed becomes overloaded and fluid is excreted into the lungs, whilst the heart becomes dilated and weakened as the cardiac reserve is exceeded. Death follows from pulmonary oedema and heart failure.

Electrolytes

Electrolytes such as normal saline, Ringer's solution and Ringer-Locke solution have all been used as plasma volume expanders, either by intravenous drip, by intravenous injection, or by subcutaneous administration. There is evidence to show that, in all but perhaps early shock, intravenous administration of electrolyte solutions can turn reversible into irreversible shock after an initial increase in blood pressure and plasma volume. It would seem to be the case that infusions of such solutions remain in the circulatory system for only a short time and, as they diffuse across the capillary membrane into the interstitial fluid, they remove some of the remaining plasma protein and severely prejudice an animal's recovery. It is probable, however, that intravenous saline may be useful when given before the onset of shock and as a prophylactic blood volume expander. These electrolytes may be used in conjunction with 5 per cent dextrose solution which will maintain blood sugar levels and provide a readily available source of energy (see also Chapter 10).

Gum Acacia

Gum acacia has been discarded as a colloid plasma expander in human shock therapy owing to the danger of allergic reactions and because of prolonged tissue storage; it also interferes with plasma protein regeneration. The last two effects occur in dogs and acacia should therefore not be used. **Pectin** has been discarded for similar reasons.

Gelatin

Gelatin is obtained from collagen. It is normally used in a non-gelling form when infused as a plasma-expanding colloid. The material has a molecular weight of around 35 000 and can be injected as a 6 per cent solution. This concentration will achieve its maximum haemodilution in 3 to 4 hr after intravenous injection. Gelatin can be degraded in the body, but is largely excreted unchanged in 24 hr. The toxicity of the product is low and it is non-antigenic.

Dextran

Solutions of 6 per cent and 10 per cent concentration are available, the former either with saline or 5 per cent dextrose and the latter only with dextrose. Dextran itself is a complex polysaccharide prepared by growing a bacterium, *Leuconostoc mesenteroides*, in a medium containing sucrose: the dextran for infusion is further hydrolysed to reduce the length of the saccharide chain and the molecular weight, which can be as high as 450 000 to 30 000–150 000. Infusion produces a volume expansion which is slightly greater than the volume injected, but is equal to the infusion volume after 1 hr: after 12 hr the increased volume is about 50 per cent of the infusion volume. Commercial solutions are claimed to be effective for up to 48 hr, over 60 per cent remaining in the circulation after 24 hr. It would seem that the dextran molecule is slowly broken down into glucose, though some is excreted unchanged in the urine, mainly during the first 6 hr: it is probably those molecules in the lower weight range (less than 10 000) which are thus excreted. Dextran solutions may be infused at the rate of 10 to 25 ml per min, about 500 ml being the average initial dose for an adult human, with a further

500 ml ½ hr later, given at a rate determined by clinical response. Doses varying from 4 ml up up to 40 ml per kg body weight have been suggested in dogs but the normal range varies between 75 and 250 ml. Reactions have occasionally been noticed in dogs and in no case should the volume injected exceed one-third of the plasma volume which may be taken as approximately 7·2 per cent of the body weight: if this proportion is exceeded the dextran has an heparin-like action and severe widespread haemorrhages may occur, leading to the animal's death. On occasion, anaphylactic reactions have occurred.

Polyvinylpyrrolidone (PVP)

A complex organic polymer which was developed by Germany during the 1939–45 war. It is water soluble and has macromolecules varying in molecular weight from 30 000 to 50 000. Following infusion of 3·5 per cent solutions there is an initial increase of blood volume equal to about 90 per cent of the infused volume. As much as 50 per cent is excreted, mainly unchanged in the urine, within 3 or 4 hr and by 12 hr the increase in volume falls to about 30 per cent of the infusion volume. Some PVP becomes stored in the cells of the reticuloendothelial system. Volumes used should be much the same as for dextran solutions.

Whole Blood

Whole blood transfusions are more frequently administered than formerly. The blood may be collected immediately before use or, if properly citrated, it may be kept refrigerated for up to 3 weeks. For blood to be used immediately, 100 ml of a 3 per cent sodium citrate solution for every 400 ml of blood should be adequate, but for prolonged storage 75 ml of acid sodium citrate should be used for every 500 ml of blood. This solution contains 1·6 g sodium citrate, 0·56 g citric acid and 1·5 g dextrose.

Although more information on blood groups in animals appears annually in the literature so far little attention is paid to them in practice, for single infusions rarely cause trouble and repeated infusions are uncommon. However, cross matching of citrated samples can be carried out on a slide, any agglutination being an indication of incompatibility.

Dosage

Horse and Cow: 1·25–2·5 litres (2–4 pt.) or more
Calf: 400 ml
Dog: 80–100 ml given at the rate of 7–14 ml per min

Blood Plasma

Blood plasma may be obtained by centrifuging citrated blood or decanting off the plasma after sedimentation. The plasma will keep for up to 2 years if refrigerated. Doses of up to 40 ml per kg body weight have been suggested, given either intravenously or subcutaneously.

Blood Serum

Blood serum is obtained by allowing blood to clot naturally and pouring off the serum. It may be used in the same way as plasma and in the same dosage.

The difficulty associated with the use of homologous blood products in veterinary medicine and surgery is the lack of blood banks and the scarcity of suitable donors at the right time. This has encouraged the search for a blood product which is suitable for use in all species and capable of being stored.

Despecified Bovine Albumin

This has been used clinically and was obtained from bovine blood by refrigeration techniques and was sterilized by heat. The advantage of despecified albumin is that it will keep almost indefinitely, it rarely induces allergic reactions, but, above all, it can be given intermittently as a concentrated solution in small volume intravenously, the blood volume being made up by oral intake or subcutaneous injection of electrolyte solutions. There is thus little chance of overloading the vascular system and the intermittent method of administration is eminently suitable for veterinary practice.

The recommended dosage for dogs is at the rate of 1 g per kg. An initial intravenous dose of 1·25 ml of a 20 per cent solution per kg followed by half this at 2 or 3 hourly intervals up to a maximum of 1 g albumin (equivalent to 5 ml) per kg was recommended. As a rule, far less than this was required to return the blood volume to normal when coupled with oral or systemic electrolyte solutions. The product has now been withdrawn.

In all intravenous injections in shocked animals, raising the vein and introducing the needle offers difficulties. If a limb is used a tourniquet is required above the injection site and the limb should be alternately flexed and extended until the vein becomes engorged. In small animals, it may be found that the sublingual veins are easy to enter, as they lie superficially below the mucous membrane.

Bibliography

BORN, G. V. R. (1968) Drugs in relation to blood coagulation, haemostasis and thrombosis. In: *Procedings of the Third International Pharmacological Meeting, 1966*, vol. 6. Oxford: Pergamon.

FEARNLEY, G. R. (1973) Fibrinolysis. *Adv. Drug Res.*, **7**, 107–65.

FORTH, W. & RUMMEL, W. (1973) Iron absorption. *Pharmac. Rev.*, **53**, 724.

HALL, D. E. (1972) *Blood Coagulation and its Disorders in the Dog*. London: Baillière Tindall.

JACQUES, L. B. (1967) The pharmacology of heparin and heparinoids. In: *Progress in Medicinal Chemistry* (ed. G. P. Ellis & G. B. West), vol. 5, pp. 139–88. London: Butterworths.

KOLB, E. (1963) The metabolism of iron in farm animals under normal and pathologic conditions. *Adv. vet. Sci.*, **8**, 50–108.

MOYER, G. A. & BUTCHER, H. R. (1967) *Burns, Shock and Plasma Volume Regulation*. St Louis: Mosby.

O'REILLY, R. A. & AGGELER, P. M. (1970) Determinants of the response to oral anticoagulants in man. *Pharmac. Rev.*, **22**, 35.

ROWSELL, H. C. (1968) The haemostatic mechanisms of mammals and birds in health and disease. *Adv. vet. Sci.*, **12**, 337–97.

SCOTT, D. B. (1969) Shock and its treatment. In: *Modern Trends in Gynaecology*, vol. 4, pp. 235–63. London: Butterworths.

10

The Urinary System

The urinary system provides a most important route by which the body can excrete substances absorbed in excess of current requirements and substances no longer required by the body, some of which are the non-volatile products of tissue metabolism. These are all filtered at the glomerular membrane and some are added to the ultrafiltrate by active transport through the wall of the proximal convoluted tubule. Much filtered solute is re-absorbed in the proximal tubule, some actively, e.g. sodium, some passively, e.g. chloride, some in a facilitated manner, e.g. glucose, and some to 100 per cent, e.g. amino acids. The non-absorbed solute provides the osmotic pressure which retains within the urinary tract that 1 per cent of the filtered plasma water which eventually appears as urine. Glomerular, proximal tubular and loop functions are obligatory, but distal tubule function is subject to regulation by aldosterone, while collecting duct permeability to water is governed by vasopressin. The volume of urine passed can therefore be increased by any technique which increases the concentration of solute which remains in the ultrafiltrate, or by interfering with the action or availability of aldosterone or vasopressin.

The degree of hydration of the body is in part governed by the resorption of water without electrolyte through the walls of the collecting ducts. Increase in plasma osmolality or decrease in plasma volume trigger the release of vasopressin which facilitates water resorption. The diuresis of diabetes insipidus demonstrates the degree of control which vasopressin normally exercises over urine volume, i.e. a 15-fold reduction of ultrafiltrate volume.

The kidney contributes significantly to the regulation of acid–base balance via the active resorption of sodium and potassium and the active secretion of hydrogen and ammonium ions.

Drugs capable of modifying renal function can therefore modify body electrolyte concentrations, pH and hydration and usually do so simultaneously (see below).

DIURETICS

These are substances which increase the quantity of urine and usually of electrolyte excreted in a given period.

The normal kidney glomerulus produces a filtrate containing a number of physiologically essential components as well as metabolites and unwanted substances. The essential substances, such as electrolytes (e.g. sodium, potassium, chloride, HPO_4 and HCO_3 ions) and glucose are selectively re-absorbed as the filtrate passes successively through the proximal convoluted tubule, the loop of Henle, the distal convoluted tubule and the collecting tubules. The urine reaching the kidney pelvis contains only excess electrolytes, exceptionally small amounts of essential non-electrolytes such as glucose, and waste products such as urea, uric acid and creatinine. In all 98 per cent of the water passing into Bowman's capsule is resorbed, a considerable part of it under the influence of antidiuretic hormone in the final concentration processes subsequent to the fluid leaving the loop of Henle. During tubular passage resorption of sodium and bicarbonate ions, in conjunction with the passage of hydrogen ions from the cells into the tubule lumen to take part in the formation of dihydrogen sodium phosphate, provides a means of rectifying imbalances of the acid base equilibrium created by the production of acids during normal metabolism. The hydrogen ions are made available by the association of metabolic and resorbed carbon dioxide and water into carbonic acid under the influence of the enzyme *carbonic anhydrase*—this carbonic acid ionizes into hydrogen and bicarbonate ions which are directed into the tubule lumen and the plasma, respectively. Water of solution accompanies these exchanged ions. These processes can be represented diagrammatically (see Fig. 12).

Heat loss during heat stroke can also be encouraged by giving diuretics together with fluid. Where cardiac, liver or kidney incompetence results in the retention of fluid in the tissues and body cavities, diuretics can be most useful in bringing about the resorption and elimination of the fluid.

Diuretics can be divided into: (1) hormone inhibitors; (2) cardiovascular diuretics; (3) miscellaneous diuretics; (4) osmotic diuretics; and (5) reducers of electrolyte resorption.

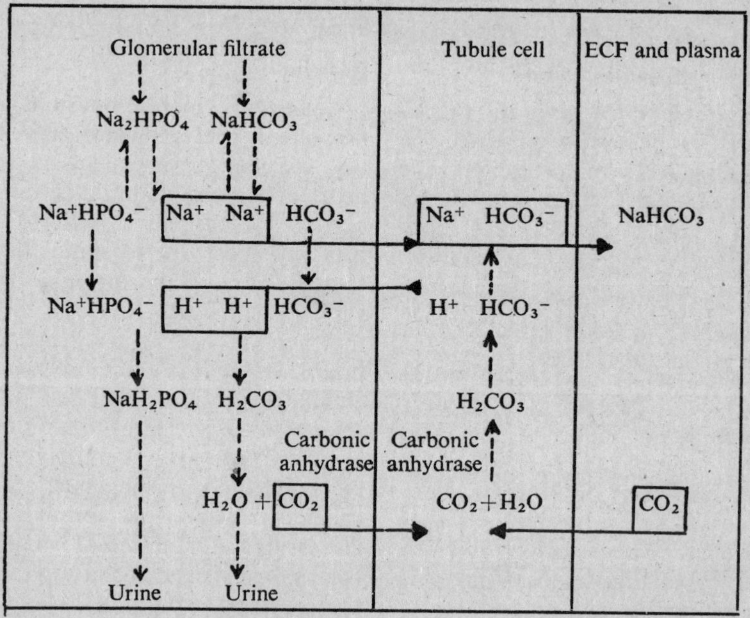

Fig. 12. A schematic representation of the resorption of sodium bicarbonate from and the acidification of the ultra-filtrate. Movement across cell walls is shown by the solid arrows.

Most diuretic drugs increase urine volume at least in part by reducing the efficiency of the carbonic anhydrase or other sodium resorbing processes.

By increasing the volume of urine output, bacterial infections can be influenced by dilution and by a mechanical 'flushing' of bacteria, débris and toxins from the kidney and urinary tract. When tissue metabolite and toxin levels build up during, for example, bacterial and viral infections, or when reduced excretion occurs as in nephritis, diuretics can be valuable both in reducing the metabolite and toxin levels but also in helping to dissipate the heat generated during the time the body temperature is elevated.

Hormone Inhibitors

Vasopressin

The production of this hormone (antidiuretic hormone, ADH) is stimulated by nerve impulses from the supra-optic nucleus initiated by osmoreceptors which are sensitive to changes in the water content of the blood. Vasopressin acts on the collecting tubules permitting an increased resorption of water unaccompanied by electrolytes. However, if the secretion of antidiuretic hormone is suppressed, either by administration of ethyl alcohol or large amounts of water, water diuresis occurs without a corresponding loss of electrolytes, until the increasing osmotic pressure of the blood once again causes the release of antidiuretic hormone.

When ADH production is inadequate, there occurs a condition characterized by the production of large volumes of dilute urine and known as diabetes insipidus. Preparations of ADH are available for treating this condition.

Aldosterone Antagonists and Inhibitors

These inhibitors act by preventing the operation of the sodium for potassium ion exchange system normally activated by aldosterone in the distal tubule. Although only the last few per cent of filtered sodium is subject to its control, aldosterone is still vitally important to sodium conservation. The most potent antagonist, **spironolactone,** is a steroid compound which acts by competing with aldosterone. It follows that the greater the effect of aldosterone on the kidney the greater will be the effect of spironolactone; the compound is therefore of most value in cases of hyperaldosteronism. Human dosage is 100 to 200 mg three or four times daily. Inhibitors of aldosterone are compounds such as amphenone and metyrapone which interfere with the biosynthesis of aldosterone. They have as yet no veterinary significance.

Cardiovascular Diuretics

The members of this section are so classified because their diuretic effect is accompanied by actions on the cardiovascular system.

In part, their effect is indirect and related to increased cardiac output and circulatory pressure. The volume and pressure of blood passing through the kidney is increased, which results in increased volume and speed of glomerular filtration, with a proportional increase in volume and rate of urine production. In many cases the urine excretion is so markedly increased that tissue oedema and fluid within body cavities is resorbed into the blood stream and then excreted by the kidney.

There are two main groups of cardiovascular diuretics: (1) the *cardiac glycosides* of digitalis, strophanthus, and squills; and (2) the *xanthine derivatives*, caffeine, theobromine and theophylline.

1. The former belief that the *cardiac glycosides* acted mainly on the kidney was largely responsible for the naming of the 'cardiovascular' diuretics. These compounds owe their diuretic effect to improved heart function and are discussed in Chapter 8.

2. The *xanthine derivatives* are rarely used as diuretics in veterinary medicine.

Caffeine

Caffeine is more important for its central nervous system stimulant action and, through this stimulation, its effect on respiration and heart action. In addition, arteriolar dilatation occurs by peripheral action which counteracts any effect due to stimulation of the vasomotor centre in the medulla. The presence of very few constrictor fibres in the kidney means that here, as well as in other areas such as the brain, the dilator effect predominates, and the circulatory volume of the kidney increases. As a result, increased filtration occurs. Increase in urine output is not entirely accounted for by increased blood throughput; some believe that caffeine is responsible for mobilizing dormant glomeruli and more recent work has shown that there is interference within the tubule cells with the enzymes responsible for electrolyte resorption. The urine produced has a relatively high water content and therefore a lower specific gravity.

Theobromine

Theobromine is suggested as the best xanthine diuretic as it has less tendency to cause gastric secretion and a more prolonged action than theophylline. It has no significant stimulatory effect in the central nervous system and can therefore be given in larger doses than caffeine.

Theophylline

Theophylline is not used on its own. Though more powerful as a diuretic than its two relatives, its action is short lived and it is therefore combined with other drugs such as mercurial diuretics. This both increases the diuretic potency and decreases the toxicity of the mercurial.

All xanthine derivatives are administered by oral or parenteral routes as double salts with organic sodium compounds such as sodium benzoate, salicylate or acetate. These salts increase solubility, and so improve intestinal absorption and make solution for injection much easier. Dosages are given on page 282.

Miscellaneous Diuretics

Volatile Oils

One of the remote local effects of volatile oils and similar compounds is the irritation of the kidney and the rest of the urinary system. Their action is said to be two-fold in that capillary dilatation occurs allowing greater filtration, whilst tubular irritation results in less resorption

Volatile oils still used for this purpose are **camphor** and **turpentine,** whilst colophony (resin) and benzoin have a similar effect but are less popular. **Chelidonium extracts,** which contain alkaloids and an irritant principle are popular in small animal practice and are used occasionally for large animal diuresis.

Most volatile oils also act as urinary antiseptics but their action is relatively slight when compared with hexamine and chemotherapeutic agents. They have the advantage of acting irrespective of urine pH and are less irritant than the mercurial diuretics.

Osmotic Diuretics

The administration of electrolyte solutions, particularly hypertonic solutions, or solutions of certain non-electrolytes, especially mannitol and urea, results in the appearance and persistance in the ultra-filtrate of osmotically active solute in increased concentration. The excess electrolytes or the unwanted non-electrolytes therefore pass through the kidney and into the bladder carrying with them the water of solution necessary to produce normal osmolarity.

The substances most frequently used for this purpose are salts of sodium and potassium, but those of ammonium, calcium and lithium are also used. These salts give the group its name, but other substances such as dextrose and urea will also exert a similar action.

Of the salts which can be used, the potassium ones are of most value, for the renal threshold of the potassium ion is relatively low (potassium extracellular concentration is, in most animals, 15 to 20 mg per 100 ml as against 316 to 340 mg per 100 ml for sodium (man)). Of the potassium salts the nitrate was very widely used in cattle practice and to a lesser extent in other farm animals. The acetate and citrate are used as diuretics in dogs.

Potassium Nitrate, B.Vet.C.

This is a white crystalline, odourless solid with a saline taste. It is very soluble in water and when dissolved with ammonium chloride in a ratio of one part of each in sixteen parts of water it gives a refrigerating solution.

Actions and Uses

Potassium nitrate ionizes freely in body fluids and is rapidly absorbed from the gut. The potassium ion is excreted, mainly via the kidney, as the bicarbonate and takes water with it. The nitrate is excreted partly as the potassium and partly as the sodium salt, some of which may be converted within the tubule to ammonium nitrate. The sodium ion can then be reabsorbed in order to conserve this essential substance. In excess, the nitrate is irritant and the preparation is contraindicated where nephritis is present. The urine is strongly alkalized.

Potassium nitrate was widely used in cattle practice as a diuretic and febrifuge. It was frequently used compounded with ammonium carbonate and acetanilide, but this type of drench was made obsolete by the advent of sulphonamides and antibiotics. The salt is not commonly used in other species, less irritant organic acid radicals being preferred to the nitrate.

Dosage and Administration:

Horse and Cow: 2–12 g (30–180 gr.)
Sheep, Pig and Goat: 1–2 g (15–30 gr.)

The nitrates are best avoided owing to the possibility of their bacterial reduction to nitrites in the rumen and intestine and the consequent production of methaemoglobinaemia.

Potassium Citrate and Acetate

Both these are used in the dog when saline diuretics are required; both ionize to give potassium and organic acid radicals. The potassium ion is excreted as the bicarbonate and takes with it water of solution. The citrate and acetate radicals are partly excreted unchanged while the greater part is metabolized to bicarbonate.

Ammonium Salts

Ammonium salts can also be used as diuretics. The most popular compound used for this purpose is **Ammonium Chloride**, B.Vet.C. which is used in horses and small animals. It acts as a urinary acidifier as well as a diuretic and is used in conjunction with organic mercurial diuretics, with which it seems to be synergistic. Part of the absorbed ammonium chloride is excreted by the lungs, and by thinning and loosening the bronchial mucous secretions it acts as an **expectorant.** The largest part of the ammonium radical is converted into urea, mainly by the liver and kidney. This urea is excreted along with its water of solution. The chloride ion depletes plasma sodium bicarbonate so causing hydrogen ion excretion and urinary acidification.

Carbohydrates

Mannitol exemplifies a non-electrolyte osmotic diuretic. Such compounds are infused intravenously in man to promote diuresis in refractory oedema and to prevent incipient acute renal failure. Sucrose and dextrose are alternatives for intravenous use. Glycerol and, recently, isosorbide have been used as oral osmotic diuretics.

Urea, B.Vet.C.

This has been used as a diuretic in doses of 60 to 250 g (2 to 8 oz.) for the horse, and 0·5 to 12 g (15 to 180 gr.) for the dog. In cattle it has been used as a feed supplement for ruminal synthesis of amino acids. Because of this ruminal utilization it cannot be used as a diuretic in cattle, sheep and goats.

Urea is a metabolite and is therefore not favoured as a systemic drug because of the danger of raising concentrations in the blood to toxic levels. This danger determines the contra-indications of hepatitis and nephritis.

Reducers of Electrolyte Resorption

These compounds all act by interfering with the efficiency of the active transport of sodium and perhaps chloride ions. Unfortunately the resorption of other electrolytes, but particularly potassium ions, is also sometimes affected; protracted use of many diuretics of this group necessitates the replacement of potassium by oral administration. Those compounds previously classified as 'irritant diuretics' belong to this group, for example, the mercurial diuretics. Although sodium resorption is reduced by all types of diuretic, the compounds included in this section are those whose main action is the impairment of sodium resorption from various sites in the renal tubule.

The Mercurials

Any compound of mercury, if absorbed, will cause diuresis during excretion through the kidney. This action is attributed to the interference with metabolism in the tubular epithelial cells, principally by combination of mercury ions with the SH groups of the enzymes controlling the active resorption of sodium. This results in the committant loss of chloride and water as these would normally passively accompany the sodium. As some sodium is resorbed in the distal tubule, there is a loss of chloride in excess of sodium and this eventually leads to a hypochloraemic alkalosis and a reduction in diuresis. This may be restored temporarily by the administration of a urinary acidifyer, such as ammonium chloride.

In therapeutic concentrations only mild irritation is caused in the kidney, but in high dosage severe nephritis may result with blood and albumin appearing in the urine. Mercurial preparations are consequently contra-indicated where nephritis is already present.

Formerly, inorganic salts of mercury were used as diuretics but the toxicity of the mercury ion is such that this practice has been abandoned.

Organic mercurials have replaced calomel for parenteral injections, for they ionize only slightly and are consequently less irritant and safer.

Mersalyl, B.Vet.C.

This white deliquescent powder containing

approximately 40 per cent metallic mercury is used as an injection or in tablet form. It is normally used in combination with theophylline, a cardiovascular diuretic, which appears to combine with the mercury compounds, making them less toxic and more active. Injection of Mersalyl, B.Vet.C. contains 10 per cent of mersalyl and 5 per cent theophylline.

The injection usually causes drowsiness and the animal may sleep for several hours, subsequent to which copious amounts of urine are passed. Excretion reaches a peak in about 6 to 8 hr but may continue at a decreasing rate for up to 24 hr. Oral medication with an ammonium chloride 0·3 g (5 gr.) tablet three times daily for the dog, and half this quantity for the cat, will increase the effect.

Dosage and Administration

Oral:

Dog and Cat: 160–240 mg

Intravenous or intramuscular:

Dog: 1–2 ml

Cat: 0·5–1 ml

Small trial doses will help in preventing accidents, for some animals show a sensitivity to these compounds. Where no ill effects result from the test dose normal therapeutic doses can be given 24 hr later. The therapeutic doses can be repeated at intervals of 3 to 7 days depending on the individual reactions.

Oral dosage, though sometimes more convenient is less effective owing to the low absorption rate of mercurial compounds. Tablets normally contain 80·0 mg mersalyl and 40 mg theophylline.

Parenteral dosing should be carried out with care. Deep intramuscular injection is the best route, but the intravenous route can be used providing the injection is given slowly, for cardiac damage may follow too rapid an injection rate. The subcutaneous route should not be used as sloughs may result from mercurial irritation.

Carbonic Anhydrase Inhibitors

These are typified by **acetazolamide** which is a complex sulphonamide. Members of this group of compounds act by inhibiting carbonic anhydrase activity in the kidney. They thus interfere with the ability of the tubule cells to make available hydrogen ions to take part in the essential acid–base interchanges. This results in increases in the excretion of sodium, potassium and bicarbonate ions in alkaline solution. This process leads to the development of acidosis which nullifies the action of the sulphonamide. Potassium losses can also be marked. It would seem that these disadvantages make the carbonic anhydrase inhibitors less satisfactory than the thiazide and hydrothiazide diuretics. They have, however, contributed most usefully toward the understanding of normal kidney function.

The Benzothiadiazine (Thiazide)

The discovery of this group of diuretics has provided the clinician with a number of safe and almost universally applicable diuretics of moderate potency.

At least three of the group are now used in the veterinary field: **Chlorothiazide** B.Vet.C., U.S.P. **Hydrochlorothiazide,** B.Vet.C., U.S.P. and **Bendrofluoazide,** B.P., B.P.C.

The thiazide and derived diuretics are the result of intensive pharmaceutical investigation of the sulphonamides and their diuretic derivatives which act by carbonic anhydrase inhibition. During these investigations disulphonamides were synthesized which were diuretic in effect but not by virtue of carbonic anhydrase inhibiting effects. It is from this latter group that the thiazide compounds were developed:

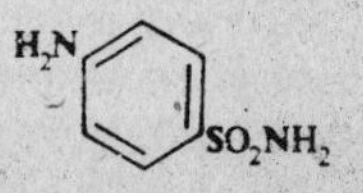

Sulphanilamide
(*p*-aminobenzene sulphonamide)

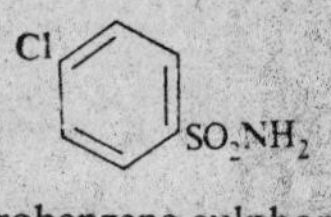

p-Chlorobenzene sulphonamide

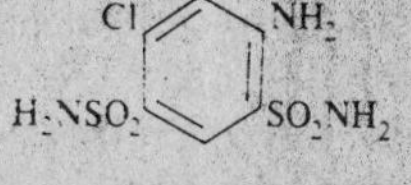

5,Chloro-2,4,disulphamylaniline
(a disulphonamide)

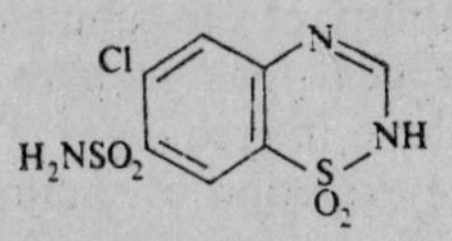

Chlorothiazide

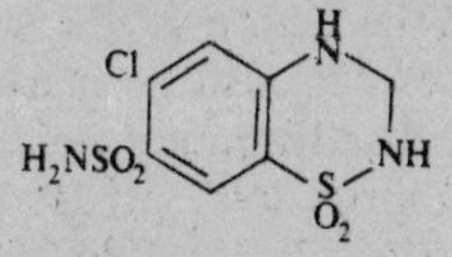

Hydrochlorothiazide

Numbers of thiazide and hydrothiazide substitution and addition products have been investigated and a structural/activity correlation has been worked out along with a comparative evaluation. The compounds are white or almost white crystalline powders. The chlorine derivatives usually have a slightly bitter taste whilst the fluorine derivatives are tasteless. In general the group has a very low water solubility and only a slightly greater solubility in organic solvents although a few are reasonably soluble in alcohol and acetone. The parent compounds are thus only available in tablet form for oral administration and it is the sodium salts which are available in solution for parenteral administration.

Actions and Uses

After absorption, which is rapid and complete following oral dosing the thiazide diuretics are dispersed throughout the extracellular fluids and do not concentrate in any tissue or organ other than the kidney. Excretion is rapid and takes place both via glomerular filtration and by tubular secretion. They act as diuretics by inhibiting the tubular resorption of sodium and chloride ions (*saluretic* action), the main effect being exerted on the proximal convoluted tubules. The two ions are excreted in approximately equi-molecular proportions when normal doses are given. At ordinary dosage levels over short periods chlorothiazide and hydrochlorothiazide have little effect on potassium excretion. In high dosage there is a disproportionate increase in chloride excretion and some effect on potassium and bicarbonate ion excretion. These actions are not complicated by a major disturbance of body acid–base balance, which is a considerable advance on most other diuretics. The compounds are active within 1 hr of oral administration but hydrochlorothiazide and bendrofluazide persist much longer, making single daily dosage possible.

The main advantage of this group of diuretics apart from their low toxicity lies in their duration of efficacy. With the older diuretics, loss of effect or refractoriness occurred after a short period of use due to the considerable powers of accommodation of the kidney.

In the veterinary field the compounds have been used for oedema of most types and in most species—parturient oedema, cardiac and nephritic oedema, bowel oedema of pigs, post-operative and non-specific oedema.

The saluretic effect also makes these compounds directly indicated for the treatment of salt poisoning and for the inhibition of lactation, especially that of pseudo-cyesis (pseudo-pregnancy).

Dosage and Administration

Hydrochlorothiazide:
- Parenteral:
  - Horse: 50–150 mg } intravenously or
  - Cow: 100–250 mg } intramuscularly
  - Pig (8–12 weeks): 50–75 mg intramuscularly
  - Small Animals: 12·5–25 mg intramuscularly
- Oral:
  - Horse and Cow: 500 mg initially followed by 250 mg daily
  - Dog and Cat: 1·25–2·5 mg/kg body weight

Chlorothiazide doses should be approximately ten times those of hydrochlorothiazide. All intravenous doses should be given slowly. As most of any intravenous dose is excreted within a few hours therapy by this route must be repeated at least twice daily although intramuscular dosage every 48 hr is reported to be satisfactory. Excretion following oral dosing covers a longer period and single daily dosing is adequate for all except chlorothiazide which should be given twice daily to obtain the best results. Potassium chloride supplements may be needed if therapy is prolonged.

Minimum effective doses of other members of

the group can be determined by reference to the following relative dosage table derived from human dosage ranges quoted by de Stevens (*Diuretics*, 1st edn, 1963, p. 116, Academic Press).

Hydrochlorothiazide	=	1·0
Chlorothiazide	=	10·0
Flumethiazide	=	10·0
Benzthiazide	=	2·0
Hydroflumethiazide	=	1·0
Thiabutazide	=	0·75
Benzhydroflumethiazide	=	0·2
Trichloromethiazide	=	0·2
Methylcothiazide	=	0·08
Polythiazide	=	0·04
Cyclothiazide	=	0·04
Cyclopenthiazide	=	0·01

Toxicity. The thiazide diuretics are remarkably non-toxic. Oral daily doses of chlorothiazide and hydrochlorothiazide of 1000 mg per kg have been given to dogs for months without structural or functional ill effects, and the intravenous LD_{50} of the sodium salts for mice exceeds 800 mg per kg.

Modern Diuretics

In recent years a number of potent diuretics have been introduced into human medicine, but as yet are not widely used in the veterinary field. Chlorthalidone exemplifies the many heterocyclic variants of the benzothiadiazines, compounds developed for increased duration of effect. Ethacrynic acid, frusemide and bumetanide are very potent, short-acting diuretics whose major action is in the ascending limb of the loop of Henle but which also act at other sites of sodium resorption in the nephron. The need for potassium supplementation with the foregoing is avoided with the so-called potassium-sparing diuretics, triamterene and amiloride, whose effects quantitatively and qualitatively resemble those of spironolactone.

URINARY ACIDIFIERS

The substances causing acidification of the urine are salts which will release into the urine an excess of acid radicals. They are all saline diuretics.

In general, they are combinations of acid radicals with basic radicals which will either be metabolized releasing an excess of acid radicals, or will be absorbed into the alkaline reserve allowing the acid radical to be excreted. Examples of the first group are the **salts of ammonia** such as ammonium acid phosphate, ammonium mandelate or ammonium chloride. The second group includes **sodium acid phosphate.** Salts of sulphuric acid are not used as the sulphate radical is only slowly absorbed, and though the nitrates are used, they tend to be more irritant than the weaker acid radicals.

Ammonium salts ionize to release ammonium and acid radicals. The ammonium is changed into urea and the excess chloride or the acid phosphate radicals are excreted, causing an acidification of the urine.

Sodium acid phosphate ionizes and the sodium ion is normally accepted into the alkali reserve whilst the acid phosphate is excreted giving an acid pH to the urine.

Urinary acidifiers are used in cats, dogs and pigs. The urine of herbivores is normally alkaline and can be rendered acid only with large doses of ammonium chloride.

It is generally found that bacterial growth is favoured by alkaline urine and, frequently, acidification will rapidly clear up an infection. This applies particularly to cystitis in dogs and cats.

The dosage of **ammonium chloride** for horses and cattle is from 4 to 30 g (60 gr. to 1 oz.) and for dogs and cats about 0·025 g (⅓ gr.) per lb. body weight per day in divided doses: 0·3 g (5 gr.) given three times daily is a common dose. The administration in large animals mainly achieves a saline diuresis with a slight shift in pH towards acidity.

Sodium acid phosphate is used extensively in small animals as a urinary acidifier in doses of 0·15 to 0·3 g (2½ to 5 gr.) given orally three times daily. It is valuable in neuter cats which are prone to urine retention; this retention, either from laziness or imprisonment, results in the urine becoming alkaline, which allows bacterial growth and cystitis (inflammation of the bladder) and also increases the incidence of the soft cream-white deposits in the urethra which are the common cause of urethral obstruction in the cat. Once this obstruction has occurred and been cleared, the animal should be dosed with acid phosphate tablets—0·15 g (2½ gr.) tablets once or twice daily are usually sufficient to prevent further obstruction. The acid phosphate is most frequently used as a urinary acidifier in conjunction with hexamine, either combined in tablet form, or as two separate solutions, when the hexamine dose is given 15 to 20 min after the acid phosphate dose. Ammonium chloride is also used occasionally for this purpose.

URINARY ALKALIZERS

These also are saline diuretics. The salts are combinations of strong basic radicals and weak acid radicals and the best therefore are **sodium bicarbonate, acetate** and **citrate. Potassium salts** are equally effective, but the acetate and citrate are deliquescent and are consequently of less practical value.

The acetate and citrate radicals are metabolized, and the base released is excreted as bicarbonate which gives a strong alkaline reaction to the urine.

The main value of urinary alkalizers is during the administration of sulphonamides whenever there is the danger of crystalluria and damage to the kidneys. The sulphonamides themselves are more soluble in alkaline media, and the formation of the less soluble acetyl derivatives is decreased. Bicarbonates are also able to increase the excretion of salicylates.

The alkalizers are not very often required for the urine of herbivores is normally alkaline and the incidence of crystalluria in carnivores is not very great.

URINARY ANTISEPTICS

The historical urinary antiseptics were mainly volatile oils and oleoresins such as the oils of sandalwood, cubebs and buchu. Of these, only buchu is now used in veterinary medicine. In doses of 0·3 to 4 ml (5 to 60 m.), **Tincture of Buchu**, B.P.C., 1949, has been combined with similar doses of **Tincture of Hyoscyamus**, B.Vet.C. in urinary antiseptic and sedative mixtures for the dog.

These urinary antiseptics have very little antibacterial action compared with **Hexamine,** B.Vet.C. which is still a very widely used urinary antiseptic, though superseded to a large extent by the sulphonamides and antibiotics, which are the most efficient of all urinary antiseptics.

In man, these compounds have been entirely replaced by the nitrofuran derivative, **nitrofurantoin** and, more recently, by **nalidixic acid.** Nitrofurantoin is active against a wide range of bacteria but often causes nausea and vomiting. Nalidixic acid is effective only against Gram-negative organisms. Both compounds are used in small animal practice.

Hexamine, B.Vet.C.

This drug is known by a number of different names; hexamethylene-tetramine, otropinure and methenamine. An odourless, colourless crystal or white crystalline powder, hexamine has a characteristic warm, sweet taste, which becomes bitter. It burns with a blue non-luminous flame and is the main constituent of many proprietary fire lighters. It is very soluble in water and alcohol.

Actions and Uses

The action of hexamine is firstly one of diuresis and secondly of urinary antisepsis, provided the pH of the urine is acid. This is achieved in small animals by the administration of sodium acid phosphate at least 20 min before the first dose, but subsequently at the same time as the hexamine. Combined tablets with 0·3 g (5 gr.) hexamine and 0·15 g (2½ gr.) sodium acid phosphate are available. In acid urine, hexamine slowly decomposes releasing ammonia and formaldehyde; the latter exerts an antibacterial action: the effect in no way compares with that of more modern treatments.

Dosage and Administration

Hexamine:

Horse and Cow: 4·8–8 g (60–120 gr.)
Sheep and Pig: 1·3–2·6 g (10–40 gr.)
Dog: 0·3–1 g (5–10 gr.)
Cat: 0·15–0·3 g (2–5 gr.)

Hexamine is administered orally as a drench in large and small animals, but, provided adequate water is available, tablets provide the easiest way of giving the drug.

Toxicity. Hexamine is non-irritant, but even slight acid concentrations release formalin which is highly irritant and will cause marked tissue reactions.

If large quantities are administered to small animals, the toxic effects will vary according to the urine pH. Thus, animals with strongly acid urine will rapidly develop nephritis with large quantities of albumin rendering the urine milky in appearance. In very severe cases urine may be stained with blood, indicating severe tissue damage and the ulceration of blood vessels. In animals with alkaline or neutral urine similar doses will cause relatively little damage and few symptoms.

Mandelic Acid

Mandelic acid has been used in human medicine in cases of urinary infections which do not respond to sulphonamide therapy. In veterinary medicine, urinary infections in dogs and cats have occasionally been treated with the drug although the newer antibiotics are safer and more effective. The acid itself is a deliquescent crystalline solid, and **ammonium mandelate** or **Calcium Mandelate,** B.Vet.C. are preferred for their greater stability. The calcium salt is particularly stable being insoluble in water and has only a slight saline taste; it must be used with a wetting and suspending agent.

Actions and Uses

Mandelic acid is absorbed following its release from the salts in the stomach. It circulates and is excreted unchanged through the kidney. In the urine, it exerts a bacteriostatic effect in acid urine but relatively little in alkaline and neutral urine. The optimum effect is exerted in urine at a pH below 5·5, and should methyl red indicator not confirm this, ammonium chloride or acid sodium phosphate should be given. The ammonium and calcium salts act as self-acidifying compounds and are all the more favoured because of this. Further intensification of the action can be achieved by restricting water intake to a minimum. This is an exception to the rule that patients with urinary infections should be encouraged to drink as much fluid as possible.

Dosage and Administration

Calcium Mandelate, B.Vet.C.
Dog: 0·5–3 g (7½–45 gr.)
Cat: 0·3–1·5 g (5–25 gr.)

Toxicity. Mandelic acid is irritant, mainly because of the extreme acidity required for optimum effect. It should therefore be administered for a maximum of 14 days, and if further treatment seems indicated, mandelic acid must not be used until 12 to 14 days have elapsed after the end of the 14 day course. In the majority of indicated cases bacteriological cure is achieved in 6 to 7 days.

URINARY SEDATIVES

These drugs are used mainly to relieve irritation in the bladder, but are also of value in depressing the irritability of the whole urinary tract.

The sedation may be achieved by: (a) removing the cause of the irritation, and (b) anaesthetizing the mucous membrane.

The first group includes antibacterial agents such as sulphonamides and antibiotics, particularly streptomycin and the 'broad spectrum' compounds. Occasionally urinary acidifiers and alkalizers may relieve irritation by changing the pH, and may also arrest bacterial growth. This first group should be used in conjunction with the members of the second group if the most satisfactory results are to be obtained.

The second group consists of preparations of belladonna and hyoscyamus, but especially the latter. Most compounds containing atropine will have a peripheral sedative action, but preparations of **Hyoscyamus,** B.Vet.C. contain a high proportion of hyoscyamine which has a stronger anaesthetic effect on the mucous membrane. The hyoscine also contained assists in countering the central nervous system stimulant action of the small quantity of atropine and the hyoscyamine.

Toxicity. Similar to atropine poisoning and the treatment is therefore the same. Hyperexcitability in the earlier stages requires short acting sedatives, and the later depression should be countered by medullary stimulants such as picrotoxin, leptazol and nikethamide. Unabsorbed alkaloid can be removed either by emesis if treatment commences before excitability is shown, otherwise tannic acid and potassium permanganate may be used to precipitate the alkaloids. The precipitated alkaloids can then be expelled by laxatives.

FLUID BALANCE, ACID–BASE BALANCE AND ELECTROLYTES

Approximately 70 per cent of the body weight of an adult animal consists of water, of which about 50 per cent is intracellular and 20 per cent extracellular. Of the extracellular water, about 5 per cent is present in the plasma and 15 per cent is interstitial fluid. The total volume normally changes very little, although the body fluids in the different 'compartments' are constantly interchanging as metabolism proceeds: water is produced during intracellular oxidation (40, 55 and 107 g of water are produced by the oxidation of 100 g of protein, carbohydrate and fat, respectively), and diffuses under the influence of osmotic pressure out of the cell and into the interstitial fluid. Diffusion continues down concentration gradients via the capillaries and into the plasma compartment. The blood then carries excess water and metabolites to the organs of excretion—the kidney and lungs. The interstitial fluid therefore acts as a cushion between the largest fluid compartment and the smallest, enabling intake and excretion of water to proceed without sudden stresses and strains on the limited capacity of the plasma volume.

The interchange of water between blood and tissues and the concentration of the electrolytes of the body are interdependent; an increase or decrease of one, although it may cause a temporary imbalance, is followed very soon by an equalizing increase or decrease of the other, although the concentration of the electrolytes varies only within fairly restricted limits. As with water, the electrolytes are constantly interchanging between the various body fluid compartments.

The electrolytes may be divided into cations, the most important being sodium, potassium, calcium and magnesium, and anions, principally chloride, bicarbonate, biphosphate and sulphate ions. There are also a number of organic acid anions, and proteins also ionize acidically, producing more anions. In plasma and interstitial fluids the predominant cation is sodium whilst the two main anions are chloride and bicarbonate, the former being in greater amounts. In the intracellular fluid, however, potassium, and to a lesser extent magnesium, are the only cations of significance whilst organic phosphate, protein, sulphate, and bicarbonate are the anions, the first predominating.

The method of interdiffusion of these electrolytes is not fully understood. However, it is known that the permeability of cell membranes is selective and that the permeability in one direction may be different from that in the other direction. Red blood cells are almost impermeable to sodium and calcium ions, fats and sugars, slightly permeable to amino acids, but readily permeable to anionic radicals, phosphates, bicarbonates, chlorides and sulphates as well as bile salts, ammonia and urea. Sodium and potassium ions increase, while calcium ions decrease, the permeability. Tissue cells vary in their permeability, but in general potassium and phosphate ions are constantly interchanging between intra- and extracellular fluids, whilst magnesium is almost static within cells, any magnesium acquired by the body or mobilized from the bone being distributed exclusively throughout the extracellular fluids.

The role of the electrolytes is complex for they influence the metabolism of the whole body. They are the major factor in the creation of osmotic pressures and take part in the buffer systems and pH mechanisms of the body; their concentrations regulate the enzyme actions and determine tissue irritability; they are essential for the transmission of nervous impulses through the polarity of cell membranes and take part in all chemical and physical reactions within the body.

It is apparent therefore that these electrolytes and the fluid necessary for their solution are absolutely essential and that any imbalance or losses will exert an influence on the whole body metabolism. This applies particularly to the young animal, in which the surface area (to which metabolic needs are directly related) is considerably greater in proportion to the extracellular water reserves: as an illustration, a cow weighing 454 kg (1000 lb.) has an extracellular fluid component of about 91 kg (200 lb.) (20 per cent) and a 45 kg (100 lb.) calf, 9 kg (20 lb.) extracellular fluid, thus the adult has a fluid reserve ten times greater than the calf and yet has a metabolic requirement (or surface area) only three or four times greater than that of the calf. All young animals are similarly placed which accounts for their greater susceptibility to the effects of scouring and vomiting as well as to poisoning with electrolytes (salt poisoning, etc.).

Dehydration. The loss of electrolytes and fluid by an animal affected with severe diarrhoea and, or, vomiting can be a critical factor in deciding between life and death. Provision of water and electrolyte nowadays forms a standard component in the management of persistent vomiting and/or diarrhoea. The poor risk surgical patient is also greatly benefited by preoperative attention to electrolyte, acid–base or fluid imbalance; e.g. a horse prior to surgery for high intestinal obstruction would be suffering additionally from severe metabolic acidosis and the sequestration of secretions anterior to the obstruction. It would benefit from a litre of plasma, 8–10 litres Ringer lactate and 2–3 litres of 5 per cent sodium bicarbonate. Whenever possible, administration by the oral route is to be preferred. In general, up to 6 per cent of the body weight may be administered daily. Ideally the adequacy and accuracy of electrolyte replacements should be monitored by clinical biochemical analyses.

Electrolyte Replacement Solutions

The simplest electrolyte solution available to combat electrolyte and fluid loss is Injection of **Sodium Chloride,** B.Vet.C., which consists of a solution of sodium chloride in water to the isotonic concentration of 0·9 per cent w/v. A more complicated solution which is of greater value in electrolyte loss during gastro-intestinal disturbances is Compound Injection of Sodium Chloride, B.Vet.C., which contains 0·86 per cent of sodium chloride, 0·03 per cent of potassium chloride and 0·04 per cent of hydrated calcium chloride; this is **Ringer's Solution. Ringer Lactate** solution contains sodium chloride 0·6 per cent, potassium chloride 0·03 per cent, calcium chloride 0·022 per cent and sodium lactate 0·03 per cent. The lactate yields bicarbonate in the body and this reduces the risk of inducing the metabolic acidosis which sometimes follows the simpler formulations. Where debility and starvation accompany dehydration these injections may be supplemented with 5 per cent dextrose. The quantities of these injections used will vary with the species but the maximum daily quantity should not exceed the 6 per cent of body weight as mentioned previously. This represents about 0·86 litre for a 14 kg dog (1½ pt. per 30 lb.) or about 3·6 litre per 50 kg body weight (6 pt. per cwt.) for farm animals. In practice a maximum single dose for horses and cattle is 3 litres (4¼ pt.) and for dogs 100 ml: these quantities can only be given subcutaneously with the assistance of hyaluronidase at a concentration of 1000 turbidity units per litre. With hyaluronidase, 10 ml per min can be injected with no discomfort or lifting of the skin from the underlying tissues. The absorption rate is also increased.

The foregoing section has described the replacement of fluid and electrolyte loss using infusion solutions whose components are present in concentrations which parallel their concentration in plasma. Such solutions were originally designed as perfusion solutions for physiological experimentation. In conditions which require fluid and electrolyte therapy there is frequently a loss of one or more electrolytes in excess of others. In such situations, it is obviously preferable to use a solution designed to replace lost electrolytes. This is known as deficit repair. In diarrhoea there is, because of the abnormal loss of intestinal secretions and unabsorbed foodstuff, a loss of fluid, sodium, potassium, bicarbonate and chloride

Intestinal fluid replacement:

Potassium chloride, 2 g
Sodium chloride, 3 g
Molar sodium lactate, 40 ml
Water, 710 ml

In persistent vomiting, on the other hand, the major electrolyte loss is chloride in excess of sodium. The replacement fluid contains potassium chloride in a concentration matching that of gastric juice. As the ammonium ion is metabolized to urea, the chloride replacement is in excess of cations.

Gastric fluid replacement:

Ammonium chloride, 1·875 g
Sodium chloride, 1·85 g
Potassium chloride, 0·65 g
Water, 500 ml

A further approach is to administer a solution whose composition reflects the daily turnover of its constituent electrolytes rather than their plasma concentrations. Such a solution is intended for use when there is no severe deficit and it will then be able to make good any form of loss, e.g. Talbots solution.

Talbots solution:

Sodium chloride, 0·117 g
Sodium lactate, 0·224 g
Potassium chloride, 0·150 g
Potassium phosphate, 0·135 g
Water to, 100 ml

This brief survey of electrolyte solutions will underline their value in therapy, but also indicates difficulties in their use. Ideally, because of their differing compositions, it is important to know whether the loss is of extracellular fluid (shock) or intracellular fluid (salt poisoning). Further the nature of the dehydration is important, i.e. has there been a proportionate loss of electrolyte and water or has one been lost in excess of the other? The only way to be certain is by carrying out plasma electrolyte determinations before and during therapy, but haematocrit determinations provide a useful guide. In general, the preliminary objective is to prevent circulatory collapse by rehydrating the extracellular space with roughly isotonic, low potassium solution. The dosage can be assessed by observing the regression of the clinical signs of dehydration and to continue therapy until the normal pattern of urination is re-established.

Parenteral alimentation refers to the administration of foodstuff by other than the oral route. This technique is of considerable importance in man in maintaining life in patients unable to eat, e.g. in coma or after oesophageal surgery, but is little used in veterinary medicine. The criteria applied to the selection of a medicament are those applied to a normal diet, i.e. it should be calorically adequate and provide the essential amino acids, fatty acids, vitamins and minerals.

Carbohydrate is commonly provided as glucose although it is easy to exceed the renal threshold and induce a diuresis. Intravenously administered *emulsions of oils* are better sources of energy as there is no threshold problem and weight for weight fat is the most calorigenic of the dietary constituents.

Protein can be given as amino acids, but is usually in the form of a digest. **Protein Hydrolysate,** B.Vet.C. is a solution of the products of enzymic digestion of first-class muscle protein of animal origin to which 10 per cent dextrose is subsequently added in order to minimize gluconeogenetic deamination. This solution may be administered by intravenous or intramuscular injection at a dose rate of 10 ml per kg.

Acid–Base Balance

The importance of the kidney in maintaining the pH of the body within the narrow limits compatible with life has already been emphasized by the frequency with which diuretic drugs cause pH shifts while the regulatory capabilities of the kidney have been exemplified by the rapidity with which the acidifying diuretics lose their efficacy.

There are circumstances under which body pH is altered. A primary rise in H^+ ions is known as metabolic acidosis. When a H^+ ion rise is accompanied by a HCO_3^- rise, as is the case when the bicarbonate buffering system reacts to an increase in P_{CO_2}, the change is called respiratory acidosis. In general, changes of this type are best controlled by attending to the exciting cause. However, there are occasions where the administration of drugs which influence pH is indicated, e.g. when cattle overeat carbohydrate rich concentrates there can be considerable production of lactic acid. This can exceed the buffering capacity of saliva and will require the administration of alkalinizing drugs, e.g. sodium bicarbonate by mouth or sodium lactate or acetate intravenously. A related problem can occur in cattle which overindulge in protein-rich concentrates. In this case there is an over-production of ammonia in the rumen and the administration of acidifying drugs is indicated, e.g. dilute acetic or hydrochloric acids, or ammonium chloride by mouth.

It will be noticed that some of these alkalinizing and acidifying drugs are the same as the compounds mentioned as alkalinizing or acidifying diuretics. Some of the salts of sodium and potassium used for these purposes and for electrolye replacement are discussed. The salts of sodium, the least toxic of the metals, are all white crystalline solids which are soluble in water, usually forming neutral or alkaline solutions.

Sodium Bicarbonate, B.Vet.C.

After absorption, the bicarbonate radical joins the alkali reserve of the plasma, lymph and intracellular buffer mechanisms. In excess, alkalosis may occur, but the fact that the bicarbonate radical is capable of being cycled via the formation of water and carbon dioxide makes elimination via the lungs relatively easy. Sodium bicarbonate may be used in the treatment of the acidosis which frequently accompanies acetonaemia in cattle. Some of any excess bicarbonate is eliminated in the urine which is thereby rendered alkaline in reaction: this phenomenon is utilized in increasing the solubility and excretion rate of sulphonamides and uric acid.

Dosage for horses is up to 60 g (2 oz.), for cattle up to 120 g (4 oz.), for sheep and pigs up to 8 g (120 gr.), and for dogs and cats up to 1·5 g (25 gr.).

Sodium Chloride, B.Vet.C.

This has many other uses besides that of maintaining electrolyte and tissue fluid balance. Externally, isotonic or slightly hypertonic solutions make useful eye and wound lotions, oral and nasal washes: in markedly hypertonic solutions sodium chloride exerts a lymph lavage (lymphagogue) action by which the excessive osmotic pressure draws fluid through the wound giving it a mechanical cleansing as well as concentrating antibodies and macrophages in the area of the wound. Repeated washing with solutions of up to 3 per cent are very effective where gross contamination is absent.

After absorption, both the sodium and the chloride ions enter the electrolyte system displacing equal amounts if the system is in balance but being retained along with water of solution if there is a negative imbalance. The total amount and the excretion of the salt and its water are controlled by a complex system which involves the posterior pituitary gland and the adrenal hormones. Excretion takes place through the kidney, the salt acting as a saline diuretic.

An interesting application of saline therapy concerned with the electrolyte balance is in the treatment of heat stroke and exertion cramps. During periods of excessive heat or exercise salt loss is proportionally greater than water loss through sweating and urine excretion and a negative chloride imbalance ensues which results in cramps, convulsions in severe cases, and a generalized alkalosis. Isotonic saline and concentrations of up to 10 per cent of compound salts have been used in quantities of up to 1·5 litres for large animals; they are best given by intravenous injection. Salt is an essential dietary factor for all herbivores and rock salt should be freely available.

Toxicity following the ingestion of excessive quantities of salt by pigs and poultry has been reported, but provided water is supplied ad libitum these animals will tolerate relatively large quantities of salt.

Dosage

Horses and cattle as a laxative, diuretic and antipyretic is up to 120 g (4 oz.).

Sodium Citrate B.Vet.C.

This is an alkalizing diuretic and may be used in place of sodium bicarbonate when 1·5 times higher doses are required to achieve an effect equal to that of the bicarbonate.

Sodium Acid Phosphate, B.Vet.C.

A urinary acidifier and diuretic. In this respect it differs from other salts of sodium. The excess phosphate is excreted as sodium orthophosphate which is converted in the tubules to the acid phosphate, making sodium ions available for resorption. Sodium acid phosphate has been used as a soluble salt which can be injected intravenously to make phosphate ions rapidly available in the treatment of the hypophosphataemia of the milk fever complex of cattle. It is also administered by mouth, often as maintenance doses after an initial dose by injection. This therapy is less satisfactory than proprietary injections which usually contain hypophosphites.

Dosage as a urinary acidifier for use in conjunction with hexamine in dogs and cats is 0·3 to 1 g (5 to 15 gr.) given three times daily. In cattle, 30 g (1 oz.) may be given initially followed by half this dose twice daily.

For small animals, sodium acid phosphate is available in tablet form combined with an equal quantity of hexamine. In cattle, the dose may be dissolved in 500 ml for injection and for oral dosing.

Potassium Salts

Although potassium ions are essential as intracellular electrolyte ions and are present in the blood to the extent of about 20 mg per 100 ml of serum, it must be remembered that an excess of potassium ions will first slow and disturb the rhythm of the heart and then cause a progressive ventricular block and finally cardiac arrest: if injections are given slowly, very little signs of impending heart arrest are shown until the critical blood level is reached when the cardiac effect suddenly becomes pronounced.

Potassium Bicarbonate, B.Vet.C.

Used in the same way and in the same circumstances as sodium bicarbonate, which is usually preferred because it is less expensive. As a urinary alkalizer and diuretic for carnivores potassium bicarbonate, as with all potassium compounds, is superior in action to the sodium salt because of its lower renal threshold.

Dosage

For dogs it is from 0·3 to 1 g (5–15 gr.).

Potassium Citrate, B.Vet.C.

This has an anticoagulant action but may not be used in blood transfusion work. It has been widely used in man as an oral alkalizing diuretic, for it is readily absorbed from the intestines and excreted as the bicarbonate via the kidneys. In large doses it is a saline laxative.

Dosage

For the dog it is up to 2 g (30 gr.) two or three times daily.

Bibliography

Fisher, J. W. (1971) *Renal Pharmacology*. London: Butterworths.

Lant, A. F. & Wilson, G. M. (ed.) (1973) *Modern Diuretic Therapy in the Treatment of Cardiovascular and Renal Disease*. Amsterdam: Exerpta Medica.

Mason, E. E. (1974) *Fluid, Electrolyte and Nutrient Therapy in Surgery*. Philadelphia: Lea & Febiger.

Stevenson, D. E. & Wilson, A. A. (1963) *Metabolic Disorders of Domestic Animals*. Oxford: Blackwell Scientific.

Watt, J. G. (1962) Fluid replacement therapy in the surgical patient. *Adv. small Anim. Pract.*, 3, 76–100.

Wilkinson, A. W. (ed.) (1972) *Parenteral Nutrition*. Edinburgh: Churchill Livingstone.

11

The Hormones

I. The Adrenal Cortex

A hormone may be defined as an active material discharged in very small amounts into the blood stream by an endocrine or ductless gland. Hormones which possess a target organ or tissue, have commonly been described as chemical messengers but their function is usually more autocratic than that of messengers, for they dictate to, control, and coordinate the functions of all the organs and tissues of the body. They may have relatively specific actions as with insulin, or act in several ways as with adrenaline: they may be functional hormones such as the oestrogens and progestins which are directly responsible for the changes in an organ, or they may be hormones which cause the production of other hormones and control and coordinate their actions: of this latter type the the anterior pituitary hormones are the most important. Most functional hormones work in equilibrium with their controlling hormones in the normal body, and an excess of a functional hormone will in turn depress the production of its controller hormone: thus an excess of an oestrogen injected into a normal female animal will depress the production of follicle stimulating hormone by the anterior pituitary gland. In some cases one hormone will antagonize the action of another as with insulin and adrenaline, the first of which causes glycogenesis with the removal of sugar from the circulation whilst adrenaline, which has the function of preparing for and maintaining the body during strenuous action, causes amongst other effects, glycolysis with the mobilization of glucose into the blood. In other cases two hormones will act together as with oxytocin and the oestrogens, the first hormone causing uterine contractions only when that organ is 'sensitized' by the action of oestrogens. Still other hormones will act synergistically in some ways and antagonistically in other ways: thus progesterone will continue the preparation of the mammary gland for lactation after the oestrogens have initiated the preparation, and yet will antagonize the effect of oestrogens in depressing the sensitivity of the uterus to the action of oxytocin: androgens will also antagonize the feminizing effect of oestrogens and yet will act in the same way as the oestrogens in causing growth of the duct system in the mammary glands. Many hormones have general metabolic effects in addition to the hormonal action on their target organ.

Some hormones are essential for the continuation of life whilst others, though their absence causes modification of body functions, are not essential. The parathyroid glands are essential, for extirpation results in the development of hypocalcaemic convulsions and death: extirpation of the adrenal glands also leads to death within 1 to 2 weeks: castration, with the removal of ovaries or testes, causes comparatively little effect on the body even though the organs are the main sources of oestrogens and androgens.

HORMONE CLASSIFICATION

Reproductive Hormones:

Female:

Ovarian: Follicle:

Natural:	Oestradiol
	Oestrone
	Oestriol
Semi-synthetic:	Ethinyloestradiol
Synthetic:	Stilboestrol
	Dienoestrol
	Hexoestrol

Corpus Luteum:
Natural: Progesterone
Semi-synthetic: Ethisterone
Male:
Testicular: Interstitial cells:
Natural: Testosterone
Androsterone
Semi-synthetic: Methyltestosterone
Male and female gonadotrophins:
Follicle and sperm stimulating trophin:
Serum gonadotrophin
Luteinizing and interstitial cell stimulating trophin:
Chorionic gonadotrophin

Pituitary hormones: Natural
Anterior pituitary trophins:
Gonadotrophins: FSH and LH
Thyrotrophin: TSH
Adrenocorticotrophin: ACTH
Somatotrophin or growth trophin: GH
Lactogenic trophin: Prolactin
Melanocyte-stimulating hormones: MSH
Lipotropins
Posterior pituitary:
Oxytocic factor: Oxytocin
Pressor factor:
Antidiuretic factor: Vasopressin

Adrenal hormones: Natural
Cortical:
Mineralocorticoids:
Desoxycorticosterone (DOC)
Aldosterone
Glucocorticoids:
Corticosterone
Cortisone
Hydrocortisone
Synthetic
Prednisolone
6α-Methylprednisolone
9α-Fluoroprednisolone
Triamcinolone
Betamethasone
Dexamethasone
Medullary: Natural
Adrenaline

'Metabolic' hormones: Natural
Parathyroid gland: Parathyroid extract
Thyroid gland: Thyroid extract
Thyroxine
Synthetic: di-Iodotyrosine
Pancreatic gland: Insulin
Glucagon

Miscellaneous hormones:
Alimentary hormones: Gastrin, secretin
Thymus, prostatic and allantoic hormones.

THE PITUITARY GLAND

Fractions of only the anterior and posterior portions of the pituitary gland are available for clinical use. Those fractions obtained from the posterior lobe will be mentioned in Chapter 12.

The Anterior Pituitary Hormones

Ten different hormones have been isolated in almost pure forms, and analysis has shown them to be of a protein nature: the somatotrophic or growth hormone of man contains 191 amino acid residues and has a molecular weight of 22 000; the ovine lactogenic hormone (prolactin) is also a sulphur containing protein and has a molecular weight of 23 000. However, work carried out on the corticotrophic fraction indicates that corticotrophic activity is located in a 39-amino acid polypeptide. Thyrotrophin, follicle stimulating hormone and luteinizing hormone are glycoproteins.

The trophins which have been closely investigated are the somatotrophin or growth hormone; two gonadotrophic hormones (similar in action to the chorionic and serum gonadotrophins), although complete molecular structures are not yet elucidated; the lactogenic hormone also known as prolactin or luteotrophin; the thyrotrophin also referred to as TSH (thyroid stimulating hormone); and corticotrophin or adrenocorticotrophic hormone (ACTH).

Twenty or more further hormones have been claimed to occur in anterior pituitary secretions, including the so-called 'metabolic' factors—a diabetogenic, a ketogenic and insulin producing trophins. It has yet to be proved that these factors are distinct entities, and at present they are therefore not of any clinical significance.

In veterinary medicine ACTH is the trophin of greatest significance. Whilst preparations of total gonadotrophins of pituitary origin are available commercially gonadotrophins from other sources are mainly used. Luteotrophin, thyrotrophin, somatotrophin, liptrophin and MSH are in experimental use.

Adrenocorticotrophic Hormone: Corticotrophin, ACTH

Corticotrophin, B.Vet.C. is a white or almost white powder, unless prepared by freeze-drying methods when it exists as fawn flakes or masses. It is a glacial acetic acid/acetone extract which is soluble in water or normal saline and contains not less than 24 units per mg when of ox or sheep origin, or 45 i.u. if of pig origin. Long-acting preparations are either zinc complexes or gelatin suspensions. A 24 amino acid synthetic analogue is also available.

Actions

The actions of corticotrophin are essentially those of adrenocortical stimulation, for the hormone has no known actions apart from the stimulation of the adrenal cortex. Following intramuscular or subcutaneous administration the action becomes apparent within a few hours, but intravenous injection produces effects within a few minutes. Non-depot forms lose their effects after a few hours whereas depot injections, though their effects are slower in onset and never reach the maximum achieved by non-conjugated forms, their action is prolonged over a period of 24 to 48 hr depending on dosage. It has been found in man that intravenous infusion of 20 units over a period of 6 to 8 hr will equal in adrenal stimulation as much as 150 units given over a 24-hr period in equal, divided doses at 4 to 6 hr intervals.

The effect of corticotrophin on the adrenal cortex is to produce a rapid decrease in the ascorbic acid and cholesterol levels within the gland. The significance of the loss of ascorbic acid is unknown, but it has been noticed in man that there is a complementary increase in circulating and urinary ascorbic acid. Concurrent with the decrease in cholesterol, there is a rise in the synthesis and output of corticosteroids, and ACTH is now known to speed the rate-limiting cleavage of cholesterol's side chain (see Table VII). Amongst these steroids are, cortisone and hydrocortisone, corticosterone, 11-desoxycorticosterone and aldosterone, various androgens and oestrogens, some progesterone, and precursors of these.

In the rest of the body, the reaction of corticotrophin is the same as would be expected from the action of mineralocorticoids and glucocorticoids. This includes a decrease in circulating eosinophils which is so marked as to have been used as a means of estimating adrenal cortex responsiveness (Thorn Test), as well as a method of corticotrophin assay, and modification of renal, thyroid, skin, connective tissue, mineral and carbohydrate metabolism. Early proteolytic inactivation gives ACTH a short half-life, but the compound markedly affects the excretion of steroids, causing an increase in the excretion of reducing corticosteroids including hydrocortisone and cortisone. There is also an increase in urinary 17-ketosteroids including androsterone, and also in oestrogen and pregnanolone (an excretory product of progesterone) excretion.

Uses

Corticotrophin has been widely used in the treatment of bovine primary or post-parturient ketosis. It is suggested that *acetonaemia* is due to temporary exhaustion of the anterior pituitary gland following the prolonged stress during pregnancy, and the acute stress of parturition, and the institution of lactation: this pituitary exhaustion results in a lack of glucocorticoid production by the adrenal cortex, a consequent drop in circulating glucose, and an inability to mobilize or synthesize more. Other sources of glucose are then tapped including fat breakdown which is incomplete, possibly due to lack of available glucose and therefore energy, although the true reason is unknown. The total result is an animal with a low milk yield, a loss of appetite and condition, low blood glucose levels and high blood, urine, milk, and breath ketone levels. The aetiology and treatment of *pregnancy toxaemia* in sheep has not received as thorough an investigation as acetonaemia of cattle, but as the clinical picture has many of the features associated with ketosis, it was anticipated that the toxaemia would respond in a similar manner. However, although cortisone has been reported to give good results ACTH has not proved as successful in sheep as had been hoped; blood glucose precursors such as propionates, proplyene glycol, acetates, glycerin, and glucose, or mixtures of these, often combined with cobalt and other minerals, are as effective. It is now apparent that the condition is analogous to starvation ketosis, associated with liver glycogen depletion and invasion by pathogens. While glucocorticoids appear to be helpful in early cases their use in the later stages is almost contra-indicated: in all cases they should only be used along with broad-spectrum antibiotic and carbohydrate administration.

Corticotrophin has also been used in the treatment of *skin conditions* in cattle and small animals. In cattle a dose of 40 to 80 i.u. was useful in the treatment of fifty cases with skin lesions due to photosensitization: whilst 20 i.u. were of assistance in treating, in conjunction with antibiotics, twenty calves affected with arthritis. In non-specific dermatitis in cats and dogs ACTH is useful, particularly in cases of 'fish', 'flea' or miliary eczema in cats. Pruritus ceases within 24 to 48 hr and hair begins to grow within 7 days. After a course of ACTH the condition can be expected to return after a period of 6 to 9 months or less. However, it has been found that testosterone implants, and progesterone implants, in doses of 25 to 50 mg, are as effective but more prolonged in action.

The effects of the ACTH are, of course, those of the corticosteroids produced as a result of adrenal cortex stimulation. It is more usual to use corticosteroid therapy.

In the field of *mastitis*, corticotrophin has been used as an adjunct to normal therapy to prevent the induration expected to occur in cases of staphylococcal and similar infections. Doses of 100 i.u. given intramuscularly, followed by 50 i.u. 24 hr later, were claimed to assist recovery without induration in seventeen out of twenty cases.

ACTH has been frequently used in the treatment of *arthritis* but has been largely superseded by hydrocortisone, prednisolone and methylfluoroprednisolones. In the long term view it is probable that stimulation, followed by hypertrophy of the patient's adrenal glands, should be far more beneficial, and the effects longer lasting than intermittent steroid therapy which may be expected to reduce intrinsic steroid production.

It was reported that 40 i.u. for sows under 125 kg (275 lb.) and 70 i.u. for those over that weight will, in conjunction with standard antibiotic therapy, drastically increase the recovery rate from agalactia and post-parturient septicaemia. ACTH administration and subsequent plasma cortisol analysis provides a specific and accurate adrenal cortex function test.

Dosage

In *cattle* the doses used in the United States for the treatment of ketosis have varied from 100 to 3000 units given intramuscularly, but the lack of standardization of unit designation has confused the issue and made comparison difficult. Dutch work indicated that three 50-unit doses given at 12 hr intervals were insufficient. Some American reports refer to the so-called 'crude' corticotrophin injection. Extensive use of the depot or long-acting forms in Great Britain indicate that doses of 80 to 160 or 200 i.u. of these compounds are sufficient for all cases, although many respond to a single 80 i.u. dose.

In *sheep*, for pregnancy toxaemia, approximately 40 i.u. intramuscularly should be adequate but may be repeated after 24 to 48 hr if necessary.

For *dogs* and *cats*, 10 to 20 i.u. of the long-acting compounds should be given daily to smaller individuals, but in severe cases of adrenal suppression or in larger breeds, 40 i.u. may be given twice or three times weekly until symptoms regress.

The Thorn Test consists of the intravenous or sometimes intramuscular administration of ACTH and the subsequent observation of the effect on the blood eosinophil count. A drop of 50 per cent or more in the number counted prior to injection is considered to indicate a satisfactory response of the cortex. Counts should be carried out between 2 and 4 hr after injection following intravenous administration but 2 hr or so later following intramuscular injection. Dosages of ACTH for intravenous use are; cattle: 100 i.u.; dogs: 25 i.u.

Toxicity. The toxicity of corticotrophin is essentially that of the steroids which its administration mobilizes into the system. It is unlikely that trouble will be experienced with intermittent dosing but prolonged administration may result in the delayed healing of wounds, lowered resistance to bacterial invasion, with a possible flare up of chronic infections. It should be remembered that natural ACTH, a protein extract, can be antigenic. Electrolyte balance may also be disturbed with sodium and fluid retention, and potassium loss.

THE ADRENAL GLAND AND ITS HORMONES

The adrenal or suprarenal glands, located in close apposition to the anterior poles of the kidneys, have two clearly demarcated zones, an internal or medullary zone which is dark brown in colour and an external or cortical zone which is much lighter in colour. These two zones elaborate hormones which differ considerably in character and action but which, because they are essential for the continued existence of the animal, makes this gland one of the most important in the body.

The adrenal *medulla* produces adrenaline and noradrenaline which are responsible for control of the heart and circulatory system and, by virtue of their release by adrenergic nerve fibres, affect all the structures innervated by the sympathetic portion of the autonomic nervous system. Because of its prime concern with the heart and circulation, adrenaline has been detailed in connection with that system.

The adrenal cortex is the site for the synthesis of the corticosteroids. As there is no storage of these hormones their production is continuous under the stimulus applied by ACTH. Cortical extracts have yielded a great many steroids, many of which are biosynthesis intermediates. The last true hormone was isolated in 1952 and is called aldosterone. Several androgens and oestrogens are included in the isolates, but it is the two groups of steroids known respectively as the *mineralocorticoids* and the *glucocorticoids* that exert the activity specific to the adrenal cortex. Though arbitrarily divided into these two groups it must be understood that each member of one group will possess the actions of the other group but to a varying and lesser degree. Thus the mineralocorticoids available for clinical use, desoxycorticosterone and aldosterone, have very little glucocorticoid action, whereas corticosterone has both electrolyte and glucose mobilizing effects. The potent synthetic glucocorticoid prednisolone has slight mineralocorticoid effect whilst the 9α-fluoro-16α-methyl-prednisolones have even less.

Chemistry. It must be remembered that steroids of the adrenal cortex are very closely allied to the steroids produced by the ovary and testes, and also that oestrogens and androgens will be produced by the adrenal cortex which may continue sexual activity after removal of the ovaries or testes. These corticosteroids have, therefore, the cyclopentenophenanthrene nucleus (i).

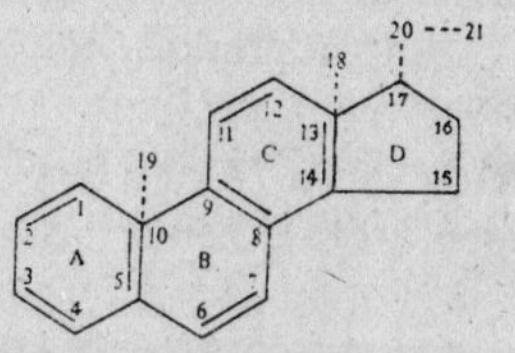

(i) Cyclopentenophenanthrene ($C_{17}H_{14}$)

Any substitution and changes which occur and any additions to this nucleus will therefore be described in the same way as are the variations affecting the sex hormones (*q.v.*). The following notes will enable a better understanding of the nomenclature:

1. Any compound with the suffix '-ene' is unsaturated and will possess double bonds in its construction. The number of double bonds is indicated by '-ene', 'di-ene', 'tri-ene' designation, and the site of the linkage is indicated by the number allocated to the first carbon atom in the bond as (ii) and (iia):

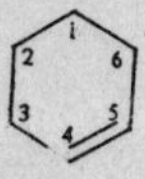

(ii) -'4-ene'

(iia) -'1, 3-di-ene'

This double bond position may also be indicated by the Greek Delta (Δ) and the position indicated as before, the figure(s) being added as in (iii) and (iiia). This is the origin of the Delta- of the many proprietary names for the adreno-cortical hormones.

(iii) -4-ene, or Δ^4-ene, or $\Delta^{4,5}$-ene

(iiia) -1,3,5-triene, or $\Delta^{1,3,5}$-triene

2. The suffix '-*a*ne' indicates a saturated compound.

3. '-ol' indicates a hydroxy-derivative, while '-diol' and 'triol' represent the di- and tri-hydroxy derivatives respectively.

4. '-one' indicates a ketone derivative (double bonded oxygen atom); hence '-dione' and '-trione'.

5. '-ol-one' will then indicate a mixed hydroxy-ketone derivative.

6. Numbers are used to indicate the positions of these groups as for examples: 3, 17-diol indicating hydroxyl groups attached at both carbons 3 and 17 of the nucleus.

7. Where a particular isomer is designated it may be referred to as 'cis-' or 'α' when the group is orientated to the rear, and 'trans-' or 'β' when the configuration is to the front. The alpha configuration confers a lower activity and these forms are more frequently found as excretory products.

8. It is also important to remember that the prefix 'pregn-' or 'pregna-' indicates progesterone origin: progesterone may be referred to as Δ^4pregnene-3, 20-dione.

Beginning with *progesterone* (iv) the substitution of a hydroxyl group for a hydrogen atom of the methyl group centred on carbon 21 produces *desoxycorticosterone* (v) also designated 21-hydroxyprogesterone. The addition of a further hydroxyl group for the hydrogen atom at carbon 11 of desoxycorticosterone produces *corticosterone* (vi) which may therefore be designated 11, 21-dihydroxyprogesterone.

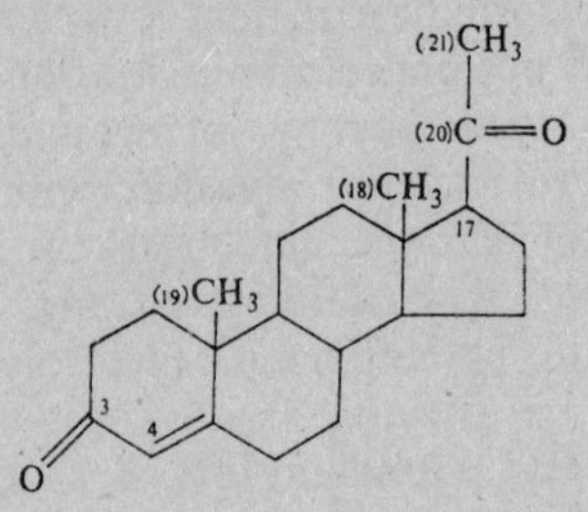

(iv) Progesterone, or Δ^4pregnene-3,20-dione

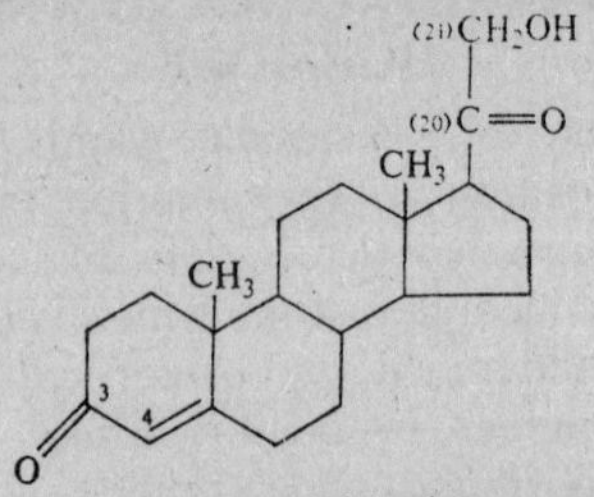

(v) Desoxycorticosterone, or 21-hydroxyprogesterone, or Δ^4pregnene-21-ol-3,20-dione

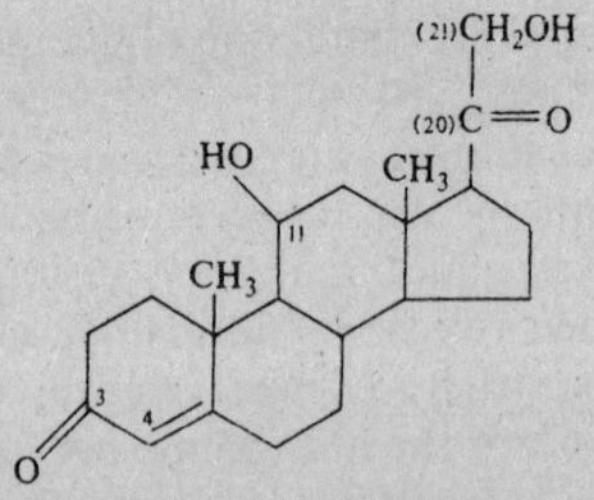

(vi) Corticosterone, or 11,21-dihydroxyprogesterone, or Δ^4pregnene-11,21-diol-3,20-dione

If we now change the hydroxyl group at carbon 11 of corticosterone to an oxygen atom (dehydrogenate and therefore '-dehydro-'), and add a hydroxyl group inclined to the rear (α) at carbon 17 we produce *cortisone* (vii).

If, instead of removing the hydrogen from the hydroxyl group at carbon 11 of the corticosterone, we merely add the hydroxyl group at carbon 17, we have hydrogenated cortisone and produced *hydrocortisone* (viii), that is, a hydrogen atom added to cortisone.

Now if we take cortisone and dehydrogenate it to make another double bond between carbons 1 and 2 in ring A of the steroid nucleus, we produce *prednisone* (ix), and similarly if we dehydrogenate hydrocortisone in the same place we produce *prednisolone* (x). These two new

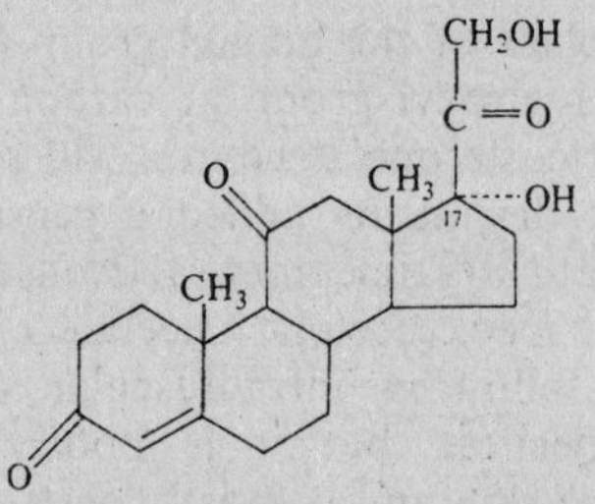

(vii) Cortisone, or 17α-hydroxy-11-dehydrocorticosterone, or 11-oxy-17α,21-dihydroxyprogesterone, or Δ^4-pregnene-17α,21-diol-3,11,20-trione

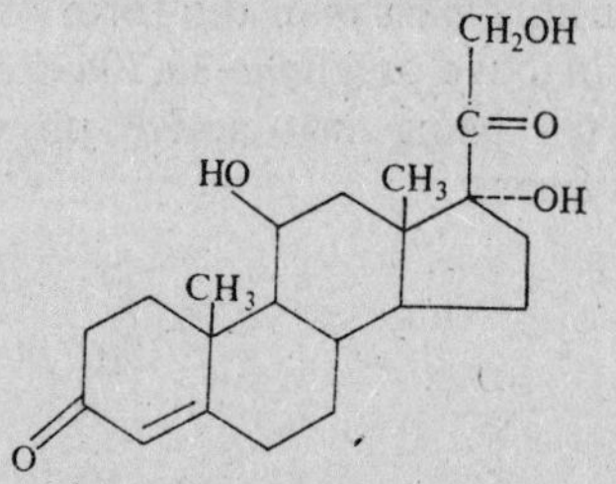

(viii) Hydrocortisone, or 17α-hydroxycorticosterone, or 11,17α,21-trihydroxyprogesterone, or Δ^4pregnene-11β,17α,21-triol-3,20,dione

compounds therefore are related to each other in the same way that cortisone and hydrocortisone are related.

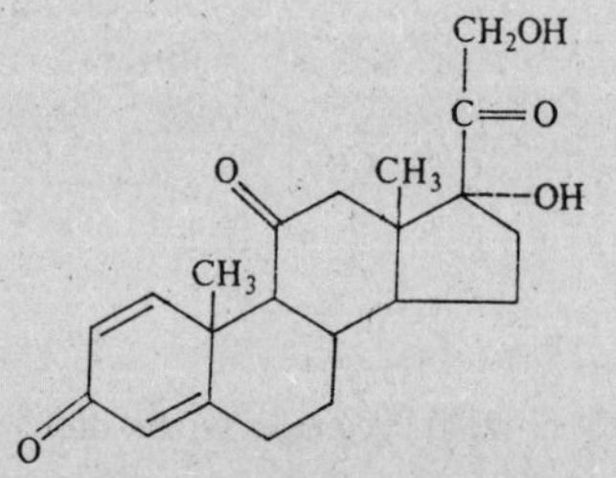

(ix) Prednisone, or $\Delta^{1,4}$pregna-diene-17α,21-diol-3,11,20-trione

It will be seen that as the steroid nucleus now has two unsaturated links the indicator in the written formula changes from '*-ene*' or one to '*-diene*' or two double bonds, i.e. $\Delta^{1,4}$, delta-one-four.

These corticosteroids can thus be classified by structure into three groups—the 11-desoxysteroids, the 11-oxysteroids and the 11,17-oxysteroids. The actions of the compounds are closely related to this presence or absence of oxygen atoms at carbons 11 and 17.

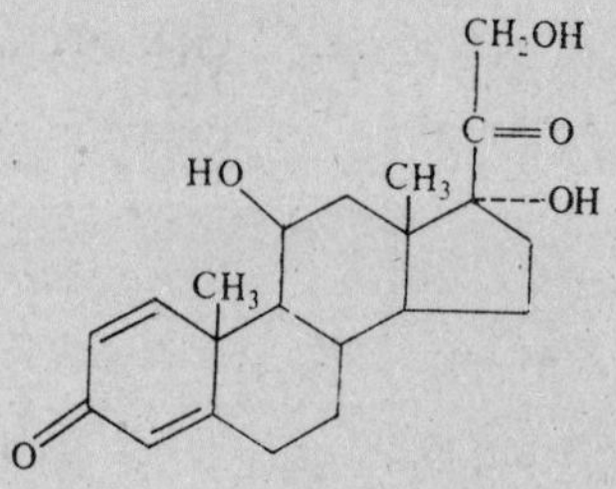

(x) Prednisolone, or $\Delta^{1,4}$pregnadiene-11β,17,21-triol-3,20-dione

Physical characters. Though many of these compounds may be extracted from the adrenal cortex of cattle, sheep and pigs, the yield is too small for economic use and all are prepared synthetically some of the structural modifications being achieved biosynthetically. (The biosynthetic processes used are the enzyme reactions which take place during bacterial fermentation, basic steroids being fed to the fermenting cultures and subsequently recovered in a modified form. An example of this is the introduction of an oxygen atom at carbon 11 of the steroid nucleus to form the hydroxyl group of hydrocortisone and its near relatives.) All are white or off-white crystalline powders, generally insoluble or almost so in water but slightly more soluble in organic solvents including propylene glycol and fixed oils. They may be used either as esters such as acetates (xi) or as free alcohols. Veterinary injections are mainly aqueous suspensions (prednisolone, cortisone acetate) and aqueous solutions of salts (betamethasone and dexamethasone disodium phosphate), but oily solutions such as those of desoxycorticosterone acetate and aqueous alcoholic solutions as those of hydrocortisone are also used. Ointments with and without broad-spectrum or narrow spectrum antibiotics are also available for topical and ophthalmic application.

Actions

The range of actions possessed by individual corticosteroids is much influenced by molecular configuration and substitutions. Thus the absence of any oxygen at carbon 11 is related to mineralocorticoid action in deoxycorticosterone (v): the addition of an oxygen atom, either in the form of a hydroxyl group or a double-bond

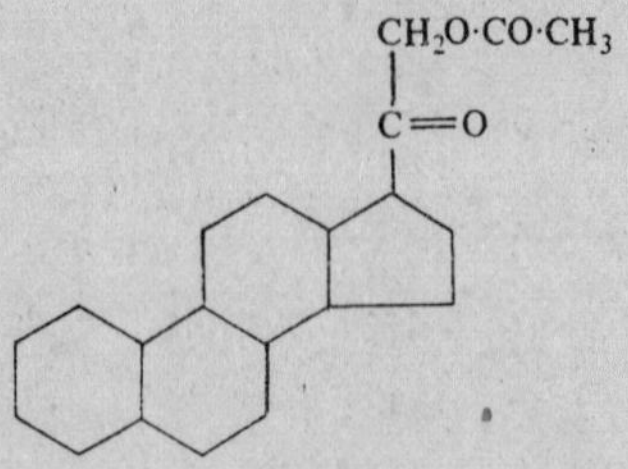

(xi) All corticosteroid acetates

oxygen confers, additionally, glucocorticoid activity—corticosterone, (vi), whilst the addition of a further oxygen, usually in the form of a hydroxyl group, at carbon 17 provides a considerable increase in the proportion of glucocorticoid activity—cortisone (vii). Desaturation of ring A by creating a further double-bond between carbons 1 and 2 greatly increases the glucocorticoid activity and reduces the mineralocorticoid effect—prednisolone (x). If fluorine atoms are introduced into the nucleus further proportional and absolute increases in glucocorticoid activity is achieved—9-fluoro, 16α-methyl-prednisolones (xiii, xiv): this particular change, however, results in a much greater increase in anti-inflammatory action. Although two distinct groups of actions mineralo- and glucocorticoid are described and the adrenocorticoids classified according to these actions, it must be remembered that the glucocorticoids all possess some active effect on the electrolyte balance, although hydrocortisone is not so active in this direction as cortisone, and their analogues prednisone and prednisolone are even less active; the fluoromethylprednisolones have mineralocorticoid effects which are virtually negligible at normal dose levels. There is therefore a constant danger that protracted treatment with glucocorticoids may cause disturbances of the electrolyte balance.

The Mineralocorticoids

Deoxycortone Acetate, B.P.

Although this hormone is synthesized in the adrenal cortex but is not the major mineralocorticoid it still remains an active and economic steroid available in the veterinary field for influencing electrolyte balance. It is likely that **aldosterone,** which has a '-CHO' aldehyde group in place of the methyl group at position 18 and a hydroxyl group at carbon 11 of the desoxycorticosterone structure, will replace this compound despite its added expense for it is twenty-five to fifty times more active as a mineralocorticoid. Deoxycortone acetate is rapidly absorbed following intramuscular or subcutaneous injection, but it is poorly absorbed through the skin and is mostly destroyed in the alimentary tract so that oral doses need to be about thirty-five times those for intramuscular injection. Excretion is typical of that of all steroids in that part is in the form of the saturated and inactive pregn*ane*-3α,20α-diol (xiia) the alpha forms being less active physiologically than beta forms.

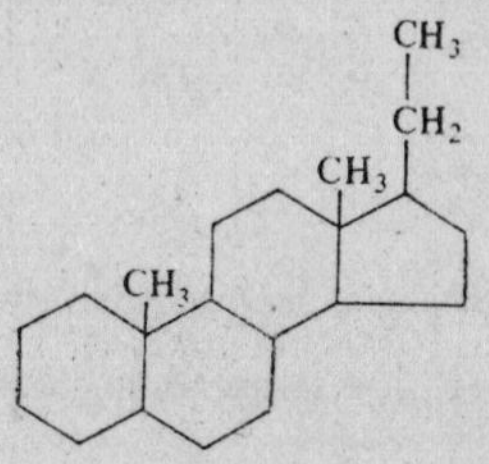

(xii) Pregn*ane*

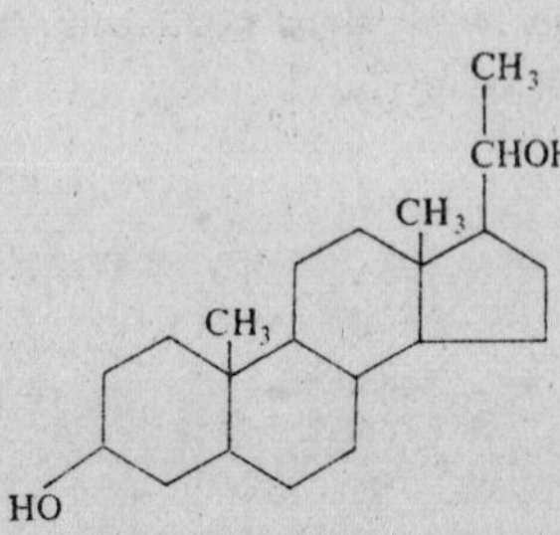

(xiia) Pregn*ane*-3α,20α-diol

Pregnane-3,20-diol is also the form in which progesterone is excreted. Some is also excreted as 17-ketosteroids, a common urinary form of most steroids in which all groups attached at carbon 17 are replaced by a single oxygen atom, thus forming a '>C=O' or ketone group. The excretion of steroids in ruminants is not identical with that of non-ruminants.

The mineralocorticoid action is the major part of the effect of deoxycortone. This action following injection into the normal animal is, in brief, a retention of sodium and chloride ions with water, an excretion of potassium ions and hydrogen ions with a tendency to alkalosis, i.e. hypercorticism. If, however, the sodium intake is restricted there is no effect on either sodium and water retention or potassium loss. Part of the effects of the compound are due to increased tubular resorption of sodium chloride and water and excretion of potassium, but there are similar effects on turnover in all the cells of the body. Using radioactive sodium (^{24}Na isotope) it has been shown that there is an increase of intracellular sodium and a decrease in potassium. There is also an increase in extracellular fluid due both to water retention and the passage of water from the cells into the extracellular spaces.

Clinically, the effects will be shown by a decrease in urine production, an increase in weight, and a 'filling out' of superficial body hollows.

In normal animals prolonged use of small doses has only an initial transient effect, but large doses may cause an increased water intake and output reminiscent of diabetes insipidus.

Deoxycortone has practically no effect on glucose metabolism, blood eosinophil counts, inflammatory reactions, or allergic oedema, but some progesterone-like activity has been reported.

Apart from experimental situations desoxycorticosterone and aldosterone are used only in adrenocortical insufficiency.

The Glucocorticoids

The glucocorticoids of clinical interest comprise cortisone, hydrocortisone, prednisone and prednisolone. The activity of these compounds is similar in quality, the only differences being related to degree of effect. In general, prednisone and prednisolone are from three to five times more effective than cortisone or hydrocortisone, but this refers only to the glucocorticoid, anti-inflammatory, eosinophil and lymphocyte suppressive actions: the effects on the electrolyte balance are considerably less than those of cortisone or hydrocortisone. Betamethasone and dexamethasone are considerably more potent than prednisolone, the factor varying between five and ten depending on the criteria used: the mineralocorticoid effects are proportionately low and are reduced to almost negligible proportions in these potent anti-inflammatory drugs.

Cortisone Acetate, B.Vet.C.

This, with a few exceptions, may be taken as typical of the glucocorticoids. Originally known as Kendall's Compound E, cortisone can be obtained by extraction from bovine adrenal glands, but is probably not the main glucocorticoid produced naturally, a position known to be held by hydrocortisone. The material used commercially is not isolated from biological sources, but is prepared by synthetic modification of naturally occurring steroids.

Absorption of cortisone, and hydrocortisone, is poor following subcutaneous or intramuscular injection, and in small animals at any rate the oral route provides a more rapid means of obtaining a clinical effect. In man, intramuscular injections give peak levels in the blood after 8 to 12 hr but with oral administration of doses no more than twice as large as parenteral doses, peak levels are achieved in 4 to 8 hr. For maximum effect doses should be repeated every 8 to 12 hr with parenteral dosing, and every 6 hr with oral dosing. Topical application achieves only a superficial action and the same applies to ophthalmic use, in which only structures associated with the anterior chamber are affected. Intra-articular injection provides early local effect. Intravenous injection and subcutaneous implants have also been used for administering cortisone in man but veterinary clinical application via these routes is rare.

After absorption cortisone disappears rapidly from the blood and is converted into hydrocortisone the greatest part of which becomes bound to a glycoprotein called transcortin and to plasma albumin.

Excretion follows the inactivation of the steroid mainly in the liver. The breakdown products are found in the urine amongst the steroid débris which forms the 17-hydroxycorticosteroids, the 17-ketogenic steroids and the 17-ketosteroids.

Cortisone exerts actions on the metabolism of carbohydrates, fats, proteins, enzymes, water and electrolytes. In addition, its effects influence the kidneys, adrenal glands, anterior pituitary gland, the skin, joints, connective tissue, and the processes of repair (anti-inflammatory action).

The effects on *carbohydrate, fat and protein metabolism* are not completely understood, but cortisone increases the formation of glycogen, increases stored glycogen, increases blood sugar levels and normally reduces the metabolic use of carbohydrates. Urinary excretion of nitrogen is increased, partly because of increased renal efficiency, but also due to increased tissue and protein breakdown.

Enzyme concentrations in the blood and liver increase and the activity of many other enzymes is enhanced, although this action is thought not to be a direct effect but an indirect one due to effects on related substrates and co-factors. Other enzymes such as hyaluronidase are inhibited and the degree of hyaluronic acid polymerization is increased following intra-articular injection: this produces a more viscous intra-articular fluid which has a greater 'cushioning' effect.

Water metabolism is influenced by cortisone in much the same way as it is by desoxycorticosterone, but in the normal animal the effects are variable and inconstant. Extracellular fluid is increased and intracellular fluid is decreased, but the total fluid content of the body may not be changed. In some cases, however, water retention may occur in the first few days of a prolonged course of treatment and be followed later by spontaneous diuresis.

Mineral metabolism is likewise variably affected in animals with normal adrenal function, but usually some tubular resorption and retention of sodium takes place, accompanied by loss of potassium. This effect decreases as administration is continued. Also there is usually some increase in phosphate and calcium excretion and a decrease in magnesium excretion. The total effect is, therefore, a decrease in intracellular phosphorus and potassium, an increase in intracellular sodium, and a decrease in blood calcium and phosphate. This would seem to discount and contra-indicate the use of these steroids in milk fever which has been advocated on the continent. As these losses of potassium, phosphorus and calcium are countered by testosterone, the use of this compound would be much more logical.

These effects on mineral balance are considerably less than those caused by deoxycortone, for cortisone is estimated as possessing about one-thirtieth to one-fiftieth of the activity of the mineralocorticoid.

The influence of cortisone on the *kidneys* of normal animals results in an increase in the rate of glomerular filtration; tubular resorption of sodium also increases and an equivalent decrease in potassium retention occurs. Uric acid and phosphate excretion is also increased.

One of the most important side actions of cortisone therapy is its depressant effect on the production of many of the important *anterior pituitary* trophins and because of their suppression, the decrease in activity of the glands controlled by these trophins.

Of primary veterinary importance is the *depression of adrenocorticotrophin* production (ACTH), for acetonaemia and skin diseases have been correlated with decreased adrenal cortex function. The steroid treatment of these conditions would therefore seem contra-indicated, in that such treatment would tend to depress natural corticosteroid function. However, it is probable that large doses or protracted therapy are more dangerous in this respect (50 to 75 mg daily in man), and it is only in such cases that caution should be exercised—periodic dosing with corticotrophin and gradual withdrawal at the end of a course of treatment are the recommended procedures.

There would seem to be some suppression of thyrotrophin but the action of cortisone in this respect is not clearly understood. There would appear to be no effect on the diabetogenic or growth trophins of the anterior pituitary gland.

The actions on *the skin* are also very important and its use in skin diseases constitutes a major indication. Cortisone induces an increased blood flow through the skin which results in increased warmth and skin moisture. There is a decrease in sebaceous secretions: the increased blood supply also provides better nutrition and removal of waste products. Where inflammatory lesions exist the fibroplasia, exudation, and other processes of inflammation are arrested, and irritation decreases; this in turn reduces self-inflicted tissue damage. To offset these advantages there is a slowing of repair processes and a decreased resistance to infection. Prolonged application causes a decrease in epidermal and dermal thickness with an increased collagen density, atrophy of sebaceous glands and a reduction in hair growth.

On the *eye*, the effects are analogous to those on the skin but are of greater importance, for exudation and fibroplasia may result in permanent blindness following corneal scar formation. This can be restricted by the use of cortisone, but antibacterial therapy must be applied concurrently. Only the anterior chamber including the ciliary body, iris, cornea and conjunctiva are affected by local application: systemic administration must be used if deeper structures are to be affected. Care should be taken if ulceration is deep, for penetration may be encouraged by the use of corticosteroids in such cases.

Muscle and bone metabolism are normally unaffected, but during prolonged treatment osteoporosis may develop due to excessive calcium and phosphate excretion: at the same time there occurs a decrease in the number of osteoblasts and cartilage cells so that the formation of new bone is slowed; this, with the calcium and phosphate mobilization, results in an increase in the proportion of intramedullary connective tissue: in very severe cases spontaneous fractures may follow. Such conditions are only likely to be met in small animals treated continuously over long periods, as in the systemic therapy of arthritis.

The effects on arthritis have achieved great prominence in human orthopaedics and in the veterinary field. The effect on joints following intra-articular or systemic administration is a non-specific anti-inflammatory action, but also an increased polymerization of the hyaluronic acid in the synovial fluid results in increased viscosity within the first few days of treatment which, in true arthritis, is accompanied by decreased pain and increased mobility. In rheumatoid arthritis in man, it has been observed during histological examination of the synovial membrane that, as a response to cortisone treatment, fewer lymphocytes and plasma cells, less oedema, less 'papillary tufting', less fibrin deposition and necrosis occur in or on the membrane. Aqueous suspensions are used and the crystals of these are rapidly adsorbed onto the synovial membrane. Following a single injection the results may persist for from 3 days to several weeks. Joints may be injected repeatedly but any infection in the region of the joint contra-indicates this route of administration.

The effects on *connective tissue* and the *anti-inflammatory* actions of cortisone are very important.

Cortisone depresses the function and production of all connective tissue cells including the reticular embryonic mesenchymal cells, but excluding elastic tissue. As the cells themselves are depressed the fibres and interfibrillar tissues are also affected, and the production of tissue exudates is decreased, particularly following trauma. Thus, granulation with its high vascularization is inhibited, fibrin production is decreased, leucocytes are fewer and the resistance to infection therefore less, fibrosis is delayed and scar formation is weak and minimal. Epithelial growth is unaffected. This weakening of wound healing can be minimized by increased protein intake.

Systemic and local reactions to all types of injury are suppressed, including febrile, antibacterial and allergic responses. In *bacterial infections* mobilization of neutrophils at the site(s) of infection does not occur, there is little thrombosis and vascular obliteration with necrosis; liquefaction of caseous material is slowed; capillary dilatation is antagonized and fluid extravasation is minimal; more viable bacteria are found both at the infection site and in the blood stream, yet the febrile and other reactions to bacteraemia, septicaemia and toxaemia are absent. This factor is of prime importance and must always be considered when contemplating corticosteroid therapy. The effect of the steroids on connective tissue (see above) also slows up the processes of combating the infection and repairing the damage, so that, whilst local reactions are decreased and marked tissue changes avoided, these advantages are offset by the dangers of bacterial invasion with death from septicaemia. This situation has occurred in people infected with tuberculosis.

Of equal importance in the veterinary field is the antagonism of allergic reactions by the corticosteroids. This includes allergies of natural origin and induced allergic responses, including serum allergy and the reactions to diagnostic allergens such as tuberculin and mallein. There is considerable confusion as to how this action is achieved, but though some inhibition of antibody formation occurs the major part of the action appears to be related not to abolition of tissue sensitivity but to the control of tissue reaction. Thus though skin sensitivity to tuberculin remains, the cortisone prevents the extravasation of fluid, dilation of capillaries, and increased blood flow, so that no appreciable local reaction occurs. The inhibition of allergic response is therefore a part of the anti-inflammatory action of cortisone.

Cortisone also exerts marked effects on the *blood and circulatory system*. The most obvious action of cortisone on the blood is the rapid decrease in the numbers of eosinophils in the blood. This is frequently used as an indication of the degree of action of the corticosteroid. The action appears to be related to the concentration of cortisone achieved in the blood, for the action is negligible following intramuscular injection whereas it is most marked following intravenous injection or oral administration, when high concentrations will be achieved. In those glucocorticoids which are absorbed rapidly after intramuscular injection an eosinopenic response will follow their administration. This applies particularly to prednisolone and later synthetic steroids.

Lymphocytopenia may occur initially following cortisone administration but continued use does not maintain the decrease. This lymphocytopenia has been used to treat lymphosarcomata and leukaemias but the response is only temporary.

An increase in circulating neutrophils occurs consistently. The effects on coagulation of blood are variable.

The effects on blood pressure are variable and probably relate to the changes in electrolyte balance, and the effect on kidney and peripheral circulation following prolonged usage.

It must not be forgotten that these actions of cortisone are only palliative and not curative. Unless spontaneous resolution occurs, or unless concurrent curative treatment is applied, suppressed symptoms will recur on cessation of treatment. It should also not be forgotten that depression of the anterior pituitary and the adrenal cortex occurs, and sudden withdrawal after prolonged treatment may precipitate severe symptoms of adrenal deficiency.

Hydrocortisone

Hydrocortisone has the same actions as cortisone but it is more effective following oral, intravenous and topical application. On intramuscular administration it is much less active possibly because the compound is less soluble in body fluids. The free alcohol is usually more active than the acetate.

Prednisone and Prednisolone

Both have actions which are similar to those of cortisone and hydrocortisone, but both are more active (about five times) in all respects except the mineralocorticoid effect, which is far weaker than that of cortisone. Prednisolone is slightly more active than prednisone is respect of glucocorticoid effects.

Triamcinolone (9α-fluoro,16α-hydroxyprednisolone) and **6α-methylprednisolone** are slightly more active than prednisolone and the former lacks the sodium retaining tendency of prednisolone although its effects on potassium metabolism are inconsistent and losses can occur. The increased glucocorticoid and anti-inflammatory activity of these two compounds allows their dosage to be reduced by one fifth compared with prednisolone.

9α-Fluoroprednisolone is widely used in the United States as a potent hyperglycaemic agent in the treatment of ketosis. It is more than twice as active as prednisolone in respect of glucocorticoid effects but appreciable mineralocorticoid effect has been reported in man and, in large doses, in cattle.

Betamethasone (xiii) and **dexamethasone** (xiv) are 9α-fluoro,16α-methyl-derivatives of prednisolone; the only structural difference between the two compounds is that the methyl group is in the beta position in betamethasone and in the alpha position for dexamethasone.

For all clinical purposes these two compounds may be taken as being about five times more active than prednisolone. The activity ratio in respect of anti-inflammatory activity actually lies between seven and ten. The beta compound is slightly more active than the alpha compound, as would be expected, but the difference is insufficient to be reflected in dosages. The side actions of the beta compound, especially in respect of thirst and polyuria are also slightly more noticeable, although of little significance at normal dose levels.

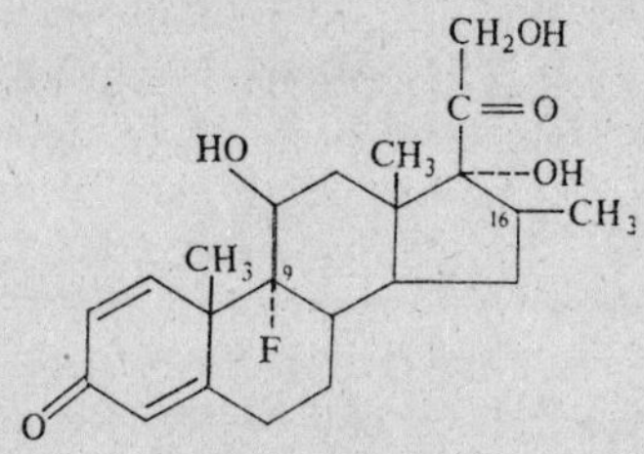

(xiii) Betamethasone

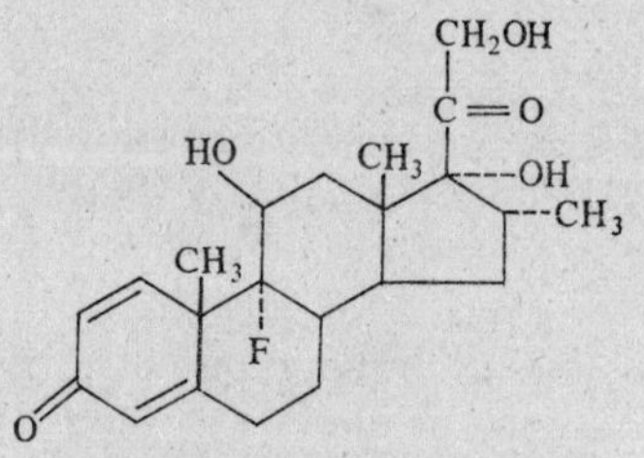

(xiv) Dexamethasone

Uses

The glucocorticoids are used in metabolic diseases (acetonaemia and pregnancy toxaemia), arthritic, reproductive and ophthalmic conditions, skin diseases, and to a lesser degree nervous diseases.

Ketosis has been treated with the glucocorticoids since the use of cortisone was first reported in 1950, following an earlier report in 1947 on the effect of adrenocortical extracts.

The dose of *cortisone* for cattle has been estimated as lying between 0·5 and 1·5 g of the acetate injected intramuscularly. The usual response of the primary, uncomplicated case, is a rapid rise of the blood glucose to normal (within 8 to 10 hr) and beyond, with a return to normal after about 2 days. Eosinophils respond almost in an inversely proportional manner, whilst acetone bodies decrease gradually over a period of about 4 days. The appetite shows improvement within about 24 hr, and milk yields improve from about the second day after injection. Relapses requiring further doses have occurred.

Clinically, hydrocortisone had little to recommend it over cortisone, and ACTH and cortisone were most widely used until the results of trials of *prednisone* appeared. This was shown to have effects far superior to those of cortisone and hydrocortisone. Prednisone is between three and five times more active than cortisone and doses from 0·3 to 0·5 g were effective.

At the same time a halogen derivative of hydrocortisone, known as 9α-fluoro-hydrocortisone or fludrocortisone, was reported to be effective in doses of 0·1 g of the acetate. This compound, however, is more active in causing sodium chloride and water retention and potassium loss than even desoxycorticosterone, and is therefore unsuitable as a glucose mobilizer in the treatment of ketosis.

It has been shown that 50 to 150 mg of *prednisolone*, as an aqueous suspension of the alcohol injected intramuscularly, gives over 80 per cent recoveries. Glucose levels rise very rapidly and strongly whilst the eosinophil reduction is equally rapid: in this latter respect prednisolone, and prednisone, are from three to five times more active than either hydrocortisone or cortisone. The anti-inflammatory and lymphocytolytic actions are proportionately greater but the effect on the electrolyte balance is less than that of the earlier corticosteroids. Prednisolone is now widely used. Betamethasone and dexamethasone are now, to a large extent, replacing prednisolone and other earlier corticosteroids. Doses are about one-fifth those of prednisolone, 10 to 30 mg for ketosis in cattle; the blood glucose levels usually rise higher and persist longer than with other corticosteroids. The anti-inflammatory activity at these accepted dose levels is considerably greater and the side effects are considerably less than with other glucocorticoids.

Although glucocorticoids have now become established as an effective means of mobilizing glucose into the blood, there is no doubt that this is not the end of the ketosis story, nor is the theory of pituitary-adrenal exhaustion entirely accepted as the true explanation of the aetiology of ketosis.

Pregnancy toxaemia in sheep is similar in many respects, including the incidence of ketonaemia and low blood glucose levels, to acetonaemia in cattle. Other factors, however, serve to show that the diseases are not identical: amongst these are the relationship of the aetiology to a low nutritional plane during late pregnancy, particularly low protein intake, the coincidence of this with the rapid growth of the foetus, the intensification of symptoms and the increased incidence in twin conceptions (hence 'twin lamb disease'), and the frequent rapid recovery of the ewe following abortion.

Despite these differences the glucose mobilization effect of cortisone has been reported of value, but not any more so than less expensive types of treatment and the results are not wholly satisfactory. It would seem, as might be expected, that the response of the animal is conditioned mainly by the stage at which treatment is commenced, and hence the level of the glycogen reserves. The earlier the corticosteroid is administered the more rapid and successful the response. Other factors involved are the level of nutrition, the stage of pregnancy, climatic conditions and general husbandry. Dosages of prednisolone are variable: in some cases 50 mg have resulted in abortion, whereas the same doses have been satisfactory elsewhere and in other cases. In the main, 25 to 40 mg given intramuscularly, and repeated as necessary, is the dosage routine.

Arthritis in dogs, pigs, cattle and horses has been treated with cortisone which gives temporary improvement often followed by relapse on cessation of treatment if repair has not taken place in the meantime. Doses are 25 to 200 mg intramuscularly, or 25 to 75 mg orally, per day in the dog, 500 mg twice daily intramuscularly in the horse, 2 g followed by 1 g daily in the bull, and 0·1 g intramuscularly per day in pigs. Cortisone is less effective than *hydrocortisone* in the treatment of arthritis and this latter compound is given orally to dogs in initial doses of 20 to 80 mg. When the symptoms regress these doses are reduced to daily maintenance doses, which will vary in each individual case. The intra-articular use of hydrocortisone is effective. In horses, intra-articular injections of about 100 mg every 4 to 7 days have been successful in controlling symptoms. Prednisolone has proved superior to these compounds: for generalized arthritis, doses of 100 to 150 mg intramuscularly each day for large animals and 5·5 to 11 mg/kg (0·25 to 0·5 mg/lb.) body weight intramuscularly each day for small animals, have been recommended commercially as initial doses, followed by gradual reduction to a maintenance dose which will control symptoms. For localized non-infected inflammations of joints, tendon sheaths, and bursae, injection directly into the cavity of the structure after removal of an equal volume of fluid has been recommended in doses of 20 to 80 mg in large animals and 10 to 20 mg in small animals: these doses can be repeated as necessary, but this mode of treatment is contra-indicated where infections exist near or in the joint. It goes without saying that scrupulously aseptic techniques must be used. In dogs, stifle lameness and ossification of the intervertebral discs have been the conditions most frequently treated. The more potent anti-inflammatory drugs such as betamethasone and dexamethasone should prove even more effective.

Reproductive disturbances can also be associated with adrenal misfunction. Such conditions as cystic ovaries, irregular and unexpected oestrus and abortion have all been correlated with disturbances of adrenal function. Unexpected oestrus, irregular oestrus and nymphomania which do not respond to standard therapy —progesterone and/or luteinizing hormone— are probably related to excessive sex steroid production by the adrenal cortex. In the mare, cow, bitch and queen cat the conditions may respond to large doses of corticosteroids, preferably betamethasone or dexamethasone which have a strong ACTH suppressant effect; initial doses should be two or three times the standard doses. Once symptoms are controlled doses should be reduced to maintenance levels. In entire animals single doses are often sufficient but in castrates the condition may return at varying periods after withdrawing treatment and intermittent courses of treatment may be required.

Ophthalmic conditions have been treated with corticosteroids since cortisone was first available, and have proved to be sight saving on occasions. Topical application is effective only in conditions affecting the structures of the anterior chamber. Deeper structures can only be reached by systemic or subconjunctival injection. The steroids inhibit fluid effusion, opacity due to cellular infiltration, and scar formation. It must be emphasized that concurrent antibiotic therapy is necessary to control infection. Buffered aqueous suspensions containing 0·5 to 2·5 per cent cortisone are applied every 1 to 2 hr but ointments, usually of 1 to 2 per cent concentrations, have a more prolonged action, and should be applied only every 3 to 4 hr.

Hydrocortisone is superior to cortisone for ophthalmic use, ophthalmic suspensions and ointments of hydrocortisone, some containing antibiotics, are available and should be applied every 2 hr or so, and two or three times a day, respectively. The fluoromethylprednisolones, beta- and dexamethasone, have almost entirely replaced hydrocortisone for ophthalmic as well as aural use. Concentrations of 0·1 per cent combined with 0·5 per cent neomycin are used both as ointments and solutions. Subconjunctival injections of solutions of these two corticosteroids are preferred to suspensions of prednisolone and other earlier corticosteroids.

Corticosteroids should not be used when deep corneal ulceration is present as penetration is encouraged. In most other eye conditions corticosteroids are extremely useful in limiting damage caused by inflammation and pressures exerted by fluid effusions which are themselves reduced or restricted.

Skin conditions of non-specific or allergic origin respond favourably to oral or topical application of the corticosteroids; pruritus is relieved within 24 hr and inflammatory reactions subside. It is usual to apply the treatment directly to the skin except in severe and generalized conditions. Eczema and otitis externa of small animals are the main indications for steroid therapy, but 'sweet itch' in horses and similar eczematous and seborrhoeic conditions will respond favourably. Ointments of 0·5 to 2·5 per cent strengths should be applied two or three times daily. Cortisone preparations, again, are inferior to those of hydrocortisone but the more potent anti-inflammatory agents beta- and dexamethasone are the compounds of choice when either systemic or local administration is required.

Among the conditions which benefit most from the treatment with corticosteroids are skin diseases and those associated with acute oedema where vascular permeability is increased: such diseases as fog fever and lung oedema in other species, acute mastitis in cattle, sheep, goats and pigs, serum allergy and other allergic conditions, should all benefit, though simultaneous treatment of specific pathogens must also be carried out. Diseases in which rapid fibrosis interferes with treatment, such as staphylococcal and corynebacterial mastitis, will also benefit from steroid therapy in conjunction with antibiotics.

The palliative effect of the corticosteroids has been stressed repeatedly, but it is very important to remember that relapses will occur at varying periods after treatment has ceased if no attempts are made to destroy or remove the aetiological factors.

Nervous conditions have been treated by oral or parenteral steroid therapy. The inflammatory reactions and damage caused by the viruses of the distemper complex in dogs (hard pad etc.) are reduced and the prognosis improved, but care must be exercised, for experimental evidence indicates that the convulsion threshold is lowered by those steroids and acute attacks of convulsions may be precipitated by steroid therapy.

Neoplasia in man is an indication for the glucocorticoids, especially in tumours where their lympholytic effect is required, e.g. lymphomas and acute lymphocytic leukaemia. Steroids are used to induce remission in a therapeutic regimen the success of which depends on the contributions of additional anti-tumour agents given in a rigorously controlled programme.

The indications described so far exemplify the use of steroids for pharmacological effect, i.e. to elicit a response not usually associated with the supposed physiological role of these hormones. The two following used are examples of their physiological indications.

Adrenocortical insufficiency causes Addison's disease in man and a parallel syndrome of hypocortism has been recorded in dogs. With adequate diet such animals remain reasonably well when supplied with mineralocorticoid supplementation. However, a greater resistance to stresses is imparted when glucocorticoids are supplied also. This may be an indication for the full spectrum of actions of 9α-fluorocortisol or 9α-fluoroprednisolone. Long-term maintenance can be effected by the implantation of one or two pellets containing 125 mg of DOCA subcutaneously. Such a pellet will liberate about 0·5 mg DOCA per day.

The **induction of parturition** in ruminants has been shown to be influenced by the secretions of the foetal adrenal by means as yet unestablished. Various workers have induced parturition in cattle at least 245 days pregnant by a single injection of 10–30 mg of dexamethasone or 5–10 mg of flumethasone. Such inductions were, however, frequently accompanied by retention of the placenta.

Dosage

Prednisolone and products with similar activity.

Large animals:
  100–200 mg daily by intramuscularly
Small animals:
  2–20 mg intramuscularly initially followed by oral therapy:
Dogs (orally):
  2–7 kg (5–15 lb.): 1–2 mg daily
  7–14 kg (15–30 lb.): 2–5 mg daily
  14 kg (30 lb.) and over: 5–15 mg daily

Doses may be trebled in severe cases until symptoms are controlled.

Cats (orally):
  half × 1 mg tablet twice daily to 1 mg tablet twice daily.

Betamethasone and dexamethasone.

Parenteral:
  Horse, Cattle: 10–30 mg
  Foal, Calf, Sheep, Goat, Pig: 2–5 mg
  Dog: 0·25–2 mg
  Cat: 0·25–0·5 mg
Intra-articular:
  Large animals: 2–10 mg
Peri-articular:
  Small animals: 0·25–5 mg
Oral:
  Small animals: 0·25–2 mg

These daily dosages should be divided whenever possible. Initial doses may be maximum or near maximum but once symptoms are controlled doses should be reduced to maintenance levels.

Where bacterial infection is present or suspected concurrent specific antibiotic therapy is imperative. Following lengthy corticosteroid therapy during the previous 2 years, animals may become, because of residual pituitary–adrenal incompetence, extremely susceptible to the effects of anaesthetics, blood loss and other causes of surgical shock. Protecting doses of two or three times standard doses should be given just before, immediately after operation and for several days post-operatively. This procedure is standard practice in human surgery.

Toxicity. It must be remembered that continued use of corticosteroids can, and in small animals does produce a Cushing-like syndrome, the major symptoms of which are excessive thirst and appetite, polyuria, oedema, often muscular atrophy and alopecia. Treatment consists of gradual withdrawal of the corticosteroids, stimulation of the adrenal cortex with ACTH two or three times weekly, and the administration of anabolics.

Contra-indications. Untoward effects are practically unknown in farm animals, but side effects are not infrequently apparent during small animal therapy. The following conditions preclude the use of corticosteroids:

Bacterial infections uncontrolled by antibiotics, but particularly chronic infections with acid fast organisms, Johne's disease in cattle, sheep and goats being the main organism involved.
Diabetes mellitus.
Recent abdominal surgery.
Cardiac insufficiencies.
Deep corneal ulceration.

Use in pregnancy should be avoided if possible, for it has been shown that large doses in the rabbit cause foetal degeneration and abortion: abortion following the use of prednisolone in pregnancy toxaemia cases has also been reported.

Bibliography

LEUNG, K. & MUNCK, A. (1975) Peripheral actions of glucocorticoids. *A. Rev. Physiol.*, **37**, 245–73.

MYLES, A. B. & DALY, J. R. (1974) *Corticosteroid and ACTH Treatment*. London: Edward Arnold.

NACHT, S. & GARZÓN, P. (1974) Effects of corticosteroids on connective tissue and fibroblasts. In: *Steroid Biochemistry and Pharmacology* (ed. M. H. Briggs & G. A. Christie), vol. 4, pp. 157–82. London: Academic.

SCHRIEFERS, H. (1967) Factors regulating the metabolism of steroids. *Vitams Horm.*, **25**, 271–311.

VEILLEUX, R. (1963) The stress concept as we see it today. *Adv. vet. Sci.*, **8**, 189–212.

12

The Hormones

II. The Control of Reproductive Function

The drugs which affect the reproductive system more or less selectively can be divided into three groups for ease in discussion:

1. Aphrodisiacs and Anaphrodisiacs
2. Ecbolics
3. Hormones

These groupings are by no means rigid. It will, for example be apparent that some of the hormones have aphrodisiac action, while oxytocin is a hormone with powerful ecbolic effects.

APHRODISIACS AND ANAPHRODISIACS

The aphrodisiacs are drugs which increase sexual desire and libido. Conversely, the anaphrodisiacs are drugs which decrease sexual desire and libido. In veterinary medicine there is little need for the use of aphrodisiacs providing the hygiene and husbandry of animals is correct.

In the past, aphrodisiacs were resorted to if a stud male was slow in serving any of the females presented to him in oestrus. Little cognizance was taken of the fact that perhaps he was overweight from lack of exercise, that his feet were overgrown and service exerted a strain on tendons and joints which caused pain, that perhaps he was deficient in vitamins due to bad housing and feeding, or merely that he was overworked. The female was not considered.

To-day there is little indication for the use of the old type of aphrodisiac, the general opinion being that if animals refuse to mate there must be some physiological, and perhaps pathological, cause. Before any drug is resorted to, therefore, a full investigation of the feeding, hygiene, and husbandry should be carried out. Following this, and if no possible cause has been found, a thorough physical examination and observation of mating behaviour is the second stage of investigation. If no pathological features such as arthritis, either of the limb joints, but more important, of the thoraco-lumbar vertebral joints, are found then, and only then, may stimulants be considered. Such stimulants may be of a general type such as strychnine therapy, but, particularly in the old male, modern therapy inclines toward the use of androgens or stimulators of androgen production (chorionic gonadotrophin). It should be remembered, however, that while such hormones may increase libido they tend to depress spermatozoa production and should not be used indiscriminately as a treatment for reproductive derangements.

In the female, suboestrus—blind bulling in the cow or silent heat, can similarly be treated with hormones.

Anaphrodisiacs have a small but definite part to play. Their use is restricted almost entirely to suppressing excessive libido in puppies at puberty, when sexual excitement and masturbation may become a problem. They may be used empirically in nymphomania in large animals. Sedatives may control these symptoms. Delmadinone acetate is a steroidal anti-androgen recently introduced to control hyper-sexuality in dogs. Where such measures fail oestrogens may be resorted to in the male, and in the female gonadotrophin (LH) with progesterone if the problem is of ovarian origin or large doses of glucocorticoids if adrenal imbalance is implicated.

Old Type Aphrodisiacs

The non-hormonal aphrodisiacs are of two types: (1) those which exert their action through their effects on the nervous system, and (2) those which irritate the urinogenital system during excretion. Strychnine, alcohol, and possibly yohimbine, are examples of group (1), whereas cantharides is representative of group (2).

Strychnine

This is only aphrodisiac by virtue of general toning up of spinal reflexes. Nevertheless it can be of use in the lazy, sluggish male when lack of inclination to serve is due to overweight and need of exercise. Without adjustment of management and feeding strychnine is not very effective.

Alcohol

In humans this has been considered as an aphrodisiac and is reputed to act by the removal of social and sexual inhibitions during intoxication. It has no application in the veterinary field.

Yohimbine

Has been used in man and animals to stimulate 'sex drive', though it is reported to be ineffective in creating sexual desire and certainly ineffective in therapeutic doses. It is reputed to act in one of two ways or possibly both: (a) by causing a general excitation of the CNS, and (b) by general vasodilatation in the skin and mucous membranes, but particularly in the genital organs, which results in erection. The vasodilatation is due to blockade of the α-adrenergic receptor by yohimbine, which also blocks 5-HT. Oral doses used have been 30 to 90 mg ($\frac{1}{2}$ to $1\frac{1}{2}$ gr.) for horses and cattle, and 0·6 to 3 mg ($\frac{1}{100}$ to $\frac{1}{20}$ gr.) for the dog.

Cantharides, B.P.C.

This has also been used as an aphrodisiac but is now rarely used for this purpose. Its effects, however, may occasionally be observed following the injudicious use of cantharides blister in the vicinity of the prepuce, when preputial contamination causes irritation and penile erection, or when a large area has been dressed and sufficient drug has been absorbed through the skin to cause irritation during excretion: trouble may also result from animals licking an area blistered with cantharides. Most of the cantharidin absorbed is excreted via the kidneys and, during excretion, irritation of the urethra occurs. This results in penile erection. Copious fluid intake and hyoscyamus will help to counteract renal and sexual symptoms of poisoning, the remaining symptoms being treated symptomatically.

ECBOLICS

An ecbolic is an agent which stimulates muscle contraction in the uterus and is frequently referred to as an oxytocic agent. The use of ecbolics in the pregnant animal may cause abortion, but though they are often referred to as abortificients, this is not their constant therapeutic effect nor their indicated use in animals, indeed their use is contra-indicated in the pregnant animal.

The use of ecbolics is indicated: (a) In parturition where position and presentation are normal and where the cervix is adequately dilated. They should not be used unless parturition is prolonged.

(b) In a flaccid post-parturient uterus which is not contracting (involuting).

(c) In post-parturient haemorrhage.

In all cases of uterine inertia or slow delivery, doses should be small and repeated.

Many substances have an ecbolic action, including all types of purgatives but particularly the 'drastics', the anthracenes, and the neuromuscular purgatives: some poisons such as the salts of lead and some volatile oils including those of ragwort are also ecbolics and abortifacients. Despite the number of substances with ecbolic actions, few are used therapeutically—ergot and its preparations, extracts of the posterior pituitary gland and the prostaglandins.

Ergot Alkaloids

Ergot is a stage in the life cycle of a fungus, *Claviceps purpurea*. It consists of a spur-like body or sclerotium derived from the ovary of rye, and is black on the outside but has a cream interior which shows dark veining when the sclerotium is fractured transversely. At one time almost unknown in Great Britain it is now fairly widespread. The active principles can survive in silage and outbreaks of poisoning occur sporadically in cattle. In those countries where rye is eaten by the human population poisoning characterized by hallucination occasionally reaches epidemic proportions.

The main active constituents of ergot are six pairs of stereoisomeric alkaloids. The naturally occurring *laevo*-rotatory alkaloids are the most important and active, whilst the *dextro*-rotatory stereoisomers are relatively inactive. These alkaloids have been classed chemically into three groups as indicated in Table VI.

Table VI. Chemical classification of the stereoisomeric alkaloids

	Laevo-rotatory	*Dextro*-rotatory
1. The ergotamine groups:	Ergotamine Ergosine	Ergotaminine Ergosinine
2. The ergotoxine groups:	Ergocrystine Ergocryptine Ergocornine	Ergocrystinine Ergocryptinine Ergocorninine
3. The ergometrine group:	Ergometrine	Ergometrinine

The discovery of these alkaloids has occurred piecemeal and of the alkaloids listed above a number are still referred to by other names both officially and unofficially. Thus ergocrystinine is also known as ergotinine, ergocorninine as ergotinine, and ergometrine as Ergonovine, U.S.P., ergobasine, ergostetrine and ergotocine. Ergotoxine is now recognized as being not a single compound but a mixture of the three *laevo*-rotatory alkaloids listed above. All are derivatives of lysergic acid.

Besides these and other alkaloids, ergot also contains various amino acids, amines, organic carboxylic acids and vitamin D, as well as about 40 per cent fixed oil, sugars and glycosides. These do not materially affect the action of ergot.

Action

Although, historically, the main alkaloids are classified under three heads, the amino acid alkaloids (Groups 1 and 2 above) are grouped together therapeutically, whilst the less complex ergometrine is more important for use in obstetrics.

Both groups of alkaloids stimulate smooth muscle though ergometrine has very little effect on smooth muscle other than that of the uterus.

The members of **ergotamine**–'ergotoxine' group have a delayed onset of their uterine effect following oral administration or intramuscular injection (15 to 30 min). Following absorption, small doses affect mainly the uterine muscle by direct stimulation, causing an increase in power and extent of the rhythmic contractions of the non-pregnant, and especially the parturient and immediately post-parturient, uterus. Larger doses will stimulate the bladder, intestine, and blood vessels when there may be a rise in blood pressure due to arteriolar constriction, though a slowing of the heart tends to mask this effect. In very large, supertherapeutic, doses, the group causes the 'adrenaline reversal' action whereby the blood pressure falls if adrenaline is injected after a large dose of 'ergotoxine' or to a lesser degree, ergotamine. This action is due to selective blockade of the α-adrenergic receptors after which adrenaline will elicit vasodilation only, due to β-site activation.

Ergometrine (ergonovine) is readily soluble in water and is rapidly absorbed and rapidly effective (5 to 8 min after oral dosing, 2 min and 1 min after intramuscular and intravenous injections, respectively). The action persists maximally for about 1 hr and gradually lessens over a period of several hours. It affects the uterus fairly selectively and is the ergot derivative most preferred for this action.

Uses

Ergot, as the prepared powder or the liquid extract, and ergometrine are useful for causing expulsion of the placenta and involution of the post-parturient uterus. They are also useful in stopping haemorrhage from the uterus after parturition, the effect being achieved mainly by rapid and considerable reduction in uterine size which exerts pressure on the vessels as the tissues become compacted. The cruder preparations are used more in the large animals whereas ergometrine, in the form of a solution of the maleate, is preferred in small animals. 'Ergotoxine' and ergotamine are rarely used.

In man, the central effects of ergot has resulted in the abuse of these compounds and especially

lysergic acid diethylamide, which is a potent hallucinogen. The ability of this compound to induce severe disorientation and, sometimes, changes resembling psychosis has lead to the use of the term psychotomimetic to describe its effects.

Note. Great care is required in the administration of ergot alkaloids, for large doses, or use during foetal expulsion, may cause spasmodic contraction of the uterus and imprisonment of the young, with considerable danger to both mother and offspring. These effects are also very painful.

Dosage

Owing to the extended duration of effect, and the catastrophic results of overdosage, it is better to err on the side of underdosage if any doubt exists and rely on repetition to achieve the desired effect.

Ergometrine Maleate, B.Vet.C. (as tablets or injection)

Horse and Cow: 10–20 mg
Sheep and Pig: 0·5–1 mg
Dog: 0·2–1 mg
Cat: up to 0·125 mg

Toxicology. Two types of poisoning occur. The acute type takes the form of hyperexcitability, muscular incoordination, and tremors, with blindness, and possible convulsions. The pupils of the eyes are often tightly constricted as in morphine poisoning. The chronic form follows vasoconstriction and endothelial damage by the amino acid alkaloids and takes the form of end-artery infarction with necrosis and sloughing particularly of areas in the legs and the feet. These symptoms are usually associated with a drop in milk yield, dullness, lethargy, inappetance and wasting. Treatment is symptomatic, the main benefit being derived from removal of the source of poisoning. In man, the treatment consists in keeping the limbs cool and the body warm: Methacholine Chloride, B.P. (acetyl-b-methylcholine), and glyceryl trinitrate (nitroglycerin) have been used to counteract the arteriolar constriction effects of the ergot alkaloids.

Injection of Pituitary (Posterior Lobe), B.Vet.C.

This injection is a sterile aqueous extract of the posterior lobe of pigs' and cattle pituitary glands. It is standardized to contain 10 oxytocic units per ml. When subjected to fractionation, almost pure preparations of the two main therapeutic constituents, oxytocin and vasopressin, can be obtained. These fractions are official as **Injection of Oxytocin,** B.Vet.C. which contains 10 oxytocic units and 0·5 units of pressor activity; and **Injection of Vasopressin,** B.P., which contains 20 units vasopressor activity and not more than 1 oxytocic unit per ml. Synthetic oxytocin is now manufactured commercially and is available as official injections in both the human and veterinary fields.

Actions

The active principles of this injection are polypeptide hormones and have three main actions of therapeutic significance and a fourth of no clinical importance. The actions of these hormones are shared by synthetic analogues, the main effects of which are oxytocic or uterine stimulant effects, and vasoconstrictor actions; these actions are due to the oxytocic and vasopressor principles respectively. Some authorities also recognized the presence of a third principle which has an antidiuretic effect, and a fourth principle which causes the dilatation of melanophores: the antidiuretic hormone has since been shown to be identical with vasopressin.

Oxytocin. Oxytocin which is an octapeptide, acts specifically on the uterine musculature but is only completely effective when the uterus is under the influence of oestrogens. It is therefore strongly active on the non-pregnant and late pregnant uterus: in the latter case, pituitary secretions are suggested to be responsible for the induction of parturition when ovarian oestrogen production reaches a 'trigger' level in the tissues. Pituitary injections have little effect during early and middle pregnancy when the uterus is under the protection of luteal and, later, placental hormones. Oxytocin is also involved to a major extent in 'letting down' of milk. This action results from stimulation and contraction of the smooth muscle fibres located around the mammary alveoli and results in the expulsion of preformed milk. Injection of Oxytocin, B.Vet.C., particularly that made from synthetic oxytocin, contains little or no pressor activity; absorption and uterine effects therefore occur much more rapidly than with the natural extract, containing, as it does, an appreciable but variable amount of the pressor factor. For this reason dosage rates must be considerably reduced. This lack of pressor activity also means that local reactions due to the restriction of local blood supply and the resultant tissue anoxia do not occur.

Vasopressin. A polypeptide similar to oxytocin, it is a general stimulant of smooth muscle with strong vasopressor activity and an antidiuretic effect. The rise in blood pressure is elicited only by doses in excess of those required for autidiuretic effect and is a direct action on smooth muscle in the vascular bed. As even the coronary vessels are constricted, this action should not be exploited in therapeutics. Propulsive activity of the large gut is also enhanced by large doses of vasopressin. Stimulation of urinary bladder contractions also results from vasopressin activity.

Uses

Injections of pituitary extracts and of oxytocin are primarily used for speeding expulsion of the foetus or foetuses when uterine inertia is diagnosed. They should not be used for this purpose unless position and presentation are normal. They are occasionally useful in obtaining the expulsion of the last puppy, piglet or kitten, and are probably of more use in these multi-foetal species than those producing only one offspring at a time. After parturition, pituitary extract will aid the expulsion of placental débris and involution of the uterus, and it is especially useful in cattle to inject 3 ml or more directly into the uterine muscle after the removal of the foetus during Caesarian section: this also controls uterine bleeding. Following uterine prolapse the administration of injection of pituitary extracts at several places in the organ will reduce engorgement and contract the everted uterus, which can then be replaced more easily.

Agalactia in sows responds well to the injection of 2 to 3 ml, and occasionally milk will spurt from the teats within a few minutes of intramuscular administration and a few seconds of intravenous injection. It is not indicated in mastitis or in inability to suckle a litter due to other causes. In examination of milk during investigation of 'baby pig disease' due to haemolytic antibodies in milk related to rhesus factor differences between offspring and sow, the collection of milk samples has been facilitated by the intravenous injection of about 1·0 ml of pituitary extract. The milk is vigorously expressed from the teats and saves the laborious task of manual expression.

Oxytocin intravenously in doses of 10 to 40 units has been used in the treatment of mastitis in cattle, as an aid to expressing all infected milk from the quarters prior to introducing chemotherapeutics, but in controlled experiments on goats and cows, no therapeutic advantage has been demonstrated.

Dosage and Administration

Injection of Pituitary (Posterior Lobe), B.Vet.C.
- Horse and Cow: 5–10 ml (30–100 units)
- Sheep and Pig: 1–5 ml (10–50 units)
- Dog: 0·25–1·5 ml (2½–20 units)
- Cat: 0·25–0·5 ml (2½–5 units)

Injection of Oxytocin, B.Vet.C.

Intramuscular:
- Horse and Cow: 10–40 i.u.
- Sheep, Goat and Pig: 2·5–10 i.u.
- Dog: 1–10 i.u.
- Cat: 0·5–5 i.u.

Intravenous:
- Horse and Cow: 2·5–10 i.u.
- Sheep, Goat and Pig: 0·5–2·5 i.u.
- Dog: up to 0·5 i.u.

The injections should be given by deep intramuscular injection. Subcutaneous injection can cause severe 'bruising', with discoloration of the area around the site of injection up to a diameter of 15 cm (6 in.) or so, depending on the dose; the centre of the area may slough out. Intravenous injection can be used, but care should be taken in small animals. Injections should be about one-third of the intramuscular doses and should preferably be diluted to 1 : 10 with sterile Water for Injection, B.Vet.C.

It must be emphasized, as with ergot preparations, that parturition cases are consistently overdosed in small animal practice, and that it is far better to give repeated small doses rather than one single large dose. A large dose, besides causing uterine spasm, also increases the danger of sloughs and causes the animal unnecessary pain and discomfort.

The physiological situation in parturition is more properly mimicked when ecbolics are administered by intravenous infusion. This technique is used in the induction of parturition in man.

PROSTAGLANDINS

These tissue hormones, which are widely distributed in the animal body, exist in several forms and can elicit a wide variety of responses. They are biosynthesized from unsaturated essential fatty acids incorporated in cell membranes and are derivatives of a 20-carbon parent compound, prostanoic acid.

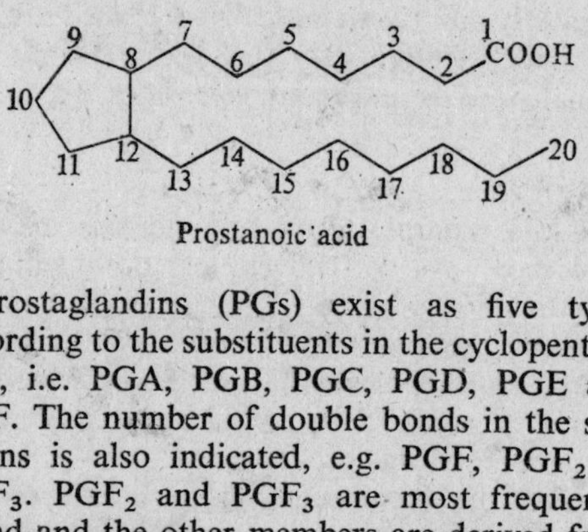

Prostanoic acid

Prostaglandins (PGs) exist as five types according to the substituents in the cyclopentane ring, i.e. PGA, PGB, PGC, PGD, PGE and PGF. The number of double bonds in the side chains is also indicated, e.g. PGF, PGF_2 or PGF_3. PGF_2 and PGF_3 are most frequently found and the other members are derived from these two primary prostaglandins. A specific dehydrogenase inactivates PGs E and F in blood to the extent of 80–90 per cent in one passage through liver or lung.

The manifold effects exerted by the PGs vary with the species of PG, the tissue and the location in the body, but many are consistent with an involvement in the inflammatory response, e.g. vascular smooth muscle contraction or dilation, increase in vascular permeability, leucocytotaxis, pyrexia and pain. In general, smooth muscle of uterus, lung and gut is contracted by PGFs and relaxed by PGEs. An ability to influence secretory function in both endocrine and exocrine glands is present. A possible CNS neurotransmitter role is postulated, as is a capacity to modulate autonomic neurotransmitter function. Metabolic effects on lipolysis, glucose fate and calcium turnover have also been reported.

Prostaglandins can be released by physical, neural and endocrine stimuli and their primary action seems to be a modification of cyclic AMP accumulation in cells, an increase of which leads to a stimulatory response. Calcium fluxes are also known to be involved in some PG responses.

The physiological functions suggested for the PGs are as numerous as its pharmacological effects and include reproductive, renal, smooth muscle tone, inflammation and neural transmission modulating effects. Its roles in reproductive function include male fertility by stimulating sperm transport by the female genital tract, induction of luteolysis and so regulating the oestrous cycle, a contribution to uterine contractile ability at parturition and a post-partum role in contracting the umbilical vessels.

Such potent agents will clearly have many potential applications in therapeutics and some of these have already been investigated. The reproductive functions are applicable to this chapter.

Both PGE_2 and $PGF_{2}\alpha$ are powerful inducers of uterine contractions in the last two-thirds of pregnancy in man and the clinical response is indistinguishable from that to oxytocin at term. These agents are also capable of inducing abortion when injected into the uterus or amnion but effective doses increase inversely with the duration of pregnancy. Side-effects occur more especially with the higher doses and are nausea, vomiting, colic and diarrhoea.

Owing to their rapid inactivation and to the incidence of side-effects, PGs are not suitable for use as drugs. Analogues with extended duration of action and with more restricted activity

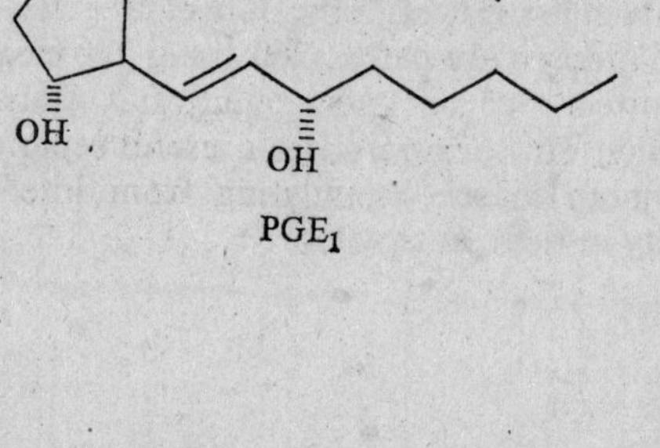

PGE_1

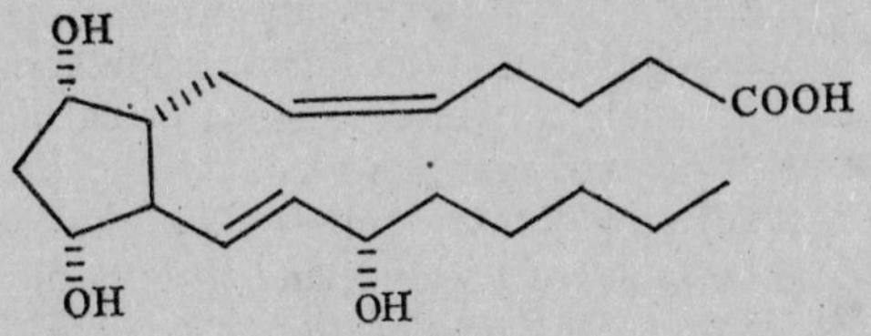

$PGF_2\alpha$

spectra are available. Their potential in veterinary obstetrics for the induction of parturition is being evaluated.

Already established, however, is the value of the two analogues *cloprostenol* and *fluprostenol* for control of the oestrus cycle in the cow and the mare respectively. These substances are powerfully luteolytic, i.e. they cause rapid regression of the corpus luteum with a dramatically rapid arrest of its secretory activity. If they are administered after the fifth day of the cycle, luteolysis is rapid and an ovulatory oestrus follows in 2–4 days. PGs are without effect in the follicular phase of the cycle. The same pattern occurs in the mare.

In the cow cloprostenol 500 μg intramuscular can be used to induce a known oestrus in suboestrus lactating cows which have a corpus luteum; can terminate by luteolysis an unwanted 1–150 day pregnancy; can cause by luteolysis expulsion of a mummified foetus up to mid-term or resolution of pyometra. Two treatments at an interval of 11 days will result in synchronization of oestrus in cows or heifers and confers managemental advantages.

In the mare fluprostenol is administered as a single intramuscular injection of 250 μg. It is of value in correcting infertility states caused by persistence of a corpus luteum, i.e. following early foetal death and resorption; in the termination of persistent dioestrus or pseudopregnancy; in the treatment of lactational anoestrus; and in correction of anoestrus due to corpus luteum persistence. Fluprostenol can also be used to time oestrus in stud management.

In the mare 4–6 times the dose causes sweating and diarrhoea. In cattle, 200 times the dose of cloprostenol causes only mild and transient scouring. This demonstrates a useful separation of smooth muscle stimulating from luteolytic activity in these analogues.

THE SEX STEROIDS

All naturally produced androgens, oestrogens and progestogens are steroids, that is, compounds related to that group of organic alcohols known as sterols. Both sterols and steroids have been classified as sterids.

The sterols are saturated and non-saturated alcohols which occur in fixed oils originating from animals (zoosterols), higher plants (phytosterols) and fungi (mycosterols). They may occur as free alcohols, or as esters and glycosides. The principle zoosterol is *cholesterol*, and the principal phytosterol is *stigmasterol*, both of which are widely used as starting points in the manufacture of steroid hormones.

The steroids are very similar in structure to the sterols and occur widely throughout the plant and animal kingdom. They are very active and frequently highly poisonous substances; consequently they are very important both pharmacologically and therapeutically. Included are the bile acids, vitamin D, the sex steroids, the adrenal steroids and many glycosides.

Chemistry

All the natural sterols and steroids may be considered as derivatives of cyclopentanoperhydrophenanthrene (i). Changes in nomenclature follow a set international pattern—see p. 121. Whilst cholesterol (ii) is the basic steroid, all sex steroids and corticosteroids have a common root in progesterone (see p. 152). However by reference to the structure of the steroids (iii), (iv), (v) (vi), (vii) and (viii), it will be seen that certain structural characteristics are associated with specific hormonal activity. Thus *androgenicity* is associated mainly with methyl groups at C_{18} and C_{19} and no further carbons attached at C_{17}, *oestrogenic action* with the absence of a methyl group at C_{19} and aromatization of ring A, and *progestogenic action* with methyl groups at C_{18} and C_{19} but also with the addition of further carbon atoms at positions 20 and 21. It will also be seen, on reference to Chapter 11, that corticosteroid action is related to hydroxylation of C_{21} in the progesterone molecule, although other possible changes including hydroxyl groups at C_{17} and C_{11} increase the corticosteroid activity.

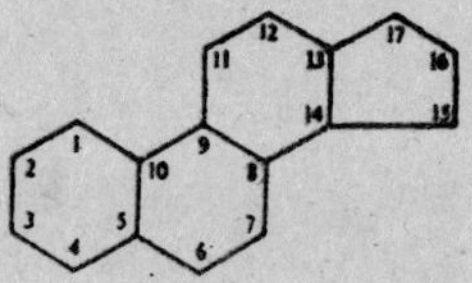

(i) Cyclopentano-perhydrophenanthrene

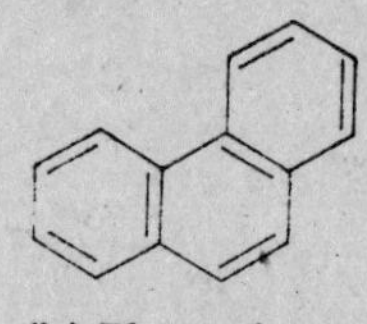

(ia) Phenanthrene

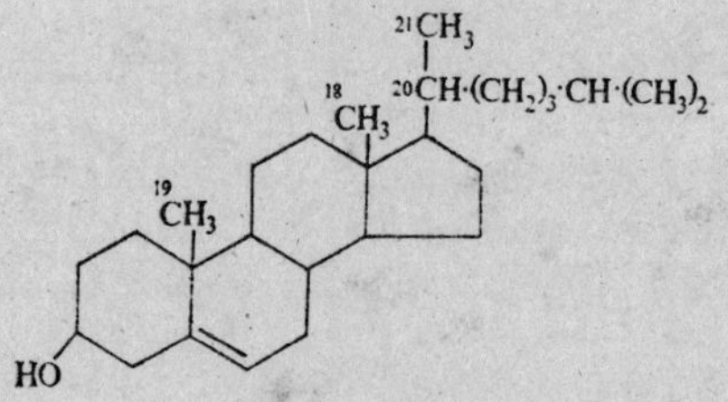

(ii) Cholesterol

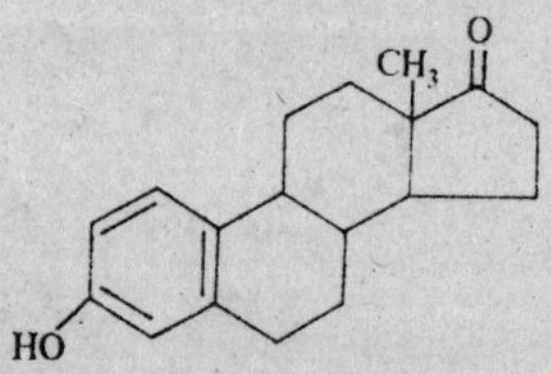

(iii) Oestrone (the single oxygen at carbon 17, giving it the '-one' structure and distinguishing it from the di-alcohol ('-diol') structure of oestradiol)

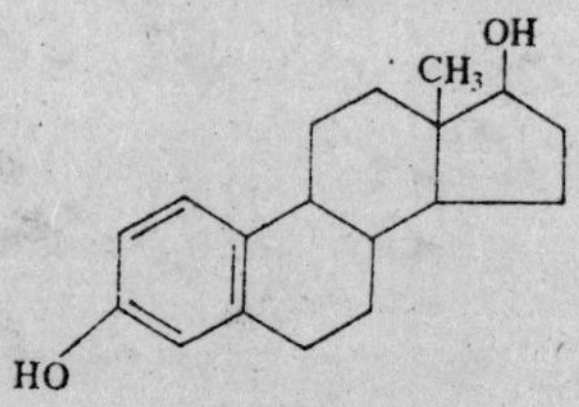

(iv) Oestradiol

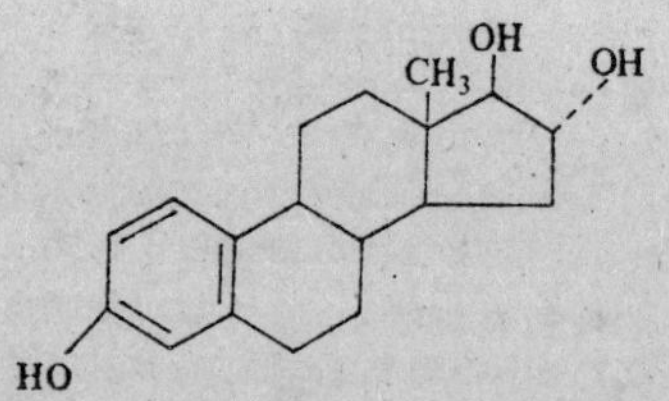

(v) Oestriol ('-triol' instead of '-one')

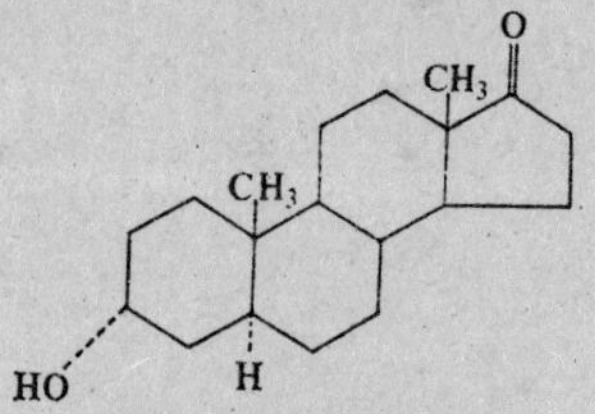

(vi) Androsterone

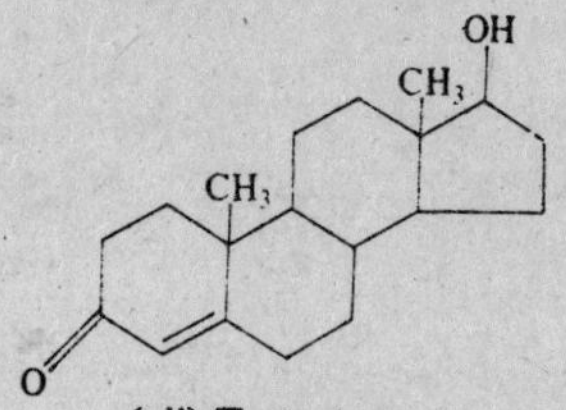

(vii) Testosterone

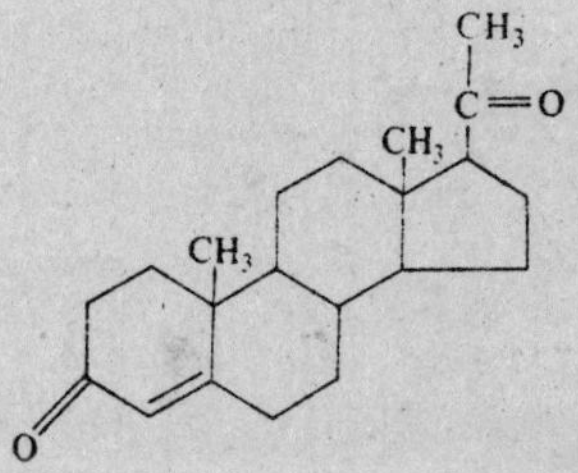

(viii) Progesterone (pregn-4-ene-3, 20-dione)

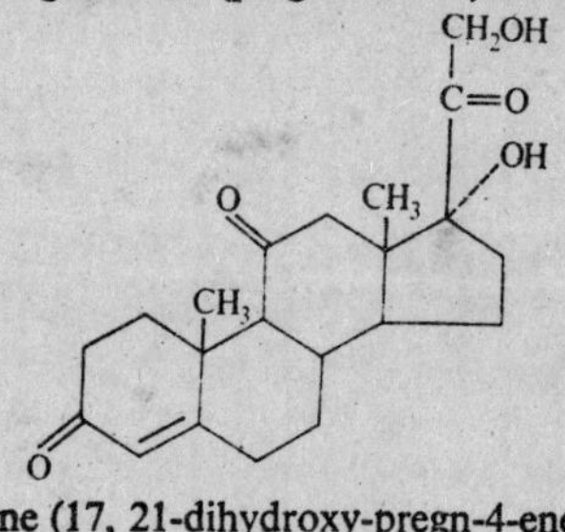

(ix) Cortisone (17, 21-dihydroxy-pregn-4-ene-3, 11, 20-trione)

Note: '-ene' indicates a double bond and is related to the first carbon atom in the link, e.g.

 is 1-ene,  is 3-ene,  is 4-ene and

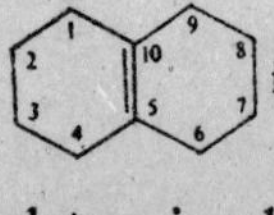 is 5(10)-ene.

The nomenclature is related to the reduction of hydrogen atoms attached to the carbons and the formation of an unsaturated compound in the same manner that cyclohex*ane* becomes benz*ene*

with the removal of hydrogen atoms and the creation of double bonds. Similarly with saturation cyclopent*eno*phenanthrene becomes cyclopent*ano*perhydrophenanthrene and vice versa.

'-one' indicates an oxygen linked by a double bond; thus two, '-dione' and three, '-trione'.

'-ol' indicates a hydroxyl group and from this '-diol' with two and '-triol' compounds with three hydroxyl groups.

'Oestr-' indicates a steroid with oestrogenic properties containing only one methyl group attached at carbon 13 in the nucleus.

'Androst-' indicates a steroid with androgenic properties containing two methyl groups—one attached to carbon 10 and one at carbon 13 in the nucleus.

'Pregn-' indicates a further stage, in which a methyl-ketone group is added to the androgen formation at carbon 17 (carbons 20 and 21).

A comparison of these formulae will show how closely related are the androgens, oestrogens, progestogens and the adrenocorticoids, of which group cortisone (ix) is a member. The androgens, oestrogens and corticosteroids are, physiologically, derived from progesterone. It is therefore less surprising that the sex steroids can be replaced in certain circumstances by each other in either sex. It is, however, remarkable that the substitution in the progesterone molecule (viii) of an oxygen for two hydrogen atoms at carbon 11, and of hydroxyl groups for hydrogen atoms on carbons 17 and 21, should create the vast pharmacological and physiological differences that exist between that compound and the cortisone (ix) which is formed by these changes.

The synthetic oestrogenic stilbenes, diethylstilboestrol (xi), dienoestrol (xii) and hexoestrol (xiii), derived from the basic stilbene nucleus (x), do not at first sight appear to be in any way structurally related to the natural compounds, but if diethylstilboestrol is graphically represented as in (xiv) it will be seen to bear some resemblance to oestradiol (xv).

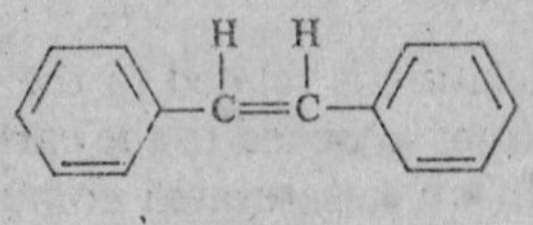

(x) Stilbene nucleus

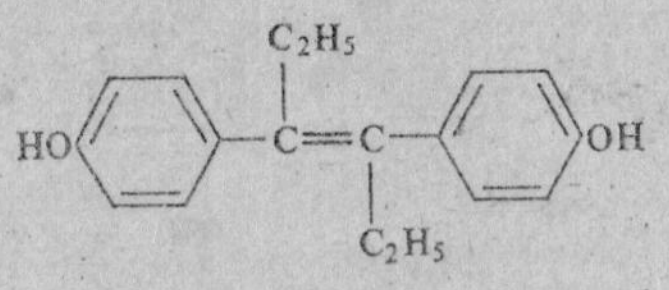

(xi) Stilboestrol

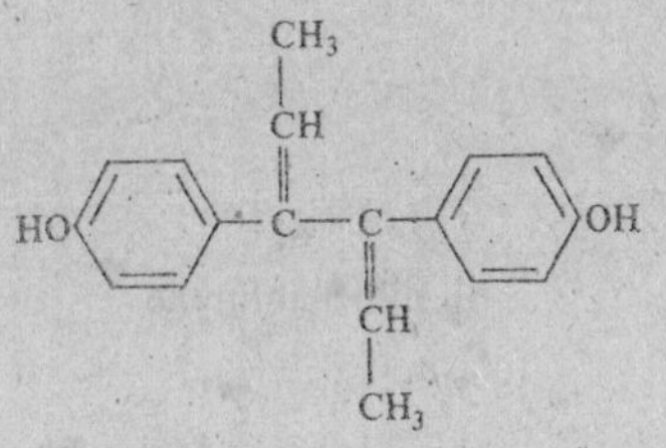

(xii) Dienoestrol

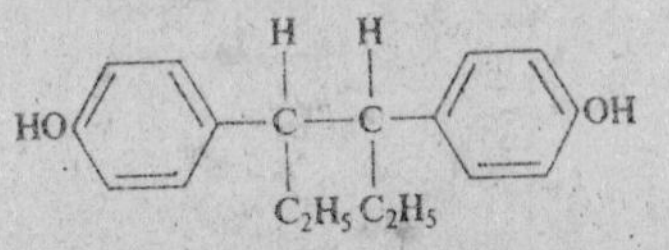

(xiii) Hexoestrol

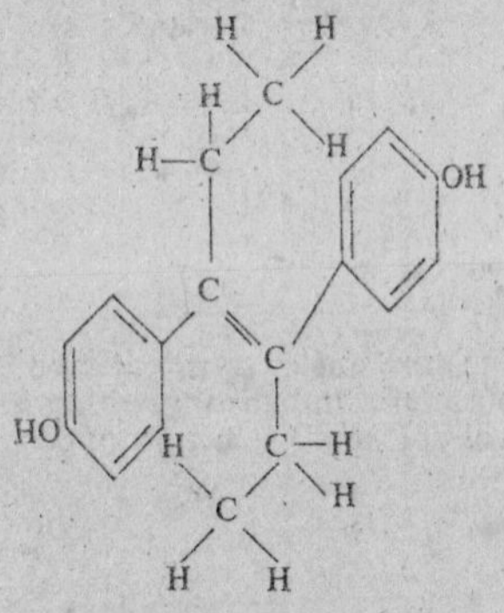

(xiv) Stilboestrol

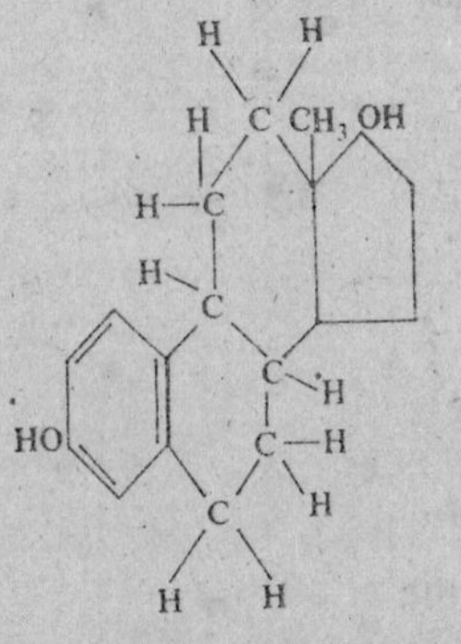

(xv) Oestradiol

The semi-synthetic compounds consist of natural compounds which have been manipulated chemically to increase or modify the action of the original compound. Thus methyltestosterone (xvi), has a methyl group added at C_{17} to increase the stability of the compound for oral administration. The addition of an ethinyl group at C_{17} whilst increasing the oral stability also changes the activity from an androgenic to a progestogenic one because the carbons added make up C_{20} and C_{21} of a progestogenic compound (ethisterone, xvii). Ethinyloestradiol (xviii) on the other hand retains oestrogenicity because of the aromatization of ring A and the absence of the methyl group at C_{19}; this compound is stable on oral administration and is still the most active oestrogen available for clinical use.

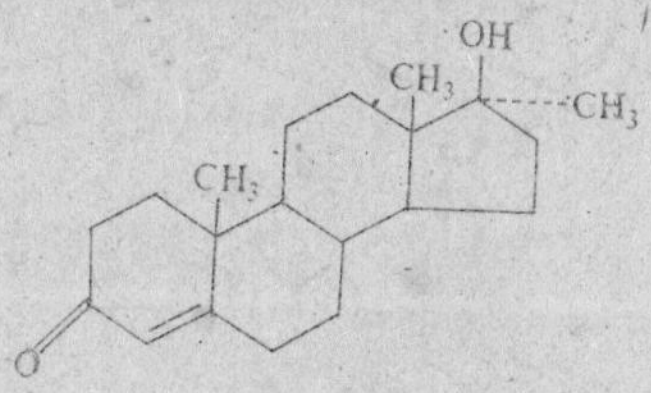

(xvi) Methyltestosterone

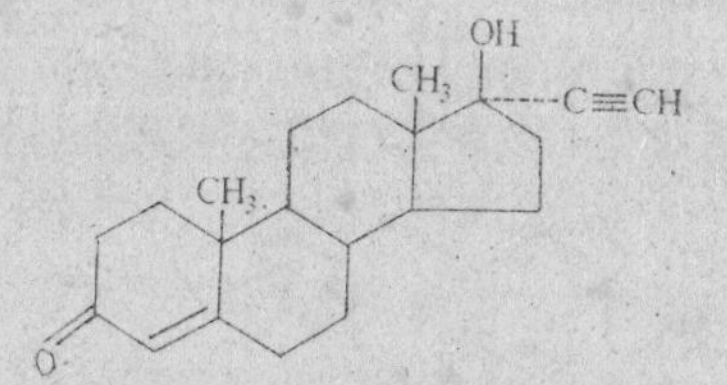

(xvii) Ethisterone (ethinyltestosterone)

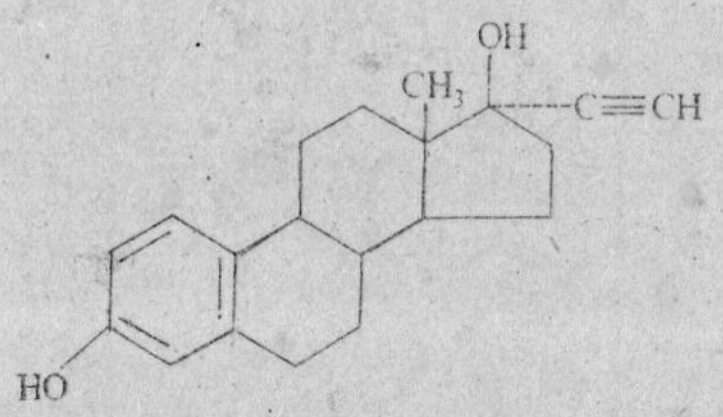

(xviii) Ethinyloestradiol

Metabolic Pathways

In the body all steroids are derived from cholesterol via progesterone which thus provides a common root (Table VII). The progressive changes are brought about by enzyme activity, each enzyme effecting a specific change at a specific point; e.g. 17α-hydroxylase which is responsible for the formation of 17α-hydroxyprogesterone as a stage in the conversion of progesterone through a testosterone. That all sex steroids and the corticosteroids are derived from progesterone is extremely important to remember for a derangement of enzyme activity in ovary, testis or adrenal cortex can result in eccentric hormone production and marked changes in the physiological status of the animal. In the ovary the major steroid produced in the follicular phase is oestradiol; small amounts of other oestrogens and androgens are also formed: in the luteal phase progesterone is produced in large amounts. In the testicle, whilst it is thought that small amounts of oestrogens are required at certain stages of sperm production, the major steroid produced is testosterone.

Steroid Esters

Few of the natural steroid alcohols are present and active in the tissues for long when administered intramuscularly or subcutaneously as solutions in oil and a series of esters have therefore been developed. These esters are formed by the attachment of various types of fatty acid chains at points where hydroxyl groups occur in the steroid nucleus. These esters prolong the activity of the compound following intramuscular or subcutaneous injection as oil solutions. The length of activity is roughly proportional to the length and/or complexity of the fatty acid side chain, reflecting the time taken to split off the side chain. Thus up to a certain length the longer the chain the longer the action. Whereas pure testosterone is active over a period of a few hours, the propionate is active for a day or two, the phenylpropionate for about 10 days, the decanoate about 21 days and the oenanthate for about the same length of time.

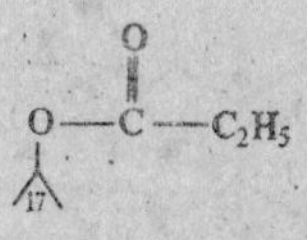

Propionate

Table VII. Metabolic pathway of the sex hormones

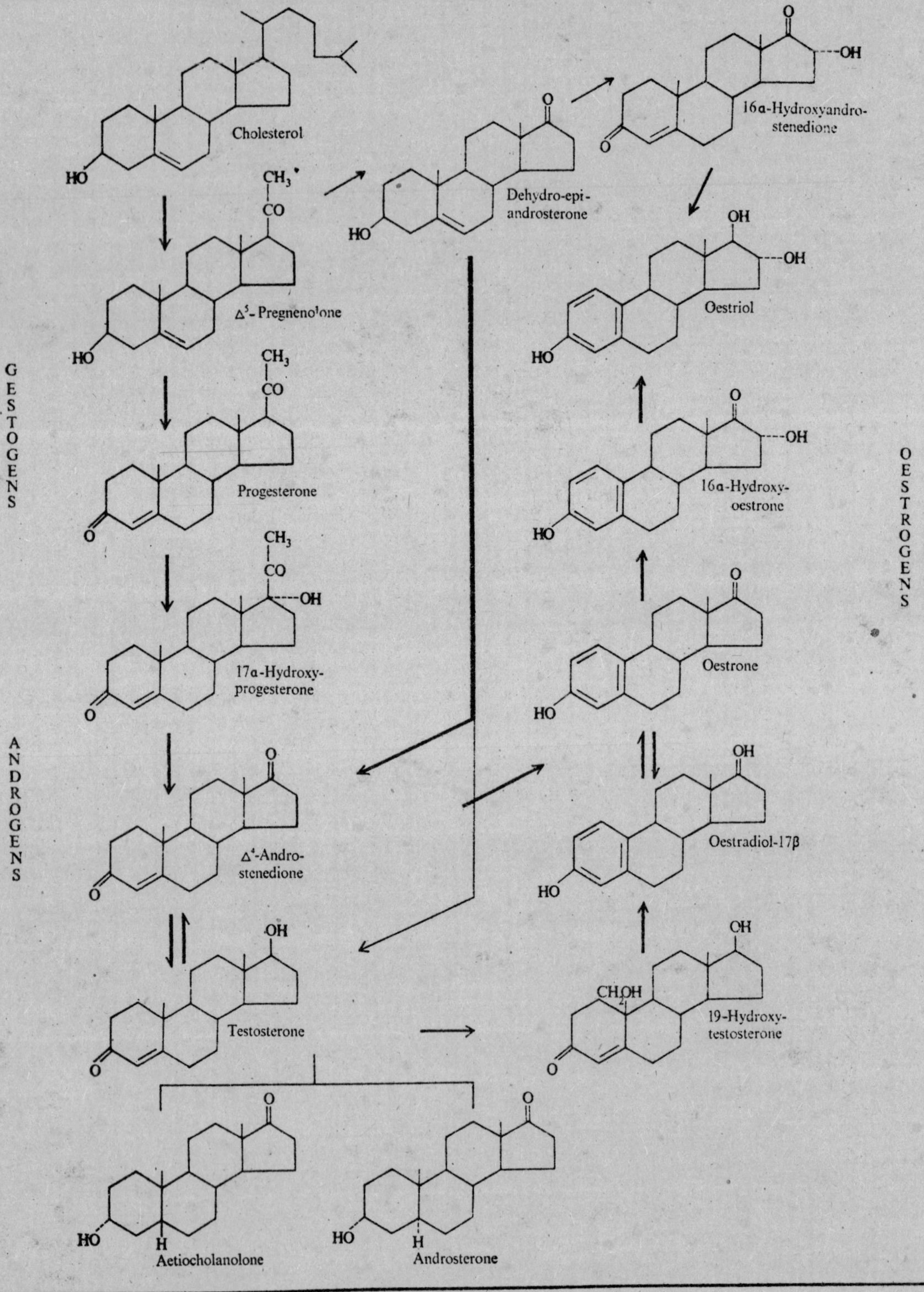

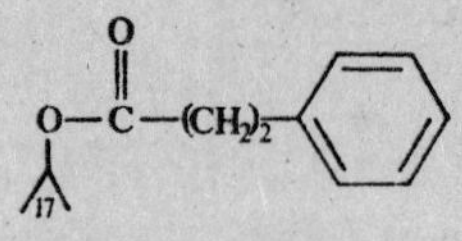

Phenylpropionate

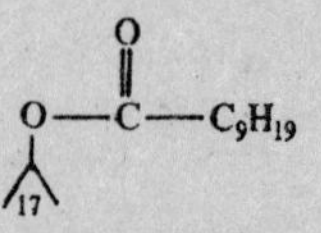

Decanoate

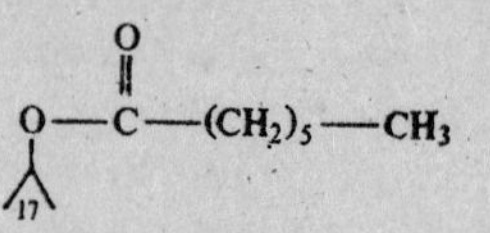

Oenanthate

Similar esters of oestradiol can likewise prolong the action but progesterone, a ketone compound with no alcohol groupings cannot be so esterified and radical manipulation of the molecule is required to obtain longer acting compounds: e.g. medroxyprogesterone acetate.

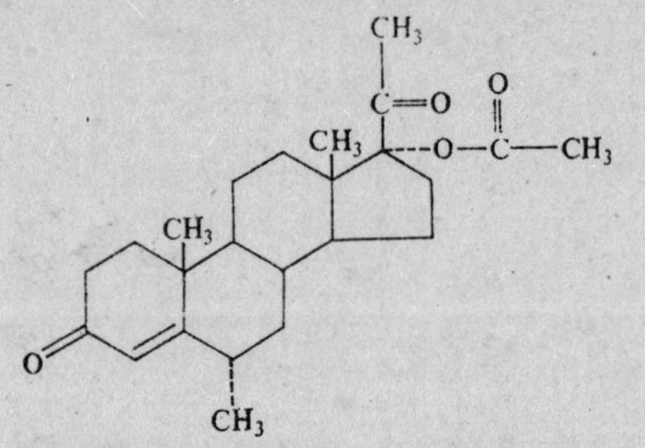

(6α-Methyl-17α-hydroxyprogesterone acetate)

However, progestogens with alcohol groups in the molecule are available and can be esterified to produce longer action of lynestrenol (ethinylestrenol) and lynestrenol phenylpropionate.

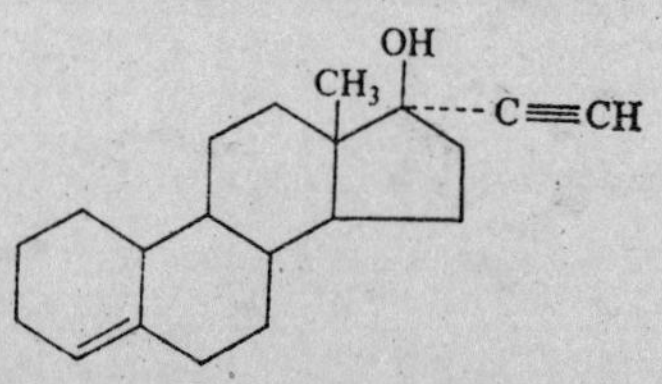

Lynestrenol (ethinylestrenol)

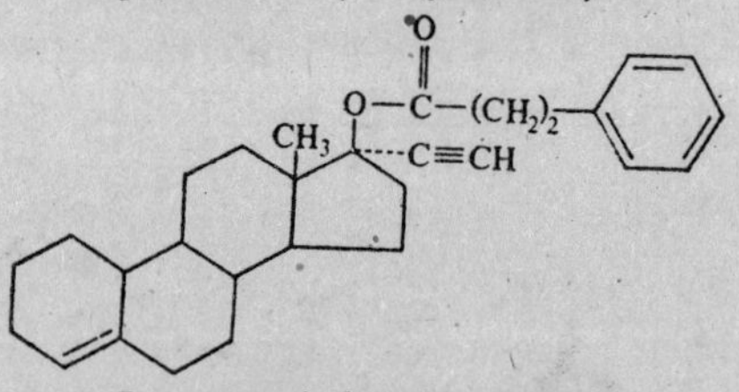

Lynestrenol phenylpropionate

17α Alkyl Derivatives

When steroids are administered orally the major part of the steroid is rendered innocuous in the liver and possibly the intestinal mucosa; very little survives in an active form for distribution to the body tissues. It is therefore usual to add an α alkyl group at C_{17}. This considerably increases the resistance of the compound to inactivation by the liver and ensures that a major portion of any oral dose reaches the general circulation in an active form. The principal alkyl groups are the methyl and ethyl radicals but ethers including ethinyl (also written ethynyl) and allyl unsaturated groupings are also used: these latter, however, by adding two *active* carbon atoms at C_{17}, modify the actions of androgens and related compounds to produce progestogenic effects.

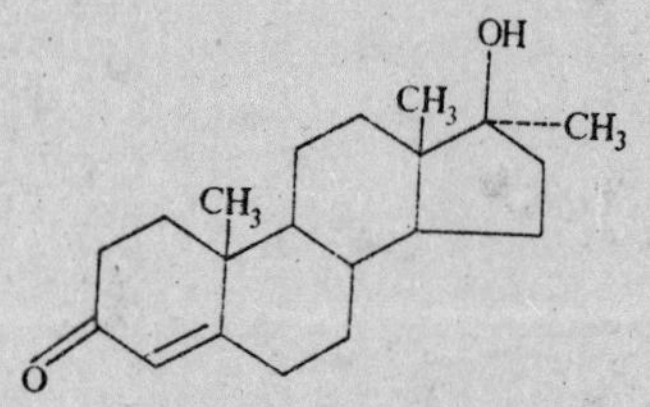

Methyltestosterone (androgen)

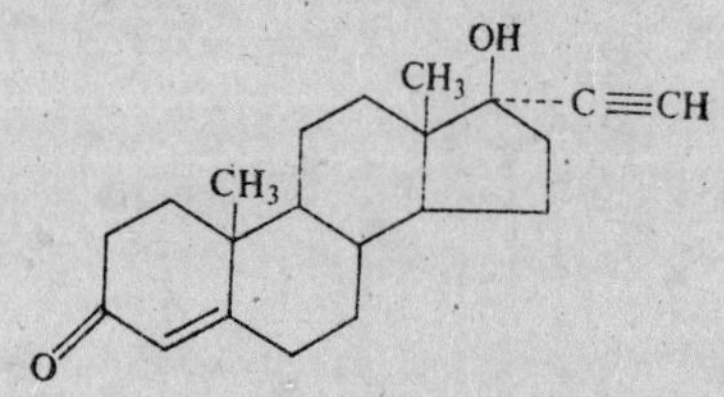

Ethinyltestosterone—ethisterone (progestogen)

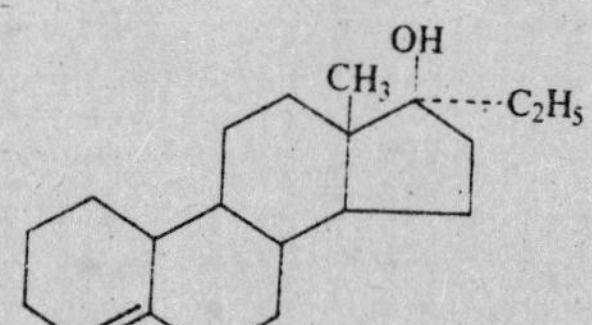

17α-Ethylestrenol (anabolic)

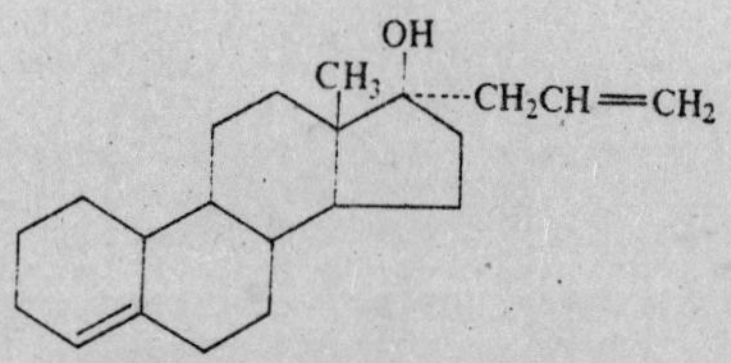

Allylestrenol (progestogen)

Whilst alkylation at C_{17} considerably increases the amount of active compound reaching target tissues from any one dose it is still necessary to increase oral daily dosage three or four times above parenteral dosages. Alkylation also increases the hepatoxicity of any given steroid.

General Properties

The steroid hormones and synthetic oestrogen compounds are white or near white crystalline compounds. In the main they are fat and oil soluble compounds which can be dissolved in ether or acetone and similar vehicles: dienoestrol and ethisterone are exceptions. In general these compounds are insoluble in water but soluble in propylene glycol and this solvent is used where a water miscible vehicle is required, especially for intravenous injection.

Most compounds are now manufactured synthetically and biological assay is therefore made unnecessary by the availability of chemical assay methods. Where the hormones are derived from natural sources biological assay methods may still be used.

Absorption of all steroids from the intestinal tract is relatively rapid and complete but, because of their susceptibility to hepatic degradation and conjugation are not suitable for oral therapy unless alkylated at C_{17}. If administered parenterally in aqueous suspension in a finely divided crystalline form absorption is rapid and complete within a few hours unless specially deposited in large crystals intramuscularly when absorption may take several days. If administered in oil the parent alcohols are rapidly absorbed but if esterified absorption takes place over a period determined by the length and complexity of the side chain (see Steroid Esters, p. 143). When administered in the form of solid, cylindrical pure crystalline implants absorption rates are related to the steroid and the implant surface area (approximately proportional to the weight of the implant) as well as the vascularity of the site of implantation. The approximate absorption rates are as in Table VIII.

The *oestrogens* in the female are entirely comparable with the effects of *androgens* in the male in so far as primary and secondary sex characteristics are concerned. These hormones are responsible for the correct growth and functioning of the primary and secondary sex characters but in the female the oestrogens collaborate with progesterone in the functional control of the uterus and other reproductive structures. They antagonize any tendency to sex reversal which may occur due to simultaneous production of either androgenic (in the female) or oestrogenic (in the male) substances in the ovary, testis and adrenal cortex. These antagonistic actions are utilized in the chemical castration of both male and female animals.

Both oestrogens and androgens increase the blood supply to certain tissues, increasing the nutrition and oxygen supply and improving tone and tissue activity; smooth and striated muscle, the brain, heart, kidneys, sex organs and the middle and inner ear are also affected.

Oestrogens and *progestogens* take part in mammary gland development and milk production but the importance of each substance varies in different stages.

Table VIII. Approximate average daily absorption and duration of steroid implants

Implant weight (mg)	Oestradiol		Progesterone		Stilboestrol		Testosterone	
	Daily intake (mg)	Duration (days)	Daily intake (mg)	Duration (days)	Daily intake (mg)	Duration (days)	Daily intake (mg)	Duration (days)
10	0·05	200						
15	0·07	210						
20	0·085	235						
25	0·1	250	0·16	150	0·3	80	0·40	60
50			0·25	200	0·5	100	0·70	70
100			0·40	250			1·10	90
500					1·25	400		

Relative Activity

The *oestrogens* are very difficult to compare, for each has activities greater in one respect than another and one hormone may therefore be more efficient for, say, suppressing lactation and yet relatively inefficient in initiating oestrus. A second hormone may have exactly the reverse potency. However, their relative activities when given orally in women have been generally assessed as: ethinyloestradiol 0·4, stilboestrol 3·0, oestradiol 13·4, and oestrone 31·5. Relative efficacies by the intramuscular routes were: oestradiol dipropionate 0·86, monobenzoate 1·06 and oestrone 3·0. When given orally, 1·0 mg stilboestrol is approximately equivalent in oestrogenic activity to 0·5 mg oestradiol monobenzoate given by intramuscular injection. By injection, stilboestrol dipropionate is slightly more active than oestrone given by the same route. Eight eostrogens were tested in spayed ewes using vaginal and behavioural changes as indicators. In both types of test oestradiol benzoate proved almost twice as active as stilboestrol, more than twice as active as hexoestrol, and about seven times more than dienoestrol.

Of the *androgens*, only testosterone or its esters and derivatives are sufficiently active for clinical use. The androgen unit is equivalent to the activity of 0·1 mg androsterone, and pure testosterone has a potency of 67 units per mg: it is thus nearly seven times more active than androsterone, which is one of its excretion products. Testosterone propionate has 50 units activity per mg. Methyl testosterone when given orally requires doses three to five times greater than those of testosterone propionate given intramuscularly.

When *progesterone* is given parenterally (the hormone is almost inactive when given orally) it is from five to ten times more active than ethisterone administered orally.

The Excretion of Steroids

Whilst a great deal of work has been carried out on the urinary excretion in man, relatively little has been carried out in the domestic animals. Little of clinical value has resulted from the work in either man or animals, although modern investigations using new techniques are gradually clarifying the situation.

All three groups of sex steroids are represented in the urine in one water-soluble form or another—sulphates and glucuronides mainly but in most species urinary excretion is of minor significance. The *oestrogens* are represented in most species by varying proportions of oestradiol (α and β forms) oestrone and oestriol in both male and female; other oestrogenic compounds occur in some species, e.g. equilin and equilenin in the mare. In the female of most species the amounts of oestrogen excreted increases progressively as pregnancy advances but exceptions occur; the sow shows a biphasic excretion curve over the pregnancy period. *Androgens* occur in the urine as 'neutral ketosteroids' typified by a double bond oxygen at C_{17} and the lack an aromatic ring A. They are derived from adrenal, ovarian and testicular androgens, as well as certain corticosteroids. A proportion of administered androgens is excreted as oestrogens. *Progestogens* are mainly represented by pregnanediol glucuronide in some species (woman and mare) but numerous related compounds which differ according to the species are also excreted.

The role of biliary and faecal excretion is now receiving a great deal more investigation and may prove of great significance in at least the dog and cattle.

Uses, Dosages and Administration

Although our knowledge of the physiological activities of hormones grows slowly, owing to the complexity of their interaction, the value of these substances in selected conditions becomes daily more apparent. Likewise, the penalties of injudicious use also become more evident.

Androgens

In the female androgens find little use except in the bitch where certain *mammary tumours* can be controlled and sometimes reduced in size. The dose of testosterone propionate is up to 5 mg every 2 days, of **testosterone propionate** 10 mg every 7 days. Where testosterone implants are preferred up to three implants of 100 mg. size may be required; these have an approximate duration of 3 months, but residual effects may persist much longer necessitating less frequent replacement than the duration of absorption infers.

Pseudopregnancy (*pseudocyesis*) in the bitch frequently responds to testosterone either as the propionate—up to 5 mg every 2 or 3 days—or the phenylpropionate—5 to 10 mg every 7 days; occasionally up to two implants of 50 mg size may be used for the 70 day duration of activity. It would appear, however, that a combination of androgen and oestrogen provide a greater anterior pituitary suppressant effect: a combination of **methyltestosterone** and ethinyloestradiol in tablet form has been very successful when given in doses of up to two tablets together three times daily for about five days. The dosage will depend on the size of the animal but more especially on the severity of the symptoms. Oestrus inductions by this combination occur only rarely.

Oestrus suppression in the bitch and queen cat can be achieved by testosterone propionate (up to 5 mg every 2 to 3 days) by **testosterone phenylpropionate** (up to 10 mg every 7 days) or by testosterone implants (up to two implants of 50 mg) but is probably less satisfactory for this purpose than progesterone. In the *male* androgens are of much greater use.

Ageing and debility associated with ageing, can be effectively countered in all species by the administration of the short or long acting injections or by the use of implants. In small animals methyltestosterone may be used in a daily dosage of up to 30 mg either to establish diagnoses or as standard treatment. Whilst the propionate can be used in the stallion, gelding and bull in doses of about 50 mg every 2 or 3 days and in the ram, goat and boar in doses of up to 20 mg every 2 or 3 days, it is more usual to reserve this preparation for dogs and cats in which the dosage is up to 5 mg every 2 or 3 days. Testosterone phenylpropionate dosage for the stallion, gelding and bull is up to 50 mg every 7 to 10 days: dosage for the ram, boar, billy-goat, dog and cat is up to 10 mg every 7 to 10 days. Testosterone implants are perhaps more suitable a method of treatment as the 25 mg implants have approximately a 60-day, 50 mg a 70-day and 100 mg a 90-day absorption period. Residual effects persist for several weeks after the implant has all been absorbed. Stallions, geldings and bulls require up to five implants of 100 mg, rams, goats, boars and dogs up to two of 50 mg and cats up to two of 25 mg. These are maximum doses which should be reduced according to the size of the animal and severity of the condition.

Alopecia of hormonal origin in the male of small animals may be treated with the same preparation and at the same dosage rates as used for ageing and debility but, additionally, where large bare skin areas exist testosterone ointments and creams can be used. This topical therapy may sometimes be used alone as absorption.

Balanitis in castrated lambs and yearling sheep has been extensively treated in Australia as a palliative measure and as a preventive with implants of testosterone propionate (25 to 50 mg).

Cryptorchidism which has failed to respond to chorionic gonadotrophin will sometimes respond to testosterone therapy. This is especially so when *hypogonadism* is also present. However, as testosterone speeds up the rate of maturation processes—closure of epiphyseal plates, constriction of the inguinal ring, etc., the possible trapping of the undescended testicle in the canal, amongst other dangers, e.g. stunted growth, must be considered before resorting to testosterone. Dosages are less than for most other conditions but must be continued for longer periods. Up to 25 mg twice weekly of the propionate and 20 mg weekly of the phenylpropionate are usual in the foal and young bull. If implants are used, up to two of 100 mg can be used. In the ram lamb, young goat and boar up to 10 mg of the propionate twice weekly or up to 20 mg every 10 to 14 days should be sufficient and up to two implants of 50 mg. In the dog 2·5 mg twice weekly of the propionate and up to 10 mg of the phenyl propionate every 14 days is adequate; no more than two implants each of 2·5 mg should be used.

Deficient sex drive in adult stud animals can be countered by testosterone. All non-hormonal causes should first be eliminated and the more physiological approach of chorionic gonadotrophin tried first. This latter therapy mainly works by stimulating the testicular production of testosterone. Dosages similar to those for ageing and debility should be tried. Whilst some depression of sperm production may occur this is rarely of practical significance and recovery, often to levels above those prior to treatment, follows cessation of therapy.

Feminization, sometimes seen at puberty in dogs but more frequently in those affected with testicular tumours later in life, testosterone in doses similar to those used in cryptorchidism can be used.

Urethal calculi of a sabulous type in the castrate male cat may be prevented from recurring by testosterone therapy. Oral therapy with up to 5 mg daily of methyltestosterone or parenteral therapy with implants in doses of one implant of 25 mg once every 3 to 5 months is usually adequate.

Dosage

Female:

Bitch
- Methyltestosterone, 5–30 mg daily in divided doses
- Testosterone Propionate (Test. P.P.), 2·5–5 mg every 2–3 days
- Test.P.P., 5–10 mg every 7–14 days
- Implants, 25 mg to two × 50 mg or exceptionally (mammary tumours) 100 mg to three × 100 mg

Queen Cat
- Methyltestosterone, 5–10 mg daily or on alternate days
- Testosterone Propionate, 5 mg every 2–3 days
- Test.P.P., 10 mg every 10–14 days
- Implants, 50 mg to two × 50 mg

Male:

Stallion and Bull
- Testosterone Propionate, 25–50 mg twice weekly
- Test.P.P., 20–50 mg every 7–14 days
- Implants, two × 100 mg to five × 100 mg

Ram, Goat and Boar
- Testosterone Propionate, 10–20 mg twice weekly
- Test.P.P., 10–20 mg every 7–14 days
- Implants, 25 mg to two × 50 mg

Dog
- Methyltestosterone, 5–30 mg daily or alternative days in divided doses.
- Testosterone Propionate, 2·5–5 mg twice weekly
- Test.P.P., 5–10 mg every 7–14 days
- Implants, 25 mg to two × 50 mg

Cat
- Methyltestosterone, 5–10 mg daily or alternate days
- Testosterone Propionate, 2·5–5 mg twice weekly
- Test.P.P., 5–10 mg every 7–14 days
- Implants, 25 mg to two × 25 mg

Oestrogens

Oestrogen sensitivity varies according to species, the order of sensitivity being from the most sensitive to the least: cow, sheep, pig and horse. It follows that whilst oestrogens should be used with care in all species, they should be used with particular care in those species which head the sensitivity list.

Oestradiol monobenzoate, stilboestrol and **stilboestrol dipropionate,** are the most commonly used oestrogens in veterinary medicine, although **dienoestrol** is increasingly used. **Hexoestrol** is also used, but caponization of poultry and fattening of cattle are the only purposes for which it has found popularity.

In the mare, oestrogens have been little used. No doubt the danger of the induction of nymphomania has been responsible. Nevertheless, where follicular activity is present but the *signs of oestrus* are slight or absent, oestradiol monobenzoate in single doses of 5 to 10 mg for maiden mares, and 10 to 15 mg for barren mares, given intramuscularly as the follicle increases to its maximum size, may induce *ovulation* and oestrus. The subsequent oestrus period is more likely to result in ovulation. Dienoestrol has been used in single 5 mg doses subcutaneously, and stilboestrol, as its dipropionate ester, has also been used. Service should follow within 12 hr of the injection and should be repeated in 24 hr if possible. In general, mares respond more effectively to the administration of follicular stimulating and luteinizing hormones, and warm, boiled water flushing of the uterus is effective, causing oestrus with ovulation in 3 to 5 days.

Two 0·5 g or 1 g implants of stilboestrol have been used in the treatment of *vaginal discharges* associated with 'bunch of grape ovaries' and, usually, irregular cycles. The implants have been left in situ for 2 to 2½ months. The same sized implants have also been utilized to reduce enlarged ovaries to a size convenient for removal *per vaginam.*

In cattle, oestrogens have been far more widely used. *Anoestrus* or *suboestrus* may be treated with small doses of oestrogens in the same way as described for the mare, but, as with the mare, the hormones which stimulate the ovary to increased function are far superior and are less likely to result in cystic ovaries and nymphomania. Large doses of oestrogens can produce a prolonged anoestrus period once the initial oestrus symptoms due to the treatment have regressed.

The relaxing effect on the cervix and the sensitizing of the uterus to the ecbolic effects of posterior pituitary extracts has led to the use of oestrogens as an aid in removing *retained placentae* and for *mummified foetus* expulsion. It has similarly been used in metritis and pyometra. Doses in these cases have been 15 and 25 mg of oestradiol monobenzoate given intramuscularly, once in the first two conditions, but repeated weekly for 2 or 3 weeks in the last condition. Stilboestrol has been used in similar doses. Controversy still exists with regard to the use of oestrogens in these conditions and their use for these purposes is far less frequent than formerly. It is probable that the large doses by producing prolonged anoestrus may do more harm than good.

In fattening steers, stilboestrol has been used for *growth promotion* Hexoestrol is similarly. active in about the same dosage. If a 50 to 60 mg implant of either of these compounds is inserted subcutaneously at the base of the ear using aseptic precautions some 3 months before the anticipated slaughter date, up to 25 per cent increase in killing-out weight can be achieved but not entirely without impairment of quality. Because of the public health danger, implants should not be inserted in any place other than the ear, for they are normally discarded with the ear at the time of slaughter. Implantation of heifers and cows is not recommended because of side actions, particularly related to nymphomania, vaginal prolapse, faulty calcium metabolism, which, with the nymphomania—and sometimes without—results in pelvic and vertebral fractures. The doses recommended in the UK and elsewhere in earlier days (1 g) are now regarded as too high and should not be used. In the United States, stilboestrol was used as a feed supplement for fattening beef cattle. The use of oestrogen implants and feed supplements has now been banned in some countries, but deleterious effects of meat from treated animals have still to be demonstrated.

In sheep, perhaps the most important use of oestrogens is the *induction of early oestrus*, with lambing antedated by as much as 2 months. By this means, and by inducing oestrus in the post-lambing anoestrus period, two crops of lambs per year have been obtained. The financial returns, however, are not great when set against the added workload. The doses used for this purpose have been 1 mg of oestradiol monobenzoate, preferably given as four daily 0·25 mg intramuscular doses, and no doubt dienoestrol would be similarly effective. Previous attempts to induce early oestrus which utilized progesterone in 25 mg doses on alternate days for three to five doses followed by up to 1000 units serum gonadotrophin (PMS) on the 9th day after the first injection and again 3 days later gave good results without any deleterious side actions. More recent work has shown that three doses of 50 mg progesterone (*q.v.*) followed by one of 750 i.u. PMS is one of the most satisfactory methods of inducing oestus in anoestrus sheep so far devised.

Oestrogens have also been used for fattening sheep, but have proved less successful than in poultry and steers and are not recommended for this purpose. The implant has been of 12 to 15 mg size.

Goats may be given oestrogens for the same purposes and in the same dosages as for sheep. Oestrus induced by oestrogens is not usually fertile, but oestrus subsequent to the artificially induced heat should be fertile.

In sows, *oestrus induced* by 5 to 10 mg oestradiol monobenzoate or similar doses of stilboestrol dipropionate, both given intramuscularly, is usually fertile and the animal should be served as soon as she comes into season. Double service with an interval of 24 hr between the two is advisable.

In the bitch, the main indication for oestrogens is in the treatment of '*mesalliance*'—unintentional mating. For this purpose, stilboestrol dipropionate or oestradiol monobenzoate, given subcutaneously or intra-muscularly in doses of 10 to 15 mg, will prevent pregnancy. The maximum interval after mating frequently stated is 4 days, but experience indicates that a delay of up to 10 days is not detrimental to the effect. A great disadvantage is that bitches may remain in oestrus for up to 6 weeks, and pregnancy has been known to follow service late on in this period. It may be that injected androgens can be used instead, as with the suppression of pseudopregnancy. This would be of great value in small animal practice. Orally administered oestrogens are of no value in averting pregnancy, even in 30 mg doses (stilboestrol).

Oestrogens in doses of 1 to 3 mg daily by mouth or 0·5 to 1 mg injected daily—can be used in fat bitches where the fat is laid down due to *hypogonadism*. It is also a value for *urinary incontinence* in old or spayed bitches. Oestradiol monobenzoate in doses of 1 mg injected daily for 3 days followed by 1 mg every third day as required can be used.

In the dog, oestrogens are of value for the treatment of *anal adenomata*, especially where ulceration has occurred, and for causing a regression of *hypertrophied prostrate glands*. For anal adenomata 5 to 10 mg stilboestrol dipropionate subcutaneously once a week is effective, but causes some increase in thirst and urination: frequently an injection is used initially and is followed by 1 to 5 mg daily by mouth. This latter course is better in the frail old terrier, where overdosing can cause rapid collapse, and death within 24 to 48 hr. For hypertrophied prostrate glands, 3 to 5 mg stilboestrol by mouth daily is usually adequate. This treatment can be successfully replaced by the use of 15 mg stilboestrol implants or preferably 10 to 20 mg of implanted oestradiol.

Excessive libido in dogs, particularly young dogs, can be treated with oral oestrogens in daily doses of 1 to 3 mg, by injection of daily doses of 0·5 to 1 mg or, in persistent cases by implants of 10 mg, when some feminization may occur.

Lactation in bitches and cats can be diminished by the oral use of oestrogens—10 mg twice with a 12 hr interval in the bitch, and 3 to 5 mg similarly in the cat. Androgens are to be preferred.

Poultry were frequently *caponized* (young and old cockerels) by implantation of 15 mg stilboestrol or combinations of 7·5 mg each of stilboestrol and hexoestrol. The implants were introduced at 8 weeks or later (3 to 4 weeks before anticipated slaughter dates is said to give optimal results), subcutaneously in the neck just below the head so that at slaughter any remaining portion of the implant was discarded with the head.

Dosage

Mare:
- Oestradiol and Stilboestrol injections, 5–15 mg
- Dienoestrol injection, 5 mg
- Stilboestrol implants, one to two × 500 mg

Cow:
- Oestradiol and Stilboestrol injections, 10–25 mg

Steers:
- Oestradiol and Stilboestrol implants, up to 60 mg

Sheep and Goat:
- Oestradiol and Stilboestrol injection, 0·25 mg

Sow:
- Oestradiol and Stilboestrol injection, 5–10 mg

Bitch:
- Oestradiol and Stilboestrol injection, 0·5–15 mg

Dog:
- Oestradiol and Stilboestrol injection, 0·5–10 mg
- Oestradiol and Stilboestrol implants, 10–20 mg

Poultry:
- Stilboestrol implants, 15 mg
- Stilboestrol/Hexoestrol, 7·5–15 mg

Progestogens

The use of progesterone and related compounds in veterinary medicine is not as yet very extensive but some novel applications have been developed in recent years.

Abortion of a habitual nature and threatened abortion, where infective agents and other non-hormonal causes have been excluded, are indications for the use of progesterone and similar compounds. In the cow 100 mg, the ewe, goat and the sow about 10 mg daily are required to maintain pregnancy in the absence of the ovaries. However, it is unlikely that ovarian progesterone is entirely absent in cases where abortion occurs and all that is required is supplementation. In the mare and cow up to four 100 mg progesterone impants are required in the first 6 months and up to three of 100 mg in the last 3 months of pregnancy. In repeat-breeder cows one to two 100 mg implants inserted during the first week after service can be very successful in preventing early abortion. In the bitch up to four 25 mg progesterone implants may be used but 2·2 to 3·3 mg per kg (1 to 1·5 mg per lb.) body weight daily is often effective in threatened abortion if continued for 10 days or so after the disappearance of symptoms. All implants should be removed about 2 weeks before foaling or calving or about 3 days before whelping.

The *deferment of oestrus* has been widely investigated. Daily injections of progesterone in doses of 100 mg for the cow and sow, and 10 mg for the sheep and goat can be used to defer oestrus. This is used to *synchronize* oestrus in these species where controlled breeding programmes are required, especially when artificial insemination is in use in large herds and flocks. Oestrus follows in the majority of animals within 2 to 4 days of cessation of therapy. Injections should commence on or before the fifteenth day of the cycle during the luteal phase following the previous ovulation. Fertility is depressed in the sow with a 100 mg daily dose; less produces a high proportion of cystic ovaries. The application of these techniques to animal production nowadays rests on long-acting or orally administrable progesterone analogues. In sheep an easily absorbed progestogen administered by vaginal tampon is available as a very effective and relatively convenient method of synchronizing oestrus. In the cat oestrus can also be deferred indefinitely as long as injections of 2·5 to 5 mg progesterone are given every 2 to 3 days starting 5 days or so before oestrus is anticipated; this is not an acceptable method for extended deferment because of the necessity to repeat the injections.

Suppression of oestrus presents greater problems than temporary deferment. These have been receiving increasing attention over the years. In cattle the fattening of heifers is complicated by the repeated incidence of oestrus; mixtures of long acting testosterone esters, progesterone derivatives and long acting oestradiol esters have been used successfully in fattening heifers but such preparations have not been accepted to any degree outside the United States. In the bitch the problem is not satisfactorily settled although preparations are now marketed for this purpose; high doses of progesterone over a period will result in pyometra and it would seem that at least some parenteral long acting progestogens produce similar changes in the uterus. However, there are indications that certain oral progestogens given in small daily doses can achieve safe oestrus suppression as long as therapy is continued. It is, however, impossible to forecast the reaction of any individual to such therapy and the effects vary according to the substance used. Residual effects may persist for some time —extended anoestrus due to anterior pituitary suppression is common. There seems to be no ill effect on post therapy breeding in the bitch, but in cats similarly treated the first litters post therapy contain many weakling and stillborn kittens. Oestrus in bitches and cats can be terminated when present by injections of progesterone; 25 to 50 mg for the bitch repeated the next day if necessary and 12·5 mg for the cat repeated in a similar manner. Cats will return to oestrus after a normal interoestrus period but it is not easy to forecast when the next oestrus period will occur in bitches. It should be remembered that the uterus during the luteal phase is perhaps more susceptible to infection and that continuous progesterone therapy may predispose animals to uterine infection although *pyometra* in cattle following stilboestrol therapy may respond to 100-mg doses on 3 successive days. Oestrus in spayed animals which may be due to aberrant ovarian tissue becoming active can be terminated with progesterone; those which do not respond to progesterone should be given triple doses of corticosteroid as many are the result of sex steroid production by the adrenal gland.

Induction of oestrus with ovulation in the ewe is not possible with follicular stimulating hormone alone but requires 'priming' doses of progesterone. Gordon followed previous extensive work, with the results of some elegant trials which indicated that 50 mg progesterone injected on three occasions at 3-day intervals followed by 750 i.u. of serum gonadotrophin 2 days after the last progesterone dose give the best chance of success in anoestrus ewes. A further approach also in sheep used an early absorbed progestogen administered by vaginal tampon to achieve the 'priming' effect and simplify the handling problems.

Dosage

Mare:
  Implants 100 mg to four × 100 mg
Cow:
  Injection Progesterone: 100 mg
Sheep:
  Injection Progsterone: 10–50 mg
Bitch:
  Injection Progesterone: 25–50 mg (2·2–3·3 mg/kg, 1·0–1·5 mg/lb. body weight)
  Implants: 25 mg to four × 25 mg
Cat:
  Injection: 2·5–5 mg exceptionally 12·5 mg

Choice of Pharmaceutical Formulation

With such potent substances as the sex steroids it must be borne in mind that once injected the substances may have residual effects which persist long after the steroids have been excreted. Also with implants it is often difficult to find them after subcutaneous insertion. It is therefore advisable, where the diagnosis is in doubt, to use oral or short acting injection therapy before proceeding to use long acting injections and implants.

Toxicity

Androgens rarely give rise to toxic symptoms and it is unlikely that signs other than increased libido will be noticed in the male. In the female suppression of oestrus may occur and because of the danger of a masculinizing of the foetus it is unwise to use androgens in pregnant females. *Oestrogens* on the other hand can cause severe acute and chronic toxicity in the male: collapse and death within 24 to 48 hr may occur but when animals live longer the initial signs are intense thirst and polyuria along with lethargy followed by collapse and death; erythema with intense irritation and feminization follow polydipsia and polyuria in sub-acute toxicity and chronic toxicity. In the female the major signs of overdosage are persistent oestrus, followed by anoestrus, the former often associated with cystic ovaries, the latter with ovarian hypoplasia. *Progestogen* overdosage effects in the male are unknown but overdosage in the female results in masculinization from which recovery takes place over a period if dosage is discontinued.

Occasionally individuals may become narcotized for several hours. This has been seen in dogs with nandrolone and progesterone compounds.

The Gonadotrophins

The anterior pituitary gland in the normal female is responsible for the cyclic changes of the ovary from follicular phase to luteal phase and back again. In the male the cyclic changes are not apparent but specific effects on androgen production and on sperm formation are exerted by analogous hormones. The two hormones involved are the *follicle-stimulating hormone*, which stimulates follicular development in the female and sperm development in the male, and the *luteinizing hormone* (female) or *interstitial cell stimulating hormone* (male), which is responsible for final maturation of the ovum and for ovulation and development of the corpus luteum in the female and the production of testosterone by the interstitial cells of the testes in the male.

These two hormones have been isolated from pig and sheep pituitary glands, but this source is not economically feasible. However, it has been found that similar hormones are produced by the chorion and excreted in the urine of pregnant women (luteinizing hormone or chorionic gonadotrophin) and by the mare, arising from special areas in the endometrium, and obtained from the serum between the days 50 and 80 of pregnancy (follicle-stimulating hormone, pregnant mare's serum, or serum gonadotrophin). Both preparations possess mixed activities, i.e. LH and FSH activity. Attempts to separate these activities remain unsuccessful and it is now believed that mixed activity is inherent in the large glycoprotein molecules of FSH and LH.

Both these gonadotrophins are white, though the chorionic one may be fawn, amorphous powders, which are soluble in water. They are kept in sealed ampoules as powders for solution in water immediately before use. The powders are assayed biologically: **Serum Gonadotrophin,** B.Vet.C. should contain not less than 100 units of follicle-stimulating activity per mg, while the **Chorionic Gonadotrophin,** B.Vet.C. should contain not less than 400 units luteinizing activity per mg. Once reconstituted the less stable PMS should be used within a few hours, but LH is usually stable at room temperatures for several weeks.

Indications

In horses, the gonadotrophins are used mainly for achieving ovulation in the female, for the treatment of cryptorchidism in the foal and yearling, and for deficiencies in sperm production and sex drive in the stallion.

In mares which fail to come into oestrus and yet on examination have follicles present in one ovary, oestrogens, in the form of stilboestrol, oestradiol monobenzoate, or dienoestrol, in doses up to 15 mg intramuscularly, can be used and sometimes cause spontaneous *ovulation* with an oestrus period of 24 to 48 hr. This is uncertain but the administration of chorionic gonadotrophin in doses of 1500 i.u. given subcutaneously or intramuscularly, repeated if necessary 2 or 3 days later, is a more reliable treatment; oestrus usually follows, with ovulation about 24 hr after the second injection. An alternative is to give serum gonadotrophin in doses of 2000 i.u. intramuscularly daily for 4 days, repeating the treatment 14 days later if necessary. Ovulation may not be accompanied by oestrus if the intravenous route is used for the administration of chorionic gonadotrophin, but follows after a normal inter-oestrous period of 16 to 21 days.

In *prolonged oestrus*, usually met in the early part of the season, and in mares where early ovulation is required to give the best chance of conception, chorionic gonadotrophin may be given intravenously on the day of service, but it is essential to ensure that a follicle of not less than 3 cm in diameter is present in the ovary. Doses may be varied from a minimum of 1500 i.u., but where a number of follicles are present making a grape-like ovary, from 3000 to 5000 units should be given. Ovulation should follow within 40 hr.

Nymphomania in mares may be treated with chorionic gonadotrophin given intravenously in doses of 2000 to 3000 i.u. and repeated 1 month later. The rationale of this treatment is not apparent, for it has been pointed out that nymphomania in the mare is not associated with cystic ovaries.

Where *lactation* has failed after parturition, luteinizing hormone in doses of 2000 to 3000 units may be given and repeated 24 hr later when necessary.

In the male, *cryptorchidism* in the foal or yearling can be treated with luteinizing hormone given intramuscularly twice weekly for 4 to 6 weeks, and where this is associated with *genital hypoplasia* 200 mg of testosterone implanted subcutaneously should be of help. It is important to make sure that there is no mechanical obstruction to testicular descent before commencing treatment. There are, of course, ethical and genetic reasons why treatment of any description, short of castration, should be avoided.

In the stallion which is '*slow*', luteinizing hormone may be used to stimulate testosterone production by the interstitial cells of the testicle. By using this hormone there is less likelihood of injuring spermatogenesis, for the small contaminating amounts of follicle-stimulating hormone should likewise stimulate *spermatogenesis*. The use of testosterone implants in this context has been recommended and it is probable that small implants would be satisfactory. A 100-mg implant, for example, should provide a daily intake of 1·1 mg testosterone over a period of about 90 days. These figures will vary with the individual and site of implantation, but they may be taken as an approximate guide. Up to 500 mg have been suggested as a dose. The possible effect on sperm production is not exactly known, but it has been stated (B.P.C., 1954) that small amounts of testosterone will stimulate sperm production. Too much reliance should not be placed on this statement, for it is definitely known that reasonable doses have reduced sperm production.

Impaired spermatogenesis can be treated with serum gonadotrophin in doses of up to 1500 units twice weekly by intramuscular injection over a period of 3 to 6 weeks.

In the gelding, where *premature ageing* can be troublesome, testosterone implants in doses up to 400 mg will prove adequate for periods of about 3 months.

In cattle, the indications for the use of gonadotrophins are similar to those in the horse.

In the cow luteinizing hormone (chorionic gonadotrophin) may be used in *suboestrus*—blind bulling or silent heat—by intramuscular injection, in ovulatory failure or to ensure ovulation given by intravenous injection during oestrus, and in place of progesterone in early *non-specific abortion*, when weekly intramuscular injections are required for the first month or so after service. Doses of 1000 to 1500 i.u. are required in these cases. In *nymphomania*, associated as it usually is with cystic ovaries, 1500 to 3000 units given intramuscularly and repeated after a month may help, especially if assisted by expressing or puncturing the larger cysts. However, a combination of an aqueous suspension of progesterone (100 to 120 mg) and a solution of luteinizing hormone (1500 i.u.) administered intravenously has achieved good, and often spectacular results in returning the animals to normal and fertile cycling. A second injection a few days later may sometimes be needed but the proportion of cases needing a second dose is small. Where infertility is a herd problem and not that affecting a few individuals only this type of treatment is often less than satisfactory and it may be assumed that these herd problems are related to genetic, nutritional or similar non-hormonal factors. Chorionic gonadotrophin may also prove effective in the treatment of *metritis* subsequent to parturition and complicated by failure to involute. Intramuscular injections daily, or on alternate days, of 1000 to 1500 units should be continued until the discharge clears.

In the bull chorionic gonadotrophin in doses of up to 5000 i.u. given by intramuscular injection twice weekly will promote the *descent of the testicles*. Treatment should continue for 4 to 6 weeks, when a short period of a few weeks should be allowed before repeating the treatment. This is necessary for observation of results and to allow the stimulatory effects on the rest of the hormonal system to be dissipated. In *genital hypoplasia*, up to 200 mg testosterone in the form of implants can be used. In lazy bulls, where other causes have been eliminated, chorionic gonadotrophin in doses of up to 5000 i.u. can be given, or up to 500 mg of testosterone may be implanted but the possible effects on spermatogenesis must be considered, especially in the latter case. Where *infertility* is related to the quantity and quality of spermatozoa produced, up to 1500 units of serum gonadotrophin (PMS) possibly in conjunction with equal doses of chorionic gonadotrophin (LH) may be given twice weekly by intramuscular injection for up to 6 weeks.

In sheep, where *twinning* and triplets are, apart from the possibility of pregnancy toxaemia, less dangerous and sometimes even desired by the farmer, the use of serum gonadotrophin will increase the production of follicles and improve the conception rate. Single doses of 500 to 1000 i.u. should be given intramuscularly between days 12 and 15 of a 3 week cycle. *Cryptorchidism* may be treated as in other species, using doses of 500 i.u. chorionic gonadotrophin, whilst deficient sex drive requires similar doses of the same compound given in the same way. Testosterone implants used for this latter purpose should not exceed 100 mg in size.

Billy-goats may be similarly treated. In the nanny out of season, induced oestrus will follow the injection of two intramuscular doses of 1000 i.u. serum gonadotrophin given with an interval of 24 hr—oestrogens have also been mentioned in this context. Where non-ovulation is suspected as the cause of returning to service, up to 500 i.u. of chorionic gonadotrophin may be given on the day of service.

In pigs the details are similar to those for sheep

and goats. *Anoestrus* in the sow may be terminated with 1000 units of serum gonadotrophin intramuscularly (or oestrogens); *non-ovulation* can be treated with 500 i.u. on the day of service, and *failure of lactation* after farrowing may be countered by two 500 i.u. subcutaneous doses of chorionic gonadotrophin given with a 24-hr interval: pituitary injection (posterior lobe) will give more rapid response but the effect may not be so sustained. In the boar, chorionic gonadotrophin is suitable for treating *cryptorchidism* and deficiency in sex drive, using the doses given for the ram, whilst *spermatogenesis* may be stimulated by using alternate 500 i.u. doses of chorionic and serum gonadotrophins at weekly intervals.

In the bitch, ovulation may be induced by two intramuscular doses of chorionic gonadotrophin given with an interval of 2 days between doses. The usual maximum dose is 500 i.u. *Oestrus* may be induced by the injection of up to 200 i.u. of serum gonadotrophin daily for 3 weeks, followed by up to 300 i.u. of the chorionic gonadotrophin on the day of service. *Failure of lactation* after parturition may be treated with up to 500 i.u. of chorionic gonadotrophin given intramuscularly and repeated 24 hr later if necessary. Chorionic gonadotrophin may also be used in *pyometra*, but the effects are not sufficient to merit an unreserved recommendation: doses used have not exceeded 500 i.u. daily. Chorionic gonadotrophin can be used to *terminate* or *suppress oestrus*, using up to 250 i.u. intramuscularly: progestational agents are probably superior.

In the dog, *cryptorchidism* responds very well to chorionic gonadotrophin in doses up to 500 i.u. twice weekly for up to 6 weeks, but not infrequently spectacular results have occurred after the first two or three injections. If *genital hypoplasia* also exists, testosterone propionate in 5 mg doses intramuscularly twice a week will be beneficial. Stud dogs which are sluggish and when non-hormonal causes have been eliminated may, as in other species, benefit from chorionic gonadotrophin in 200 i.u. doses twice weekly for up to 6 weeks; testosterone implants of from 25 to 100 mg sizes may also be used, but it is likely that implants of 100 mg will be too large and 50 mg should normally not be exceeded.

Pituitary Inhibitors

In addition to those drugs which reproduce the effects of anterior pituitary gonadotrophins, various compounds are used to control animal breeding because they reduce the secretory activity of the pituitary gland. Progesterone and long-acting progestogens such as medroxyprogesterone or megestrol are able to defer or suppress oestrus because they limit FSH and LH release via the normal feed-back mechanism which controls the oestrus cycle. Although successful in the bitch and in ruminants, synchronization of detectable oestrus in the sow is more difficult with steroids, as it requires phased administration of progestogen, FSH and oestrogen.

Clomiphene

This compound can be shown to be both mildly oestrogenic and more strongly anti-oestrogenic. More importantly, it also modifies the secretion of pituitary gonadotrophins. It is believed to compete with oestrogens for the hypothalamic binding site or receptor via which the oestrogen feed-back mechanisms functions. This interaction is believed to modify LH–RH, FSH–RH and pituitary gonadotrophin release. In man this action is used in the treatment of anovulatory infertility.

Gonadotrophin Releasing Hormones

The chain of command in the hormonal regulation of the oestrous cycle contains several links. In a normal animal, it is possible to modify reproductive function by manipulating the contribution made by each link. The steroid hormones are the final specific links although even their action may be shown to be mediated via an intracellular 'second messenger'. The steroids are released from the primary sex organs in response to stimuli provided by glycoprotein endocrine hormones released by the anterior pituitary. These trophins are available as therapeutic agents in the form of PMS (FSH) and HCG (LH). It is now known that anterior pituitary gonadotrophins are themselves synthesized and released in response to polypeptide releaser hormones which arise in the hypothalamus.

These are neurohormones with a variety of activities about which knowledge is as yet incomplete. There appears to be just one regulatory factor for both FSH and LH which, in sheep and pigs, is a decapeptide. No release-inhibiting factor such as exists for growth hormone and prolactin has yet been isolated. However, synthetic LH–RH/FSH–RH analogues are already available and have undergone evaluation. Some evidence is available to support the proposition that such a substance will be of value in inducing ovulation and normal luteinization of the follicle and also that by intravenous or intramuscular injection it will be of value in bringing about a physiological luteinization of cystic follicles in cattle.

Bibliography

BRENNER, R. M. & WEST, A. B. (1975) Hormonal regulation of the reproductive tract in female mammals. *A. Rev. Physiol.*, **37**, 273–303.

COLE, H. M. & CUPPS, P. T. (1969) *Reproduction in Domestic Animals*, 2nd ed. New York: Academic.

HAWKINS, D. F. (1974) The pharmacology of the pregnant human uterus. In: *Obstetric Therapeutics*, pp. 1–71. London: Baillière Tindall.

JÖCHLE, W. (1973) Corticosteroid-induced parturition in domestic animals. *A. Rev. Pharmac.*, **13**, 33–57.

KADOWITZ, P. J., JOINER, P. D. & HYMAN, A. L. (1975) Physiological and pharmacological roles of prostaglandins. *A. Rev. Pharmac.*, **15**, 285–307.

SCHWARTZ, N. B. & MCCORMACK, C. E. (1972) Reproduction: gonadal function and its regulation. *A. Rev. Physiol.*, **34**, 425–73.

SHORT, R. V. (1967) Reproduction. *A. Rev. Physiol.*, **29**, 373–401.

WILSON, C. A. (1974) Hypothalamic amines and the release of gonadotrophins and other anterior pituitary hormones. *Adv. Drug Res.*, **8**, 119–205.

13

The Hormones

III. Anabolic Steroids

The physiological terms anabolism and catabolism which are applied to the tissue building and tissue breakdown components of metabolism are familiar to all. The term *anabolic* when applied to a steroid has a special significance; it indicates that the particular steroid is a derivative of testosterone, in which the androgenic hormonal activity of that compound has been reduced relative to the tissue building activity. Testosterone is the strongest anabolic steroid available but its androgenicity precludes its routine use for anabolic effect.

Chemistry

Methylandrostenediol (i) the first true anabolic while superior to methyltestosterone which it replaced clinically, was still too androgenic, but by further manipulation of the testosterone molecule a whole series of valuable anabolic steroids has been produced for both human and veterinary therapy.

Chemically, the anabolics at present in use fall into approximately two groups—those in which the methyl group at carbon 19 is absent and those which are more closely associated with testosterone in having this group present, but possessing greater or less modification in the region of ring A in the steroid molecule. Nandrolone and its esters (phenylpropionate, decanoate and laurate) norethandrolone and ethylestrenol fall into the first category, methandienone, oxymetholone, boldenone and stanozolol in the second.

Nandrolone (ii) this differs from testosterone only in the absence of the methyl group at carbon 19. The compound acts chemically as a base forming esters with acids which are attached in the 17β position as illustrated by the **phenyl propionate** (iii), and the **laurate** (iv), both of which are available commercially in the European veterinary field (as Nandrolin and Laurabolin, respectively).

Norethandrolone (v) differs from nandrolone in the attachment of an alkyl group (ethyl) in the α position at carbon 17. The compound is not used extensively in the veterinary field.

Ethylestrenol (vi) also available in the veterinary field in Great Britain as Nandoral, represents a further step from testosterone towards an oestrogen by the removal of the oxygen atom at carbon 3; this effectively removes all androgenic activity.

Methandienone (viii) also available in the veterinary field (Vetanabol) differs from methyl testosterone (vii) in respect of a further double bond in ring A of the steroid nucleus between carbons 1 and 2.

Oxymetholone (ix) and **stanozolol** (x) however, possess a saturated ring A and further differ from methyl testosterone in having a hydroxymethylene group at carbon 2 (oxymetholone) and a pyrazole ring attached at carbons 2 and 3 (stanozolol) whilst stanozolol also has the hydrogen atom at carbon 5 directed backwards in the β position. Stanozolol is used in the veterinary field in the United States but neither oxymetholone nor stanozolol are much used by veterinary surgeons elsewhere.

The physical characters are similar to those of the parent sex steroid.

Actions

The activity spectrum of the anabolic steroids can be considered under two headings. The first group are those actions associated with anabolism and anti-catabolism, nitrogen retention with protein anabolism, calcium and

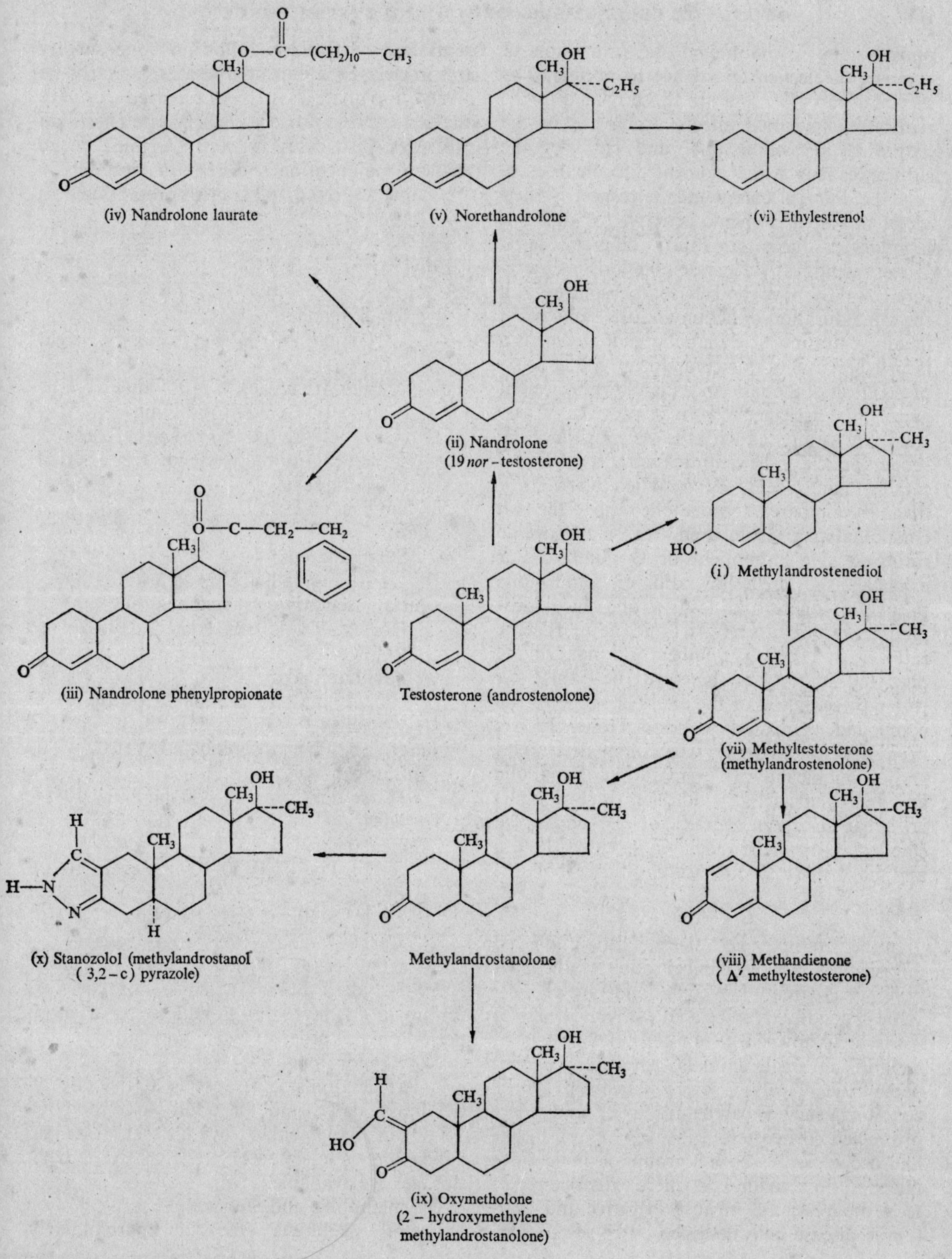
(iv) Nandrolone laurate
(v) Norethandrolone
(vi) Ethylestrenol
(ii) Nandrolone
(19 *nor* - testosterone)
(i) Methylandrostenediol
(iii) Nandrolone phenylpropionate
Testosterone (androstenolone)
(vii) Methyltestosterone
(methylandrostenolone)
(x) Stanozolol (methylandrostanol
(3,2 - c) pyrazole)
Methylandrostanolone
(viii) Methandienone
(Δ′ methyltestosterone)
(ix) Oxymetholone
(2 - hydroxymethylene
methylandrostanolone)

phosphorous retention and the promotion of a return of electrolyte balance to normal. The second group of actions are those related to the remaining hormonal effects which are in the nature of side actions (*q.v.*) and vary with the substance. It is at the moment difficult to compare the different compounds in respect of their anabolic and androgenic activity, for different laboratories have developed different tests. Direct scientifically accurate comparisons therefore can only be made between substances with the same duration of action and the same route of administration. The anabolic/androgenic ratio is an expression of this comparison and consists of the effect of the substance (as shown by weight gain) of a muscle, usually the levator ani of castrate rats, compared with the weight gain of the seminal vesicle or the ventral prostrate of the same rat due to androgenic effects. When the two products are compared, the anabolic ratio of the two compounds can be compared with the androgenic ratios to give a quotient or Q number. For example, the anabolic ratio of nandrolone phenylpropionate and testosterone phenylpropionate is approximately 4 : 1 and the androgenic ratio 1 : 4. The Q number for nandrolone phenyl propionate is therefore 16 : 1 this is a true representation of the anabolic value of the compound compared to testosterone. It has been suggested that only nandrolone derivatives show a significantly wide separation of anabolic and androgenic activity although marked separation has also been claimed for methandienone stanozolol, oxymetholone and 1-methyl-Δ' 4, 5 and dihydrotestosterone acetate—Primobolan.

Uses

In the veterinary field there is little published information but the clinical value of the anabolics is apparent. The compounds are of most value in conditions associated with a high rate of tissue breakdown, e.g. non-terminal phases of nephritis in small animals, parasitic bronchopneumonia of calves, severe parasitism, febrile conditions such as canine distemper, and where protracted corticosteroid therapy or overdosage has occurred. A second group of indications includes those conditions where extensive repair is in progress; e.g. in post-operative and post-febrile disease convalescence, after severe accidents and particularly following bone surgery and in delayed union of fractures: nandrolone phenylpropionate has been reported to have analgesic and improved recovery rate effects on prolapsed disc cases in dogs. A third group includes malnutrition and debility especially in the young and aged and growth promotion.

Dosages

Ethylestrenol
- Oral:
  - Small animals:
    - 0·05 mg/kg (0·034 mg/lb.) body weight daily.
    - up to 4·5 kg (10 lb.): 0·5 mg
    - up to 14 kg (30 lb.): 1·0 mg
    - up to 27 kg (60 lb.) and over: 2 mg
  - Treatment normally extends over a period of 10–14 days

Methandienone B.P.
- Oral:
  - Small animals:
    - 5–10 mg daily for up to 3 weeks
- Intramuscularly or subcutaneously:
  - Horses: 100–150 mg every 3 days for up to 6 weeks.
  - Small animals:
    - 12–25 mg every 3 days for up to 6 weeks

Nandrolone Phenylpropionate B.Vet.C.
- Intramuscularly:
  - Horse and Cow: 200–400 mg every 10–14 days
- Subcutaneously:
  - Sheep, Pig, Calf: 50–100 mg every 10–14 days
  - Dog: 25–50 mg every 10–14 days
  - Cat: 12·5–25 mg every 10–14 days

In acute cases injections may be given every 3–4 days but as a single dose persists for up to 14 days, this should be for a short period only in order to avoid cumulation.

Norethandrolone, B.P.

The childrens' oral dose is at the rate of 0·5 mg/kg daily, and the adult dose is about 25–50 mg daily either by intramuscular or subcutaneous injection or orally. No veterinary dosages are available.

Oxymetholone and Stanozolol

Oral: Childrens' dosage is approximately

2 mg daily for a period not exceeding 30 days. Veterinary dosages are not available, but small animals might be given up to this dosage. In the treatment of hypoplastic anaemia, oxymetholone has been shown to be of value in man.

The recent advent of cheaper anabolic hormones, e.g. trienbolone acetate, has opened up the field of growth promotion to these agents. Trienbolone is implanted either alone (300 mg) or combined (200 mg) with 17β oestradiol (40 mg) and is more effective in steers, heifers or cows than in intact bulls. The combined preparation is better than either hormone singly in achieving increased live weight gain over a period of 6–8 weeks. At slaughter 12 weeks after implanting, tissue residues were very low and the drug is in any case destroyed by heating.

Side effects. These are related to hormonal activity. In the case of methandienone and nandrolone phenylpropionate, oxymetholone and stanozolol, androgenic activity is the main side action. This shows as increased libido in males and male castrates, and in neutered male cats or female cats the passage of strong smelling urine. The nandrolone compound has insignificant effects at recommended dosage levels, but signs may develop if high doses are continued for several weeks. Methandienone has slightly greater androgenicity but oxymetholone and stanozolol are both appreciably androgenic, the former showing marked effect even in dosages insufficient to produce growth acceleration. Norethandrolone has appreciable progestational activity. Early closure of epiphyseal discs may be promoted in growing animals.

Toxicity. All steroids with alkyl groups attached in the alpha position at C_{17} possess some hepatotoxic effects in man. In animals, however, such an effect is difficult to produce experimentally particularly with the nandrolone esters: many hundred times the normal dose of ethylestrenol is needed to produce signs of cholestatic hepatitis in dogs.

Bibliography

KRÜSKEMPER, H. L. (1968) *Anabolic Steroids.* New York: Academic.

VIDA, J. A. (1969) *Androgens and Anabolic Agents.* New York: Academic.

WYNN, V. (1967) Anabolic steroids and protein metabolism. In: *Modern Trends in Endocrinology* (Ed. by H. Gardiner-Hill), vol. 3, pp. 254–88. London: Butterworths.

14

The Hormones

IV. The Thyroid, Parathyroid and Pancreas

The so called 'metabolic hormones' are the only remaining hormones of clinical significance. These are the hormones of the thyroid, the parathyroid and the pancreatic glands.

THE THYROID GLAND

This endocrine gland consists in most species of two masses of tissue situated on either side of the larynx and upper trachea, with an 'isthmus' joining them anteriorly. From the dried glands of cattle, sheep and pigs, a buff-coloured, amorphous powder with a meat-like odour and saline taste is prepared by defatting with ether and drying (Thyroid, B.Vet.C.). The B.P. stipulates that the powder is standardized to contain 0·1 per cent iodine as thyroxine but the U.S.P. merely requires 0·2 per cent in organic combination, which is not necessarily iodine as thyroxine.

Thyroxine Sodium, B.Vet.C.

Thyroxine sodium is the *laevo*-rotary sodium salt of thyroxine which is twice as active as the racemic (*dl*) form. It is a cream coloured powder which is slightly soluble in water but more so in dilute acids or alkalis. *l*-Thyroxine sodium possesses the activity of thyroid extracts qualitatively, but quantitatively 0·1 mg is approximately equal in activity to 60 mg of the dried extracts. Its main advantages are chemical synthesis, lack of need for biological assays and stability.

The active constituent of the thyroid gland was originally thought to be thyroxine, an iodine-containing substance isolated in 1914 and synthesized in 1927. A second principle more potent than thyroxine, known as tri-iodothyronine, was isolated and synthesized in 1952. The thyroid hormones are synthesized then stored extracellularly incorporated within the molecule of the glycoprotein, thyroglobulin. Mono-iodotyrosine (MIT) and di-iodotyrosine (DIT) are biologically inactive precursors of thyroid hormones extractable from gland homogenates.

Metabolism. The metabolism and action of the thyroid hormones and iodine are more or less identical so that an iodine deficiency automatically constitutes a thyroid deficiency. Reference to the formula of *l*-Thyroxine Sodium, B.Vet.C. (i) will make this readily understandable, for it will be seen that four iodine atoms are contained in each molecule of thyroxine and comprise 504 parts of the molecular weight of thyroxine which is 777; that is, over five-eighths of the total molecule by weight is iodine.

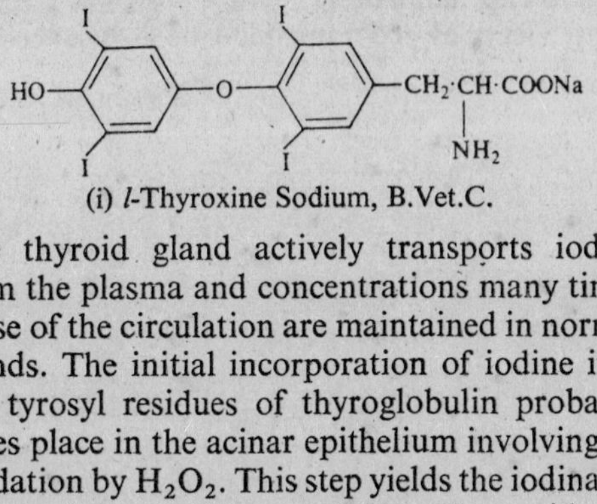

(i) *l*-Thyroxine Sodium, B.Vet.C.

The thyroid gland actively transports iodine from the plasma and concentrations many times those of the circulation are maintained in normal glands. The initial incorporation of iodine into the tyrosyl residues of thyroglobulin probably takes place in the acinar epithelium involving an oxidation by H_2O_2. This step yields the iodinated intermediates MIT and DIT. The oxidative coupling of a mono-iodotyrosyl residue with a di-iodotyrosyl residue yields tri-iodothyronine (T_3). The coupling of two DIT residues yields thyroxine (T_4). Secretion of thyroid hormones is preceded by endocytosis of colloid into follicular cells followed by lysosomal proteolysis of the thyroglobulin. Thyrotrophic hormone (TSH) regulates I_2 uptake and the final discharge of T_3

and T_4, both of which become highly protein-bound. In thyroid insufficiency the plasma content of this trophic hormone increases. Excess of circulating trophin is possibly responsible for the increase in size of the thyroid which occurs in iodine deficiencies and other goitrogenic circumstances. With excessive thyroid secretion thyrotrophic hormone production is suppressed: if this excess is therapeutic and temporary, reactivation of the pituitary gland and consequent stimulation of the thyroid gland follow cessation of the treatment. Following administration of thyroid extracts or *l*-thyroxine sodium about half the oral dose is absorbed, but the effects are not apparent for several days and maximum effect from single doses may not be achieved for as long as 10 days; residual effects may continue for as long as a fortnight. Because of this, thyroid hormone must be considered as potentially cumulative.

After absorption thyroxine causes an increase in metabolic rate with increased combustion of carbohydrate which results in depletion of liver glycogen; there is also increased combustion of fat with a mobilization of fat from all depots, and of protein, with increased urinary nitrogen, salt and water. This increased oxidation rate, which is the principal effect of thyroid substances is not a direct effect but is probably due to the effect on coenzymes and other intermediate factors involved in intracellular energy transforming processes. These effects are accompanied by increased irritability of the vasomotor centre, increased heart rate and output, and an increased circulatory rate. In addition to these metabolic effects the thyroid hormone works in conjunction with other hormones such as the parathyroid, pancreatic, adrenal and anterior pituitary hormones, playing an important part in normal mental and physical growth, and sexual maturation.

Iodine deficiency. Deficiencies of thyroid hormone and iodine are interrelated and have been shown to occur widely throughout the world causing considerable economic losses in herbivorous species, carnivores and birds. Clinical manifestations take the form of *goitre*, and are most apparent in young animals. In the severest forms young animals may be born dead after prolonged gestation, completely without hair, oedematous and with enlarged thyroid glands—hairless calves and piglets are perhaps the most common instances of the condition in Great Britain, but sheep, goats, horses, dogs, cats and birds may all be similarly affected.

Adults are affected but the condition is less spectacular and takes the form of unthriftiness, loss of coat colour and character, and sometimes sterility in male and female with loss of sex drive in the male. Goitre may be due directly to iodine deficiency, or be associated with the goitrogenic effect of substances in the diet—thiocyanates, possibly from linseed or clover, substances in kale, and possibly goitrogens in other foods. Besides the incidental effects of such drugs as phenothiazine: this last possibility was receiving much attention when this drug was being supplied in mineral lick forms which resulted in prolonged intake and therefore a greater exposure to possible goitrogenic effects. In some instances the goitre can be due to both iodine lack and goitrogenic substances intake.

Toxicity. In excessive dosage *thyrotoxicosis* may result with intensification of all these actions Increased pulse rates and cardiac irregularities, restlessness, and sleeplessness occur: if this dosage is continued osteoporosis may follow excessive mobilization and excretion of calcium and phosphorous. Early manifestations of overdosage are increased and irregular heart action, loss of weight, a tendency towards a higher body temperature and restlessness. In all therapy with thyroid compounds these signs should be watched for and animals checked weekly until a satisfactory maintenance dose has been established.

Use

Both iodine and thyroid substances may be used in the treatment of thyroid deficiency in all species, but it must be remembered that thyroid deficiencies are not invariably associated with an insufficiency of iodine and in cases where the two are not associated iodine is not a substitute for thyroid compounds. These deficiencies are known under various names, including 'hairless calves', and 'hairless piglets', 'big neck', goitre and myxoedema. These are the names accorded due to the signs associated with advanced thyroid and iodine deficiencies.

Less obvious conditions for which iodine and thyroid treatment is indicated are unthriftiness, slow growth, lack of 'wire' in wire-haired breeds or straightness in curly coats, poor colour in the coat, lethargy, obesity, oedema, possibly nymphomania in female cattle, loss of libido and depressed spermatogenesis in rams, boars, bulls, etc., where they occur in known iodine deficient areas. The clue to the thyroid and iodine deficiencies may be given by occasional births of goitrous young.

In dogs, a bilateral alopecia, most commonly associated with wire haired breeds, responds well to thryoid treatment. Obesity and urinary incontinence in dogs have also been effectively treated with thyroid, but other hormones and diet regulation are probably more beneficial and safer.

Thyroid extract is unstable: purchased samples have been examined and have been shown to vary considerably in their thyroxine content. In some cases activity has been as low as one-third of the original levels. Thyroxine should, therefore, be used in preference to the natural extract.

Dosage

In acute foetal iodine and thyroid deficiencies treatment is often successful in animals which survive for 2 or 3 days after birth. Both iodine and thyroid substances may be given but thyroid on its own is too slow in action. Dosage of iodine may be instituted by intravenous injection of sodium iodide in calves, but oral dosage with either sodium or potassium iodide from the onset is usually satisfactory in all species. Doses will vary from a few milligrams daily in puppies and kittens to a few grains daily in calves and foals. Thyroid may be given additionally by mouth in doses of approximately 1 mg ($\frac{1}{60}$ gr.) per lb. body weight (1 to 2 mg per kg) in all species. The average dose for calves is 0·3 g (5 gr.) twice daily, and for dogs 30 mg to 0·3 g ($\frac{1}{2}$ to 5 gr.)

Deficiencies in adults, which are of less acute nature, may be treated initially with thyroid at the rate of 1 to 2 mg per kg (approximately $\frac{1}{60}$ gr. per lb.), and this should be supplemented by iodine feed supplements and salt licks. Adult cattle have been treated—especially bulls with lowered fertility and impaired sex-drive—with *l*-thyroxine sodium, using a dosage of 5 to 8 mg daily given by subcutaneous injection or 50 to 100 mg daily by mouth: this has supplanted the use of iodinated casein (the original use of which was in the attempted improvement of milk yield in cows: it proved uneconomical). In small animals doses of thyroxine sodium should be in the region of 0·05 mg daily by mouth, the dosage being modified according to response. Up to 30 μg daily of *l*-tri-iodothyronine by mouth is recommended commercially for small animals and up to 400 μg for large animals. In all cases of thyroid administration, a careful check on the heart should be made weekly until a maintenance dose is established. Any signs of cardiac irregularity should be followed by reduction or temporary cessation of dosage in view of the known cumulative action of thyroid extracts.

Mineral supplements usually contain about 0·03 per cent of potassium iodide, or the more stable iodate especially in tropical climates; this is represented by 15 g in 50 kg of supplement ($\frac{1}{2}$ oz. in 1 cwt). Daily doses of such a supplement should be:

- Milking cattle: 45–140 g ($1\frac{1}{2}$–5 oz.)
- Other adult cattle: 60–180 g (2–3 oz.)
- Calves: 30–45 g (1–$1\frac{1}{2}$ oz.)
- Brood mares: 60–120 g (2–4 oz.)
- Other horses: 60–90 g (2–3 oz.)
- Ewes: 15–22·5 g ($\frac{1}{2}$–$\frac{3}{4}$ oz.)
- Lambs: 5–10 g ($\frac{1}{6}$–$\frac{1}{3}$ oz.)
- Sows: 60–120 g (2–4 oz.)
- Store pigs and others: 30–60 g (1–2 oz.)
- Laying hens: 220–400 g (8–15 oz.) per 100 birds
- Chicks: 30–60 g (1–2 oz.) per 100 birds
- Older chickens: 90–160 g (3–6 oz.) per 100 birds

Salt licks should contain about 0·05 per cent, or 30 g of potassium iodide in 60 g per 50 kg (2 oz. per 100 lb.) salt but up to 50 kg (1 oz. in 100 lb.) have been recommended for dairy cattle.

Antithyroid Substances

Though the development of tolerance to thyroid administered over a period can occur due to the development of antibodies, this is not considered to be of clinical significance.

Apart from the naturally occurring goitrogens already mentioned previously, a group of substances known as the **thiouracil derivatives** have a specific depressant effect on thyroid activity and thyroxine production. They induce hypothyroidism, with a slight goitre, a lowering of the metabolic rate, and improved fattening rates, although the inferior carcase quality produced by this technique rendered it commercially unattractive. The compounds act by interfering with the iodination of tyrosine.

Antithyroid substances find occasional use in the treatment of hyperthyroidism. Approximate daily doses of thiouracil are: Cattle, 1-4 g (15–60 gr.); Sheep 0·2–0·5 g (3–7½ gr.); Pigs 0·2–0·6 g (3–9 gr.); Dogs 50–100 mg (1–1½ gr.); Cats, 20–50 mg (⅓–1 gr). Doses of propyl thiouracil should be about a quarter of these doses.

THE PARATHYROID GLAND

There are four small parathyroid glands which are located at the poles of the thyroid lobes. In some species they are embedded in the thyroid gland but in others they are distinct entities.

A desiccated form of the gland is available and has a pale yellow amorphous appearance; it has a faint meaty odour. A lipid extract of parathyroid gland is claimed to counteract the acute symptoms of parathyroid removal. Very little of any oral dose of parathyroid escapes destruction in the gastro-intestinal tract. The active principle is a single chain polypeptide with a molecular weight of about 9500: this is now referred to as parathormone.

Calcitonin, secreted by the thyroid parafollicular C cells, has actions opposite to those of parathormone.

Parathyroid Injection, U.S.P. is a sterile water soluble extract of parathyroid gland for subcutaneous or intramuscular administration: it contains not less than 100 U.S.P. parathyroid units per ml, and the unit is described as '1/100th of the amount necessary to raise the calcium content of 100 ml of the blood serum of normal dogs 1 mg within 16 to 18 hr after administration'. A synthetic parathormone (PTH) analogue is now also available.

Actions and Uses

The function of the parathyroid is to respond to low levels of ionic calcium in plasma by secreting pre-formed parathormone (PTH) which promotes active inorganic phosphate and calcium absorption from the intestine, their mobilization from bone and calcium and phosphate excretion into the urine and milk. PTH stimulates osteoclastic bone resorption and so elevates blood calcium levels while calcitonin inhibits bone resorption and so corrects a hypercalcaemia. An injection of parathyroid hormone in normal animals causes an increase in urinary phosphate excretion, a decrease in the inorganic phosphorous blood level, an increase in blood calcium, and an increased calcium excretion. The effect on the calcium blood level is seen in about 4 hr and should reach a peak in about 18 hr, followed by a return to normal levels approximately 24 hr after the injection.

Lack of sufficient parathyroid activity causes the reverse of the above effects, and as the level of blood calcium falls muscle and general nervous irritability increases until the convulsive threshold is reached.

In excess, parathyroid hormone causes a considerable increase in calcium and phosphate excretion, a low serum phosphate and a continuous calcium mobilization. In extreme and chronic cases, decalcification of the bones occurs with loss of bone rigidity, fractures and malformations. In acute excess the blood calcium levels are approximately doubled and the blood phosphorus rises. Soft tissue metastatic calcification occurs with eventual death from renal failure. Treatment is surgical.

Parathyroid extracts have frequently been given orally in tablet form combined with calcium lactate, but as digestive enzymes destroy the majority of any dose, this practice should be abandoned in favour of subcutaneous or intravenous injections of 2–15 U.S.P. units for a dog. The chemical identity of the active moiety of the PTH molecule remains to be confirmed but a synthetic analogue is already available. Owing to its effects on phosphate, and its side-effects, use of PTH in therapy should be confined to hypoparathyroidism. Even then, vitamin D, through which the action of PTH appears to be mediated, offers an acceptable alternative.

In the treatment of experimentally induced hypoparathyroidism it has been found that antibodies against parathyroid extracts develop and the effect of a standard dose is progressively reduced.

THE PANCREATIC GLAND

The pancreas, which lies in the loop of the duodenum, contains two vastly different types of glandular tissue. The greater part of the gland consists of an exocrine gland, which secretes a fluid rich in digestive enzymes via the common bile duct into the small intestine. Dispersed throughout the mass of exocrine gland are the islets of Langerhans whose β cells secrete the polypeptide hormone insulin. **Glucagon** is a smaller, single chain polypeptide secreted by the islet α-cells. In that it has glycogenolytic and gluconeogenetic effects its actions oppose those of insulin.

Injection of Insulin, B.Vet.C.

The *British Veterinary Codex*, quoting the *British Pharmacopoeia*, describes this injection as 'a sterile solution of the specific anti-diabetogenic principle of the mammalian pancreas'. It is a clear, colourless liquid containing 20, 40 or 80 units in each millilitre: the solvent consists of water acidified with hydrochloric acid, and about 1·5 per cent glycerin.

Actions

Insulin is rapidly absorbed following injection and produces maximum effects within about 3 hr of administration. More rapid effects are achieved by intravenous or intramuscular injection than by subcutaneous therapy, but the hormone is inactivated in the gastrointestinal canal and is useless when administered orally. The Protamine Zinc Insulins and Globin Zinc Insulins are effective for about 48 and 24 hr, respectively, but the plain injection is effective for a period of about 8 hr only, and requires injection twice a day.

Following absorption, insulin exerts a powerful effect on carbohydrate metabolism, the general effect being one of removal of glucose from the circulation. The primary effect is to increase the phosphorylation of glucose, with the formation of glucose-6-phosphate: the means by which it achieves this is undecided, but may be one of releasing the enzyme catalyst hexokinase inside the cells from the control of corticosteroids and other hormones; another observation is that insulin increases the permeability of the cell membrane to glucose which, following rapid entry into the cell, can be phosphorylated by hexokinase present in the cell. These biochemical effects make phosphorylated glucose available for rapid metabolization and results in the removal of glucose from the blood. Some of the absorbed glucose is used to supply energy for cell metabolism, and the rest is converted into glycogen and deposited in the liver and possibly the muscles. Fat is formed in the liver and fat depots. Protein synthesis from amino acids is also assisted by insulin which aids the entry of amino acids into cells and inhibits protein catabolism.

In the complex and, as yet, incompletely understood hormonal regulation of intermediary metabolism, insulin and growth hormone are anabolic in effect. As a generalization, the glucocorticoids, thyroid hormones and catecholamines are catabolic and exert actions which oppose those of insulin on glycogen synthesis, on nitrogen balance and on the deposition of fat.

Uses

In true diabetes mellitus in the dog, which is very similar to that which occurs in man, hyperglycaemia and glycosuria are the dominant clinical features: ketones are also found in the blood and urine. The sugar, present in the blood in large quantities, is obtained from various sources—absorption from the intestines, mobilization of carbohydrate reserves, and breakdown of fat and protein. The breakdown of fat is incomplete and results in the formation of ketones and the provision of glycerol. The availability of amino acids is increased, partly from protein catabolism, and gives rise to glucose following deamination in the liver. Pituitary factors and glucagon drive this production of glucose which in the absence of insulin remains extracellular and metabolically unavailable.

This unproductive drain on carbohydrate, protein and fat reserves is responsible for other diagnostic symptoms: increased appetite and thirst, emesis and diarrhoea, increased urine production and a progressive emaciation with marked muscular wasting and lethargy. Bilateral cataract occurs occasionally. Frequently the condition is complicated by pancreatic enzyme deficiency due to generalization of the pancreatitis: insulin, of course, has no effect on such deficiencies and the symptoms they cause.

Insulin injections will reduce the blood glucose level rapidly by both hepatic glycogenesis and by increasing tissue breakdown of sugar with the release of energy. Glycosuria automatically ceases when the blood level decreases below the renal threshold. Ketone production ceases and the breakdown of proteins and fats is reversed.

In acetonaemia in cattle and pregnancy toxaemia in sheep, insulin has been used in conjunction with glucose therapy.

Dosage and Administration

Dog: 5–100 units or more according to requirements

Cow: 50–200 units

The optimum dose of insulin for a diabetic dog is decided by one factor and that is the degree of glycosuria: an adequate dose is one which is sufficient to reduce the urine sugar to only a trace when tested at the time of maximum effect. This time will vary according to the type of insulin used. The sugar should not be completely absent from the urine.

When commencing treatment it is wise to admit the dog into kennels, commence with a dose of 5 units and increase the dose by steps of 2 units. Urine should be tested at the appropriate time after each increase in dosage. Larger increases may be used provided a close watch is kept and glucose solutions are available to counteract any tendency to the development of convulsions and coma.

If 'regular' insulin is used twice daily injections are advisable, but mild cases will respond to restriction of diet and single daily injections. In severe cases combinations of 'regular' and protamine zinc injections are helpful.

A feature of many cases of diabetes appears to be the progressive increase in insulin requirements.

The official Injection of Insulin, B.Vet.C. is available with 20, 40 and 80 units per ml, whilst Injection of Globin Zinc Insulin, B.P. and Injection of Protamine Zinc Insulin, B.P. may contain either 40 or 80 i.u.

The **sulphonylureas** are hypoglycaemic sulphonamides and are capable of supporting life in diabetics which retain some functional β cells, but are without effect in pancreatectomized animals. Their oral administration is followed by hypoglycaemia in normal animals. It is believed that the mode of action of these drugs is a direct stimulation of the secretory activity of pancreatic β cells. **Tolbutamide** is a short-acting member of the group and **chlorpropamide** (iii), which is not metabolized, is long-acting.

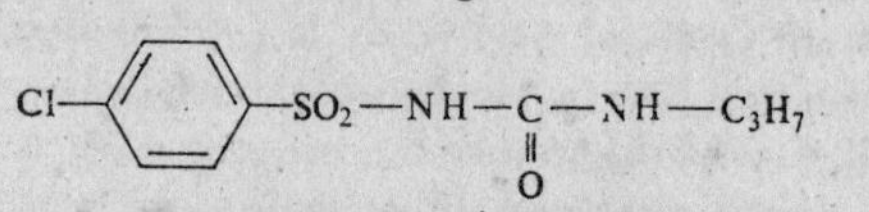

(iii) Chlorpropamide

The exocrine secretion of the pancreatic gland is a thin watery fluid with the consistency of saliva and contains a number of enzymes which are responsible for the breakdown of protein, fat, and carbohydrate (starches) in the small intestine. These enzymes are active in slightly alkaline pH (7 to 9); acidity destroys the enzymes. The main proteolytic enzyme is *trypsin*, excreted in the inactive form known as *trypsinogen*. Activation is mediated by the enzyme *enterokinase*. A related enzyme chymotrypsin, excreted as chymotrypsinogen is activated by the trypsin and both are responsible for splitting the proteins as far as the proteosepeptone stage. Optimum pH is 8 to 9. A third enzyme *carboxypeptidase* is responsible for hydrolysis of proteoses and peptones, at the end of chains where the free carboxyl groups occur.

The lipolytic enzyme in its inactive form is known as *pro-lipase* but the means of activation to give *lipase* (*steapsin*) is unknown.

The starches are hydrolysed by *amylase* (*amylopsin*) which has an optimum working pH of 7·1.

Pancreatin, B.Vet.C.

This is a standardized, buff amorphous powder with a meat-like odour. It is slightly hygroscopic and is soluble in water but not in the organic solvents: the solution is turbid. It is available as an official compound tablet containing sodium bicarbonate as a buffer to limit gastric inactivation of the enzymes. The official solution is similarly buffered, but is nevertheless unstable and must be prepared not more than 24 hr or so before use. Enteric coated tablets or fresh pancreatic gland are the usual means of administration of the enzymes.

Actions and Uses

The enzymes are essential for normal intestinal digestion. Their absence results in a typical train of events. Some starch and carbohydrate digestion is effected by oral and gastric juices, and similarly pepsin is responsible for some protein digestion. Beyond this, little digestion occurs and voided faeces will, therefore, contain undigested protein and fat—shreds of meat, and fat which congeals on cooling, are readily recognized in canine or feline faeces. This loss of essential food substances promotes a nutritional crisis, the animal has a ravenous appetite yet continues to lose weight rapidly as its fat reserves, glycogen reserves, and muscle proteins are drawn on to provide the energy necessary to maintain the vital processes. Acute cases of pancreatitis have been described but the condition is not ordinarily amenable to replacement therapy owing to tissue destruction and toxaemia. Where chronic atrophy without acute systemic disturbance occurs replacement therapy with the enzymes is quite successful. A test to determine the degree of enzyme production using gelatin film (photographic negative) has been described, and other tests include test-feeding with lean meat given in small cotton bags, recovered following defaecation and examined. Normally the meat is digested but in pancreatic deficiencies shreds of meat remain in the bag.

Trypsin is gaining favour as a means of liquefying dead tissue—enzymytic debridement—in abscesses, in otitis, haematomata, 'foul in the foot' and mastitis. Other enzymes, including pancreatic dornase (desoxyribonuclease), streptokinase and streptodornase have also been used. These enzymes have been used locally or parenterally, but some are too expensive for routine use. They should be used in conjunction with antibiotic therapy and liquefied material removed periodically. For example, pancreatic dornase in purulent otitis in the dog has been flushed out after 5 to 10 min following the application of 5 ml of an aqueous solution containing 25 000 units: 20 ml of solution containing 100 000 units introduced into the sinuses of horses is flushed out after 24 hr, and the treatment repeated daily as long as required: similar solutions, used daily in the treatment of mastitis, should be milked out with the liquefied material at least once, but preferably twice, daily prior to antibiotic administration.

Pancreatic enzymes may also be used in the preparation of protein hydrolysates and predigested foods. Two of the official tablets are sufficient when dissolved in 150 ml (5 fl.oz.) tepid water to peptonize a pint (0·6 ml) of milk. Body temperature is the optimum for peptonizing and about 15 min should be sufficient time for adequate peptonization. Enzyme action can be stopped by boiling. Other foods can be treated similarly and given to sick or convalescent animals.

Dosage and Administration

In dogs and cats the dose of pancreatin will vary according to the amount of functional pancreatic tissue remaining in each individual. Doses of up to 6 g (90 gr.) daily have been satisfactory. Trypsin has been injected via the teat canal in mastitis in doses up to 400 000 units daily dissolved in about 150 ml water, or preferably in phosphate buffer solution as this maintains pH and prolongs activity: normally 75 per cent activity is lost in 3 hr at room temperature. The fluid should be milked out after about 4 hr and the treatment repeated daily as required.

THE THYMUS GLAND

Though it is generally believed that the thymus gland has an endocrine function associated with growth and maturation and that it is influenced by the anterior pituitary gland secretions, no positive proof has yet materialized. The thymus gland produces the so-called T-lymphocytes which mediate cellular immunity and process antigens enabling them to initiate elaboration of immunoglobulins by the B-lymphocytes.

GASTRO-INTESTINAL HORMONES

A hormone given the name of *gastrin* was postulated as long ago as 1906 and is now known to be a heptadecapeptide secreted by the G-cells of the pyloric mucosa. It stimulates the secretion of gastric acid, pepsin, intrinsic factor and also increases the flow of pancreatic juice. **Pentagastrin** is a synthetic pentapeptide with gastrin-like activity. It is used as an alternative to histamine in tests designed to measure the secretory capability of the gastric mucosa.

A further digestive hormone, secretin, elaborated in the intestinal mucosa, is used in a diagnostic procedure. This produces an increase of pancreatic secretion but only on intravenous injection. It has been used as a test for pancreatic function. Bile secretion is also influenced.

Other substances also stimulate or inhibit the activity of the gastro-intestinal tract but are dealt with in other chapters.

Bibliography

FOSTER, G. V. & MACINTYRE, I. (1968) The endocrine control of calcium metabolism. In: *Recent Advances in Endocrinology* (ed. V. H. T. James), London: Churchill.

HAYMOVITS, A. & ROSEN, J. F. (1972) Calcitonin in metabolic disorders. *Adv. metab. Disorders*, **6**.

IRVINE, W. J. (1967) Immunobiology of the thymus and its relation to autoimmune disease. *Mod. Trends Immunol.*, **2**, 250.

LEVINE, R. & LUFT, R. (ed.) (1974) Proceedings of a symposium on insulin. *Adv. metab. Disorders*, **7**, 141–257.

PRANGE, A. J. (1974) *Thyroid Axis, Drugs and Behaviour*. New York: Raven.

WILKINSON, J. S. (1963) Diseases of the pancreas of the dog. *Adv. small Anim. Pract.*, **4**, 31–41.

15

The Hormones

V. The Mediators of Allergy, Anaphylaxis and Shock

The endocrine glands release their secretions into the circulation and the hormones are carried by the blood to their sites of action. In most cases it is possible to extract the hormone both from the gland or from a blood sample. The transmitter substances released by peripheral voluntary and autonomic fibres are less readily isolated. These are also known as neuro-hormones and while adrenaline and noradrenaline released by the adrenal medulla may be compared with the endocrines, acetylcholine, which, because it is liberated in the immediate vicinity of its site of action and is rapidly inactivated, is not found in significant amounts in the blood, is more closely comparable with the tissue hormones. Tissue hormones, or local hormones, are released within tissue and are usually vasoactive, i.e. they exert an effect on the walls of blood vessels. The release of tissue hormones is triggered by any stimulus which leads to cellular damage or destruction, e.g. chemical irritants, heat, ultraviolet light, ionizing radiations, trauma, bacterial toxins or immune reactions. It is now thought that some tissue hormones are the chemical mediators of the inflammatory response and shock. While it has been possible to reproduce the changes of acute inflammation with histamine, the mediation of the chronic phase remains an area for debate.

Many active substances have been extracted from the CNS and while it is known that some can function as transmitters, it is suggested that others may function as local hormones, possibly exerting a widespread facilitation or depression of function. Although many of the tissue hormones are amines (histamine, 5-hydroxytryptamine) or peptides (bradykinin, kallidin) other structures are also represented (amino acids, fatty acids).

In this chapter will be discussed those agents whose actions in relation to disease processes are to some extent understood and for which antagonists are available. The more controversial members will be described only briefly.

HISTAMINE AND THE ANTIHISTAMINICS

Histamine is widely distributed in plants and is found in animal foodstuffs (silage) and digesta. When administered by intradermal injection, it elicits what is called the triple-response. The small vessels at the site of injection dilate, then the neighbouring vessels dilate via an axon reflex of a sensory nerve. Finally, as plasma escapes through the walls of the most widely dilated vessels, the central area of the reaction becomes oedematous and appears raised and blanched. This sequence of flush, flare and wheal occur following a nettle sting (the sting contains histamine) and has given us the medical name *urticaria* to identify the phenomenon. That histamine can stimulate nerve-endings is well-remembered from the itch and pain of a nettle sting and it has been suggested that histamine is the peripheral mediator for sensations of itch and pain.

The importance of histamine as a factor in the causation of certain pathological conditions was first suggested in 1910 by Dale and Laidlaw, who pointed out a possible relationship between histamine and the aetiology of anaphylaxis and allergy: since that time it has been established with reasonable certainty that histamine release is a major factor in the development of these two phenomena. It has also been suggested that wherever tissue is damaged there is a production of histamine which causes either local or general shock: this postulation is debated, but there is no doubt that histamine administration is followed in most species by a train of reactions very similar to those seen during the development of traumatic and surgical shock. There is also no doubt that other factors besides histamine release are involved.

Histamine

This occurs in an inactive form in all tissues, but especially in the lungs, skin, intestinal and muscular tissues. It is formed from the amino acid histidine by decarboxylation mediated by an enzyme *histidine decarboxylase*: bacterial decomposition of proteins also produces histamine and fairly high concentrations exist in septic foci where tissue destruction is in progress. Histamine is destroyed by enzymic transmethylation or oxidation by diamine oxidase which is concentrated mainly in the intestinal mucosa, liver and kidneys, but occurs elsewhere in lower concentrations.

The mode of action and physiological purpose of histamine are unknown: even its effects have not been identified in all species although enough is known to indicate that species variations occur. In general, man, monkey, dog, cat and fowl respond in a similar manner, showing a marked fall in blood pressure, whereas rabbits and guinea-pigs show a rise, probably due to arteriolar constriction. The effects in the larger domestic animals are presently under investigation.

The main actions of histamine, and those against which antihistaminics are of most use, are a marked fall in blood pressure, an increase in capillary permeability, and spasm of all smooth muscle including that of the bronchi, intestines and uterus. An action against which H_1-antagonists are of no value, is the marked increase in gastric secretion: this secretion has a high acidity and low pepsin content.

The fall in arterial blood pressure is mainly caused by a dilatation of venules and capillaries, excluding the pulmonary vessels which contract in concert with the bronchial muscles, but including, in the dog, the coronary vessels. An increased heart rate with weakened muscle power accompanies the fall in blood pressure.

The clinical effects consist of salivation, vomiting in those species capable of this function, abdominal pain (colic) and diarrhoea, dilatation of skin capillaries and increased heat loss, a rapid weak pulse, a decreased circulating blood volume with loss of fluid and possibly protein into the tissues, and an increased respiratory rate with acute dyspnoea.

The effects of histamine may be antagonized by:

1. *Physiological antagonists.* These consist mainly of the sympathomimetic amines, adrenaline, noradrenaline, ephedrine, amphetamine and others. These drugs are still used quite frequently to counter the cardiovascular effects of allergic reations.

2. *Destructive antagonists.* Histaminase has already been mentioned as the enzyme which destroys histamine produced naturally in the body. It has been used clinically but is too slow in acting to be of much value.

3. *Pharmacological antagonists.* The classical antihistaminics are competitive antagonists of histamine's actions at what are called the H_1 receptors. They fail to antagonize certain elements in histamine's spectrum of activity, e.g. its ability to stimulate gastric secretions. Such actions are now ascribed to activation of a second species of histamine receptors simply called the H_2-receptors.

The Classical Antihistaminics

Interest in antihistaminics was stimulated by a report, in 1937 by Bovet and Staub of the Pasteur Institute in Paris, on two compounds designated 929F and 157F by their discoverer Fourneau, an eminent French organic chemist. These two compounds were shown to have antihistaminic activity but proved too toxic for clinical use. Since then many compounds with antihistaminic activity have been developed and marketed.

The antihistaminics show some chemical resemblance to each other and the relationship of chemical structure and clinical activity is now becoming clearer, although in both these respects exceptions exist.

The majority of the antihistaminics conform to a basic structure which is very similar to that of histamine.

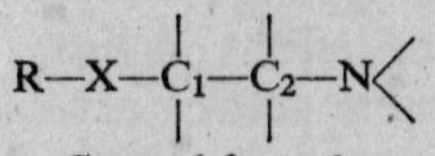

General formula

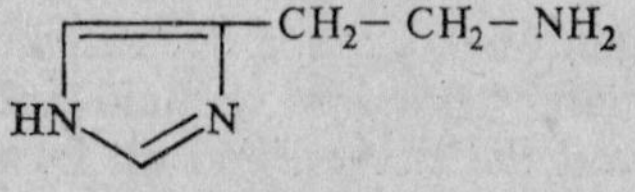

Histamine

1. In this formula R is often known as the nucleus whilst the rest is referred to as the side chain, but in many cases the atom X is often incorporated in the ring structure which comprises the nucleus (see Promethazine). The nucleus is usually composed of two or more cyclic groups which may be aromatic or heterocyclic or both.

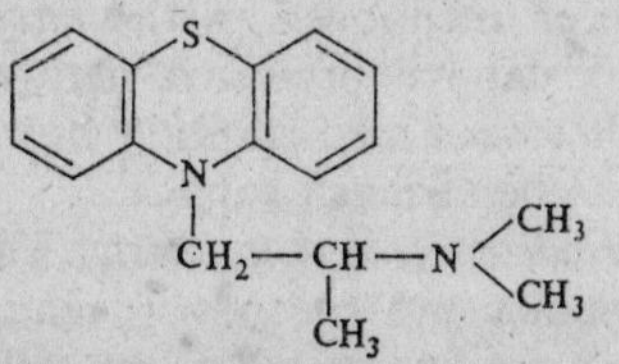

Promethazine (*N*-(2-dimethylamino-2-methyl) ethyl phenothiazine)

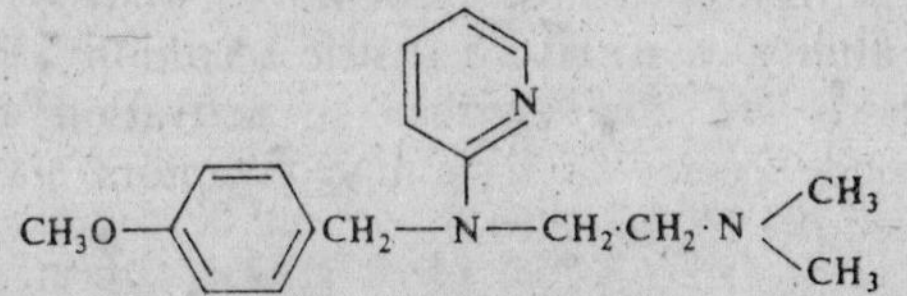

Mepyramine (2-[(2-dimethylaminoethyl) (*p*-methoxybenzyl)amino]pyridine)

2. The place of 'X' in the general formula may be taken by one of three atoms, nitrogen, oxygen or carbon. Antihistaminics representative of these three types are, respectively, tripelennamine (or mepyramine), diphenhydramine (also dimenhydrinate), and prophenpyridamine. These chemical differences are of considerable clinical significance, for compounds containing nitrogen at 'X' are usually far more powerful than those containing either oxygen or carbon atoms in this position; those containing carbon, though they have a decreased activity also possess fewer side actions and a lower toxicity, whilst those compounds with oxygen in place of 'X' have sedative side effects, though such effects are not by any means confined to oxygen-containing compounds as reference to chlorpromazine will show. (Chlorpromazine may be looked upon as an antihistaminic in which the sedative side effects are developed almost to the exclusion of the antihistaminic action.)

3. The structure of the remainder of the side chain also has a marked effect on the antihistaminic action for maximum effect is achieved when this chain consists of two carbon atoms and a dimethylated nitrogen atom. The carbon atom valencies are normally completed by hydrogen atoms although attachment of a methyl group to carbon 2 of the side chain as in promethazine appears to increase antihistaminic activity, whilst similar substitution of one of the hydrogen atoms on carbon 1 reduces activity. Increase of the carbon atoms in the side chain in any other manner reduces activity. Antazoline (iv) constitutes a partial exception to this side chain

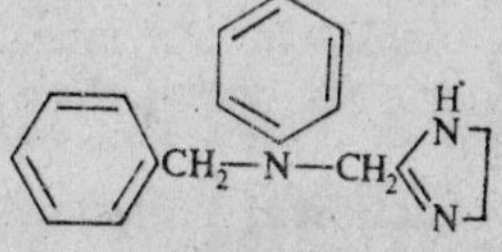

Antazoline

theory for the dimethyl-aminoethyl side chain is replaced by a methyl-iminazoline group. Nevertheless, in support of the theory the potency of the compound is less than that of more orthodox preparations and the side actions are also weaker, the lack of local irritation being of clinical significance.

The physical characters are very similar. All are white or creamy white crystalline powders which are either odourless or have only a slight odour, most have a bitter taste and may cause a local numbing of the tongue due to local anaesthetic effects. All are basic compounds which form salts, the solubilities of which in water, alcohol and chloroform, vary from less than 1 : 1 up to 1 : 6, with the exception of antazoline which is soluble only to the extent of 1 : 50 in water, 1 : 65 in alcohol and is only slightly soluble in other solvents. Many of the salts are unstable and darken on exposure to air and light.

Actions

Antihistaminics are satisfactorily absorbed from the gastro-intestinal tract of simple-stomached animals but not of ruminants.

Although satisfactory, absorption following oral administration in small animals is relatively slow and a delay of up to 30 min can be expected before any effects become apparent. The delay is shorter following intramuscular administration, while immediate effects follow administration by the intravenous route which, however, is not a satisfactory route for the administration of antihistaminics because of the danger of causing cardiovascular disturbances. The duration of effect varies from as little as 3 hr for diphenhydramine hydrochloride up to 24 hr or more with promethazine hydrochloride; this duration of action is proportional to, and is an indication of, the rate of destruction.

External application is suitable only for the treatment of skin lesions in which the surface is broken or congestion and inflammation are present. Little absorption takes place through intact skin.

The method by which the antihistaminics counter the action of histamine is not known, but the general accepted theory suggests that, because of structural similarities, the antihistaminics act as antagonists and compete with histamine for the specific 'receptors' in the tissue cells, thereby blocking the entry of histamine into the cells. This means that the antihistaminics must be available to the cells in greater quantity than the histamine or must have a greater affinity for the cell receptors. The action is therefore analogous to those of the various antagonists which compete with acetylcholine for attachment to the differing subpopulations of its receptors on cell membranes of receptive tissues.

The antihistaminics counter the actions of histamine in varying degrees. The most easily countered effects are the contractions of the smooth muscle in the bronchi, intestines and uterus. The vascular effects are less easily countered but the action of the antihistaminics in this respect is of greatest clinical importance for they prevent capillary dilatation and permeability—the effects of histamine which cause marked falls in blood pressure and oedema of the tissues, especially the subcutaneous plaques and swellings so typical of serum reactions.

Significant side actions occur but the nature and intensity of these vary with each individual drug and patient. The side effects of most importance are:

1. Central nervous system depression (cf. chlorpromazine), which results in drowsiness and lethargy during treatment.

2. Local anaesthetic effects, which may be three or four times as great as those of procaine and are of value in external therapy, although irritant effects on tissues prevent their specific use as local anaesthetics.

3. Gastro-intestinal disturbances in simple-stomached animals when oral administration is prolonged.

4. Intravenous administration may cause prolonged periods of excitement which may necessitate barbiturate sedation.

5. Marked interference with the allergic diagnostic tests, particularly the tuberculin test. Antihistaminics must be avoided while such tests are in progress.

Uses

It must be emphasized that antihistaminics have no effect on the causal factors of any of the conditions mentioned below and administration must therefore be continued until the causes have been removed either by concurrent therapy or by the natural body processes. Should the aetiological factors not be removed then symptoms will recur following withdrawal of antihistaminic treatment. Antihistamine therapy is thus symptomatic treatment only.

If allergies are acute and severe, antihistamine compounds are usually too slow in action to be of immediate use unless given by slow intravenous injection. As administration by this route is attended by some danger of causing hyperexcitability, it is advisable to give adrenaline initially, followed by oral or intramuscular dosing with an antihistaminic. In all cases, systemic therapy should be continued for about 3 days unless dramatic recovery follows within 24 hr of dosing. Oral dosing in small animals should be carried out either at meal times or shortly after feeding; gastric irritation is thereby decreased and absorption time prolonged.

The conditions in which antihistamines are of most value are those associated with allergic congestion, oedema, pruritus and any condition usually associated with traumatic histamine release.

In the first group occur all types of urticaria, serum allergies and allergic dermatitis cases, whilst in the second group may be placed moist eczema in dogs, especially when associated with otitis, 'sweet itch' or 'summer itch' in horses, plant photosensitization cases, insect bites, burns and scalds.

A number of conditions commonly believed to be allergic in origin have also been treated with antihistamines although the character of these conditions makes assessment of results difficult. Purpura, lymphangitis, 'blue nose' and pulmonary emphysema (heaves or broken wind) in horses respond to varying degrees. Some cases of bloat in ruminants (those due to histamine atony of the rumen following over-eating), have been treated with antihistaminics and good results have been claimed.

In addition, conditions associated with gross tissue destruction are reported to benefit from antihistamine treatment. The rationale for this use of antihistaminics is that in the process of protein destruction large amounts of histamine are produced and absorbed to the detriment of the sick animal: neutralization of this histamine confers benefit proportional to the amount present. Acute septic and gangrenous mastitis, septic metritis and retained placentae, myoglobinuria and azoturia, as well as acetonaemia, in which protein conversion to sugar involves protein breakdown, have all responded to varying degrees.

An illicit use of antihistaminics has been discovered in recent years; this is the use of their sedative effects for the doping of greyhounds.

The use of antihistamines to combat surgical shock has given disappointing results, although clinical assessment of any benefit is extremely difficult. The corticosteroids are valuable alternatives to the antihistaminics as anti-inflammatory and anti-oedema therapeutics.

Toxicity. The main symptoms of toxicity are the hyperexcitability and possibly convulsions which result from overdosage. Treatment is symptomatic and usually consists of the injection of sedative doses of short acting barbiturates.

Antazoline Hydrochloride, B.P.C.

Better known as Antistin (commercial name), antazoline is less active than many of the other antihistaminics but has the advantage of causing less residual irritation, which makes it more acceptable for local application. Parenteral administration is accompanied by fewer untoward reactions. It may be used as a 0·5 per cent ophthalmic solution, as a 2 per cent cream for topical use, and as ampoules or tablets containing 100 mg for parenteral and oral administration respectively. Applications need to be repeated every 6 to 8 hr to achieve maximum success.

Dosage

The drug has not been used extensively for systemic action, but the dose for small animals would be about 100 mg given three to four times daily.

Diphenydramine Hydrochloride, B.Vet.C.

This compound is widely known as Benadryl. It is a potent and effective antihistaminic and is more sedative than other common members of the group. Onset of effects is fairly rapid and the duration of effect is somewhat longer than that of antazoline. The strong local anaesthetic effect makes the compound very useful for local application in urticarial pruritus and other skin conditions. It has been used orally in capsule form or as a solution for the treatment of travel sickness and for urticaria in small animals. It may be given parenterally and is also applied externally as a 1 per cent lotion or 2 per cent cream.

Dosage

Dosage for large animals is from 0·2 to 0·3 g for the adult horse and cow repeated in 12 to 24 hr if necessary, but in the dog the dosage rate is about 2 mg per kg body weight by mouth every 12 hr and 0·5 to 1 mg per kg body weight parenterally.

Mepyramine Maleate, B.Vet. C.

Known in America as pyrilamine maleate, this medium duration antihistaminic was the first of the group available in Great Britain and is known widely as Anthisan. The effects of mepyramine maleate persist from 4 to 6 hr. The compound is a good general antihistaminic which is fairly well tolerated and may be given by oral or intramuscular administration for systemic action; it can also be applied topically for local effect. Because of the irritant action following parenteral use it is advisable to use 5 per cent solutions only in large animals: the 2·5 per cent solutions are suitable for small animal treatment.

Dosage

Intramuscular dosage of the 5 per cent solution is 10 to 30 ml in cattle and horses, 5 to 10 ml in sheep and pigs; 1 to 5 ml of the 2·5 per cent solution is satisfactory for dogs. Oral doses in dogs vary from 50 to 100 mg. Doses should be repeated every 8 to 12 hr. Topical applications of a 2 per cent cream should be applied twice daily.

Promethazine Hydrochloride, B.Vet.C.

This compound, the second antihistaminic to be introduced into veterinary practice in Great Britain, is known commercially as Phenergan. Promethazine is one of the most active of the group but its side actions are also more pronounced including anticholinergic and hypnotic activity. Its duration of effect is in excess of 24 hr which reduces the need for repeated injections—single daily doses are sufficient. Intramuscular and oral administration follow the same pattern as with other antihistaminics. Local applications usually contain 2 per cent of promethazine hydrochloride and require application once or twice daily.

Dosage

Large animals are injected intramuscularly with a 5 per cent solution at the rate of 5 to 20 ml for cattle and horses, 2·5 to 10 ml for sheep and pigs: small animals (dogs) require from 1 to 5 ml of a 2·5 per cent solution. Single daily injections should be given on three occasions.

H_2-Blockers

One of the few clinical uses to which histamine has been put in man is in a diagnostic test in which the ability of the gastric mucosa to produce an acid secretion is measured following the subcutaneous injection of histamine. This response to histamine is not blocked by the classical antihistaminics and it is now believed that there are two types of histamine receptors. Increased gastric acid secretion, stimulation of other exocrine secretions, some instances of smooth muscle contraction inhibition, enhancement of cardiac function and a contribution to dilation and increased permeability of small blood vessels are mediated via H_2-receptors. The other actions of histamine, i.e. all those blockable by classical antihistaminics, are mediated via H_1-receptors. Drugs possessing the ability to prevent the H_2 effects of histamine were introduced in 1972 and have come to be called the H_2-blockers. The first of these, burimamide, was poorly absorbed from the gut and has been replaced by a closely similar structure, metiamide. Both drugs bear molecular similarity to histamine itself.

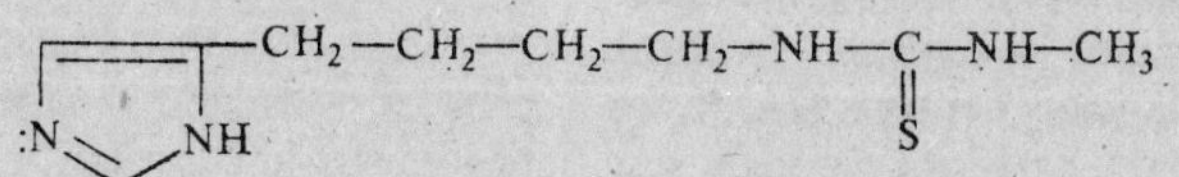

Burimamide

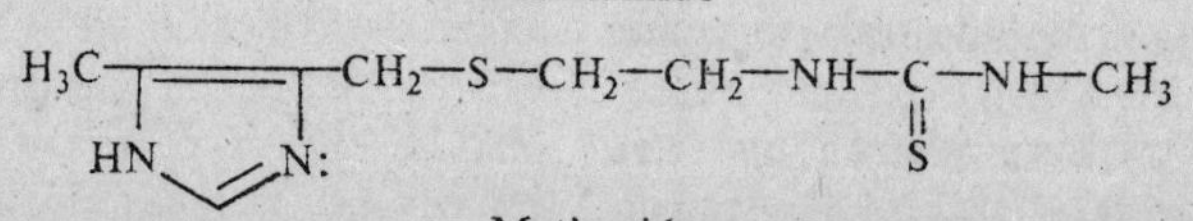

Metiamide

In man these drugs are of potential value in controlling conditions where over-production of gastric acid is a problem. Duodenal ulceration is an obvious example. H_2-blocking drugs are currently being used in many studies aimed at a fuller understanding of histamine's role in physiological and pathological processes and undoubtedly therapeutic applications will flow from this work.

ALLERGY AND ANAPHYLAXIS

The normal response of the reticuloendothelial system (RES) to a foreign protein or to a complex of foreign material plus endogenous protein is the production of antibody. Antibodies are found in the globulin fraction of plasma protein in which are several distinct immunoglobulins (IGs). A fraction distinct from normal IG has been shown to sensitize host tissues for systemic or local anaphylaxis and is described as reaginic when it combines with target cells, e.g. mast cells. A subsequent exposure to the same antigen leads to a reaction with the cell-fixed antibody which results in the disruption of the cell with consequent liberation of its vasoactive constituents. Histamine is in this way liberated from sensitized mast cells. Anaphylaxis is the term used to identify the resultant shock-like state which supervenes in the animal. However, a process indistinguishable from anaphylaxis can follow the reaction of soluble antibody with antigen. In the blood, histamine is carried in the granules of basophils and platelets. Anaphylactic shock was frequently elicited by repeat administrations of antisera produced in a different species (heterologous antisera), but is nowadays more frequently encountered following certain drugs. In the dog, anaphylaxis is characterized by constriction of the hepatic veins which dams back venous blood in the liver and portal vessels. Intestinal haemorrhage may occur. In the guinea-pig, bronchoconstriction is the major symptom. In both species, the symptoms can be reproduced by histamine administration and anaphylaxis can be prevented by the previous injection of an antihistaminic. Certain drugs are able to cause the release of stored histamine, e.g. morphine, curare, stilbamidine, trypsin, dextran and the polymer Compound 48/80. The administration of these drugs may be accompanied by the symptoms of histamine release.

Allergy can be regarded as a localized hypersensitivity reaction whereas anaphylaxis is usually a generalized response. In man and cattle, penicillin allergy results in skin lesions, whereas sensitivity to inhaled allergen results in nasal congestion (hay fever) and/or bronchoconstriction (asthma). Should the allergen be ingested, then the hypersensitivity reaction is often expressed as a digestive system disorder. Although many of the symptoms of anaphylaxis can be reproduced by histamine not all are and it is now known that other substances contribute especially in delayed (e.g. contact dermatitis) rather than immediate hypersensitivity reactions (e.g. urticaria, rhinitis, broncho-constriction). In relation to all hypersensitivity reactions, it can be said that the size of the reaction will depend on the amount of antigen and the amount and type of antibody participating. The amount and number of active substances released will vary with the kind of cell damaged, the enzymes consequently activated and substrates accessible to them, and factors which influence the activity of the enzyme, e.g. pH. The pharmacological or clinical effects produced will be determined by the distance between the site of action and point of release of the active substance, by factors which effect the activity of the active substance, e.g. rate of inactivation and by the type of cell influenced by the active substance. The interaction of these variables is responsible for species' differences and individual variation in anaphylaxis and allergy. Mediators known to be involved in immediate type hypersensitivity reactions and which cause increase in secretory activity, smooth muscle contraction, increased permeability of blood vessels, itch and pain are discussed in the following pages.

Histamine Liberation

It is known that histamine is formed in those cells in which it is stored (i.e. mast cells and basophils) in association with heparin in the so-called secretory granules. From these sites the amine can be released by a variety of drugs, often as a side-effect of the drug and also during immediate type hypersensitivity reactions involving cell-bound, reaginic antibody. The discharge of histamine does not involve cell disruption but is an active, secretory process perhaps initiated by the entry of Ca^{++} ions after the permeability of the cell membrane has been increased by the antigen–antibody reaction on its surface.

Compound 48/80 is a mixture of basic polymers capable of causing mast cells to liberate rapidly, by the usual process of exocytosis, their stored histamine. It is of experimental interest only.

Chelators of calcium can arrest the degranulation of mast cells even after the antigen–antibody reaction has taken place, but it is very unlikely that they will be able to be used therapeutically because widespread side-effects could occur.

Sodium cromoglycate is a drug introduced in recent years for the prevention of asthma in man. It prevents the release of histamine and SRSA during pulmonary allergic reactions. This valuable effect seems to be mediated via the pulmonary mast cells which following exposure to the drug fail to discharge their granules when the sensitized cell is exposed to antigen. The powdered drug is administered by inhalation and it has no value other than in prophylaxis. Indeed, its effect is relatively specific to the lung although it is now undergoing evaluation for gut allergies in man.

5-Hydroxytryptamine (5-HT)

This base, also known as serotonin or vasotonin, was first identified as a vasoconstrictor released from platelets lysed in the coagulation process. It is now established that considerable quantities of 5-HT occur in the argentaffin cells of the gastro-intestinal tract. It also occurs in the mast cells of rats, mice and bovines in which species it is released during anaphylaxis. It is present in other tissues and is believed to be a transmitter substance in the CNS.

5-HT is produced in the body by the hydroxylation of tryptophan to 5-hydroxytryptophan which is then decarboxylated to 5-HT. The blood 5-HT is stored in the platelets and these, just like the granules of argentaffin and mast cells, can accumulate 5-HT against a concentration gradient. Elimination of 5-HT is by a monoamine oxidase catalysed oxidation to 5-hydroxyindolacetic acid or by acetylation to yield *N*-acetyl-5-hydroxy tryptamine and excretion in the urine. The noradrenaline-releaser drug, reserpine, also liberates stored 5-HT and it is suggested that the 5-HT released in the brain is responsible for the depression produced by reserpine. 5-HT is released in allergic reactions in rabbit, mouse, rat and possibly in ruminants, but not in man.

The actions of 5-HT are constriction of blood vessels, especially in the lungs, and contraction of the smooth muscle of intestine, uterus and bronchial tree. Nerves and ganglion cells may be stimulated and adrenaline is released from the adrenal medulla. Many of the effects of 5-HT are the consequence of reflex actions following sensory ending stimulation, e.g. the peristaltic reflex of intestine and reflex slowing of the heart. The interplay between the various possible actions is probably the basis for species and individual variation in response to 5-HT.

Competitive antagonists of 5-HT are available, and may help to control anaphylaxis in the bovine. They are members of the ergot series, e.g. dihydroergotamine, or are derivatives containing lysergic acid, e.g. lysergic acid diethylamide (LSD), 2-bromo-lysergic acid and methysergide. In man, these drugs are used as hallucinogens in psychotherapy, in the treatment of migraine and in the pre-operative control of tumour of argentaffin tissue.

Bradykinin

This is a nonapeptide which is produced from a circulating, inactive, globulin precursor called bradykininogen by enzymic action. Such enzymes (kallikreins), or enzyme precursors (kallikreininogens) are to be found in blood and certain body secretions and they can be released in some antigen–antibody (Ag–Ab) reactions. Apart from these mechanisms the proteolytic enzymes trypsin and fibrinolysin can also split bradykinin from its precursor. Precursor concentration is much reduced after severe or prolonged allergy, as in asthma in man. As with other mediators, it is also possible that much bradykinin is formed, exerts its action and is inactivated by a kininase within the tissues and this means that blood levels may not be indicative of body turnover of bradykinin.

Bradykinin is a very potent vasodilator and causes a fall in blood pressure and an increase in capillary permeability. It owes its name to its ability to induce slow contractions in intestinal smooth muscle and also causes uterine and bronchial musculature to contract. Its occurrence in association with exocrine glands suggests a functional role in secretion. Additionally bradykinin causes adrenaline release from the adrenal medulla and stimulates autonomic ganglia. Its most prominent actions are those which also characterize early inflammation, i.e. vasodilation, increased vascular permeability, pain production and accumulation of leucocytes.

Kallidin, a decapeptide, is lysyl bradykinin. It may be liberated with bradykinin and has the same effects.

Eledoisin, physalaemin and caerulein are further examples of peptides, which although derived from lower animals, have bradykinin-like action in mammals.

Substance P is a substance first extracted from horse intestine and brain and whose molecule is known to be an endecapeptide. Again, it has bradykinin-like actions and may function in maintaining peristalsis or as an inhibitor in CNS.

Angiotensin. An enzyme, renin, can be released by ischaemic juxtaglomerular cells in the kidney. It separates a decapeptide (angiotensin I) from a plasma α-globulin precursor (angiotensinogen). A second enzyme present in plasma converts angiotensin I to the octapeptide angiotensin II. Angiotensin is a powerful vasconstrictor and its release is followed by a rise in blood pressure due to a direct action on vascular smooth muscle. This action is accompanied by a release of adrenaline. Angiotensin also causes smooth muscle of other systems to contract. This polypeptide is not a mediator of inflammation and is only included at this point because of its chemical relationship to the kinins. Its function may well be the maintenance of an adequate kidney blood flow. Another physiological role is the stimulation of aldosterone secretion which causes increased plasma volume. The renin–angiotensin–aldosterone mechanism is involved in the aetiology of chronic hypertension in man.

SRSA

This is one of a group of fatty acids known to possess pharmacological activity. It is released from lungs during an anaphylactic reaction and owes its name to its slow onset of action (slow reacting substance of anaphylaxis). It is as yet not identified owing to its complexing abilities. The most pure form examined is a vasodilator, but has no gut contracting ability.

The contribution of the kinins and SRSA to bronchoconstriction in experimental allergy under both in vivo and in vitro conditions can be antagonized by the **salicylate** antipyretics and, more powerfully, by the related compound, **sodium meclofenamate.** These drugs appear to act at the level of the released mediator rather than by preventing the release of mediators via an immune reaction. The antihelmintic drug **diethylcarbamazine** can inhibit the elaboration of SRSA during anaphylaxis and may therefore be of value in bovine anaphylaxis.

Prostaglandins

The contribution of various prostaglandins to inflammatory processes is complex and variable according to the site. In general the effect is vasodilator of all elements of the microcirculation (PGE, PGA) and permeability is increased (PGE_1, PGE_2, $PGF_1\alpha$, $PGF_2\alpha$). Pulmonary arteries and veins are constricted by $PGF_2\alpha$ and dilated by PGEs and PGAs. Similarly bronchial smooth muscle is contracted by $PGF_2\alpha$ and relaxed by PGE. Prostaglandins, in common with the other tissue hormones, are believed to act via specific receptors, the activation of which probably causes a change in membrane permeability with resultant influx of Ca^{++} ions. Additionally PGs can increase or decrease the levels of cyclic AMP in target tissue. These agents are believed to be associated with the slow aspects of inflammation, are present in synovial fluid in some instances of arthritis and have been detected during allergic reactions. Various anti-inflammatory drugs, e.g. aspirin, phenylbutazone and indomethacin, have been shown to inhibit prostaglandin synthesis and this is believed to be the basis of their anti-inflammatory effect.

Permeability Factors

Lymph node permeability factor (LNPF) is a recently discovered principle which plays a part in the inflammatory process of delayed allergy, e.g. the tuberculin reaction. It is released by lymphocytes and increases the permeability of blood vessels, attracts leucocytes and causes the perivascular deposition of an eosinophilic material. Interestingly, this factor is antagonized by salicylates. This material has not been characterized as yet and it is possible that it acts indirectly, e.g. by the local induction of histamine or bradykinin.

Bibliography

COCHRANE, C. G. & KOEFFLER, D. (1973) Immune complex disease in experimental animals and man. *Adv. Immunol.*, **16**, 186–248.

FERREIRA, S. H. & VANE, J. R. (1974) New aspects of the mode of action of nonsteroid anti-inflammatory drugs. *A. Rev. Pharmac.*, **14**, 57–75.

Gell, P. G. H. & Coombs, R. R. H. (1975) *Clinical Aspects of Immunology*, 3rd ed. Oxford: Blackwell Scientific.

Kaliner, M. & Austen, K. F. (1975) Immunologic release of chemical mediators from human tissues. *A. Rev. Pharmac.*, **15**, 177–91.

Schachter, M. (ed.) (1973) *International Encyclopaedia of Pharmacology and Therapeutics, Section 74, Histamine and Antihistamines.* Oxford: Pergamon.

Schild, H. O. (ed.) (1968) Immunopharmacology. In: *Proceedings of the Third International Pharmacological Meeting 1966*, vol. 2. Oxford: Pergamon.

Walton, G. S. (1971) Allergic responses involving the skin of domestic animals. *Adv. vet. Sci. comp. Med.*, **15**, 201–30.

16

Drugs for the Management of Metabolic Disease in the Ruminant

The term metabolic disease in its narrower context is used in veterinary medicine to describe a range of conditions in which the major biochemical symptom is a reduced blood concentration of calcium, magnesium, phosphorus or glucose. This low blood level is not always associated with an absolute deficiency, but is believed to reflect a temporarily impaired ability to mobilize reserves within the animal. Although straightforward dietary deficiencies of these substances can and do occur, such problems are not the primary concern in this chapter. The roles of sodium, potassium, chloride, copper, cobalt, iron, iodine, zinc and selenium are dealt with elsewhere.

CALCIUM, MAGNESIUM AND PHOSPHATES

Although all three ions have important electrolyte functions, especially in the acid–base buffer systems, the part these ions play in tissue metabolism is more important clinically, as a reference to pathological conditions in which the three are involved will show: rickets and osteodystrophia, milk fever, milk tetany in calves and hypomagnesaemia in adults, uterine inertia, and sterility are some of the more important ones. In many of their functions the three ions are interrelated, although this is not always the case, for in hypomagnesaemia the calcium and phosphate blood levels are unaffected whilst the magnesium levels may be almost undetectable: likewise in ordinary milk fever the calcium and phosphate levels may be extremely low whilst the magnesium levels may be normal or higher than normal.

Calcium, magnesium and phosphates are present in the extracellular fluids in relatively small amounts when compared with sodium, bicarbonate and chloride ion concentrations, but the intracellular fluids contain large amounts of magnesium and phosphate ions as well as of potassium.

Should the 'mineral' concentrations in the blood fall below minimal levels (which are slightly variable according to seasonal and individual fluctuations), a condition is precipitated which is known in cows as milk fever but in most other species as eclampsia. If the magnesium level alone is affected the condition resulting is known as hypomagnesaemia.

THE MILK FEVER SYNDROME

The variety of symptoms which collectively constitute the syndrome are related to acute pathological decreases in the blood levels of calcium, magnesium and the phosphates: the deficiencies are referred to respectively as hypocalcaemia, hypomagnesaemia and hypophosphataemia. The aetiology of the condition is still largely a matter of conjecture but, summarizing the present theories, it would appear that the main exciting cause is the drain of calcium and of phosphates into the colostrum at the time of calving, and though this occurs in normal cows at calving, the milk fever cow is unable to compensate for this drain. The inability to cope with this sudden call for minerals may be due to lack of mobilizable reserves or to disorganization of the process of mobilization either through an enzyme or a hormone imbalance, or a combination of both.

Analysis of the incidence of the condition shows that it occurs in both dairy and beef cattle, but that in the former the disease is predominantly a calcium and phosphate imbalance whereas in beef breeds it is the hypomagnesaemic type which seems more frequent. The majority of cases occur within 2 days before and 3 days after calving. Age, and therefore the number of pregnancies undergone, is a predisposing factor—most cases appear to occur with or after the third pregnancy, presumably as the mineral reserves become depleted. Climatic conditions appear to affect the incidence and 'magnesium' cases appear to be precipitated by cold weather, especially if such spells occur intermittently and alternate with mild wet periods. This may be partly correlated with increases in the growth rate of grass but probably also with a lowering of the animals resistance or even an induced metabolic effect on mineral levels.

The excitability of the tissues is calcium dependent. Calcium, magnesium and phosphate ions are all vitally linked in blood, bone, muscle and nerve metabolism. They work essentially in equilibrium and, should the concentration of any one member of the three-cornered equation be changed, the function of the other two, and various related coenzyme systems, are also disturbed. Calcium and magnesium are both able to depress nerve and muscle membrane irritability: in the presence of excess calcium or magnesium ions, the membrane does not become permeable to sodium until its depolarization crosses a higher than normal threshold. This action is known as membrane stabilization. Calcium is believed to 'couple' the depolarization of a cell membrane with the response of the cell to the stimulus. In voluntary, smooth and cardiac muscle, the presence of calcium is essential for normal function of the contractile elements. Calcium ions are intimately involved in the release of acetylcholine. The coupling, contractile and releaser actions of calcium are antagonized by magnesium. These functional relationships mean that absolute or relative changes in calcium and magnesium levels will be accompanied by marked abnormality of nerve and muscle excitability.

Magnesium is also necessary in the coenzyme system of *phosphatase 1*, which is involved in calcium and phosphate deposition in the bone, and also the coenzyme system concerned in glycogen, adenosine triphosphate and creatine phosphate reactions which are essential for producing the energy required for muscle contraction. Phosphate ions also play a large part in these enzyme systems concerned with producing energy for the muscles:

Adenosine triphosphate (ATP)→adenosine diphosphate (ADP)+inorganic phosphate→ energy for muscle contraction. The ADP formed reacts with creatine phosphate to form creatine and ATP, while the creatine is reconverted into creatine phosphate, using energy supplied by the tri-carboxylic acid cycle in which phosphates also play a part. If phosphate ions are not freely available the production of energy is curtailed and muscular contraction is weakened proportionally. This is reflected in the symptoms of hypophosphataemia which are typically those of a recumbent cow unable to rise or stand.

The blood picture in milk fever is by no means constant. The normal values of serum calcium, magnesium, and whole blood inorganic phosphate of from 8·5 to 11·5 mg, 1·8 to 3·2 mg and 3·5 to 6 mg respectively (averages 10, 2·5 and 5 mg) per 100 ml show significant changes at normal parturition, with considerable falls in calcium and phosphate and a rise in magnesium. These falls of calcium and phosphate are exaggerated in milk fever animals but the magnesium levels may fall, rise or stay normal. Levels below 7 mg calcium, 1·6 mg magnesium and 2·5 mg phosphate may be considered pathological, although exceptions do occur and in Channel Island cattle normal magnesium levels are higher than in other breeds.

The symptoms of milk fever cannot easily be correlated with the blood picture. Uncomplicated milk fever is accompanied by dullness, a dry nose, a dilated pupil, a constricted anal sphincter and a disinclination to eat. 'Paddling' of the hind legs, with stiff extended joints and nervousness due to lessened muscle control, precede collapse, and a period of narcosis in which the neck may be typically kinked may follow. Finally, coma supervenes in which the head may be turned onto the flank. This sequence of symptoms is found in 'classical' milk-fever.

Acute hypophosphataemia occurs very commonly in the modern form of milk fever. The lack of blood phosphate is correlated with muscle weakness: a typical case will show acute lumbar pain, parturient oedema of the brisket and udder, and be unable to rise. In many cases calcium injections allay the more acute signs but the animal, though apparently normal, is unable to rise; in other cases, there is a temporary improvement but the animal relapses a few hours later and rarely responds to a second injection. These cases usually become chronic if treated first with calcium and may stay recumbent for several days, even if subsequently treated with phosphorus injections.

The typical hypomagnesaemic milk fever case shows what is known in some areas as 'the blinks'—a fluttering of the eyelids: the animal is extremely hypersensitive, shows muscular tremors and may move in circles, occasionally appearing to 'attack' attendants: the pupil is often constricted and the anal sphincter flaccid. Convulsions may occur and the animal falls to the ground: the stimulus of inserting a needle may be sufficient to induce a fatal spasm.

Treatment of the uncomplicated hypocalcaemia case is fairly straightforward and failure to respond to plain calcium borogluconate injections is not frequent. It is common practice to inject 400 ml of a 40 per cent calcium borogluconate solution intravenously and provide a maintenance reservoir by injecting 400 ml of a 20 per cent solution subcutaneously.

Acetonaemia is a complicating factor in milk fever. Complex calcium, magnesium, phosphorus and dextrose solutions should be used in these cases, but where acetonaemia persists it should be dealt with by normal methods once the mineral levels are corrected. Cortisone derivatives, ACTH, glycerin and propylene glycol, propionates, acetates and lactates, as well as glucose injections, may be used.

Prophylactic measures have been investigated in recent years. Various workers have shown that high protein and calcium diets during pregnancy predispose cows to attacks of milk fever. Low calcium and high phosphorus diets during the dry period before calving have therefore been suggested: this is provided by hay and concentrate mixtures supplemented by the addition of monosodium phosphate in the proportion of 1·5 to 5 per cent of the basic ration. Absorption of Ca^{++} is enhanced and PTH secretion is increased. It has also been shown that daily doses of 10 to 30×10^6 units of vitamin D_2 for up to 7 days before calving will reduce the incidence of milk fever. It should be remembered that a danger attaching to overdosing is the widespread deposition of calcium in the blood vessels and organs. For Ca^{++}, PTH and vitamin D interrelations, see Chapter 18.

Dihydrotachysterol, a steroid related to the D vitamins has been used to prevent milk fever. It is administered by injection close to the expected day of parturition and has been shown to cause an increase in plasma calcium concentration. Unfortunately, this is accompanied by a fall in plasma inorganic phosphate concentration and would therefore appear to be of value only in instances where milk-fever is not complicated by a concurrent hypophosphataemia.

Calcium Salts

The salts of this alkali metal are white or off white amorphous or crystalline powders, many of which have a low water solubility. The insoluble salts are relatively inert externally and act as skin sedatives; internally they also tend to have a gastro-intestinal sedative and anti-diarrhoeic action. Absorption of calcium is conditioned by three main factors:

1. *The intestinal pH.* An acid pH tends to promote absorption by encouraging ionization, whereas an alkaline pH is detrimental to absorption by encouraging precipitation of insoluble salts.

2. *The calcium–phosphorus ratio* in the diet is an important factor in calcium and, of course, phosphate absorption. The optimum ratio is about 1·5 : 1.

3. *Adequate amounts of vitamin D* are essential for normal calcium metabolism, including absorption. If vitamin D concentrations are maintained, fairly wide calcium–phosphorus ratios can be tolerated: ratios of from 1·2 : 1 and 3·3 : 1 have been tolerated by calves when vitamin D was freely available (by sunlight). In general, vitamin D deficiency, a high phosphorus diet, excess fatty acids in the diet, alkali therapy and chronic diarrhoea are all detrimental to calcium absorption. It should be remembered that normally about two-thirds of any oral dose will be excreted in the faeces, though a propor-

tion of the faecal calcium represents that excreted from the blood into the intestine: the remainder is excreted in the urine. The urinary route is of less importance in the ruminant than the other domesticated species.

In the blood, calcium exists partly as ionized calcium and partly as protein bound calcium: it is only the ionized calcium which is the active portion. The amount of calcium in the blood (9 to 15 mg per cent) is regulated by the parathyroid hormone calcitonin mechanism (see Chapter 14).

The principal functions of calcium are:

1. The provision of a rigid skeleton and teeth. About 99 per cent of the body's calcium is found in bone and teeth.

2. The maintenance of normal function of the nervous and muscle systems; it is also essential for normal cardiac function.

3. Calcium is one of the basic ions which play an important part in the regulation of the acid–base balance: it is required in a definite ratio to potassium and sodium in the maintenance of membrane permeability.

4. The clotting mechanisms of both blood and milk are dependent on the presence of adequate concentrations of calcium.

Calcium Gluconate, B.Vet.C., and Calcium Borogluconate, B.Vet.C.

These two salts are perhaps the most widely used salts of calcium. Calcium gluconate is only slowly soluble in cold water (1 : 30) but is slightly more soluble in boiling water (1 : 5). By the addition of boric acid in the proportion of about 1 : 5 (17 : 83) the gluconate can be changed into borogluconate, with a solubility in cold water 1 : 5 (20 per cent) and in hot water of over 50 per cent; such solutions remain stable on cooling.

Calcium borogluconate is used for the treatment of hypocalcaemia in all species and by all parenteral routes. It is also useful as an antidote to overdosage with magnesium salts and chlorinated hydrocarbons. In the early days of calcium therapy of milk fever, intravenous injections of calcium chloride were used but, though the salt was suitably soluble in water, any extravascular leakage caused extensive sloughing. Attention then turned to the use of calcium lactate, which, though less soluble was also less irritant: it nevertheless frequently caused sloughs following perivascular injection and in 1930 Greig reported on the use of calcium gluconate. This too caused some irritation and occasional sloughs following subcutaneous injection, so that the more soluble and therefore more rapidly absorbed borogluconate superseded it. The search for more suitable calcium preparations continues and within recent years the calcium salt of levulinic acid, itself a derivative of hexose sugars, has been tried experimentally. Advantages of calcium levulinate are its greater solubility and higher calcium content (13 per cent w/v against 9 per cent of the gluconate). Even so, its clinical effects appear to be no better than those of the borogluconate and certainly do not justify preference over the cheaper gluconate.

Dosages of calcium borogluconate are:

Horse and Cow: 60–180 g (2–6 oz.)
Sheep and Pig: 6–20 g (90–300 gr.)
Dog and Cat: 1–5 g (15–75 gr.)

Aqueous solutions of up to 20 per cent w/v strength are suitable for subcutaneous injection (10 per cent w/v is better for small animals), but stronger solutions should be administered intravenously. All injections should be given at blood heat, especially subcutaneous injections which should also be well dispersed by massage or the use of hyaluronidase at the rate of about 75 units per 400 to 500 ml. It is now common practice in cattle to give 400 ml of a 40 per cent w/v solution intravenously and an equal volume of 20 per cent w/v subcutaneously, the latter injection ensuring a fairly sustained blood level. This technique has been found to be necessary in practice to reduce the incidence of relapses and has been validated by studies using labelled calcium. It was shown that calcium leaves the circulation very rapidly following intravenous injection and most is deposited in the skeleton. It is now known that the movement of calcium between bone, blood and tissue is much more rapid and quantitatively important than was believed previously. Intravenous injections of calcium solutions should be made slowly to minimize the risk of ventricular fibrillation or heart block. In eclampsia of the bitch, about

20 ml of a 20 per cent w/v solution is usually adequate: for lambing sickness in sheep and parturient eclampsia of the sow, 100 ml of a 20 per cent w/v solution are sufficient.

Calcium Hydrogen Phosphate, B.Vet.C.

This is an insoluble calcium salt, is used as an alternative to steam bone flour as a calcium and phosphate supplement for the treatment of calcium and phosphorus deficiencies (acalcicosis and aphosphorosis) but is not of any value in acute conditions.

Dosages of up to 12 g (180 gr.) in horses and cattle, up to 2 g (30 gr.) in sheep and pigs, and up to 0·6 g (10 gr.) in dogs and cats may be used in the therapy of rickets. Attention should be paid to the calcium–phosphorus ratio in dry cows in which a low calcium–high phosphorus ratio is advisable as a prophylactic against milk fever (*q.v.*).

Calcium Lactate, B.Vet.C.

This is used mainly in conjunction with vitamin D for the treatment of rickets in small animals in doses of 0·3 to 0·6 g (5 to 10 gr.) two to three times daily. Calcium Sodium Lactate, B.Vet.C., is used for the same purpose but is more soluble and more easily absorbed. In severe cases of rickets it is also advisable to provide bilateral support with plaster bandaging to prevent 'bowing': both legs must be bandaged, for otherwise the major part of the weight will be taken by the unbandaged leg on the opposite side and severe bowing of that leg may then occur, even when not apparently affected in the first place.

HYPOPHOSPHATAEMIA

If the cow is recumbent and suspected of suffering from hypophosphataemia, with or without hypocalcaemia it is advisable to inject phosphorus solutions intravenously immediately—either 30 g (1 oz.) of sodium acid phosphate in approximately 400 ml of water or preferably one of the commercial phosphorous injections which usually contain small amounts of calcium as well as phosphorus, usually in the form of hypophosphites. Oral dosing with 90 g (3 oz.) of acid sodium phosphate every 8 hr has also been practised. Where there is acute lumbar pain and it is desirable to have the animal on her feet as quickly as 'normal' milk fever cases, analgesics may be given. The old practice of udder inflation may be tried in cases which do not respond well to phosphate administration.

In many parts of the world, there is a soil phosphate deficiency and stock require supplementation in order to prevent the appearance of chronic deficiency i.e. anorexia and pica, impaired weight gains, infertility and lameness. Calcium, phosphorus and vitamin D are intricately involved in bone calcification and their deficiency results in rickets in growing animals and osteomalacia in adults. In pseudo-rickets, a vitamin A deficiency is also present. The phosphate ion is an important component of body buffer and the form of its excretion (i.e. as hydrogen phosphate or dihydrogen phosphate) in urine is a significant factor in adjusting body pH. In the cell phosphate is found in nucleic acids, phosphatides and is most important in energy transfer via the energy rich phosphate bond.

Phosphate is absorbed best from an acid environment and in order to prevent the formation of insoluble salts, needs to be present in a correct concentration relative to calcium. Absorption is increased by vitamin D and parathormone and excretion is via the kidney and intestinal mucosa.

Phosphorus Compounds

It is usual to supply phosphate along with calcium in the ratio of 1·5 or 2 parts calcium to 1 part phosphorus. Steam bone flour is the obvious substance to supply both calcium and phosphorus in the correct ratio.

Acid sodium phosphate (*q.v.*) is a soluble form of phosphorus for use in acute hypophosphataemia and may also be used orally in more chronic conditions and as a phosphate supplement for milk fever prophylaxis (*q.v.*). Most commercial injections for hypophosphataemia contain hypophosphites.

HYPOMAGNESAEMIA

Hypomagnesaemia, which is a complicating factor in the diagnosis of milk fever, is less related to calving than hypocalcaemia and hypophosphataemia. Magnesium blood levels show a seasonal fluctuation, being highest at the end of a normal summer and lowest at the end of winter just before the spring flush of grass. Two types of clinical hypomagnesaemia occur—and acute type which usually follows abrupt changes of diet, especially from indoor feeding to outdoor grazing on fresh leys. Aftermath grazing may sometimes precipitate an outbreak. The subacute type is usually recurrent and indicates a borderline low blood magnesium level. Cases may be precipitated by cold, wet periods, and often follow oestrus. Most cases occur between November and early March, but if the cold weather persists, the season may be prolonged. When low calcium levels also occur and the animal calves, milk fever may occur in which calcium, phosphate and magnesium are all implicated.

The acute or subacute types may occur in bucket-fed calves. In this instance, the condition represents a straightforward dietary deficiency of magnesium.

Hypomagnesaemia cases can be treated by combined calcium, phosphorus and magnesium injections, magnesium–dextrose injections, calcium borogluconate with magnesium injections, or by plain 25 per cent injections of magnesium sulphate. Magnesium sulphate should normally be given by subcutaneous injection but the others may be given by any parenteral route. Owners should be warned before commencing treatment that hypomagnesaemic animals may die as soon as the needle is inserted—explanations are less likely to be believed after the event. Prophylaxis against hypomagnesaemia is now being practiced on a large scale by the inclusion of calcined magnesite in the food at the rate of 30 to 90 g (1 to 3 oz.) per head daily. More recently it has been shown that hypomagnesaemia in grazing cattle can be controlled by the application of magnesian limestone or calcined magnesite in quantities of 5 tonnes and 1 tonne per ha (2½ tons and 10 cwt per acre), respectively—these quantities supply the same amount of magnesium.

Magnesium Compounds

The salts of this alkali metal which are of clinical value may be divided into the amorphous insoluble salts and a single crystalline soluble compound. The carbonates, oxides, and the trisilicate are the insoluble compounds, and magnesium sulphate the soluble substance. When administered orally these magnesium salts may be divided into three groups according to their clinical effects which are dependent on their individual solubilities in the gastro-intestinal fluids: the relatively insoluble compounds are unreactive and have an antidiarrhoeic action due to mechanical coating (magnesium trisilicate); the slightly soluble salts are laxative (magnesium carbonate and magnesium oxide), whilst the soluble compounds have a strong purgative action (magnesium sulphate). Magnesium is not very easily absorbed, though a restricted water supply following the administration of magnesium sulphate to a dehydrated animal may result in the absorption of sufficient magnesium sulphate to cause severe nervous system depression and anaesthesia. The absorption of magnesium is from the proximal small intestine and is by active transport. Uptake interference occurs in the presence of excessive amounts of calcium and phosphates or fats—the calcium takes up the bicarbonate radicals, the phosphate forms insoluble triple phosphates of magnesium, and fats form dibasic soaps which are insoluble: alkaline pHs are detrimental to magnesium absorption. The magnesium ion is essential for normal bone metabolism, normal nerve and muscle irritability (low magnesium blood levels are accompanied by tetany and convulsions—milk tetany in calves, grass tetany, stomach staggers or Hereford disease in adult cattle), for normal carbohydrate metabolism, in which it links up with phosphate metabolism and the provision of energy for muscle contraction via the coenzyme system in which magnesium plays an essential part. Around 50 per cent of the body magnesium is found in bone, 45 per cent is in the intracellular fluid, while 5 per cent is present in the ECF. Excretion is renal.

Magnesium Sulphate, B.Vet.C.

In the treatment of hypomagnesaemia the salt of choice is the soluble sulphate. It is prepared as a 25 per cent w/v solution and is routinely administered by subcutaneous injection. The local anaesthetic effect of this preparation is of some value as the solution is hypertonic and is often administered in large volume. Intravenous infusion is hazardous but sometimes indicated in severe cases. It should be performed quietly and slowly after the owner of the animal has been informed of the risk of precipitating cardiac arrest. The magnesium ion, like the potassium ion, is depressant to the SA node, the conducting system and the musculature. It also reduces blood pressure by a direct action on vascular smooth muscle supplemented by ganglionic blockade.

Elevation of Mg^{++} concentration causes a reduction in the release of acetylcholine both in the neuromuscular junctions and in the CNS. The relaxant and depressant consequences of these actions have made attractive the use of magnesium salts in anaesthesia. However, the intensity of effect related to dose is rather unpredictable and safer alternatives are available. These effects no doubt contribute to the clinical recovery of cases of hypomagnesaemic tetany. They have been further exploited in the use of saturated solutions of magnesium sulphate for the induction of euthanasia in dogs. Such solutions are usually administered by intracardiac injection and are dramatically effective. Dogs destroyed by this means sometimes make a distressing noise. Death should be confirmed after some minutes as immediate signs of death are probably due to peripheral rather than central actions of the drug.

Dosage for hypomagnesaemia by subcutaneous injection.

Cattle: 200 to 400 ml 25 to 50 per cent w/v solution

Calcined Magnesite, B.Vet.C.

This is prepared by heat treating the natural mineral, magnesite (magnesium carbonate). It is a white or off-white powder which is principally magnesium oxide. Although it seems rather unpalatable to stock, it is the most suitable magnesium compound for prophylactic purposes and its consumption can be encouraged by incorporation with concentrates. Where there is a history of hypomagnesaemia, calcined magnesite should be offered throughout the period of risk at a dosage of 60 g per day for cattle and 15 g per day for calves and sheep. Increasing the dosage promotes purgation and this would reduce magnesium absorption.

KETOSIS

Just as the successful treatment of milk fever has become more difficult in recent years so has the management of ketosis. While it is possible to cure cattle with a single intravenous infusion of dextrose, it is nowadays often necessary to combine dietary and managemental considerations with injections of potent glucocorticoids.

A full discussion of the aetiology of the two conditions of bovine acetonaemia and ovine pregnancy toxaemia is beyond the scope of this book. However, it can be said that both are diseases of a disordered metabolism and would appear to be induced by the stress of overproduction. In cows the stress is that of coping with the rise in metabolic loading consequent upon a rapidly increasing milk production in early lactation while in sheep the predisposing cause is that of the demands of twin lambs in utero at a time when the ewe is likely to be in negative nutritional balance due to the declining nutritive value, or lack, of forage. Despite the different precipitating causes, the two conditions share certain biochemical symptoms, the main ones being an elevation of plasma ketone body levels and a depression of plasma glucose concentration. In both, the liver is depleted of glycogen and, especially in sheep, undergoes fatty change. Spontaneous recovery occurs on occasion and this makes the evaluation of treatments difficult. In cattle, recovery follows a fall in milk yield and in sheep recovery often follows the death of one or more lambs in utero, caesarean section or abortion.

Treatment of both species is aimed at correcting the hypoglycaemia with glucose or glucogenic agents and thus providing the animal with a source of readily available energy. An alternate approach is to stimulate gluconeogenesis (the provison of glucose from non-carbohydrate precursors) within the animal. This is the rationale behind the use of the glucocorticoids. As the body carbohydrate deficiency is believed to impair the functioning of the citric acid cycle, many of the agents used in therapy are given in the belief that they will either stimulate the cycle (e.g. thiamine or cobalt salts) or will provide intermediates that are lacking (e.g. fumarate, propionate). The 2-carbon fragments resulting from fat mobilization and arising from ruminal fermentation are not further metabolized to carbon dioxide and water by the citric acid cycle and give rise to the ketone bodies—acetone, acetoacetate and β-hydroxybutyrate. The following agents are those commonly used in the treatment of ketosis.

Dextrose B.Vet.C.

The correction of a hypoglycaemia by an intravenous infusion of glucose solution is attractively simple. Unfortunately, the renal threshold for glucose is easily exceeded and much of the dose is lost in the urine. Despite the drawback of this transient action, cases frequently respond to one or more injections both biochemically and clinically. Better results can be obtained if the glucose is administered by intravenous drip. This technique is rather troublesome and, not surprisingly, is therefore not popular.

Dosage for cattle: 500 ml of 50 per cent w/v Dextrose solution. In the mistaken belief that sugars given by mouth would be absorbed as such by a ruminant, treacle or molasses was often used. These would rapidly be fermented in the rumen to yield volatile fatty acids which would only add to the animal's biochemical embarrassment.

Glycerol, B.Vet.C.

Glycerol has many uses in veterinary medicine, but by far the most important use of glycerol discovered in recent years is its glycogenic action when given orally to cattle and sheep. It is administered as a warm 5 per cent aqueous solution as a treatment for pregnancy toxaemia of the ewe and acetonaemia of cattle. Doses are up to 120 ml (4 fl. oz.) in sheep; given as single daily oral doses at the rate of 3 ml per kg body weight, glycerin reduced pregnancy toxaemia (artificially induced) mortality from 83 per cent in controls to 25 per cent. In cattle, doses of 500 ml (about 1 pt.) in food or as a drench should be given twice daily for 2 or 3 days.

Propylene Glycol, B.Vet.C.

Like glycerol, propylene glycol is a water-miscible, clear, colourless, oily liquid with a slightly sweet taste. It is glucogenic, and it has been stated that it is an intermediate in the synthesis of glycogen. Various workers have reported the glycogenic effect of propylene glycol in acetonaemia. They recommended doses of from 200 to 400 ml (6 to 12 fl. oz.) given twice daily for 4 days. Toxicity is not great, though large amounts can cause incoordination and narcosis.

Sodium Propionate, B.Vet.C.

A white, deliquescent powder which is administered for the treatment of acetonaemia in cattle. The rationale of this therapy is related to the fact that soluble carbohydrates, including glucose and treacle, are not absorbed as such following oral administration to ruminants. In acetonaemia, increases in fatty acid absorption are to be discouraged and the absorption of the immediate precursors of glucose encouraged: glycerol, propylene glycol, lactates and propionates including sodium propionate, are such precursors; acetates are glucose sparing substances. Doses vary from 60 to 240 g per day (2 to 8 oz.) for 4 to 10 days: a common method is to give 8 oz. as an initial dose and follow it with 4 oz. twice daily for 4 days. Mineralized mixtures are available commercially, the addition of cobalt in particular is useful in promoting a healthy appetite following bacterial synthesis of vitamin B_{12}. As sodium propionate is unpalatable, it is often unwise to rely solely upon voluntary intake of this rather slow-acting drug. Propionate enters the tricarboxylic acid cycle as succinyl-CoA. Propionic acid is the only fatty acid produced in the rumen which is glucogenic.

The Glucocorticoids

The adrenal cortex hormones and the anterior pituitary trophic hormone ACTH have been fully discussed in Chapter 11. Although their use is also described there, the importance of this group of potent steroids is such that their relevant actions may be summarized here in relation to various beliefs concerning the means by which they are of benefit in ketosis.

Initially it was believed that the adrenal cortex became exhausted by the stress of lactation or pregnancy and that replacement of the missing corticosteroids should correct the condition. This theory was modified from an absolute to a relative deficiency when it became apparent that the adrenal cortex showed evidence of hypertrophy and that the plasma glucocorticoid level was often higher than normal, i.e. although steroid production was increased, the increase was still not enough to meet the body's demands. It has been suggested that in some sheep cases, the steroid level is so high that it interferes with the peripheral utilization of glucose—the so-called diabetogenic effect of cortisol.

What is established is that the administration of glucorticoids to ketotic cattle and sheep is followed by a rise in plasma glucose concentration and a fall in the level of ketone bodies. These changes are usually accompanied by an early return of appetite and a rapid improvement in clinical condition. The principal mechanism by which the hypoglycaemia is corrected is by the stimulation of gluconeogenesis. Corticosteroids share with the other steroid hormones the ability to form steroid/receptor protein complexes in the cytoplasm of target cells. These complexes migrate into the nucleus and there stimulate protein synthesis. In the liver the glucocorticoids cause an increased presence of those enzymes involved in the elaboration of glucose from amino acids. The production of glucose is accompanied by protein breakdown and increased excretion of nitrogen. It is assumed that the improvement in the animal is consequent upon the rise in blood sugar concentration. The extent to which a possible interference with the peripheral utilization of glucose contributes to the rise in plasma glucose concentration is not known.

Although investigations into the regulation of carbohydrate metabolism in ruminants have continued apace in recent years, the precise interrelationships of pituitary, adrenal and pancreatic hormones have not yet been established. For this reason, it is not yet possible to allocate a therapeutic role to insulin, glucagon or growth hormone in the management of ketosis.

Bibliography

CARE, A. D. (1971) The effect of diet on calcium homeostasis. *Rep. Rowett Inst.*, **27**, 106–11.

KRONFELD, D. S. (1971) Parturient hypocalcaemia in dairy cows. *Adv. vet. Sci. comp. Med.*, **15**, 133–52.

LENG, R. A. (1970) Glucose synthesis in ruminants. *Adv. vet. Sci. comp. Med.*, **14**, 209–52.

PHILLIPSON, A. T. (ed.) (1970) *Physiology of Digestion and Metabolism in the Ruminant*. Newcastle-upon-Tyne: Oriel Press.

REID, R. L. (1968) The physiopathology of undernourishment in pregnant sheep, with particular reference to pregnancy toxaemia. *Adv. vet. Sci.*, **12**, 164–226.

STEVENSON, D. E. & WILSON, A. A. (1963) *Metabolic Disorders of Domestic Animals*. Oxford: Blackwell Scientific.

UNDERWOOD, E. J. (1966) *The Mineral Nutrition of Animals*. Rome: FAO.

17

The Vitamins

I. Water Soluble Vitamins

A vitamin is an organic compound which, in very small quantities, is essential to normal tissue metabolism. In the absence of vitamins a syndrome often develops that is specific to each vitamin. Some vitamins need not be included in the diet of some species, for they are either synthesized by the intestinal flora (vitamin B complex in ruminants) or in some body tissue as yet unknown (vitamin C in most domestic animals—guinea-pig, monkeys and man do not synthesize vitamin C and must therefore obtain all their requirements from outside sources).

Vitamin deficiencies were at one time very common but the specific symptoms and the dangers of such deficiencies are now so widely recognized that acute deficiencies are rare. Chronic deficiencies do not always cause characteristic symptoms and may exist unrecognized: intensively reared animals, especially those reared on synthetic or near synthetic diets, are more likely to suffer from such deficiencies. This possibility is also well recognized and reputable manufacturers of animal feedstuffs are extremely careful when formulating and compounding rations, to ensure that wherever possible adequate and stable vitamins are included.

Because some species do not normally require certain vitamins in their diet, it must not be assumed that such species cannot suffer from deficiencies or insufficiencies of those vitamins. For example, vitamin B_{12} is a cobalt compound; should a ruminant be kept on a cobalt free diet, and vitamin B_{12} is not synthesized by the ruminal flora a vitamin B_{12} deficiency will develop.

The presence of a vitamin in a foodstuff is also no guarantee that a deficiency of that vitamin will not occur, for other factors may be present in the diet which prevent assimilation. It is well known that continuous therapy with liquid paraffin can prevent the absorption of carotene because it dissolves the compound and forms a solution which is not absorbed by the intestinal mucosa: vitamin A deficiencies may result. It is also an established fact that excess cod-liver oil in the diet will quickly produce a vitamin E deficiency with the onset of muscular dystrophy. Intestinal infections interfere with vitamin A absorption and its utilization. The disturbance of fat hydrolysis and absorption automatically influences the absorption of all fat soluble vitamins: the absence of bile salts or pancreatic enzymes prevents the absorption of both fats and fat soluble vitamins.

The assay of vitamins has been carried out by biological methods and these tests must still be used for estimation of the natural vitamin content of foodstuffs. Nevertheless, most vitamins can now be produced synthetically and chemical methods of estimation are increasingly used.

The nomenclature of vitamins is undergoing metamorphosis as more of their chemistry becomes known. Initially the name *vit-amine* was adopted for an essential nutritional and growth factor, but it was later found that all such compounds were not proteins or amino acids, and the name was reduced to *vitamin*. As different forms of vitamins were recognized they were designated by numbers, as, for example, vitamins D_1, D_2 etc. later it was found that vitamin D_1 was not one but two different compounds so that no vitamin D_1 is now recognized. This nomenclature has led to confusion and the vitamins are now more frequently referred to by their chemical names: calciferol or vitamin D_2; pyridoxine or vitamin B_6; cyanocobalamin or vitamin B_{12}; the tocopherols or vitamin E.

Classification of the vitamins is an arbitrary

one which divides the vitamins into two groups according to their solubility in water, or fats and oils. As absorption is better when substances are water soluble or water miscible, compounds exerting the actions of fat soluble vitamins but having water solubility as a characteristic have been devised: these are referred to as water-soluble analogues, e.g. Menaphthone Sodium Bisulphite, B.Vet.C. which is a water-soluble salt. Water miscible dispersions of vitamin A are available.

1. *Water soluble vitamins*
   The vitamin B complex:
   - Aneurine—vitamin B_1
   - Riboflavine—vitamin B_2 or G, lactoflavin
   - Pyridoxine—vitamin B_6
   - Nicotinic acid or pellagra preventing (P.P.) factor, niacin
   - Cyanocobalamin—vitamin B_{12}
   - Biotin—vitamin H, coenzyme R, bios II
   - *para*-Aminobenzoic acid—PABA, vitamin H
   - Inositol—bios I
   - Folic acid—*L. casei* factor
   - Pantothenic acid—filtrate factor
   - Choline

   Vitamin C:
   - Ascorbic acid

   (Vitamin P):

2. *Fat soluble vitamins*
   Vitamin A:
   - Precursor: α-, β- and γ-carotenes and cryptoxanthine
   - Vitamin A_1
   - Vitamin A_2

   Vitamin D:
   - Calciferol: vitamin D_2
   - 7, dehydrocholesterol: vitamin D_3

   Vitamin E:
   - α-, β-, γ- and δ-tocopherol; and α-, β-, γ- and δ-toctrienol

   Vitamin K:
   - Vitamin K_3

Two other vitamins have been postulated but so far the evidence advanced has been inconclusive. These are vitamin P which has been considered an associate of vitamin C concerned with vascular permeability, and vitamin U reputed to be associated with vitamin K and responsible for the prevention of gizzard ulceration in chickens.

The *stability* of vitamins varies considerably but certain principles can be applied to all. The majority of vitamins are sensitive to light and should be stored in light proof containers. Some, but not all, are susceptible to oxidation, and it is wise to store them in airtight containers: calciferol for example is kept under nitrogen in sealed ampoules. The majority of water soluble vitamins are less stable in solution, whilst the fat soluble vitamins are generally more stable in solution. Most vitamins are heat stable, the main exceptions being vitamins A, D and E; calcium pantothenate is also heat labile but vitamin C, once thought to be heat labile has been shown to be heat stable in the absence of oxygen.

WATER SOLUBLE VITAMINS

The water soluble vitamins comprise the vitamin B complex, vitamin C and the vitamin P flavonoids. They may all be obtained from either plant or animal sources; in addition vitamin C is manufactured in the tissues of domestic animals, possibly in the gastro-intestinal mucosa. The vitamin B complex is available to ruminants following synthesis by ruminal bacteria: in other species the vitamin B complex is synthesized by caecal bacteria, especially in the horse, but very little is available to the animal short of coprophagia, which is not an uncommon source of vitamin B in the animal kingdom.

THE VITAMIN B COMPLEX

This complex includes a number of familiar names but there is also a number of vitamins the functions and importance of which are imperfectly understood:

- Aneurine or vitamin B_1
- Riboflavine or vitamin B_2
- Pyridoxine or vitamin B_6
- Nicotinic acid
- Cyanocobalamin or vitamin B_{12}

Subsidiary, but nevertheless quite important members of the group, are: biotin or vitamin H; *para*-aminobenzoic acid or vitamin H^1; pantothenic acid; choline; inositol and folic acid.

Aneurine Hydrochloride, B.Vet.C.

This vitamin is also known as Thiamine (U.S.P.), and vitamin B_1. Aneurine hydrochloride is a white crystalline powder with a slight but characteristic odour. It is very soluble in water (1:1), and in glycerin, but not in ethyl alcohol, ether, acetone, or similar organic solvents.

Sources. The majority of aneurine used therapeutically is a synthetic product, but it is present in most plants, the most concentrated sources being yeast and wheat germ: ruminant requirements are met by bacterial synthesis in the rumen. Milk, meat, eggs and fish, are the main sources for carnivores.

Stability. Aneurine is one of the least stable of the B complex members, but in the dry state it will withstand 100°C for as long as 24 hr without appreciable deterioration. At pH 5 or less solutions may be sterilized by steaming or autoclaving, but above this pH the vitamin is very susceptible to high temperatures. The powder absorbs moisture on exposure and is affected by light: it should therefore be stored in air- and light-tight containers. Solutions also deteriorate when exposed to air.

Assay. The assay is chemical, relying either on the fluorescence produced on conversion to thiochrome by oxidation, on gravimetric estimation following precipitation of the silicotungstate, or on estimation of the chlorine content. Biological assays are also available.

Action

Aneurine is readily absorbed from the intestines; very little is stored, the liver, kidney, brain and heart muscle being the main depots. Excess is excreted in the urine as a pyrimidine or as unchanged material.

When absorbed, aneurine is converted, probably in the liver, into aneurine pyrophosphate (co-carboxylase). As such, it acts as a coenzyme for several decarboxylating enzyme systems, the most important of which is probably *decarboxylase*. This enzyme is necessary for the decarboxylation of pyruvic acid, well known as an intermediate stage in carbohydrate build up or breakdown. When carbohydrates are a major source of energy the body requirements of aneurine increase.

Deficiency results in a loss of energy, a build up of pyruvic and lactic acids in the tissues, and a decreased intestinal absorption of glucose.

Clinical symptoms are associated with brain lesions, which may be partly responsible for the cardiac and intestinal abnormalities that also occur. The main symptoms are muscular incoordination, especially of the hind legs, paralysis with the head turned over the back (opisthotonus), convulsions, possibly vomiting where possible, and diarrhoea. The symptoms and lesions have been reported in the fox, the cat, pigs, non-ruminating calves, horses and lambs. The workers reporting on the chronic deficiency in pigs, and those experimenting on cats, indicated that the polyneuritic syndrome in man did not apply to these animals; nor was any effect on the peripheral nerves observed. The work so far recorded on other species has not commented on this aspect.

The aetiology of these deficiencies has been commented on by various authors when reviewing aneurine deficiencies, but though diets that include species of raw fish which contain thiaminase (an anti-aneurine enzyme), diets containing horsetail (*Equisetum* species), and various synthetic experimental diets were mentioned, the effect of bracken which, like *Equisetum* also contains thiaminase, is not mentioned. Bracken poisoning in horses is probably the main form in which practitioners are likely to meet pure acute aneurine deficiencies. Aneurine deficiency in pigs following the ingestion of bracken rhizomes has been reported. The condition in cattle is not one of pure thiamine deficiency.

Deficiencies in aneurine can be induced in a very short time in carnivores, birds, non-ruminating calves, lambs and goats, and also horses. Dogs will show symptoms in about 4 weeks, and pigeons in about 2 weeks, when kept on an aneurine free diet. Adult ruminants and horses will not show deficiency symptoms on normal diets but bracken poisoning will induce acute deficiencies in horses. Intensively reared fur bearing animals and poultry are particularly open to aneurine deficiencies, usually associated with deficiencies of others of the vitamin B group, although poultry obtain adequate amounts if on a whole grain diet. Dogs on a meat free diet may also suffer from such deficiencies. Necrotic enteritis in pigs involves similar deficiencies. Cerebrospinal necrosis in sheep is believed to be caused by a thiamine deficiency, due to sheep eating fungi containing a thiaminase.

Uses

Aneurine is specific for the treatment of bracken poisoning in horses and pigs but cattle and sheep do not respond. Up to 80 per cent recoveries have been achieved in cattle by the slow intravenous injection or intramuscular injection of about 1 g *dl*-batyl alcohol in an emulsified form. Oxytetracycline 0·5 g intravenously and procaine penicillin 3 mega-units intramuscularly were also administered; doses of about 100 mg should be given daily for 4 days. The thiaminase induced deficiencies in other species respond favourably. Aneurine injections have also been used empirically, in common with many other vitamins, for the treatment of nutritional hysteria in dogs, nutritional gastroenteritis particularly of pigs, in nervous diseases, and in general debility of all species. Known *nutritional requirements* are (Nat. (Amer.) Res. Council. Comm. Animal Nutrit. No. 1, 11, VII, VII (1944–1954)):

Dog:
16 μg/kg body weight daily
Fox and Mink:
0·5 mg/lb. dry food
Chicken:
0·8 mg/lb. of feed
Pig:
2·5 mg/50 kg body weight
(0·5 mg per score live weight) daily

Dosage

By mouth or by subcutaneous or intramuscular injection:
Horse: 100 mg
Calf: 10 mg
Pig: 2·5–15 mg
Dog: 1–10 mg
Cat: 1–5 mg

Most frequently yeast, brewers' grains, or vitamin extracts such as Marmite, are used for the treatment of vitamin B deficiencies irrespective of any particular requirements, for a specific deficiency is uncommon.

Riboflavine, B.Vet.C.

Synonyms for this member of the B complex are *lactoflavin* and *vitamin G*, and, more frequently used, *vitamin* B_2. It is an orange-yellow crystalline powder with a slightly bitter taste and slight odour. It is more soluble in normal saline and 10 per cent urea solutions than in water, and is almost insoluble in organic solvents.

Sources. As with other B vitamins, riboflavine is present in yeast and may be extracted commercially, but by far the most important source is synthetic manufacture. Riboflavine is present in milk and milk products, meat, eggs, fish roes, malt and, to a lesser degree, in fruit and vegetables.

Stability. When crystalline and dry, riboflavine is light- and air-stable—it is a strong oxidizing agent—but solutions deteriorate on exposure to light, especially alkaline solutions. Correctly compounded solutions will withstand autoclaving.

Assay. This is now a chemical test, relying on the fluorescence or on the nitrogen content of the compound. As with aneurine, biological tests may also be used.

Action

Riboflavine is rapidly absorbed from the intestines or injection site and is rapidly distributed throughout the body.

Excretion via the urine commences within about ½ hr of administration. Little storage occurs in the body and most of what is stored may be found in the liver, kidney and heart, though small quantities exist in practically all organs and tissues.

Following absorption riboflavine is phosphorylated, forming two coenzymes, riboflavine phosphate (riboflavine mononucleotide) which consists of one molecule each of riboflavine and phosphoric acid, and riboflavine adenine dinucleotide which is one molecule each of riboflavine and adenine but two molecules of phosphoric acid. Both are esters. The simple phosphate coenzyme is associated with, amongst other flavoprotein enzymes, cytochrome-C reductase and *l*-amino-acid oxidase. The more complex compound is coenzyme to *d*-amino-oxidase, xanthine oxidase, aldehyde oxidase and others. All these flavoproteins promote alternate oxidation and reduction reactions, often in conjunction with other enzymes such as those containing nicotinic acid. Riboflavine thus has a direct effect on the metabolism of carbohydrates, amino acids and aldehydes.

Deficiency which is normally confined to simple stomached animals and pre-ruminating ruminants, results in a slowing of growth; abnormal haemopoiesis with anaemia; decreased fertility; ocular disorders with pain, irritation and ocular discharges, in severe cases vascularization of the cornea and in most cases photophobia; generalized dermatitis with ulceration and deep skin fissures. The clinical effects of deficiency are as follows:

Chickens:
    Curl toe paralysis of young chicks and reduced egg yield in adults
Turkeys:
    Dermatitis at angles of beak, scaly foot pads and inflamed vents
Pigs:
    Vomiting, skin ulcers, abnormal gait and still births
Calves:
    Scours, salivation, dermatitis

In horses, it has been suggested that periodic ophthalmia may be prevented but not cured by riboflavine. In dogs, partial blindness, especially at night, and dermatitis are reported. Of these conditions the 'curl toe paralysis' of chicks is the only riboflavine deficiency in which symptoms are pathognomonic (permitting a definite diagnosis). Young chicks may die suddenly without characteristic symptoms but in those of 2 weeks of age or more, more usually between 3 and 6 weeks, the birds rest on their hocks and the claws curl inwards in a characteristic manner. Chicks may have diarrhoea, they are ruffled and untidy, with delayed feathering and a poor growth rate. Spontaneous recovery occurs in all birds after a time, but losses from smothering occur. Coprophagia in birds kept on solid floors makes the incidence lower than in birds kept on wire. (This is one of the advantages of the deep-litter system of rearing chickens.) Enlargement of the sciatic nerves is found on post mortem. In adult birds, lowering of egg yields and hatchability are the main features of the deficiency.

Riboflavine is one of the vitamins which may not be provided by the major components of a ration and must therefore be added as a supplement.

Daily nutritional requirements for non-ruminants are approximated in the following table:

Horse:
    2–4 mg/50 kg body weight
Pig:
    3–4 mg/50 kg body weight or 2–3 g per tonne of feed
Dog:
    40 mg/kg body weight
Fox and Mink:
    1·8 mg/kg. of dry food
Fowl:
    1·6–3·4 mg/kg. of feed or 3–4 g per tonne of feed
Turkey:
    4·5 g per tonne of feed

Dosage

In prophylaxis of periodic ophthalmia in the horse, a tentative dose of 40 mg daily in the food has been suggested, but the efficacy of such therapy has not been established. In dogs, 2 to 10 mg is normally satisafctory.

Pyridoxine Hydrochloride, B.P.C.

Vitamin B_6 or pyridoxine when in the form of the hydrochloride is a white crystalline powder, which is soluble in water but only slightly so in alcohol, and even less soluble in the other common organic solvents.

Sources. Pyridoxine exists naturally and also as its aldehyde, pyridoxal, as an amine form, pyridoxamine, all of which have vitamin B_6 activity. The sources are typical of the vitamin B group and include yeast, grains, potatoes, cabbage, bananas, legumes, meat, fish, eggs and milk. Synthetic pyridoxine is widely used.

Stability. This vitamin is one of the most stable of the vitamin B complex. Acid solutions keep indefinitely at room temperature and may be autoclaved: they are also light stable unless at a neutral or alkaline pH (solutions are normally about pH 3). The powder is air stable but is slowly affected by light.

Assay. The assay has been biological, but no international units are established and dosage is in terms of weight of pure substance.

Action

Absorption, storage and excretion are similar to those of other members of the group.

After absorption pyridoxine is utilized, probably as the aldehyde, in the formation of co-enzymes. The ester, pyridoxal-5-phosphoric acid, is formed which is the coenzyme for decarboxylase (hence its name co-decarboxylase), and also for transaminase. These enzymes are involved in the metabolism of proteins and amino acids.

Deficiency results in *acrodynia* which is a dermatitis, characterized by hyperkeratosis and acanthosis of the skin. These features are particularly noticeable on the nose, paws and ears. The muscle action is weak, and nervous symptoms occur. In the dog and pig, microcytic and hypochromic anaemias are seen, and in pigs degeneration of the dorsal root ganglion cells of the spinal cord and axis cylinders of the peripheral nerves, with demyelination, have been reported.

Such lesions and symptoms are rarely met clinically, but pyridoxine has been recommended for vomiting due to excessive stilboestrol dosage, and to X-ray irradiation in man. This vitamin is also the antidote to cyanacethydrazide (Dictycide) overdosage and should be given on an equal weight for weight basis. This would mean a maximum of 5 g pyridoxine for cattle, and 1 g for sheep, given intravenously. *Dietary* requirements and *dosage* are said to approximate with those of aneurine (*q.v.*).

Nicotinic Acid, B.Vet.C., and Nicotinamide, B.Vet.C.

Niacin has been suggested as an alternative name for the acid, and the term *pellagra preventing factor* is applied to both. These two substances are white crystalline, odourless, or almost odourless, powders, the former with a slight acid taste and the latter with an intensely bitter taste. The acid is soluble 1:55 in water, but less soluble in alcohol, and almost insoluble in ether: the amide is much more soluble in water (1:1) and alcohol (1:1·5), but is only slightly soluble in ether.

The formulae are relatively simple and, in common with pyridoxine, are based on a pyridine nucleus:

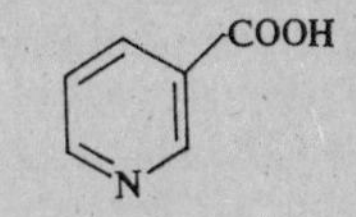

(i) Nicotinic Acid

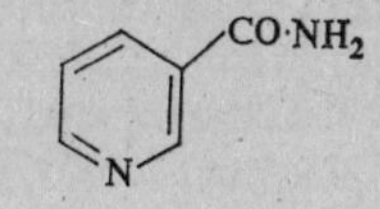

(ii) Nicotinamide

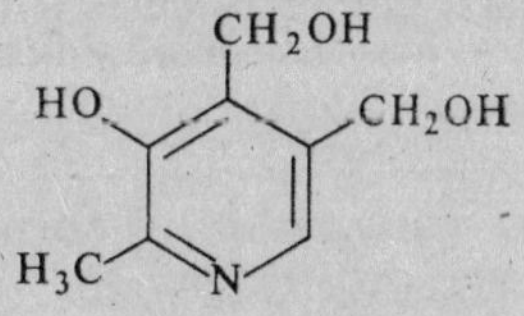

(iii) Pyridoxine

Sources. The majority of these compounds are of synthetic origin often derived from nicotine but the vitamins are also present in meat, whole grains and their products, yeast, mushrooms and legumes, particularly soya beans. There are relatively small amounts in milk, green foods and fresh fruit.

Stability. The amide is a little more stable than the acid, but both are reasonably stable, for solutions may be sterilized by autoclaving. Nevertheless, solutions and powders should be kept in airtight and light proof containers.

Assay. Both compounds are assayed chemically, the acid by a straight acid/alkali titration, and the amide by ammonia production.

Action

Nicotinic acid and nicotinamide are both absorbed rapidly when given orally, subcutaneously, or intramuscularly, and in the case of nicotinic acid vascular reactions may occur within 5 or 10 min of oral administration. Little of the vitamin is stored but the liver usually contains appreciable amounts. The vitamin undergoes methylation in the liver, and some is excreted in the urine by glomerular filtration and tubular secretion as *N*-methylnicotinamide.

After the vitamin has been assimilated, the

acid is converted into nicotinamide. Nicotinamide enters into combination in the tissues with the protein adenine, the sugar ribose, and phosphoric acid, to form two coenzyme systems, coenzyme I (diphosphopyridine nucleotide, coenzymase, co-dehydrogenase I, or DPN; now known as NAD), and coenzyme II (triphosphopyridine nucleotide, co-dehydrogenase II, or TPN; now known as NADP). Coenzyme I consists of one molecule each of adenine and nicotinamide with two of *d*-ribose and phosphoric acid, whilst coenzyme II contains an extra molecule of phosphoric acid. These two coenzymes are associated with a number of enzymes which are involved in oxidation processes. The coenzymes appear to function by taking up two hydrogen atoms in the pyridine ring of nicotinamide, with the loss of a double bond. They give up these hydrogen atoms on oxidation and revert to the form which has a double bond associated with the nitrogen atom. This reversible process may be graphically represented by:

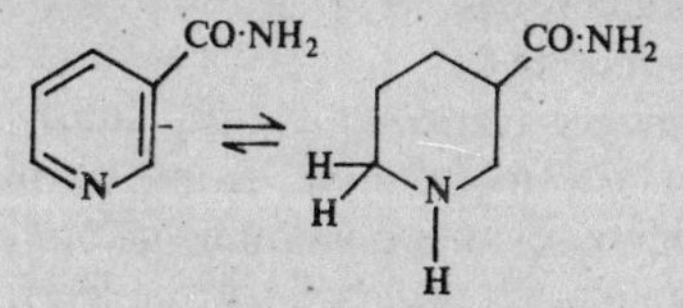

The coenzymes containing nicotinamide therefore act as hydrogen carriers in the oxidation–reduction processes of cell metabolism, for which they appear to be essential. These nicotinic enzymes often work in conjunction with those containing riboflavine. They are particularly important in carbohydrate metabolism.

Additional actions of nicotinic acid only are those of causing peripheral vasodilation and depression of plasma cholesterol levels. These have little or no clinical application in veterinary medicine.

The essential amino acid tryptophane is a precursor of nicotinic acid, to which it is converted in the body, and nicotinic acid is not required in the diet if sufficient tryptophane is present. Pyridoxine and vitamin E are essential for this conversion.

Deficiency results in *pellagra* in man; a rough scaly skin, oral and gastro-intestinal ulceration with diarrhoea, in pigs; perosis, inflammation of the tongue, and dermatitis, in chicks; enlarged hocks in turkeys; 'black tongue' or 'brown mouth' in dogs; and a 'feline pellagra' has been described. Decreased growth rates and fertility are shown in chronic, subclinical, nicotinic acid deficiencies.

Maize (corn) does not contain much available nicotinic acid or tryptophane, and diets consisting mainly of this grain must be fortified with the vitamin.

Dietary requirements:

Pig:
12·5 mg/50 kg body weight daily or 11–18 g per tonne of feed

Dog:
0·22 mg/kg body weight daily

Fox and Mink:
9 mg/kg. dry feed

Fowl:
12–20 g per tonne of feed

Turkey:
17–20 g per tonne of feed

Dosage

Calf:
25 mg daily

Pig:
0·1–0·3 g subcutaneously, 0·2–0·9 g orally

Dog and Cat:
5–10 mg/kg body weight given intramuscularly
10–30 mg/kg body weight given orally

These doses are tentative. Usually 100 mg two or three times daily by mouth in the pig, and 50 mg three times daily for the dog and cat are sufficient.

The amide should be used on all occasions, for it provides the vitamin in an immediately utilizable form and eliminates any vascular side actions.

Cyanocobalamin, B.Vet.C.

This compound, known as vitamin B_{12}, was isolated simultaneously in Great Britain and America in 1948. It is a dark red, crystalline, hygroscopic powder, which is soluble in water (1:80) and alcohol, but insoluble in the normal organic solvents.

Chemically, cyanocobalamin is a complex polyacidic organic base, and has a molecular weight of approximately 1300. It contains, beside carbon hydrogen and oxygen, the metal cobalt (4·5 per cent), nitrogen and phosphorus. Cyanocobalamin is one of a series; other members are formed when the cyano-group is replaced by other organic groups such as a nitrito-, a thiocyanato-, or similar groups. This substitution produces analogues of vitamin B_{12} which have been designated B_{12a}, vitamin B_{12b}, etc.

Vitamin B_{12a} has been shown to be a hydroxy-analogue of cyanocobalamin, in which—CN is replaced by—OH; it is known as hydroxocobalamin.

Vitamin B_{12b} is aquocobalamin, of which hydroxocobalamin is the hydrated form.

Vitamin B_{12c} is nitritocobalamin, whilst **vitamin B_{12d}**, derived by removal of the nitrite group of B_{12c}, is apparently identical with vitamin B_{12b}. The portion of the molecule minus these specific groups is commonly referred to as cobalamin, though confusion is added by the generic use of cobalamins as a comprehensive name including the whole group of analogues.

Sources. Cyanocobalamin is present in liver (20 tonnes produces only 1 g of the vitamin), kidney, meat, in fish meals, milk, whey, eggs, ruminant and rabbit faeces, and sewage. Little is present in green food, fruit or yeast.

The majority of the vitamin produced commercially comes as a by-product of streptomycin and other antibiotic manufacture. Vitamin B_{12} is synthesized by growing moulds and is extracted from the residue remaining after removal of the antibiotic.

Stability. Cyanocobalamin is relatively stable and neutral solutions will withstand autoclaving for 20 min at 120°C. Alkalis, strong acids, reducing and oxidizing agents, and heavy metals cause decomposition. The compound is susceptible to light and will decompose in air.

Assay. The compound is assayed electrochemically.

Action

Cyanocobalamin is absorbed rapidly following parenteral administration, but gastro-intestinal absorption in man, at least, is dependent on an 'intrinsic stomach factor'. After absorption the vitamin circulates in the blood for a period during which appreciable amounts are excreted in the urine. A proportion becomes bound to protein. Small quantities are stored in all tissues but most is found in the liver.

The metabolism of cyanocobalamin is complex and is closely associated with that of *folic acid* (another member of the vitamin B complex), and vitamin C, ascorbic acid.

The release of folic acid from conjugated forms, and also the conversion of folic acid to the metabolically active folinic acid, are probably brought about by cyanocobalamin, possibly in conjunction with ascorbic acid. These three are important factors in carbohydrate, protein and nucleoprotein, and fat metabolism.

The synthesis of nucleoproteins is a complex ladder of chemical actions in which cyanocobalamin is required at various stages. The probable sequence of events in the light of present knowledge may be represented as follows:

Conjugated folic acid + B_{12} → folic acid + B_{12} and C → folinic acid + organic precursors (e.g. uracil) → purines and pyrimidines (e.g. thymine) + B_{12} → nucleosides (e.g. thymidine) → nucleotides → nucleic acids.

Through its transmethylating action in the formation of methionine from choline and homocysteine, cyanocobalamin has an effect on fat metabolism, particularly in the liver.

Adequate vitamin B_{12} is also necessary for maintaining normal levels of sulphydryl groups in the reduced state.

These metabolic actions are reflected in the importance of cyanocobalamin in maintaining normal haemopoiesis in the bone marrow and elsewhere, in the formation and maintenance of the myelin sheaths of nerve fibres, in the protection of the liver in some species from the fatty changes normally caused by such compounds as the chlorinated hydrocarbons (e.g. chloroform, carbon tetrachloride, tetrachloroethylene), in the maintenance of ovum and sperm production and therefore fertility, in the maintenance of all types of muscle tissue, in the maintenance of healthy skin, brain and pancreatic metabolism; in general, cyanocobalamin plays an important part in the maintenance of all tissues in which protein and nucleoprotein synthesis occurs and where fat conversion proceeds.

It is only to be expected that a compound

which affects tissue regeneration will have a marked effect on growth. Even before it was recognized, vitamin B_{12} was referred to as the 'animal protein factor'. This effect has been well recognized in the use of antibiotic manufacturing residues which contain high concentrations of vitamin B_{12}, and also of fish meals and meat and bone meals, in the feeding of pigs and poultry.

The haemopoietic action was likewise well known before vitamin B_{12} was isolated, when liver extracts were used in the treatment of pernicious anaemia in man: the factor referred to as the 'antipernicious anaemia factor' is now known to be vitamin B_{12}.

Deficiencies of vitamin B_{12} do not occur in ruminants unless cobalt is deficient, when a condition known as 'pine' (Bush Sickness, Nakuruitis, Grand Traverse Disease) occurs. In pigs, skin lesions, hind leg incoordination, anaemia and loss of 'voice' occur; in dogs and cats similar effects might be expected. In all species a loss in appetite occurs which is recovered following administration of the vitamin.

Pine is mainly a disease of sheep in Great Britain, but adult cattle are sporadically affected in established pine areas. The symptoms are similar in the two species and are typified by progressive debility, loss of wool or hair giving the animal a bedraggled, ragged appearance, the appetite decreases progressively and emaciation becomes apparent. Lambs fail to grow and have a stunted, unthrifty appearance. The animals are usually markedly anaemic. Incoordination, particularly in the hind limbs, is followed by collapse. In marginal deficiencies the symptoms are less acute and may be difficult to diagnose. Test dosing is followed by a marked improvement within a week to 10 days and soil analysis will normally confirm the condition: for cattle a test dose would be 0·5 to 1 g in a single dose, and for sheep about one-eighth of this should be sufficient: pasture cobalt content is approximately 0·03 to 0·04 ppm in deficient areas. Animals may be given parenteral cyanocobalamin (0·3 mg per sheep), but this is uneconomic and oral cobalt (8 mg cobalt chloride per week per sheep) or pasture top dressing with 1·8 kg (4 lb.) per ha are superior. Mineral licks containing cobalt are available.

It should be remembered that in some areas—Western Australia (coast disease) and Florida (salt sickness)—other minerals, especially copper, may also be deficient.

Nutritional requirements have been suggested:

Pig:

Up to 50 mg for weaners and up to 20 mg for pigs over 25 kg body weight in each tonne of feed

Dog:

20 μg/kg dry diet

Fox and Mink:

26 μg/kg dry food

Poultry:

Up to 10 mg per tonne of food, but this figure may be ignored for chicks in deep litter houses where adequate vitamin B_{12} is obtained from this litter

Dosage

No exact dosage can be given, for requirements vary in each individual case. As a general tonic, which is the main use of the vitamin 2 to 4 μg intramuscularly per kg body weight per day may be used as a guide. The normal dose for a dog or cat is 100 μg.

Biotin

This vitamin has been known under a number of names—vitamin H, bios IIb, and coenzyme R. It is a white crystalline powder which is soluble in water and alcohol, but not in other solvents. Biotin is present in all animal tissues—liver, meat, in most plants, but is highest in egg yolk; small quantities are found in milk. It is heat stable but is unstable in alkaline solution and in the presence of oxidizing agents: it should therefore be kept in light and air tight containers. The metabolic role of biotin is of a coenzyme nature taking part in several enzyme carboxylation reactions by virtue of its carbon dioxide binding capacity. Precise requirements are unknown and no definite clinical effects have been proved, althought it is frequently claimed to have a beneficial effect on non-specific skin conditions in dogs and pigs. High dosing with egg white which contains avidin, an antagonist, may cause a biotin deficiency, but this is not of much clinical importance.

Doses of 150 to 300 mg daily have been used experimentally in man.

para-Aminobenzoic Acid, N.F.

Also referred to as PABA and vitamin H[1], this member of the vitamin B complex is a white crystalline powder which is slightly soluble in water (1:170), soluble in alcohol (1:8), and in ether (1:50). It is stable for all normal purposes but discolours on exposure to light.

This compound is present in the highest concentration in liver and yeast, but is widely found in animal and plant tissues. It is a component of the folic acid vitamin structure, but otherwise its present importance lies mainly in its antagonism of the sulphonamides (*q.v.*), in that it is an essential growth factor for many bacteria and is displaced by the sulphonamides. It is also an essential growth factor for chickens, and is reported to be useful in countering organic arsenical poisoning.

Dosage and *nutritional requirements* are unknown, but in man up to 8 g have been given by mouth initially, followed by up to 3 g every 2 h, in the treatment of rickettsial infections: antibiotics are more effective and less toxic. These doses however, give some indication as to dosages in arsenical poisoning following tri- or penta-valent compound administration to small animals (e.g. arecoline-acetarsol overdosage during tapeworm eradication).

Inositol

This vitamin, also known as bios I, is a fine white crystalline powder with a sweet taste. It is soluble in water, but only slightly soluble or insoluble in organic solvents. It is found in quantity in all tissues of animal or plant origin and is manufactured synthetically. It is assayed chemically.

Biochemically, inositol appears to be esterified with phosphoric acid in the tissues, and is associated with fat metabolism, although in a different manner from choline which forms lecithins, amongst other compounds, whereas inositol is found as cephaelins.

Clinically its significance is undetermined, but it has a curative action on alopecia in mice and protects the liver of rats from fatty changes. All experimental rations for mice contain 10 mg inositol in 100 g.

Doses of 1 to 3 g by mouth daily in divided doses have been used in man for the treatment of fatty livers and liver cirrhosis, but it is doubtful whether inositol deficiencies will ever occur clinically, in view of the widespread distribution of the compound.

Folic Acid, B.P.

Also known as pteroylglutamic acid, and identical with the *Lactobacillus casei* factor, folic acid has been shown to be the basis of many similarly acting compounds variously designated vitamin B_c, vitamin B_c conjugate, and fermentation *L. casei* factor. These compounds are conjugates of pteroylglutamic acid with extra molecules of glutamic acid attached through peptide linkages

$$(R'(NH_2)COOH + NH_2R^2COOH = R'(NH_2)CO.NH.R^2COOH + H_2O)$$

Thus vitamin B_c conjugate is a heptaglutamic acid conjugate, and the fermentation *L. casei* factor is a triglutamic acid conjugate.

Folic acid is a yellow or orange-yellow odourless crystalline powder, soluble in acid or alkaline aqueous solutions, but practically insoluble in water and organic solvents. It is stable as a dry powder but in solutions below pH 6 stability is considerably decreased. Alkaline solutions are stable under normal conditions but are decomposed in the presence of some sulphonamides. Light destroys the compound.

The vitamin is available for injection as sodium folate but is more commonly given orally. The natural sources are similar to those of other members of the B complex—yeast, wheat, green foods, meat, including liver and kidneys. Cooking and storage will destroy the vitamin. Intestinal bacteria in all species are probably responsible for supplying some of the needs of healthy animals. The vitamin is also manufactured synthetically.

Folic acid is assayed chemically.

Action

Folic acid is absorbed fairly well from the intestines, for most species seem to absorb sufficient for normal purposes after synthesis by intestinal bacteria. Folic acid takes part in many metabolic processes, particularly where single carbon unit transfer takes place, acting as a coenzyme precursor. The interrelation of folic acid, folinic acid, vitamin C and cyanocobalamin in nucleic acid synthesis has been mentioned under cyanocobalamin (*q.v.*). The acid is also involved in transmethylating actions along with cyanocobalamin, both in nucleoprotein formation and in fat metabolism through methylation in the synthesis of choline and methionine. Through this latter action folic acid and cyanocobalamin reduce nutritional requirements of choline and methionine.

Clinically the symptoms shown in experimentally induced deficiencies are unlikely to occur. These symptoms include macrocytic anaemias with reduction in the numbers of all cellular elements of the blood, diarrhoea, skin lesions, reduced growth rates in pigs, dogs, foxes, mink, experimental animals and birds, which also may show poor feathering and perosis. It should be noted, however, that, as with *para*-aminobenzoic acid, which is a precursor of folic acid, small quantities of the gut 'active' sulphonamides will promote a deficiency, probably by reducing bacterial numbers and therefore the synthesis of this and other vitamins.

Dietary requirements have not been ascertained in detail, but it is suggested that 10 to 20 g per ton of feed should be adequate for poultry and fur bearing animals.

Doses used in man for therapy in anaemia are up to 1 mg daily by injection: as much as 100 mg have been given intravenously without harm, though administration was slow and solutions were well diluted. Use in canine anaemia has not given very satisfactory results.

Pantothenic Acid

This complex organic acid is used mainly in the form of calcium pantothenate, a water soluble, white, crystalline, hygroscopic, bitter tasting powder.

Sources. The vitamin is widely distributed in plant and animal tissues, the richest source being royal jelly of the honey bee: yeast, milk, liver, meat, eggs, peanuts and wheat, also contain appreciable quantities. The calcium salt and pantothenyl alcohol are manufactured synthetically.

Stability. Calcium pantothenate is unstable in solutions outside pH 5 to 7, and is destroyed by heat. It is probably the least stable of all the members of the B complex of vitamins—stewing meat reduces the pantothenic acid content by two-thirds. The alcohol is slightly more stable than other forms.

Action

Pantothenic acid is similar to other members of the B complex in occurring universally throughout all tissues. It is absorbed from the gastro-intestinal tract and the alcohol can also be absorbed percutaneously in quantities sufficient to cure deficiencies. Its subsequent metabolism is not well understood. It is a component of coenzyme A and as such is concerned in acetylation processes throughout the body. In this coenzyme it is probably combined as a phosphorylated ester with cysteine. Examples of the reactions in which it is involved are the conversion of acetates to sterols (hence the probable relationship with adrenal cortex necrosis observed in deficiencies), the conversion of choline to acetylcholine, the acetylation of PABA and the sulphonamides, and the conversion of acetate to acetoacetate and oxaloacetate to citrate. Pantothenic acid is also concerned through its incorporation in coenzyme A with carbohydrate and fat metabolism. The conversion of α-ketoglutaric to succinic acid results in the formation of phosphorylcoenzyme A which then cooperates, by transferring the phosphoryl group, in the formation of adenosinetriphosphate, a compound essential for the release of the energy necessary for muscle contraction and other vital processes. The vitamin may also be associated with peptide metabolism.

Deficiencies result in dermatitis and loss of hair in all species, particularly in the region of the mouth and eyes. In pigs, ulcerative dermatitis and intestinal necrosis, an exaggerated high stepping leg action, and foetal resorption have been observed; in calves diarrhoea and weakness are the main symptoms.

Dietary requirements in non-herbivores are:

Pig:

18·5 mg/50 kg body weight daily or

12·5 mg/kg of feed (11–13 g per tonne of feed)

Dog:
55 μg/kg (25 μg/lb.) body weight daily

Fox and Mink:
7·2 mg/kg dry food

Fowl:
10–11 g per tonne of feed

Turkey:
11–16 g per tonne of feed

Dosage

No adequate dosage scheme has been drawn up. Human dosage indicates a wide margin of safety, the usual range being from 10 to 50 mg, but doses of up to 1 g have been considered safe. Intramuscularly, 100 mg have been given three times daily, and ointments containing up to 5 per cent have been applied topically. In chickens the requirement of pantothenic acid increases with the fat content of the diet, and with cyanocobalamin deficiency.

Choline

This vitamin B complex member is a colourless viscid, strongly alkaline liquid, which is very hygroscopic and is soluble in water, in alcohol, but not in other organic solvents. It is used as the bitartrate, the dihydrogen citrate, the chloride or the gluconate.

Sources. Choline is a component of lecithin phospholipides, which are found in high concentration in nervous tissue and are therefore widely distributed throughout animal tissues: egg yolk, liver, yeast, brain, kidney, heart, wheat and soya beans, all contain appreciable amounts.

Stability. Choline tends to adsorb carbon dioxide when exposed to air. It is also hygroscopic but withstands autoclaving.

Assay. The assay is a chemical one.

Action

Absorption is fairly rapid from the intestines or following parenteral injection. After absorption the compound is distributed widely, a large amount localizing in the liver. Some of the administered choline is excreted unchanged but a proportion is converted into trimethylamine and excreted as such by all normal excretory routes.

Choline is concerned in methylation processes and increases the phospholipid metabolism in the liver. It will prevent fatty changes in the liver when the animal is on an excessively fatty diet, and is also thought to mobilize the fat from an already fatty liver. It has been used for this latter purpose in animals and man, but its clinical efficacy has not been fully assessed. Cyanocobalamin and folic acid are also associated with transmethylation and choline will, to a certain degree, 'spare' these vitamins.

Deficiency results in perosis (slipped tendon) in fowls and turkeys. Piglets with a choline deficiency suffer from ataxia, fatty livers, and the mortality rate is high.

Nutritional requirements of choline are high, but methionine, of which it is a precursor, and betaine will offset a choline deficiency.

Pig:
900–1000 g per tonne of feed for piglets up to 22·6 kg (50 lb.)

Fowl:
1500 g (optimum) per tonne of feed for chickens up to 18 weeks.

Turkeys:
Up to 1700 g per tonne of feed for turkeys up to 8 weeks.

Choline has been used in the treatment of milk fever and acetonaemia, in both of which conditions fatty livers are probably responsible for a number of cases which annually resist normal treatment. It has also been used in the treatment of fatty livers caused by chlorinated hydrocarbons and in the treatment of liver conditions in dogs.

Dosage

Cow:
25–50 g by mouth or subcutaneous injection.

Dog:
44 mg/kg body weight daily by mouth

The members of the vitamin B complex are rarely administered individually for as has been seen in the foregoing monographs many of them are interdependent. For this reason many commercial vitamin preparations contain most of the series in varying amounts. These are usually expensive, and for normal purposes *yeast* and yeast extracts such as Marmite are quite useful.

Dried Yeast, B.Vet.C. is a buff granular powder with a characteristic odour and taste. It contains in each gramme not less than 0·1 mg aneurine hydrochloride, 0·3 mg nicotinic acid and 0·04 mg riboflavine: in many instances the content is nearly twice these figures. Small amounts of pyridoxine and pantothenic acid are present, with appreciable amounts of choline. The aneurine and nicotinic acid are usually present in phosphorylated coenzyme form. Besides the vitamins, about 45 per cent of yeast consists of protein, 36 per cent of carbohydrate and in the remainder occurs a number of enzymes. Fresh bakers yeast apparently is less active as a vitamin supplement. Dried yeast and Marmite have been used in piglet scours, in poultry food, and as general 'tonics' for small animals.

Dosage

Dog and Cat:

8–15 g (120–240 gr.) daily. This represents 2·4–4·8 mg nicotinic acid, 0·8–1·6 mg aneurine hydrochloride and 0·32–0·64 mg riboflavine

In pigs and poultry concentrations of 3 to 5 per cent have been incorporated in food rations.

Marmite has been given to piglets at the rate of about one teaspoonful per day for each piglet.

VITAMIN C

Ascorbic Acid, B.Vet.C.

This is not an essential nutritional factor in the domestic animals, for it is synthesized by all species. Guinea-pigs, man and monkeys are, however, unable to synthesize the vitamin, and it is consequently an essential nutritional as well as an essential metabolic factor. The vitamin is a white crystalline powder without odour but with a sour taste. It is soluble in water (1:3) and alcohol (1:30), but insoluble in other organic solvents.

Sources. The citrus fruits contain large quantities of vitamin C, but all green foods, meat, especially liver, root crops, tomatoes and blackcurrants, contain appreciable amounts. Commercial ascorbic acid is usually synthetic.

Stability. Ascorbic acid in the dry powder form is fairly stable, though it slowly darkens on exposure to light. In solution, however, it is rapidly destroyed by oxidation and by heating in the presence of oxygen. Solutions may be phosphate buffered at pH 6–6·5, stabilized with 0·005 per cent of thiourea, or with 3 parts sodium chloride to each 10 parts of ascorbic acid.

Asssay. Ascorbic acid has a chemical assay.

Action

Ascorbic acid is destroyed in the rumen but is readily absorbed from the intestines of single stomached animals. Little is stored, though the liver localizes more than other tissues, with the exception perhaps of the adrenal glands. Excretion of any excess occurs in an unchanged form via the kidneys.

The metabolic effects of ascorbic acid are not entirely known. It is essential as a coenzyme in certain oxidative processes, including the oxidation of tyrosine and phenylalanine. It is necessary for normal folic acid function (*q.v.*). Ascorbic acid is also essential for normal wound healing, possibly by inhibiting hyaluronidase-hyaluronic acid activity in the collagen and ground substance of connective tissue: this is also the possible mechanism in maintaining normal capillary function—capillary fragility is one of the main features of scurvy. The role of ascorbic acid in adrenocortical metabolism is unknown, but in times of stress or shock the level within the gland decreases rapidly: following ACTH administration the concentration within the gland increases along with an increase in cholesterol. The significance of these facts is unknown.

The *clinical indications* for the use of ascorbic acid are not determined though the vitamin has been used in the treatment of infertility in cattle. Lowered blood levels in horses, cattle, and dogs, have been reported, though the significance of these findings is not clear. As the site of synthesis in the body is unknown it is difficult to predict the circumstances in which vitamin C deficiencies might occur. However, it is known that vitamin C concentrations are reduced in fever, that the compound is useful in the detoxication of toxins and chemicals, including arsenicals, sulphonamides, salicylates, stilboestrol and barbiturates: these facts serve to suggest various circumstances in which vitamin C therapy might be useful.

Dosage

Horse:
  2–4 g by subcutaneous injection.
Bull:
  1–2 g per 450 kg (1000 lb.) subcutaneously every 3–4 days for up to 6 weeks
Cow:
  1–2 g intravenously and 2 g subcutaneously before mating, or 2 g subcutaneously once or twice a week for up to six doses
Dog:
  25–75 mg orally or subcutaneously each day

Vitamin P. This is not an established vitamin. It is suggested as a factor responsible for maintaining normal capillary permeability and is said to occur naturally along with vitamin C. Two chemicals of a group of compounds collectively referred to as flavonoids and extracted from plants and citrus fruits, *rutin* and *hesperidin*, are claimed to have similar actions. There is yet no clinical indication for the use of vitamin P analogues.

Bibliography

BECK, W. S. (1968) Deoxyribonucleotide synthesis and the role of vitamin B_{12} in erythropoiesis. *Vitams Horm.*, **26**, 413–40.
HÖTZEL, D. (1966) Contributions of the intestinal microflora to the nutrition of the host. *Vitams Horm.*, **24**, 115–66.
SEBRELL, W. H. & HARRIS, R. S. (ed.) (1967) *The Vitamins*, vols II and IV. New York: Academic.
STOKSTAD, E. C. R. (1968) Experimental anaemias in animals resulting from folic acid and vitamin B_{12} deficiencies. *Vitams Horm.*, **26**, 443–60.

18

The Vitamins

II. Fat Soluble Vitamins

This group comprises vitamins A, D, E and K. A further member was known as vitamin F but this has been shown not to be a true vitamin, and linoleic and linolenic acids, hitherto thought to comprise this vitamin, are now classified with the essential unsaturated fatty acids.

The fat soluble vitamins are more stable as a group than the water soluble ones but they have some instabilities, the one perhaps of most importance being the destruction of vitamin E in rancid oils.

Other than stability, various factors may influence the intake of fat soluble vitamins; the interference of carotene absorption by prolonged liquid paraffin taken orally, and the destruction of vitamin E by excessive quantities of fish oils in the diet are of prime importance. Other circumstances such as excessive amounts of calcium in the diet may also influence fat and fat soluble vitamin absorption. These factors are assuming greater importance as more becomes known. For example there is a tendency to decrease the regularity of cod-liver oil, replacing it either by intermittent supplementation of the food or by the use of concentrated preparations which avoid the feeding of large volumes of unsaturated fatty acids. Greater use is also being made of vitamin 'stabilizers' which are usually antioxidants: these are of particular importance to foodstuff manufacturers, who must ensure that the vitamin content of meals and cakes is adequate, even after months of storage.

VITAMIN A

This is a term applied to several compounds which have 'vitamin A activity'. Pure vitamin A is a complex organic alcohol which exists in two main forms, vitamin A_1 (retinol) and vitamin A_2 (3-dehydroretinol), the latter occurring in freshwater fish livers, whilst the former comes from salt water fish livers. There also exist several less active isomers of vitamin A, quantities of which usually accompany the main compounds. Vitamin A_1 differs from A_2 only in having one less double bond in its molecule and yet is three times as active as the freshwater fish oil extractive. The vitamins are usually found not as the free alcohol, but as the acetate or the palmitate: the palmitate ester occurs in greater quantities than the acetate in the liver oils.

The pure alcohol can be obtained as yellow prisms, the vitamin A acetate as pale yellow prismatic crystals, and the palmitate either as yellow crystals or an amorphous powder. Each has a characteristic melting point and is soluble in fixed oils, which are the usual vehicles for the purposes of administration.

Sources. Vitamin A_1, commonly referred to simply as vitamin A, is derived mainly from saltwater fish livers. The vitamin is extracted in the liver oils, the commonest fish being cod and halibut: tuna and other lesser known fish livers are also used, and whale-liver oil is a potent source of vitamin A. The less active *vitamin A_2* is obtained from freshwater fish liver but is of little clinical importance. The livers of food animals also contain appreciable amounts, but vitamin A does not occur as such in plants. One of the major sources of vitamin A in the diet of the domestic animals and the only source in herbivores is the group of compounds known as the *vitamin A precursors*. These compounds are *carotenoids*, but only four of the group are important as sources of vitamin A. They are primarily plant pigments known as α-, β- and γ-carotenes and cryptoxanthine: they are present in fresh green foods (associated with chlorophyll) good quality hay, carrots, tomatoes, egg yolk, milk, butter and cheese. These carotenoids, but mainly β-carotene, are converted into vitamin A by enzyme hydrolysis, probably in the liver and intestinal mucosa. Under optimum conditions β-carotene will form two molecules of vitamin A, but, carotene utilization varies according to the form in which the carotene is presented, the quantities of vitamin A preserving antioxidants present (vitamin E, vitamin C, etc.), and the concentration of the antagonists, such as unsaturated fatty acids, which occur in the food. The carotenes are dark yellow or orange crystalline solids which have a slight odour and little taste. They are fat soluble and dissolve easily in ether but more easily in chloroform and benzene.

Stability. In the absence of oxygen and oxidizing agents vitamin A and its precursors are heat stable, and ordinary cooking does not destroy them. Frozen foods maintain their vitamin content, but rancidity of an oil and exposure to air and light will destroy the vitamin: vitamin C (ascorbic acid), citric acid, and vitamin E (tocopherols), amongst many antioxidant compounds, can be used to maintain stability. In all circumstances the esters—acetate and palmitate—are more stable than the alcohol, and solutions in oil are more stable than the pure, dry compound. Cod-liver oil will keep satisfactorily for many months in a closed, light proof container.

Assay. This was initially carried out by biological methods but modern spectrographic methods have replaced these.

Action

Vitamin A and the carotenes are absorbed from the intestines only when adequate fat, bile salts, and pancreatic enzymes are present. However, too high a concentration of unsaturated fatty acids or liquid paraffin will decrease the amount of vitamin A absorbed, the former by promoting oxidative destruction of the vitamin, whilst the latter dissolves the vitamin and sequestrates it, preventing absorption. Enteritis will also interfere with absorption. Under normal conditions absorption is rapid from both the intestines and sites of parenteral injection. Following absorption the vitamin passes into the lymph and so into the circulation where it may be found as esters of the fatty acids, the free alcohol, and protein bound carotene. The majority of any one dose will be stored in the liver, which normally contains about 95 per cent of the vitamin A in the body: other tissues including the kidneys, adrenals, lungs, and fat, all contain some vitamin A and/or carotene. Following administration, this storage commences within a few hours, and it has been shown by several groups of workers that rats, for example, will store enough vitamin A from one feed on cod-liver oil or carotene to meet its requirements for the subsequent 2 months.

The *metabolic effect* of vitamin A is mainly on epithelial tissues, including the retina, skin, and mucosa of the respiratory and urinary systems. The retinal effect relates to the fact that vitamin A is a part of the conjugated protein *visual purple* (rhodopsin), and also of visual violet (iodopsin). The former is concerned with accommodation to light (hence night blindness in deficiencies), whilst the latter is concerned in colour vision: the visual purple is obviously of greater veterinary significance. Young animals are normally born with low reserves, though these reserves are usually proportional to the vitamin A levels in the blood and tissues of the dam, and can therefore be increased by adequate vitamin therapy during pregnancy; after birth, these reserves are usually supplemented by the vitamin in colostrum and milk.

Deficiency symptoms, where the avitaminosis is marked, are mainly those of hyperkeratosis of the skin, including the cornea (xerophthalmia, which once established is almost impossible to cure), blindness, nervous symptoms, decreased growth rates and a lowered fertility. Details of effect vary in the different species:

The *horse* does not convert carotene to vitamin A very easily, and plasma vitamin A levels are normally lower than those of the cow. Symptoms of deficiency include night blindness, corneal keratinization which results in photophobia, closed lids and lacrimation. The appetite is capricious and, as the condition progresses, respiratory embarrassment, as well as interference with normal mating processes and fertility occur. Interference in hoof formation, with a scaly and encrusted secretion from the perioplic band, has been reported.

Cattle provide perhaps the most spectacular symptoms of vitamin A deficiency, calves showing blindness due to abnormal bone growth around the optic foramina which constricts, and finally causes pressure necrosis of the optic nerve. The calves grow slowly, scouring is common, hyperkeratinization may occur producing a rough, scaly, dry skin which is particularly obvious in the region of the neck, withers and tail head: the coat is staring. In older animals, night blindness may result from retinal abnormalities; abortion sometimes occurs, and oedema of the legs and lower parts of the body (anasarca) may be seen. In all cases increased cerebrospinal fluid pressure may cause derangement of motor reflexes, with muscular incoordination, a staggering gait and convulsions. Sterility in young bulls has been attributed to lack of vitamin A during the prepubertal period. Response to treatment is slow, and prophylactic feeding with cod-liver oil is therefore widely practiced. It is now common practice to include high vitamin A levels in diets with a high cereal content, to counteract hypovitaminosis.

In *sheep* and *goats* vitamin A deficiency has received little attention, and efforts to identify specific disease syndromes associated with such a deficiency have not had much success. It would seem that, in sheep at any rate, symptoms of vitamin A deficiency follows a pattern similar to that in cattle, but of a less acute nature.

Vitamin A deficiency in *pigs* may occur in intensively reared herds where provision of vitamin supplements is omitted. If large amounts of swill are fed vitamin A deficiencies are more likely. Symptoms may include unthriftiness, slow growth, poor appetite, keratinization of skin and cornea, posterior incoordination and paralysis, often with convulsions: adults may be infertile, but where conception is achieved, piglets may be aborted or suffer from weakness and various deformities, including microphthalmia. Resistance to secondary pathogens is low and enteritis and pneumonia occur frequently. The condition must be differentiated from swine fever.

Carotene conversion is relatively effective in pigs and access to green foods normally provides sufficient vitamin A. This can be supplemented by cod-liver oil or synthetic preparations, but overdosage with these compounds can be harmful.

Small animals, including the dog, cat, rabbit, fox, mink, etc., apparently require vitamin A in amounts similar to those required by the pig on a weight for weight basis. Symptoms are typical, but deafness due to errors in bone growth similar to those which cause blindness in calves has been reported in dogs: the eighth cranial nerve may also be constricted. Lesions of the oral mucosa, and inadequate kidney function have also been reported in dogs.

Poultry require relatively large amounts of vitamin A, and turkeys more than fowls. Symptoms result after a few weeks on a deficient diet, ducks and turkey poults showing symptoms 3 to 4 weeks after hatching. These symptoms follow much the same pattern as in mammals and form a syndrome referred to as 'nutritional roup'. Young birds become lethargic, stand about with ruffled feathers, and when they do move show incoordination with a staggering gait. The birds stop growing, become progressively emaciated and, if the condition is untreated they die, usually from secondary infections: the mortality rate in turkeys is very high unless treatment commences within a fortnight or so of symptoms becoming obvious; with chickens the condition is not so acute. Specific lesions occur in association with the eyes and respiratory system: in the early stages there are watery discharges from the eyes and nose, but later these discharges become inspissated, and cheesy exudates cover mucous membranes and block sinus exits, tear ducts and possibly the bronchi. These birds then show facial swellings as the sinuses fill and swell, and respiratory embarrassment becomes progressively worse. Xerophthalmia (corneal keratosis) is occasionally apparent. Adults will show most of these symptoms, but also a decreased egg yield and poor hatchability. Uric acid crystalline deposits in the kidneys are very common.

Vitamin A is available as crystalline vitamin A alcohol or vitamin A acetate; 1 unit of vitamin A equals 0·3 μg of crystalline vitamin A alcohol or 0·344 μg of vitamin A acetate.

Nutritional requirements are not standard, either for species or individuals, but the following will serve as a guide:

Horse, Cow, Sheep and Pig:
3000 units/50 kg body weight daily

Small animals:
about 66 units/kg body weight daily
Poultry:
4000–8000 units daily per kg of feed.

Pigs may be fed at the rate of 2 500 000 units per tonne of feed for all but breeding stock, which require between three and four times this quantity. Fowls require about 3 000 000 units per tonne of feed, though laying and breeding stock require 5 000 000 and 4 000 000 units per tonne respectively. Turkeys require about 6 000 000 units per tonne.

Dosage

Vitamin A is commonly given by mouth along with vitamin D as Cod-Liver Oil (Veterinary), B.Vet.C. Less commonly it is given as Concentrated Solution of Vitamin A, B.Vet.C., or as the Concentrated Solutions of Vitamins A and D: parenteral injections of these are available commercially. Halibut-Liver Oil, B.Vet.C. is also used.

Cod-Liver Oil (Veterinary), B.Vet.C.

Cod-liver oil has characteristics known to all —a clear, yellow, oily liquid with a fishy odour and taste: it contains not less than 600 i.u. of vitamin A, and not less than 85 i.u. per g of vitamin D. The dose is:

Horse and Cow: 15–60 ml (½–2 fl. oz.)
Sheep and Pig: 4–15 ml (60 m.–½ fl. oz.)
Dog and Cat, etc.: 0·5–8 ml (8–120 m.)
Poultry: ½–1 per cent in the food (6–12 litres or 10–20 pt./ton)

It should not be given to food animals for at least 2 weeks before slaughter. It is advisable, where food medication is practised, to add the oil as near to the time of use as possible so that vitamin losses from oxidation may be minimal.

Cod-liver oil is used as eye drops and also as a vehicle for skin therapeutics.

A cod-liver oil emulsion and a combination with extract of malt are also detailed in the B.Vet.C., but these are used less frequently than the plain oil.

Halibut-Liver Oil, B.Vet.C.

This is similar in character to the cod-liver oil but its vitamin content is vastly different: it contains not less than 30 000 units vitamin A per gram, and though no specific limits are prescribed for vitamin D this usually approximates to 3000 units per gram. This oil is obviously preferable for the treatment of vitamin A deficiencies, but it is expensive and is rarely used other than in capsules for small animal treatment.

The dose for horses, cattle, sheep and pigs, is 0·2–0·3 ml per 50 kg (3–5 m./cwt) daily, and for dogs and cats: 0·075 to 0·3 ml (1–5 m.)—usually one or two capsules daily.

The oil in extract of malt is also official but is not commonly used.

Concentrated Solution of Vitamin A, B.Vet.C.

A fortified fish-liver oil, or a solution of vitamin A in a bland oil, usually arachis oil: it may or may not have a fishy taste and odour, and is normally a golden colour. It contains between 45 000 and 55 000 units of vitamin A per gram and is therefore superior to halibut-liver oil in its potency.

The dose for all farm animals varies widely but it is usual to give the equivalent of about 100 to 200 units per kg (45 to 90 units per lb.) body weight daily—this is approximately 1 to 2 g per day for a 450 kg (1000 lb.) cow. Fowls are given 0·07 g and turkeys 0·22 g per kg (0·1 g per lb.) food—these are equivalent to 3000 and 4000 units, respectively.

VITAMIN D

A number of sterols possess antirachitic activity, i.e. the ability to prevent and cure rickets. Two such substances, vitamin D_2 or *calciferol* and vitamin D_3 or *cholecalciferol*, are commonly used in therapeutics. Vitamin D_3 occurs naturally in fish liver oils along with vitamin A while calciferol is the best antirachitic sterol produced by the irradiation of the provitamin *ergosterol* in yeasts and fungi. Vitamin D's common nickname of the sunshine vitamin derives from the ability of ultra-violet light to convert the provitamin dehydrocholesterol into cholecalciferol in the skin. Both processes are utilized in the natural and synthetic production of vitamins D_2 and D_3. Vitamin D status is determined by dietary availability and sunlight. In that it is synthesized in the body, it is scarcely a true vitamin. Its intestinal absorption requires adequate bile secretion and fat absorption and is impaired by enzyme lack, excess of fat or calcium and scouring. It is transported in blood bound to an α-globulin prior to its activation or storage in the liver. It is now known that cholecalciferol is a prohormone for factors which control calcium turnover in the body.

Calciferol, B.Vet.C.

This is produced by irradiation of ergosterol, which occurs in yeast and moulds in up to 2 per cent concentrations of the dried substances. It is a white crystalline, odourless and tasteless powder, which is soluble in organic solvents including fixed oils, and is affected by air and light. Each milligram contains 40 000 units of antirachitic activity as estimated biologically using the effect on bone calcification in rats as the indicator—other methods of assessment may be used.

Activated 7-Dehydrocholesterol, U.S.P.

Vitamin D_3 or cholecalciferol is very similar in all characteristics to calciferol. It occurs in cod-liver and other fish-liver oils, and may be manufactured synthetically from cholesterol by way of dehydrocholesterol. Vitamin D_3 is used as the international reference standard for assays of antirachitic compounds, and is considered to be more effective than calciferol in the same species.

Cod-Liver Oil (Veterinary), B.Vet.C., contains not less than 85 i.u. antirachitic activity, the majority of which is due to vitamin D_3. Halibut-Liver Oil, B.Vet.C. contains about 3000 units of antirachitic activity (see vitamin A—dosage).

Action and Uses

The need to maintain the plasma Ca^{++} level within the narrow limits compatible with health has probably led to the evolution of its complex and delicate control mechanism. Parathormone is secreted in response to hypocalcaemia which it then corrects by moving calcium from bone into the ECF, by increasing the active absorption of calcium from the gut and by increasing the resorption of calcium from renal ultrafiltrate. The calcitonin release is believed to be triggered by hypercalcaemia which it probably corrects by diminishing bone resorption. These hormonal effects are modulated by activated vitamin D whose biosynthesis is itself regulated by plasma Ca^{++} concentration.

Cholecalciferol (CC) undergoes 25-hydroxylation in the liver, then re-enters the plasma and binds to an α-globulin. D_3 in excess of requirements remains bound to α-globulin or is stored in body fat. In the kidney, under the influence of hypocalcaemia, hypophosphataemia and PTH, 25-OH CC is converted to 1,25-dihydroxy CC, now known to be the most active form for vitamin D_3 and responsible for its effects in increasing the rate of Ca^{++} absorption and transport from the intestine and so promoting calcification of bone. Under normocalcaemia or hypercalcaemia, and in the absence of PTH, the kidney instead converts 25-OH CC to the less active 24,25-dihydroxy CC. Further hydroxylation to 1,24,25-trihydroxy CC yields a metabolite whose major action is to facilitate the intestinal absorption of calcium.

Vitamin D deficiency is one cause of rickets in young animals and osteomalacia in adults. Frank deficiencies are rare nowadays and vitamin D is used more in the prevention than the treatment of deficiency states dietary intake and biosynthesis are adequate. Hypoparathyroidism can often be managed safely with vitamin D and attention to diet. Dihydrotachysterol is an analogue of low antirachitic but considerable hypercalcaemic potency. It is used in hypoparathyroidism and in the prophylaxis of bovine hypocalcaemia.

Rickets is a disease of the immature growing animal in which calcium and phosphate metabolism are disturbed and incomplete ossification of the bones occurs. This lack of ossification leads to enlarged painful joints, bowed long bones, and 'rickety rosary', which is an enlargement of the costochondral junctions of the ribs:

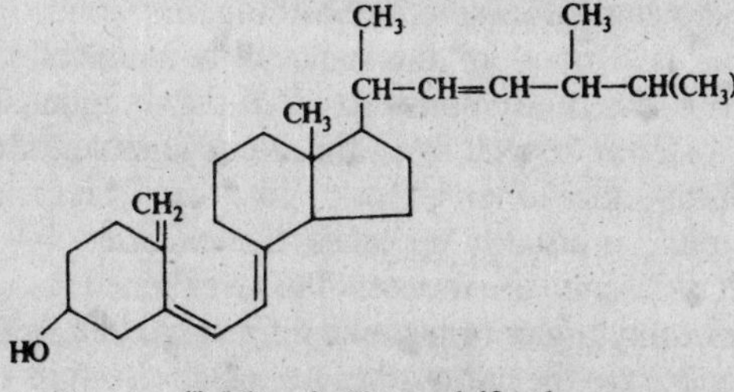

(i) Vitamin D_2—calciferol

fractures may occur spontaneously. These lesions result in marked lameness. In all cases the appetite becomes capricious: this may cause some temporary recession in symptoms for the growth rate decreases and the demand for vitamin D and, or, minerals remains static—it is always the larger breeds and rapidly growing animals which suffer most.

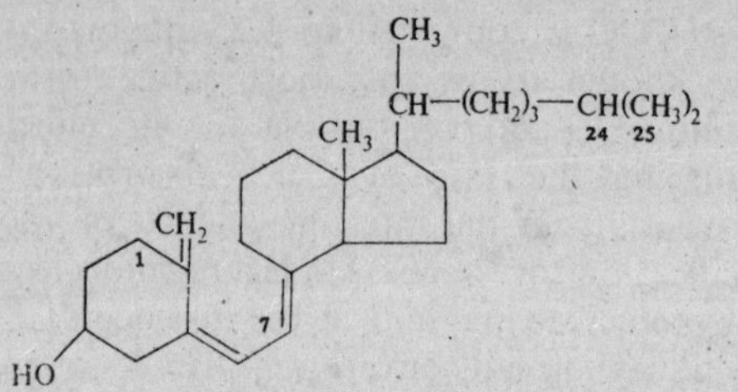

(ii) Vitamin D_3—activated 7-dehydrocholesterol

In *calves* and *lambs*, rickets is commonly due to deficiencies of either calcium, or phosphorus, or both, rather than vitamin D, for irradiation of sterols in the skin plus whatever occurs in the diet will usually provide sufficient vitamin D for normal purposes. It has, however, been reported that vitamin A and its precursors may be the cause of rickets in lambs on pastures adequate in calcium and phosphorus, and where vitamin D deficiency is unlikely: this action is suspected tobe due to an effect on phosphorus metabolism.

Pigs are dependent to a greater extent on vitamin D supplies: dogs are similarly dependent, but it is *poultry* which suffer most markedly from vitamin D deficiencies.

Osteomalacia is the equivalent disease in adults. It is characterized by the formation of excessive amounts of osteoid tissue which fails to calcify: this osteoid tissue is mainly confined to the diaphyses of the bones, for it is formed during the normal process of rebuilding the bone, but some is formed in the cancellous bone of the epiphyses of long bones: osteoid tissue under the periosteum, especially at the site of ligament and tendon attachments, and in the Haversian systems, gradually replaces the calcified tissue and weakens the bones. The weakened bones eventually break or become deformed, and in the former case the callus does not calcify so that the bone is unable to support weight.

Recovery of young stock will follow readjustment of the diet although malformations will remain. In adults the prognosis is rarely satisfactory, and where recovery does take place progress is slow and malformations persist.

Nutritional requirements vary widely. There are, for example, species differences, for foxes and mink require very little vitamin D if calcium and phosphorus are present in the diet in correct proportions: other species, dogs, calves, foals and especially poultry, have far greater vitamin D requirements even when the calcium and phosphorus elements are present in the correct ratio. Fast growing breeds of animals require more of the vitamin than the slower growing breeds: this is very well shown in the larger breeds of dog—the greyhound, St Bernard and Great Dane. Besides these differences it has been shown that breeds with pigmented skins have greater requirements than light skinned breeds, that pregnant and lactating females require more, that animals in northern or southern climates require more than those in temperate or equatorial areas, and that greater quantities of vitamin D are required in the winter than in summer. Of considerable importance is the fact that vitamin D requirements in all species and individuals is minimal when calcium and phosphorus are present in the diet in optimal proportions (a ratio of approximately 2:1). As the ratio changes the vitamin D requirements increase and the vitamin becomes of greater importance, for in many species it tends to overcome the effects of the imbalance and promotes normal utilization. However, though vitamin D will improve the assimilation and use of any calcium or phosphorus present in the food it will not replace an absolute mineral deficiency.

Average *daily minimum requirements* have been assessed as follows:

Horse, Cow, Sheep and Pig:

300–500 units/50 kg (1 cwt.).

Dog:

10–20 units/kg (5–10 units/lb.) body weight.

Normally about three times these quantities should be given daily if full requirements are to be met.

Poultry requirements concern vitamin D_3, for vitamin D_2 is relatively inactive. Fowls need about 1 000 000 units vitamin D_3 per tonne of food, turkeys 2 000 000 units. Where vitamin D is given to pigs in food up to 8 000 000 units per ton of food has been recommended for piglets of up to 100 kg (50 lb.) weight, and 500 000 units for older pigs though breeding stock may be given 1 000 000 units per tonne of food.

Dosages for treatment of deficiencies should approach ten to twenty times the minimum daily requirements.

VITAMIN E

This is a collective term applied to a group of eight alcohols known as the α-, β-, γ- and δ-*tocopherols* and α-, β-, γ-, δ-*tocotrienol*, of which *α-tocopherol* is by far the most active. Alpha-tocopherol is a yellow, almost odourless, viscous oil: it oxidizes very easily and darkens with exposure to air and light. The tocopherols as a group are insoluble in water but soluble in organic solvents, including fixed oils. The acetate esters, which have the same solubilities, are preferred for therapeutic use because of their greater stability.

Sources. Vitamin E is present in green foods, crude vegetable oils and animal fat: its most potent source is wheat germ oil which has been the main therapeutic source of the vitamin until DL alpha tocopherol was produced synthetically.

Stability. The vitamin E alcohols are rapidly oxidized and destroyed when exposed to air and light. The acetates, however, are less affected by atmospheric oxidation, but are still susceptible to the effects of light, though to a lesser degree than their parent alcohols. Any process which involves oxidation automatically involves the tocopherols which are antioxidants. Thus any oil or fat which is rancid will contain less active vitamin E, for the oxidative processes which result in rancidity will have been restrained by the antioxidant effects of the tocopherols and marked rancidity will only have developed when some of the tocopherols have been destroyed.

Assay. The tocopherols are assayed chemically.

Action

Under normal conditions the tocopherols are readily absorbed from the intestines, and excess is either stored in the fat depots of the body or excreted, mostly in the bile but partly in the urine. Where fat absorption is restricted the absorption of vitamin E will be likewise affected.

The predominant action of the tocopherols would appear to be that of regulating the oxidation processes of the body. This anti-oxidant action links its physiological effect with vitamin A metabolism (it is used for the stabilizing of vitamin A in foodstuffs), and with that of the unsaturated fatty acids. It is probable that it is this antioxidant 'protecting' action on unsaturated fatty acid which is responsible for the 'conditioned' vitamin E deficiency that has occurred fairly widely due to the regular feeding of excessive quantities of cod-liver oil, although it has been pointed out that various other factors, including mineral imbalances, influence the course of this type of deficiency. In this 'conditioned' deficiency of vitamin E, the tocopherol content of even severely affected muscles is higher than in normal muscles. Interference with energy exchanges of cells may occur, perhaps due to the formation of abnormal phospholipids on cell interfaces by the excess unsaturated fatty acids. It is suggested that tocopherols either prevent the action of such compounds or facilitate their metabolic neutralization. Increased requirements of vitamin E would therefore follow an increased intake of unsaturated fatty acids. More recent workers have indicated that excessive iron intake can induce severe myodystrophy. It has also been shown that selenium can have a vitamin E sparing action so that under certain circumstances inadequate vitamin E in the diet can be offset by the provision of selenium compounds. Sodium selenite should be given to ewes 3 to 4 weeks before lambing in doses of 10 mg per head and followed up with the same dose to each lamb when 3 to 4 weeks old. In cattle doses at the rate of 0·05 mg per kg body weight should be combined with vitamin E therapy.

These details add emphasis to the fact that the biochemical status of vitamin E is as yet incompletely known and is probably not simply one of controlling oxidation processes, although investigations now proceeding suggest that the vitamin may be involved in the cytochrome-C reductase system. Animals can be reared without vitamin E and it has been suggested that the role of vitamin E can be taken by one of the co-enzyme Q group.

Deficiency syndromes which respond to tocopherol therapy have been reported in all domestic species.

It is probable that vitamin E deficiency in all species is accompanied by muscular disorders. These are most marked in calves and lambs, in which extensive investigations have related muscular dystrophy, white muscle disease, and stiff lamb disease (some types), to tocopherol deficiencies. In horses, pigs and other species, vitamin E deficiencies have not been examined so intensively. The muscle damage results in changes of stance, movement, and general behaviour as the affected animals attempt to transfer the function of affected muscles to other unaffected muscle groups. Lameness is bilateral, and usually numbers of individuals are affected at the same time. The symptoms will vary considerably according to the groups of muscles which are affected, but three main types are described: a skeletal type, in which locomotor and supporting muscles and symptoms are involved; a thoracic type, in which intercostal and diaphragmatic muscles are affected: and a cardiac type, in which the heart muscles are affected.

In addition to muscular involvement other effects such as diminished fertility, oedema, and steatitis, have been demonstrated in some species. In *pigs*, for example, embryonic death occurred in sows maintained on vitamin E deficient diets. In *poultry*, fertility is also affected—cockerels develop testicular degenerations and hatchability of eggs is affected. In *mink* and *pigs*, an inflammation of fatty tissues and a yellow-brown discoloration of the fat, steatitis or 'yellow fat', have been widely reported. In *poultry*, exudative diathesis, which is a generalized oedematous condition resulting from increased vascular permeability (also reported in pigs), and a condition known as nutritional encephalomalacia or 'crazy chick disease', which usually occurs in chicks from 3 to 6 weeks of age but has been reported shortly after hatching, are two other ways in which vitamin E deficiency may show itself.

Apart from pigs and poultry no fertility effects have been reliably reported in domestic animals, though such effects in rats and mice have been fully established and were the reason for the early references to vitamin E as the 'anti-sterility' or 'fertility' vitamin.

In lambs and calves the incidence of the disease is mainly in spring and early summer, if animals are housed during winter. Increased exercise precipitates an attack of the disease and symptoms occur within 2 days or so of turning the animals out to grass. Cardiac cases are very acutely affected and are often found dead. With less acutely affected cases a large proportion will recover if treatment is instituted before gross, irreversible muscular damage is caused. The appreciation of the relationship of poor winter feeding and vitamin E deficiencies with muscular dystrophy has now greatly reduced the incidence and mortality rate, but prior to this in some areas of Scotland, Norway, Finland, Canada, and the United States, the morbidity rate was as high as 25 per cent and the mortality rate was also very high. Annual losses are still appreciable.

Vitamin E and its Importance in Large Animal Nutrition

Vitamin E activity is shared by a number of related compounds, of which α-tocopherol and its derivatives are the most important. This activity is still defined in terms of the prevention of resorption of foetuses in the rat ('the gestation-resorption test'). One International Unit represents the activity of 1 mg DL-α-tocopheryl acetate in this test. Vitamin E is available commercially both as the synthetic racemic (DL) form and the natural, optically active (D) form. The natural forms have a higher activity than the synthetic: generally the potency ratio of natural to synthetic is about 1·3:1. The tocopherols are generally fat-soluble, though water-soluble derivatives can be prepared.

Specific factors have also been discovered in some feedstuffs, e.g. lucerne and raw kidney beans, which interfere with tocopherol uptake.

The role of selenium. During the last 10 years, it has gradually become clear that selenium deficiency plays an important part in many diseases which had previously been thought to be due purely to lack of vitamin E. Thus the muscular dystrophies of sheep and cattle appear naturally in areas where soil is low in selenium.

Sometimes the selenium deficiency appears to be more important, as on the lush pastures of New Zealand where there is probably adequate vitamin E but low selenium. Sometimes both factors are probably deficient, as in parts of Scotland where cattle pastured on land low in selenium are fed during the winter on roots and hay, both poor sources of vitamin E.

Deficiency diseases. The principal vitamin E-deficiency disease of cattle and sheep is muscular dystrophy ('white muscle disease' or 'stiff lamb disease'). This occurs especially in young animals, and the clinical syndrome characteristically develops after some sort of stress, e.g. after a cold, windy night in Scotland, or after a flock has had to walk some miles to new pastures in New Zealand. The affected animals have a typical stiff gait, and may die suddenly of cardiac failure. The disease has been reported from Scotland, the Rocky Mountain States, New Zealand and many other parts of the world. Confirmation of the clinical diagnosis is usually by examination of the muscle for the typical white areas (Zenker's degeneration) which are found in the skeletal and sometimes also in the cardiac musculature. The syndrome also occurs occasionally in calves fed milk substitutes if the ration is not adequately fortified with vitamin E.

In pigs there are a number of conditions which seem to be due to a deficiency of vitamin E and/or selenium. Hepatosis dietetica, steatosis ('yellow fat disease'), dietetic micro-angiopathy ('mulberry heart disease') and muscular dystrophy have all been described in pigs. In the case of the cardiac disorders, the first sign that something is amiss is often deaths in the herd. The exact aetiology of these diseases is by no means fully understood: it is not certain, for instance, that the 'dietetic micro-angiopathy' described in Scandinavia is exactly the same as the disease seen in the United Kingdom. Nor is it certain what is the factor or factors responsible for triggering off outbreaks—'new grain', rancid cereals and various other things have been blamed from time to time, but it is often difficult to correlate these factors with the occurrence or severity of disease.

In the horse, 'yellow fat disease' has been recorded in New Zealand, but clinical vitamin E deficiency is rare in this species.

Prophylaxis and therapy. The provision of adequate levels of selenium or vitamin E in the diet is the most effective prophylaxis. What is an 'adequate level' depends partly on the composition of the rest of the diet (e.g. if unsaturated oils, such as cod liver oil, are being given, more vitamin E will be needed). Selenium salts can be given by injection, but the possibility of overdose must always be borne in mind, since selenium is a comparatively toxic element. It has been suggested in the United States that, since relatively low levels of selenium are needed to prevent these disorders, steps should be taken to exchange feedstuffs between areas of selenium adequacy or excess and areas of selenium deficiency.

Therapy when an outbreak of deficiency disease is suspected usually consists of immediate administration of vitamin E and/or selenium in whatever form is appropriate and convenient.

Nutritional requirements are not defined. This is possibly because the availability of the vitamin is conditioned by the composition of the diet. Where the level of unsaturated fatty acids in the diet is high, the vitamin requirements will also be high, but it should be remembered that other compounds and conditions besides these acids will depress absorption of vitamin E, including tri-orthocresylphosphate (in some species), possibly sulphaguanidine and gastroenteritis: it must also be borne in mind that variations in mineral and carbohydrate concentrations in a diet will modify the symptoms resulting from a vitamin E deficiency or insufficiency. Deficiencies other vitamins such as choline may predispose muscles to the effects of vitamin E deficiencies.

Dosage

Therapeutic doses must be increased if the exciting factor is unknown. Should the factor be cod-liver oil, it is advisable to discontinue the administration of this, and supply vitamins A and D in some other form in which unsaturated fatty acids are not present in such high proportions—if at all. Daily doses of **Tocopheryl Acetate,** B.Vet.C. are:

Calves: 0·3–2 g
Lambs: 0·1–0·5 g
Pigs (weaned): 0·5 g minimum
Dogs: 30–100 mg

These doses may be given orally, by subcutaneous, or by intramuscular injection.

The B.Vet.C. (1965) quotes initial doses of

0·6 g and 0·1 g for calves and lambs, respectively, followed by daily doses of 0·2 g and 35 mg.

Wheat Germ Oil, B.Vet.C. (1953)

Wheat Germ Oil has been used but, because of its variable vitamin E content, it is now being largely replaced by synthetic α-tocopherol preparations. The average daily doses recommended are:

Calves: 10 ml
Lambs: 1 ml
Dogs: 1–2 ml

Toxicity. Extremely large doses of tocopherols have produced neither acute nor chronic toxicity and consequently dosage is conditioned more by size of effective dose and expense than by any other consideration.

VITAMIN K

The name of this vitamin was derived from *koagulation*. As with other vitamins, the specific vitamin K activity is possessed by a group of compounds, three of which are of significance. All are naphthaquinones. The first to be discovered—that occurring naturally in green foods—was 2-methyl-3-phytyl-1,4 naphthaquinone, and was designated vitamin K_1. The second, discovered in putrefied fish meal, and designated vitamin K_2 has since been shown to be a series of nine closely related compounds. Vitamin K_3 is used to designate a synthetic naphthaquinone found after an intensive investigation of a large number of naphthaquinone compounds; it is simply 2-methyl-1,4-naphthaquinone. Vitamin K_1 is recognized as **Phytonadione** by the U.S.P., and vitamin K_3 as **Menaphthone,** B.Vet.C. or as **Menadione** in the U.S.P. This source of vitamin K_1 is 3·3 times more active than natural vitamin K_1.

Menaphthone, B.Vet.C. (1953)

This synthetic vitamin is a bright yellow crystalline powder with a faint but characteristic smell. It is insoluble in water, slightly soluble in alcohol and fixed oils, but more soluble in other organic solvents. The powder is very irritant to all mucous membranes, and alcoholic solutions are vesicant. Menaphthone Sodium Bisulphite has been included in the *British Veterinary Codex* 1965 and is a white crystalline, odourless hygroscopic powder. Various menaphthone derivatives are available—acetomenaphthone for oral administration, and the water soluble compound, menaphthone sodium bisulphite for parenteral administration, are the more important ones.

Sources. Menaphthone is synthetic, but vitamin K_1 is present in green leaves—lucerne and horse chestnut leaves being credited with the greatest concentrations—but species of *Brassica* contain appreciable quantities. Vitamin K_2 is derived mainly from rotted fish, but in common with vitamin K_1 can be manufactured synthetically. Meat and animal products contain little vitamin K.

Stability. The powder is said to decompose to some extent on autoclaving at 15 lb. pressure for 30 min, but solutions in oil may be sterilized by heating at 150°C for 1 hr. Menaphthone is decomposed by light which results in a darkening in the colour of both powder and solutions.

Assay. The assay is chemical.

Action

Ruminants are independent of dietary vitamin K once rumination has commenced, and it is possible that most other species derive some of their vitamin K from bacterial synthesis in the intestines. However, in young ruminants prior to the onset of rumination and in simple stomached species, at least some vitamin K must be supplied in the food. Like all fat soluble vitamins the absorption of vitamin K from the intestines is dependent on the presence of bile and fat in the intestines. Storage is mainly in the liver and excretion is via the urinary system.

Menaphthone and the other vitamin K compounds are essential for the formation of prothrombin in the liver, and lack of the vitamin results in hypoprothrombinaemia. Lack of prothrombin in the blood breaks the chain of events which results in the formation of blood clots, so that when small contusions occur prolonged bleeding follows with the formation of large haematomata, and the development of severe haemorrhagic anaemia and haemolytic jaundice. It is only when animals are kept under intensive artificial conditions that deficiencies are likely to occur and, with poultry in particular, it is essential to ensure an adequate level of vitamin K in the diet.

A conditioned *deficiency* exists in cattle and is known as 'sweet clover' disease, in which dicoumarol or bis-hyroxycoumarin elaborated in the sweet clover is absorbed and counteracts the formation of prothrombin in the liver. The same principle applies to the action of Warfarin, an anticoagulant rat poison. Vitamin K is the physiological antidote, but in Warfarin poisoning (see p. 96) absorption of the vitamin is usually too slow following oral or intramuscular administration, and slow intravenous injection of menaphthone sodium bisulphite should be tried. If possible, menaphthone treatment should be associated with blood transfusions. In the treatment of Warfarin poisoning, vitamin K_1 has been shown to be more effective than the water-soluble analogues. The recommended dosage rate is 1·1 to 2·2 mg per kg body weight by oral or parenteral administration. If the intravenous route is used for the initial dose some shock might be anticipated with certain commercial preparations of vitamin K_1 and antihistaminic premedication is advisable. Dosing should continue every 24 hr for at least 4 days.

Vitamin K is of some value for the treatment of the epistaxis in racehorses associated with the excitement prior to and during races, but the perfect treatment for this condition has yet to be found. Given with vitamin C, menaphthone and other vitamin K analogues are useful in salicylate poisoning; they have also been used beneficially in quinine poisoning and in diseases such as hepatitis and gastro-enteritis when vitamin K absorption is restricted, or prothrombin formation impaired.

Dietary requirements have not been generally assessed but it has been suggested that 0·4 g per ton of food is adequate for chickens of up to 18 weeks of age.

Dosage

No established doses are available but 50 or 60 mg have been given intravenously in dogs without any toxic effects. It is probable that in Warfarin poisoning the safe maximum human dose of menaphthone sodium bisulphite, which is 100 mg daily, may be taken as a guide for the treatment of dogs, cats and pigs. Intramuscular injections of up to 5 mg of menaphthone have been suggested by the *British Veterinary Codex*. Commercially, acetomenaphthone has been recommended in daily oral doses of 2 to 10 mg for the dog, and 100 to 400 mg for large animals. For the treatment of vitamin K deficiencies in poultry quantities of 1 to 2 g menaphthone sodium bisulphite per ton of feed are recommended by the 1965 B.Vet.C.

The Polyunsaturated Fatty Acids

These 'essential' fatty acids were formally referred to as vitamin F, but this name has now been discarded. There are two main groups of unsaturated fatty acids conforming to the general formulae of $C_nH_{2n-4}O_2$ (linoleic acid), and $C_nH_{2n-6}O_2$ (linolenic acid). Members of the first group contain two double bonds, and of the second three double bonds; the degree of unsaturation naturally increases with the number of double bonds. One or two acids with higher degrees of unsaturation are of some importance.

Linoleic Acid

One of the two main essential unsaturated fatty acids, belongs to the second group and contains within its structure two double bonds: it has the empirical formula $C_{17}H_{31}COOH$.

Arachidonic Acid

Arachidonic acid $C_{19}H_{31}COOH$, is the important acid. It is more highly unsaturated than any of the three main groups mentioned, and contains four double bonds.

The main natural *sources* of these fatty acids are soya bean oil, corn embryo oil (maize germ oil) and lard. The first contains about 50 per cent linoleic and between 5 and 10 per cent linolenic acid, the second about the same amount of linoleic but only about 2 per cent linolenic acid, but lard contains only 10 per cent or so of linoleic acid.

These acids are relatively unstable but the addition of tocopherols provides a certain degree of protection by virtue of their antioxidant action.

Deficiencies have resulted in stunted growth, skin lesions and tail necrosis in rats but so far little of clinical significance in domestic animals has been demonstrated, though beneficial effects in human skin lesions have been claimed. Linolenic acid is far less important than either linoleic or arachidonic acids in the experimental work so far carried out. Linoleic acid prevents the onset of certain of the lesions associated with pyridoxine deficiencies in rats (acrodynia).

These acids have recently been shown to be the starting point in the in vivo synthesis of a pharmacologically very active and widely distributed group of compounds known as the prostaglandins (see Chapter 15).

Bibliography

DeLuca, H. F. (1967) Mechanism of action and metabolic fate of vitamin D. *Vitams Horm.*, **25**, 316–62.

Martins, C. (1966) Mode of action of vitamin K in animals. *Vitams Horm.*, **24**, 441–6.

Sebrell, W. H. & Harris, R. S. (1967) *The Vitamins*, vols I, III and V. New York: Academic.

19

Agents Affecting the Skin

I. Stimulants

Compounds which have an action on the skin can be applied in various forms, as powders, aqueous solutions and suspensions, alcoholic solutions and suspensions, mineral oil and fat solutions and suspensions, and vegetable and animal oil or fat solutions and suspensions.

In general, their degree of action is related to their position in this scale, for drugs applied as powders have only a superficial action unless applied to wounds, whereas the same compounds incorporated in animal fats could have a deep penetrating action and might be absorbed in sufficient quantities to have a systemic and even a toxic effect.

Iodine compounds provide an illustration. Iodine is not normally applied to the skin in powder form but Aqueous Solution of Iodine B.Vet.C., Lugol's iodine, which contains 5 per cent of iodine is of very little use externally. The Weak Solution of Iodine, B.Vet.C., which is a 2½ per cent alcoholic solution, is sufficiently active to be used as a counter-irritant. A 10 per cent solution is frequently used as a penetrating counter-irritant in both human and veterinary medicine in the treatment of chronic inflammations of structures below the skin. Swollen lymph glands and chronic, suppressed abscesses, which may occur in strangles in the horse, bursitis and synovitis, have all benefited from such treatment. Iodine dissolved in mineral oil, and ointments with bases incorporating animal fats, have a stronger effect still. The mineral oil solutions, if applied to large areas or too frequently, will produce iodism—iodine poisoning—within a few days.

It follows from this that the form and method of applying drugs externally must be carefully chosen according to the degree and type of action required.

The effect of concentration is, of course, additional to these two factors. In most cases, higher concentrations will produce actions differing from those of lower concentrations in degree only, but in some cases the clinical results may appear totally different. The application of mercuric iodide in a mineral fat base at a concentration of 1:64 will cause a slight reddening of the skin (rubefacient action) and will also have a fungicidal effect. The same drug in a concentration of 1:4 in the same base, or an animal fat base, will produce severe blisters and, in certain circumstances, sloughing of the skin and scar formation. Pathologically however, this vesication is only an advanced stage of an action, the early stage of which is dilatation of the superficial blood vessels and congestion—a rubefacient action. Another example of actions apparently varying according to concentration is that of mercuric chloride or corrosive sublimate. In concentrations of 1:10 000 mercuric chloride acts as an antiseptic and bactericide; in concentrations higher than 1:500 the compound destroys tissues and will act as a deeply penetrating caustic. These two effects, however, are different degrees of a single action of mercuric chloride, namely, that of poisoning cell protoplasm.

There are exceptions to this rule of concentrations. Thus glycerin in dilute concentrations exerts an emollient or softening and soothing action on tissue, whereas in strong concentrations it exerts a high osmotic pressure and will have a 'lymph lavage' effect, particularly in wounds and on abraded surfaces. This 'lymph lavage' action describes the osmotically induced increased flow of tissue fluids to the area of application.

It should be realised that counter-irritation can be induced efficiently and more safely with physical agents, e.g. heat from pads, packs and lamps or short-wave diathermy. The persistence of some of the following dangerous agents of doubtful efficacy in veterinary use probably reflects conservatism among equine practitioners.

CAUSTICS

These are agents which destroy tissue. They may be used to destroy excessive granulation, to destroy indolent granulation with the aim of stimulating normal growth, to destroy minor superficial tumours—warts, and to destroy horn buds in calves. They were once extensively used in conjunction with wooden clams for castration of cattle and horses, but this practice is almost entirely discontinued in favour of surgical methods.

Silver nitrate, phenol, and antimony trichloride, are in frequent use, whilst podophyllin resin and glacial acetic acid are also occasionally used for the removal of warts (carbon dioxide is rarely used for this purpose in veterinary medicine.)

Nitric acid has been used for acid firing, but though there are modern advocates, this method of counter-irritation is not liked because of the difficulty of limiting the action. The possibility of a severe caustic effect with sloughing and scarring is always present.

Caustic soda and **caustic potash** are used for their caustic effect only for the destruction of horn buds of calves. The caustic is often made into a paste, of which Vienna Paste is an example this usually consists of equal parts of potassium hydroxide and quicklime, but other pastes are also used in which the quicklime is replaced by chalk and similar diluents.

Mercuric chloride, though a caustic, is not used as such because of its dangerous corrosive action. (A corrosive action is one where cell protein is destroyed but either remains in solution or is redissolved in an excess of the agent thus allowing penetration of the corrosive substances to the underlying tissue. Penetration of the corrosive will continue almost indefinitely.

Phenol is used occasionally for cautery of corneal ulcers, but the difficulty of restraining the action has limited its popularity in the veterinary field, where silver nitrate is normally the drug of choice. **Liquefied Phenol**, B.Vet.C., consisting of 80 per cent phenol in distilled water, is applied by touching the ulcer with a swab moistened in the solution. Applied to the skin, phenol has some corrosive action, forming a white eschar which turns red and then brown. Anaesthesia of nerve endings occurs which can disguise the fact that necrosis of the tissue is taking place. The dead tissue sloughs without pus formation. Phenol burns are best treated with glycerin which considerably reduces the caustic effect.

Antimony Trichloride, B.Vet.C. is a very deliquescent, colourless crystalline substance which is incompatible with water, giving an insoluble white powder, the oxychloride. It is soluble in dilute hydrochloric acid, absolute alcohol, chloroform, and ether.

The compound is an ingredient of disbudding preparations for calves. It is combined usually with salicyclic acid to soften the keratin layers of the skin in concentrations of about 28 per cent chloride with 5 to 7 per cent acid. Solutions in oil are available, but these fluid preparations are difficult to control and the more suitable collodion vehicle is most frequently used. Collodion of Antimony Trichloride, B.Vet.C. is the official example. The volatile components of the collodion evaporate following application, leaving the active agents applied to the area with a thin film of nitrated cellulose (pyroxylin) covering and ensuring adherence of the caustic. Three things are important to remember with these caustic collodia:

(1) The material will not adhere to wet or greasy surfaces. It is therefore necessary, before applying the collodion, to clip, clean and defat the horn bud area with ether or chloroform. Application is best achieved by a short stiff bristled brush of the type used for stencil painting.

(2) Normal skin surrounding the horn bud should be protected by the application of a protective greasy ointment such as zinc oxide ointment.

(3) Disbudding, to achieve the best results, should be carried out within 48 hr of birth, and treatment should be thorough.

Podophyllum Resin, B.Vet.C. is obsolete as a drastic purgative but it is occasionally used in an oily vehicle as a paint for warts. The 1965 edition of the B.Vet.C. includes Compound Paint of Podophyllin consisting of 14·63 per cent resin in Compound Benzoin Tincture B.Vet.C. Care should be taken: (1) to see that the material is not applied to normal skin of either the operator or patient, (2) to prevent licking of the area dressed, and (3) to prevent contact with the eyes.

Silver Preparations

The preparations of silver used in veterinary practice consist of Silver Nitrate, B.Vet.C. and various proteinates which depend for their activity on the ionization of the compounds. There are two proteinates in the Codex—**Silver Protein,** B.Vet.C. and **Mild Silver Protein,** B.Vet.C., the former contains 8 per cent w/v and the latter about 20 per cent w/v metallic silver. 'Lunar caustic' is a caustic pencil formed by fusing silver nitrate with a proportion of potassium nitrate.

These silver compounds are mostly used for the treatment of ocular conditions. The nitrate is used for cauterization of wounds and ulcers, including those of the cornea, as well as for conjunctivitis and keratitis, although these last conditions require less drastic treatment and the organic compounds are preferred.

Silver Nitrate, B.Vet.C.

Consisting of colourless sheet crystals with no odour and a bitter metallic taste, silver nitrate is very soluble in distilled water but less soluble in the organic solvents. It is incompatible particularly with halogens, proteins and tannic acid.

Actions and Uses

The action of silver nitrate is related to its incompatibilities, in that it precipitates protein as silver proteinate. When applied to tissue not protected by keratin it will therefore precipitate the protein in all the superficial cells. These superficial cells will be destroyed, but those cells in the deeper layers will be protected by this film of almost insoluble proteinate. Some of the proteinate will slowly redissolve and ionize to release minute amounts of silver ions, which cause a certain amount of counter-irritation and stimulate healing. At the same time these slowly released silver ions have an antibacterial or antiseptic effect. Any silver compound thus has an effect which is proportional to the concentrations of the dissociated silver ions.

Strong silver nitrate, usually in crystal or 'lunar caustic' form, is used to cauterize corneal ulcers and small superficial wounds; also to remove small warts, although occasionally the nitrate appears to stimulate the return of the wart. In concentrations of 1 per cent in distilled, chloride-free water, sometimes combined with about ½ per cent zinc sulphate, silver nitrate has an astringent effect and is used as an eye lotion or eye drops for the treatment of general keratoconjunctivitis.

Silver Protein Compounds

In many cases silver nitrate is too irritant and caustic. Following the realization that silver nitrate when combined with protein became less caustic and corrosive, compounds of protein and silver were prepared.

These protein salts of silver are all relatively non-irritant brown powders which are slowly soluble in water, but insoluble in the organic solvents. They contain various quantities of silver —the concentrations of the official preparations have already been mentioned—and their activity is inversely proportional to this concentration, but closely and directly related to the concentrations of dissociated silver ions.

Though the action of these compounds is weaker than that of silver nitrate, they are valuable in that they are not only less irritant but also more persistent, for the protein compounds will not be neutralized to any great extent by tissues and tissues fluids, particularly proteins and chlorides.

Of the official preparations the strong proteinate is used in all cases where dilute silver nitrate could be used but where irritation must be avoided; concentrations of 2 to 10 per cent are used in aqueous solutions and ointments containing up to 5 per cent are available. The mild compound is used in solutions of up to 25 per cent for general mucous membrane antisepsis and up to 50 per cent for corneal ulcers; 10 per cent ointments are also used. The action is similar to that of the stronger compound but the antiseptic effect is weaker.

All solutions should be freshly prepared and stored in containers which exclude light.

Toxicity. Local argyria is occasionally seen following repeated treatment of ocular conditions. It is a black pigmentation of the sclera and cornea as the silver is converted into black compounds by the action of light.

Keratolysis is the name which describes the ability of some drugs to soften keratin and cause desquamation of the superficial layers of keratinized epithelia. Substances with keratolytic effect are exemplified by salicylic and benzoic acids. Such agents have been used in the removal of small warts and excrescences and in ointments for application in chronic eczema or ringworm infestation.

COUNTER-IRRITANTS

Counter-irritants are substances which are applied to the skin to cause local irritation and inflammation. The reasoning which prompts their use is that, where a chronic inflammation exists, a counter-irritant will cause an over-riding acute inflammation with a great increase in vascularity and consequently an increased blood supply to the area. This increased blood supply results in a greater concentration of blood borne antidisease factors such as leucocytes, increased nutrition, and an improved removal of waste products by the venous and lymphatic systems. It is hoped that when this acute inflammation subsides the cause and some of the results of the chronic inflammation will also have been removed.

Counter-irritants are therefore applied to the skin covering the area where chronic inflammatory lesions exist just below the surface. These chronic inflammations may affect bone, ligaments, tendons or glandular tissues. Chronic 'arthritis', which is frequently not a true joint lesion but a strain and ossification of ligamentous attachments, ruptured flexor tendons in the horse, and glandular infections such as strangles of the horse, are commonly treated by counter-irritation. The use of a vesicant on an animal enforces rest for a period of up to three months, depending on the severity of blistering, but also on the psychological use the veterinary surgeon makes of the treatment to enforce the resting of an overworked animal.

Three stages of counter-irritation are described:

1. A **rubefacient action,** where the irritant applied causes only mild irritation and a small increase in congestion—literally translated the term means 'to make red'. Rubefacients often have pain relieving or anodyne properties which may be due either to some pharmacological action of the drug or to the creation of a diversionary sensation. Iodine, camphor, methyl salicylate, turpentine and ammonia, are frequently used rubefacients.

2. **Vesication,** which occurs when the irritation is severe; this damages the capillary system and results is serous exudates collecting in the superficial skin layers to form blisters. This is the origin of the common name for vesicant dressings—'blisters'. Mercuric iodide and cantharides ointments are the preparations most frequently used for blistering.

3. **Pustulation** is the third stage where the irritation is so great that the deep layers of the skin are damaged and the blisters become filled, not with serous exudate, but with the products of cellular destruction—pus. As the deeper layers of skin are damaged, including the Malpighian layers, the ability of regeneration is lost and scar tissue is formed which, when it contracts, may limit the action of joints or tendons and at the best cause permanent blemishes.

It follows that no substance should normally be applied in a manner or form capable of causing irritation beyond the stage of vesication.

The counter-irritants in use are preparations of: (1) iodine, (2) ammonia, (3) volatile oils such as camphor, turpentine and mustard, (4) cantharides, and (5) mercuric iodide.

Most caustics, if applied in dilute concentrations, will exert a counter-irritant action, and no doubt part of the beneficial action of cauterization is the irritation and stimulation of the tissue not destroyed by the caustic.

The application of counter-irritants should be in accordance with the principles involved in the absorption of substances applied to the skin. Choice of the base can assist in varying the effect of most counter-irritants from a mild rubefacient action to one of vessication. Thus *mercuric iodide* in soft paraffin at a concentration of 1:40 will have a very mild effect, whereas the same salt at a strength of 1:4 in lard will have an effect only just short of pustulation. A change in concentration alone will not achieve such a wide variation.

When volatile drugs are in use bandaging will, by preventing evaporation, intensify the effect of the counter-irritant. This can be demonstrated by bandaging an area immediately after the application of **White Liniment,** B.Vet.C., which is a commonly used stimulating embrocation. The ammonia and turpentine cannot evaporate and blistering of the area can result.

Further factors involved in the use of this group of compounds are the amount of friction used and the duration of application. Apart from an increased circulation which normally follows, vigorous rubbing will massage the dressing into the skin. This penetration will increase proportionally to the vigour and duration of massage, and the degree of effect will similarly increase.

Iodine, B.Vet.C.

Iodine and its preparations have been used as rubefacients for many years by both the medical and veterinary professions, but more recently the popularity of these compounds with the veterinary profession has declined. Iodine possesses some advantages, for its action is persistent due to its take up by the cornified layers of skin and its slow release. Its action is also mild and vesication only follows repeated applications of strong alcoholic solutions.

For strong counter-irritation the accepted method is to paint the area with a 10 per cent alcoholic solution, or Churchill's iodine (a 6·5 per cent alcoholic solution) on alternate days until signs of desquamation appear.

For weak counter-irritation the **Weak Solution of Iodine**, B.Vet.C., containing 2·5 per cent iodine in alcoholic solution can be painted on daily or on alternate days, but **Non-staining Iodine Ointment,** B.Vet.C., is less irritating, This ointment is an organic iodine complex prepared by dissolving 5 per cent iodine in arachis oil and making up to 100 per cent with yellow soft paraffin. A similar ointment containing 5 per cent methyl salicylate is described in the *Veterinary Codex*. This will be slightly more active as a counter-irritant. The ointments are less irritant than the alcoholic solutions and should be applied at least once daily.

On occasions, **Aqueous Solution of Iodine,** B.Vet.C., known as Lugol's iodine, which is a 5 per cent iodine solution in water with potassium iodide to assist solution, has occasionally been injected, following drainage, as a counter-irritant into the lesion in cattle known as 'hygroma'. This is a chronic bursitis of the carpus due to repeated knocking on the manger during rising and lying down. No doubt counter-irritation plays some part when this preparation after dilution is injected into the uterus or vagina in the treatment of infertility in the cow.

Ammonia

Ammonia and its preparations are probably the compounds most frequently used for achieving a rubefacient effect. They are non-corrosive and non-persistent, but nevertheless are strongly irritant and if bandaged on can cause severe blistering. The compounds, however, are rarely bandaged on but are applied in an emulsified form as a liniment—for external application with friction. **Ammonium hydroxide,** in the form of **Strong Solution of Ammonia** B.Vet.C., **ammonium carbonate,** and **ammonium chloride,** are all used, sometimes together, frequently with other drugs. Volatile oils such as camphor, methyl salicylate (oil of wintergreen), and turpentine, are commonly used with ammonium compounds.

The following are representative of the many formulae in common use:

℞

Ammonium carbonate	8 g (120 gr.)
Oil of turpentine	30 ml (1 fl. oz.)
Soft soap	15 g (½ oz.)
Water	to 0·25 litre (to 8 fl. oz.)

or,

℞

Strong solution of ammonia	30 ml. (1 fl. oz.)
Rape oil	to 0·25 litre (to 8 fl. oz.)

To this latter prescription may be added 1·5 ml of turpentine should a stronger action be required.

A camphorated liniment, containing 12·5 per cent w/v of camphor with 25 per cent w/v of strong solution of ammonia in 90 per cent industrial methylated spirits, is also frequently used.

These liniments are best applied to areas clipped of hair, but clipping is not essential.

Strong Solution of Ammonia, B.Vet.C.

This is a 33⅓ per cent solution of ammonia gas in water. It is colourless, with a pungent, characteristic odour and taste, and is strongly irritant.

Volatile Oils

Most volatile oils will cause a dilation of cutaneous blood vessels due to local irritation, and also will cause a slight local anaesthesia which gives them an anodyne (pain killing) effect. Not all volatile oils are suitable rubefacients but camphor, turpentine and mustard oil are frequently used with good effect.

Camphor, B.Vet.C.

This stearoptene is a white crystalline substance obtained from a Sino-Japanese tree and purified by sublimation. It is also prepared synthetically. It has all the properties of volatile oils and is slightly soluble in water but freely soluble in normal organic solvents and fixed oils. The odour is characteristic and the taste is pungent followed by a cool feeling due to slight anaesthesia.

Actions and Uses

Externally, camphor is a mild rubefacient and anodyne. It is applied to the skin as:

(1) **Liniment of Camphor, B.Vet.C.,** commonly known as 'camphorated oil', which is a 20 per cent solution of camphor in arachis oil.

(2) **Ammoniated Camphor Liniment, B.Vet.C.,** already mentioned as containing 12½ per cent camphor.

(3) **Liniment of Soap, B.Vet.C.** (Lin. Saponis) which contains 4 per cent of camphor with 8 per cent soft soap in alcohol and water.

These liniments should be applied only to chronic conditions, with vigorous friction.

Methyl Salicylate, B.Vet.C.

This salicylate is a colourless or pale yellow liquid with a characteristic odour and taste. It has the solubilities of a volatile oil.

Methyl salicylate forms the greater part of oil of wintergreen or gaultheria and is dispensed wherever this oil is prescribed.

Actions and Uses

Though methyl salicylate possesses most of the properties of salicylates, it is not normally given by mouth. Externally, it has a mild rubefacient and anodyne action and has been used as such in ointments at a concentration of up to 10 per cent. The *Veterinary Codex* details a liniment which contains 25 per cent methyl salicylate in a bland oil, a 5 per cent methyl salicylate ointment and a compound ointment of 5 per cent methyl salicylate in non-staining iodine ointment. **Kaolin Poultice,** B.Vet.C. contains 0·2 per cent methyl salicylate, more to give the preparation an odour than to exert any pharmacological action; it is more for this purpose that methyl salicylate is now used.

Capsicum or Cayenne pepper has also been used as a counter-irritant, but the irritant principle capsicin causes pain and severe prolonged irritation. The compound is now rarely used.

Black Mustard, B.Vet.C. (1953) (*Sinapis nigra*)

This, when ground, is used in varying proportions with white mustard to make mustard flour.

Actions and Uses

If made into a paste with luke-warm water (NOT HOT, or the enzyme may be inactivated), an enzyme *myrosin* reacts with a glycoside *sinigrin* to form a volatile oil, allyl isothiocyanate, **Volatile Mustard Oil,** B.Vet.C. This oil is a very rapidly acting irritant and should not be left in contact with the skin for longer than 1 hr, otherwise vesicles and pustules may be produced. If the paste is applied to skin between two layers of gauze the application may be removed after each 15 min or so to examine the degree of reddening. Blisters produced by mustard are slow in healing and mustard should therefore only be used as a rubefacient. It has been used in pneumonic conditions, but its use in this context is obsolete.

Turpentine, B.Vet.C.

This volatile oils is usually incorporated with other volatile rubefacients, particularly ammonium compounds, to give the preparations greater effect. It is usually applied in liniment form, **White Liniment (Linimentum Alba),** B.Vet.C. being a typical example.

The preparations of turpentine include a sulphonation product terebene which is used primarily as an inhalant, Liniment of Turpentine, B.P., or 'compound camphor liniment', and White Liniment, B.Vet.C., or white oils. The first liniment contains turpentine 65 per cent, camphor 5 per cent, soft soap and water to 100 per cent, whilst the second consists of 8·33 per cent oleic acid, 4·5 per cent dilute solution of ammonia, 1·25 per cent ammonium chloride, 25 per cent turpentine oil, and 62·5 per cent water. An alternative but unofficial formula consists of turpentine oil 30 ml (1 fl. oz.), ammonium carbonate 8 g (120 gr.), soft soap 15 g (½ oz.), and water to 180 to 240 ml (6–8 oz.).

Actions

The two degrees of counter-irritation can be achieved: a rubefacient or reddening effect by application in liniment form with friction, and vesication or blistering by preventing evaporation, which is achieved by bandaging. Severity of action in both cases can be varied by shortening or lengthening the period of rubbing. It is well known that horses are, as a species, susceptible to turpentine and ammonia vesication so that extra care must be taken with these animals, particularly thin skinned blood stock.

Cantharides

Widely known as 'Spanish fly', these dried beetles are crushed to prepare a brown powder with iridescent green particles. This powder contains an irritant active principle which is chemically a lactone of cantharidic acid, and is officially recognized under the name of **Cantharidin,** B.Vet.C. This compound is a colourless, odourless, crystalline substance.

Cantharides ointments are standardized by their cantharidin content and commercially available samples contain from 0·01 per cent to 0·16 per cent of cantharidin. **Liquor Epispasticus** or **Blistering Liquid,** B.Vet.C., contains 0·4 per cent cantharidin in oil, acetone and resin, but the concentration and difficulty of controlling fluid blisters makes this preparation unpopular.

Actions and Uses

Widely known as an aphrodisiac, its action as such is not very efficient and though it causes engorgement of erectile tissue in the male and female, there is no increase in sexual desire. There is considerable irritation of the urethral mucosa.

Externally, cantharides preparations cause intense irritation, with a rubefacient action when dilute, and blistering effect in stronger concentrations. The blisters formed are large in area but penetration is relatively slight and inflammatory reaction is therefore not very deep. Cantharides and its derivatives are often combined with Red Mercuric Iodide, B.Vet.C., to make use of the diffuse superficial effect of the former with the more circumscribed but penetrating effect of the latter. These combined blisters are very powerful and should be used circumspectly.

Administration. As with the application of all blisters, the affected area should be clipped and a protecting and limiting application of a bland zinc ointment smeared around the area. Special attention should be shown to protecting the inner angles of joints where misapplication may cause severe damage in the creased skin areas. The blistering ointment should be rubbed in for 5 or 10 min, depending on the degree of action required, and then any excess of ointment removed by rubbing with cotton wool or a cloth. Blistering will occur after a minimum of 3 hr and will appear more rapidly in thin skinned animals and those where strong concentrations are applied with vigour. Animals should be prevented from licking and biting the blistered area. Cantharides blisters should not be applied to sites where direct contamination of mucous membranes, particularly of the urogenital system, can occur.

Toxicity. Cantharidin can be absorbed through the skin in sufficient quantities to cause trouble if large areas are dressed. Should this happen, symptoms of acute gastro-intestinal enteritis and nephritis will be seen, possibly with penile erection in the male and difficulty in urination. Treatment consists of washing off the local application and the oral administration of demulcents of a non-oleaginous character. Hyoscyamus tincture may help in counteracting the urinary system irritation.

Mercury Compounds

The pharmacology and usage of this metal and its derivatives are governed mainly by their individual solubilities and the ionization of each compound. Those which are soluble and highly ionized will be the most active, whilst the least soluble and un-ionized compounds will be relatively inactive. This principle, of course, is not confined to mercury compounds alone but applies to all simple inorganic salts. As mercury ions, but particularly mercuric ions, are protoplasmic poisons, it follows that soluble and ionized compounds will be highly toxic and therefore only used externally, if at all. The less soluble forms will be safer and available for internal use.

The mercury compounds in veterinary use, arranged according to activity in descending order, are: mercuric chloride, mercuric iodide, ammoniated mercury, mercuric oxide, mercurous chloride, metallic mercury, and organic mercurials. Of these compounds only mercurous chloride and metallic mercury are used internally. All have external actions and uses, although mercuric chloride is so dangerous that it should not appear on the shelves of any veterinary pharmacy.

Mercuric Chloride, B.Vet.C. (1953) (Corrosive Sublimate, Hydrarg. Perchlor)

This is a white crystalline solid with no odour; it is soluble in water and very soluble in alcohol, giving water-white solutions; solutions should be coloured with methylene blue or crystal violet to protect against accidental drinking. Metal instruments should not be used in contact with the salt because of its strong corrosive action.

Actions and Uses

Mercuric chloride precipitates protein and poisons protoplasm. In tissues, the mercury–protein compound is soluble in excess protein and precipitation does not occur, thus no protective proteinate pellicle is formed and the mercury can affect successive tissue layers until all the free mercuric ions are taken up. This may take a considerable time, requiring a large amount of tissue, and very deep penetration can occur; this is a corrosive action. A penetrating action would not be so dangerous but for the protoplasmic poisoning action. Every cell is destroyed in which the mercuric ion reacts with the protein.

These corrosive and protoplasmic poisoning actions have been utilized when mercuric chloride solutions were used for antiseptic and disinfectant purposes. There is now no justification for its use in this way.

The corrosive effect was made use of when the salt was incorporated in caustic pastes; counter-irritant solutions containing mercuric chloride have been recommended for the treatment of traumatic arthritis. These uses are also unjustifiable.

Toxicity. Mercuric chloride poisoning is usually very acute and severe. The acute toxicity, which follows mistaken or accidental administration, is typified by acute and extremely severe gastroenteritis. If the animal revives from these conditions there follows stomatitis and, from the chloride absorbed, acute hepatitis but more especially nephritis, within 24 to 48 hr. Death results from cardiac and circulatory collapse subsequent to renal failure and concomitant uraemia with convulsions.

Treatment consists of gastric lavage where possible and the administration of sodium phosphite or hypophosphite (10 g per cwt.) by mouth. Raw white of egg has been used as a very temporary measure to bind the mercury with the protein, even though in a soluble form; charcoal is also very useful. These and other treatments are aimed at preventing absorption; if they are to be effective, they must be administered immediately and the gastric contents extracted within 10 min. Unless these measures are applied quickly recovery from toxic doses is unlikely. Once absorption has taken place treatment is problematical. It is possible that the chelation compounds such as Sodium Calciumedetate, B.Vet.C., calcium disodium ethylene-diamine-tetracetate (calcium disodium versenate), would be of value and should be tried.

Chronic toxicity follows repeated applications of skin dressings, repeated dosings with calomel, grey powder, etc., and vapour inhalation. Mercuric chloride is rarely responsible for this type of poisoning now that it is no longer used for dressing wounds. The chronic toxicity is called *mercurism* and results from cumulation. The symptoms are those of stomatitis and nephritis which regress when the source of the poison is eliminated and symptomatic treatment applied.

Mercuric Iodide, B.Vet.C. (Biniodide of Mercury, Hydrarg. Iod. Rub.)

This is the most important inorganic mercury salt in veterinary use. It is a bright vermilion heavy powder, amorphous, odourless and tasteless. Almost insoluble in water, mercuric iodide can be dissolved with the aid of an equal weight of potassium iodide. It is slightly soluble in alcohol and ether.

Actions and Uses

The external actions of mercuric iodide are counter-irritant, fungicide, and antiseptic. The iodide of mercury is not markedly corrosive but it is strongly irritant to the tissues, particularly if applied in alcoholic solution. In a soft paraffin or animal fat base mercuric iodide has been and is widely used both as a mild and a strong counter-irritant. Concentrations of 1:40 or 1:50 have a mild rubefacient action, but at strengths of 1:8 a strong vesicant effect is given. Exceptionally, concentrations of 1:4 in lard have been rubbed in vigorously for 5 to 10 min to be followed after several hours by severe blistering. There is always a danger at this strength of causing pustulation.

The biniodide blister penetrates deeply, producing many small blisters which, after a period, will coalesce. Alcoholic solutions are even more effective than animal fat ointments but, because of the difficulty of limiting the action of fluid blisters and the danger of pustulation, they are not frequently used.

Strengths of 1:40 to 1:50 have been effectively used for ringworm dressings. Alcoholic solutions of 1 per cent of the iodide have also been used.

Internally, mercuric iodide is sometimes used as a supplementary or interim treatment for actinobacillosis when iodism has followed the administration of either sodium or potassium iodide. It seems that the mercuric compound in small doses (0·3 to 0·5 g or 5 to 8 gr.), with an equal quantity of potassium iodide, in water given twice daily will often maintain the disease in equilibrium until the iodism symptoms subside and the potassium or sodium salt can be used again.

Bibliography

ADAMS, O. R. (1974) Methods of therapy. In: *Lameness in Horses*, pp. 429–62. Philadelphia: Lea & Febiger.

KRAL, F. & SCHWARTZMAN, R. M. (1964) *Veterinary and Comparative Dermatology*. Philadelphia: Lippincott.

SCHILLING, J. H. (1968) Wound healing. *Physiol. Rev.*, **48**, 374.

20

Agents Affecting the Skin

II. Sedatives

The depression of skin irritability and the relief of irritation or pruritis may be achieved by the use of a drug or drugs from one or more of the following groups: (1) emollients, (2) sedative and protective powders, (3) astringents, (4) local and general anaesthetics (analgesics), (5) corticosteroids, and (6) antihistaminics.

In addition, where there is an exciting cause for the irritation, then a suitable antiseptic, antibiotic, antifungal or parasiticide should be applied locally. On occasion systemic administration is to be preferred and may be reinforced by concurrent dosing with tranquillizing drugs.

EMOLLIENTS

Emollients are agents which soften, soothe, lubricate and, protect a skin surface. By this definition they will therefore be fats, fixed oils, waxes, both natural and synthetic, hydrocarbons, and derivatives of these.

The *oils* used are olive, castor, rape, arachis, almond, theobroma and cottonseed oils. Linseed oil is not used alone because of its property of oxidizing to produce varnishes when exposed to the atmosphere. Fish oils such as cod-liver oil may also be used. Derivatives include the oleates and stearates of metals such as zinc, magnesium, copper and aluminium. Glycerol (glycerin) and other polyhydric alcohols such as propylene glycol are also emollient derivatives of oils.

The *fats* in common use are lard (pig-fat), lanolin (wool fat) and wool alcohols derived from lanolin.

The *waxes* used are beeswax, a natural wax, and synthetic ones such as the polyethylene glycols and Emulsifying Wax, B.Vet.C.

The *hydrocarbons* obtained as a by-product in oil refining are all well known—liquid paraffin, soft paraffin and hard paraffin.

Fixed Oils

The fixed plant oils are all colourless to amber liquids depending on their purity, except oil of theobroma which is solid at normal room temperature but fluid at body temperature; some, such as olive oil and arachis oil, partially solidify at winter temperatures. They each have a slight but characteristic odour and in pure forms, a bland taste—impure forms frequently have a disagreeable taste; fish oils usually have a strong fish odour and taste. On keeping, all oils tend to go rancid, but fish oils are much worse than plant oils which may keep several years without becoming unusable. Physical recognition of fixed oils depends on testing for insolubility in water, lubricant properties, the ability to make a permanent grease stain on brown paper, a black stain with osmic acid, and saponification with alkalis.

Actions and Uses

The plant oils are widely used as vehicles for the external, oral and parenteral administration of drugs. Externally, rape oil, castor, and cottonseed oils are frequently used for preparing counter-irritant liniments, parasiticidal dressings, and bland emollient ointments such as the well known Ointment of Zinc and Castor Oil, B.Vet.C. (zinc and castor oil ointment). If mixed in equal parts with lime water various liniments are formed which were used extensively at one time for the treatment of burns and scalds: olive oil forms linimentum calcis and linseed oil forms carron oil. These are still useful for extensive burns in large animals if sodium bicarbonate paste is not available. Arachis oil is rarely used externally, being an 'edible oil' and almond oil, though it has similar external actions to olive oil and rape oil, is not frequently used because of its cost. Oil of theobroma is a solid oil used mainly as a vehicle for pessaries and suppositories, but it is being replaced by glycogelatin which is more stable and of a standard quality. Internally, fixed plant oils are laxative and purgative (*q.v.*), for which actions castor oil and linseed oil are the best, as their respective soaps, ricinoleates and linoleates, are more irritant than those of olive oil and rape oil.

The fish oils can be used externally in the same way as plant oils but their odour is objectionable and restricts their use. The vitamin A and D content is widely believed to impart better healing properties when applied to the skin, and there would appear to be some clinical justification for the belief. Cod-liver oil is very effective as a vehicle in oily eye drops and lotions.

These natural oils are frequently replaced by synthetic agents having no rancidity problems in modern formulations, e.g. isopropyl myristate.

Derivatives of Fixed Oils

The **oleates** and **stearates** are soaps of the fatty acids present in fixed oils and fats. The sodium and potassium soaps are well known as detergents, but the insoluble oleates and stearates have other uses, mainly of skin sedation. The oleates in general penetrate the skin very easily and the attached metallic ion is readily absorbed: mercury oleate was at one time applied to the skin specifically as a means of systemic mercury therapy in syphilis: 5 per cent copper oleate is still used in parasitic and antifungal dressings. The stearates, which are white or yellowish powders, possess strong skin sedative properties—zinc stearate is used in concentrations of up to 50 per cent in ointments and can be applied as a dusting powder in any concentration, occasionally being substituted for talc; it is claimed to have mild astringent and antiseptic properties. Magnesium and aluminium stearates are used pharmaceutically as lubricants, water repellants, and gelling agents. Aluminium stearate is used as a suspending agent for parenteral injections and intramammary injections in strengths of approximately 2 per cent. Lithium stearate is also used in intramammary injections.

Some **polyhydric alcohols** possess strong emollient properties. Glycerol (glycerin) and propylene glycol are used extensively for their emollient and solvent properties.

Glycerol, B.Vet.C.

Sometimes known as glycerin, this compound is a thick, syrupy, colourless, odourless liquid with a sweet taste.

Actions and Uses

The majority of the effects of glycerin are due to the high osmotic pressure of the compound and its aqueous or alcoholic solutions. These high osmotic pressures cause slight pain when pure glycerin is applied to abraded skin and to the conjunctiva, but the pain is transient.

Externally, glycerin applied to unbroken skin is a very useful emollient. Applied to wounds, the osmotic pressure results in a lymphagogue or lymph lavage action, which is defined as the bathing of a wound by lymph drawn into it in the process of neutralizing the osmotic pressure exerted by the glycerin or other lavage agents (salines, sugars). Besides mechanical cleansing by passage of the lymph, there is a concentration of the body's defence agents—antibodies, phagocytes, etc., which help to overcome infection and aid healing. The combination of these two actions makes glycerin a very useful skin dressing for chapped, dry areas and for wounds; **Glycerin of Starch**, B.Vet.C. (1953) is especially useful for this purpose. A wound dressing, Morison's Paste—**Paste of Magnesium Sulphate,** B.Vet.C., containing approximately equal parts of exciccated magnesium sulphate and glycerin, is a very useful large animal wound dressing, particularly if wounds are dirty and pocketing.

By virtue of its high osmotic pressure, glycerin is also mildly antiseptic and is sometimes used as a preservative for vaccines, enzymes and pathological specimens, but tends to shrink specimens and does not always control moulds.

Glycerin is also used in conjunction with gelatin to make the glycogelatin bases of pessaries and suppositories, as well as those of lamellae, capsules and throat pastilles.

Propylene Glycol, B.Vet.C.

Actions and Uses

Externally, propylene glycol has emollient and humectant properties, but these properties are not so marked as with glycerin. It is, however, used in ointments and lotions mainly because of its greater solvent properties: most small animal ear dressings have propylene glycol as a base. It is also used as an area disinfectant when dispersed as an aerosol: 1 part in 26 kills staphylococci but not spores.

Fats

This group of emollients differs from fixed oils only in that its members are solid at normal temperatures and consist of glycerides of palmitic and stearic acids (oils are mainly glycerides of oleic and palmitic acids). Beef tallow or suet, pig fat or lard, are fats which are used in ointments to aid penetration and to act as emollients but their use is not so frequent as formally owing to their tendency to go rancid on keeping. Two per cent benzoin was often incorporated to prevent or defer rancidity.

Wool Fat, B.Vet.C.

Described as the purified, anhydrous, fat-like substance obtained from sheep wool, wool fat is a pale yellow thick tenacious fat-like substance with a characteristic odour. It is insoluble in water but will absorb about 50 per cent of its weight of water. Sparingly soluble in alcohol, it dissolves freely in ether, chloroform, other organic solvents and fixed oils. **Hydrous Wool Fat,** B.Vet.C. (lanolin), is produced by causing the anhydrous fat to take up 30 per cent of water. This compound is more easily incorporated into ointment bases than is the anhydrous compound and will mix more readily with aqueous solutions and adsorb them in quantities up to two or three times its own volume, forming emulsions.

Chemically, wool fat differs from other animal fats in that the fatty acids are combined not with glycerin but with higher alcohols such as cholesterol, oxycholesterol and others. **Wool Alcohols,**B. Vet.C., contains a number of alcohols, including not less than 28 per cent cholesterol, obtained by saponification of the crude grease from sheep wool. It is, at normal temperatures, a hard, brittle, dark amber coloured solid which fractures to give a smooth shiny surface. It is plastic when warm. Its solubility is similar to the other wool fat compounds but it is more soluble in alcohol. These compounds do not go rancid very quickly.

Actions and Uses

The ability of these compounds to act as emulsifying agents, and their relationship to normal skin secretions, has made them popular for incorporation into emollient ointment bases, usually in concentrations of 5 or 6 per cent. Examples are **Simple Ointment,** B.Vet.C. (unguentum simplex), and **Ointment of Wool Alcohols,** B.Vet.C., which contain 5 per cent of wool fat and 6 per cent of wool alcohols, respectively. Recently, some of the alcohols have been shown by patch tests to promote allergic reactions and wool fat was omitted from the ointments of the U.S.P. fourteenth revision, but was reinstated in the fifteenth revision. Though the newer emulsifying waxes now enable aqueous solutions to be incorporated easily into ointments, lanolin is still very popular and in many cases is a better emollient. Its property of aiding penetration varies according to the nature of the incorporated drugs and the state of the skin, but in general it is the best animal fat available for this purpose.

Waxes

Natural waxes resemble the fats physically, but chemically are more nearly related to wool fat, being composed of esters of the higher fatty acids with high molecular weight alcohols. Beeswax, the main example, contains a high percentage of myricyl palmitate ($C_{39}H_{61}$. O . CO . C_5H_{31}) with some cholesteryl esters, hydrocarbons, and other complex organic compounds. Beeswax is available in the natural yellow form, **Yellow Beeswax,** B.Vet.C., and also a bleached white form **White Beeswax,** B.Vet.C. The solubilities are approximately those of wool fat. These compounds are used as hardeners for emollient ointment and suppository bases, and they are the main ingredients of plasters (or charges) mentioned under counter-irritants (*q.v.*). They are still used as suspending agents in some intramammary injections though superseded by various stearates.

Synthetic Waxes. The manufacture of synthetic waxes has become a vast industry and the number of such waxes is countless. Two types have become official and represent the two main groups:

1. Those which combine sulphated and non-sulphated fatty alcohols, the former often as sodium salts. **Emulsifying Wax,** B.Vet.C., is the example.

2. Those which are combinations of fatty alcohols and condensates of fatty alcohols with ethylene oxide. **Cetomacrogol Emulsifying Wax,** B.Vet.C., is the example.

The first group are anionic emulsifying waxes and the second group are non-ionic. Because of the incompatibility of anionic preparations with cationic substances (detergents and antiseptics, particularly quarternary ammonium compounds and basic dyes), the non-ionic compounds of the second group are becoming more popular.

Emulsifying Wax, B.Vet.C.

The compound is a white or pale yellow solid wax which has a faint but characteristic odour and is insoluble in water, slightly soluble in alcohol, and miscible with oils and organic solvents. It forms emulsions with water.

Chemically, the compound is a mixture of 90 per cent cetostearyl alcohol and 10 per cent of sodium lauryl sulphate, an anionic detergent.

Actions and Uses

Thirty per cent emulsifying wax is incorporated in **Emulsifying Ointment,** B.Vet.C., along with white soft paraffin and liquid paraffin. This ointment is miscible in all proportions with water. Thirty per cent of emulsifying ointment is in turn incorporated in **Simple Cream,** B.Vet.C., a water miscible cream base.

As some difficulty was experienced due to incompatibility with cationic agents, non-ionic condensates were developed.

Cetomacrogol Emulsifying Wax, B.Vet.C.

This is also known as Non-ionic Emulsifying Wax, and belongs to the group of non-ionic waxes prepared by combining a fatty alcohol with a polyethylene derivative of a fatty alcohol. This particular wax consists of 20 per cent **Cetomacrogol 1000,** B.Vet.C., and 80 per cent **Cetostearyl Alcohol,** B.Vet.C. This wax can be used to replace emulsifying wax in any preparation, but especially in those containing cationic drugs. It is therefore used in creams containing quaternary ammonium antiseptics, and is part of the official bases for the two gamma benzene hexachloride creams in the Veterinary Codex.

Polyethylene glycols, also known as the polyoxyethylene glycols, and in the case of some solid condensates as Carbowaxes, are now widely used either as bases or part bases for many ointments, creams, and suspensions. They consist of long chain, high molecular compounds, made by subjecting ethylene oxide and water to high pressures in the presence of a suitable catalyst. The ethylene oxide and water first produce ethylene glycol

$(CH_2 . O . CH_2 + H_2O = HO . CH_2 . CH_2OH)$

and as the condensation is continued polymers are made which conform to the formula $H(OCH_2 . CH_2)_nOH$: '*n*' may be as high as 185. The highest of these compounds used normally—polyethylene glycol 4000—contains polymers with '*n*' varying from 70 to 85. The lowest is polyethylene glycol 300 with '*n*' ranging from 5 to 5·75. Between these two there are a number of condensates which have as a suffix the average molecular weight of the polymers. Those with average molecular weights below 700 are liquids, while those above 1000 are wax-like solids. The viscosity varies directly according to the molecular weight. The liquids are clear, colourless, and viscid, miscible with water, alcohol, acetone and other glycols, but insoluble in ether. They are also miscible with aromatic hydrocarbons (benzene derivatives), but insoluble in aliphatic hydrocarbons (paraffin hydrocarbon derivatives). They do not, therefore, mix with oils and fats. They are hygroscopic according to molecular weight, varying inversely; polyethylene glycol 4000 is almost non-hygroscopic. The wax-like solids are, at lower molecular weights, similar in consistency to soft paraffin, whereas the high molecular weight compounds are like paraffin wax.

Actions and Uses

The liquid polyethylene glycols act as emollients and humectants in much the same way as glycerin and propylene glycol, and are used in similar formulae. The solid preparations are likewise used as part of the water miscible section of cream bases. In many cases, mixtures of the polymers are made so that a cream base is entirely made up of polyethylene glycols. A number of antibiotic formulations are now being made with these compounds as bases, including intramammary creams.

Polyethylene glycols are also used in preparing a series of suspending and emulsifying agents known by various names, of which Polysorbate of the *United States Pharmacopoeia* is one. If ethers or anhydrides derived from sorbitol (a sugar alcohol, isomeric with mannitol), and called sorbitans, are combined with fatty acids by partial esterification, oil soluble but water miscible dispersing and emulsifying agents are produced. If these are then combined with polyoxyethylene glycols, etc., to give polyoxyethylene sorbitan esters, there is formed a series of powerful water soluble emulsifying agents of which polyoxyethylene sorbitan mono-oleate is one. Depending on the fatty acid and the length of the polyoxyethylene side chains used, a series of emulsifying agents can be formed which vary from predominantly lipophilic at one extreme to predominantly hydrophilic at the other. There are in this series, emulsifiers suitable for all types of emulsions and dispersals.

The mono-oleate mentioned is used in many procaine penicillin and streptomycin buffered aqueous suspensions.

Hydrocarbons

The hydrocarbon 'oils and fats' used in veterinary pharmacy are Light Liquid Paraffin, B.P., Liquid Paraffin, White and Yellow Soft Paraffins, and Hard Paraffin, all of the *British Veterinary Codex*. There are also 'mineral oils' of viscosity lower than the official liquid paraffin which have a wide variety of veterinary uses—from pig rubbing oils to the manufacture of water imscible creams and liniments.

The soft paraffins. Unpurified soft paraffins obtained during petroleum distillation are a deep amber colour, but as purification proceeds stage by stage the colour is lightened until the **Yellow Soft Paraffin,** B.Vet.C., is achieved. This product is a soft oily mass with the consistency of butter and a slightly fluorescent pale amber-yellow colour. It is miscible with fixed oils and organic solvents, but not alcohol. It is insoluble in water. **White Soft Paraffin,** B.Vet.C., has the same general characters as the yellow compound but the hydrocarbon has been further purified and bleached until white.

Actions and Uses

The soft paraffins form the major part of all veterinary ointments and are emollient, inert, non-irritant compounds. They act as bulking, suspending vehicles and do not assist penetration in any way. Should absorption of drugs be required, animal fats or emulsifying waxes must be incorporated with the hydrocarbons. It is usual to have 85 to 90 per cent soft paraffin, 5 per cent wool fat and 5 per cent hard paraffin as a routine ointment base—**Simple Ointment,** B.Vet. C., contains 5 per cent cetostearyl alcohol also. The paraffins exclude air and water from the skin and restrict fluid evaporation. This is advantageous in some circumstances and 'tulle' dressings for burns consist of soft paraffin impregnated gauze; penicillin and sulphonamide are commonly incorporated as bacteriostatics. It is usual to include white paraffin in ointments with white ingredients and yellow paraffin with coloured ingredients but this rule is rarely adhered to in veterinary practice, the yellow substance being used almost universally.

Hard Paraffin, B.Vet.C., is only of significance as a stiffening agent for ointments, particularly if intended for tropical use. The common concentration in Great Britain and similar climates is 5 per cent. In character and consistency it is similar to candle wax but contains hydrocarbons of a lower melting point. Outside pharmacy the compound has numerous uses—as a depilator, and, with various melting points, as an embedding agent for preparing histological and pathological sections.

Liquid Paraffin, B.Vet.C. As an emollient, liquid paraffins of both high and low viscosities are used as the main ingredients of the oil phase of water-miscible cream bases for both pharmaceutical and cosmetic use. The light mineral oils are useful as bases for liniments, emulsions, and, in conjunction with small proportions of essential oils such as citronella, are used both as pig rubbing oils to prevent thickening and scaling of the skin, so maintaining suppleness, and also as engineering penetrating and lubricating oils. Perhaps one of the most important uses is as the base of many intramammary injections. Parenteral injections should not contain liquid paraffin as the substance is not easily absorbed, and in mammals causes foreign body reactions with severe local oedema and swelling if given subcutaneously: intramuscular injections appear to cause less trouble.

SKIN SEDATIVE AND PROTECTIVE POWDERS

The use of emollients is not always indicated in the relief of pruritus, nor are ointments and creams the best treatment for all cutaneous lesions and wounds. In cases of pedal eczema in the dog and clean wounds which are not suitable for stitching, dusting powder applications are most suitable. The ingredients of such dusting powders may be inert protective powders such as chalk and talc, astringent and protective such as oxide or carbonate of zinc, adsorbent such as kaolin, antiseptic such as proflavine, antifungal such as dichlorophen, or local anaesthetics such as chlorbutol, amethocaine and benzocaine. Many of these agents such as dichlorophen and chlorbutol have multiple actions: many are antiseptic, but all are antipruritic.

Zinc Oxide, B.Vet.C.

The compound is a white or yellow-shadowed, smooth, non-gritty amorphous powder, odourless, tasteless, and insoluble except in acid or alkaline solutions. On exposure it absorbs carbon dioxide and moisture, becoming gritty: if this happens the compound should be discarded as the gritty particles are irritant.

Zinc carbonate is very similar in character to zinc oxide, for which it can be used as a substitute externally: it does not become gritty on keeping.

Calamine, B.Vet.C.

This was originally prepared by calcining native, pink zinc carbonate, but the official compounds are synthetic, consisting in the United States of zinc oxide coloured with red, ferric iron oxide, and in Great Britain of zinc carbonate so coloured. The actual depth of colour varies according to each manufacturer's standards.

Actions and Uses

Externally, these three compounds react similarly. They are all mildly antiseptic, astringent, and protective. They are applied in dusting powders, lotions and liniments, ointments and creams. As dusting powders they are usually combined with other protectives, antiseptics, etc., such as starch, boric acid, talc, or iodoform; various combinations of zinc oxide, starch, boric acid, and talc are official in the *Veterinary Codex*, the powders being combined in equal parts. If more active drugs such as iodoform are used they are diluted to concentrations of 1:8 by the less active compounds. For pedal eczema in the dog, a powder containing 25 per cent sulphanilamide and 25 per cent each of calamine, boric acid and starch is very useful if a bandage 'boot' is made. For 'wet eczema' a mixture of 4 parts calamine or zinc carbonate, with 1 part each of sulphanilamide, zinc sulphate, copper sulphate, and starch, will cake on the patches, relieve irritation and encourage healing under the scab. There are many such formulae.

Incorporated in ointments, zinc oxide, carbonate and calamine are widely used as bland protective dressings. Zinc and castor oil ointment is widely known and is recognized in the *Veterinary Codex;* an ointment of 15 per cent zinc oxide in simple ointment base is also official. Similarly a compound ointment and a simple ointment of calamine are official, the latter consisting of approximately 83 per cent white soft paraffin and 17 per cent calamine. An eye lotion or ointment made with 10 per cent zinc oxide in cod-liver oil is very useful in severe conjunctivitis and corneal ulceration, relieving pain and allowing the animal and the eye to rest.

Creams and lotions are also available, though perhaps less popular than in human medicine. Lotion of Calamine, B.Vet.C. contains 15 per cent calamine, 5 per cent zinc oxide, 3 per cent bentonite and 5 per cent glycerin preserved with 0·5 per cent phenol, and made up to volume with distilled water.

All these preparations are useful on abraded and irritated skin areas and should be used in preference to sulphonamide and antibiotic dressings unless gross infection has occurred. The zinc oxide ointment is very useful in protecting joint folds and in limiting the spread of counter-irritants and caustics, as in blistering and disbudding or in the caustic removal of warts.

Zinc stearate, though official in the *Veterinary Codex*, is not frequently used by the practitioner except as part of commercially manufactured ointments.

Internally, the only compound of zinc used is the carbonate, which has recently become popular as a treatment for parakeratosis of pigs —caused by an insufficiency of zinc.

Zinc Sulphate, B.Vet.C.

Actions and Uses

Zinc sulphate is strongly astringent, precipitating superficial protein but not penetrating the barrier so formed. It is only used externally. Compound Eye Lotion of Zinc Sulphate, B.Vet.C., is very useful in mild chronic conjunctivitis. It contains 0·34 per cent zinc sulphate and 2·29 per cent boric acid in sterile distilled water. A similar but more active preparation which is very useful as eyedrops contains 0·4 g (2 gr.) zinc sulphate and 1·0 g (4 gr.) silver nitrate in 100 ml (1 fl. oz.) sterile distilled water. This is very effective for corneal ulcers.

Lead, or White Lotion (Lotion of Zinc Sulphate and Lead Acetate, B.Vet.C.) contains 4 per cent of the zinc compound and 5 per cent of lead acetate dissolved in water. It is very widely and effectively used as an astringent and sedative lotion for the treatment of superficial strains, bruises, inflammations, and pruritus in the horse and cow. It should not be used on abraded skin because of the danger of lead absorption.

Starch, B.Vet.C.

Actions and Uses

Externally, in powder form, it is absorbent and protective and, as a glycerin gel, emollient. It is used frequently in dusting powders as a sedative and absorbent in all proportions: the official preparations vary in the starch content from 25 to 75 per cent. The **Glycerin of Starch,** B.Vet.C. (1953), contains about 8·5 per cent starch, and 74·5 per cent glycerin in distilled water brought up to a gelling temperature. It is one of the finest dressings for chapped and fissured skin when applied once or twice daily. Starch, specially treated to remove its gelling properties, is used as a lubricant for surgical gloves instead of talc, since the latter has resulted in internal foreign body reactions and adhesions when introduced on gloves into wounds and body cavities. Starch is completely absorbed.

Internally, **Mucilage of Starch,** B.Vet.C., containing 2·5 per cent starch, is a demulcent and an easily digested carbohydrate food—flour gruel is sometimes used instead for this purpose. Starch is also an antidote for iodine overdosage.

Talc, B.Vet.C., and Purified Talc, B.Vet.C.

These are products of a native hydrated magnesium silicate. Also known as 'French chalk', they are fine, white, odourless, tasteless, insoluble powders, sometimes containing aluminium silicate (kaolin).

Actions and Uses

Talc is perhaps more widely used in human medicine and cosmetics than in veterinary medicine where kaolin is more often available. However, these native silicates are extremely good skin sedatives and protectives when incorporated in dusting powders, acting as lubricants on the skin. They should be sterilized if used in wounds or on abraded skin, for they may contain spores, particularly tetanus spores.

LOCAL ANAESTHETICS

Local anaesthetics, particularly amethocaine, benzocaine and chlorbutol, act as skin sedatives, but have been mentioned in the chapter dealing with local anaesthetics (Chapter 21).

ASTRINGENTS AND HAEMOSTATICS

The action of skin astringents is the same as that

of intestinal astringents, many of the compounds being used in both contexts. The astringent, when applied to the skin, coagulates protein, forming a protective pellicle or scab under which healing can take place. At the same time, should there be any slight haemorrhage not of arterial origin, the astringent will, by coagulating and precipitating blood proteins, hasten blood clotting. This process is referred to as a *haemostatic* or *styptic action.*

The metallic sulphates, particularly of iron, copper, aluminium, and zinc, act as external astringents and are used in dusting powders in strengths of up to 5 per cent. On occasions, crystals of these salts can be used but their action then tends to be irritant and caustic, destroying the superficial cells rather than coagulating their protein as a reversible reaction.

Perhaps the best external astringent is **Ferric Chloride,** B.Vet.C., which, as Strong Solution of Ferric Chloride, B.Vet.C., a dark brown clear fluid, is a safe, effective, astringent and haemostatic for application to small superficial wounds and abraded skin. It is also used as a haemostatic in amputating tails, digits, and small warts of puppies and dogs. For this purpose, the bleeding should be temporarily stopped by pressure, the wound swabbed clean and the chloride immediately applied with a swab made by twisting cotton wool round the tips of a pair of forceps or round a swab 'stick'. The swab should be wetted but not saturated with the chloride, especially if the wound is on an eyelid because of the pain caused if the astringent runs into the eye. If blood is welling from a wound, application of the astringent merely precipitates blood protein in the extravasated blood and not, as required, in the broken end of the vessel; bleeding therefore continues under the mass of precipitated blood, and harm rather than good results. A wound properly treated should be almost dry, clean, but brown from the precipitated chloride.

Lead Acetate, B.Vet.C.

Lead acetate, the only salt of lead used therapeutically, is a white crystalline substance with a sweet astringent taste and an acetic odour. It is very soluble in water—giving a clear solution—and is consequently dangerous—soluble in alcohol up to approximately 2 per cent, and freely soluble in glycerin.

Lead acetate, besides being used in the preparation of lead lotion as described under zinc sulphate, is also used to make Strong Solution of Lead Subacetate, B.Vet.C., which consists of a mixture of the acetate with litharge (lead monoxide) and water.

Actions and Uses

Lead lotions are astringent and sedative. Astringency is typical in that lead albuminates are precipitated to form an impenetrable pellicle. The sedative action of lead lotions is due in part to the astringency but also to the local anaesthesia of nerve endings. Because of the danger of chronic lead poisoning from cutaneous absorption and acute lead poisoning from licking, lead preparations should not be used internally, should not be applied to abraded skin and wounds, should not be applied for long periods and should not be used on small animals. Mucous membranes should not be dressed with lead lotions and in particular the eye should not be washed because of the danger of precipitation and localized permanent white opacities following the healing of corneal ulcers.

Toxicity. Lead poisoning is one of the main hazards of calf rearing. Toxicity very rarely results from maladministration or overdosage of lead salts, but frequently due to the licking of paint from walls, wood-work or from discarded tins, from grazing herbage contaminated by discharges into the atmosphere from smelting works and similar industrial concerns, and also from lead arsenate crop sprays.

Symptoms of acute cases are indicative of acute indigestion and nervous system derangement—grunting, grinding the teeth, sweating, convulsions, leaping high into the air, excessively high temperatures and blindness. Chronic cases also show signs of indigestion, blindness, walking in circles or pressing the head against walls or posts.

Treatment consists of intravenous injections of Sodium-Calciumedetate, B.Vet.C., available commercially as Calcium Disodium-Versenate. The B.Vet.C. dosage rate is 75 mg per kg body weight intravenously or subcutaneously, but initial doses of 3 to 6 g intravenously in 500 ml sterile distilled water, followed by an equal second dose 2 or 3 days later if recovery is slow, appear to be most effective. This treatment should be combined with oral doses of magnesium sulphate to precipitate and evacuate any lead remaining in the intestines.

ANTISEPTICS AND DISINFECTANTS

Agents which are classified under this heading are discussed in Chapter 38.

ANTIHISTAMINICS

These drugs are described in Chapter 15. It is relevant to state here that in use they have failed to live up to the optimism with which they were introduced. The lack of effect may indicate that these compounds do not achieve sufficient concentration at the site of action of histamine following topical application, but also serves to underline the importance of inflammatory response mediators other than histamine.

CORTICOSTEROIDS

The anti-inflammatory action of the glucocorticoids has provided a most significant advance in the treatment of skin diseases. These drugs by virtue of their ability to decrease macrophage accumulation and their direct effect of inhibiting the vasodilation, increased permeability and exudate formation of acute inflammation irrespective of exciting cause, clearly have a very wide range of application. In addition they inhibit fibroblasts and diminish the fibrosis of chronic inflammation. These drugs are described in Chapter 11.

DIMETHYLSULPHOXIDE

The remarkable solvent properties of dimethylsulphoxide (DMSO) have been exploited in many ways to the advantage of drug action. It is used as a highly penetrating solvent vehicle which improves the topical efficiency of the dissolved drug, e.g. corticosteroids. In addition, however, it acts in an unknown way to reduce pain, pruritus and inflammation and to benefit the process of healing when applied topically. It is therefore of value in eczema and dermatitis and for symptomatic relief of strains, sprains, bruises and damage to muscles, tendons and bones. The material is applied gently to clean dry areas with a brush two or three times daily and is absorbed rapidly. For a horse 10–15 ml and for a dog 2–5 ml per application is recommended for a 5–10-day period with an absolute maximum of 14 days.

The drug is contra-indicated in food-producing animals and in chronic conditions. Its topical use may be followed by signs of local irritation which occasionally progress to vesication. Long-term high-dosage experiments have caused changes in the refractive index of the lens of the eye in rabbits, dogs and pigs.

Bibliography

David, A. A. (1972) The pharmacology of dimethylsulphoxide. *A. Rev. Pharmac.*, **12**, 353–75.

Kral, F. (1962) Skin diseases. *Adv. vet. Sci.*, **7**, 183–219.

Muller, G. H. & Kirk, R. W. (1969) *Small Animal Dermatology*, Philadelphia: Saunders.

Polano, M. R. (1973) Topical therapy. *Rec. Adv. Derm.*, **3**, 372–411.

Vickers, C. F. M. (1966) Percutaneous absorption. *Mod. Trends Derm.*, **3**, 84–110.

21

Depressants of the Central Nervous System

The definitions of terms used to describe the drugs which cause depression of the central and peripheral nervous systems overlap and can be confusing. The following definitions will serve as a general guide and provide a reasonable understanding of terms used to describe the type and degree of neuronal depression caused by a particular drug.

1. *Sedatives.* These are agents which will cause a lowering of cerebral perception and reaction, whilst not preventing the patient from taking part in the normal life surrounding it. Until recently this term has been applied to such compounds as ethyl alcohol, potassium, sodium and ammonium bromides, but these are now being displaced by a group of drugs known collectively as *tranquillizers* or *ataractic drugs.* Such compounds do not induce the extreme lethargy and apathy associated with the older sedatives.

2. *Hypnotics or soporifics.* This series of nervous system depressants produces a state of central nervous system depression equivalent to sleep, and in some cases very similar to sleep. Most of the non-volatile anaesthetics will produce this state when given in suitable doses. The barbiturates are perhaps the most useful.

3. *Narcotics.* The stage of CNS depression called narcosis is one of profound sleep from which the animal can only be roused with difficulty: as soon as the rousing stimulus ceases the patient immediately relapses into its former state. Such drugs are usually, but not always, analgesic. Where an animal under the influence of a narcotic is later subjected to complete anaesthesia by the application of an anaesthetic, the term *basal narcotic* is often used to describe the drug. Light, medium and deep narcosis are terms used to describe the depth of narcotic depression. In some cases the level of light narcosis is accompanied by excitement and struggling, which may result in self-inflicted injuries: the incidence of narcotic excitement is highest during the induction of, and recovery from, general anaesthesia, but may also occur when narcotized animals are hurt or stimulated by noise and movement.

4. *Analgesics or anodynes.* These are simply described as 'pain killers'. They may act centrally and systemically, as do salicyclic acid derivatives, chlorbutol, morphine and pethidine, or they may act locally. Local analgesia of a higher order can be induced by the local anaesthetics.

5. *Anaesthetics.* These are agents which are used to render a patient insensible to all sensation either locally, as with local anaesthetics, or completely, as with general anaesthetics. The local anaesthetics are described in Chapter 25 and the general anaesthetics in Chapter 22.

From these definitions it will be seen that simple sedatives, hypnotics, narcotics and anaesthetics are successive stages of increased nervous system depression and progressive loss of consciousness. Analgesia, though closely allied to, and a natural corollary of this scale of depression, is a distinct phenomenon. This may perhaps be illustrated by referring to nitrous oxide which is a particularly good analgesic and yet has a relatively low anaesthetic potency; and also by reference to the barbiturates which are powerful sedatives, hypnotics, narcotics and anaesthetics and yet relatively poor analgesics.

From the realization of this division has arisen the concept of modern anaesthesia in which the control of a patient during an operation is achieved by the employment of drugs having three different pharmacological actions: (1) analgesia, (2) narcosis, used in the sense of the elimination of consciousness, and (3) muscular relaxation. These three actions may be effected unselectively by the use of one drug only, as with chloral hydrate or barbiturate intravenous injections, but in most cases the doses which will achieve all these effects are so large that undesirable side effects such as depression of respiration, body temperature, or blood pressure, may also occur. *Balanced anaesthesia*, which involves the use of a different drug for each of the three effects mentioned, makes it possible to ensure a complete loss of pain sensation, both from a surgical point of view and from conscious appreciation, by the use of an analgesic such as pethidine; makes it possible also to maintain a stage of *light narcosis* so that medullary depression is minimal—the general anaesthetics used in small doses are indicated for this purpose; and makes it possible to obtain the complete muscular relaxation which cannot be produced by anaesthetics used in narcotic doses. Muscle relaxants, such as the curare-like and the succinylcholine-like drugs, are used for this purpose. When balanced anaesthesia is used it is essential that all three phases should be correctly applied, for the use of narcotics and muscle relaxants with incomplete analgesia will cause severe surgical shock when the operation begins. Similarly, if either of the other two phases is incomplete considerable distress and shock will result.

Many of the indications which make the CNS depressants such an important group in human medicine have no veterinary parallel. Their major role in veterinary medicine is as anaesthetic adjuncts.

SEDATIVES AND HYPNOTICS

The drugs which cause sedation will usually, in larger doses, cause hypnosis.

The *barbiturates* of the long, medium and short acting types will all exert sedative and hypnotic effects when given in small doses either orally or by parenteral injection: the oral route is the most suitable one for the routine administration of all sedatives and hypnotics. Phenobarbitone, methylphenobarbitone and butobarbitone are the most commonly used of the barbiturates, but small doses of pentobarbitone will have sedative effects. The dosage of the first three compounds varies from about 30 mg (½ gr.) three times daily for small breeds of dogs or for light sedation in larger breeds, to 90 mg (1½ gr.) three times a day for heavy sedation in the larger breeds. Higher or more frequent doses may cause motor incoordination. In large animals barbiturates are not used commonly.

In general, the barbiturates are not satisfactory when pain is present and must be combined with analgesics and indications for true sedative or hypnotic effects occur rarely in veterinary practice.

Traditionally, chloral hydrate is the large animal sedative. Other details concerning barbiturates and chloral hydrate may be found in Chapter 22.

Bromides

The bromide salts of potassium, sodium and ammonium are effective sedatives for dogs and have occasionally been used in pigs, although they have now fallen into disuse.

The bromides have sedative and anticonvulsant actions on the brain and are administered to dogs in doses of 0·3 to 1 g (5 to 15 gr.) two or three times daily: 0·6 g (10 gr.) doses are adequate and should normally not be exceeded. The salts are absorbed rapidly, are metabolized in a manner similar to chlorides, and have actions identical with chlorides in all respects other than effects on the CNS: bromide ions are depressant to the CNS and, as they displace chloride ions and are more slowly excreted, there is a danger of cumulation, indicated clinically by progressive depression and incoordination. Both motor and sensory nervous activities are depressed, but side effects relating to respiratory and circulatory depression are insignificant.

Toxicity. This is uncommon and can be discounted providing the bromides are not used for longer than a week. Local gastric irritation may occur following tablet administration, but emesis subsequent to administration is uncommon.

Chlorbutol, B.Vet.C.

This white crystalline solid with a characteristic odour is more soluble in organic solvents than in water, and is particularly soluble in alcohol, which could be anticipated from its formula—trichlorodimethyl ethyl alcohol.

Actions and Uses

The actions of chlorbutol are similar to those of chloral hydrate but it is far less irritant to tissues. Its action following topical administration is sufficient to ensure its mention as a local anaesthetic and antiseptic, but once absorbed it has a central depressant effect on both cerebrum and medulla. Amongst the medullary centres, it depresses the vomiting and respiratory centres.

Chlorbutol is used externally in dusting powders (1 to 2 per cent) for its analgesic and antiseptic effects: it is also used at a 0·5 per cent concentration as a preservative for certain parenteral injections, both as an antioxidant (adrenaline injections) and as a bactericide. Internally, chlorbutol has a gastric sedative action both by local analgesic effects and by central depression of the vomiting centre after absorption: it is incorporated in arecoline worm tablets and capsules for this purpose [160 to 195 mg ($2\frac{1}{2}$ to 3 gr.) chlorbutol with 15 mg ($\frac{1}{4}$ gr.) arecoline hydrobromide]. Chlorbutol is also a useful prophylactic against travel or motion sickness in dogs: for this purpose, capsules containing 130 mg to 0·6 g (2 to 10 gr.) chlorbutol are used. Chlorbutol is usually given in capsule form mainly because of the very slow rate of disintegration of tablets containing this compound.

Dosage. Dogs 0·13 to 0·6 g (2 to 10 gr.). It is rarely used in other species.

ANTICONVULSANTS

This term is applied to a group of drugs which possess an enhanced ability to depress the convulsive threshold of the motor cortex. They are of less importance in veterinary medicine than in human medicine, and the study of these drugs for veterinary purposes is not extensive. They are nevertheless widely used to control the epileptiform convulsions which occur as a symptom of the neurotropic virus syndrome of dogs and which are also associated with digestive disturbances during teething.

The *barbiturates* are widely used in the control of epileptiform convulsions, but usually hypnosis and incoordination are seen before convulsions are controlled and the newer compounds are preferred.

Phenytoin Sodium, B.Vet.C.

Also known as Diphenylhydantoin Sodium, U.S.P., has less hypnotic and greater motor effect than the barbiturates and may be used on its own or combined with phenobarbitone.

Dosage

The dose for dogs is 0·1 to 0·3 g given by mouth three to four times daily, followed by a decrease when the convulsions have been controlled.

Troxidone, B.P. (Trimethadione, U.S.P.)

This oxazolidine derivative has been used in veterinary medicine to control epileptiform convulsions in dogs. The dosage is 0·3 to 1 g (5 to 15 gr.) by mouth three times daily, the dosage being varied in the range according to reaction. Doses are increased in size until the convulsions are controlled, or until signs of toxicity are seen: once control has been established doses can be reduced to a maintenance level. The compound is available in capsules and is relatively expensive.

Toxicity. The main sign of toxicity is photophobia. As this is a common symptom of distemper or hard-pad the only way of determining the presence of toxicity is to withdraw the drug and see if the phobia remains.

Primidone, B.Vet.C.

This deoxybarbiturate is used in doses of 12 to 50 mg per kg body weight daily in divided doses. The usual method is that doses are progressively increased until convulsions are controlled and are then reduced until a maintenance dose is established. By such a method dosing has been continued for months without any deleterious effect. Recent work has suggested that primidone is metabolized to phenobarbitone in the body.

Toxicity is very low and appears clinically as a progressive sedation and hypnosis. Symptoms usually disappear on decreasing the doses, though they can also regress spontaneously, perhaps indicating the onset of tolerance.

TRANQUILLIZERS

This is a term which has been applied comprehensively to a group of drugs which will, by action on the central nervous system, alter the temperament of a patient, thereby making it less responsive to outside stimulation. Another term, 'ataractic' or 'ataraxic', is being increasingly used to describe these drugs.

The application of such drugs in the sedation of temperamentally unstable and nervous patients provides an obvious and logical use, but the side actions of many of the group are numerous and in some cases are more important therapeutically than the tranquillizing effect. These side effects include antihistaminic; general and local anaesthetic effects; 'sparing' effects or synergism with narcotics, analgesics and both general and local anaesthetics; antispasmodic effects; antipyretic and hypothermic properties, antemetic and anti-autonomic actions. The comprehensive nature of these side actions and the varying individual susceptibility to each and all of them makes clinical prescribing difficult and clinical assessment very protracted and involved. Much still remains to be learnt of the pharmacology of these tranquillizing drugs.

Meprobamate

This is 2-methyl, 2 n-propyl, 1, 3-propanediol dicarbonate. It is administered orally and has an *action* similar to that of mephenesin (*q.v.*) on the transmission of impulses between association neurones in the spinal cord, causing a depression of all but monosynaptic motor transmissions in the spinal cord. Meprobamate's mild central depressant effect and increased duration of action over mephenesin were responsible for its launch as one of the first anti-anxiety drugs in human medicine. Although classed as a minor tranquillizer, the central effects of meprobamate are closer to those of a small dose of a barbiturate than of a phenothiazine. In some respects its insolubility and restriction to the oral route of administration are a disadvantage, for dosing is followed by a delay of about 1 hr before any tranquillizing effects are apparent.

Carisprodal

This is a further example of the propanediol derivative tranquillizers. It is used in man to treat painful conditions associated with increased tonus of voluntary muscles. Its properties of tranquillization and centrally-induced muscle relaxation would make it especially useful in cases where protrusion of the intervertebral disc is incriminated.

Phenothiazine Derivatives

The CNS depressant side-effect of the antihistaminic promethazine led to investigation of other derivatives of phenothiazine with a view to finding compounds with more intense central activity. This has led to the discovery of a number of compounds of important veterinary as well as human medical significance. In human use, the minor tranquillizers (e.g. promazine) are used to control neurotic disorders while the major tranquillizers (e.g. chlorpromazine) are indicated for the control of psychoses. The actions of all phenothiazine derivatives including promethazine, acetylpromazine, chlorpromazine, pecazine, promazine and trimeprazine differ quantitatively rather than qualitatively. In veterinary practice they have been used as antiemetics, antipruritics and anaesthetic premedicants. Behavioural modification is rarely sought but may be exemplified in introducing animals to new environments, e.g. dogs to quarantine kennels, heifers to milking parlours.

Chlorpromazine Hydrochloride, B.Vet.C.

This synthetic organic compound is susceptible to light and oxidation, and solutions should contain antioxidants such as hydroquinone, ascorbic acid or sodium metabisulphite.

Chemically the compound is a relatively simple amino-alkyl derivative of halogenated phenothiazine; it will be seen that chlorpromazine differs from promethazine (10-(2-dimethylaminopropyl)-phenothiazine) the potent antihistaminic, only in the attachment of the chlorine atom at carbon 2 in the nucleus and in the attachment of the dimethylamino group at carbon 3 of the side chain instead of carbon 2.

Actions

Chlorpromazine is fairly rapidly and completely absorbed by whatever route the compound is administered. Maximum blood levels are obtained about 3 hr after oral dosing and somewhat more than 1½ hr after intramuscular injection. Effects occur within 5 min of intravenous administration and become maximal shortly after, remaining so for about 1 hr; the duration of effect depends upon the indicator used, for most effects may persist as long as 6 hr, whereas hypothermic effects may be evident even after 24 hr. The 6-hr limit for the main action suggests a dosage rate of three times daily where repeated oral medication is indicated.

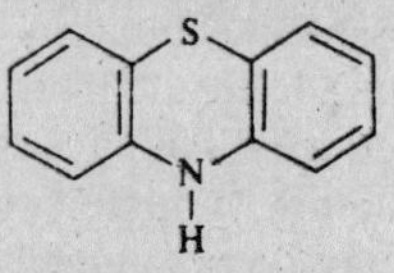

(i) Phenothiazine

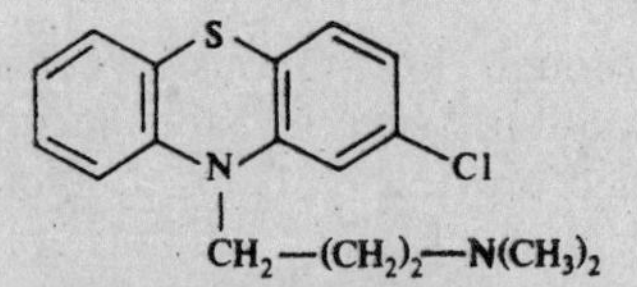

(ii) Chlropromazine base or 2-chloro-10-(3-dimethyl aminopropyl)-phenothiazine

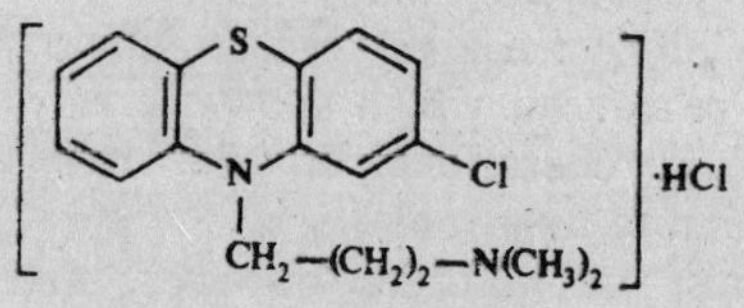

(iii) Chlorpromazine hydrochloride

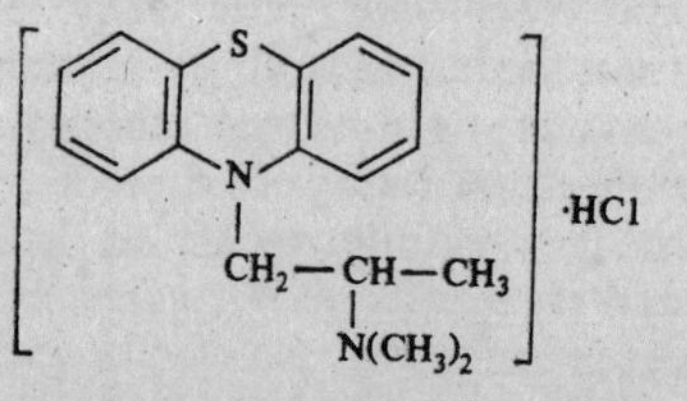

(iv) Promethazine hydrochloride antihistaminic

The *metabolic fate* of chlorpromazine in the domesticated species is not completely known, but part of the dose is excreted in the urine as chlorpromazine sulphoxide and conjugates.

The *pharmacological actions* of chlorpromazine are still incompletely known is spite of the wide use and valuable therapeutic effects achieved. The major action is a marked depressant effect on the brain stem; very little action seems to be exerted on the cerebral cortex and spinal cord. Subsidiary but important actions include the potentiating effect on general and local anaesthetics and analgesics, slight antihistaminic effects, adrenolytic and anti-acetylcholine actions, hypothermic and antipyretic effects, and a marked antemetic action where emesis is due to stimulation of the vomiting centre by certain chemicals.

Central sedative effects result in drowsiness and an inclination to sleep, and it has been shown by electro-encephalography that brain activity approximates more closely to that of normal sleep than the activity present during barbiturate depression. In spite of the drowsiness induced, mental acuity is little impaired. From this central sedation arises the 'sparing' or potentiating action which chlorpromazine exerts when administered prior to general anaesthesia, local anaesthesia, or analgesia: the anaesthetic dose of a barbiturate is cut by as much as 50 per cent when chlorpromazine premedication is practised. Although it has been shown that chlorpromazine has a local anaesthetic potentiating action it is not usual to reduce the amount of local anaesthetic administered. Also associated with the sedative action, and closely integrated with the analgesia potentiating effect, is the tranquillizing action: apprehension is reduced or abolished and the reaction to pain decreased.

The *antemetic* effect of chlorpromazine, which is associated with depression of the vomiting centre, makes the compound remarkable in that it was the first drug discovered to have an antagonistic action to the emetic effect of apomorphine. It has been found to antagonize the emetic effect of many chemicals which act centrally, but has no limiting effect on the action of local emetics.

The *antihistaminic* action is relatively weak (over 100 times less) when compared with that of promethazine, in spite of the very small difference in chemical structure.

The *autonomic* effects of chlorpromazine on the other hand are quite significant. It has α-adrenoceptor and uptake-blocking actions and also blocks 5-HT and acetylcholine.

The *adrenolytic* action is of major importance in anaesthesia, for in addition to causing some peripheral vasodilatation, chlorpromazine causes an inhibition of the pressor response to adrenaline. If, during the course of anaesthesia, shock supervenes, with vasodilatation and a drop in blood pressure, the autonomic reflex vasoconstriction cannot occur when the pressor effect is abolished by premedication with chlorpromazine; the fatal tissue ischaemia which might result is therefore avoided. It has been pointed out that administration of chlorpromazine after the onset of shock can prove fatal, probably because a further and fatal drop in blood pressure may result. In the cerebral cortex, chlorpromazine inhibits the neuronal re-uptake of the putative transmitter agents noradrenaline and 5-HT. It can also block dopaminergic receptors in the brain.

The *parasympatholytic* effect of chlorpromazine is less prominent than the adrenolytic effect; it is also less than that of atropine. Nevertheless, the action is sufficiently great to reduce intestinal spasm and peristalsis. The flow of bronchial and salivary secretions is reduced.

On respiratory rate, chlorpromazine has been demonstrated to have a stimulant effect when administered to anaesthetized rabbits, but this action is not established in other species and, in fact, respiratory depression is a clinical feature of sedation, though not a marked one.

The *thermo-regulatory* centre is depressed by chlorpromazine. This means that the body will tend to adopt the temperature level of the immediate environment, and chlorpromazine is therefore useful in assisting the achievement of induced hypothermia as an anaesthetic technique. The antipyretic action stems from this effect on the thermo-regulatory centre, and excessively high temperatures can be reduced, making the compound of value in the treatment of heat stroke.

Chlorpromazine has other less well defined actions which will no doubt be clarified as experience increases.

Uses

A considerable number of publications have appeared, and still are appearing, concerned with the veterinary applications of chlorpromazine and other tranquillizers.

1. *Pre-anaesthetic medication.* The use of chlorpromazine as a pre-anaesthetic sedative has been reported in all the domesticated species. The advantages of premedication with chlorpromazine are numerous and include:

(a) A tranquillizing effect, which besides reducing reaction to pain and irritation, and therefore allowing intubation to be achieved without causing obvious discomfort, also reduces by as much as 50 per cent the quantity of anaesthetic and analgesic required. The sedation allows uneventful anaesthetic induction and recovery, particularly when barbiturates are used as anaesthetics.

(b) Since less anaesthetic is used, the toxic effects which might result from anaesthesia are considerably reduced. This is especially valuable in poor risk patients, though it should be realized that sedation may be prolonged in such animals.

(c) Bronchial and salivary secretions are diminished.

(d) Although intravenous injection usually causes blood pressure to fall, chlorpromazine, as explained previously, helps to avoid irreversible surgical shock when administered prior to the induction of a shocked state.

(e) The anti-emetic effect assists in cases showing emesis due to central action (e.g. pyometra cases in bitches).

(f) Chlorpromazine causes some skeletal muscle relaxation which assists in carrying out abdominal operations, in the reduction of fractures and luxations etc.

It must be realized, however, that chlorpromazine sedation is uncertain, particularly in horses and cases will occur where excitement rather than sedation follows administration. Chlorpromazine is now not recommended for pre-anaesthetic sedation in horses and cattle because of its variability of action in the former species and the prolonged post-operative recovery period in cattle.

2. *General restraint.* In all species small doses will produce tranquillization and so facilitate the handling of otherwise intractable animals. In *small animals*, chlorpromazine administration will make the scaling of teeth, dressing of eczema patches, abdominal palpation, general handling, restraint for X-ray, opening of abscess, injection of local anaesthetics and many similar manipulations, relatively simple in cases where such processes were previously difficult or impossible.

In *large animals*, the handling of feet, clipping, shoeing, transportation including loading and unloading, teeth rasping, the drawing of the penis in the bull and similar operations, will all be facilitated by chlorpromazine.

3. *Treatment of psychoneurotic conditions.* Although such affections have not received the same attention in the veterinary as in the human medical field, according to various reports certain emotional disturbances have benefited from chlorpromazine therapy.

Many authors agree on the benefits conferred by chlorpromazine on kennelled and confined animals. Restlessness and noisiness decrease, and the animals eat, drink, and sleep normally, with no sign of fretting or apprehension. In season bitches which refused to stand, cases of pseudo-pregnancy, and similar conditions which may be associated with psychotic disturbances all benefit from chlorpromazine therapy.

4. *Use in specific diseases.* Those *skin diseases* of small animals which are associated with pruritis respond very well to chlorpromazine therapy combined with specific agents, for the phenothiazine derivative appears to remove the animals incentive to scratch and nibble so that the skin is not continually damaged by self-inflicted trauma.

In large animals *tetanus* is perhaps the most important disease to respond to repeated chlorpromazine therapy.

Colic cases in horses, particularly of the spasmodic type, have responded excellently to 1·5 mg per kg (75 mg per cwt.) given intramuscularly, symptoms disappearing within 30 min.

The relief of *choke* in cattle is greatly facilitated by chlorpromazine. The use of chlorpromazine as a muscle relaxant in the reduction of shoulder luxations in cattle, using a 500 mg dose for a 450-kg. animal and in sheep prior to Caesarian section under local anaesthetic was found to be very effective.

Toxicity. In itself chlorpromazine is not highly toxic but it is somewhat unpredictable in its sedative effects, especially in horses, and following intramuscular administration. Excitement and struggling during the onset of action and during post-anaesthetic recovery are the most important possibilities. Barbiturates or other sedatives may be used to control these idiosyncratic responses. Should nephritis or hepatitis be present animals will show an exaggerated and prolonged response to chlorpromazine. Chlorpromazine is contra-indicated in deeply anaesthetized animals. In man, the occurrence of obstructive jaundice and blood dyscrasias is its more serious toxicity problem.

Dosage

For pre-anaesthetic medication and general sedation, the dose for all species lies between 0·5 and 2 mg per kg body weight (50 to 100 mg per cwt.) when administered intramuscularly. The oral dose rate for the dog is 5 mg per kg. Large animals are dosed at the lower end of the range and the intravenous route is rarely used nowadays so as to minimize idiosyncratic reactions.

Administration in large animals is normally by intramuscular injection of a 5 per cent solution. This is sometimes followed by painful but temporary reactions at the sites of injection. In small animals intravenous injection of 2·5 per cent solutions can be given, but this concentration is irritant when given intramuscularly and should be diluted to 0·5 per cent concentrations whenever this is possible.

In small animals collapse of the third eyelid and wrinkling of the face provides an indication of the point when further procedures may commence following administration of chlorpromazine. In all species it is usual to allow about 5 min to elapse after intravenous injection and 40 min or more (up to 90 min has been suggested) after intramuscular injection: oral dosing may not give maximum effects for up to 3 hr after dosing.

Duration of effect will vary with the individual and dose. Normal dosage gives sedation periods of about 6 to 8 hr, but the higher doses mentioned may provide obvious sedation for up to 24 hr or longer. For sedative purposes doses may be repeated from two to four times daily depending on results and requirements.

Acepromazine

Acepromazine differs chemically from promazine and chlorpromazine in having an acetyl group at carbon 2 in the nucleus.

Actions and Uses

Similar to those of its near relatives although the delay before onset of effect is shorter—only 10 min after oral administration and 5 min following intramuscular injection. It is claimed that the incidence of aberrant reaction in the horse is less with this compound and it has found wide acceptance in the United Kingdom.

Dosage

Farm Animals:
- Parenteral: 0·05–0·1 mg/kg (approximately 0·25–0·5 mg/10 lb.) body weight

Dogs and Cats:
- Oral: 1–3 mg/kg (approximately 0·5–1·4 mg/lb.) body weight
- Parenteral: 0·125–0·25 mg/kg (approximately 0·5–1 mg/10 lb.) body weight

Toxicity. Nor-adrenaline should be used in place of adrenaline should a circulatory crisis arise in an animal sedated with acetylpromazine. In this it differs little from the other members of the promazine series.

Perphenazine

This highly active, piperazine derivative of phenothiazine has been used extensively as an ataractic in human medicine under the names of Fentazin and Trilafon. It has been used in veterinary practice at the dose rate of 0·22 mg per kg (0·1 mg per lb.) body weight in cattle, pigs, dogs and cats. Sedation following intravenous injection lasts about 7 hr: results are more satisfactory when this route is used. Tachycardia and hypotension have been reported in pigs, a high proportion of which (16 per cent) do not tolerate the drug very well. Other side-actions reported include excitement and excessive salivation.

Promazine Hydrochloride, B.Vet.C.

This compound differs from chlorpromazine in having no chlorine atom at carbon 2 of the nucleus; it can be considered as the root compound of the whole series of the phenothiazine derived ataractics.

Actions and Uses

The product has had wide acceptance and is useful in horses and cattle, especially since chlorpromazine is no longer recommended for anaesthetic premedication in these species. It is recommended for all farm animals, dogs and cats. Sedation lasts for periods of up to 6 hr, but the actions in other respects are similar to those of chlorpromazine. It is less potent and less toxic than chlorpromazine.

Dosages

Small animals:
- 2–6 mg/kg intramuscularly or intravenously.

Large animals:
- 0·4–2 mg/kg intramuscularly.

Trimeprazine Tartrate, B.Vet.C.

This phenothiazine derivative is 10-(3,dimethylamino-2,methylpropyl) phenothiazine. Solutions are light sensitive and should not be used when discoloured.

Actions and Uses

Trimeprazine is qualitatively similar to chlorpromazine but differs quantitatively in being only two-thirds as potent a sedative and having a shorter duration of activity; effects persist for about 1 hr after intravenous injection. It has a strong antihistaminic action but virtually no anti-adrenaline activity. As with chlorpromazine, barbiturate doses should be reduced by about one-third. It has been particularly recommended for the control of pruritus in skin conditions in dogs.

Trimeprazine is recommended as a sedative for all species of domestic animals, but is particularly indicated for horses and cattle where chlorpromazine is no longer used.

Dosage

All species: 1 to 4 mg per kg by all routes. As with all phenothiazines, intravenous injections should be made slowly, preferably using the lower end of the dosage scale.

Neuroleptics

The state of depression induced by most tranquillizers or antipsychotics is relatively minor and it is one from which the animal can readily be roused by noise or pain, so that it is not of great value in the field of chemical restraint. The neuroleptics are agents capable of greater CNS depression with an enhanced ability to reduce the animal's tendency to voluntary motion.

Xylazine

This agent reliably induces in ruminants a CNS depression ranging from tranquillization to narcosis and is also of value in the horse, but not in other domesticated species. It is a solution for intravenous or intramuscular injection. Most cattle go down after xylazine and remain deeply sedated for about half an hour. Depression is greater if the intravenous route is used. Increasing the dosage increases the duration rather than the intensity of depression and side-effects become apparent, e.g. muscle tremors, bradycardia which may be accompanied by A–V block and depressed respiratory rate. Sedation is accompanied by analgesia which may require local fortification for surgery.

The Butyrophenones

Examples of this group are the compounds dehydrobenzperidol, and azaperone. Other members have been used as major tranquillizers in man, e.g. haloperidol and triperidol.

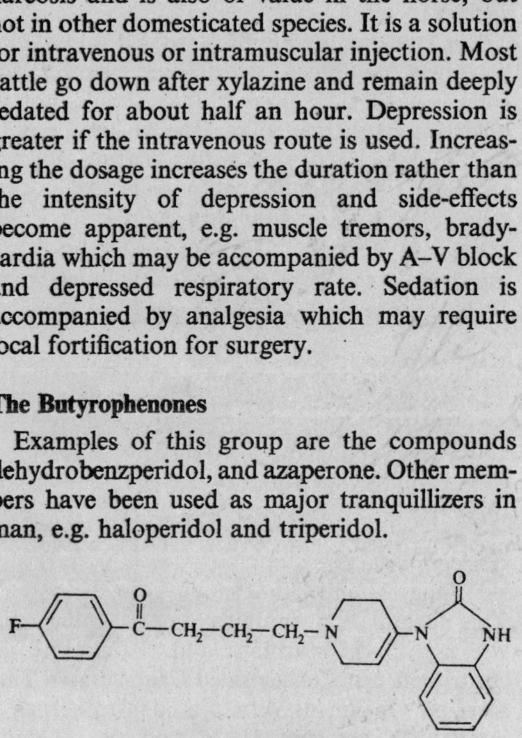

Dehydrobenzperidol

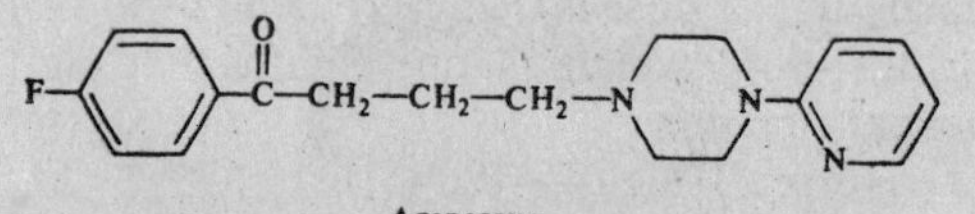

Azaperone

A considerable advance in the technique of chemical restraint was achieved with these substances. Azaperone has been evaluated in the pig and horse and fluanisone in the dog.

They are potent psychomotor sedatives and their veterinary examples although being hypothermic are notably less adrenolytic and interfere less with other autonomic functions than the phenothiazines. Additionally, with dosage variation, their effect varies from sedation to immobilization. Analgesia is not pronounced nor is the tendency to narcosis.

The drugs are administered by the intravenous or intramuscular routes and can be used in conjunction with other agents to induce neuroleptanalgesia and anaesthesia. Dehydrobenzperidol and azaperone have been widely used for sedation and tranquillization in the pig in which the phenothiazines are not very effective. The butyrophenones have been administered intramuscularly and shown to be of value as anaesthetic pre-medicants, in the facilitation of surgical interference and in the control of fighting when mixing batches of pigs.

Phencyclidine, B.Vet.C.

This is another neuroleptic agent and it is a piperidine derivative. It was introduced for ataractic use in the domesticated species, but was later withdrawn owing to the high incidence of aberrant reactions. At present, it is used for the control of primates and finds routine use in making monkeys more amenable to handling and experimentation. It can produce degrees of CNS depression from sedation to anaesthesia depending on dosage (0·5–5 mg per kg) but the usual effect desired approaches catalepsy.

NARCOTICS

Narcosis is a state of CNS depression which borders on anaesthesia and may be achieved by the use of subanaesthetic doses of barbiturates or chloral hydrate or by the administration of morphine or opium extracts such as papaveretum. Morphine and its many synthetic analogues are known as narcotic analgesics and in subnarcotic dosage can induce the most profound analgesia. Thiambutene hydrochloride is a synthetic drug which has analgesic, hypnotic and narcotic actions: its action is reversed by nalorphine hydrobromide.

These narcotics may be used alone for restraining hysterical animals or for general sedation but may also be administered prior to anaesthesia by general or local anaesthetics. In view of the increasing use of combinations of non-volatile basal narcotics and non-volatile anaesthetics it should be stressed that, though this practice is safe in normal animals, the liver and kidneys of the toxic animal may be unable to tolerate the increased work entailed in detoxicating and excreting these compounds. It is therefore advisable in these cases to use volatile gaseous anaesthetics in conjunction with the non-volatile narcotics.

Among the older narcotics there is an established order of species suitability. Chloral hydrate is suitable for adult farm animals; the barbiturates can be used in farm animals and household pets; morphine and its analogues are more suitable for dogs: pethidine (meperidine) may be used in cats, but its sedative action in this species is not as powerful as that of morphine in the dog.

The barbiturates and chloral hydrate have been described elsewhere (see Chapter 22) but morphine, opium and the synthetics may best be mentioned here.

Opium, B.Vet.C. (D.D.A.)

Unprocessed or raw opium consists of the dried exudate obtained by the incision of the growing unripe seed capsules of a poppy, *Papaver somniferum* L., cultivated for this purpose in the Middle East, the Indian continent, and China. It is imported as brown resin-like masses which have an aromatic and characteristic odour. Opium should contain at least 9·5 per cent anhydrous morphine, but a great number of other alkaloids are present: these may be classified into *phenanthrene* derivatives, of which morphine and its esters, both natural and synthetic, such as methyl-morphine (codeine), dimethyl-morphine (thebaine) and diacetyl-morphine (heroin), are examples, and the *isoquinoline* derivatives such as papaverine and narcotine. The former group are predominantly sedative in their effects but the latter group tend to cause relaxation of smooth muscle and have relatively low narcotic effects.

Actions and Uses

The greatest part of the action of opium is related to its morphine content which provides analgesic, sedative and hypnotic actions. The isoquinoline derivatives add the smooth muscle relaxant effects, giving a broader application to a combination of alkaloids than is possible with the individual alkaloids. It is for this reason that opium has a greater constipatory effect than pure morphine: there is little local anaesthetic action, but direct action on the intestinal muscle combined with central action causes closure of the sphincters and a slowing of peristaltic waves.

Opium is used mainly as an intestinal sedative in gastroenteritis of all species other than the cat. It is also a useful cough sedative in the same species.

For intestinal sedation, powdered opium may be given combined with chalk, kaolin and catechu, but such preparations should not be given in obstructive colics. The tincture may be used in drenches of similar formulation. In large animals, these opium compounds are not as reliable as in man or the dog.

Dosage

Tincture of Opium, B.Vet.C. (Laudanum)

Horse: 30–80 ml (1–2½ fl. oz.)
Cow: 60–120 ml (2–4 fl. oz.)
Sheep: 6–20 ml (90–320 m.)
Dog: 0·3–2 ml (5–30 m.)

This compound contains 1 per cent anhydrous morphine.

Camphorated Tincture of Opium, B.Vet.C. (Paregoric)

Dog: 2–6 ml (30–90 m.)

This formulation contains only 0·05 per cent w/v anhydrous morphine and is therefore not controlled, but **Concentrated Camphorated Tincture of Opium,** B.Vet.C. is controlled as it contains 40 per cent w/v of tincture of opium and consequently 0·4 per cent w/v of anhydrous morphine: its dose is 0·25 to 0·75 ml (4 to 12)m. for the dog, which is one-eighth that of the normal tincture.

Horse: 3–8 g (45–120 gr.)
Cow: 6–12 g (90–180 gr.)

Powdered Opium, B.Vet.C.

Sheep: 0·5–2 g (8–30 gr.)
Pig: 0·3–1 g (5–15 gr.)

Powdered opium is standardized to contain 10 per cent w/w anhydrous morphine and is controlled.

Morphine

This alkaloid, extracted from opium, is

mainly used in the form of either the hydrochloride or the sulphate. These two salts are white crystalline powders which are odourless but have the typical bitter taste and the water solubility of alkaloidal salts. They are far less soluble in alcohol, although the hydrochloride, with a solubility of 1 : 100 is about ten times more soluble than the sulphate.

Actions

There are no external actions. Following oral or parental administration, absorption is rapid and the actions are clinically evident within 2 or 3 min. In all species there is at least an initial period of stimulation but this soon passes in the dog: in the cat, however, stimulation predominates throughout, whilst in other species restlessness may give way to depression. This irregularity of response makes morphine unpopular in all species but the dog.

In the dog, clinical signs following subcutaneous injection consist of restlessness with panting, salivation, and finally emesis and defaecation. Subsequently, depression deepens to narcosis and a drop in body temperature occurs due to peripheral vasodilatation. Narcosis is achieved in about 20 min and will last 2 or 3 hr unless the animal is toxic, when it may persist for as long as 24 hr.

1. The *cerebral cortex* is initially stimulated, but depression of both motor and sensory cortex soon follows. Brain morphine receptors have been isolated and recently, a polypeptide having morphine-like effects also.

2. The *medullary centres* are affected in two ways: (a) The vagal, oculomotor and chemosensitive trigger zone of the vomiting centre are initially stimulated and later depressed. This selective stimulation is mainly responsible for the salivation, emesis, defaecation and a constricted pupil (this effect is irregular). The heart rate initially increases due to nausea, but is then slowed by the stimulated vagus. (b) Group 2 comprising the respiratory, vasomotor and cough centres which are depressed without any initial stimulation. This depression is responsible for peripheral vasodilatation and loss of heat but there is little effect on the blood pressure; respiration is slowed but may be deeper than normal, and the response to elevated P_{CO_2} is lowered: the cough reflex is materially affected even by small doses. These effects on respiration are very important if morphine is used as a basal narcotic and may affect the choice of anaesthetic.

3. The *alimentary tract* is quite strongly affected, the action being both central and local. The sphincters are closed and the intersphincter muscle remains in tonic spasticity so that food does not pass and constipation results. These effects follow the initial period of stimulation which results in emesis and defaecation. Opium, due to the muscle relaxant effect of the isoquinoline derivatives, has a much stronger constipating effect.

Tolerance. An acquired tolerance to morphine and dependence can be induced following repeated dosing. However, morphine is rarely administered chronically to animals.

Idiosyncrasy. Individual animals of all species will show unusual reactions to morphine. In large animals, restlessness and excitement may result when sedation is expected. In the dog, the most common idiosyncractic reaction could be classed as inherent tolerance, for doses as large as 100 mg have been given with little effect. The stored histamine released by morphine causes vasodilatation, itching and hypotension.

Excretion. The major part of a morphine dose is excreted via the urine in the 24 hr following administration. It is present in the urine mainly in a conjugated form, but partly unchanged. A small quantity is excreted into the intestines.

Uses

Morphine has three main uses:

1. *Premedication.* Its powerful analgesic effect makes it very suitable for this purpose and the narcotic effect enables smooth and easy induction of anaesthesia with either a non-volatile or an inhalant anaesthetic. Further advantages are the reduction of the amount of anaesthetic required and an uneventful recovery period. Morphine remains one of the more potent narcotic analgesics and may therefore be used to manage severe pain, e.g. post-surgery.

Morphine is not the ideal basal narcotic. Its disadvantages include the incidence of vomiting and salivation, the constipatory effect, elimination of pupil reflexes and respiratory depression. In toxic patients the relatively slow excretion rate makes full recovery very slow and provides a problem in maintaining body temperature and freedom from hypostatic pneumonia.

2. *Intestinal sedation.* Morphine is a constituent of many scour drenches including Tincture of Chloroform and Morphine, B.Vet.C. (chlorodyne) which contains 0·229 per cent morphine hydrochloride but is exempted from control under the Misuse of Drugs Acts. Chlorodyne is frequently combined with chalk, kaolin, catechu and ginger, as a treatment for diarrhoea in cattle, horses and dogs, and morphine as the hydrochloride, acetate, or sulphate is similarly used. These preparations are especially useful in spasmodic colic and where diarrhoea is accompanied by painful straining (tenesmus).

3. *Cough sedation.* Codeine is superior to morphine for this purpose, but traditionally morphine was a constituent of many cough mixtures, pills, and tablets, used in veterinary practice in both large and small animals. It depresses the cough reflex and is therefore indicated in the 'non-productive', harsh, dry cough usually occurring at the beginning of bronchopneumonia.

Dosage

Morphine Salts by subcutaneous injection.
- Narcosis:
  - Dog: 5–10 mg ($\frac{1}{12}$–$\frac{1}{6}$ gr.)
- Sedative:
  - Horse and Cow 0·14–0·6 g (2–10 gr.) (normally 0·14–0·3 g or 2–4 gr.)
  - Sheep and Pig: 30–60 mg ($\frac{1}{2}$–1 gr.)

Solution of Morphine Hydrochloride, B.Vet.C.
- By mouth:
  - Horse and Cow: 15–60 ml ($\frac{1}{2}$–2 fl. oz.)
  - Sheep and Pig: 2–8 ml (30–120 m.)
  - Dog: 0·3–1·3 ml (5–20 m.)

This solution contains 1 per cent w/v morphine hydrochloride.

Tincture of Chloroform and Morphine, B.Vet.C.
- Horse and Cow: 30–60 ml (1–2 fl. oz.)
- Foal and Calf: 15–30 ml ($\frac{1}{2}$–1 fl. oz.)
- Dog: 0·3–1·5 ml (5–25 m.)

The tincture contains 0·229 per cent w/v of morphine hydrochloride.

Note. These compounds are not recommended for use in cats.

Nalorphine Hydrobromide, B.Vet.C.

In doses of 10 to 40 mg ($\frac{1}{6}$ to $\frac{2}{3}$ gr.) nalorphine is a specific antagonist for use in opiate respiratory depression; it may be given intravenously or subcutaneously and acts within 20–30 sec. Nalorphine itself has some CNS depressant activity and should only be used in an attempt to reverse opioid-induced CNS depression.

Codeine

This natural opium alkaloid has been described already in relation to its valuable cough suppressant action. Its advantage over morphine is a much reduced tendency to induce dependence. Moreover, its oral effective dose is only one-third greater than its equi-analgesic subcutaneous dose, while that of morphine is about eight times higher. Codeine is therefore well suited pharmacokinetically for its role as an orally administrable cough or pain remedy. Codeine's analgesic potency is about one-twelfth that of morphine and about ten times greater than that of aspirin, with which it is frequently combined to enhance the analgesic potency of the combination. This is preferable to simply increasing the dose of aspirin because of aspirin's side-effects.

Diethylthiambutene Hydrochloride, B.Vet.C.

This compound known is a member of a group of narcotic and analgesic drugs classed as amino-alkyl-tertiary-carbinols. It has the formula 3-diethylamino-1:1-di(2′-thienyl)but-1-ene hydrochloride and is a white, water soluble, bitter tasting powder that gives unstable solutions which, though they can be sterilized by boiling at 100°C, turn brown on keeping for a few days. Solutions are prepared for injection by dissolving a hypodermic tablet in sterile water immediately prior to administration.

Actions

Thiambutene is used only in dogs. Following subcutaneous administration a local anaesthetic effect prevents pain. Intravenous injection is also suitable. Given by these parenteral routes, thiambutene causes analgesia and sedation, hypnosis or narcosis, the degree depending on the size of the dose and the route of administration. Small doses will cause mild analgesia and sedation, whereas doses may be increased to give strong analgesia and deep narcosis: subcutaneous administration will tend to give a slower onset but more prolonged duration of effect and lighter level of CNS depression, whereas intravenous injection will give stronger analgesia and deeper narcosis in a shorter time and for a shorter period.

Normally no emesis follows parenteral injection but salivation may occur which, however, can be prevented by the use of atropine or hyoscine. Defaecation sometimes occurs during the recovery stages.

Antagonism. The morphine derivative nalorphine hydrobromide acts by competitive antagonism as a stimulant for animals depressed by morphine and morphine-like substances, including thiambutene hydrochloride. In some cases full recovery occurs but in others the animal relapses into a hypnotic state and only recovers fully several hours later. This happens because nalorphine has a half-life much shorter than that of thiambutene.

Uses

The main indication for the use of thiambutene is in place of morphine, as a basal narcotic. Barbiturates and inhalant anaesthetics may be used in conjunction with thiambutene. It has also been used as a sedative and analgesic for minor manipulations and surgery.

Toxicity. Thiambutene must not be used in cats, because in this species a CNS stimulant effect predominates just as with morphine. In the dog, a considerable range of response to a standard dose exists, extending from convulsions to coma. The drug is a powerful and long-acting respiratory depressant but also elicits tachycardia and increased blood pressure. Nalorphine can reverse thiambutene's central depression, but convulsions are better controlled with barbiturates.

Dosage

As with most narcotics and sedatives, the highly nervous animal will require more thiambutene than the placid animal. Small dogs will require slightly more in proportion to their weight than the large dog. Solutions containing from 25 to 100 mg per ml may be used.

Subcutaneous injection: 5–20 mg/kg.

Intravenous injection: 1–4 mg/kg.

Fentanyl Citrate, B.Vet.C.

This is a synthetic compound with morphine-like actions but with an analgesic potency about 80 times that of morphine. It is usually presented in combination with a butyrophenone neuroleptic (dehydrobenzperidol or fluanisone) in a formulation designed for producing the state of neuroleptanalgesia. This alternative to surgical anaesthesia is produced by a combination of a neuroleptic and a narcotic analgesic and is characterized by stupor, catalepsy (immobilization) and analgesia. The side-effects are bradycardia, fall in blood pressure due to droperidol's α-adrenergic blockade, respiratory depression, salivation and defecation. The product is administered intramuscularly, achieves maximal effect in about 15 min and has a rather short duration of analgesia. Voluntary movements sometimes occur and surgery may require supplementary local anaesthesia. The technique is, however, of value for minor surgery, orthopaedic problems and dental manipulations and has the advantage of rapid termination by an intravenously injected narcotic antagonist. Alternatively, full anaesthesia could be induced by a suitably reduced dose of a barbiturate should this seem desirable.

Etorphine

Thebaine is a phenanthrene alkaloid of opium with a relatively low order of potency. From thebaine have been developed the oripavines of which etorphine is presently available as an injectable product in combination with acepromazine for large animals or methotrimeprazine for small animals. The tranquillizer is incorporated in an attempt to overcome some of etorphine's shortcomings, e.g. its tendency to cause excitement, especially in horses. In this species the violent muscle tremblings, staring eyes, considerable tachycardia and elevated blood pressure are distressing. A further problem is that enterohepatic cycling of the drug occurs and sometimes causes relapse into depression or excitement about 8 hr after initial recovery. Despite these problems, the drug is a valuable asset when faced with having to castrate a horse single-handed and has seen extensive use in dart-gun capturing of wild animals, its considerable potency allowing the use of conveniently small injection volumes. In conventional situations, the drug is administered by intramuscular or intravenous injection. It is likely that the incidence of side-effects will limit the use of this drug to the dog, in which species they are minimal, and to other species only in situations where there is no alternative. It is possible that additions of further drugs to the schedule will allow the realization of etorphine's potential for chemical restraint, e.g. it has been shown that prior administration of propranolol to a horse greatly reduces the etorphine tachycardia. Etorphine narcosis can readily be deepened to full surgical anaesthesia with a volatile anaesthetic. As with thiambutene, the ready reversibility of etorphine's effects with the antagonist diprenorphine makes it a drug of great convenience for use in the small animal clinic.

ANALGESICS

Drugs of the opium series are used mainly for their narcotic or constipant effects in veterinary medicine. In human practice the use of these drugs for the control of severe and long-lasting pain is of major importance. Unfortunately, this pattern of use creates problems of dependence. This has stimulated many attempts to find a derivative or analogue which retains analgesic potency but which is free of addiction liability and respiratory depression. Sadly, although compounds of enhanced potency are available their side-effects are similarly enhanced. Nonetheless, some have found their way into veterinary use, e.g. pethidine as an analgesic, etorphine for its narcotic–cataleptic properties and fentanyl in the production of neuroleptanalgesia. It has, however, been found possible to separate cough suppression from the narcotic effects, e.g. in dextromethorphan.

Pethidine Hydrochloride, B.Vet.C.

Known in America as Meperidine Hydrochloride, U.S.P., pethidine hydrochloride is a synthetic white crystalline powder with little odour and a bitter taste. It is soluble in water, less soluble in alcohol and chloroform, and almost insoluble in other organic solvents.

Actions

Pethidine was originally developed as a smooth muscle relaxant. Although it is structurally unlike morphine, it has about one-tenth of its analgesic potency and shares morphine's other effects. Absorption is rapid. Following oral administration in dogs, the maximum analgesic effect is given in approximately 45 min and lasts 4 to 6 hr; after subcutaneous injection, maximal effects follow in about 20 min and persist for 3 to 4 hr. The duration of effect varies according to the severity of the pain—the greater the pain the shorter the period between doses.

The *antispasmodic or spasmolytic* action of pethidine is due to a relaxant effect on smooth muscle of the large intestine, the uterus, bladder and the bronchi, partly by an atropine-like action at the parasympathetic nerve endings but mainly by a direct action on muscle fibres similar to that of papaverine or narcotine. Through these two effects it will be seen that the greater the tonicity of the muscles the more marked will be the action. This is amply supported by clinical observations. The effect on the small intestine is one of decreased tonus which results in a delay in the passage of food though this is less marked than with morphine. Diphenoxylate is a synthetic drug structurally related to pethidine and selected for its useful separation of constipant from narcotic effects.

The *sedative* action is slight when compared with that of morphine, but hypnosis will follow the administration of large doses, especially if hyoscine is administered simultaneously.

Excretion is mainly via the kidneys after inactivation in the tissues either by hydrolysis or methylation. Traces will be found in saliva within 30 min of intramuscular injection and some may be found in the faeces. Foetal respiration is less depressed by pethidine so making it preferable to morphine around parturition.

Uses

The main advantage of pethidine over morphine is its usefulness as a sedative analgesic in cats: other advantages lie in its greater freedom from emetic and constipatory side-effects and its shorter duration of action. In horses the action is a little irregular, the sedative action being absent and transient excitement occurring occasionally, but it use in spasmodic colic has been recommended. As a premedication drug in dogs and cats it has been found quite useful.

Dosage

Intravenous injection is not recommended for animals owing to the vasomotor effects which may result.

All species: 3 to 5 mg per kg by mouth or by intramuscular injection.

Toxicity. This is relatively insignificant. Overdosage may cause a drop in blood pressure, and convulsions which can be counteracted by pentobarbitone sodium injections. Nalorphine hydrobromide will counteract respiratory and generalized depression in cases where overdosage causes sedative effects only.

HEAT REGULATION

The normal oxidation processes of the body are constantly producing heat: it is therefore in those tissues where oxidation is a more or less continuous process that most heat is produced. It has been shown that skeletal muscle is the most important site of heat production, whilst the liver is the next most important structure. During infection with bacteria, viruses, etc., the body tissues react strongly to the attack of the organisms and the increase in metabolism results in increased heat production; an interference with central control of temperature may occur and body temperature rises—this constitutes a fever or pyrexia. The blood is responsible for the removal of heat from the points of production and for its distribution to the various sites of elimination—the skin, lungs, kidneys, etc.

The normal methods of heat loss are:

1. Radiation, conduction and convection, from the body surface, which together account for about 60 per cent of the heat loss in man. These processes function only when the external temperature is below that of the body; equal temperatures produce equilibrium and higher temperatures reverse the flow of heat and warm the body. The skin circulation is the controlling factor in heat loss by these three means for a greater amount of blood circulating through the skin means more heat brought to the surface and therefore a greater chance of heat loss. As the dilatation of the vessels is controlled by the vasomotor mechanisms (vasomotor centre, adrenaline activity, etc.) this means of heat loss is directly related to nervous system reaction. In most animals the piloerector system is also involved, for erection of the hairs tends to restrict airflow over the skin and slows down the rate of heat loss.

2. The vaporization of water from the lungs and skin. About 25 per cent of heat generated during rest is lost by this method. In cows and the dog, which have poorly developed sweat gland systems, the diffusion of water through the skin at low temperatures is just as important as that lost by animals such as the horse, which have a well developed sweat gland system, but at high temperatures the cow, the sheep and the dog lose far more heat by loss of water through the lungs, and these species therefore develop polypnoea (a rapid respiratory rate) during hot weather.

3. The urine and faeces are also a means by which heat is lost.

Apart from the mechanisms controlling heat loss and heat conservation, there are centres in the hypothalamus which exert a controlling influence on the total production of heat, its subsequent distribution, and the dissipation of any heat in excess of requirements. Nuclei in the anterior portion of the hypothalamus are concerned with heat loss, whereas nuclei in the posterior hypothalamus appear to be associated with heat conservation and production. These areas are very sensitive to slight changes in the blood temperature and excessive heat will stimulate the anterior hypothalamus, while a fall in blood temperature will stimulate the posterior section: these two areas act reciprocally one being inhibited whilst the other is stimulated.

ANTIPYRETIC ANALGESICS

These non-steroidal anti-inflamatory agents

were originally introduced for the symptomatic control of pyrexia: they will not reduce a temperature rise due to physical conditions (e.g. heat stroke). They are also referred to as *febrifuges*. Such drugs also possess mild analgesic action.

Pyrexia or fever is the natural tissue reaction to infection and it is a doubtful practice arbitrarily to reduce a high temperature without at the same time reducing the infection. However, where the temperature is very high and persistent it may be advisable to reduce the temperature in order to conserve the body's energy reserves and prevent exhaustion.

Antipyretics act in the following ways:

1. By reducing heat production. This may be achieved either by inhibiting the heat production centre in the hypothalamus or by reducing the rate of oxidation in the tissues.

2. By increasing heat loss. The main methods of losing body heat are involved.

Increased losses follow dilatation of the cutaneous blood vessels; increased sweating causes a further loss as latent heat of vaporization; an increase in respiratory rate is followed by an increase in evaporation losses via the lungs. These reactions may result from stimulation of the temperature lowering, and inhibition of the heat production and conservation, areas of the hypothalamus: it is probable that antipyretics can compete with an endogenously liberated leucocyte pyrogen for receptors involved in thermoregulation.

The Salicylates

Sodium Salicylate, B.Vet.C.

This salt may be obtained either as pearly crystalline flakes or a white crystalline powder, both of which are odourless and have an unpleasant sweetish taste. Both forms are very soluble in water, though the flakes take much longer to dissolve in cold water.

Acetylsalicylic Acid, B.Vet.C.

This is aspirin and is available as tablets or a white crystalline powder which is only slightly soluble in water but fairly soluble in alcohol. Calcium salts are available commercially as Soluble Aspirin—**calcium acetylsalicylate.**

Both the sodium salt and the acetic ester of salicyclic acid react with the hydrochloric acid of the stomach and release a proportion of salicylic acid, which, though somewhat antiseptic in effect, is also irritant and may cause emesis in dogs. After absorption, sodium salicylate circulates as such whilst aspirin may circulate partly as sodium salicylate (the portion which is absorbed as salicylic acid) and probably also as acetyl-salicyclic acid. Both have an antipyretic and analgesic action, although aspirin is the more effective of the two. Both also have an effect on rheumatic conditions including joint and muscular types: in this respect the sodium salt appears to be more effective. The mechanism of this action remains in debate. However, there is a reduction in inflammatory exudate formation and swelling. An increase of glucocorticoid secretion, reduction in glucocorticoid binding and inhibition of hyaluronidase have been claimed as the basis for the peripheral component of the antipyretic's ability to reduce inflamation and control mild to moderate pain. More recently several of these drugs have been shown to inhibit the synthesis of prostaglandins. A mild central analgesic action is also present.

Excretion in the form of salicyluric acid is rapid and may be further encouraged by the simultaneous administration of sodium bicarbonate which will also reduce gastro-intestinal dissociation. During excretion, the salicylates make the urine positive to reducing tests for sugars.

Dosage

Dosage of sodium salicylate is 15 to 120 g (½ to 4 oz.) for the horse and cow, 1 to 4 g (15 to 60 gr.) for the sheep and pig, 0·3 to 1 g (5 to 15 gr.) for dog, 3 mg to 0·3 g (½ to 5 gr.) for the cat and up to 120 mg (2 gr.) for p poultry.

Dosage with aspirin is usually confined to small animals: Dog: 0·3 to 1 g (5 to 15 gr.), Cat: 0·1 to 0·3 g (1½ to 5 gr.). It is frequently administered as Compound Tablets of Codeine,

B.Vet.C., which also contain phenacetin and codeine phosphate.

Toxicity. Recently, salicylates have been shown to be toxic in the cat, perhaps due to its inability to form glucuronides. The use of salicylates is requuently accompanied by gastric irritation, ftinnitas, hyperpnoea and hypoprothrombinaemia.

Aniline Derivatives

This group contains three antipyretics of veterinary interest: acetanilide, phenacetin and paracetamol; only paracetamol is now used to any great degree.

Acetanilide, before the advent of sulphonamides, was widely used as an antipyretic for farm animals, mainly cattle, in doses of about 4 g (60 gr.) two or three times daily.

Phenacetin, B.Vet.C., with similar physical properties and dosage to acetanilide, was used mainly as an antipyretic with analgesic actions for small animals in doses of up to 0·3 g (5 gr.) three times daily for dogs, and up to 0·1 g (2 gr.) administered at the same rate for cats. It was incorporated in Compound Tablets of Codeine, B.Vet.C., which contained 0·25 g (4 gr.) each of aspirin and phenacetin and 8·1 mg ($\frac{1}{8}$ gr.) of codeine per tablet. Both acetanilide and phenacetin cause damage to liver and kidney and are now obsolete. Neither possess aspirin's anti-inflamatory potency and both yield hydroxylated compounds which cause haemolysis and methaemoglobinaemia.

Paracetamol is the active metabolite of phenacetin in man. It is a useful analgesic and is the best member of this series as it does not yield phenylhydroxylamine which is the metabolite responsible for methaemoglobin production. It is well absorbed from, and of a lower order of irritancy to, the gastro-intestinal tract.

All three compounds are central nervous system depressants; from this there stem analgesic and antipyretic actions. The latter effect is exerted entirely through depression of the heat regulating centres in the hypothalamus.

The main disadvantage of these phenolic compounds is their effect on the blood if dosing is prolonged. Overdosage can cause nephro- and hepatotoxicity. Paracetamol is the least toxic of the three compounds.

Pyrazolone Derivatives

Phenazone, B.Vet.C., or antipyrin, differs from the aniline derivatives in its chemical structure, though it contains a phenol ring, and in its water solubility (1:1). This compound is also an antipyretic and analgesic, though rarely used now. *Dosage* for horses and cattle is up to 16 g ($\frac{1}{2}$ oz.) twice daily.

Both phenazone and the closely related aminopyrine have been abandoned because of the occurrence of agranulocytosis following their use.

Phenylbutazone is a more modern representative of this group and is widely used in the control of chronic arthritic problems. It has salicylate-like antipyretic activity, but a rather slower onset of action. Unlike the salicylates this drug is available as an injectable as well as a tabletted form. It possesses analgesic and anti-inflammatory properties and provides symptomatic relief only. In man it causes gastric irritation, nausea, salt retention and oedema, skin rashes and leucopenia. The degradation of phenylbutazone is much more rapid in the dog than in man and this may account for the lower incidence of side-effects in this species.

Oxyphenbutazone is a recent addition to this series and is claimed to be less toxic than phenylbutazone of which it is a metabolite. Both of these drugs are extensively protein-bound and can cause inter-action toxicities.

Anti-inflammatory Effects

The antipyretics provide an interesting example of a group of drugs whose therapeutic indications have changed down the years as a result of continuing advances in the understanding of their properties.

Originally these drugs were synthesized in the days of antiseptic surgery as derivatives of phenol which might be capable of exerting internal antisepsis; it was later realized that their beneficial effect was not related to control of infecting organisms but to their non-specific antipyretic action. Then followed a long period of use in the management of febrile states in the era before the development of antibacterial drugs. When the non-specific control of fever became unnecessary with the availability of drugs with the capability of eliminating invading organisms, the antipyretics remained of interest for their mild analgesic effect and for their ability to increase the rate of excretion of uric acid (uricosuric action) which is of value in the management of gout. Their long-term use in the management of arthritic pain led to the realization that the drugs were also modifying the course of the disease. Similarly they could reduce the incidence of heart valve damage in rheumatic fever in man. Attention was thus focused on the anti-inflammatory action of the salicylates and the other antipyretic analgesics and it is now recognized that these compounds owe part of their analgesic effect to a direct peripheral anti-inflammatory action. The problem of systemic side-effects which accompany the long-term use of the anti-inflammatory glucocorticoids has caused research attention to be directed to the so-called non-steroidal anti-inflammatory drugs (NSAIDs). The outcome of these researches is the recognition that many NSAIDs owe their anti-inflammatory and perhaps their antipyretic action to their ability to modify the genesis or fate of the chemical mediators of the inflammatory response (see Chapter 15). Many impair the synthesis of prostaglandins from arachidonic acid by inhibiting the enzyme prostaglandin synthetase, e.g. salicylates, indomethacin, phenylbutazone, fenamates. The generation of kinins is impaired via an anti-plasmin action of salicylates, aminopyrine and phenylbutazone. Hyaluronidase is inhibited, so limiting the spread of inflammation, while inhibition of collagen formation will limit fibrosis.

Indomethacin represents one of the many rather unsuccessful attempts to replace the salicylates with a drug of superior efficacy and less toxicity. It is used in the management of gout and arthritis in man, but causes gastro-intestinal irritation and blood dyscrasias.

Fenamates are derivatives of anthranilic acid, e.g. flufenamic acid, sodium meclofenamate. Gut irritation, severe diarrhoea and blood dyscrasias occur in man where these agents are used if aspirin cannot be tolerated. Their ability to protect against anaphylaxis in the bovine is claimed to be due to suppression of SRS–A and kinin release.

Ibuprofen exemplifies a further group of anti-inflammatory analgesics various members of which are claimed to have fewer side-effects than aspirin. Nonetheless, severe gastro-intestinal bleeding has occurred following therapeutic dosage in man.

Nitrite Antipyretics

This group of antipyretics is now also rarely used in veterinary practice. The group consists of **amyl nitrite, sodium nitrite, ethyl nitrite** in the form of Spirit of Nitrous Ether, B.Vet.C. 1953, or Concentrated Solution of Nitrous Ether, B.Vet.C. 1953, and glyceryl trinitrate.

The actions of individual members of the group differ only in speed of onset and duration of effect. The action is one of dilatation of all peripheral blood vessels and also the coronary vessels. Amyl nitrite, which is administered by inhalation, exerts its action within a few seconds and effects pass off within 10 min, whereas the other preparations are effective within 10 min and continue their action for up to 2 hr. The vasodilatation causes a drop in blood pressure and an increased heart rate. Such drugs are used in man for the relief of angina pectoris. Vasodilation will relieve the myocardial ischaemia and also reduces peripheral resistance and, thus, the myocardial work load.

Dosage

Dosage of amyl nitrite for the dog is up to 0·3 ml (5 m.) by inhalation. It is rarely used.

Dosage of Spirit of Nitrous Ether, is up to 90 ml (3 fl. oz.) for the horse, up to 120 ml (4 fl. oz.) for the cow, and up to 15 ml (½ fl. oz.) for the sheep and pig. The concentrated solution should be diluted with seven times its volume of water and is then equivalent in strength to the spirit.

Dosage of sodium nitrite for use in conjunction with sodium thiosulphate (*q.v.*) is up to 2 g (30 gr.) for the horse and the cow, and up to 1 g (15 gr.) for the sheep, by intravenous injection. For its diaphoretic (sweat producing) and antipyretic effects it may be given in doses of up to 2 g (30 gr.) for horses and cattle, and up to 0·1 g (2 gr.) for dogs.

Dosage of glyceryl trinitrate for horses and cattle is 10 to 20 mg ($\frac{1}{6}$ to $\frac{1}{3}$ gr.) and for the dog is up to 0·6 mg ($\frac{1}{100}$ gr.).

Central Voluntary Muscle Relaxants

Mephenesin, B.Vet.C.

This compound is chemically 3-(2-methylphenoxy) propane-1-2-diol and is entirely different from the previous muscle relaxants mentioned but is related to the propanediol derivative tranquillizers. It is a white crystalline compound almost odourless but with a bitter taste followed by local anaesthesia. It is soluble 1:100 in water but freely soluble in organic solvents.

Actions

Mephenesin produces muscular relaxation following oral or intravenous administration. In normal doses facilitation and spreading of impulses up and down the spinal cord are reduced, but monosynaptic transmission is unaffected. This would account for the fact that muscular relaxation is neither so marked nor so complete as with tubocurarine and similar relaxants. In addition to muscular relaxation mephenesin has sedative, hypnotic, and ataractic effects: it also has some local anaesthetic action. Respiratory depression is unlikely unless large doses are given, when a fall in blood pressure and heart block may also be anticipated.

Subsequent to intravenous injection the compound takes effect quickly and lasts for up to 30 min, but the duration may be extended up to 2 hr by giving the drug orally.

Uses

Mephenesin may be used in cats and dogs as a muscle relaxant during anaesthesia, but is less satisfactory for this purpose than the agents already described.

Although the need for frequent administrations is inconvenient, mephenesin will abolish the convulsions of strychnine poisoning and it has been used in tetanus in the horse in oral doses of 50 to 70 g per 450 kg (1000 lb).

Toxicity. Side actions when given orally are nausea and emesis. Solutions given intravenously at concentrations greater than 5 per cent may cause intravascular haemolysis, haemoglobinuria, and venous thrombosis at the site of injection. In the dog overdosage, causing respiratory collapse, heart block, and a fall in blood pressure, should be treated by artificial respiration and the slow intravenous injection of methylamphetamine in doses of about 15 mg.

Dosage

Dog and cat:
- 22 mg/kg (10 mg or $\frac{1}{6}$ gr./lb.) given intravenously to effect
- 0·5–1 g ($7\frac{1}{2}$–15 gr.) orally three times daily.

Bibliography

BORISON, H. I. & CLARK, W. G. (1967) Drug actions on thermoregulatory mechanisms. *Adv. Pharmac.*, **5**, 129–303.

BRANCKER, W. M. (1963) Sedation of cats. *Adv. small Anim. Pract.*, **4**, 9–13.

CROSSLAND, J. (1967) Psychotropic drugs and neurohumoral substances in the central nervous system. *Prog. med. Chem.*, **5**, 251–309.

LIM, R. K. S., ARMSTRONG, D. & PARDO, E. G. (ed.) (1968) Pharmacology of pain. In: *Proceedings of the Third International Pharmacological Meeting 1966*, vol. 9. Oxford: Pergamon.

LUDWIG, B. J. & POTTERFIELD, J. R. (1971) The pharmacology of the propanediol carbamates. *Adv. Pharmac. Chemother.*, **9**, 173–232.

MERCIER, J. (ed.) (1973) *International Encyclopaedia of Pharmacology and Therapeutics, Section 19, Anticonvulsant Drugs*, vols 1 and 2. Oxford: Pergamon.

MINDHAM, R. H. S. & SHEPHERD, M. (1973) Clinical psychopharmacology. *Rec. Adv. Med.*, **16**, 293–333.

OSWALD, I. (1968) Drugs and sleep. *Pharmac. Rev.*, **20**, 274–96.

22

Non-Volatile Anaesthetics

Veterinary anaesthesia has received much attention during the last three decades and great advances have been made. For further detailed study of anaesthesia and anaesthetic techniques reference should be made to *Wright's Veterinary Anaesthesia and Analgesia*, 7th ed. (Ed. by L. W. Hall) Baillière Tindall (1971).

Anaesthetics are agents which eliminate body sensation. The term anaesthesia is now broadly applied in the human surgical field to a complex procedure by which the patient is prepared psychologically for the operation and then rendered insensible to all feeling in the operation area; by which blood pressure and oxygenation of the blood are maintained at their optimum levels, by which physiological means are used to keep the field of surgery clear of blood, and by which muscle tonus is decreased so that manipulations may be carried out easily and quickly. This procedure involves extensive premedication and the administration of ancillary therapeutics after anaesthesia has been induced. Drugs are used which aim at achieving the following:

1. A reduction in the likelihood of the incidence of shock: (a) By lowering the blood pressure and preventing the physiological constriction of the peripheral blood vessels: this maintains normal capillary function under conditions which would otherwise result in peripheral vasoconstriction followed by capillary anoxia and an increased vascular permeability—the normal reaction to traumatic shock. Concurrently with the use of these hypotensive drugs the blood volume must be maintained by the injection of blood-volume expanders should a state of shock appear imminent. *Note.* These hypotensive agents should not be administered after the onset of shock, for they will then only serve further to reduce the blood pressure, possibly to a fatal level. (b) By reducing apprehension and voluntary struggling due to fear.

2. A reduction in the amount of anaesthetic required: (a) by lowering the level of brain activity, and (b) by the use of muscle relaxants.

3. A reduction in salivary and respiratory secretions.

4. A reduction of haemorrhage in the operation field. This is achieved by the use of the hypotensive drugs mentioned under 1(a) above.

Hexamethonium bromide and chlorpromazine have hypotensive actions, the long and medium acting barbiturates, chloral hydrate, morphine, pethidine, and also chlorpromazine, have general sedative effects, and in addition some have analgesic actions. Muscle relaxants include tubocurarine chloride and succinylcholine, whilst atropine and hyoscine (scopolamine) have a depressant action on secretions.

This system of balanced anaesthesia, while offering many advantages over simple anaesthesia, is technically cumbersome and therefore not widely used under general practice conditions.

There are two main groups of anaesthetics:

1. *General anaesthetics* which, besides causing the loss of sensation and the abolition of voluntary motor response, also induce sleep, or a level of central nervous system (CNS) depression equivalent to sleep.

2. *Local anaesthetics* which merely desensitize one area of the body and have no direct effect on the CNS.

GENERAL ANAESTHETICS

An ideal general anaesthetic would possess a number of essential characteristics. It must: (1) be easily and painlessly administered; (2) induce a rapid loss of consciousness without causing voluntary or involuntary struggling; (3) not cause any physiological changes such as a drop in blood pressure, depression of respiration and heart action, or an increase in respiratory and salivary secretions; (4) give adequate analgesia and muscular relaxation at the minimum level of dosage required to cause loss of consciousness; (5) be completely non-toxic; (6) be easily neutralized by a non-toxic antidote so that the duration of anaesthesia can be shortened at will; (7) have a short recovery period without excitement; (8) be compatible with premedication and other ancillary therapeutics; and (9) be reasonably priced and non-explosive.

No single anaesthetic is known which possesses all these attributes but by combining drugs with complementary pharmacological actions, an approximation to the ideal can be achieved. This combination of complementary drugs is now spoken of as 'balanced anaesthesia'. For example, *premedication* with relaxing, analgesic and CNS sedative drugs such as pethidine or morphine, along with hyoscine, *a light anaesthesia* produced by some general anaesthetic such as pentobarbitone sodium, and a *muscle relaxant* such as tubocurarine chloride or suxamethonium chloride, will together produce something approaching perfect anaesthesia.

In this suggested combination for balanced anaesthesia the pethidine and morphine provide basal narcosis (CNS sedation at the level of narcosis prior to the induction of anaesthesia by another substance), provide analgesia and also lower the amount of pentobarbitone sodium required. This lower dosage of anaesthetic means a duration of anaesthesia and period of recovery shorter than that which follows the use of the barbiturate alone. The hyoscine, besides taking part in the CNS sedation also, by its atropine-like effect, reduces respiratory and salivary secretions. The muscle relaxant ensures relaxation of abdominal muscles with only light barbiturate anaesthesia and greatly facilitates intra-abdominal manipulation or visualization.

There are still great imperfections attaching to the use of these combinations. In most cases, for example, paralysis of the respiratory muscles occurs with muscle relaxants and positive pressure breathing apparatus is required: this, even in its simplest form, requires special apparatus, There are also dangers inherent in the induced muscle relaxation, including that of over-extension of joints and the stretching of nerves, ligaments, and tendons, when the animal is tied to the operating table.

However, in small animal surgery these methods of anaesthesia are becoming established as the best that have so far been devised. In large animals, the use of muscle relaxants is still under discussion.

The general anaesthetics now available each possess one or more of the attributes of a perfect anaesthetic. They are classified into three types according to their physical characters:

1. Non-volatile anaesthetics which are mostly water soluble solids and must be given orally or by parenteral injection:
   (1) The barbiturates.
   (2) Chloral hydrate.
2. Volatile anaesthetics which are liquids. They give off vapours which have anaesthetic properties and are therefore administered by inhalation:
   (1) Chloroform.
   (2) Ether and vinyl ether.
   (3) Trichloroethylene.
   (4) Halothane.
   (5) Methoxyflurane.
3. Gaseous anaesthetics, which are stored under pressure in some cases as liquids, but which are administered as gases at normal temperature and pressure:
   (1) Nitrous oxide.
   (2) Cyclopropane.

Control of Anaesthesia

During the *induction* of anaesthesia and *recovery* the depth of effect is determined by repeated tests of the animal's reflex response to stimuli and the degree of relaxation of appendages such as the tail.

During *anaesthesia* the depth of anaesthesia and the circulatory condition are determined by frequent examination of mucous membranes, pupillary and other reflexes, respiratory rate, depth and regularity, pulse character and rate.

Reflexes in Anaesthesia

The majority of anaesthetics are spoken of as 'descending anaesthetics', indicating that the reflexes of the head and forepart of the body are eliminated before those of the hind-quarters.

Anaesthetic reflexes in the approximate order of disappearance:

1. (a) Ear twitch, and (b) jaw and tongue.
2. Front pedal.
3. Body reflexes—scratch and panniculus response.
4. Hind pedal.
5. Tail relaxation.
6. Palpebral.
7. Corneal.
8. Pupillary.

The stimuli which initiate these reflexes are touch in the cases of (1a), (6) and (7), forcible opening of the mouth in (1b), light in (8) and pain in (2), (3) and (4). The jaw reflex is not readily applicable in large animals.

The most important are the corneal and pupillary reflexes for these give the final warnings against overdosage. In light anaesthesia the eyelids are open, the corneal response (closure of lids on touching the cornea) is rapid and the pupil is usually dilated, but will frequently respond to light. In deep anaesthesia the pupil is tightly constricted and the lids are closed: in dogs and cats the corneal reflex may be absent, but in horses the corneal reflex may persist until death is imminent. Sudden dilatation of the pupil is an indication of overdosage and that death is imminent if not already a fact.

Control of Oxygenation

The state of oxygenation and circulatory well being are normally estimated by examination of the oral and ocular mucous membranes, the pulse and respirations. The membranes should be light pink in colour, not dark red or blue; the vessels should not be engorged and pressure on the gum should produce a white mark which disappears within a few seconds. It must be remembered, however, that where hypotensive agents such as the hexamethonium salts are used, the blood pressure will be low and mucous membrane recovery will be relatively sluggish. Precautions relevant to oxygenation are intubation, whereby a tube is introduced into the trachea to ensure a clear air flow, premedication with atropine or similar drug to reduce salivary and respiratory secretions, and positioning the animal with its head lower than the body so that fluid, whether of gastro-intestinal or salivary and respiratory origin, will drain out of the mouth and not back into the lungs.

NON-VOLATILE ANAESTHETICS

The use of non-volatile anaesthetics is sharply divided between the species, chloral hydrate being used exclusively in the large animal field and the barbiturates being used almost entirely in dogs and cats, although their use in farm animals and exotic species is increasing. Nevertheless, these anaesthetics have certain group characteristics, though each individual anaesthetic has attributes which are not shared by the rest of the group.

The non-volatile anaesthetics can be given by the oral or parenteral routes. The intravenous route is that most commonly used because it alone allows the maximal exploitation of one of these anaesthetics' greatest advantages, i.e. a very fine control of dosage and the ability accurately to adjust the degree of response to the required level. The other routes introduce their customary problems of slow and perhaps irregular absorption. Additionally, induction of anaesthesia can be rapidly achieved without sympathetic nervous system activation. Nor is there stimulation of bronchial and salivary secretion. There is no fire or explosion hazard with these compounds nor do they yield toxic decomposition products. Postoperative vomiting is less prevalent and the drugs are not specifically damaging to liver, kidney or heart.

A further attribute common to chloral hydrate and most of the barbiturates is their suitability as basal narcotics prior to the induction of anaesthesia by a volatile, gaseous or local anaesthetic. Chloral hydrate has been widely used for this purpose in cattle and horses and pentobarbitone sodium is useful in small animals, whilst thiopentone sodium is widely used for inducing anaesthesia which is then maintained with nitrous oxide, ether, cyclopropane or halothane.

With the barbiturates, the induction of anaesthesia is usually rapid and unaccompanied by struggling, providing the correct techniques are used, but with chloral hydrate struggling occasionally occurs. A disadvantage of considerable importance which applies to chloral hydrate and the short and medium acting barbiturates is the length of the recovery period, during which excitement and voluntary struggling may cause considerable distress and even death in some species: the convalescent period may be lengthened and damage to, and breakdown of, the operation wound may occur. In some cases the recovery period may be so long that frequent turning over may be required to prevent the occurrence of hypostatic pneumonia.

Neither chloral hydrate nor the barbiturates are satisfactory as analgesics and in balanced anaesthesia pethidine or morphine are used for their additional analgesic effects. Similarly, neither of them produces good muscular relaxation and for this reason they are frequently used only as narcotics, volatile or gaseous anaesthetics being used to bring about anaesthesia.

All non-volatile anaesthetics are detoxicated and eliminated by the liver and kidneys, and where these organs are functioning inadequately prolonged anaesthesia and slow recovery may result. This is not a problem with anaesthetics given by inhalation which are also excreted via the lungs.

Foetal anaesthesia due to placental diffusion makes all non-volatile anaesthetics unsuitable either as narcotics or anaesthetics for Caesarian section.

Greater skill is required in the administration of the non-volatile anaesthetics but the injections can be given with the help of unskilled assistance and once the required depth of anaesthesia is reached no further help is needed. In emergencies, anaesthesia can be induced without any assistance whatsoever if non-volatile anaesthetics are used.

Chloral Hydrate, B.Vet.C.

This white crystalline substance has a characteristic pungent odour sometimes compared with that of melons. The crystals gradually turn yellow if kept in clear glass containers for long periods: they are very soluble in water, alcohol and other organic solvents: exposure of solutions to light over a period may result in some decomposition, and alkalis such as the barbiturate sodium salts cause rapid decomposition with the formation of chloroform and a formate. Solutions should therefore be prepared immediately before they are needed.

Actions

The anaesthetic effect following the administration of chloral hydrate is attributed to trichlorethyl alcohol produced by reduction following absorption. It was thought that chloral hydrate liberated chloroform in the body and that the anaesthetic effect was due to the chloroform. Because of this is has long been believed despite evidence to the contrary that chloral hydrate causes liver damage.

The *systemic* effect of chloral hydrate is one of CNS depression affecting the motor impulses of the brain and spinal cord. The sensory routes are only mildly affected short of anaesthesia. This selectivity results in a loss of motor function before sensory perception ceases, and incoordination due to loss of muscle control may result in fright and struggling, with possibly serious consequences. It is for this reason that, in horses at any rate, there should be a minimum of noise and the eyes should be covered during induction and the early recovery stages. When the stage of anaesthesia is reached dosage is such that the medullary centres and particularly the respiratory and vasomotor centres are depressed: this results in a slowing of respiration and peripheral vasodilatation with a drop in blood pressure. There is no direct action on the heart except where toxic doses are given, and occasionally during the recovery stages (see 'Toxicity'). In doses short of those required to give anaesthesia there is little analgesic (pain killing) effect except in so far as cerebral perception is dulled. The *local* effect of chloral hydrate is one of irritation, whether given by mouth, applied externally in liniments or inadvertently injected extra-vascularly. External applications and oral dosage make use of this action as well as the slight local anaesthetic effect.

Uses

Chloral hydrate is used mainly in the horse for: (1) a *light sedative* action, as in the treatment of colic, (2) a *basal narcotic* effect prior to anaesthesia with volatile or local anaesthetics, and (3) *general anaesthesia* by intravenous injection either alone, with magnesium sulphate, or with magnesium sulphate and pentobarbitone sodium.

It is also used in the cow as a treatment for acetonaemia but only rarely for other purposes. Chloral hydrate should not be used during Caesarian section, for it passes the placental barrier and depresses foetal respiration.

For *sedative effects*, chloral hydrate may be administered as a drench, as a bolus, in drinking water after a period of water starvation (18 to 48 hr), or given by stomach tube. It is useful in colic for its central and local sedative, carminative and antizymotic actions: frequently it is combined with chlorodyne in doses of ½ fl. oz. of each (15 g and 15 ml), given every 6 hr, or given as a single dose with aloes and turpentine in a vehicle of linseed oil. As a sedative for fractious horses which object to clipping, or a fractious bull which must be handled 15 to 45 g (½ to 1½ oz.) chloral hydrate should be given orally in water about half an hour before, or by intravenous injection immediately before handling. In the treatment of acetonaemia 30 g (1 oz.) has been given initially followed by 15 g (½ oz.) twice daily, further dosing at reduced rates following according to reaction: the use of chloral hydrate is only indicated in the 'nervous type' of acetonanaemia.

For *basal narcosis*, chloral hydrate may be given in drinking water after starvation, by stomach tube, or by intravenous injection; when the first two methods are used, dosing should take place about 20 min prior to anaesthesia, and the dosage should be at the rate of 3 to 6 g per 50 kg (45 to 90 gr. per cwt.) body weight in all species. Narcosis varies when the oral route is used, and intravenous dosage to effect is preferred. Subsequently, anaesthesia may be induced by a volatile or local anaesthetic. In light narcosis the animal will remain standing, but will become cast in medium and deep narcosis: the last of these approaches light anaesthesia. Narcotic effects will persist for up to 2 hr, depending on dosage and route of administration.

For *anaesthesia*, chloral hydrate should be administered only by the intravenous route. It has been used for this purpose in horses, cattle, sheep and pigs, but is rarely used for any species other than horses and cattle. For intravenous injection a 10 per cent solution [60 g (2 oz.) in 600 ml] in water or saline should be used and it is usual to include 1 per cent sodium citrate (6 g per 600 ml or 90 gr. per 20 fl. oz.) as an anticoagulant. On all occasions it is advisable to make up more solution than is required according to the estimated dose: this is the normal practice in all cases of non-volatile anaesthetic administration; it is intended as an emergency reserve for use in animals which are idiosyncratic and tolerate doses greater than the average.

Injection should be carried out by gravity apparatus in the normal manner over a 5 to 10 min period, but extra care must be taken to see that the needle is in the vein and that its point is some distance from the point of entry. These precautions, aimed at preventing perivascular leakage, are necessary because of the strong irritant effect of the hydrate on the tissues.

A further important point which must be observed during chloral hydrate administration is that injection should cease at a point when the depth of anaesthesia is just short of that required, as anaesthesia continues to deepen for 10 to 15 min after injection has stopped.

Following anaesthesia the animal may be recumbent for 3 hr or more.

Excretion is via the kidneys in the form of urochloralic acid which is formed in the liver by the conjugation of chloral hydrate with glucuronic acid; some is excreted unchanged. The presence of urochloralic acid in the urine causes a strong Fehling's or Benedict's reduction which should not be interpreted as a sign of glycosuria.

If the drug is administered by the oral route the animal should be starved for up to 24 hr in order to empty the stomach and small intestine so that absorption, and therefore nervous system depression, may not be affected by dilution and dispersion in the food.

Whether orally administered, injected intravenously or intraperitoneally (this route has been used in pigs but it is not the one of choice), chloral hydrate should be well diluted in order to avoid irritation.

Dosage and Administration

1. *Oral*
   Sedative:
   Horse and Cow: 15–45 g (½–1 oz.)

Sheep and Pig: 2–4 g (30–60 gr.)
Dog: 0·3–1 g (5–15 gr.)
Cat: 0·12–0·6 g (2–10 gr.)

Narcosis (Wright):

Horses

Light	Hunter	37 g ($1\frac{1}{4}$ oz.)
	Vanner	45 g ($1\frac{1}{2}$ oz.)
	Heavy draught	52 g ($1\frac{3}{4}$ oz.)
Medium	Hunter	52 g ($1\frac{3}{4}$ oz.)
	Vanner	60 g (2 oz.)
	Heavy draught	75 g ($2\frac{1}{2}$ oz.)
Deep	Hunter,	67 g ($2\frac{1}{4}$ oz.)
	Vanner	75 g ($2\frac{1}{2}$ oz.)
	Heavy draught	90 g (3 oz.)

2. *Intravenous injection*

Anaesthesia:

Farm animals: 6–9 g/50 kg (90–120 gr./cwt.)

Narcosis:

Light, one-half	of the anaesthetic dose
Medium, two-thirds	
Deep, five-sixths	

In all cases doses will be judged according to effect.

Toxicity. Chloral hydrate anaesthesia has a narrow safety margin, a disadvantage made the more sinister when its poor analgesic and muscle relaxant properties are recalled. Existing heart, liver or kidney disease increases the anaesthetic risk.

Apart from overdosage, and death from respiratory failure or occasionally circulatory collapse, the main danger in chloral anaesthesia results from cardiac fibrillation and sudden death during the recovery period. This cardiac fibrillation and heart block has been attributed to vagal action. The vagus reacts reflexly to the increase in blood pressure which follows the release of adrenaline by fright and struggling during the recovery period. Because of this *adrenaline is contra-indicated as a stimulant in chloral hydrate anaesthesia* and animals must not be encouraged to rise until it is reasonably certain that they can do so and remain on their feet.

In an endeavour to overcome the disadvantages of chloral hydrate various combinations are used, including mixtures with magnesium sulphate alone or along with pentobarbitone sodium.

Combined Formulations

Chloral Hydrate–Magnesium Sulphate

There are several suggested formulations containing various proportions of chloral hydrate and magnesium sulphate. These proportions vary from 3:1 to equal parts, but probably the most popular is the proportion of 2:1 in a solution containing 12 per cent of chloral hydrate and 6 per cent magnesium sulphate. This solution is used only in the horse and must be given slowly (30 ml per min) if overdosage is not to occur, for the progressive deepening of the depression found with chloral hydrate also occurs with this mixture, and there is the added danger of magnesium depression of the heart and respiration if too rapid an injection is given.

The magnesium in this formulation gives a more rapid induction of anaesthesia with less excitement, and the depth of anaesthesia is increased if equal volumes of the plain solution and the combined formulation are compared. As magnesium sulphate is excreted rapidly, duration of anaesthesia is less (30 min) and recovery is more rapid, incoordination being less and of shorter duration because of the reduced dose of chloral hydrate.

Dosage

Dosage is according to reaction, allowance being made for the post-injection deepening of anaesthesia. The quantity of a 12 to 6 per cent solution required varies from 200 to 400 ml (7 to 14 fl. oz.) according to size and condition of the horse. Injection should take a minimum of 10 min.

This mixture of chloral and magnesium sulphate is not as safe as one would wish and the addition of pentobarbitone sodium has been used as a means of reducing further the proportions of each component of the anaesthetic and therefore reducing the toxic effect of the combination.

Chloral–Magnesium–Barbiturate

The original solution of Millenbruck and his colleagues contained 30 g (1 oz.) chloral hydrate, 15 g ($\frac{1}{2}$ oz.) magnesium sulphate and 6·4 g (100 gr.) pentobarbitone sodium, per litre of aqueous solution. A further modification doubled and quadrupled the amount of chloral hydrate and magnesium sulphate (60 g with 30 g or 120 g with 60 g, respectively, per litre), retained the same dose of pentobarbitone salt, and added 200 ml ethyl alcohol per litre. The addition of alcohol made the product more stable and the more highly concentrated solution made the volume required very much lower, which in turn decreased the injection period. This solution must be prepared and used within an hour or so because of the hydrolysis of the pentobarbitone sodium and its reaction with the chloral. More stable solutions containing the acid barbiturate in place of the sodium salt, and using high proportions of propylene glycol and ethyl alcohol, are marketed commercially, usually in concentrations similar to those above.

This combination of drugs combines the rapid induction properties and muscular relaxation of the barbiturate with the short anaesthetic period (30 min) and rapid recovery rate of the chloral–magnesium mixture. Return of consciousness is approximately paralleled by return of muscle control.

Dosage and Administration

Horses and cattle: approximately 650 ml of the Millenbruck solution. Proportionately less will be required of the more concentrated solutions.

Administration follows the pattern of barbiturate anaesthesia in small animals in which half the dose is given fairly rapidly (5 min) and the remainder slowly until the required depth of anaesthesia is reached. Half the dose is usually sufficient to approach medium narcosis. In cows it is advisable to cast the animal by Reuff's method before attempting anaesthesia. Where animals are cast prior to anaesthesia injections can be made slowly, as the initial 'knock-down' effect is not required.

THE BARBITURATES

This group of organic CNS depressants are either white or yellow (sulphur containing compounds), odourless, bitter tasting crystalline powders. The acids are only slightly soluble in water, but are soluble in alcohol and other organic solvents: their sodium derivatives are very soluble in water, usually soluble in alcohols, but insoluble in other organic solvents; these sodium salt solutions are frequently highly alkaline. All powders are hygroscopic and hydrolyze in aqueous solution; they are also decomposed by air, heat and light.

Traditionally, the barbiturates are classified according to their duration of effect, for in most other respects they have few distinguishing features, their mode of action in particular being the same for all members of the group. There are four classes, as shown in the table at the foot of the page.

The long and medium acting compounds are usually given as the parent acid but the short and ultra-short acting compounds are normally administered as sodium salts by injection.

Chemistry

The whole group are based chemically on barbituric acid (malonyl urea) which has a pyrimidine ring as a nucleus. Its formula may be written in two ways: (i) by the '*keto*', and (ii) by the '*enol*' method:

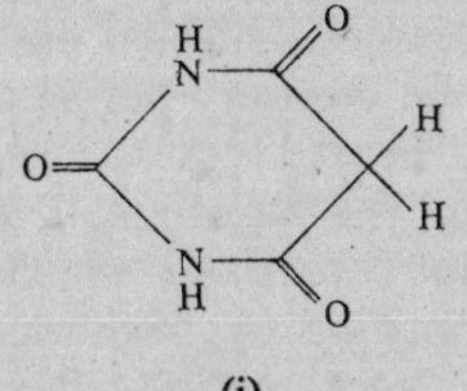

(i)

Class	Compound	N_1	C_2	R_1	R_2
1. Long acting	Phenobarbitone	H	O	C_2H_5	C_6H_5
	Methylphenobarbitone	CH_3	O	C_2H_5	C_6H_5
2. Medium acting	Butobarbitone	H	O	C_2H_5	C_4H_9
3. Short acting	Pentobarbitone	H	O	C_2H_5	$(CH_3)C_4H_8$
4. Ultra-short acting	Methohexitone Sodium	CH_3	O	C_3H_5	$(CH_3)C_4H_6$
	Thialbarbitone	H	S	C_3H_5	C_6H_9
	Thiamylal Sodium	H	S	C_3H_5	$(CH_3)C_4H_8$
	Thiopentone	H	S	C_2H_5	$(CH_3)C_4H_8$

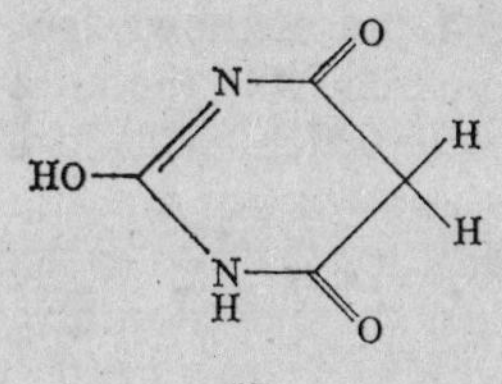

(ii)

The enol tautomer predominates in an alkaline environment. In the British usage all barbiturate names end in '-one' while the American equivalent name ends with 'al'. The basal radical of the salts is usually shown at the C_2 oxygen.

The chemical structure of the barbiturates is varied by means of substitution of organic or combined inorganic and organic groups at one, two or all of three sites in the acid structure: (a) nitrogen 1; (b) carbon 2; and (c) carbon 5. The whole group may therefore be represented by the graphical formula (iii).

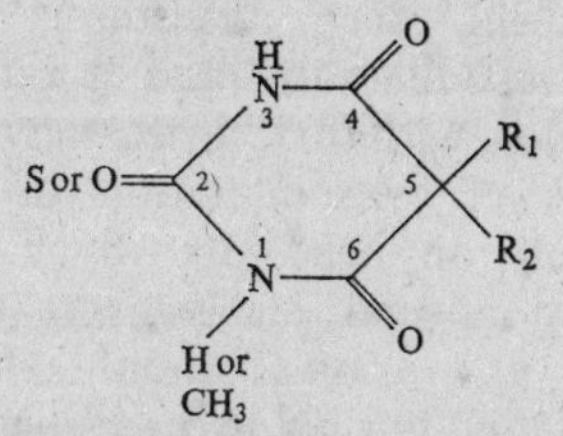

(iii)

Action. The action of the barbiturates can be divided into two distinct parts, each of which is closely identified with a certain part of the chemical structure.

1. The CNS depressant action, which is common to the whole group, is related to the common barbituric acid nucleus, and if this nucleus is disrupted the depressant action is lost. It is interesting that barbituric acid alone has no depressant action, but, when aliphatic and/or cyclic groups are substituted for the hydrogen atoms at R_1 and R_2, the resulting compound has a CNS depressant action.

2. Just as the character of the group is linked with the nucleus, so are the individual differences linked with the side-chains that may be substituted in any or all of the sites indicated above. These side chains are responsible for individual differences in metabolism and therefore determine rates of absorption, detoxication, and excretion, and hence duration of depression, dosage and indications, etc.: they have very little influence on the nature of the action, affecting only the degree.

For example, a barbiturate with the shortest and most stable side chains commensurate with activity will be long acting, as with Phenobarbitone, B.Vet.C. (iv) where hydrogen atoms are attached at nitrogens 1 and 3, oxygen at carbon 2, and ethyl and phenyl groups at carbon 5.

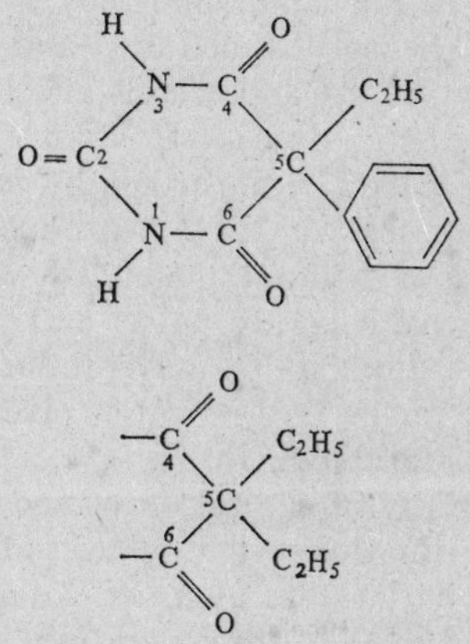

(iv) (a) Phenobarbitone (b) Barbitone

Conversely, if nitrogen 1 has a methyl group, carbon 5 having one relatively long unsaturated or branched chain, and/or carbon 2 has sulphur substituted for oxygen, the resulting compound will be metabolically unstable and therefore short acting, as with thiopentone (v).

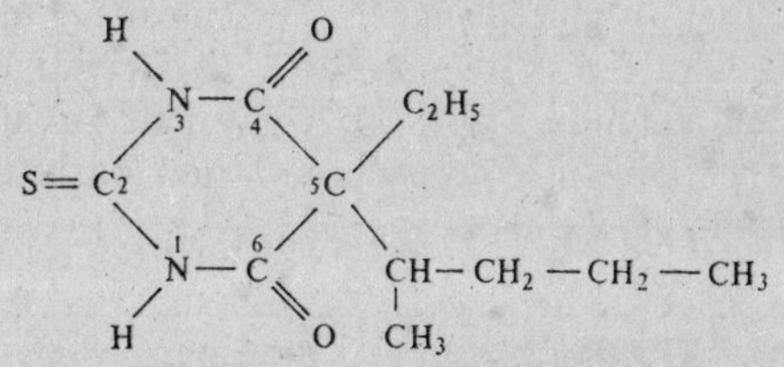

(v) Thiopentone

Between these two extremes, by variation of the side chains, a whole scale of compounds can be produced with a more or less predictable and graduated degree of effect.

CNS Depressant Action

The action of the barbiturates on the brain cells is generally accepted as one of interference with oxidation processes, probably in connection with the breakdown of glucose, but the precise details are unknown. Alternatively, barbiturates have been shown capable of interfering with axonal, polysynaptic and ganglionic transmission and of reducing transmitter release.

The clinical action is one of CNS depression, mainly of the motor functions of the brain and spinal cord, preceded, in the case of the medium and short acting compounds, by a period of excitation. Sensory functions, and therefore pain perception, are appreciably affected only by near anaesthetic doses (cf. chloral hydrate). These details may be illustrated by reference to phenobarbitone, which will produce marked incoordination without materially dulling the animal's sensitivity to pain, and to pentobarbitone sodium which may produce narcotic excitement in the dog or cat if given by slow intravenous injection.

The intensity of clinical effect and its rate of onset and duration will vary according to the individual barbiturate used, its dosage, and its method and route of administration. Individuals of the same species, of approximately the same body weight and under identical conditions may require appreciably different doses to achieve the same level of CNS depression. Moreover, the same individual can vary in its response from one occasion to another.

The central nervous system actions are prolonged by the administration of many derivatives of glucose, especially phosphate derivatives, and occasionally glucose itself. This means that the administration of glucose and its derivatives in plasma extenders for the treatment of shock must be carefully considered when barbiturate anaesthetics are used.

Other Systems

The barbiturates have no appreciable effect on systems other than the respiratory and cardiovascular systems, except in near toxic doses. During the induction of anaesthesia, the injection of barbiturates is almost invariably accompanied by a temporary cessation of respiration (apnoea); this is an acknowledged fact and is particularly apparent with the ultra-short acting barbiturates, with which the induction period is of only one or two seconds duration. In the normal animal this apnoea is only temporary and respiration restarts spontaneously: where the apnoeic period is prolonged artificial respiration should be applied until the breathing becomes spontaneous once more. The phenomenon is due to a temporarily high concentration of barbiturate in the brain which affects the respiratory centre, but as soon as the drug is adequately circulated the temporary high concentration becomes reduced to a sub-paralytic level. The cessation of respiration is not the catastrophe that one normally associates with respiratory paralysis, for in barbiturate over-dosage the heart will continue to function satisfactorily for sufficient time to allow the institution of normal respiratory resuscitation—artificial respiration and analeptics.

Depression of the vasomotor centre in the medulla during deep anaesthesia results in peripheral vasodilatation and a fall in blood pressure which, with toxic doses, may be reduced to shock level. Vasodilatation of peripheral vessels results in heat loss which cannot be replaced by normal mechanisms because of anaesthesia and animals should therefore be kept reasonably, but not excessively, warm.

General Pharmacology

The individual metabolic fate determines whether the compound is a long, medium, or a short acting barbiturate. This duration bears an inversely proportional relationship to the reversible binding of the barbiturate to plasma albumin or to its solubility in fat which means that long acting compounds are adsorbed onto protein in the lowest degree, whilst the ultra-short acting compounds are those with the highest fat solubility. *Long acting barbiturates*, of which phenobarbitone is probably the most widely used, are readily and completely absorbed from the intestines but, whether given orally or parenterally, the onset of effect is slow and in general the depth of CNS depression is not so great as that caused by the shorter acting compounds. The duration of effect is prolonged due to the slow rate of renal clearance, for no disruption of the molecule occurs in vivo and the compounds are excreted via the kidneys unchanged. It takes from 7 to 10 days for a single dose of barbitone to be eliminated; consequently, this and the other long acting compounds are cumulative and should be used circumspectly when dosing should be continued for long periods. If any signs of incoordination are shown the daily dosage rate should be cut immediately. Any renal blockage will seriously impair excretion, and prolonged action will result.

Medium and short acting barbiturates are rapidly and well absorbed when given by any route. Their effects are relatively rapid in onset and their duration limited to a period of a few hours. CNS depression can be varied from light sedation to deep anaesthesia by varying the dosage or route of administration. Unlike the long acting group, these barbiturates undergo detoxication mainly in the liver and are excreted neither completely disintegrated nor unchanged: the metabolic products vary according to the individual barbiturate, but from radioactive isotope investigation it would appear that oxidation of the longer chains at carbon 5 is the most important method of detoxication. These medium and short duration barbiturates are therefore not mainly dependent, as are the long acting compounds, on renal clearance for a limitation of the duration of effect: this responsibility is shared with the liver. Consequently any liver damage will slow up the detoxication of these barbiturates and activity will be prolonged.

The *ultra-short acting* barbiturates can be absorbed from the gut but the intravenous route is the only one for which these drugs are intended. As with all barbiturates, the drug distributes readily throughout total body water, binds to tissue proteins and dissolves in fat cells. Their special virtue, however, is the speed with which they achieve entry to brain tissue, highest initial concentrations being found in the most vascular areas. Other highly perfused tissues, e.g. heart, liver, kidney and spleen, achieve peak concentrations within seconds of administration.

The very short (10–20 min) duration of action of these barbiturates is not due to their rapid metabolic inactivation but to the much slower rate at which the less vascular tissues achieve equilibrium with plasma concentrations, e.g. skin, fat gut and muscle. After a single injection of thiopentone, the only source for drug to replace that lost from plasma into less vascular tissues is the more vascular, more easily saturable tissue, especially the brain. By this means drug is leached out of the brain and clinical recovery takes place.

Toxicity

Normally these drugs are not very toxic but cumulation of the long acting compounds may occur: this is indicated by increased general depression and the onset of incoordination. With the short and ultra-short acting barbiturates, toxicity is mainly that of CNS depression with respiratory and circulatory collapse due to temporary or permanent overdosage, the former resulting from too rapid an injection rate. Treatment consists in the first place of artificial respiration, if the heart is still beating, and the administration of analeptics (convulsants) to stimulate the respiratory and vasomotor centres in the medulla.

Individual uses, dosages and administration. In general, the long and medium duration barbiturates are used orally as anticonvulsants and routine CNS sedatives: the short acting compounds may be used by all routes as anticonvulsants and sedatives, but their main indication is for intravenous injection as basal narcotics and general anaesthetics: the ultra-short duration barbiturates are of no use other than as general anaesthetics for intravenous administration.

Long Acting Barbiturates

These comprise phenobarbitone, methyl phenobarbitone, and their sodium salts; barbitone is also available but is now rarely used in veterinary medicine.

Phenobarbitone, B.Vet.C., Phenobarbital, U.S.P. (5-ethyl, 5-phenyl barbituric acid)

This barbiturate is a typical example of a long acting barbiturate. It is therefore absorbed slowly and excreted slowly; between these pharmacological mile-posts is a long period of CNS sedation, for excretion of a single dose takes several days.

Uses

The main use of phenobarbitone is as an anticonvulsant in the treatment of the neurotropic virus syndrome of dogs (distemper, hard-pad, etc.), and in the sedation of puppies suffering from convulsions and hysteria associated with teething and the digestive upsets related to teething or other causes. Its power in these respects is greater than that of other barbiturates, with perhaps the exception of methyl phenobarbitone. Where pain or severe skin irritation is present, the barbiturates are relatively useless and must be combined with analgesics which may be the aspirin type of compound or the more active Pethidine, B.Vet.C. (Meperidine, U.S.P.).

Sodium phenobarbitone has been used by intravenous injection as a sedative in the horse in doses up to 12·5 g, but its effect is of long duration and it is therefore infrequently used.

Dosage

Dogs: orally 30 to 130 mg ($\frac{1}{2}$ to 2 gr.) given three times daily. This dosage may occasionally result in cumulation, especially if liver or kidney malfunctions are involved: indications of this are the development of incoordination, with an inability to negotiate furniture or steps. Cessation or reduction of the dosage is usually followed by a return to normal and no other measures are required.

Methylphenobarbitone, B.Vet.C. (Mephobarbital, N.F., 5-ethyl-l-methyl-5-phenyl barbituric acid), in which a methyl group is substituted for the hydrogen atom at nitrogen 1, may be used instead of phenobarbitone, as its central sedative action is less than that of its parent and its anticonvulsant action is superior. Dosage is the same.

Medium Duration Barbiturates

These sedative/hypnotics are exemplified by amylo-, seco- and butobarbitone.

Butobarbitone, B.Vet.C., Butabarbital, N.F. (5-ethyl-5-n-butylbarbituric acid)

This medium duration barbiturate is a slightly more effective central sedative than phenobarbitone but its anticonvulsant action is weaker. The indications and dosages are similar to those for, phenobarbitone. It has the advantage of being eliminated in about half the time of phenobarbitone.

Short Duration Barbiturates

In reality, this group differs little from the medium duration barbiturates.

Pentobarbitone Sodium, B.Vet.C., Pentobarbital, U.S.P. (5-ethyl-5-methylbutyl barbituric acid)

This white, bitter tasting, water soluble barbituric acid derivative is typical of the short acting group of compounds. It is therefore reasonably quickly absorbed when given by all but the subcutaneous route and can be used to induce all stages of cerebral, medullary and spinal depression. The degree of effect is related to route of administration and dosage.

The major use of pentobarbitone in the veterinary field is as a surgical anaesthetic. It may, however, be used as an antidote to eliminate the tonic spasms associated with strychnine poisoning, for which purpose more than the normal anaesthetic dose may be required: the affected animal is kept anaesthetized until the alkaloid is metabolized and excreted. It is also widely used for euthanasia of dogs and cats, being injected intravenously or intraperitoneally as an aqueous–alcohol–glycol solution containing 195 mg (3 gr.) per ml: the dose used varies but is approximately 0·7 ml per kg (1 ml per 3 lb.) body weight given intravenously.

Oral administration should be used routinely only if the barbiturate is required for general sedation and hypnotic (sleep-producing) effect in small animals. Narcosis in small animals can be achieved by this route, but if there is a large amount of food in the gastro-intestinal tract this will affect the absorption rate and the maximum blood and tissue levels. There is always the danger of producing a state of almost unmanageable narcotic frenzy, which though not of common occurrence is never pleasant to meet. Anaesthesia in cats is frequently achieved in single handed practice by administering an estimated dose (twice the intravenous dose, i.e. 60 mg per kg or approximately $\frac{2}{5}$ gr. per lb.) in capsular form, keeping the cat in a small cat basket until anaesthetized. Euthanasia can be conveniently performed in this manner. Anaesthesia in dogs should not be attempted by this route unless all other routes of administration are disallowed. In these cases the animal may be starved overnight and the pentobarbitone sodium given in solution: anaesthesia should follow in 15 to 20 min. This method can be used for euthanasia where help is not available and the animal will not allow injection. The application of a tape-muzzle and the spraying or pouring of a solution into the mouth can prove an easy, humane, and reasonably quick method of anaesthetizing a dog prior to destruction.

In general, narcosis or anaesthesia should not be induced by the oral route, because it is impossible to compute a dose which will provide a predictable level of CNS depression.

Parenteral administration, excluding the use of intravenous injections, gives results almost as unpredictable as the oral route. The *subcutaneous route* may be suitable for small doses intended for prolonged cerebral sedation, but for other purposes, such as the induction of anaesthesia or narcosis, absorption is too slow; though narcosis or anaesthesia may eventually be achieved, providing the dose is large enough, a period of excitement may occur lasting as long as 30 min.

The *intramuscular route* is not practicable because of the volume of solution required.

The *intraperitoneal route* is very useful where intravenous injection is not possible, for narcosis or anaesthesia may be obtained within 15 min. The usual disadvantages of intraperitoneal injection apply.

The *intrathoracic route* can be used to induce rapid anaesthesia in hysterical dogs where intravenous injection is not possible. Anaesthesia can be obtained within 5 min, but great care must be taken in computing the dosage, for absorption is rapid, and the time lag which occurs following intraperitoneal injection is not present to act as a buffer and prevent temporary overdosage. A fine needle should be used and the injection made fairly high on the chest wall, behind the shoulder and just anterior to a rib.

The *intravenous route*, including intracardiac injection, is used routinely for inducing anaesthesia and performing euthanasia. All the precautions normal to intravenous injections must be observed scrupulously, but there are a number of points concerning procedure which are peculiar to pentobarbitone sodium: unless certain principles are adhered to, serious consequences may follow.

The induction of anaesthesia should be performed as follows: Prepare the animal in the normal manner. Dosage should be computed according to weight (20 to 33 mg per kg or $\frac{1}{4}$ to $\frac{1}{7}$ gr. per lb.) and an extra 65 mg (1 gr.) or more in the case of large doses, included to allow for idiosyncrasy. When the needle is in the vein inject about half the computed dose taking about 10 sec: the animal will tense and then relax, possibly extruding the tongue, curling it up over the nose. Lie the animal on its side making sure that the needle remains in the vein. Proceed to inject the solution in 1-ml fractions, allowing about 20 sec to elapse between injections, during which time the various reflexes should be tested: this second part of the induction period should take 5 to 10 min and proceed until the required depth of anaesthesia is reached.

This method is both suitable and advisable in healthy dogs and cats but should not be used for toxic animals, sheep or pigs, for which the normal method is a slow continuous injection until narcosis is achieved, when slow intermittent injection should continue until the required level of CNS depression is reached.

Where light anaesthesia has been induced, a period of 15 to 20 min will be available for operational purposes. Medium anaesthesia will provide a 'safe' period of 15 to 30 min, and deep anaesthesia will normally last for up to 45 min. The recovery period will last for up to 8 hr, by which time the animal should be able to stand. Incoordination may persist for 24 hr and in extreme cases as long as 48 hr. During this recovery period animals should be kept quiet and warm. Should the animals be stimulated during the medium and light narcosis recovery stages, appreciable narcotic excitement can result and self-inflicted damage may be considerable. Pentobarbitone sodium alone is not suitable for inducing anaesthesia in horses, is of doubtful value in large adult cattle, should not be used in young calves which may remain anaesthetized for up to 48 hr, but has been used successfully in sheep and pigs, although thiopentone is preferred in view of its short post-anaesthetic recovery period. The compound β methyl β ethylglutarimide sodium makes it possible to antagonize pentobarbitone and cut short the recovery period. Pentobarbitone sodium with chloral hydrate (*q.v.*) and magnesium sulphate has proved to be useful combination for inducing surgical anaesthesia in horses and cattle.

In *balanced anaesthesia* pentobarbitone anaesthesia may be combined with morphine or, better still, pethidine narcosis. This reduces the quantity of pentobarbitone required by as much as two thirds, but usually one half, of the normal requirements, and therefore increases the chances of survival, as it gives a lighter level of anaesthesia and a shorter recovery period. Another type of premedication drug which will reduce the amount of anaesthetic is chlorpromazine; this potentiates the barbiturate effect, reducing dosage and recovery period: it should not be used for premedication of shock cases, however, for it may cause a further, and fatal, drop in blood pressure. Muscle relaxants are useful in conjunction with pentobarbitone sodium for, as with all barbiturates, muscle relaxation is not marked and may make prolonged and involved abdominal surgery difficult.

Dosage

Dosage by intravenous injection to give anaesthesia: 20 to 35 mg per kg ($\frac{1}{4}$ to $\frac{1}{7}$ gr. per lb.). In general, large animals will require about 15 mg per kg ($\frac{1}{9}$ gr. per lb.) body weight and small dogs and cats may require as much as $\frac{1}{4}$ gr. per lb. body weight. The dosages for intraperitoneal and intrapleural injection are the same.

Doses given orally for sedative purposes are much the same as those of other barbiturates. Doses injected intramuscularly for sedative purposes should be about half those given for anaesthesia. Pentobarbitone sodium is commercially available as stable solutions containing 65 mg (1 gr.) per ml for anaesthetic purposes and 195 mg (3 gr.) per ml for euthanasia and large animal anaesthesia. Capsules are also available.

Ultra-Short Duration Barbiturates

Numbers of this type of barbiturate have been synthesized, including methohexitone, thiamylal sodium, thialbarbitone and thiopentone, but the latter is the one used most frequently in Great Britain.

Methohexitone Sodium, B.Vet.C.

This ultra-short acting barbiturate differs from the other members of the sub-group in that it has an oxygen and not a sulphur atom at carbon 2 in the barbiturate nucleus. It is the sodium salt of 1-methyl,5-allyl,5-(1-methylpentynyl)barbituric acid. The molecule contains two points of unsaturation and is therefore readily metabolized in the body. It is, however, about three times more potent than thiopentone. It is available as a powder or a ready-to-use solution.

Actions and Uses

The duration of anaesthesia in dogs and cats may be as short as 3 min or as long as 8 min following a single dose. *Dosage* for household pets is at 11 mg per kg body weight in the form of either a 25 mg or a 50 mg per ml aqueous solution. Half the dose should be given in the first 30 to 60 sec and the remainder at the rate of 25 mg per min. Too cautious an administration rate can result in hyperexcitability and too rapid, apnoea. The duration of action is rather shorter than that of thiopentone and unlike thiopentone where recovery is due to redistribution of the drug, recovery from methohexitone is aided by the rapid inactivation of the oxybarbiturate. Muscle tremors may occur during induction and recovery stages.

Thialbarbitone Sodium, B.Vet.C. (Kemithal Sodium)

This is another thiobarbiturate with an ultrashort anaesthetic duration. It is a pale yellow hygroscopic powder with a characteristic odour. It is soluble in water and solutions are stable for only 4 or 5 hr. As an intravenous anaesthetic it has an effect similar to that of thiopentone, although its potency is approximately half that of thiopentone.

Thiamylal Sodium, N.F.

This thiobarbiturate differs from thialbarbitone in having a (methyl-butyl) side chain at carbon 5 in place of the (cyclo-hexenyl) ring. It is therefore also closely related to thiopentone sodium.

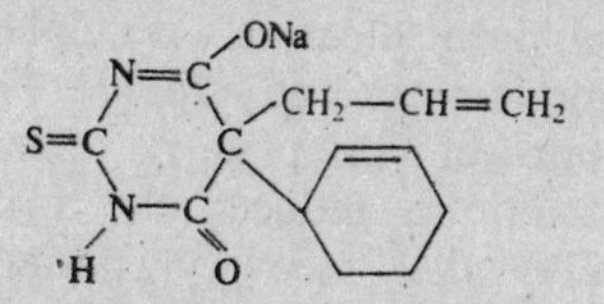

(vi) Thialbarbitone sodium

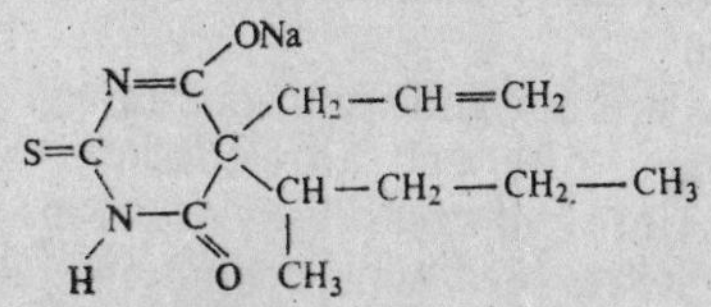

(vii) Thiamylal sodium

This anaesthetic, closely related to thiopentone sodium, is a pale yellow hygroscopic crystalline substance which, in aqueous solution (it is freely soluble), has a strong sulphurous odour. It may be used as a 2·5 per cent solution (pH about 10·8) or a 4 per cent solution.

The anaesthetic should be given intravenously in the same way as thiopentone sodium; intraperitoneal administration gives irregular results. It has a potency about 50 per cent greater than thiopentone, has a slightly less depressant effect on the respiratory centre, but seems to cause greater salivation and requires atropine or hyoscine premedication. Excitement during induction or recovery is less common than with thiopentone sodium.

Toxicity is not great, up to 2·25 times the anaesthetic dose being tolerated in the majority of dogs. The heart is usually unaffected by doses which cause apnoea.

Dosage

Dosage depends on the method of administration, the more rapid the injection the smaller the dose and the shorter the duration of anaesthesia.

Dog and Cat:

17·8–23·2 mg/kg (8·1–11 mg or $\frac{1}{8}$–$\frac{1}{6}$ gr. per lb.)

Pig:

approximately 18·26 mg/kg (8·3 mg or $\frac{1}{8}$ gr. per lb.) by slow intravenous injection

Where medium anaesthesia is induced, the duration is approximately 10 min, but by continuing to deep anaesthesia duration may be extended to about 20 min. Morphine, atropine, hyoscine and tubocurarine have all been used safely in conjunction with thiamylal sodium, and ether is satisfactory for continuing anaesthesia induced by this barbiturate.

Thiopentone Sodium, B.Vet.C. (Thiopental Sodium U.S.P.)

This is a yellow, unstable, water soluble salt, marketed in ampoules sealed under nitrogen. When made up with water for injection, the solution should normally be used within a few hours, for hydrolysis occurs fairly rapidly. No solution should be used after 3 days at room temperature. Chemically the compound is identical with pentobarbitone sodium except for the substitution of sulphur for the oxygen at carbon 2.

The action is typical of the ultra-short duration barbiturates, and though it has anticonvulsant properties which are superior to those of pentobarbitone, it can only be used as an intravenous injection for the induction and maintenance of anaesthesia. To control the spasms of tetanus in man, it has been used successfully over a period of 10 days by intravenous drip techniques in conjunction with intramuscular pethidine.

Thiopentone sodium is used in all species of domestic animals for the induction of anaesthesia which may then be maintained by further intermittent or continuous flow injections of thiopentone or by changing to volatile or gaseous anaesthetics. Thiopentone may be used with or without prior sedation and analgesia (with chlorpromazine, morphine or pethidine), and may also be combined with muscle relaxants. It is compatible with atropine and hyoscine which may be administered in a combined solution with pethidine or morphine.

Intravenous administration follows much the same pattern as pentobarbitone sodium, but it should be remembered that the ultra-short acting barbiturates exert their effects immediately following injection and the impact of a temporarily excessive dose is immediate. The most frequent consequence of this is temporary (15–60 sec duration) apnoea owing to the transient depression of the respiratory centre. Assistance is rarely necessary. Laryngospasm or bronchospasm are sometimes seen and can be fatal in the absence of determined intervention, e.g. intubation.

In dogs and cats it is usual to give the preparation as a 5 per cent solution, ampoules of 0·25 g being dissolved in 5 ml water for injection, 0·5 g in 10 ml, and 1 g in 20 ml of water. Half the dose is given fairly quickly (about 10 sec), followed by a wait of up to 1 min, during which the breathing should become even and regular. If apnoea is protracted, artificial respiration should be given by pressure on the chest every 3 or 4 sec or by other suitable methods. After assessing the depth of CNS depression, injection is continued intermittently until the required depth of anaesthesia is reached. This may take up to 5 min. Duration of anaesthesia will depend on the dose given: with light anaesthesia, usually requiring a dose of approximately 13·2 mg per kg body weight (60 mg or $\frac{1}{10}$ gr. per lb.), anaesthesia lasts about 10 min, but 26·4 mg per kg ($\frac{1}{5}$ gr. per lb.) will normally give deep anaesthesia with an anaesthetic duration of up to 25 min. Recovery is usually complete within 1 to 2 hr and struggling is not common. Old animals may show idiosyncrasy with protracted incoordination.

It is now common practice, when an operation is scheduled to take 30 to 60 min, to strap the syringe onto the leg with the needle in the vein, using elastic or zinc adhesive tape, so that an assistant may reinforce the initial injection by further intermittent injections. By this means, anaesthetic duration similar to that obtained with pentobarbitone is achieved without the prolonged recovery period attaching to the latter drug, though some extension of the recovery period must be expected.

By varying the methods of injection, thiopentone sodium can thus be used for operations and manipulations of short duration, such as X-ray procedures, dental surgery, minor surgery and abdominal palpation: it may also be used for protracted abdominal or thoracic surgery, surgical repair of fractures, and similar procedures which require prolonged anaesthesia but in which it is desirable to avoid the prolonged recovery period which follows anaesthesia induced by pentobarbitone sodium or chloral hydrate.

Thiopentone has been used to anaesthetize all the domesticated species by the intravenous route and is satisfactory for short duration interferences. It is widely used in small animal practice and premedication with a phenothiazine tranquillizer is a routine adjunct. This reduces the amount of thiopentone necessary for induction and so reduces the toxic hazard. Additionally, the recovery period is then relatively free from the convulsive struggling which frequently accompanies barbiturate anaesthesia.

Horses are not good subjects for thiopentone anaesthesia. Although anaesthesia is achieved in the majority of cases, heavy falls may occur, Cheyne-Stokes breathing is common and violent struggling during the recovery period is frequent. It is therefore not to be considered as an anaesthetic for routine use in horses. However, when combined with the relaxant suxamethonium or with a suitable phenothiazine type ataractic, it can provide a very satisfactory means of inducing anaesthesia in the horse.

For this purpose, the rapid intravenous technique is used. This differs from the slow intravenous technique used in small animals in that the computed dose is injected as rapidly as possible. For this method, dosage is computed on the basis of 11 mg per kg (1 g per 200 lb.) body weight. If the skin overlying the jugular vein has previously been anaesthetized with a local anaesthetic, minimal restraint is required. Anaesthesia is produced within 20 to 30 sec and it will last 3 to 4 min. It is characterized by an initial apnoea and a rather violent recovery. This latter undesirable feature is much reduced or eliminated in animals premedicated with a tranquillizing drug. In this instance, the barbiturate dosage is not reduced as the technique is effective only because the dose administered reaches the brain in a relatively undiluted 'slug'. This high concentration produces almost immediate unconsciousness. In the meantime, barbiturate has become well mixed with the blood and is leaving the blood to diffuse into the tissue fluids. When this barbiturate-depleted blood returns to the brain, the drug diffuses out of the brain, and back into the blood. As brain levels fall, so consciousness returns. This is usually avoided as the animal will within this time have become exposed to a maintenance agent, e.g. chloroform from a face-mask or halothane via an endotracheal tube. It is pharmacologically interesting to note that while clinical recovery from thiopentone anaesthesia depends on the transfer of the drug into less vascular tissues, this sequestered anaesthetic is still potentially biologically active. Its elimination is effected over a period of days by the usual processes of metabolism and excretion.

When thiopentone is injected slowly, as in the standard technique in the dog, larger doses are needed to induce anaesthesia, but the anaesthesia is of longer duration. This is because the slower administration allows a considerable diffusion of the drug throughout body water and entry into body fat and, consequently, the rate of back-diffusion from the brain is much reduced.

When using thiopentone it is important to ensure that none of the solution is injected into the peri-venous tissues. Premedication and local analgesia of the injection site will much reduce the chances of the animal's moving during the injection of the barbiturate. However, should peri-venous leakage occur, it is important to inject large volumes (0·5 to 1 litre) of saline containing hyaluronidase as soon as possible. This precaution will significantly reduce the tissue reaction to the barbiturate and will eliminate the possibility of skin sloughing. Tissue damage will follow the peri-vascular injection of solutions of a strength of 5 per cent or greater. Tissue necrosis is not caused by 2·5 per cent solutions but 5 per cent solutions can cause spasm of the vein and thrombosis. Even stronger solutions (10 per cent or more) can be used safely via the jugular vein of large animals owing to the rapid dilution of the anaesthetic.

Dosage

Large animals: 10 to 15 mg per kg; small animals: 20 to 35 mg per kg.

Thiopentone sodium is available commercially in 0·25-, 0·5-, and 1-g ampoules. The suitable sizes of ampoules of water for injection (5, 10 and 20 ml, respectively) are usually included in the packs.

Ketamine Hydrochloride

This is a product introduced as a cataleptoid anaesthetic for cats. It is *dl*-2(*o*-chlorophenyl)-2-(methylamino) cyclohexanone hydrochloride and is a short-acting derivative of phencyclidine. In man ketamine induces a state of dissociative anaesthesia. In this state analgesia and amnesia are pronounced, but muscle relaxation is poor and the patient remains capable of responding to instructions. Respiratory function is little altered but there is some increase in heart rate and blood pressure. Salivation occurs but is not a problem as swallowing is not impaired. Eyes remain open and it is recommended that they be protected from drying and retinal damage.

Dosage is in the range 10–40 mg per kg intramuscularly depending on condition, age and intended interference. The duration of anaesthesia is 30–45 min. Increasing the dosage tends to increase duration of effect rather than intensity. Intravenous injection of 5–10 mg per kg is also satisfactory for small animals. Atropine may be used to control unwanted secretions.

The persistence of the eructation and swallowing reflexes is claimed in pregnant sheep undergoing experimental surgery. This is a considerable advantage for general anaesthesia in ruminants. Induction is achieved by a dose of 2 mg per kg and thereafter is maintained by an intravenous infusion of a 2 mg per ml solution in 5 per cent glucose at a rate of 4 ml per min. A similar technique has been successfully employed in cattle, differing only in that infusion was carried out at a rate of 10 ml per min.

The difficulty in assessing the degree of unconsciousness coupled with the low order of muscle relaxation make it doubtful whether this drug should be used alone for major surgery, although its ease of administration and wide safety margin make it attractive for routine minor use. Return to full consciousness occurs slowly over a period of several hours, making the drug more suitable for in-patient use in small animals.

Alphaxolone

CNS depression and anaesthesia was described as a steroid side-effect some time before the first steroidal anaesthetic, hydroxydione, was introduced in 1955. It had a wide safety margin but unfortunately all except very dilute solutions tended to cause damage to the vein at the point of infusion. With such solutions induction of anaesthesia was very slow.

Alphaxolone (3α-hydroxy-5α-pregnane-11,20-dione) is an anaesthetic steroid of high potency which induces rapid anaesthesia, has a low order of irritancy and a wide safety margin. The compound is without significant hormonal activity. Its low water solubility is increased 100-fold by a surfactant and by a further factor of three by the related anaesthetic steroid alphadolone acetate which is the 21-acetoxy derivative of alphaxolone. The final product contains 9 mg alphaxolone and 3 mg alphadolone per ml and may be administered by intravenous or intramuscular injection as it has a pH of about 7 and is isotonic with blood. Unfortunately, the surfactant is a potent histamine releaser in the dog and the product is therefore contra-indicated in this species. In the cat recovery occurs within 30–60 min and appears to be without hangover effects. Because the anaesthetic effect is terminated by degradation of the drug in the liver, rather than by redistribution, there is no cumulative tendency as is the case with most other injectable anaesthetics. This, coupled with its advantageous safety margin and its compatibility with other anaesthetics and anaesthetic adjunct drugs, suggests that alphaxolone is a significant advance in anaesthesia of the cat.

Metomidate

This imidazole derivative has been used for the capture of wild birds by ingestion on bait, for anaesthesia of birds including hawks by intramuscular injection and following intramuscular azaperone (2 mg per kg) for intravenous (4 mg per kg) or intraperitoneal (10 mg per kg) induction of neuroleptanalgesia in pigs. This state was often accompanied by pallor and trembling, but in general sleep, analgesia and muscle relaxation were satisfactory and the pigs regained their feet between 2½ and 4 hr.

Metomidate (3·5 mg per kg intravenously) has been evaluated for the induction of brief periods of anaesthesia in horses premedicated with azaperone and appears capable of inducing anaesthesia as rapidly as thiopentone but without some of thiopentone's hazards, e.g. apnoea, tachycardia and fall in cardiac output. Recovery from anaesthesia was sometimes abrupt and accompanied by motor excitement. Unfortunately, metomidate causes haemolysis, the extent of which increases with the rate of intravenous infusion and this is not an acceptable side-effect.

α-Chloralose

This is a condensation product of glucose and chloral hydrate and has anaesthetic properties which are not, however, suitable for clinical exploitation as the injection volume is large, induction is slow, anaesthesia is shallow, duration is lengthy and recovery is slow. These characteristics resemble those of chloral hydrate to which the drug is converted in the body.

The material is, however, used in the control of rats and pigeons and the baited grain may be consumed by other animals. Chloralose is by this means an occasional cause of a poisoning characterized by CNS depression and hypothermia. Its prevalence may be increased by the illicit baiting of carcases or eggs in attempts to control other species, e.g. birds of prey. In poisoning cases, evidence of excitation usually precedes prostration. Treatment involves emesis and isolation in a warm environment. Gastric lavage is necessary in comatose subjects and artificial respiration may also be required.

Bibliography

ADRIANI, J. (1963) General anaesthetics: absorption, distribution and elimination. In: *Physiological Pharmacology* (ed. W. S. Root & F. G. Hofmann), vol. 1. New York: Academic.

EVANS, F. T. & GRAY, T. C. (ed.) (1958) *Modern Trends in Anaesthesia*. London: Butterworths.

GRAY, T. C. & NUNN, J. F. (ed.) (1971) *General Anaesthesia*, 3rd ed., vols 1 and 2. London: Butterworths.

HUNTER, A. R. (1972) Drugs for producing dissociative states. In: *Recent Advances in Anaesthesia and Analgesia*, 11th ed., pp. 1–15. London: Churchill Livingstone.

LUMB, W. V. & JONES, E. W. (1973) *Veterinary Anaesthesia*. Philadelphia: Lea & Febiger.

SCURR, C. & FELDMAN, S. (1974) *Scientific Foundations of Anaesthesia*. London: Heinemann Medical.

SHEPHERD, N. W. (1964) *The Application of Neuroleptanalgesia in Anaesthetic and Other Practice*. Oxford: Pergamon.

23

The Volatile and Gaseous Anaesthetics

VOLATILE ANAESTHETICS

The members of this group of anaesthetics are fluids with low boiling points which volatilize freely at normal temperature and pressure, the gases produced having anaesthetic effects when inhaled. They are:

1. Ether and vinyl ether.
2. Chloroform.
3. Trichloroethylene.
4. Halothane.
5. Methoxyflurane.

Volatile anaesthetics may be used to anaesthetize any species of animal: chloroform is widely used in farm animals and was used in the dog but because of toxicity considerations pure chloroform is not used in the cat: ether is valuable in small animals but its use in large animal anaesthesia is limited by the need for special apparatus, of which there are various types (this problem is common to the volatile anaesthetics); vinyl ether is not used commonly but has been recommended for dogs and cats: trichloroethylene has not made its mark as a veterinary anaesthetic in this country: halothane is now established as a safe, effective and reliable volatile anaesthetic, whilst methoxyflurane is becoming increasingly used, particularly in the United States.

Induction

The volatile anaesthetics are administered, using various types of masks or apparatus utilizing intratracheal tubes, in one of two ways:

1. By auto-inhalation in which air is inhaled through or over the anaesthetic, or through a pad soaked in the anaesthetic, thereby volatilizing the liquid or,

2. By positive feed methods whereby air, oxygen, or oxygen combined with 5 per cent carbon dioxide, is bubbled through or passed over the anaesthetic and carries with it into the mask or intratracheal tube a quantity of anaesthetic vapour.

In the *open method* of administering anaesthetics the anaesthetic is applied to a pad of lint or cotton wool held in the region of the mouth and nostrils either on a metal frame, as with the Schimmelbusch mask, or in the bottom of a container such as a plastic tumbler or jar: this method may be used in emergency. The *semi-open method* involves the use of special masks which fit closely round the animals muzzle; air is breathed in through a sponge or cotton wool wad soaked in anaesthetic and in most cases exhaled gases also pass out through the anaesthetic pad. Such auto-inhalation methods are wasteful of anaesthetic and unpleasant for the anaesthetist but, nevertheless, they are useful where more sophisticated equipment is not available. In a *closed circuit*, anaesthetics are vapourized by positive feed oxygen and after being breathed by the animal all exhaled gases are retained, the carbon dioxide absorbed by soda-lime and the remainder rebreathed after the addition of more oxygen. The *semi-closed* circuit is similar to the closed circuit method in that anaesthetic is supplied in a stream of oxygen, but the exhaled gases for the most part are lost into the atmosphere through non-return expiratory valves. Soda-lime canisters are sometimes used in conjunction with semi-closed circuits.

The **stages of induction** are common to all inhalation anaesthetics. They are well defined and it is essential that they should be known and recognized. They are:

Stage 1: Voluntary struggling and excitement due to fear.
Stage 2: Involuntary struggling with narcosis.
Stage 3: Anaesthesia.

A fourth stage usually due to overdosage, that of bulbar paralysis and death, is frequently added to this list.

The indications vary slightly between the species, but in general they follow the same pattern.

In the *horse* stage one is typified by all the signs of fear—spasmodic struggling, rapid deep respirations, dilated pupils and a frequent, strong pulse. Voluntary holding of the breath may occur. Struggling is more violent if the air supply is so low as to verge on suffocation and the anaesthetic concentration high.

As stage one passes to stage two the periods between struggling get longer and movements become less concerted and sluggish, although occasionally relatively continuous but less powerful leg movements may occur. The eyelids open, the pupils remain dilated, nystagmus occurs and breath-holding of an involuntary nature may be noticed. The third stage follows with a general relaxation of muscles and loss of limb and trunk reflexes, the tail is relaxed, respiration becomes shallower, slower and abdominal in type. The pulse should be regular and normal. In light anaesthesia the eyelids should be open and the corneal reflex brisk, but as anaesthesia deepens to medium anaesthesia the eyelids close and the corneal reflex becomes sluggish but is always present; the pupil is constricted; breathing and the pulse are normal. In deep anaesthesia the breathing becomes much slower and shallow, the pulse is normal in rate but less powerful than in the previous stages, the anal and bladder sphincters relax, occasionally allowing involuntary defaecation and urination, the corneal reflex is inconsistent, but the pupil should always be constricted.

Should anaesthesia be taken further, respiratory and circulatory depression arises and the fourth stage, in which CNS paralysis with cessation of breathing and, later, heart failure, supervenes. The onset of this stage is indicated by progressive, or sometimes sudden, dilatation of the pupil and irregular breathing.

Unless euthanasia is intended, anaesthesia should not be allowed to progress beyond the depth of medium anaesthesia, with closure of the eyelids and the development of a sluggish corneal reflex.

In *cattle* the use of volatile anaesthetics is less common than in horses, mainly because of their more phlegmatic disposition and suitability for local and regional anaesthetic techniques. The tendency of all ruminants to develop tympany when cast and the danger of respiratory failure due to the pressure exerted on the diaphragm, as well as the possibility of regurgitation of ruminal contents, allied with the profuse salivation caused by the anaesthetics, combine to make general anaesthesia relatively hazardous. Never the less the induction of narcosis and light anaesthesia using chloroform has been frequently and successfully practised, often combined with local anaesthesia. In general, the stages of induction follow a course similar to that in horses, but the struggling, so typical of inhalation anaesthesia in other species, is relatively mild and insignificant.

The *pig* is not a good subject for the induction of general anaesthesia by volatile anaesthetics, partly because of the short neck and the possibility of occlusion of the air passage, but also because of the susceptibility of the species to chloroform poisoning—ether and other volatile anaesthetics are not suitable for normal auto-inhalation methods. Induction stages are the same as for the horse, stage one and two being accompanied by struggling and squealing but induction is rapid and the stages therefore short.

In the *dog and cat* small variations occur in the signs indicating the various stages of induction.

Stage one is accompanied by the normal signs of fright—convulsive struggling, growling and biting, defaecation and urination. Vomiting is not common. Salivation usually occurs, due more to the irritant effect of the anaesthetic than the response of the animal to anaesthesia. The pupil is widely dilated. Both species will hold their breath voluntarily for as long as 10 sec, after which they inhale deeply: this may occasionally be responsible for high anaesthetic concentrations in the blood followed by respiratory and cardiac paralysis so that great care should be taken to dilute the anaesthetic during these periods either by lifting the mask or by opening air inlets, etc.

Stage two arrives imperceptibly, the character of the struggling and the noise changing; movements are less convulsive and less coordinated, whilst growling and angry noises change to whining and whimpering during the expiratory phase. The pupil is still dilated and nystagmus is usually marked in cats but not in dogs.

As stage three commences all movements and noises cease and muscular relaxation occurs, breathing becomes regular and slower, nystagmus ceases and all reflexes are lost, the corneal being the last reflex to go. At this stage blood pressure falls and temperature control is diminished; body temperature may fall as much as 5° and one may need to keep animals warm during anaesthesia. Anaesthesia should normally not be required to go beyond the stage of a sluggish corneal reflex, but its absence has not the significance that it has in the horse and cow.

The imminence of stage four can only be determined by watching the breathing and mucous membranes. The former becomes irregular and shallow, whilst the latter become progressively cyanosed, with a sluggish pressure reaction, and engorged vessels.

Excretion and Recovery

The volatile and gaseous anaesthetics are mainly excreted through the lungs. Only a minor part (8 to 20 per cent of ether) is lost in the urine or by biotransformation. Excretion begins immediately the mask is removed and about half is lost from the body within 5 min but the remaining percentage is lost only gradually, ether taking up to 4 hr and chloroform almost double this period.

The stages of recovery are similar to those of induction but more rapid in reverse. Struggling is, however, not so marked but can still be dangerous and with the horse restraint should continue until the animal is able to rise and stay on its feet; this restraint usually consists of keeping the head down only. With cattle and small animals, if they are propped on their brisket (sternum) and left free in a box, kennel, or basket, recovery quickly follows without undue excitement.

The greater danger period during inhalation anaesthesia is the stage of light narcosis, both in induction and the recovery phases. In both phases death is usually due to cardiac fibrillation and syncope (heart failure). Death occurs following excessive anaesthetic concentrations allied with reflex respiratory stimulation of the heart and adrenaline outpouring during the struggling and excitement. Of the two periods, struggling during recovery is probably more dangerous, for the venous anaesthetic concentrations are high and there is no means of reducing them rapidly, whereas during induction the intake of anaesthetic can be controlled with care. This danger is greatest with those agents that sensitize the heart to the action of adrenaline, e.g. chloroform and ethyl chloride.

Contra-indications. Inhalation anaesthesia is contra-indicated in patients with respiratory diseases, and in the case of chloroform anaesthesia, hepatic and renal diseases.

Balanced anaesthesia. Inhalation anaesthesia is now frequently used following premedication with narcotic doses of chloral hydrate, a barbiturate, pethidine or an opium derivative. Belladonna derivatives are very valuable in controlling the excessive salivary and bronchial secretions which follow the irritant action of chloroform and ether. In small animals it is now common practice to induce anaesthesia with an ultra-short acting barbiturate, insert an endotracheal tube and continue anaesthesia with a volatile or gaseous anaesthetic. In the teaching schools and more advanced practices it is now also common practice to follow this procedure in large animal anaesthesia.

Chloroform, B.Vet.C.

This important volatile anaesthetic is a heavy, clear, volatile liquid with a characteristic odour and burning taste. It is non-inflammable but when exposed to air and light, oxidation produces the poison gases, phosgene and free chlorine. To reduce this process, all anaesthetic chloroform contains suitable concentrations of recognized stabilizers. Chloroform is slightly soluble in water (0·5 per cent) but miscible with most organic solvents.

Chemically, chloroform is trichloromethane ($CHCl_3$) and is therefore a close relative of carbon tetrachloride (CCl_4) which is used as an anthelmintic. This relationship accounts for the anaesthetic properties of carbon tetrachloride.

Chloroform Anaesthesia

The method by which chloroform exerts its anaesthetic action is not known precisely, but the Overton–Meyer theory, or the 'lipoid theory', classifies chloroform, ether and other fat soluble anaesthetics together, suggesting that the compounds have an affinity for lipoid tissue and that they are therefore taken up by the brain in relatively large quantities because of the high fat content of the brain. This theory was postulated before the new era of pharmacological investigation and accounts only for the distribution of the anaesthetics. The intracellular effects are as yet unknown but are presumed to be related to the reversible blockade of enzymes. It is probable that these lipophilic anaesthetics interfere with intracellular oxidation processes and that the different groups of anaesthetics do not differ in their pharmacological actions as much as was once thought (see p. 11).

The *local* irritant action of chloroform on the respiratory and naso-pharyngeal mucosae is responsible for increased salivation and bronchial secretions in all species, but especially in ruminants. This increase of secretions, allied with the decrease in activity of tracheal and bronchiolar cilia and depression of the cough reflex, increases the danger of post-anaesthetic pneumonia. It is advisable to ensure that if operations are to be at all protracted, the animal lies with its head below the general body level so that secretions will drain towards the mouth.

The *circulatory system* effects of chloroform are important and dangerous. In general, chloroform causes circulatory depression with a progressive fall in blood pressure as the stage of anaesthesia is reached, due mainly to depression of the vasomotor centre, but splanchnic dilation also occurs. Circulatory reflexes are also affected, the inhibitory ones being intensified whilst the pressor reflexes are depressed: this includes the carotid sinus reflexes and the vagal centre reflex reaction to respiratory mucosa irritation. The *heart* muscle is directly poisoned by high concentrations of chloroform and becomes very sensitive, particularly in cats, to the stimulant effect of adrenaline.

Heart failure, which has long been acknowledged as the most frequent cause of death during chloroform anaesthesia, is therefore probably due to three major factors: (1) direct muscle toxicity due to high blood concentrations following erratic inhalation, (2) ventricular fibrillation caused by excessive adrenaline production during a period when the heart is particularly sensitive to its action; and (3) vagal arrest. The last two actions are probably intensified by the muscular weakness following the direct toxic action. It follows that anaesthesia must be conducted as quietly and smoothly as possible and that premedication with vagal inhibitors (e.g. atropine) and sedatives (e.g. pethidine and phenothiazine ataractics) should decrease the likelihood of cardiac collapse.

The effects of chloroform on the *respiratory system* are mainly those of depression but in the early induction and late recovery stages they are masked by the effects of fear and narcotic excitement. Strong concentrations of chloroform during the induction stages may produce reflex respiratory arrest. The carbon dioxide excitability threshold is elevated so that reinstitution of breathing may not be stimulated until tissue anoxia has reached a fatal stage. This situation may be responsible for death during induction of anaesthesia.

In deep anaesthesia the character of the breathing changes from predominantly thoracic to diaphragmatic or abdominal breathing, and if anaesthesia is continued further becomes irregular, laboured, and finally stops due to respiratory centre paralysis.

Death from respiratory paralysis rarely occurs except following overdosage to the stage of medullary paralysis.

The alimentary system is not markedly affected by chloroform, but relaxation and cessation of intestinal movement do occur. In the vomiting species—pig, dog and cat, vomition occasionally occurs during the induction but more commonly in the recovery stages: it is more likely to occur in toxic patients.

Chloroform has a toxic effect on the liver and kidneys but has little effect on the uterus or foetus in utero and is therefore useful as an anaesthetic for Caesarian section.

Muscular relaxation is good during chloroform anaesthesia. *Analgesia* occurs before the onset of anaesthesia and loss of consciousness. As with ether, sensory pathways are paralysed before the motor reflex components.

Toxicity. Sheep and cats are the most susceptible species of the normal domestic animals but poisoning does affect other species. The acute toxicity of chloroform on the heart and respiratory system has been mentioned, but delayed toxicity is probably as great a danger. The fear of causing delayed poisoning probably predisposes to acute toxic effects for there is a tendency to use only light anaesthesia and to cease administration at the earliest opportunity: the maintenance of a level plane of anaesthesia is difficult and lightening into the narcotic stage happens all too easily, with consequent dangers of narcotic excitement and surgical shock.

Delayed chloroform poisoning takes the form of fatty changes in all parenchymatous organs as well as the heart muscle. Toxic patients, pregnant or parturient animals, young and aged patients, are most susceptible. Protracted operations, in which chloroform anaesthesia is maintained for 1 hr or more may cause delayed chloroform poisoning in normal adults but in the susceptible group shorter periods than this may be dangerous. Generally speaking affected animals are slower to recover from anaesthesia but definite symptoms may not be seen for about 48 hr. Symptoms include depression, 'dirty' mucous membranes with cold extremities, repeated vomiting, and the development of jaundice. Depression increases and develops into a coma which is followed by death within about a week of the anaesthesia.

Treatment of acute poisoning depends on the cause. Adrenaline is contra-indicated in heart failure during the early stages of induction and the later recovery stages, but can be useful in deep anaesthesia. It is wiser in narcotic stage heart failure, to rely on artificial respiration with cardiac massage in small animals and the use of oxygen is available. Analeptics have also been used but rarely can a heart be restarted no matter what the treatment. Prophylactic treatment includes premedication with atropine or similar vagal paralytics which prevents at least one of the possible contingencies. With respiratory failure, artificial respiration following clearance of the air passages, and the use of oxygen and analeptics usually reinstitutes breathing.

Delayed chloroform poisoning should be prevented, for treatment is rarely successful. As the resistance of the liver and kidneys is related to their glycogen content, the feeding of glycogenic food and the reduction of protein for a few days before a scheduled operation can be useful. Dextrose and calcium injections have been given prophylactically before and therapeutically after chloroform anaesthesia. This is analogous to the treatment of poisoning with carbon tetrachloride.

Dosage

By inhalation: anaesthetic doses for the various species will obviously vary according to the depth and duration of anaesthesia, but in addition they will vary according to atmospheric temperature and therefore rate of volatilization. In winter it is frequently difficult to achieve anaesthetic vapour concentrations during inhalation unless the chloroform container is heated with warm water, whilst the warmth of expired air will drive off and waste anaesthetic applied to swabs or sponges used in auto-inhalation masks not fitted with non-return valves. However, for induction, concentrations of 2 to 3 per cent are normally required, and for maintenance, 1 to 2 per cent of inhaled gases.

Horse:

30–60 ml (1–2 fl. oz.) initially, followed by 15–30 ml (½–1 fl. oz.), and then 4–8 ml (60–120 m.) as and when required. These doses are usual when the horse is standing. If the animal is cast prior to anaesthesia the initial 'knock down' dose and the air restriction commonly applied at the same time should be omitted

Cow:

Up to 30 ml (1 fl. oz.) initially, followed by 15 ml (½ fl. oz.) as required, in conjunction with local or regional anaesthesia

Pig:

8–15 ml (¼–½ fl. oz.) initially, followed by similar doses when required

Dog:

1–4 ml (15–60 m.), repeated as required

At temperatures of 50°to 60°F and in a sheltered atmosphere unaffected by air currents these amounts should give narcosis in about 5 or 10 min and anaesthesia in 10 to 20 min, but where temperatures are lower than this and anaesthesia is being induced in the open these times may be greatly extended.

Anaesthetic Ether, B.Vet.C.

This volatile anaesthetic is a light, highly

mobile, 'water white' liquid with a characteristic odour and taste. It is soluble in water 1:10 and miscible with all organic solvents. Ether is highly inflammable and at certain concentrations in air, oxygen or nitrous oxide, forms an explosive mixture with a very low flash point. It is therefore unwise to use ether in the vicinity of an exposed flame, but where this is unavoidable the flame or electric fire must be above the patient as ether vapour is heavier than air. In a dry atmosphere, and where operating tables and covers are insulated, antistatic precautions must be observed. Thermocautery should not be used in conjunction with ether anaesthesia.

Actions and Uses

By inhalation, ether is a relatively non-toxic but weak general anaesthetic mainly used in small animals. Although it is no more irritant than chloroform it is less effective and vapour concentrations necessary to achieve anaesthesia are three times greater than for chloroform. Consequently the irritant effect is far greater, resulting in profuse salivation and increased secretions throughout the respiratory tract in all species. As with chloroform anaesthesia, ciliary movement decreases and great care must be taken to see that secretions neither drain back into the lungs nor block the airway.

The *circulatory system* is not adversely affected by normal anaesthetic concentrations. In the early stages of induction and during surgical anaesthesia there is a tendency for decreased vagal control and increased heart action to develop, with possibly a slight rise in blood pressure. Normal heart muscle is never affected but weak and fatty heart muscle may be weakened. If anaesthesia is taken to the stage of bulbar paralysis there is a gradual vasomotor depression with a slow fall in blood pressure and some fall in peripheral resistance. Ether does not sensitize the myocardium to adrenaline-induced arrhythmias.

The *respiratory system* is frequently stimulated in the early stages of induction, intensifying the effect of fear. This decreases the carbon dioxide content of the blood, which can be dangerous, causing respiratory arrest and death. In deep anaesthesia there is progressive depression of the respiratory centre with a raising of its reaction threshold to elevated blood CO_2 concentration (PCO_2). Respiratory paralysis usually occurs before cardiac arrest with ether overdosage.

The *alimentary system* is affected by ether in much the same way as by chloroform, though the degree is less. In man, ether causes more nausea but much of this is of psychological origin, and though vomiting does occur in cats, this can be largely avoided by overnight fasting and also by ensuring that large amounts of saliva are not swallowed. Salivation may be decreased by atropine premedication. The gastro-intestinal depression soon passes off and a period of greater activity may occur after recovery from anaesthesia.

The effect of ether on the liver and kidneys is normally slight, and the effect on the foetus in utero is negligible.

Muscular relaxation is usually very good due to central and peripheral blocking action. *Ether is known to potentiate tubocurarine* and it is advisable either to *avoid using muscle relaxants* with ether *or to reduce the dosage rates.*

Analgesia occurs before loss of consciousness and before the motor components of the reflexes are lost as in chloroform anaesthesia.

Toxicity. Ether is the least toxic of all general anaesthetics, its greatest disadvantage in this respect being its local irritant effect. Though damage to the liver and kidney of a type similar to that caused by chloroform may occur, it is only of clinical importance when liver or kidney diseases are present. Cardiac depression does not occur, and though there may be respiratory arrest due to irregular breathing and fluctuating vapour concentrations, artificial respiration is usually successful in reinstituting normal respiration. Delayed poisoning does not occur with ether.

Dosage

By inhalation: it is difficult and unnecessary to give exact doses for ether anaesthesia. Where an open mask is used in cats a commencing dose of about 2 ml is required: this is supplemented by similar doses as necessary. If semi-closed positive feed apparatus is used there is no exact dosage, but a steady stream of oxygen is bubbled through the anaesthetic, the rate being determined by individual effect. Dogs and other species cannot be adequately anaesthetized by open mask methods: by semi-closed methods dosage is varied according to individual effect.

Vapour concentrations necessary to achieve anaesthesia may be as high as 10 per cent and maintenance levels 4 to 6 per cent by volume.

Ether and balanced anaesthesia. The lack of toxicity makes ether very suitable as a maintenance anaesthetic following induction by thiopentone sodium, or as a general anaesthetic subsequent to narcosis induced by morphine or pethidine premedication. In both cases hyoscine or atropine should be used to restrict the profuse secretions induced by ether irritation.

Chloroform–Ether Anaesthetic Mixtures

Both ether and chloroform have disadvantages, and for small animals ether and chloroform mixtures have been examined. Chloroform has an acute and a delayed toxic potentiality but is non-inflammable, relatively non-irritant and a potent anaesthetic: ether is highly inflammable, is highly irritant locally in anaesthetic concentrations, is not a potent anaesthetic, but is almost non-toxic. The mixtures aim at combining the good points of both—the potency, non-inflammable nature, the low irritant effect, and the rapid induction of chloroform, with the low toxicity of ether. Alcohol is sometimes used as a diluent. The commonest mixtures used are ACE: Alcohol–Chloroform–Ether, in the proportions of 1:2:3, and a straight mixture of chloroform and ether in the proportions of 1:2.

With all mixtures of this type two points are of great importance. The first is that ether evaporates more rapidly than chloroform, and that unless the anaesthetic mixture is freshly prepared accidents will happen for the proportion of chloroform increases with keeping. The second is that the mixtures, though less inflammable than ether alone, are much more dangerous if ignited, for the ether burns the chloroform and the poison gas, phosgene, is produced.

These mixtures were used mainly with auto-inhalation open mask methods.

Halothane, B.Vet.C.

This more recent addition to the range of volatile anaesthetics is an extensively halogenated hydrocarbon ($CF_3CHClBr$) or 2-bromo-2-chloro-1,1,1-trifluoroethane. Marketed under the name Fluothane it is a volatile heavy liquid which is clear and colourless with a sweet chloroform-like odour. It is neither inflammable nor explosive, mixing safely with oxygen and nitrous oxide. As it does not react with warm or cold soda-lime it is suitable for use in closed or semi-closed anaesthetic circuits. Whilst some decomposition occurs when exposed to light it is normally stored in amber bottles and stabilized with 0·01 per cent thymol.

Actions

Comparison with other anaesthetics indicates that halothane is approximately twice as effective as chloroform and four times more potent than ether. Excitement invoked is minimal and no increase in salivation is caused. The frequency and depth of respiration are reduced although the depression of the respiratory centre is less than with other standard inhalation anaesthetics. When respiratory arrest occurs normal respiration is re-established after only a short period of artificial respiration. Blood pressure is reduced by halothane especially during the induction phase and this factor is of more significance in cats and dogs. The sensitivity of the myocardium to adrenaline is greater than with chloroform. These factors combine to make the induction phase the most dangerous. However, respiratory failure is produced long before circulatory arrest occurs and the anaesthetic is much safer than chloroform in all species.

Muscular relaxation is usually adequate but where greater relaxation is required it is better to resort to curarization than risk the circulatory collapse which is to be feared when high concentrations of halothane are used. Smooth muscle of the vascular and respiratory systems is usually somewhat relaxed.

Uses

It is usual to administer the anaesthetic by either a closed or a semi-closed circuit, vapourizing the liquid with a stream of oxygen. A finely calibrated vaporizer should be used. Whilst the anaesthetic is suitable for administration by any of the standard methods of volatile anaesthetic administration it is usual to use either a closed or a semi-closed circuit in the larger species and frequently in the smaller species. This is mainly a means of economizing on the anaesthetic. Close fitting latex face masks or endotracheal intubation may be used to administer the anaesthetic in all species. Induction requires a concentration of 2 to 4 per cent while 0·5 to 2 per cent is sufficient for maintenance purposes.

The fall in blood pressure which accompanies halothane anaesthesia was originally ascribed to its ganglion-blocking action. It is now believed that depression of the vascular smooth muscle, myocardium and S–A node is more important.

Halothane is capable of inducing anaesthesia in all the domesticated species and in this respect is about twice as safe as ether or chloroform. Recovery is rapid and uneventful. In large animal practice, halothane is more frequently used as a maintenance anaesthetic.

Dosage

Horse: 450 kg (1000 lb.):
- Induction: 30–40 ml
- Maintenance: 35–40 ml (30–35 ml if induction is by thiopentone sodium)

Cattle: 450 kg (1000 lb.):
- Induction: 30–40 ml
- Maintenance: 30 ml (25–30 ml if induction is by thiopentone sodium)

Sheep and Goats:
- Induction: 2–4 per cent
- Maintenance: 1·5–2 per cent

Dogs and Cats:
- Induction: 2–4 per cent
- Maintenance: 1–2 per cent (0·4–0·8 per cent if premedication is used)

Toxicity. A degree of respiratory depression and some lowering of blood pressure are usual, while deep anaesthesia carries the risk of circulatory crisis. Liver damage occurs rarely and halothane can induce malignant hyperpyrexia. Shivering on recovery is a frequent side-effect.

Methoxyflurane, B.Vet.C.

2,2-Dichloro-1,1-difluorethyl methyl ether is a pleasant-smelling substance with an established reputation for safety as a maintenance agent in anaesthesia. Its physical properties are such that induction is very slow. It is neither inflammable nor explosive, but is expensive. Anaesthesia and effects on organ function are very much the same as for halothane except that it is much more difficult to overdose with methoxyflurane. Respiratory depression and muscle relaxation are more apparent than with halothane and recovery is much slower. Myocardial sensitization and hypotension are much less pronounced than with halothane. The long persistence of this highly fat soluble agent in the body is accompanied by a greater percentage degradation than is usual for volatile anaesthetics. The question of liver and kidney damage after anaesthesia awaits a definitive answer.

Trichloroethylene, B.Vet.C.

The chemical formula of trichloroethylene is $CHCl:CCl_2$ and its characteristics are similar to those of chloroform, from which it may be distinguished by a blue dye which is included for this purpose. It decomposes in bright light and air, but thymol is usually included to act as a stabilizer. When combined with oxygen at a temperature of 25·5°C and over it becomes inflammable and burns to produce phosgene and hydrochloric acid in much the same way as chloroform. Interaction with soda-lime prevents its use with closed anaesthesia circuits: dichloroacetylene is produced.

Trichloroethylene is almost insoluble in water but miscible with the organic solvents.

Actions and Uses

This chlorinated ethylene is a volatile anaesthetic used for small animals, and has good analgesic properties. It has a potency below that of ether, which makes it relatively unsuitable for dogs, though following prenarcotization it can prove quite satisfactory. It has also been used in pigs and is effective in cats but can cause narcotic excitement. Pharmacologically it is less irritant and less toxic than chloroform but requires high concentrations and needs to be volatilized by bubbling air or oxygen through a warmed solution. It is not commonly used as an anaesthetic, mainly because of its low potency and tendency to cause excitement during induction and recovery.

Muscular relaxation is inadequate at levels of anaesthesia which are not accompanied by tachypnoea and cardiac arrhythmia.

Vinyl Ether, B.P.

This volatile anaesthetic has the chemical formula $(CH_2:CH)_2O$. It is a light, clear, colourless liquid with a characteristic odour. It contains 4 per cent absolute alcohol and a small proportion of a stabilizer to prevent its decomposition to formaldehyde and formic acid on exposure to air and light. The stabilizer gives the anaesthetic a purplish fluorescence. Solubility in water is 1 per cent and it is miscible with organic solvents. Inflammability is similar to diethyl ether.

Because of its rapid decomposition small containers are used and no liquid should be used from a container opened more than 48 hr previously.

Actions and Uses

An indication of the potency of vinyl ether, which is about four times greater than that of diethyl ether, is the rapidity with which anaesthesia can be induced in small animals even without the aid of prenarcotization. Recovery is likewise very rapid, and administration must be continued to the end of any operation. Reflex indications of the depth of CNS depression are reported to be unreliable in man, and the respiratory reactions are used as a gauge. Muscular relaxation is of about the same degree as with ether but it is less reliable. Analgesia is reasonably good.

Toxicity. In short duration anaesthesia toxicity is slight but there is a tendency to cause liver and renal damage beyond 20 min or so; sensitization of the myocardium occurs.

Vinyl Ether–Ether Mixtures

The use of a mixture of 1 part vinyl ether with 3 parts ethyl ether has been advocated, used and reported to possess good and safe effect in dogs and cats. The anaesthetics form an even mixture and a vapour of constant composition which may be administered by open, semi-closed or closed methods.

The small amount of vinyl ether is sufficient to give the mixture the necessary potency for rapid induction but insufficient to materially affect the safety of ether anaesthesia.

GASEOUS ANAESTHETICS

The gaseous anaesthetics that may be used are three in number: (1) cyclopropane, (2) nitrous oxide, and (3) ethyl chloride. These anaesthetics have a less restricted use than formerly in veterinary anaesthesia, with the increased utilization of expensive and complicated closed and semi-closed circuit positive feed apparatus. The anaesthetics are kept in cylinders under pressure and are relatively expensive unless used in closed circuit with total rebreathing.

Cyclopropane, B.Vet.C.

The cylinders containing cyclopropane are coloured orange in toto. They should be stored in a cool room away from inflammable material. The gas is colourless and has a characteristic odour; it is inflammable and when mixed with oxygen or air is explosive. It should therefore not be used in the presence of open flames and static electrical sparks must be avoided—intrapulmonary explosions have occurred during human anaesthesia, with fatal results.

Actions and Uses

Cyclopropane is a safe, powerful, non-irritant gaseous anaesthetic. Induction and recovery are rapid. Normally there is little effect on respiration and the anaesthetic has been preferred for thoracic surgery. Cardiac irregularities occur, and adrenaline or conditions which are conducive to adrenaline production, are contra-indicated. Blood pressure is rarely affected, though occasionally marked systolic falls have occurred. Liver and kidney toxicity is rare.

Cyclopropane is administered as a 20 per cent mixture in oxygen, preferably by way of an endotracheal tube. This entails induction with pentobarbitone or thiopentone sodium salts. The technique has been used in dogs, cats, sheep, goats, ponies and donkeys. Normally a 4 per cent concentration is sufficient to give analgesia and narcosis; 15 to 20 per cent will give anaesthesia in about 3 min. Disadvantages of cyclopropane are poor muscle relaxation and a tendency for increased superficial bleeding due to vasodilation and post-recovery vomiting.

Nitrous Oxide, B.Vet.C.

'Laughing gas' is supplied under pressure in cylinders coloured completely blue. As with all gas cylinders, those of nitrous oxide should be stored in a cool place away from inflammable material. The gas is colourless, with a characteristic odour and sweet taste, is non-inflammable, but supports combustion.

Actions and Uses

Nitrous oxide is one of the oldest and safest of the general inhalant anaesthetics. Unfortunately its potency is only about 15 per cent of that of ether, and concentrations necessary to achieve anaesthesia are so high that adequate oxygenation is not possible. The place of nitrous oxide in veterinary anaesthesia is therefore as a maintenance anaesthetic following thiopentone sodium induction or as a diluent for ether–oxygen anaesthesia. The amount of oxygen used with nitrous oxide should never fall below 20 per cent. Analgesia is very good with nitrous oxide but muscular relaxation is not marked and local anaesthetics or muscle relaxants may be used. Recovery is very rapid.

Toxicity is almost non-existent and any difficulties are usually associated with lack of oxygenation rather than anaesthetic reactions. The gas must be administered by semi-closed or closed methods.

Ethyl Chloride, B.Vet.C.

Ethyl chloride has been used for anaesthetizing kittens for spaying, and for anaesthetizing birds. The anaesthetic is supplied as a liquid under pressure in glass vials equipped with trigger release valve nozzles: the nozzles are of two types, a coarse one for general anaesthesia and a fine one for local anaesthesia. General anaesthesia can be achieved by confining the patient in a small glass chamber into which the ethyl chloride can be sprayed, or the anaesthetic can be sprayed into the end of a mask, providing no anaesthetic reaches the face in a fluid form. Induction is rapid but muscular relaxation is not good and respiratory depression occurs. It is not commonly used owing to its ability to sensitize the myocardium to adrenaline.

For local anaesthesia of short duration the chloride is sprayed on the skin. Evaporation is rapid and freezing occurs with local anaesthesia. This type of anaesthesia is suitable for removing splinters and similar foreign bodies from the skin, for opening abscesses, and for removing broken claws from small animals.

PHARMACOKINETICS OF INHALATIONAL ANAESTHETICS

Anaesthesia is achieved because inhaled gas diffuses from alveolar air into pulmonary membranes, dissolves in blood in the pulmonary vasculature and diffuses from the blood into body tissues, especially the brain. Initially the agent is flowing down a concentration gradient from air into arterial blood into brain. During maintenance these three phases are in equilibrium and during recovery, when no more vapour is added to the inspired air, the direction of flow reverses, i.e. from brain into venous blood into alveolar air. The rates of induction and recovery and the depth of anaesthesia depend on the interaction of a number of physicochemical properties of the various agents over some of which the anaesthetist can exert control.

The major controllable factor is the rate at which the anaesthetic vapour is delivered to the subject. The anaesthetist controls this by varying the concentration of vapour present in the inspired mixture. With gaseous agents this can be achieved easily to such an extent that adequate oxygenation may be jeopardized, e.g. with concentrations of N_2O greater than about 85 per cent. Some volatile agents have a boiling point which is so high that rates of vaporization adequate for induction are difficult to achieve, e.g. chloroform and methoxyflurane. The rate of delivery can also be increased during induction by forced hyperventilation in a closed-circuit system. This technique can counter the depressed respiratory function which will be present if a narcotic premedicant has been used.

The rate of entry of anaesthetic into blood is determined by its solubility in blood, e.g. ether is highly soluble, cyclopropane has a low solubility and is expressed as a blood : gas partition coefficient. This is the ratio of blood : air concentrations when diffusion has reached an equilibrium state. At equilibrium, the partial pressures exerted by the anaesthetic in both blood and air phases will, of course, be equal. The effect of this property is that the less soluble agents achieve equilibrium more rapidly than do the soluble ones because less agent has to be delivered to the blood volume in unit time in order to establish the equilibrium.

The rate at which anaesthetic enters tissues depends on similar considerations, i.e. the relative tensions (partial pressures) between blood and tissue, the volume of blood delivered to the tissue in unit time (its perfusion rate) and the partition coefficient between the blood and the tissue. Once again the less soluble agents will achieve the most rapid equilibration and will achieve it first in those tissues with the higher blood flows, e.g. brain, liver, kidney. Most anaesthetic agents have a tissue : blood coefficient of around 1 for the lean tissues of the body and this includes grey matter. Some have a high tissue : blood coefficient when the tissue is fat and this includes the white matter. As depot fat is relatively avascular, it achieves equilibrium slowly in relation to other more vascular tissues. As the tissue tension approaches that of arterial blood, the rate of diffusion from blood into the tissue decreases. During maintenance it is necessary to supply enough anaesthetic to maintain brain concentration by replacing that portion of the administered drug which is being metabolized, which is being excreted and which is continuing to enter those tissues in which equilibrium has not yet been achieved.

During recovery, no anaesthetic is added to inspired air and so diffusion from blood into alveolar gas takes place. As this equilibration occurs rapidly and alveolar air is exhaled, blood is rapidly cleared of its anaesthetic load. For as long as tissue tensions remain higher than blood levels, the anaesthetic vapours will diffuse from tissues back into blood. The more insoluble agents are cleared most rapidly and so have the shortest recovery times, e.g. cyclopropane and N_2O. The more fat-soluble agents are those with the longer recovery and elimination times because of the depot effect of anaesthetic sequestered in body fat. It was formerly believed that volatile anaesthetics were eliminated entirely unchanged. It is now known that some metabolism does occur, e.g. to the extent of 0·5 per cent for cyclopropane, 3 per cent for ether, 13 per cent for halothane and 20 per cent for methoxyflurane.

ANAESTHETIC ACTION AND POTENCY

Down the years various workers have attempted to relate anaesthetic potency to some physico-chemical parameter. One of the earliest was the Overton–Meyer proposition that potencies could be ranked in terms of increasing lipid solubility, this property allowing the agent to accumulate in the highly lipid brain tissue. The value of such a system would be that of being able to predict anaesthetic potency by making a physico-chemical measurement rather than having to conduct extensive evaluations in animals. Unfortunately, although such systems exist, they overlook the problem of anaesthetic side-effect prediction.

Similarly attempts to relate CNS depressant potency to measurable biochemical phenomena, e.g. uptake of oxygen or depression of glutamate oxidation, or to biophysical phenomena, e.g. surface tension effects, membrane permeability effects or neurotransmission interference, while adding information as to actions of anaesthetic agents still have not reached an end-point in terms of distinguishing incidental effects of the drug from actions which cause anaesthesia.

The minimal anaesthetic concentration (MAC) in the alveoli is that concentration required to abolish movement in response to a standardized stimulus in 50 per cent of subjects in a state of equilibrated anaesthesia. This parameter under such circumstances can be assumed to reflect the anaesthetic tension in the brain. Comparison of MAC values allows a numerical potency ranking for the volatile and gaseous anaesthetics. Moreover, multiples of MAC concentrations allow the comparison of any selected physiological parameters under the influence of a range of anaesthetic agents while each induces the same degree of anaesthesia.

Bibliography

Hall, L. W. (1971) *Wright's Veterinary Anaesthesia and Analgesia*, 7th ed. London: Baillière Tindall.

Hewer, C. L. (1972) *Recent Advances in Anaesthesia and Analgesia*, 11th ed. London: Churchill Livingstone.

Seeman, P. (1972) The membrane actions of anaesthetics and tranquillizers. *Pharmac. Rev.*, **24**, 583–657.

Soma, L. R. (1971) *Textbook of Veterinary Anaesthesia.* Baltimore: Williams & Wilkins.

24

Stimulants of the Central Nervous System

The drugs which stimulate the nervous system may be arbitrarily classified according to which part of the system they affect. Two groups result from this type of classification: (1) cerebral and medullary stimulants, and (2) spinal cord stimulants. Division must be arbitrary, for the members of the first group will cause stimulation of the spinal cord if given in large enough doses, and vice versa.

The main therapeutic value of the cerebral and medullary stimulants is for counteraction of anaesthetic depression. Certain specific antagonists to nervous system depressants will be included in this chapter, for though some have little or no stimulant effect on the nervous system they all aim at returning the activity of the nervous system to normal.

Group 1 comprises:
- Atropine
- The xanthines—caffeine, theophylline, theobromine
- Amphetamine and methylamphetamine
- The analeptics—picrotoxin, leptazol, nikethamide and bemegride

Group 2 comprises:
- Strychnine

Specific antagonists
- Morphine antagonist: Nalorphine

CEREBRAL AND MEDULLARY STIMULANTS

The action of the analeptics is almost specific to the medulla but the other members of the group cause cerebral as well as medullary stimulation.

Atropine has been referred to in detail in Chapter 5. Although it is unlikely that atropine will be used for its central effects, these effects are of interest from a plant toxicology point of view since atropine is contained in *Belladonna*—deadly nightshade, *Hyoscyamus*—henbane, and *Datura*—thorn apple. In small doses the main result of atropine action on the CNS is one of slight stimulation of respiration: in larger doses stimulation of the cerebrum occurs causing restlessness and excitement; very large doses may cause convulsions which are followed by depression of both medulla and cerebral cortex; death occurs from respiratory paralysis. When poisoning with any of the plants mentioned occurs, this initial stimulation followed by depression must be remembered and hypnotic doses of chloral hydrate or pentobarbitone sodium may be used to control excitement for repeated periods of perhaps 4 to 6 hr, depending on the degree of excitement and the dose given. In the depression stage long acting stimulants such as amphetamine or methylamphetamine are required.

Xanthine Derivatives

These compounds consisting of caffeine, theobromine, and theophylline, are also referred to as 'the purine derivatives'. They are methylxanthines, and xanthine is dioxypurine so that both terms are correct though the reference to xanthine is more descriptive:

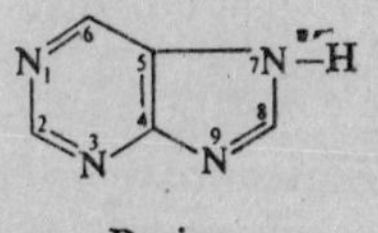

Purine

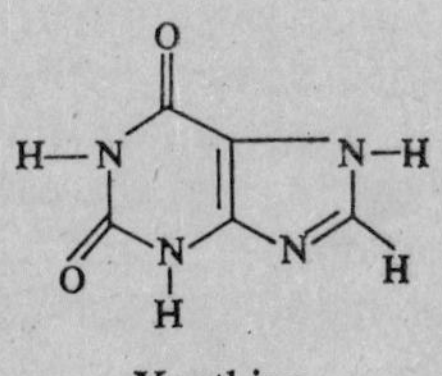

Xanthine

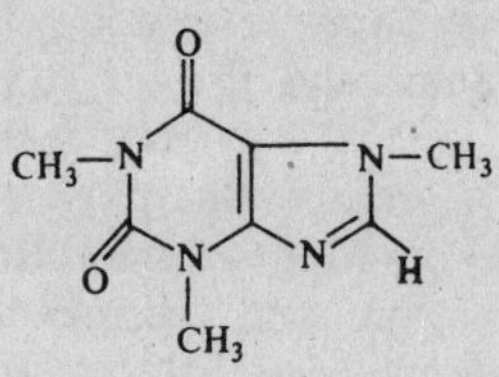

Caffeine

Caffeine is therefore 1, 3, 7-trimethylxanthine, theobromine is 3, 7-dimethylxanthine; and theophylline is 1, 3-dimethylxanthine. All three may be classified as alkaloids but their activity as bases is very slight, and theophylline forms water soluble compounds with alkalis. They are all white crystalline substances with varying solubilities: caffeine is soluble in water 1:50, in alcohol 1:75 and 1:6 in chloroform: theophylline is soluble in water 1:120, alcohol 1:80 and only slightly soluble in chloroform: theobromine is the least soluble and therefore the least useful of the three—water 1:1800, alcohol 1:2400 and chloroform 1:6000.

Actions

The xanthine derivatives are readily and completely absorbed following oral or parenteral administration. Following distribution they have a number of actions which can be divided into central and peripheral effects. Caffeine is preferred for its powerful central stimulation, whereas the other two have more powerful peripheral actions: thus caffeine has the most potent effect on the brain and medulla and theobromine the least; on the circulatory system theophylline is the most powerful and caffeine the least, whilst as diuretics, theophylline is the most powerful but is short acting and theobromine, though less active a diuretic has less central stimulation side actions and can therefore be used in greater quantities than caffeine, thus obtaining a greater effect.

Caffeine, B.Vet.C.

As a CNS stimulant caffeine in particular has powerful effects on the sensory and motor cortex, increasing mental and physical power and diminishing fatigue without causing depression later. The total amount of work done is increased without impairment of accuracy.

The medullary centres are moderately stimulated, the respiratory centre being most markedly affected. The vagus and vasomotor centres are stimulated but peripheral actions usually override these central effects.

In overdosage, the xanthines will produce restlessness, excitement, and convulsions. Increased respirations, vomiting and diarrhoea may also occur. Tachycardia and extrasystoles are prominent. The symptoms associated with CNS hyperactivity can be controlled by suitable depressant drugs such as the short-acting barbiturates.

Circulatory system effects are complex and vary according to dosage and the individual. In normal therapeutic doses the relaxant effect on smooth muscle causes vasodilatation which overrides the central vasomotor stimulation, but a direct stimulant action on the heart causes an increase in rate and amplitude and a slightly raised blood pressure. In small doses the heart may be slowed by central vagus stimulation.

Uses

Caffeine itself, as the citrate, and with sodium benzoate, is used as a respiratory and general nervous system stimulant. Occasionally it is combined with digitalis to give cardiac stimulation in the period between administration and the onset of stimulation by the digitalis. It is not of great value in barbiturate depression but may be useful in depression by other narcotics and sedatives. It is also combined with other diuretics.

Theobromine is rarely used, though one or two reports on its combinations with sodium acetate, calcium and sodium salicylate compounds, indicate its value as a diuretic when used intermittently in periods of 4 to 6 days duration. Oil of theobroma, which is solid at room temperature but liquid at body temperature, is occasionally used as a base for suppositories and pessaries.

Theophylline is rarely used on its own, being combined with mersalyl for its diuretic action, and with ethylene diamine to form aminophylline, whose bronchodilator effect relieves 'broken wind' in horses; with digitalis it has been used for congestion following cardiac decompensation in dogs.

Dosage and Administration

Caffeine, B.Vet.C.:
- Horse and Cow: 1–4 g (15–60 gr.)
- Sheep and Pig: 0·3–1·5 g (5–25 gr.)
- Dog: 50–250 mg (¾–4 gr.)

Caffeine and Sodium Benzoate, B.Vet.C.

The dosages are the same as for caffeine but this compound is usually administered by subcutaneous injection. Cats may be given about one-third the doses for the dog.

Caffeine Citrate, B.Vet.C.

All species may be given orally twice the doses of caffeine owing to the reduced caffeine content of the salt.

Theobromine

Calcium salicylate:
- Dog: 0·3–1 g (5–15 gr.)

Sodium acetate and salicylate:
- Horse and Cow: 8–15 g (120–240 gr.)
- Dog: 0·3–1 g (5–15 gr.)

Aminophylline

Horse: 2–5 g (30–75 gr.) given in food or bolus three times daily for 2–3 weeks

Dog: 50–100 mg (¾–1½ gr.) three or four times daily orally, intramuscularly or intravenously.

Amphetamine, B.P. and Methylamphetamine, B.P.C.

The basic compounds are mobile liquids with the odour of geranium leaves. The methyl compound, however, is used most commonly as the hydrochloride and amphetamine as the sulphate or mono- and dibasic-phosphates. Both exist in *dextro*- and *laevo*-rotatory forms and are available as racemic mixtures. These salts are water soluble white crystalline substances with a bitter taste but no odour.

Action

Both compounds have powerful vasopressor activity and are classified as sympathomimetic amines which stimulate all structures having an adrenergic nerve supply. They are, additionally, powerful CNS stimulants and, because of strong medullary stimulation, are used as analeptics. The respiratory and vasomotor centres are strongly stimulated and also the cerebral cortex. Clinically, the central and peripheral effects are indicated by a gradual but powerful and sustained rise in blood pressure following myocardial stimulation and peripheral vasoconstriction, increased mental activity, and improved respiration. These effects appear within 2 min of parenteral injections of methylamphetamine, and within 30 min of oral dosage, but amphetamine salts are slower in taking effect. Both compounds maintain their effects for up to 9 hr or more.

Methylamphetamine is more powerful than amphetamine but the '*l*' form of the methyl compound has a more powerful pressor action than the '*d*' form, whereas the '*d*' form of amphetamine has a more powerful central effect than the '*l*' form. In both cases the racemic mixtures have intermediate potency. Both compounds have a potentiating effect on mental and physical capacity and are therefore called psychomotor stimulants, but depression follows their use for this purpose if dosing is protracted. A further action is that of diminishing hunger, doses of amphetamine (*d*-isomer) of 0·5 to 2 mg subcutaneously have suppressed feeding in fasting dogs for at least 1 hr. This property has been used for slimming in humans and has been applied in the treatment of obesity in dogs.

Uses

The amphetamines are mainly used as analeptics, especially in cases of shock with respiratory and circulatory collapse or near collapse. They cause a lightening of anaesthetic depression, with a stimulation of rate and depth of respiration and a rise in blood pressure up to, and possibly beyond, the pre-hypotensive level. With this raised and sustained blood pressure there is an increased danger of post-operative haemorrhage, and cases where the amphetamines have been used should be closely watched for signs of bleeding during the succeeding 12 hr.

Toxicity. Toxic symptoms are rarely seen but consist of excitement, restlessness, wakefulness, and possibly convulsions. Short acting barbiturates will counteract these effects.

Dosage

Amphetamine, as the *dextro*-isomer, is mainly

used orally in man and the methyl compound for either oral or parenteral administration.

Amphetamine:

Horse and Cow: 100–300 mg subcutaneously

Dog: 1·0–5·0 mg/kg (0·5–2 mg/lb.) parenterally, the higher doses being necessary in cases of anaesthetic depression.

For stimulant and weight reducing purposes, the dose for small animals is 2·5 to 10 mg repeated according to response. For the first purpose dosing is by the parenteral routes but for the latter use the oral route is usual.

Note: The minimum lethal dose by the subcutaneous route for sheep is 15 mg per kg, for rabbits is 20 mg per kg and for monkeys 20 to 25 mg per kg. The oral minimum lethal dose for the dog is 20 mg per kg.

Methylamphetamine may be given in approximately three-quarters of the amphetamine doses, on the basis of human dosage, but should be used with caution by the intravenous route.

For counteracting anaesthetic overdosage single parenteral doses, either by slow intravenous or by intramuscular injection, are usually sufficient, but, if the initial dose does not lighten the depression sufficiently within 15 to 20 min, small supplementary injections may be given. In man 2·5 to 10 mg are given orally once or twice daily; as a parenteral analeptic 10 to 15 mg are given by slow intravenous injection and the dose repeated in 15 to 20 min if necessary: intramuscular doses are higher (up to 30 mg).

Methylphenidate represents an attempt to synthesize a sympathomimetic with relatively clean arousal effect. It is a CNS stimulant which can cause convulsions in higher dosage.

Drugs which effect the mood are much used in attempts to relieve depression in man. Such psycho-analeptics are exemplified by amphetamine and the xanthines. More modern drugs differ from the psycho-analeptics in being used mainly in depressive states associated with mental disease. Such compounds are known as the antidepressants or thymoleptics and are exemplified by **imipramine.** Although it was long believed that amphetamine owed its biological activity to its ability to inhibit monoamine oxidase (MAO), it is now known that this is not its mechanism of action. However, a group of monoamine oxidase inhibitors has been developed for use as antidepressants in man. These compounds are claimed to be relatively specific for brain MAO and many are derivatives of hydrazine, e.g. **phenelzine.** It is doubtful whether diagnostic skills will develop sufficiently to allow the rational use of these drugs in veterinary therapeutics.

The Analeptics

This term is applied to compounds which if administered to normal animals in therapeutic doses will cause convulsions. This property is used to stimulate and revive patients which are over depressed by central nervous system depressants, but particularly by anaesthetics. True analeptics have very few actions other than their central nervous system stimulant effects.

The main analeptics are **Leptazol,** B.Vet.C. (1953); **Nikethamide,** B.Vet.C.; and **Picrotoxin,** B.Vet.C. Although amphetamine and its methyl derivative are used as antisedative drugs they are not so specifically stimulating to the medullary centres as the true analeptics.

Leptazol, B.Vet.C. (1953)

Also known as Pentylenetetrazole, U.S.P., the compound is a white crystalline powder with a bitter taste, it is soluble in water, alcohol, ether and chloroform.

Actions

Leptazol is a central nervous system stimulant. In normal animals its use causes stimulation of the higher centres and results in convulsions. Normal doses affect the medullary centres, but where these centres are depressed, as in narcosis or anaesthesia, the medullary effect is intensified. The respiratory, vagal and vasomotor centres are stimulated indiscriminately so that parasympathetic and sympathetic systems are stimulated equally. Apart from increased rate and depth of respiration, there is an increase in the blood supply to the coronary arteries following vasomotor stimulation, and the heart, and hence the circulation may benefit. The drug appears to decrease neuronal recovery time.

Leptazol is more rapidly absorbed than picrotoxin and has a shorter latent period.

Uses

Leptazol is mainly used in over depression due to any anaesthetic or narcotic, but has been reported of most use in barbiturate depression. For this purpose it may be given intravenously until muscle twitchings appear, and repeated every 15 min until recovery is assured. In less acute emergencies intramuscular or subcutaneous injections every 15 min should provide adequate stimulation.

Dosage

Horse and Cow: 0·5–1 g
Dog: 50–100 mg
Cat: 25–50 mg

The official injection contains 10 per cent w/v Leptazol.

Nikethamide, B.Vet.C.

This relative of nicotinic acid and nicotinamide is a yellowish or colourless oily liquid or crystalline solid with a faintly bitter taste. It is miscible with water and is soluble in organic solvents.

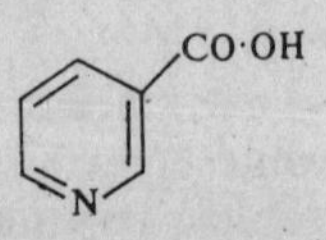

Nicotinic acid

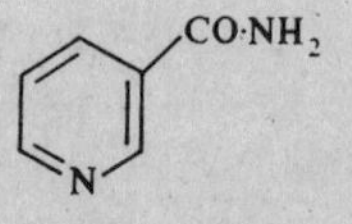

Nicotinamide

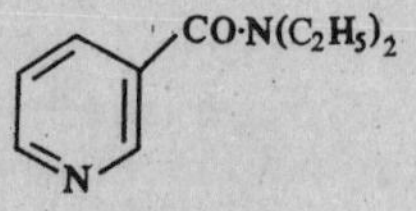

Nikethamide

Action

Nikethamide acts on the CNS non-selectively causing an increase in respiratory rate and depth by stimulation of the medullary centres and by action on the chemoreceptors of the carotid body. Some peripheral vasoconstriction is caused but the rise in blood pressure which usually follows is sometimes preceded by a sudden but transient fall. It is not such a potent analeptic as picrotoxin or leptazol.

Its mode of action is unknown but its ability to produce transient respiratory stimulation at doses below those producing more generalized effects is shared by doxapram and ethamivan.

Dosage

Horse and Cow: 10–25 ml
Dog: 1–3 ml
Cat: 1–2 ml

The official injection is a 25 per cent w/v solution in water which may be given orally or by any parenteral route. Doses must be repeated every 5 to 10 min.

The official injection is a 25 per cent w/v solution in water which may be given orally or by any parenteral route. Doses must be repeated every 5 to 10 min.

Picrotoxin, B.Vet.C.

Whereas the other two analeptics are synthetic, picrotoxin is the active principle prepared from the seeds of a shrub grown in India and the Malayan Archipelago. It is a white crystalline substance slightly soluble in water, soluble in alcohol and benzene but only slightly in other organic solvents.

Actions

Picrotoxin is a powerful stimulant of the whole central nervous system. The motor cortex, midbrain, and medullary centres, are strongly stimulated by therapeutic doses: larger doses also stimulate the spinal cord. These effects cause a marked increase in respiratory rate and depth, and a rise in blood pressure is also evident. Picrotoxin causes salivation and, occasionally, vomiting but is a reliable stimulant for barbiturate poisoning. Maximum effect is produced with doses which border on the convulsant level and twitching often accompanies the use of picrotoxin.

Picrotoxin is slow acting, taking 5 to 8 min to show results following intravenous administration, and 10 to 30 min after oral administration.

Leptazol is often preferred in barbiturate depression because of its more rapid action and the susceptibility of cats to picrotoxin in overdosage. The two may be used in conjunction during anaesthetic collapse.

Dosage

Horse and Cow: 60 mg
Dog: 1·5–6 mg

If pentobarbitone is the anaesthetic involved, dosage should be calculated at the rate of 2 to 3 mg picrotoxin for 60 mg (1 gr.) of pentobarbitone. If thiopentone is responsible for the depression the ratio is up to 2 mg per 60 mg (1 gr.) of thiobarbiturate. Intravenous or intramuscular doses may be repeated approximately every 15 min until recovery is assured. The usual strength of injections is 3 mg per ml.

In anaesthetic emergency, analeptics should never be used to the exclusion of artificial respiration, clearance of airway and support of circulatory function. In man, survival from barbiturate overdosage has improved significantly since reliance on analeptics alone has been abandoned. The mode of action of the analeptics is unknown, but at least part of picrotoxin's action is due to its ability to reduce GABA-mediated pre- and post-synaptic inhibition in the CNS.

Spinal Cord Stimulants

Although strychnine and the related alkaloid brucine will stimulate most areas of the brain, their main action is on the spinal chord.

Strychnine, B.Vet.C. (1953)

This alkaloid with a complex ring structure is derived from the seeds of a tree, *Strychnos nux vomica*, native to India, Malaya, Indo-China, and Northern Australia, but also from Ignatus beans, the seeds of a climbing shrub native to Indo-China and the Philippines. It is available in the form of the powdered seeds—Prepared Nux Vomica, B.Vet.C. (1953), which is a fine light fawn powder—and also as the liquid and dry extracts, and the tincture: the pure alkaloid is usually administered as the hydrochloride or the sulphate, which are typical alkaloidal salts with a bitter taste and about a 2 per cent water solubility.

Actions

In spite of the use of nux vomica in Europe from the fifteenth or sixteenth centuries and the isolation of strychnine in 1818 the pharmacological action of strychnine has only recently been understood.

Following oral or parenteral administration, strychnine is quickly absorbed and distributed to the tissues in the plasma and blood cells. No specific localization occurs in the nervous system.

After absorption, the main effect of strychnine is on the nervous system. The cerebral cortex may be slightly stimulated by small doses, but the medullary centres are only appreciably affected by near convulsive doses, and the use of strychnine as an analeptic in anaesthetic medullary depression is not recommended.

Within the CNS, there is considerable potential for the spreading of electrophysiological events through polysynaptic pathways. Limitation of spreading is believed to be mediated by inhibitory reflex pathways and mediators. Clinically evident stimulation may follow true excitation or the blocking of inhibition. Inhibition may be exerted at both pre- and post-synaptic membranes. Leptazol and nikethamide are believed to be excitants but picrotoxin, perhaps by competing with the endogenous inhibitory neurotransmitter GABA (α-amino-butyric acid), alleviates both pre- and post-synaptic inhibition. Strychnine interferes with the post-synaptic inhibition of spinal motoneurones by competitively blocking the inhibitory transmitter which is normally released by the Renshaw cells, probably glycine. Tetanus toxin acts by preventing the release of inhibitory transmitter.

Some strychnine is excreted in the urine but the majority is degraded in the liver.

Uses

In therapeutic doses, strychnine will act as a 'general nerve tonic' causing improved reflexes and a general speed up in the nervous response. Strychnine was widely used as a convalescent tonic in horses, cattle, and dogs, solution of Strychnine Hydrochloride, B.Vet.C. (1953) given as a subcutaneous injection in large animals, and

Easton's Syrup (Syrup of Ferrous Phosphate with Quinine and Strychnine, B.Vet.C. 1953) or similar formulae in tablet or liquid form are given to dogs.

Strychnine in the form of Nux Vomica was at one time used as a treatment for colic (Coulton–Reekes method), but gastro-intestinal stimulation by strychnine is slight and the use of these heroic methods is unjustified. Strychnine's bitter taste does, however, improve appetite and digestion when used as a drench, for it then causes reflex salivary and gastric secretion—see 'Sialagogues' (Chapter 5).

Toxicity

The dog is the species which has been most subject to poisoning, because, in addition to the species susceptibility which reputedly exists, the use of strychnine as a rodent and fox bait poison means that the dog may pick up the poison by scavenging. In Great Britain rat poisons no longer contain strychnine but occasional cases of strychnine poisoning are still seen when hounds and farm dogs gain access to poisoned fox and mole bait. *Symptoms* are unmistakable, consisting mainly of tonic spasms similar to those of tetanus in which the legs and tail are rigidly extended and the head drawn up and back. These spasms increase in severity until the death occurs from asphyxia during a spasm. Also unmistakable is the association of these spasms with any external stimulation, no matter how slight—sudden movements within the animal's sight, or slight noises, are quite sufficient. These symptoms are due to augmented reflexes, loss of the normal inhibitory mechanisms, and indiscriminate tonic contraction of all skeletal muscle: the rigid extension of limbs and appendages is due to the greater strength of extensor muscles which overrides the contraction of the flexors. Great pain results from this indiscriminate muscular contraction. Consciousness is unaffected until near the end, when anoxia due to respiratory paralysis during spasms causes cerebral depression. *Treatment* consists of the administration of local emetics only if convulsions have not commenced, followed by the induction of narcosis or anaesthesia, preferably with non-volatile short or medium duration anaesthetics. Pentobarbitone sodium is most suitable for all small species, whereas chloral hydrate would be suitable for large animals, both given intravenously. Dosage should be determined by effect and repeated as soon as muscle twitchings reappear. It will be found in dogs that from three to five times the normal doses are required to achieve anaesthesia, i.e. 88 to 143 mg per kg body weight (see also p. 243, Mephenesin).

Dosage

The doses of nux vomica powder and tincture have been detailed in the section on 'Bitters'.

Nux vomica

Dry extract

Horse and Cow: 0·3–1·5 g (5–25 gr.)
Sheep: 0·1–0·3 g ($1\frac{1}{2}$–5 gr.)
Pig: 30–150 mg ($\frac{1}{2}$–$2\frac{1}{2}$ gr.)
Dog: 3–20 mg ($\frac{1}{20}$–$\frac{1}{3}$ gr.)

This extract contains 5 per cent strychnine alkaloid.

The liquid extract contains 1·5 per cent strychnine alkaloid and the doses are therefore approximately three times those of the dry extract, but represented in fluid measures.

Strychnine salts. Oral or subcutaneous administration (hydrochloride, sulphate, nitrate and phosphate):

Horse and Cow: 15–60 mg ($\frac{1}{4}$–1 gr.)
Sheep: 5–15 mg ($\frac{1}{12}$–$\frac{1}{4}$ gr.)
Pig: 2–8 mg ($\frac{1}{30}$–$\frac{1}{8}$ gr.)
Dog: 0·3–1 mg ($\frac{1}{120}$–$\frac{1}{60}$ gr.)
Cat: 0·1–0·5 mg ($\frac{1}{600}$–$\frac{1}{120}$ gr.)

Solution of strychnine hydrochloride, B.Vet.C. (1953) contains 1 per cent w/v of the hydrochloride—about 10 mg per ml ($\frac{1}{110}$ gr. per m.).

Easton's Syrup contains 0·3 mg of strychnine hydrochloride per ml ($\frac{1}{360}$ gr. per m.). Doses can be determined by reference to those of strychnine hydrochloride, for the alkaloid content of these compounds determines their dosage.

Morphine Antagonist

In the past the analeptics, caffeine, and strychnine, have all been used in the treatment of morphine overdosage, but with doubtful effect. The superiority of *N*-allyl morphine has revolutionized the antagonism of the effects of morphine and its relatives.

Nalorphine Hydrobromide, B.Vet.C.

This compound is chemically related to morphine having a chemical name of *N*-allylnormorphine hydrobromide. The hydrochloride is also available and is similar in its details. The compounds are white crystalline powders which darken on exposure to air and light and which are soluble in water but less soluble in alcohol and other organic solvents.

Actions

In spite of its relationship to morphine, nalorphine hydrobromide has little or no analgesic or narcotic action, and when administered to patients depressed with morphine or its analgesic and narcotic relatives, nalorphine causes rapid revival. This is not an analeptic effect, for the substance has no convulsant action on normal individuals, nor does it assist in the revival of individuals depressed by barbiturates, chloral hydrate, or the volatile anaesthetics. Nalorphine is thought to act by competitive displacement of morphine and other compounds from the enzyme–coenzyme systems which they block. The effect is thought to be related to the greater affinity of nalorphine for the receptor.

Uses

Nalorphine became prominent as an antagonist to morphine, etc., for veterinary use in Great Britain with the introduction of the narcotic analgesic thiambutene hydrochloride. It has been used extensively in dogs with success. Administration to animals narcotized with thiambutene causes a dramatic recovery in under 1 min following intravenous injection, whilst recovery takes up to 20 min following subcutaneous administration. In some cases animals relapse into a state of sedation, or occasionally narcosis, but others will show no tendency to relapse. This is related to the relatively short duration of action of nalorphine. It is also of value in cases of overdosage with morphine and related depressants, e.g. pethidine, diamorphine and codeine.

Toxicity in dogs is not reported but overdosage may cause nausea and sedation. The margin of safety is very wide.

Dosage

The dose of nalorphine should be equivalent to that of the narcotic drug, but may be reduced in proportion to the time which elapses between the taking of the narcotic and administration of the antagonist. In the *dog* doses up to a total of 40 to 50 mg may be given as a routine, calculated at the rate of 2·2 mg per kg (1 mg per lb.) body weight. In deep narcosis this may be increased up to 3·3 mg per kg (1·5 mg per lb.). Intravenous, intramuscular or the subcutaneous routes may be used. Nalorphine hydrobromide is available in Great Britain in 5 ml vials with 20 mg per ml. The U.S.P. injection contains 5 mg per ml. The narcotic antagonists can reverse the narcotic component of neuroleptanalgesia.

Levallorphan, Naloxone, Diprenorphine

These drugs are further examples of competitive antagonists to morphine-series central depressants. They are all of greater potency than nalorphine, i.e. ×5 levallorphan, ×13 naloxone, ×35 diprenorphine. Whereas nalorphine and levallorphan both have some agonist activity, naloxone and diprenorphine are without depressant effect and are therefore superior. Additionally, their duration of action is usefully longer.

Bemegride, β-ethyl-β-methylglutarimide, B.Vet.C.

This compound is superficially related in structure to the barbiturates.

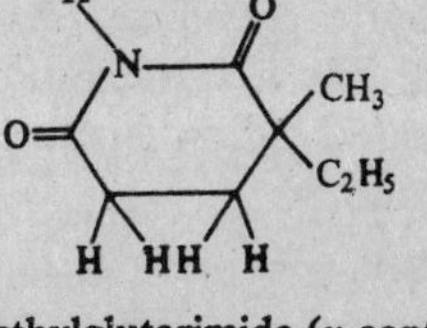

(i) Barbitone (diethyl barbituric acid)

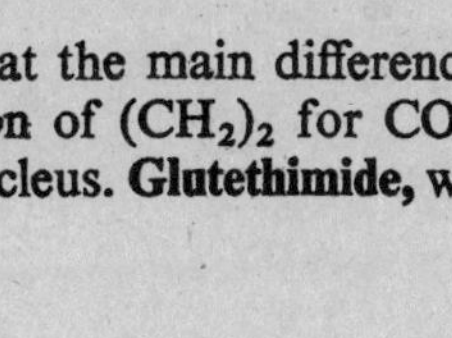

(ii) Ethyl methylglutarimide (α configuration)

This shows that the main difference consists of the substitution of $(CH_2)_2$ for CO . NH in the barbiturate nucleus. **Glutethimide,** which consists of a phenyl group substituted for the methyl group in the α configuration (ii) above, is related to phenobarbitone in the same way and is actually a sedative and hypnotic drug.

The sodium salt of this glutarimide alkyl derivative is a white crystalline compound which is soluble in water.

Actions and Uses

The pharmacological action of ethyl methylglutarimide is unknown, but it was thought that it acted with barbiturates in a similar manner to that of nalorphine with morphine. The salt, when injected intravenously into individuals anaesthetized with a barbiturate, lightens the depression immediately: in cats it does no more than this, but in dogs, calves, adult cattle, and sheep, an accelerated return to consciousness occurs. It is now accepted as a valued antagonist for use in circulatory, respiratory and general depression due to barbiturates.

Although it was originally described as a specific antagonist to barbiturates, it is nowadays more usual to regard bemegride as a general analeptic.

The drug would appear to be indicated mainly as a barbiturate antagonist following overdosage, but additionally can be used to terminate normal barbiturate anaesthesia, especially in poor risk patients.

Toxicity is not reported but overdose effects are similar to those of other analeptics—muscle twitching with convulsions in severe overdosage—these are controlled by barbiturates.

Dosage

Ethyl methylglutarimide sodium may be given by any parenteral route but in an emergency intravenous injection is required. Dosage is around 20 mg per kg but may be doubled in severe barbiturate overdose. As this analeptic, like all others, has a rather short half-life, the need for repeated administration must always be borne in mind.

Psychotomimetics

The sympathomimetics with CNS stimulant activity can elicit effects ranging from arousal of drug-sedated patients, through elevation of the mood in normal or depressed patients, to the induction of paranoid psychosis in overdose. The psychotomimetics (hallucinogens) are drugs which alter, as a primary effect, both quantitative and qualitative aspects of sensory perception and thought in a manner akin to psychosis.

LSD (lysergic acid diethylamide) is a prototype for this class of CNS stimulants and is thought to act by modifying the action or fate of 5-HT in the brain. It and certain analogues can certainly antagonise peripheral actions of 5-HT. Iontophoretic administration has shown 5-HT to be capable of stimulating certain neurones in the brain, an effect also blockable by LSD.

Mescaline is a much less potent cactus-derived hallucinogen with a molecular structure related to the sympathomimetics.

Cannabis induces sympathomimetic side-effects too, although this does not appear to be the basis of its psychedelic action. In contrast to the foregoing it produces a degree of concomitant CNS depression. At higher dosages euphoria is replaced by features associated with psychosis.

These agents are not official and are not obtainable through normal channels of supply. They have no place in veterinary medicine and are mentioned here only for the sake of completeness, as some members are used as research tools in studies of CNS function.

Bibliography

BERTELLI, A. (ed.) (1969) Pharmacological and clinical evaluation of analeptic drugs. In: *Circulatory Drugs*, part 3, pp. 232–318. Amsterdam: North-Holland.

BRIMBLECOMBE, R. W. & PINDER, R. M. (1975) *Hallucinogenic Agents*. Bristol: Wright Scientechnica.

MYERS, R. D. (1974) *Handbook of Drug and Chemical Stimulation of the Brain*. London: Van Nostrand Rheinhold.

VAN ROSSUM, J. M. (1970) Mode of action of psychomotor stimulant drugs. *Int. Rev. Neurobiol.*, **12**, 309–73.

25

Drugs which Modify Peripheral Nerve Transmission

Drugs which modify peripheral nerve transmission (local anaesthetics) may be defined as agents which, when applied to a nerve ending or trunk, block the transmission of impulses, causing loss of sensation if sensory trunks are anaesthetized and muscle paralysis if motor nerves are affected. In no case is brain activity influenced directly during normal use. The action is reversible, and with true anaesthetics no tissue damage is caused when they are used correctly.

In general the smaller diameter fibres in a mixed trunk are anaesthetized before the larger ones and the non-myelinated fibres are blocked before the myelinated ones. The smaller fibres usually recover more slowly. In sensory nerves the order of loss of sensation is pain, cold, warmth, touch and deep pressure.

The number of local anaesthetics is large but a relatively small number of these is in common use. In general, they may be of three types:

1. Substances causing anaesthesia by freezing:
   (a) Carbon dioxide spray
   (b) Ethyl chloride spray

2. Substances causing anaesthesia by chemical action:

Amethocaine Hydrochloride, B.P., B.P.C. (Tetracaine Hydrochloride, U.S.P., Anethaine, Decicain, Pantocaine)

Benzocaine, B.P., B.P.C., B.Vet.C. (Ethyl Aminobenzoate N.F., Anaesthesin)

Butacaine Sulphate, B.P., B.P.C., N.F. (Butyn)

Butyl Aminobenzoate, B.P., B.P.C., B.Vet.C., N.F. (Butesin, Planoform)

Chlorbutol, B.P., B.P.C., B.Vet.C. (Chlorbutanol, U.S.P., Chloretone)

Cocaine Hydrochloride, B.P., B.P.C., B.Vet.C., U.S.P.

Lignocaine Hydrochloride, B.P., B.P.C., B.Vet.C. (Lidocaine Hydrochloride, N.F., Xylocaine Hydrochloride)

Orthocaine, B.P., B.P.C., (Orthoform)

Procaine Hydrochloride, B.P., B.P.C., B.Vet.C., U.S.P. (Novocain, Ethocaine, Planocaine).

This list gives some idea of the complexity of nomenclature. Many of these substances have been developed by commercial houses and their proprietary trade marks have become widely known.

3. Substances causing local anaesthesia but not primarily used for this purpose, nor are they normally classified as local anaesthetics. This group includes substances that are generally quite toxic and may, and do, cause necrosis. Phenol, lead salts, belladonna, ethyl alcohol, magnesium sulphate, have all seen some use because of their anaesthetic attributes, but their value is limited.

Chemical Classification

The compounds listed in group 2 are only the more useful few of a large number of local anaesthetic compounds which have been developed, initially by modification of the cocaine molecule, and later by manipulation and modification of the procaine molecule.

The majority of the more useful compounds are derivatives of *para*-aminobenzoic acid (i) which is produced from these compounds during metabolic breakdown, and is responsible for sulphonamide antagonism. They may be:

1. Simple alkyl esters such as benzocaine (ii) and butyl-aminobenzoate (iii).

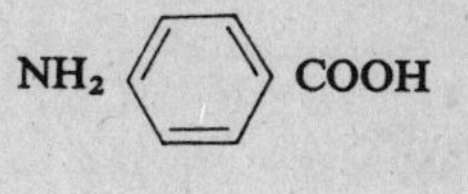

(i) *para*-Aminobenzoic acid

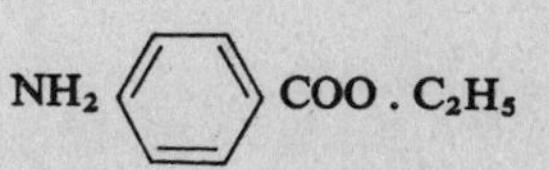

(ii) *para*-Aminobenzoic ethyl ester (Benzocaine)

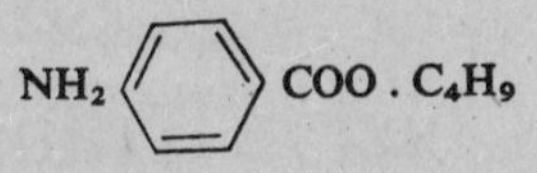

(iii) *para*-Aminobenzoic butyl ester (Butyl-aminobenzoate)

2. Alkyl amino- and simple alkyl ester combinations, such as procaine (iv) and butacaine (v):

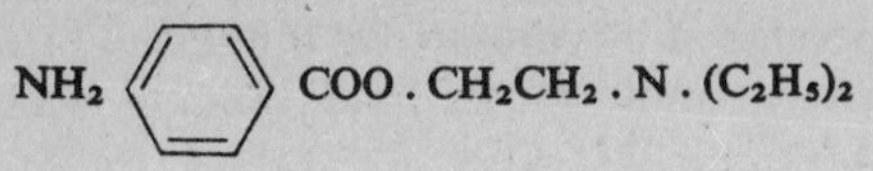

(iv) *para*-Aminobenzoic ethyl aminodiethyl ester (Procaine)

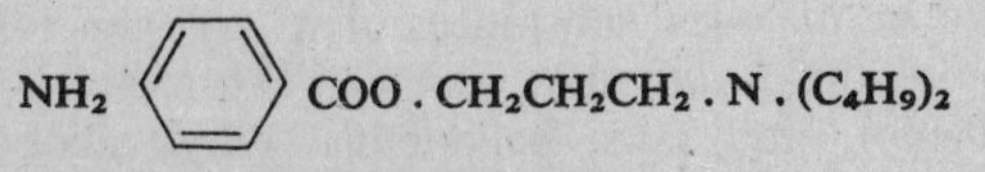

(v) *para*-Aminobenzoic propyl aminodibutyl ester (Butacaine)

Complex compounds having a basic structure similar to these form the bulk of the local anaesthetics now available and most can be grouped as substituted tertiary amino alcohol esters of substituted benzoic acids. Under this heading can be classed amylocaine, amethocaine, piperocaine (U.S.P.), hexylcaine (N.N.R.), and even cocaine (vi), which is perhaps the most complex of all, and yet contains within its structure the basic grouping (vii)

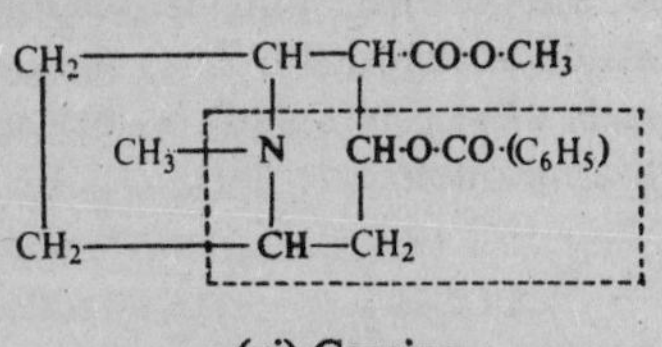

(vi) Cocaine

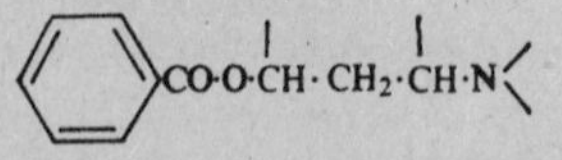

(vii)

Lignocaine (viii) may be looked upon as conforming to this general pattern but possessing a modified ester linkage:

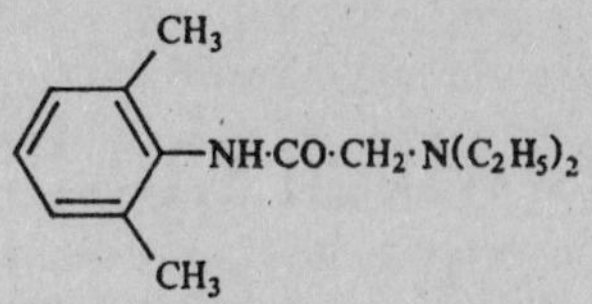

(viii) Lignocaine

A number of local anaesthetics do not fall under this general classification, but they have little veterinary significance.

Fate

The precise mechanism by which these drugs block nerve transmission is not yet known.

Absorption. Local anaesthetics are not absorbed through intact skin, but if the surface is broken or inflamed, certain anaesthetics applied in ointment form or in solution will be absorbed, the rate depending on the degree of damage to the skin and whether adrenaline is included in the formulation. If adrenaline is included the rate of absorption and onset of anaesthesia will be severely curtailed, but the duration of effect will be prolonged. This effect of adrenaline or other vasoconstrictors is also utilized, except in the case of cocaine, to diminish the anaesthetic's toxic side-effects. Following infiltration, the absorption rate depends on the vascularity of the tissues. Inflammation will therefore speed up absorption and reduce the duration of anaesthesia, whilst the inclusion of adrenaline will slow up absorption and increase the duration of effect. Too high a concentration of adrenaline will cause ischaemia, prolonged anaesthesia and, later, a delay in healing, or even necrosis and sloughing: 1:50 000 should normally be the maximum concentration to use. Hyaluronidase added in concentrations of 150 to 300 units per 100 ml anaesthetic will increase the speed of onset, enlarge the area anaesthetized by about 40 per cent but will shorten the duration owing to an increased absorption rate. It is well recognized that the toxicity of the local anaesthetics increases with the rate of their absorption.

Detoxication and elimination. The rate of breakdown and elimination varies in the different species and with different anaesthetics. Little is known of the metabolic fate of local anaesthetics other than procaine, but with this and related compounds *p*-aminobenzoic acid is released and with procaine the side chain becomes diethylaminoethyl alcohol, which is mainly excreted unchanged, but is in part metabolized. The acid is further converted to glycine or conjugated as a glucuronide, though some is excreted unchanged. These changes are achieved by the action of an enzyme possibly identical with plasma pseudo- or butyryl cholinesterase.

Sulphonamide antagonism occurs with all local anaesthetics which produce *p*-aminobenzoic acid during their metabolic destruction. This will include all the members of the procaine series.

Toxicity. This is similar with all local anaesthetics and is typified by restlessness, excitement, and convulsions, followed by depression, coma and medullary paralysis. Death is due to respiratory failure when the centre is paralysed. Treatment is similar—barbiturate sedatives in the excitement stages and stimulants in the later stages.

It should be remembered that hyaluronidase will increase the possibility of toxic symptoms if near maximal doses of anaesthetic are used.

Structure and Action

All the soluble local anaesthetics share the basic structure shown below.

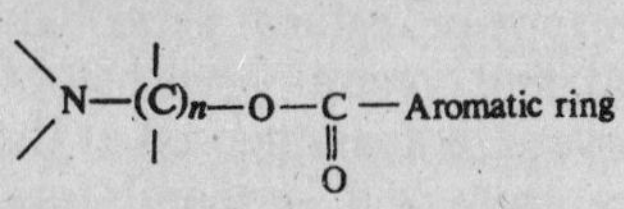

The following three chemical requirements have been postulated as basic to local anaesthetic activity:

1. A cyclic or heterocyclic ring to give the compound solubility in lipoid tissue, which is essential for local anaesthetic action.
2. An ester or isosteric group as a link.
3. A hydrophilic group to provide water solubility so that administration is easy. This is usually related to an amino alcohol or an isosteric group.

The explanation classically offered is that the local anaesthetic makes non-available the channels in the membrane through which the sodium ions flow, by occlusion, by increasing the surface pressure of the membrane or by increasing the degree of disorder in the membrane. Potassium ion fluxes are also impaired.

Local anaesthetics are weak bases which are unstable in solution and of limited solubility. Their water-soluble salts are, however, of satisfactory stability. In the body the acid salt is neutralized and the free base liberated in the pH of the tissue fluids. As the pKa of most local anaesthetics is alkaline relative to body pH, most of the base will be present in the uncharged non-ionized form. This is a property which aids the drug's ability to cross cell membranes by the usual process of diffusion and so to gain access to its site of action inside the axon. It is, howevei, likely that it is the charged cationic moiety which is of greater importance in bringing about conduction block following dissociation of the base inside the axon.

Whatever the chemical structure of these anaesthetics, they all interfere with the metabolism of nerve fibres, affecting cell metabolism probably by blocking enzyme pathways at the cytochrome C–cytochrome oxidase level. All act by deposition of lipoid soluble bases in the tissues, their salts being used because of their water solubility, which makes manufacture and administration easy. Once deposited in the vicinity of the nerve fibre, they probably diffuse into the lipoid tissue of the fibre and prevent depolarization of the nerve—at the nodes of Ranvier in myelinated nerves; they therefore prevent the transmission of nerve impulses. The smaller fibres within a nerve trunk are affected first, but with increase in time or concentration, the larger fibres are affected progressively in order of size. This means that the senses are also progressively affected beginning with vasomotor control (the finest fibres), followed by temperature appreciation and pain, touch, general and special sensation such as joint sensation, taste, deep pressure, etc. Motor nerves are affected after the sensory fibres and finally proprioceptor innervation is affected.

Ionic theory of nerve impulse transmission. It is believed that a molecular barrier to ionic diffusion exists around the axoplasmic cylinder of nerve fibres. This plasma membrane, called the axolemma, is described as being possibly only two layers in thickness and composed of lipoid and protein molecules. This is surrounded by the spirally arranged layers of lipoprotein which form the insulating myelin sheath, broken every millimeter or so by the nodes of Ranvier. This in turn is surrounded by the neurilemma.

In the resting stage, the plasma membrane provides a demarcation between a positive electrical potential on the outside and a negative potential on the inside (ix). These potentials are created by a high concentration of electro-positive sodium ions with a low concentration of electro-positive potassium ions on the outside, and the reverse on the inside with the addition of electro-negative amino acid, the electro-negative chloride ions remaining more or less static. Following the initiation of an impulse by an outside influence, the plasma membrane becomes permeable to sodium ions which pass into the cylinder of cytoplasm (by a mechanism not known), whilst potassium ions are static. This therefore reverses the potential for an excess of anions will exist on the inside (x). This change in potential passes along the nerve in a wave motion and is, of course, the nerve impulse. Potassium

(ix) Resting period

Outside fibre	+ive	$K^+Cl^-Na^+$
Inside fibre	−ive	$K^+Cl^-Na^+$
		Amino acid$^-$

(x) Active period

Outside fibre	−ive	$K^+Cl^-Na^+$
Inside fibre	+ive	$K^+Cl^-Na^{++}$
		Amino acid$^-$

diffusion takes place to the outside neutralizing the potential difference and this is later followed by a complete reversal of ion migration and a return to the resting state. At synapses and motor end-plates, acetylcholine is responsible for the changes in permeability of the plasma membrane, which is followed by the changes in potential.

Local anaesthetics are able progressively to stabilize the ion distribution and prevent the generation and passage of the action potential.

Uses and Application

The method of application determines the effect and therefore the usage of local anaesthetics. They may be applied in several ways:

1. *Topical anaesthesia*, indicating the application of a local anaesthetic to an external surface. This type of anaesthesia is useful for anaesthesia of the mucous membranes of the eye, nose, mouth, ear and perineal region. Topical anaesthesia of the skin is difficult unless abrasions or inflammatory areas exist. Some anaesthetics such as cocaine and amethocaine are very useful in this respect, and others such as procaine are relatively useless.

2. *Infiltration or field anaesthesia* in which a local anaesthetic is injected around the site of irritation or operation. Cocaine, procaine, and lignocaine, are examples of local anaesthetics suitable for field anaesthesia: the less soluble agents such as amethocaine, butacaine, and benzocaine, are unsuitable for this purpose. Concentrations should be lower than those intended for regional anaesthesia.

3. *Regional anaesthesia*, which includes:

(a) Nerve blocking or conduction anaesthesia, in which local anaesthetic solution is injected around, or near and parallel to, a nerve trunk at a superficial point remote from the operation or injured site. By this means the whole region innervated by this nerve trunk is rendered insensitive and, if the trunk incorporates the regional motor fibres, immobile. Examples of this are the blocking of the plantar and median nerves of the horse, sometimes used in the diagnosis of lameness: paravertebral nerve block in the cow is another type of regional nerve block in which the dorsal and ventral branches of the main spinal nerves are anaesthetized at their point of emergence from the vertebral foramina. The usual nerves that are blocked are the last dorsal (thirteenth), and first three lumbar nerves. Once blocked, the whole flank is completely anaesthetized.

(b) Epidural (spinal) nerve block, by which anaesthetic is introduced into the neural canal but not below the meninges covering the cord. This means that the injection is made into the lower end of the vertebral canal where there is normally space available for the introduction of anaesthetic. Low (posterior), medium, and high (anterior), epidural anaesthesia can be obtained, depending on the volume of fluid used: 'low' anaesthesia is given by the smallest volume (5 to 15 ml in the cow), as the anaesthetic tends to fill the caudal end of the space first; as the quantity is increased, the space is filled further and further forward giving medium, and with the greatest volume (up to 130 ml for the cow), high, epidural anaesthesia. Low epidurals are useful for tail and perineal operations and for stopping straining in obstetrics. The animal remains standing. Solutions of 1 to 2·5 per cent procaine hydrochloride are suitable for this purpose. High epidurals usually cause hind leg paralysis and the animals become cast: they are useful for abdominal operations, surgery of the hind legs and udder, drawing of the penis, and obstetrics. Unilateral epidural anaesthesia can be obtained by injecting the anaesthetic with the animal on its side.

Great care must be observed when carrying out high epidural anaesthesia, for bilateral paralysis of the splanchnic nerves containing the vasomotor fibres will produce splanchnic area vasodilatation and an alarming and dangerous fall in blood pressure. These nerves originate at the level of the last seven or eight dorsal, and the first one or two lumbar vertebrae. This danger is avoided to a great extent by using 1 per cent solutions only for anterior epidurals, but where it does occur, adrenaline intravenously and the use of cardiac and respiratory stimulants are indicated.

(c) Intrathecal (spinal) injection, which entails the withdrawal of cerebrospinal fluid from the subarachnoid space and its replacement with local anaesthetic, is a technique now rarely if ever used in veterinary anaesthesia.

4. Intravenous injections of local anaesthetics have been used for generalized analgesia, but there are few reports and the procedure is not yet generally accepted in the veterinary field.

An alternate intravenous technique is used to obtain regional anaesthesia of a limb distal to an occluding tourniquet. Below this the drug is injected into a superficial vein from which it diffuses to its site of action.

Topically Applied Anaesthetics

The local anaesthetics can be divided fairly rigidly into those applied topically and those which may be applied universally. The topically applied may again be divided into those causing anaesthesia by freezing and those which act by biochemical reaction.

Freezing anaesthetics. There are two of these available: they are of occasional use for superficial and transient anaesthesia. They are carbon dioxide and ethyl chloride. Both are gases stored under pressure. When released, these gases evaporate rapidly drawing heat from the surface upon which they are sprayed; this causes an intense cooling and literally a freezing of the tissues, which results in a loss of sensation. Short duration superficial surgery can be carried out during the period of anaesthesia: this applies to the removal of warts, foreign bodies, the opening of abscesses, the removal of portions of broken claws, and similar operations.

In addition carbon dioxide snow, prepared by releasing the gas rapidly through a nozzle can be moulded and used as a 'pencil' for the removal of warts: repeat applications should only be applied after the inflammation caused by the preceding application has subsided, and normal skin should always be protected by bland ointments.

Biochemically active anaesthetics. Of the list given in group 2 of the general classification on page 293, those examples which lack the hydrophilic terminal substituted tertiary amino group and are therefore of a low aqueous solubility, and consequently too slowly absorbed to be toxic, are used for topical application to wounds and raw surfaces. The slow rate of absorption of these powders extends their duration of action in comparison with the soluble compounds.

Amethocaine Hydrochloride, B.P.

Chemically, this anaesthetic is the hydrochloride of 2-dimethyl-aminoethyl *para*-butylaminobenzoate. It is a typical local anaesthetic—a white, crystalline, odourless powder with a slightly bitter taste followed by a slight numbness. The hydrochloride is very soluble in water, soluble in alcohol, but insoluble in other organic solvents. Amethocaine hydrochloride is unstable in neutral solutions, but in 0·001 per cent hydrochloric acid will withstand temperatures up to 115°C.

Actions and Uses

Amethocaine hydrochloride may be used topically or parenterally. When applied topically its action is approximately ten times greater than cocaine, and when used parenterally it is similarly ten times more active than procaine. It is mainly used for surface anaesthesia. Topically, amethocaine has a similar action to cocaine but more prolonged, and 0·25 to 1 per cent solutions are very effective in the eye, whilst 1 per cent ointments are used on the skin. The effect on intact skin is not very great but, when applied to inflamed or abraded surfaces, anaesthesia is good. Amethocaine is a useful anaesthetic for relieving the irritation caused by otitis in the dog and cat.

Where parenteral administration is intended, strengths of 0·05 to 0·25 per cent may be used for infiltration, and for epidural injection, 0·5 to 1 per cent. In both cases the onset of anaesthesia is slower than with procaine but more prolonged.

Toxicity. Amethocaine hydrochloride is approximately twelve times more toxic than procaine and though this is largely offset by the greater activity, procaine is still preferred for parenteral use.

Benzocaine, B.Vet.C.

Ethyl *para*-aminobenzoate is the chemical name for this white, odourless, crystalline powder with a slight bitter taste which is followed by numbing of local sensation. It is almost insoluble in water but soluble in alcohol, ether and chloroform (1:8, 1:4 and 1:2, respectively), and to the extent of 2 per cent in the fixed oils. Benzocaine acid salts such as benzocaine hydrochloride are irritant and of no clinical value. Solutions of benzocaine cannot be sterilized by heat. The base must be sterilized and the anaesthetic incorporated with aseptic techniques.

Actions and Uses

Benzocaine, being relatively insoluble in water and not very soluble in oil unless ether is incorporated, is not suitable for parenteral therapy, and when applied topically does not penetrate deeply. It is therefore indicated only for superficial anaesthesia applied to abraded surfaces and areas of irritation; it is of relatively little value on intact skin or mucous membranes. When applied externally, it is used in the form of 5 to 20 per cent ointments, or dusting powders of similar strengths diluted with starch, zinc oxide, calamine, or talc.

Toxicity is under one-tenth of that of cocaine and is of little significance when used externally.

Butacaine Sulphate, B.Vet.C. (1953)

This sulphate of a *para*-aminobenzoic acid ester is a white, odourless, crystalline powder with a slightly bitter taste and a mild, transient anaesthesia of the mucous membranes. It is slowly but very soluble in water, very soluble in alcohol and acetone: insoluble in ether and almost so in chloroform. Butacaine sulphate solutions may be sterilized by boiling.

Actions and Uses

Butacaine sulphate effectively penetrates mucous membranes and has an action equal to that of cocaine. The onset of anaesthesia is more rapid than that with cocaine and the effect is more prolonged. With these features, butacaine is particularly suitable for anaesthesia of the eye and does not cause drying, ischaemia and necrosis of the cornea, all of which result from repeated use of cocaine. Solutions are almost non-irritant and single applications will give superficial corneal anaesthesia in 1 min: further applications at 3-min intervals will give surgical anaesthesia of the cornea and sclera but there is no effect on the iris and pupillary action is maintained.

Solutions of 2 per cent butacaine sulphate are suitable for anaesthesia of all mucous membranes it may also be used on the skin.

Toxicity. The toxicity of butacaine sulphate when applied externally lies between that of cocaine and procaine, but parenterally it is more toxic than cocaine and should therefore not be so administered.

Butyl Aminobenzoate, B.Vet.C.

This local anaesthetic is butyl *para*-aminobenzoate and therefore closely related to benzocaine. Its characteristics are very similar to those of benzocaine although it is slightly more soluble in organic solvents. Solutions for injection may be sterilized by steaming.

Actions and Uses

The action of butyl aminobenzoate is prolonged, injections in oil providing anaesthesia for periods of up to 24 hr. The compound can be used as a dusting powder either undiluted or diluted with talc, boric acid, zinc oxide and similar diluents, but it must not be used in crystalline form for the crystals will cause mechanical irritation. A concentration of 1 per cent is usual in ointments. Injections of butyl aminobenzoate alone are uncommon but combinations of 6 per cent of the aminobenzoate with 5 per cent benzyl alcohol, as a solvent and adjuvant, often with 1·5 per cent procaine base, have been widely used for veterinary purposes. Dusting powders and ointments are useful and effective in the treatment of irritant skin lesions, wounds, and burns: solutions in oil have also been used for similar purposes, but these are probably of most value externally in allaying irritation in the ears of small animals. Parenteral injections have been used successfully for prolonged field anaesthesia and for epidural injections to control 'straining' following the return of uterine, vaginal, and rectal prolapses. A similar preparation, but with a base of polyethylene and propylene glycols, and containing procaine hydrochloride also, has been used to achieve epidural and field anaesthesia for periods of a week or more. Reports of tissue necrosis and neuritis have been reported following human therapy, though the damage has been attributed to the vehicle rather than the anaesthetic: this may be so, but it is also probable that the crystals precipitated in the tissues and slowly resorbed cause local irritation and necrosis.

Butamben picrate (A.D.I. 1974) is a combination of butyl aminobenzoate and picric acid. It has been produced commercially as Butesin Picrate. The yellow powder is soluble in water at 1:2000 and in oil at 1:100. The aqueous solution has anaesthetic effect when applied to the eye but it is most commonly used as a 1 per cent ointment both for the treatment of burns, scalds, and pruritis of the skin, and also for ocular lesions. All tissues are stained yellow by the picrate.

Toxicity. Butyl aminobenzoate is not significantly toxic, but prolonged use in man has caused dermatitis; this may be severe with the picrate due to the irritant effect of the picric acid.

Chlorbutol, B.Vet.C.

Is most commonly used as a remedy for motion sickness in small animals, but besides its hypnotic central action, chlorbutol is also an effective local anaesthetic for topical application. It is occasionally used for this purpose at a 1 to 2 per cent concentration in dusting powders for application to granulating wounds, for which purpose its antiseptic action is also useful. (A saturated solution—0·8 per cent in water—is completely bacteriostatic and the pure substance has a phenol coefficient of 1·2.) When given orally, the local sedative action on the gastric mucosa is a useful adjunct to the central action: this action is made use of in the oral administration of drugs which tend to cause emesis, particularly anthelmintics. The combination with arecoline hydrobromide in the treatment of tapeworms in the dog is an example.

Orthocaine, B.Vet.C. (1953)

The chemical name is methyl, 3-amino-4-hydroxybenzoate. It is a white, odourless, tasteless, crystalline powder which decomposes on boiling with water, especially alkaline water. Orthocaine is only slightly soluble in water, but is soluble in alcohol and ether to the extent of 1:7 and 1:50, respectively.

Actions and Uses

Orthocaine is not used very frequently. It may be substituted for benzocaine, as a dusting powder diluted or undiluted, or as ointments in strength of 5 to 20 per cent, but its action is weaker.

Toxicity. This is of little significance systemically, for its low solubility prevents absorption. Externally it may be responsible for severe irritation and sloughs.

Parenteral Local Anaesthetics

The local anaesthetics which can be classed as parenteral ones are cocaine, procaine, and lignocaine. Cocaine, though no longer recommended for parenteral use, was the first of the true local anaesthetics and was in use as recently as 1943, for field anaesthesia, in conjunction with morphine sulphate and chloral narcosis. Its efficiency is beyond doubt, but legal control and the lower toxicity of procaine and lignocaine, as well as other advantages, make it less attractive except for ophthalmic use. Procaine is entirely restricted to parenteral use but lignocaine is valuable externally as well as internally.

Cocaine Hydrochloride, B.Vet.C.

Any preparation containing more than 0·1 per cent of cocaine is subject to control under the Misuse of Drugs Act. Any preparation which is merely a dilution of cocaine is controlled in all proportions.

Although many of the oily applications contain the pure alkaloid, it is the hydrochloride which is most widely used. The alkaloid is prepared by extraction of a mixture of alkaloids from the dried leaves of shrubs of the *Erythroxylum* genus. The mixture of alkaloids is usually hydrolysed to ecgonine, from which the cocaine is synthesized. Cocaine hydrochloride is a typical alkaloidal salt: it is a white, crystalline compound with an intensely bitter taste, no odour and a solubility in water of 1:0·5, in alcohol and glycerin of 1:3, in chloroform of 1:20, but it is insoluble in fixed oils. It differs from most other alkaloids in causing temporary anaesthesia of the oral mucosa when tasted. Solutions and ointments cannot be sterilized by heat: vehicles must first be sterilized and the cocaine incorporated aseptically.

Actions

The *local* action of cocaine is one of anaesthesia, first of the nerve endings of touch and pain, and later of pressure. Still later there is a paralysis of motor fibres. Cocaine, by impeding the re-uptake of noradrenaline, potentiates the action of locally released transmitter, which causes a constriction of local capillaries, slows the absorption, and prolongs the action of the cocaine. Anaesthesia takes about 5 min to take effect and lasts for ½ hr or more depending on concentration and tissue vascularity. When anaesthesia passes off, overdilation of the capillaries occurs and post-operative bleeding should be anticipated.

When applied to the eye, absorption across the conjunctiva is fairly rapid and deep actions are soon apparent, for the adrenaline potentiation effect causes a dilatation of the pupil, although accommodation is normally unaffected. Intra-ocular pressure reacts irregularly and depends mainly on the degree of iris collapse, closure of the spaces of Fontana and restriction of the drainage. With anaesthesia, the blink reflex is lost and the cornea dries; with the drying arises the danger of conjunctivitis and ulceration. Prolonged treatment with cocaine increases the risk of corneal ulceration by a direct toxic action.

The *systemic* effects of cocaine are of little significance in veterinary medicine and surgery, for ocular anaesthesia is now almost its only use. The action is mainly one of stimulation of the CNS, circulatory system and skeletal muscle. In small quantities mental and physical fatigue are decreased, the heart reacts more powerfully and this, in conjunction with vasoconstriction due to the adrenaline potentiating action, causes a rise in blood pressure. The general clinical reaction is one of restlessness. Large doses cause convulsions followed by paralysis, coma, and death from respiratory paralysis and asphyxia.

Uses

Cocaine is now almost entirely used for ocular and mucous membrane anaesthesia. For surgical anaesthesia, 4 to 5 per cent solutions or ophthalmic ointments are satisfactory, but, for repeated use, concentrations should not be greater than 1 per cent because of the danger of corneal ulceration, and of paralysis of the blink reflex which would result in drying of the conjunctiva and the adherence of dust.

If anaesthesia of intact skin is required, oily solutions of cocaine base are more easily absorbed and are more likely to give anaesthesia.

Oily eye drops (1 per cent) with and without mercuric chloride are official, and also aqueous eye drops with (1 per cent), and without (2 per cent), homatropine (2 per cent). The B.P.C. eye ointment is of 0·25 per cent strength.

Toxicity. Cocaine is unsafe to use parenterally and large wounds or abraded surfaces should not be dressed with cocaine, for absorption is more rapid than detoxication and excretion, and toxic levels may be reached rapidly. Treatment is symptomatic. It should be remembered that rabbits are particularly susceptible to cocaine poisoning though ocular anaesthesia with cocaine is not precluded.

Procaine Hydrochloride, B.Vet.C.

This was the first of the synthetic true local

anaesthetics, and until the advent of lignocaine, was considered the best for routine veterinary use. It is still used very widely.

The systemic and local toxicity of cocaine coupled with its potential for abuse and its stringent control by law provided the impetus for the search for more acceptable analogues. Unfortunately, in many of these, it has not been possible significantly to dissociate local analgesic potency from systemic toxicity. The first valuable analogue was procaine although it is ineffective topically and short-acting.

Actions and Uses

Procaine hydrochloride is an effective local anaesthetic which has a latent period of 5 to 10 min and a duration of about 1½ hr with a 2 per cent solution containing adrenaline of 1: 100 000. *Paravertebral* injections of the same solution in cattle will give anaesthesia for about 2 hr and *epidural* administration also gives anaesthesia for 2 hr. *Topically*, procaine is about sixteen times less active than cocaine and is considered unsuitable for this purpose.

Procaine is almost non-irritant to tissues; the hydrochloride is therefore used for *field anaesthesia* in strengths of 2 and 5 per cent, but for *regional anaesthesia* it is advisable to use the 5 per cent concentration. *Epidural* injections of strengths from 1 to 5 per cent are used, but where anterior or high epidural is intended 1 per cent should be used so that disastrous falls in blood pressure may be largely avoided; for cows up to 12 to 18 ml (1200 lb.) will give good epidural anaesthesia for about 2 hr without causing the animals to lie down. Inclusion of adrenaline 1:50 000 to 1:100 000 will prolong anaesthesia to 2·4 hr but an increase in volume or concentration will cause animals to go down. If 2 per cent strengths are used, posterior or low epidural doses will be: heifers: 10 ml; medium sized cows; 12 ml; large cows: 15 ml. With concentrations of 3 to 5 per cent the maximum should be 10 ml. Anterior epidurals may require 50 to 130 ml of a 1 per cent solution.

Toxicity. Procaine is four times less toxic than cocaine when used alone, but with adrenaline it is as much as ten times less toxic. Maximal doses of solutions containing adrenaline are as quoted under lignocaine toxicity (*q.v.*) scaled up or down according to concentration. Symptoms are those of restlessness, excitement and convulsions followed by depression, coma and death from respiratory paralysis. This is typical of most local anaesthetics. Treatment is symptomatic. Procaine can cause hypersensitization and elicit allergic reactions. The toxicity of procaine depends upon its rate of absorption, i.e. no toxic signs are seen unless the rate of absorption exceeds the hydrolysing capacity of the detoxifying enzyme. By reducing the rate of absorption, adrenaline reduces the toxicity of procaine.

Lignocaine Hydrochloride, B.Vet.C.

This valuable local anaesthetic was developed and first used in Sweden. Details were published in 1946. Since then it has gained world-wide recognition in both the human and veterinary fields. It is supplanting procaine for routine veterinary local anaesthesia. Chemically, it is an alkyl amine derivative of *meta*-xylene (dimethylbenzene).

Lignocaine hydrochloride is a white crystalline powder with a slight but characteristic odour. It is soluble in water, alcohol and chloroform, but insoluble in ether, benzene, and oils. Solutions may be sterilized by heat.

Actions and Uses

Lignocaine is approximately twice as effective as procaine both in potency and area of diffusion, but it is also more toxic in concentrations greater than 0·5 per cent, being approximately 50 per cent more toxic in 2 per cent concentrations. The greater potency more than offsets this toxicity. The onset of anaesthesia is more rapid and the duration is more prolonged. Solutions are almost non-irritant. In a series of tests comparing the effects of lignocaine on cattle with those of procaine and other local anaesthetics, lignocaine was found to have the most rapid onset and the best diffusion properties. Lignocaine hydrochloride 2 per cent with adrenaline 1:100 000 was used: for *epidural* anaesthesia, 1·25 ml per 50 kg (1 cwt.) gave anaesthesia of the perineal region in as little as 3 min and a duration of 4·2 hr compared with 1·7 hr for procaine: for *paravertebral* anaesthesia 7·0 ml were infiltrated around each nerve; anaesthesia lasted for 4·3 hr and was complete, whilst procaine gave an average of 2 hr of less profound and incomplete anaesthesia with a longer period of induction: for *infiltration* anaesthesia the same solutions were used and though standardization was difficult, lignocaine gave good anaesthesia, rapid diffusion, a shorter latent period and a duration of 2·7 hr compared with procaine, which gave a duration of only 1·6 hr. *Topical* effects were not compared, but whereas procaine is relatively useless, lignocaine is suitable for ophthalmic purposes, anaesthesia of all mucous membranes, and is effective in pruritus and pain from abraded surfaces, wounds, and inflammatory lesions.

Toxicity. No exact figures of maximum doses are available, but if volumes of adrenalinated solutions similar to those used for procaine solutions are used no trouble need be anticipated (B.Vet.C., 1965). The equivalent of approximately 100 ml of a 2 per cent solution has been used in cattle without any signs of toxicity. On the basis of 2 per cent procaine and cocaine solutions, the following volumes would be maximal for 2 per cent lignocaine solutions given subcutaneously:

Horse and Cow: 400 ml
Sheep and Pig: 80 ml
Dog: 25–65 ml
Cat: 8·5 ml

In practice these volumes will rarely be required.

Availability. Lignocaine hydrochloride is available commercially as sterile injections of 0·5, 1 and 2 per cent strengths, with adrenaline 1:80 000 or 1:100 000 in some cases. There are isotonic solutions of 4 per cent and ointments of 5 per cent concentrations for topical application, including ophthalmic use, and a 2 per cent carboxymethylcellulose gel also for topical application.

Lignocaine is an amide and is not hydrolysed by pseudocholinesterase. It is detoxified much more slowly than procaine and so the rate of absorption and the concentration of the solution used are very important. For these reasons the toxicities of solutions above 0·5 per cent concentration increase more rapidly than the corresponding dilutions of procaine. Without adrenaline, it is possible that enough lignocaine would be absorbed from a regional block to produce drowsiness. Overdosage will lead to twitching, respiratory depression, hypotension and convulsions.

The amide-type local analgesics do not cross-react in ester-type hypersensitivity states. Further examples of such agents are **prilocaine, bupivacaine** and **mepivacaine.** These lignocaine analogues have been synthesized for enhanced analgesia relative to local and systemic toxicity, rapidity of onset of analgesia, more intense action and greater duration of action. Prilocaine can cause methaemoglobinaemia through a metabolite, but bupivacaine has twice the duration and four times the potency of lignocaine, while mepivacaine is also more rapid in onset.

MUSCLE RELAXANTS

Interference with the transmission of nerve impulses can take place not only at the axon but also at the endings of the motor nerves, in which case the effect produced is paralysis of striated muscle.

When the action potential arrives at the nerve terminals in the neuromuscular junction it causes a release of acetylcholine (ACh) from storage vesicles in the endings. This neurotransmitter diffuses across the synaptic cleft and causes the muscle membrane to become depolarized by modifying membrane permeability and so allowing the influx of Na^+ ions. This depolarization may be measured as the end-plate potential which, if of sufficient size, triggers the muscle action potential and thus muscle contraction. This site of ACh action, along with the ganglia and the adrenal medulla, are said to be nicotinic. The pharmacological spectrum of ACh's actions may be subdivided into muscarinic and nicotinic effects, i.e. effects reproducible by the alkaloids muscarine or nicotine. These effects may be antagonized by specific blocking drugs, e.g. atropine in the case of muscarinic sites. The nicotinic receptors are not homogeneous in that drugs which are ganglion blockers (Chapter 8) are not effective at the myoneural junction. Those drugs of clinical value as voluntary muscle relaxants have affinity for the ACh nicotinic receptors in the end-plate and so prevent the released ACh from gaining access to its receptor. The blockade produced is classically described as competitive, reversible or surmountable when it can be reversed by increasing the availability of ACh at the receptor. Drugs which induce reversible blockade do not depolarize the end-plate and are therefore known as non-depolarizing blockers.

Other drugs do depolarize the membrane before blocking transmission and muscle fasciculations are a recognizable feature of the depolarizing blockers. As the blockade is in this case made more intense by ACh, it is known as non-competitive, irreversible or insurmountable blockade.

A third group of drugs possesses a lesser ability to induce generalized voluntary muscle relaxation by an action mediated through the CNS (Chapter 21).

Botulinus toxin is derived from *Clostridium botulinum* and, in that it produces voluntary muscle paralysis by diminishing the ability of the action potential to cause the liberation of ACh, induces what is called prejunctional block. ACh release is enhanced by Na^{+} and Ca^{++} ions but is reduced by Mg^{++} ions.

In using the muscle relaxants as adjuncts to anaesthesia it is essential for humane reasons that the depth of anaesthesia is such as to ensure adequate analgesia.

Competitive Blocking Agents

The members of this group of compounds are also referred to as *curarizing agents* since the alkaloids of curare were the first of these agents to be investigated and used surgically. The important members of this group are Tubocurarine Chloride, B.Vet.C., Dimethyl Tubocurarine Iodide, N.F., and Gallamine Triethiodide, B.Vet.C.

Tubocurarine Chloride, B.Vet.C.

This white odourless, microcrystalline alkaloidal salt is soluble in water to the extent of 1:20, and in alcohol of 1:45. It is insoluble in chloroform and ether. The preparation was standardized biologically, like many other useful drugs, before its chemical identity was fully evaluated. With the preparations of improved purity now available, standardization is effected spectrophotometrically.

Action

Tubocurarine chloride affects the myoneural junction by preventing the access of acetylcholine to the motor end-plate: this may also be expressed as raising the stimulation threshold of the end-plate. This results in a paralysis of skeletal, or voluntary, muscle, commencing in the eyes and then affecting successively, muscles of the face, neck, limbs, abdomen and respiratory muscles (intercostal and diaphragm). Recovery takes effect in the reverse order. The action is rapid following intravenous injection, the first effects appear in about a minute and maximum action in 3 to 5 min: duration of effect is about 45 min, depending on individual sensitivity and dosage. Where respiratory paralysis is going to occur it may be expected 7 to 10 min subsequent to administration.

Dogs and cats appear to tolerate tubocurarine less well than other species.

Note: Tubocurarine has no sedative or analgesic effect and may induce some hypotension due to incidental ganglion block: a means of administering artificial respiration should always be at hand in case of diaphragmatic paralysis.

Excretion. Tubocurarine is excreted mainly in the urine, partly unchanged and partly metabolized.

Toxicity. The side actions of tubocurarine are mainly those associated with histamine release, which occurs in a number of cases. The chief symptoms of histamine release are salivation, a drop in blood pressure, bronchial secretion, and bronchial spasm: the administration of antihistaminics will counteract these effects. Intratracheal intubation is advisable in all cases submitted to relaxant injections.

Overdosage produces death from asphyxia; occasionally a large drop in blood pressure is an aggravating factor. Relaxation of the cardiac sphincter and oesophageal muscles may allow regurgitation of food from the stomach.

Treatment consists of artificial respiration using oxygen, combined with intravenous Edrophonium, B.P.C. and neostigmine injections. The anti-cholinesterase, neostigmine prevents acetylcholine destruction and allows it to build up to a concentration which exceeds the stimulation threshold of the motor end-plate.

Uses

Tubocurarine is chiefly indicated as a muscle relaxant for use in surgery and should always be used in conjunction with anaesthetics. It is compatible pharmacologically with all anaesthetics but should be used in reduced doses with ether, which has itself some curare-like effects; it should not be mixed with non-volatile anaesthetic solutions as precipitation may occur.

Dosage and Administration

It must be remembered that the American unit is equivalent to 0·15 mg tubocurarine chloride pentahydrate. Dosages are not satisfactorily determined and administration should be slow and continued until the desired effect is achieved. With aqueous solutions the intravenous route is preferred but with oily suspension the intramuscular route must be used. A marked fall in blood pressure also occurs in dogs and cats. It must be stressed that dosage recommendations as set out below are maximal and will probably elicit respiratory arrest.

Horse:
Up to 22–25 mg per kg.

Calf and Lamb:
Up to 0·6 mg per kg.

Pig:
Up to 0·3 mg per kg.

Dog and Cat:
Up to 0·4 mg per kg.

Note: Doses should be reduced considerably if ether is used concurrently with tubocurarine. Aminoglycoside antibiotics also potentiate curare.

Dimethyl Tubocurarine Iodide, N.F. This is similar in character and actions to tubocurarine chloride but it is about six times as potent as the chloride in rabbits and cats: in man it is only three times as active. Clinically in man it has been shown to cause little respiratory depression, little tendency to produce histamine or cardiovascular shock, and to possess a slightly longer duration of effect; it has therefore been advocated for thoracic surgery. In the dog this drug is 2–3 times as potent as tubocurarine but has a rather shorter duration of action. It is said to have less action at ganglia than tubocurarine.

Gallamine Triethiodide, B.Vet.C.

A white, crystalline, synthetic powder which is hygroscopic and therefore very soluble in water: it is slightly soluble in alcohol and acetone but insoluble in other organic solvents. Gallamine acts in a similar manner to tubocurarine but improves both blood pressure and heart rate rather than depresses them, causes little histamine release and has a potency about five times less than tubocurarine. Duration of effect is the same or slightly longer. Administration is by intravenous injection to effect.

Doses are best administered in repeated amounts of 0·25 to 0·5 mg per kg. Doses of 0·8 to 1 mg per kg usually cause an apnoea of 10 to 20 min duration in all species except the pig in which 2 mg per kg or more are required. Gallamine has also been administered by dart as an aid to the capture of wild animals.

Attempts to eliminate ganglion-blocking and histamine-releasing side effects have led to the introduction of such agents as alcuronium and pancuronium, the latter having a very rapid onset of effect. Both are effective in domesticated animals.

Muscle Relaxant Antagonists

Of the relaxants used in veterinary surgery, curare, tubocurarine and gallamine triethiodide may be reversed by anticholinesterase drugs such as physostigmine and especially neostigmine following the administration of atropine, but a substance, related to the quaternary ammonium group of relaxants, Edrophonium Chloride, is also a useful antagonist. Suxamethonium and suxethonium are potentiated by these antagonists and over-dosage cannot be countered by such means.

Edrophonium Chloride, B.P., B.P.C.

Dimethylethyl (3-hydroxyphenyl) ammonium chloride is a white, odourless, crystalline powder, readily soluble in water and alcohol, but almost insoluble in other organic solvents.

Actions and Uses

Edrophonium chloride is a competitive antagonist for curariform muscle relaxants which act by preventing acetylcholine depolarizing the motor end-plate. It is anticholinesterase and so potentiates acetylcholine and allows it to resume its normal function. It may therefore be used to terminate the action of a relaxant at the end of an operation, or to eliminate the effects of overdosage. Edrophonium chloride has fewer side actions, but is less persistent than neostigmine.

It should be remembered that circulatory collapse occasioned by overdosage with curariform relaxants is not reversed by edrophonium chloride and must be treated separately.

Toxicity. In excessive doses, edrophonium chloride may itself cause curare-like paralysis, but normally some indication, such as muscle twitching, as well as salivation and other signs of parasympathetic stimulation, should indicate impending overdosage.

Dosage

An estimated dose for the domesticated species is 0·1–0·2 mg per kg and relates to the human dose of 5–10 mg. The onset of action is rapid, ½–1 min compared with 2–4 min for neostigmine. Duration of effect is 5–8 min for neostigmine which is about twice that of edrophonium and repeat dosage may therefore be necessary. Preliminary intravenous atropine is recommended for the control of unwanted muscarinic side effects of the antagonists. Assisted respiration is also helpful.

Contra-indications. Edrophonium should not be used as an antagonist to succinylcholine derivatives (suxamethonium), for these relaxants are potentiated and not antagonized by it.

Depolarizing Agents

The members of this group of compounds all have a chemical similarity in their quaternary ammonium structure. The most important members of this group are Decamethonium Iodide, Suxamethonium Chloride and Suxethonium Bromide B.Vet.C. It must be remembered that physostigmine and neostigmine potentiate the action and may not be used as antagonists for the depolarizing agents.

Decamethonium Iodide

Decamethonium Iodide is now rarely used in anaesthesia. It is a white, bitter-tasting crystalline powder which is soluble in water (1:10) and alcohol (1:50). The *action* is typical of the group, causing depolarization of the motor end-plate. Following intravenous injection, muscle paralysis begins in 2 to 3 min and is maximal in 4 to 8 min; the duration varies from 15 to 25 min and is therefore shorter than that of tubocurarine. Muscle cramps, salivation and laryngospasm are disadvantages: no safe antidote is available. Decamethonium, however, has the advantage of not causing histamine liberation or blood pressure reduction and, because of its shorter duration, is more suitable for controlled dosing by small repeated injections. The short-acting succinic acid derivatives of choline have displaced it because of their greater suitability for intravenous drip administration and therefore more accurate control of relaxation. Toxicity is of a low order but overdosage can only be counteracted with artificial respiration and oxygen.

Decamethonium is more active in the cat than tubocurarine and *doses* should be about one-fifth of those of tubocurarine. In man, doses of 1 to 2 mg initially by intravenous injection are followed by further doses of 0·5 to 1 mg every 5 to 10 min for as long as relaxation is required

Suxamethonium Chloride, B.Vet.C.

This muscle relaxant is a white crystalline powder which is soluble in water and alcohol. Acid solutions are stable but in alkaline solution destruction is rapid even with refrigeration.

Action

Known also as succinylcholine chloride, this compound, in common with the bromide, acts by depolarization of the motor end-plate or myoneural junction. Its effects following intravenous injection are seen within 15 sec and consist of irregular uncoordinated muscular contractions which, in the absence of an analgesic, would be painful: these contractions last about ½ min and, along with the profuse salivation, are an indications of the progress of depolarization with its stimulatory effects. Respiratory paralysis frequently follows and the drug must not be used unless facilities are available for administering artificial respiration. Duration of action varies in the species according to the concentration of the enzymes cholinesterase and pseudo-cholinesterase in the red blood cells and plasma respectively. Thus, in cats 1 mg per kg causes respiratory paralysis for about 3 min, whereas a third of this dose will cause a paralysis in dogs lasting five times longer. In the horse, quantities necessary to produce muscular paralysis and casting (0·125 to 0·20 mg per kg) will give a duration of up to 8 min. Cattle are much more susceptible to the effects of suxamethonium, and doses about $\frac{1}{8}$ to $\frac{1}{10}$ of those for the horse will cause recumbency for about 15 min. The use of succinylcholine is contra-indicated in cattle. The difference between the reaction of the cow and the horse is attributed to the low pseudo-cholinesterase in the plasma of cattle. In pigs, 2 mg per kg produces only transient paralysis.

All investigators stress the need for pre-sedation before the relaxant is administered, and also the danger of aspiration of ruminal contents in cattle.

Uses

Suxamethonium may be used in small animals to obtain muscular relaxation in all types of surgery, for examination by X-ray, or manipulation, in conjunction with narcosis, but it is particularly suited for use in thoracic and orthopaedic surgery. In large animals, with prior sedation, it has been used for casting. In this situation unexpected susceptibility to the drug has on occasion been shown to be due to exposure to organophosphorus agents within the previous month or to a genetically mediated inadequacy of pseudocholinesterase.

Toxicity. The side actions of decamethonium and suxethonium are those associated with the initial depolarization, and consist of incoordinated muscular contractions, salivation, and possibly bradycardia. Dogs have survived 300 times the normal lethal dose provided artificial respiration was given. Toxicity is therefore extremely low. Serum potassium level may become elevated and autonomic stimulation can occur. These changes often affect heart rate. The occasional occurrence of malignant hyperpyrexia in man has been blamed on succinylcholine. This condition is characterized by an uncontrollable and progressive rise in temperature in the post-operative period.

Antagonists are not available but are usually not necessary if a means of providing sustained artificial respiration is available. Dexamethasone has been shown to be of value in attempts to manage episodes of malignant hyperpyrexia.

Administration

These muscle relaxants have been administered by intravenous, intramuscular and intrathoracic injections, but only the first route is entirely satisfactory. The drug may also be given as a 1:1000 solution in saline as an intravenous infusion.

Bibliography

CHEYMOL, J. (ed.) (1972) *International Encyclopaedia of Pharmacology and Therapeutics, Section 14, Neuromuscular Blocking and Stimulating Agents*, vols 1 and 2. Oxford: Pergamon.

FELDMAN, S. A. (1973) *Muscle Relaxants*. London: Saunders.

LECHAT, P. (ed.) (1971) *International Encyclopaedia of Pharmacology and Therapeutics, Section 8, Local Anaesthetics*. Oxford: Pergamon.

TUFVESON, G. (1963) *Local Anaesthesia in Veterinary Medicine*. Södertälje, Sweden: Astra.

WESTHUES, M. & FRITSCH, R. (1964) Local anaesthesia. In: *Animal Anaesthesia*, vol. 1. Edinburgh: Oliver & Boyd.

PART III

THE CONTROL OF INFECTIOUS DISEASE: CHEMOTHERAPY

26

Introduction

Chemotherapy is the treatment of a disease by the use of pure chemicals which have specific antagonistic effects on the organism causing the disease. The action is achieved by interference with the metabolic processes of the organism either directly or indirectly, by antagonizing enzymes or by preventing coenzyme formation, thus starving the organism of a substance essential for normal life. Examples of chemotherapeutics in three different fields are: sulphanilamide and other sulphonamides used in the treatment of many bacterial diseases including mastitis due to *Streptococcus agalactae*; piperazine salts used for the treatment of ascarid worm infestations; and quinuronium sulphate used against the protozoan blood parasites, the babesias or piroplasms which cause haemolytic anaemias in domestic animals.

Although chemotherapy has been practised empirically for centuries, as for example by the use of quinine against malaria since 1630, the principle of chemotherapy was first postulated about 1900 by Paul Ehrlich, who gave this name to the compounds produced by a systematic search for chemicals which would attack infecting organisms without harming their host. He stated that chemotherapeutic agents should be 'maximally parasitotropic and minimally organotropic', and thereby established the modern assessment of therapeutic value—the *Therapeutic Index* represented by the ratio of:

$$\frac{\text{Median lethal dose}}{\text{Median curative dose}}.$$

The wider the index the more valuable is the drug.

Since Ehrlich and his collaborators discovered arsphenamine, which is of value in the treatment of trypanosomiasis and syphilis, by the systematic substitution and interchanging of organic chemical nuclei and side chains, an army of scientists has been engaged in investigating the biological activity of vast numbers of organic chemicals and whole series of related compounds. While these investigators may be working individually, increasingly such research activities involve teams of workers representing several scientific disciplines. This work has continued concurrently with, and sometimes co-ordinated with, research into the metabolic minutiae of parasite and host metabolism.

It has been found, for example, that methylthionine hydrochloride (methylene blue—one of the early interests of Paul Ehrlich) and other oxidizing antiseptics acts by diffusing into the organism and there blocking essential reduction processes. Some substances combine with an essential metabolite or growth factor, forming a compound which the organism cannot utilize. Other chemotherapeutics react with enzymes, forming a combination which is similar to that normally produced, owing to the chemical similarity of essential growth factor and chemotherapeutic, and yet is dissimilar in a way that makes it useless for normal metabolism; this breaks the chain of metabolic reactions. It is this last process which is thought to be the method by which the sulphonamides achieve bacteriostasis, since the basis of all these compounds is *para*-aminobenzoic acid, which is part of the vitamin B complex and an essential growth factor for many bacteria as well as for animal life.

The aim of modern chemotherapeutic research is to discover a peculiarity of an organism's metabolism which is not common to that of the host. Having found it, a compound is sought which will utilize that peculiarity in attacking the organism, without affecting the metabolism of the host. Frequently this is only a dream. The rare event of a satisfactory end point being achieved may follow months or years of painstaking trial and error. However, with expanding knowledge and by collating all available evidence, biochemists increasingly can forecast with reasonable accuracy, that particular organic groupings when attached to a nucleus in a certain position will increase the compound's activity against a particular organism, whereas other groupings will decrease its activity or usefulness. As an illustration of this Fourneau and his colleagues refer to the effect of homologues of pamaquine against plasmodia, the blood parasite which causes malaria. They state that—'As a result of these observations it was believed that the methoxy group in position 6 in the quinoline nucleus is not essential for antimalarial action, while the substitution of (—OCH_3—) for (—OC_2H_5—) appears always to exert an unfavourable action, as does the group (—C . CH_2—O—R) placed in the diamino chain. In at least one compound replacement of (diethylamino-) by (diethylamine) reduced toxicity ten times.'

The quest for antibiotic agents has continued to play a large part in medical research. Hundreds of thousands of soil samples from all over the world have been tested for organisms which may yield products with activity against bacteria, viruses, fungi or neoplasms. Even when antibiotics have been isolated the search continues, since the original organism yielding an antibiotic may not produce as high a yield as a related organism. Thus the original *Penicillium notatum* of Fleming preferred surface growth and a higher yielding organism, which was a mutant of *Penicillium chrysogenum*, was isolated from a decaying cantaloupe melon and was shown to give up to one hundred times better yields in deep culture. This is the basis of present day commercial manufacture of penicillin.

It is interesting to note that microscopic organisms are not the only ones to yield antibacterial substances, although they are so far the only ones of commercial importance. Some plants such as the buttercup family, cabbages and algae yield active antibacterial products.

From animal sources hormones have a limited activity, and lysozyme present in various fluids including tears has an enzymatic destructive effect on certain bacteria.

Insects have been shown to produce inhibitory substances and the larvae of certain blowflies secrete such compounds. This is claimed to be one of the reasons that maggots can in certain circumstances help to clear up a badly infected purulent wound. Another substance iridomyrimecin has been shown to have considerable antibacterial potency. It is isolated from the secretions of the anal gland of the Argentine ant and although unlikely to prove a valuable commercial source of the drug, the mere fact of its isolation bears testimony to the patience of investigators and the wide area over which their searches have been directed.

By researches of this nature, tempered with patience and industry, and assisted by chance, two important groups of antibacterial chemotherapeutics, the sulphonamides and the antibiotics were discovered and new powerful anthelmintics and insecticides have been developed.

THE USE OF VETERINARY DRUGS

We have now reached the stage in the control of animal disease where more thought must be given by the veterinarian to the proper use of the range of new drugs in all fields of disease control. In every area of therapy, whether it be antibiotics, anthelmintics or coccidiostats, it is clear that pharmaceutical research has produced a very considerable reward in the form of highly active drugs which can assist in the control of the majority of the daily diseases of domestic and farm animals.

The level of daily dosage required has been reduced in the case of coccidiostats, for instance, to 0·01 per cent or less of drug in the feed. This high activity at low levels can produce problems as in many ways there is still a very great gap between the understanding of the use of drugs by the scientist and veterinarian who may have investigated a drug, and the pharmacist who dispenses it and the farmer who is the end user. Because there is often relatively poor communication between the scientist and the farmer, misuse of drugs is possible and this can lead to ill-informed criticism of the value of therapy in the veterinary field by the public health and medical authorities.

It is important now to ensure that where possible effective routines of therapy commensurate with the animal production system are defined for all the major disease areas, so that it will be clear both to the farmer and to the veterinarian how drugs should be used and how it is possible to minimize both the dangers of drug resistance in the farm environment and the likelihood of persistent tissue residues in meat and milk sold for human consumption. In subsequent chapters therefore emphasis will be placed on the importance of ensuring that drugs are used as part of control programmes, rather than as specifics for the treatment of individual cases.

Bibliography

BLOOD, D. C. & HENDERSON, J. A. (1974) *Veterinary Medicine*, 4th ed. London: Baillière Tindall.

British Pharmacopoeia (*Veterinary*) (1977) London: Pharmaceutical Press (in preparation).

British Veterinary Codex (1965) London: Pharmaceutical Press.

GRUNSELL, C. S. & HILL (1976) *The Veterinary Annual*. Bristol: John Wright.

27

Antibiotics

Introduction

The antibiotics are a group of complex organic chemicals which are produced, initially, by micro-organisms during their growth and which in minute quantities have a detrimental effect on other organisms. They are therefore metabolites produced by bacteria and moulds. These antibiotics are effective against living bacteria, some rickettsiae, some viruses, some fungi, and a few helminths. An average single dose of penicillin for injection into a cow's udder in the treatment of streptococcal mastitis is 100 000 international units. This quantity of penicillin represents approximately 62·5 mg weight, which is just under one grain. A further indication of activity is the blood level of penicillin which is effective in most systemic infections—about 0·1 i.u. per ml, which is equivalent to 0·06 mg per ml or approximately 1 gr. in 200 gallons of blood.

MANUFACTURE

The composition of many of the commonly used antibiotics is known. Some are prepared from fluids on, or in which the moulds and bacteria have been grown. Others are of synthetic or semi-synthetic origin. Chloramphenicol was the first synthetically produced antibiotic. Others which are prepared semi-synthetically include the derivatives of the penicillin nucleus such as ampicillin and cloxacillin. The usual process is known as the 'deep vat culture' or 'submerged culture fermentation' method, which is a mass production process developed during the Second World War by British and American collaboration, following the failure of the original surface culture methods to produce sufficient quantities to meet world requirements.

The method consists of inoculating cultures into a suitable fluid medium contained in vats holding as much as 20 000 gallons of medium. During the fermentation period air is blown through the culture and temperatures are maintained at the optimum for growth of the particular organism. The whole process is carried out under strictly aseptic conditions, for it can be imagined what damage could be done if the mould culture, air or vat were contaminated with an organism capable of growing more rapidly than the mould, especially if that organism produced a substance which destroyed the antibiotic.

Following fermentation, the fluid is filtered off and subjected to various extraction and filtration processes. Finally the resulting crystalline antibiotic is suitably prepared as one of the various presentations for oral, parenteral or topical use.

The correct strain of mould and optimum fermentation conditions will greatly increase yield, and the process may also be aided by the inclusion of certain medium supplements to increase production or favour certain forms of the antibiotic.

ASSAY

Once prepared, the antibiotics must be standardized by biological, chemical or physico-chemical methods.

Biological standardization is achieved by comparing the bacteriostatic effect of a standard sample of the antibiotic with that of the unknown, using a standard strain of a susceptible organism. The International Standard preparations for each antibiotic are kept at the Department of Biological Standards, National Institute for Medical Research, Mill Hill, London, but National Standards are maintained in each country and each working laboratory carries its own Working Standard which has been accurately assayed against the National Standard. The test organisms are available from the various National Collections of type cultures; the British one now being kept at the Central Public Health Laboratories, Colindale, London.

While there are numerous modifications of the methods of biological standardization, the following procedures form the basis for the majority of these tests:

1. The Serial Dilution Method

This consists essentially of inoculating a standard suspension of a susceptible bacterium type culture into a series of test tubes containing graded broth dilutions of an antibiotic of unknown potency. These tubes are incubated for a given period alongside a similar series containing known dilutions of the antibiotic standard preparation. At the end of the period the tests are read either by determining the dilution at which minimum growth occurs or by recording culture turbidities by subjecting the tubes to an absorptiometric test, if the so-called turbidimetric method is being used. Graphs are plotted, the standard and unknown preparations are compared, and the potency of the unknown calculated. Accuracy to within errors of 4 and 10 per cent have been given by the turbidimetric and broth dilution methods.

Serial dilution methods are only suitable where the activity of the preparation is known within close limits, for high dilutions must be used because of the great activity of the compounds and only a narrow range of dilutions can be tested at one time. Also, where impure preparations of an unknown antibiotic are to be standardized, potentiators or inhibitors may form part of the impurities and affect bacterial growth in a way which cannot be reproduced in the standard tubes.

Some of these disadvantages are avoided by diffusion methods of assay.

2. Diffusion Methods

There are two main methods but both are essentially the same.

The *cylinder-plate* method consists of comparing the antibacterial effect of known phosphate buffer dilutions of standard and unknown, quantities of which are pipetted into porcelain diffusion cylinders just embedded in the surface of Petri dish or sheet plates of agar. The agar sheets are previously seeded with a test suspension of a standard susceptible bacterial culture. After the filling of the cylinders the plates may either be incubated immediately or allowed to stand at room temperature for 1 to 1½ hr before incubation; this latter delay favours diffusion, enlarges the zone of inhibition and gives clearer and better results. With given dilutions, a more sensitive test with less error is achieved, for the larger zone decreases the proportion of the error but also a clearer zone with sharply defined edges is given as the antibiotic is able to repress growth from the commencement of incubation. The plates are incubated overnight at 37°C and the following day the diameters of the clear zones created by the absence of bacterial growth within the diffusion zone of the antibiotic are measured with calipers or an illuminated scale. Graphs are plotted using the diameter and the logarithm of the concentration, which should give a straight line, and comparisons are drawn between unknown and standard. The greater the number of cylinder tests for each dilution the more accurate will be the results obtained.

The *cup plate* method is very similar, differing only in that discs of agar are cut out of the seeded plate with a cork borer and removed carefully so that the edges of the circle are not lifted. Into these 'cups' are pipetted the antibiotic dilutions and it will be realized that were the edges of the circle lifted from the glass these solutions would underrun the agar and give useless diffusion zones. Greater accuracy is required in measuring the quantities of fluid placed in the cups than is needed in the cylinder method.

Accuracy of both these methods is approximately 4 per cent.

Other modifications are used, one of which consists of using filter paper discs which have been soaked in standard antibiotic solutions and dried, instead of the cups or cylinders. This method is simple, quick, and can be adapted for testing antibiotic content of body fluids. Rapid methods have been devised involving up to 3 or 4 hr incubation instead of 12 hr or more.

Differential methods are used in which various test organisms are used to identify different types of a given antibiotic.

Other tests use reverse phase techniques in which the nutrient medium is in a disc moistened with a liquid such as milk containing antibiotic activity. When placed on a seeded plate and incubated the extent of bacterial inhibition on the plate gives an indication of antibiotic activity in the milk.

Some tests depend on colour changes in sensitive reagents, brought about by the extent of bacterial activity in a test culture when incubated under standard conditions in the presence of an antibiotic. The colours indicate the presence or absence of chemicals produced by the bacterial growth. If enough antibiotic is present, growth is inhibited.

3. Chemical Methods

These consist in converting the antibiotic salt into a form which can give a colour reaction when treated. The conversion of penicillin to penicilloic acids by penicillinase is a widely used example. Gravimetric methods by which a compound containing the antibiotic is precipitated, filtered off, dried, weighed, and the amount of antibiotic calculated according to its proportion in the molecule, are also used.

A further development from the standard methods is the use of paper chromatography.

Bacterial sensitivity tests. These tests are modifications of the methods used for biological standardization. Some are laboratory diagnostic procedures aimed at determining the most suitable antibiotic for treatment of a case or a series of related cases.

Conversely, there are tests which are used to determine the susceptibility of numbers of strains or types of organisms to one antibiotic.

The *first* type of test consists of a modification of the cylinder-plate assay method. A Petri dish plate is poured, using agar seeded with the organism previously isolated in pure culture from the patient. The plate is divided into segments; these segments are suitably marked for identification. A paper disc containing a known amount of each available antibiotic is then embedded in the surface of the agar plate within each segment according to the code letter or number. Sometimes a multiple disc is used with individual antibiotics at the end of radiating spokes. The whole plate is then incubated for a given period depending on the organism and clinical requirements. A disadvantage is that in vivo responses do not always follow those in vitro.

It is important to note that disc sensitivity tests are purely qualitative and that they give only a limited indication of the value of an antibiotic when applied clinically. The conditions on a dish vary greatly from those encountered in the body and it is impossible accurately to simulate these by any simple test.

While the existence of a zone of inhibition round a disc indicates a degree of activity of the antibiotic in that disc, it is not possible to compare the probable clinical efficiency of two antibiotics in this type of test by the comparison of the diameter of their zones of inhibition. This will vary according to a number of extraneous factors including the rate of diffusion of the antibiotic, the rate of growth of the bacteria, the size of the inoculum, the quantity of antibiotic in the disc and the delay before a disc is incubated after seeding. The rate of diffusion depends on the molecular weight of the antibiotic where low molecular weight compounds diffuse through the agar more rapidly than those of higher molecular weight. Thus the zones from the former will tend to be wider than those from the latter, other circumstances being equal. This will sometimes give a false impression of greater activity of the low molecular weight compound.

The various factors are dependent also on the manipulative skills of the bacteriological worker using the tests, and much of the interpretation will depend on his experience.

While disc tests give a broad indication of usefulness of an antibiotic there is frequently the tendency to read more into the results than can be justified.

The *second* type of test is of less clinical significance, though it can be adapted to serve clinical as well as laboratory needs. Known as the 'ditch-plate' sensitivity test, it consists of a nutrient agar or, if required, a special solid medium plate from which a 'V' ditch has been cut across the diameter. Into this is run sufficient agar to fill it containing a known quantity of the antibiotic under test. When set, the plate is streaked at right angles to the ditch with various organisms. The sensitivity of each organism is shown following incubation by the linear zone of growth suppression on either side of the ditch.

Both these tests are easily carried out in any small laboratory equipped for simple bacteriological procedures using media tablets for the preparation of plates.

SAFETY TESTS

In addition to assays to determine potency, antibiotics must be sterile and free from extraneous toxic material such as pyrogens, which are toxic products of bacterial lysis. If present in toxic quantities, pyrogens will give a temperature reaction in test animals. Rabbits are customarily used for these tests. Cultural tests for sterility and pyrogen tests are therefore routine procedures carried out before batches of antibiotics are released for sale.

STABILITY

The major factors influencing the stability of antibiotics are listed below:

1. Moisture. In the early days of antibiotic therapy many antibiotic preparations had to be kept as the dried powder, due to the deterioration in potency which took place on exposure to moisture. Due to its hygroscopic qualities penicillin was particularly susceptible to degradation. It was necessary to dissolve the powder in a sterile solvent shortly before use and to store the solution in a refrigerator once prepared.

Now by using various salts of the antibiotics and special formulations, frequently with elaborate buffering systems, it is possible to make up stable liquid injections with a satisfactory shelf life of months or years in the unopened vial. In some cases slow deterioration may take place once the vial is partly used, as the air which replaces the extracted liquid may cause slow oxidation. In such preparation the space above the injection may be filled initially with nitrogen or other inert gas.

2. pH. This is probably the most important factor affecting antibiotic stability. Solutions should be buffered to the optimum pH for the individual antibiotic. A wide variety of buffering systems have been devised to produce optimal conditions of pH for a given antibiotic preparation.

3. Temperature. Stability of pure antibiotics is markedly affected by temperature, particularly when in solution, although some are more stable than others.

4. Oxidizing agents. These, as well as reducing agents, may cause rapid loss of potency. Chlorates, permanganates, peroxides, hydrogen sulphide and sulphur dioxide, are all antagonistic to antibiotics.

5. Enzymes. Some antibiotics may be destroyed by selective enzymes. Penicillin, for example, is rapidly inactivated by penicillinase, an enzyme which is elaborated by a number of bacteria of which *B. subtilis* is one.

6. Miscellaneous agents. A widely varied list of substances and conditions in addition to those already mentioned will affect antibiotics. *Heavy metals* such as copper, iron, zinc and mercury, will all inactivate penicillin; *light* affects most antibiotics, particularly when they are in solution; *cysteine* and other sulphydryl substances will rapidly destroy streptomycin and penicillin activity; *alcohols* are antagonistic to penicillin; which means that when syringes are kept in a spirit bath they should be well rinsed before injections of antibiotics are made with them.

ACTIONS AND USES

The antibiotics are all *bacteriostatic*, that is, in suitable concentrations, they prevent the growth and division of bacterial cells. It has been observed that with some bacteria, if concentrations lower than those which inhibit growth completely are used, some organisms increase in size, commence chromosome division but are unable to complete it so that abnormally large forms are produced: other bacteria join in long chains of swollen forms, whilst yet others are morphologically unaffected. These abnormal bacteria in vivo are highly susceptible to destruction by the body's natural defences.

Some antibiotics are also *bactericidal* and destroy bacteria when conditions are suitable: *bacteriolysis*, or solution of bacterial bodies occurs in some cases.

Mode of Action

Since antibiotics were first discovered in the 1950s, much knowledge has been acquired on their mode of action and the significance of this action on their relative merits in the therapy of man and animals.

Antibacterial agents can be divided into four groups as they affect the synthesis of:

1. Nucleic acid.
2. Protein.
3. The formation of the cell wall.
4. Cell membrane.

The detailed synthesis of protein in the bacterial cell is described in detail by Garrod et al. (1973).

Nucleic Acid

Replication of the nucleic acids of the bacterial cell is prevented directly by nalidixic acid and rifamycin and indirectly by the sulphonamides.

Effect on DNA. Sulphonamides ultimately deprive the cell of nucleic acid and the presence of nalidixic acid prevents its replication.

Effect on RNA. Rifamycin specifically inhibits the bacterial enzyme concerned with the replication of RNA. This is achieved by the binding of one molecule of the antibiotic to one molecule of the enzyme.

Protein

Streptomycin, tetracyclines, chloramphenicol, macrolides and lincomycin all act by interfering with the normal development of the protein combinations necessary for bacterial growth. In the growth phase, the process of linking together of amino acids to form a new bacterial cell has to proceed in a cyclical fashion and any interruption of the cycle locks the process at the point of interference and growth is halted.

Because of the continuation of processes which may have been affected by the interrupting mechanism, the overall integration of cell metabolism is disrupted and the long-term effect is so severe that the cell dies. Streptomycin affects the key step of the initiation of polysome construction and thus prevents total protein synthesis. In high concentrations it alters the protein synthesis by causing the cell messenger code to be misread and the wrong amino acid is incorporated into the growing peptide chain. The change induced is often a very small one, often the misplacement of a single amino acid, but this is sufficient to disrupt the chemical sequence.

The effect on cell growth is relatively slow, and probably a considerable amount of faulty protein accumulates before cell function ceases. In addition to protein derangement, streptomycin also affects the permeability of the cell and the respiration ability.

Tetracycline interrupts the cycle by which amino acid is carried and attached to the ribosome.

Chloramphenicol interrupts the transfer of a growing peptide chain to newly attached amino acids.

The process of translocation, i.e. the movement of an amino acid to a donor site on the ribosome, is prevented by the macrolides and lincomycin. The translocation step appears to be particularly accessible to modification as both agents, i.e. macrolides and lincomycin, are liable to give rise to resistant mutants.

Cell Wall

The cell wall of bacteria is tough and rigid and lies external to the cell membrane, giving the whole cell protection from possible osmotic damage. Such a cell wall is absent from mammalian cells, so that agents which interfere with its construction do not have a damaging effect on the mammalian host. Several agents affect cell wall synthesis, the most important being penicillins and cephalosporins.

The major constituent of the cell wall of Gram-positive organisms is a mucopeptide and that of a Gram-negative organism is a combination of a mucopeptide and a lipoprotein. The cell wall is built up of layers of acetylmuramic acid and acetylglucosamine molecules which unite in long strands with trailing polypeptides. In the final step, the glycine chains attached to the short peptides of one layer are fixed to the short peptides of the next.

Most antibiotics which act on cell walls interfere with the later stages of the cell wall synthesis. Bacitracin, for instance, interferes with the phospholipid carrier which carries wall material from the centre of the cell to the cell wall structure. Penicillin and cephalosporin inhibit the final process of cross-linking of the glycine and peptide chains.

When a cell is about to grow, the bonds which give the wall its strength have to open to allow for the insertion of fresh wall material. Low concentrations of penicillins or cephalosporins interfere with septum formation, but do not entirely prevent cell growth. Gram-negative organisms continue to grow under these conditions and instead of dividing become longer filaments. When higher concentrations of penicillins and cephalosporins are present, the cell membrane is progressively deprived of external support as the wall weakens and the cell ruptures. It is thus the osmotic force which is responsible for the death of the cell. In some circumstances, even though the bacterium is exposed to antibiotic, the cell membrane remains intact and continues to grow. Through the weakened cell wall it may expand into a spherical cell covered only by membrane, which is known as a *protoplast*; one which retains some cell wall elements is known as a *spheroplast*. As they are deficient in wall components, these cells are unaffected by the continuing presence of penicillin and may remain dormant until therapy is stopped and synthesis of the cell wall may be resumed.

Such organisms are not resistant in the ordinary sense of the term, as their progeny is susceptible to further treatment with penicillin. It is assumed that this phenomenon is related to the particular state of development of a cell at the time of penicillin therapy.

Cell Membrane

Beneath the cell wall is the lipoprotein layer of the cell membrane, which corresponds with the similar membrane found on mammalian cells. Therapeutically the most important agents which act on the cell membrane are the *peptides*. They consist of circular molecules which when they attach to the membrane modify the ion flux (magnesium ions) and bring about lysis of the cell. Subinhibitory doses of colistin make *E. coli* susceptible to penicillin and assist its entry into the cell. Although there is not a complete understanding of the action, it appears that these drugs have a combined action on cell wall and membrane.

Table XI. Kinetics of antibiotics used in the veterinary field

Antibiotic	Nature acid/base	% bound in serum	Serum half-life (hr)	% excreted unchanged in urine	Route of administration
Penicillins					
Ampicillin	A, B	25	1–2	90	oral/i.m. i.v.
Benzylpenicillin	A	50	0·5–1	60–90	i.m. i.v.
Carbenicillin	A, A	47	1–2	90	i.m. i.v.
Cloxacillin	A	95	0·5–1	30–40	oral/i.m. i.v.
Cephalosporins					
Cephalothin	A	56	0·5–1	60–90	i.m. i.v.
Cephaloridine	A	5	1–2	70	i.m. i.v.
Rifampicin	B	85	2–3	15	oral
Macrolides					
Erythromycin	B	20	1·5	15	oral/i.m.
Lincomycin	B	90	5	15	oral/i.m.
Tetracyclines					
Oxytetracycline	B	30	10	70	oral/i.m.
Tetracycline	B	50	10	60	oral/i.m.
Aminoglycosides					
Gentamycin	B	25	2·5	90	i.m./i.v.
Streptomycin	B	30	2·4	80	oral/i.m. i.v.

Table XII. Absorption pattern of various antibiotics

Antibiotic	Absorption pattern
Penicillins	
Benzyl penicillin	Freely diffusible by intramuscular route. Not suitable for oral use (destroyed by gastric juices)
Ampicillin	Freely diffusible, well absorbed when given orally
Carbenicillin	Not absorbed when given orally; well distributed when given intravenously or intramuscularly
Cloxacillin	Incompletely absorbed when given orally; well absorbed by intramuscular or intravenous routes
Cephalosporins	
Cephalothin Cephaloridine	Poorly absorbed; must be given intravenously or intramuscularly
Rifampicin	Well absorbed when given orally; enterohepatic recirculation maintains high blood level
Macrolides	
Erythromycin	Variable absorption by the oral route; lactobionate salt gives effective intravenous injection
Lincomycin	Readily absorbed by mouth and by intramuscular route
Tetracyclines	
Oxytetracycline Chlortetracycline	Absorption is incomplete by the oral route; low but continuous levels are obtained by the intramuscular route
Aminoglycosides	
Gentamycin Streptomycin Neomycin	Not absorbed by oral route; good absorption when given by intramuscular route

ANTIBIOTIC SPECTRUM

This term has been adopted to define the field of antibacterial activity of an antibiotic.

A *narrow spectrum* antibiotic is one in which the antibacterial effect is restricted to a relatively small number of organisms. Penicillin, which is active mainly against Gram-positive organisms and polymyxin B, which is active only against Gram-negative organisms, are two narrow spectrum antibiotics. The newer semi-synthetic penicillin ampicillin is active against a much wider range of pathogens than penicillin and is regarded as having broad spectrum activity.

A *broad spectrum* antibiotic is one which is effective against a wide variety of organisms—Gram-positive and Gram-negative bacteria, possibly rickettsiae and the larger viruses, and even protozoa and helminths: chloramphenicol, chlortetracycline hydrochloride, oxytetracycline hydrochloride and ampicillin are examples. Some antibiotics may have broad-spectrum activity in vitro and yet be restricted to use as a narrow spectrum antibiotic in vivo because of toxicity or other considerations.

Combined Antibiotic Therapy

Where mixed infections occur, or where larger doses of an antibiotic are required but may not be given because of toxicity, it may be useful to use two antibiotics together. In mixed infections, antibiotics which have a selective action on each particular organism may be used so that the total infection is promptly controlled. In the second case, a massive attack can be made on a particular organism without administering a toxic dose of either antibiotic: the combined use of streptomycin and dihydrostreptomycin is an example.

If two antibiotics are to be used together they must be carefully chosen, for some antibiotic combinations are actually less effective than either used singly.

A general guide has been given by Jawetz and his colleagues in several papers published between 1950 and 1952. The antibiotics fall into two groups: (1) narrow and medium-broad spectrum antibiotics, penicillin, cloxacillin, streptomycin, bacitracin and neomycin; and (2) broad spectrum antibiotics—chlortetracycline hydrochloride, chloramphenicol and oxytetracycline hydrochloride. With a few reservations these react as follows:

(a) Pairs from group 1 are frequently synergistic, occasionally additive but very rarely antagonistic. This is amply borne out in practice, where combinations of penicillin and streptomycin are used routinely in mastitis due to penicillin resistant organisms: where bacitracin and neomycin are very effective when applied topically in skin infections; and where penicillin and bacitracin have been effectively used in penicillin resistant infections.

(b) Pairs from group 2 may be additive but are rarely synergistic or antagonistic.

(c) Where pairs drawn from group 1 and group 2 are used against organisms sensitive to the group 1 member then the action of the group 1 antibiotic may be antagonized by small quantities of that from group 2. This is easily demonstrated by combinations of penicillin or streptomycin and chloramphenicol, or by combinations of penicillin and chlortetracycline. This antagonism is not always functional in reverse for:

(d) The action of members of group 2 is not usually impeded by the addition of small quantities of a group 1 member and may under certain circumstances even be synergized by it.

It is possible that a bactericidal drug may be inhibited in its usefulness by combination with one which is bacteriostatic. Since the bactericidal property of the penicillin is based on interference with the growth of the cell wall, causing it to rupture, it is likely to be most effective during a phase of rapid growth. If multiplication and cell growth are slowed down by a bacteriostatic agent such as a tetracycline, the action of the penicillin against the cell will be impeded and the cell may survive.

In all cases, synergism only occurs if both antibiotics are active against the infective organism even though bacterial inhibition by one of them may occur in concentrations higher than are therapeutically practicable. Thus chloramphenicol and oxytetracycline hydrochloride combinations with penicillin have both been used effectively in the treatment of human bacterial endocarditis where penicillin and streptomycin have failed.

Antibiotics Combined with Other Chemotherapeutics

The treatment of tuberculosis in man provides several examples for useful combinations of antibiotics and other chemotherapeutics. Streptomycin and dihydrostreptomycin, which must be used over long periods, soon loose their effect due to the development of bacterial resistance. If the antibiotics are combined either with *para*-aminosalicylic acid (PAS) or with isonicotinic hydrazide and related compounds the development of bacterial resistance is considerably delayed.

Sulphonamide and antibiotic combinations are probably the most popular examples of joint therapy but antibiotics are also combined with inorganic salts such as potassium iodide in the treatment of actinomycosis.

Factors Influencing Clinical Use

There are a number of important details of antibiotic activity concerning absorption, circulation and metabolism within the animal, that must be considered if the full benefit is to be obtained and the antibiotics used correctly.

1. The use of combined antibiotics and joint use of antibiotic and other chemotherapeutics has already been mentioned.

2. The blood and tissue levels of available antibiotics are of paramount importance when considering anti-infective activity in vivo. The attainment of early high levels and their subsequent maintenance will depend on the route or routes of application and the nature of the formulation of the antibiotic. Thus the choice of intravenous, intramuscular, subcutaneous, intraperitoneal or other parenteral route will influence the rate of attainment and duration of activity. Further therapy by oral medication may be used to continue treatment or it may be used in some cases of less severity as the sole means of treatment, particularly when an infection appears to be mainly in the gut.

3. Reaction with blood and tissue fluids. Some antibiotics such as penicillin are not decreased in potency by contact with blood, serum, pus, cerebrospinal fluid, etc., whereas others such as streptomycin are influenced unfavourably by these fluids (cf. sulphonamides).

The absorption of some antibiotics from the intestine may be inhibited by the presence of other chemical substances in the intestinal contents. Thus a high calcium level of the ingesta will reduce the systemic availability of certain tetracyclines given orally. To overcome this either a specially reduced calcium diet may be given for a short period or potentiating agents may be given with the antibiotic. These do not have a direct anti-infective function but by combination with the inhibitory substances in the ingesta, they render them unavailable to antagonize the antibiotic absorption. The final result is that a particular dose of antibiotic becomes more effective in its ultimate function.

4. Physiological barriers. There are a number of physiological barriers which some antibiotics will pass across quite easily whereas others pass only with difficulty. These barriers consist of:

(a) The blood–brain barrier, which is a physiological division between the blood and cerebrospinal fluid. Some antibiotics such as penicillin and streptomycin do not pass the blood–brain barrier except when the meninges are inflamed, and even then therapeutic levels may not be achieved in the cerebrospinal fluid. Intrathecal injections should therefore be given at least to initiate treatment. Other antibiotics such as oxytetracycline will normally achieve therapeutic concentrations in the cerebrospinal fluid only when the meninges are inflamed. A further group, of which chloramphenicol is outstanding, passes the blood–brain barrier fairly easily and achieves concentrations in the cerebrospinal fluid as high as 50 per cent of the blood levels.

(b) The placental barrier does not normally resist the passage of antibiotics.

(c) The intestinal barrier. When given orally a few antibiotics, because they are water soluble, are not absorbed into the system in therapeutic concentrations. Streptomycin and neomycin are examples. The reverse may also hold good. For example, when streptomycin is injected, very little is excreted into the intestinal lumen except through the biliary system. This is important, for in such cases only parenteral injection is suitable for treating systemic infections and only the oral route can be used for treating intestinal infections.

(d) Serous membranes. Some antibiotics such as penicillin will not readily pass either way across the pleural or peritoneal membrane, whereas others, for example chlortetracycline hydrochloride, diffuse quite readily into the pleural and peritoneal cavities.

(e) Milk. Most antibiotics diffuse into the milk in small quantities, but with one or two exceptions concentrations achieved do not warrant the systemic use of antibiotics for the treatment of mastitis. This is particularly so when staphylococci are the cause of mastitis. Effective therapy depends on the maintenance of an adequate antibiotic level in the udder.

5. Bacterial resistance. The development of antibiotic resistance is now known to be a phenomenon of the greatest importance.

6. Ecological factors. A dangerous situation has been reported quite frequently in human medicine where broad-spectrum antibiotics have eliminated bacterial flora leaving the field free for the multiplication of yeasts and fungi which have taken on pathogenic attributes. So far this danger would not appear to have occurred widely in veterinary medicine, but it is a possibility which has to be borne in mind.

A related situation arises when it is necessary to identify a pathogen and to determine its sensitivity to chemotherapeutics: antibiotic treatment may alter the morphological, cultural and biochemical reactions of a bacterium, particularly if resistance is developed, and make it impossible to achieve either of these aims. Where mastitis surveys and laboratory diagnoses are being undertaken samples must always be taken *before* treatment is commenced.

7. Antibiotics and immunity. Some diseases do not recur because natural recovery from an attack results in the creation and maintenance of adequate levels of immune bodies in the blood and tissues. If, however, large doses of antibiotics or indeed if any other anti-infective agents are given and the infection rapidly eliminated, the immunological reactions are not always properly stimulated and the body is susceptible to further infection as soon as the antibiotic is withdrawn. Such situations have arisen in the control of coccidiosis by chemotherapeutic food medication. In human medicine this situation has been overcome by using intermittent therapy whereby

massive doses of an antibiotic have been given at intervals instead of lower doses continuously over a longer period. Attention must be given to the danger of creating drug allergy or sensitization if this intermittent dosing method is used.

BACTERIAL RESISTANCE

The problem of the development of resistance among bacterial populations exposed to antibiotics is one which has received a great deal of attention, both in human and veterinery medicine. Broadly speaking there are two types of resistance to antibacterial agents.

In some cases bacteria have a natural ability to resist the effect of a particular drug because of the presence of enzymes which are capable of destroying the drug. An example of this type of resistance is found in the penicillinase producing staphylococcus which is able to break down the molecular structure of the ordinary penicillin compounds.

In other cases resistance is exhibited, which does not depend on the destruction of the drug by enzyme action, but rather on the ability of the bacterial cells to survive in the presence of the antibiotic without direct interaction.

Where bacteria are resistant they may be present in small numbers in a mixed population. If this population is then exposed to an antibiotic the sensitive strains will be eliminated, enabling the resistant strains to multiply in greater numbers and so leave a population which is largely or entirely resistant. This type of emergence of resistance is commonly demonstrated in the presence of streptomycin and may be a very rapid process.

In addition to bacterial colonies containing some individuals with a natural resistance, it has been found that cultures which are originally naturally sensitive can become resistant when in the presence of antibiotics which are sub-lethal. In such a culture of normally sensitive cells it is possible, in the vast number of reproductive divisions which take place, that a genetic mutation may arise from time to time, which confers a greater degree of antibiotic resistance upon certain strains. While this mutation may be quite random in its incidence with a low frequency of perhaps one in several million cell divisions, the resultant sub-generations will retain the characteristics as genetically heritable. Under these circumstances, in the continued presence of the antibiotic, the newly developed resistant mutant will again have an advantage over the sensitive strains in the same way as a naturally resistant organism, and will thrive to exclude the sensitive strains. This type of resistance which develops in concentrations of antibiotics which are sub-lethal, indicates the necessity for adequate levels of medication at early stages in a disease, preferably with a bactericidal drug or one with sufficient bacteriostatic activity to enable the body defences to eliminate the infection.

Some authorities consider that mutation towards an enhanced antibiotic resistance is a completely spontaneous change in the genetic structure of the cell which is similar to any other mutation and which is quite random in its incidence, without dependence on the presence of the antibiotic. Others maintain that a process of adaptation occurs in the bacterial cell which is to some extent influenced by the presence of the antibiotic in the environment. In view of the complications of the subject it is quite likely that these processes may each contribute to changes in the resistance pattern of the total cell population in a particular flora.

Once established the mutant or adapted strains will maintain the advantage of increased resistance and will offer a therapeutic problem with the antibiotics to which resistance has developed. The temporary removal of the antibiotic will not be likely to influence the resistance pattern of succeeding cell generations. If bacteria continue to reproduce over longer periods in the absence of the antibiotic there is generally a slow reversion towards sensitivity or to a balance of sensitive and resistant strains similar to that existing before antibiotic treatment.

There may be reverse mutations to the sensitive state or partial colonization with sensitive strains. If these sensitive cells are capable of more rapid growth they will establish themselves again in the population and multiply until balance is achieved.

While changes in the resistance pattern can involve problems of clinical treatment, it should be emphasized that the alterations in the cell brought about by these changes do not necessarily imply an increase or decrease in the pathogenicity of the particular cells. The problem is one of numbers rather than of altered virulence.

Transmissible resistance. while specific resistance to antibiotics has been recognized for a considerable number of years, it was initially thought to be a property which developed in bacterial strains and which was confined to the subsequent progeny of these strains. It was accepted, however, that some bacteria were capable of developing resistance to one antibiotic which may give a degree of cross-resistance to other chemically related antibiotics.

More recently it has been shown that genetically defined characteristics from one set of cells can be transmitted to other unrelated cells which are present in the same environment, and which may be of different strains or species. This is particularly evident so far among the Gram-negative enterobacteriaceae. Factors for antibiotic resistance can be involved in this type of transfer and the process has been referred to as *transmissible resistance* or *infective drug resistance*. The method of transmission of this genetically defined resistance involves the passage, from one cell to another in the same environment, of fragments of nucleic acid bearing the coded information for these heritable characteristics. This can include resistance which has developed to one or several antibiotics. The transfer of the nucleic acid can induce this resistance in previously sensitive cells which have not themselves been exposed to the antibiotic. It should be emphasized that the affected cells are not directly related to the donors by direct sub-culture.

Two processes may be involved in the transfer of the nucleic acid and these are known as conjugation or transduction. Conjugation is the process whereby two cells join and then separate with some exchange of cellular material. This may contain the nucleic acid bearing the factors for resistance. Transduction takes place where a bacterial phage, which is a virus like bacterial parasite, passes from one cell to another bearing the nucleic acid as a contaminant. Thus the bacterial phage acts as a vector.

The significance of this type of transmissible resistance is that certain individuals may acquire a population of bacteria which would resist treatment even though the host has never received antibiotic medication. While this is an important consideration when reviewing the chances of success of future antibiotic treatment, in practice so far the therapeutic difficulties in the veterinary field do not appear to be extensive. It is of course prudent to bear in mind the possibility of an increase in the problem but this has to be balanced against the existing benefits arising from treatment and the difficulty of providing an alternative solution.

Hartley and Richmond (1975) described studies which may help to establish the significance of antibiotic resistance in the bacteria found in man and animals.

It was found that relatively few of the bacterial strains that flourish in human intestines harbour the resistance factors and that none of these strains normally cause disease. There is a danger, however, that they can become reservoirs for drug resistance factors and thus increase the chance that resistance might be transmitted to bacteria that cause disease. When a patient or animal is taken off therapy with antibiotics, the resistant strains tend to disappear from the gut; this may be because resistant strains can only compete effectively with non-resistant strains when an antibiotic is present which is suppressing the sensitive strains.

It was also found that the effects of antibiotics vary from one individual to another. Of eight human volunteers tested, one had no resistant bacteria in his gut 3 days after he had ceased to take antibiotics, while in another resistant strains persisted for 9 months.

Some studies reported by Williams Smith (1975) show that amongst *E. coli* organisms in pigs, resistance may persist at a high level for several years despite the withdrawal of antibiotics as feed additives. In his findings, this occurred with tetracyclines, streptomycin and sulphonamides. The resistance to ampicillin tends to fall after withdrawal and in the case of furazolidone, neomycin and polymixin, it remains at a low level.

It is likely that the total incidence of resistant strains on a farm will also be related to the individual idiosyncrasy reported by Hartley and Richmond (1975).

PRINCIPLES OF ANTIBIOTIC THERAPY

Intelligently and judiciously used, antibiotics are invaluable, but before they are resorted to there are a number of questions that should be asked, considered and answered by the veterinary surgeon.

1. Is an antibiotic required or would another drug cure the animal, perhaps less quickly and spectacularly yet with less interference with the normal bacterial flora of the body and the development of natural immunity?

2. If an antibiotic is indicated because of the specific nature, or acuteness and severity of the disease, is it possible to use one of the narrow spectrum antibiotics which are more selective and therefore less likely to cause disruption of the normal resident bacteria. The success of such therapy is dependent on a full knowledge of the bacterial cause of the disease.

3. If an antibiotic is to be used, for how long should it be used and at what dosage rate? It is far better to give high doses for a short period *providing the organism is totally eliminated*, for there is then less likelihood of antibiotic resistant strains of bacteria developing. Should the infection not be eliminated, then a relapse may occur or the infection become chronic and the illness protracted. This is particularly important in the treatment of pyogenic infections when chronic abscesses may be formed.

4. Is it safe and economic to give an antibiotic in these doses over this period or would the longer period and less active drug be safer and, perhaps because of this, more economic in the end?

If an antibiotic is to be used, then it should be specific as possible and given immediately. It should be applied to achieve the maximum concentration in the tissues in as short a time as possible, and continued until 48 hr after disappearance of symptoms, unless signs of toxicity are shown. The reasons are:

(1) Bacteria are more effectively dealt with in their early logarithmic phase, and therefore the earlier the treatment the more likely are the infecting bacteria to be at this stage of development.

(2) As the action of the antibiotic generally depends on molecular diffusion and absorption into or adsorption onto the bacterial cell, the activity of a given dose will be inversely proportional to the number of bacteria present. This is another principle which applies to most, but not all chemotherapeutics. Because of this it is advisable to use large doses as early as possible whilst the bacterial numbers are still low. Polymyxin and the penicillins are exceptions, for their activity is not affected by the number of bacteria present.

Besides using large doses of a specific antibiotic as early in the disease process as possible, as has been mentioned the rapidity of action can also be affected by choice of route and form of administration. It is advisable in systemic infections to use parenteral administration of mega-units) of sodium penicillin, intramuscularly. (In the case of penicillin intramuscular injections promote therapeutic blood levels at a rate only very slightly below that of intravenous injections). At the same time a 'depot' or 'repository' injection in the form of procaine penicillin aqueous suspension would be given intramuscularly, the dose being about 2 000 000 units (2 mega-units). No further antibiotic need be given for about 24 hr when the repository dose would be repeated. It is probable that in such early cases only one, or at the most two, further doses would be needed. Commercial penicillin injections are available combining water soluble and soluble and easily diffusible forms for initial doses so that blood levels, upon which depend tissue levels and therefore antibacterial activity, are rapidly achieved. These blood levels can then be maintained by continuous slow absorption from 'depots' of less soluble and slowly diffusible antibiotic forms. As an illustration, the treatment with penicillin of equine strangles due to *Streptococcus equi* might be quoted. Presuming that the case is in a young animal and the infection is severe, with signs of septicaemia, the first dose of antibiotic would be about 500 000 units (0·5 insoluble salts so that immediate absorption and repository forms can be given at one injection.

Antibiotic Therapy in Animals and its Relevance to Disease Control and its Relationship to Human Health

Antibiotics have been used in the therapy of animal disease since 1950, but there is now a general realization by both registration authorities and veterinarians that their use should continually be reviewed so that they can be employed in the most effective therapeutic manner and so that their use in the animal field should in no way be harmful to their use in the control of human disease.

The Swann Report in 1969 considered the use of antibiotics in animal husbandry and veterinary medicine in the United Kingdom and in particular the effect of animal therapy with antibiotics and antibacterials on general public health and made several specific recommendations. These were:

1. Antibiotics to be used for addition to animal feed without prescription should be restricted to those that are of economic value in livestock production, those that have little or no application as therapeutic agents in man or animals and those that will not impair the efficacy of a prescribed therapeutic antibiotic or antibiotics through the development of resistant strains of organisms.

2. Permission should be withdrawn for the use of chlortetracycline, oxytetracycline, chloramphenicol, penicillin, tylosin, sulphonamides and the nitrofurans as feed additives and in the future these drugs should only be issued on prescription.

3. Chloramphenicol should only be prescribed for veterinary therapy under exceptional conditions.

4. No change should be made in the law which allows the supply of antibiotics on veterinary prescriptions, thus leaving full prescribing rights with the veterinary surgeon.

The recommendations of the Swann Committee, although based on a study of animal health conditions in the United Kingdom, have had an influence on the thinking of registration authorities throughout the world, and the U.S.A. Federal Authorities sent a task force to the United Kingdom to discuss the recommendations of the Swann Report and to consider its relevance to United States conditions.

Subsequently in April 1973 on the basis of the task force report, the Food and Drugs Administration published a Statement of Policy and Interpretation regarding Animal Drugs and Medicated Feeds and promulgated regulations. The major provisions of the order were:

1. Antibiotic-containing products would be withdrawn unless data from studies conducted by the sponsors under criteria established by the FDA task force were submitted which would resolve the issues concerning their safety to man and animals and their effectiveness.

2. Target dates were established for completion of these studies and the submission of data derived from them. On 6 August 1974 the Commissioner of Food and Drugs published a proposed amendment to 21 CFR 135.109 listing 136 antibacterial drugs for which satisfactory work is in progress. As a corollary to that proposed amendment the Commissioner proposed to amend 21 CFR Parts 121, 135e, 135g, and 144 to revoke those antibacterial products which were not in compliance with the requirements set forth in the order of 20 April 1976 because satisfactory responses were not submitted.

3. Manufacturers would be given until 20 April 1975 to provide data to establish the safety and effectiveness of their products under the guidelines recommended by the task force. Currently, 17 firms are involved in test work which will attempt to meet the FDA requirements.

4. Products would be withdrawn immediately if test results at any point indicated that a human health hazard was present.

5. Manufacturers who market tetracyclines, streptomycin, dihydrostreptomycin, penicillin and sulphonamides were required to test the effect of use of these products on the *Salmonella* reservoir in animals within one year. Responses have been received from 11 sponsors of these priority drugs. These drug products were selected for special attention because they are commonly used in animal feeds to promote growth and control disease as well as for treatment of disease in human and veterinary medicine. The *Salmonella* reservoir was monitored because *Salmonella* are disease-producing bacteria which can be transmitted to man through food and they can be easily measured. If the tests demonstrated a significant increase in the *Salmonella* reservoir, the products were to be removed from the market.

The FDA then evaluated data gathered by 11 drug companies and independent research on the effect of antibacterials on the number of *Salmonella* present in an animal, the number shed by the animal, the length of time *Salmonella* were shed and the drug resistance of the *Salmonella*.

The FDA in February 1975 concluded as follows on the data submitted:

1. Research conducted by FDA has shown that when test animals on subtherapeutic levels of antibacterial drugs are exposed to *Salmonella* which are drug-resistant, the bacteria are more likely to survive and multiply in the gastrointestinal tract of animals which are receiving the antibiotic to which that *Salmonella* is resistant. These studies, using drug levels in the upper ranges of subtherapeutic antibiotic dosage, indicate that one can expect an increase in numbers of *Salmonella* being shed. These studies were instituted in order to place the FDA in a better position from which to evaluate the data expected to be submitted by drug sponsors. In the protocol review phase of the overall programme, drug sponsors were requested to conduct at least a portion of their studies with *Salmonella* organisms which were resistant to the particular drug being tested.

2. The data submitted by drug sponsors for the most part were concerned with drug-sensitive *Salmonella*. In general, the data indicate that when animals receiving antibacterial drugs in the higher ranges of subtherapeutic levels are exposed to drug-sensitive organisms, one can expect to see a suppressive effect on the bacteria or no significant differences in quantity of *Salmonella* shed, duration of shedding, prevalence in animals, or tissue invasion, between test animals and controls.

3. *Salmonella*, with or without the presence of antibiotics, invade tissues outside the gastrointestinal tract of the test animals.

Therefore, while animals fed antibiotic drugs in subtherapeutic dosages which are exposed to infecting numbers of drug-resistant *Salmonella* organisms will be likely to shed increased numbers of *Salmonella* and animals under the same antibiotic feeding conditions which are exposed only to drug-sensitive *Salmonella* will be likely to shed decreased numbers of the organisms, these findings must be evaluated in light of FDA's current information concerning *Salmonella*. Upon assessing current information on the total population, the exact drug-sensitive versus drug-resistant status of naturally occurring *Salmonella* is unclear. The published literature contains information from surveys conducted on *Salmonella* isolated by diagnostic laboratories and these surveys indicate that a substantial number of these isolates are resistant to one or more antibacterial drugs. The full scientific value of these data cannot be determined until the drug-resistant segment of the naturally occurring *Salmonella* reservoir can be quantitatively established. Steps are currently being taken by FDA in an effort to answer questions concerning the drug-sensitive versus drug-resistant status of the naturally occurring *Salmonella* reservoir.

In 1974, in an effort to develop common thinking on drug regulations, a European Veterinary Conference was held in Dublin to consider the relationship between the use of animal medicines and human health, and the following subjects were discussed:

1. Veterinary medicine regulations within the European Economic Community.

2. Standardization of registration requirements.

3. How to arrive at a common view point taking into account conflicting scientific opinions.

4. Future harmonization of veterinary medicines regulations within the framework of the EEC.

Although this conference covered other drugs in addition to antibiotics, it became clear that there is a great need for the international harmonization of the regulations on the use of drugs in animals, so that any discussions at WHO and FAO levels which relate to human public health and animal therapy can have full international significance. In addition to considering the advantages of harmonization from an efficacy of drugs point of view, there is also an overwhelming need for harmonization to ensure that drugs remain economically viable, to both the veterinary surgeon and the animal owner.

The Use of Antibiotics in the Therapy of Mastitis

The use of antibiotics in the therapy of

mastitis is now commonplace, but it has to be appreciated that the methods of application of drugs must be continually under review, as systems of husbandry change and in some cases the disease pattern is altered. As a result of the detailed studies carried out by the National Institute for Research in Dairying in England, and the workers at Cornell University in the USA, into the control of mastitis, new concepts have evolved as to the effective therapy of mastitis during lactation and the dry period.

Ziv (1975) has put forward his pharmacokinetic concept for antibiotic treatment in lactating and dry cows. He suggests that as the prime objective of antimicrobial therapy is to eradicate the causal organisms of disease it is essential to deliver an optimal amount of drug at the site of bacterial invasion. The result of such therapy will be assessed by its clinical effectiveness. Clinical efficacy is often difficult to quantify, particularly in diseases such as mastitis where there is great variability in response from one case to another. He suggests, therefore, that properly designed pharmacokinetic studies can disclose the relative biological potency of a series of drugs, or two or more formulations of the same drugs. He considers that special consideration must be given to the relationship between the concentration of antibiotic in the milk and its potential clinical efficacy in the udder.

Intramammary Treatment

Intramammary treatment alone is accepted as the route of choice in the treatment of chronic or non-acute mastitis cases. In acute mastitis, however, it is often considered advisable to treat parenterally or to complement intramammary treatment with parenteral therapy, as there may be uneven distribution of the drug following intramammary therapy because of the intense swelling of the udder parenchyma, which may prevent antibiotic reaching the site of bacterial invasion. If parenteral therapy is to produce effective levels of antibiotic in the udder tissue and milk, it is important to choose a drug which will achieve effective levels. As a result of his investigations, Ziv defines certain properties that are necessary if a drug and formulation are to be effective for the therapy of mastitis. He considers that the following properties are important:

1. For parenteral use
   - (a) Low MIC against the majority of udder pathogens.
   - (b) High bioavailability from the site of injection.
   - (c) Low degree of serum protein binding.
   - (d) Chemically a weak base or highly non-ionized in serum.
   - (e) Sufficiently lipid soluble.
   - (f) A long half-life in the body.
2. For intramammary infusion during lactation
   - (a) Minimal irritancy to the udder tissue.
   - (b) Low MIC.
   - (c) Low degree of binding to milk and udder tissue proteins.
   - (d) Chemically a weak base or highly non-ionized in milk.
   - (e) Sufficiently lipid soluble.
   - (f) Short milk-withholding period.
3. For intramammary infusion during the dry period
   - (a) Completely non-irritant to the udder.
   - (b) Low MIC with a bactericidal mode of action to minimize the chances of drug resistance developing when low levels of antibiotics may be present in the udder in the latter part of the dry period.
   - (c) High degree of binding to dry udder secretions and udder tissue proteins.
   - (d) Chemically a strong acid or a strong base.
   - (e) Highly hydrophilic properties.
   - (f) Large molecular weight.
   - (g) Stability of microbiological activity in the dry udder for at least 3 weeks.

An essential in all types of treatment is that an antibiotic should be chosen to which the causative organisms of mastitis are sensitive.

When considering dry cow formulations, Ziv puts particular emphasis on the half-life and duration of effective antibiotic in dry udder secretions. The formulations he describes are infused after the last milking of lactation.

Table XIII gives an indication of some antibiotics at present used in 'dry cow' formulations.

This detailed study by Ziv on the behaviour of antibiotics in the udder of the cow is a very considerable advance on previous approaches to the development of intramammary formulations. Similar studies are necessary to determine the relative merits of oral and parenteral therapy with antibiotics in the treatment of coliform and *Salmonella* infections in farm animals and to establish the period of therapy necessary for an effective cure.

Table XIII. Antibiotics used in 'dry cow' formulations

Antibiotic	Dose/quarter	$T\frac{1}{2}$ (days)	Duration of effective concentration (days)
Procaine penicillin	$1{\cdot}2\times10^6$ I.U.	0·8	5–10*
Benzathine penicillin	$1{\cdot}2\times10^6$ I.U.	1·5	10–15*
Dihydrostreptomycin	1000 mg	5·7	5–10
Benzathine cloxacillin	500 mg	8·0	15–25
Neomycin	500 mg	7·4	5–10
Novobiocin	500 mg	5·0	15–20
Erythromycin	600 mg	1·0	5
Furazolidone	500 mg	1·0	5
Spiramycin	400 mg	4·8	15–20
Colistin	200 mg	7·1	5–25

* Dependent on sensitivity of the udder pathogen.

Bibliography

GARROD, L. P., LAMBERT, H. P. & O'GRADY, F. (1973) *Antibiotics and Chemotherapy*. Edinburgh: Livingstone.

GEDDES, A. M. & WILLIAMS, J. D. (1973) *Current Antibiotic Therapy*. Edinburgh: Churchill Livingstone.

HARTLEY, C. L. & RICHMOND, M. H. (1975) Antibiotic resistance and survival of *E. coli* in the alimentary tract. *Br. med. J.*, **4,** 71.

MOL, H. (1975) *Antibiotics and Milk*. Rotterdam: Balkema.

SMITH, H. WILLIAMS (1975) Persistence of tetracycline resistance in pig *E. coli*. *Nature, Lond.*, **258,** 628.

WILLIAMS, J. D. & GEDDES, A. M. (1975) *Chemotherapy*, vol. 4, *Pharmacology of Antibiotics*. London: Plenum.

ZIV, G. (1975) Pharmacokinetic concepts for systemic and intramammary antibiotic treatment in lactating and dry cows. Seminar on mastitis control. International Dairy Federation, Document 85.

28

Penicillins

I. PENICILLIN

Penicillin is a narrow-spectrum antibiotic which, in its many various forms, is officially recognized in all the pharmacopoeias of the world.

History

Penicillin was the first antibiotic discovered. It is also noteworthy in that it has yet to be superseded as the most useful and least dangerous of all antibiotics.

The first report indicating a major medical discovery appeared in 1929 when the late Sir Alexander Fleming of St Mary's Hospital, Paddington, London, described the lysis of staphylococcal colonies near a contaminant mould on a bacteriological plate. He also described the culturing of the mould on the surface of a peptone broth and his attempts to concentrate and isolate the active principle which he had shown to be present in the culture medium. He failed to isolate the substance, but continued to use the crude filtrate to depress unwanted bacterial growth of cultures of resistant organisms including those which cause whooping cough (*Haemophilus pertussis*), calling the filtrate *penicillin*, a word derived from the generic name of the original contaminant mould, *Penicillium notatum*. Fleming's efforts to isolate the substance were continued by Raistrick, Clutterbuck and Lovell at the School of Tropical Medicine, London. They were only able to demonstrate how difficult the problem was and showed that, while it was possible to make ethereal extracts of the substance, these extracts lost their activity on evaporation in air. The advent of the sulphonamides in 1933 focused attention away from penicillin until Florey and Chain at Oxford, interested themselves in the problem and published, in 1939, the initial paper on the chemotherapeutic activity of their crude penicillin powder. With this powder, probably containing no more than 1 per cent of penicillin, they were able to inhibit staphylococcal growth in vitro with a dilution of 1:500 000, and to protect mice against streptococci, staphylococci and *Clostridium septique*.

Later research has provided methods of manufacture and purification so efficient that **Sodium Benzylpenicillin,** B.Vet.C., is now a very inexpensive antibiotic.

Manufacture

Penicillin, originally made on a small scale from surface broth cultures of the mould, is now manufactured on a vast scale by deep vat culture methods using various 'medium supplements' and various strains and species of penicillin moulds, so that different forms of penicillin can be produced at will in high concentration. Purification processes are such that, instead of the originally manufactured yellow, amorphous, impure penicillin, the product now marketed is a white crystalline substance with a high degree of purity.

Chemistry

Penicillin exists both as an amorphous powder, now rarely used, and a white crystalline substance, **Benzylpenicillin,** B.Vet.C., which is the sodium or potassium salt of the acid. Benzylpenicillin contains not less than 1480 units per mg (potassium salt) or 1500 units per mg (sodium salt). It is soluble in water, but insoluble in fixed oils and the paraffins, and has a faint mould-like odour and salty taste.

The final crystalline substance produced from *Penicillium* moulds is a mixture of the salts of five main penicillins—organic acids with a common root formula represented by the following structure:

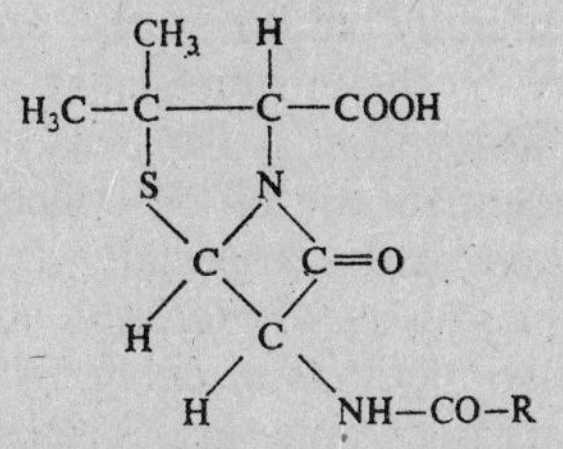

known as the β-lactam formula.

Represented in this manner it has a nucleus formed by a five membered thiazolidine ring and a four membered β-lactam ring. The five membered ring is comparatively stable, but the four membered ring is very unstable and is probably responsible for the instability of the compound.

The five acids differ only in the constitution of the side chain attached to the β-lactam ring and designated R in the graphic formula. They are named as in Table XIV.

All are active, though penicillin IV (K) is the least active and penicillin II (G) the most active.

There are many other penicillins, which are spoken of as biosynthetic penicillins because their formation by the mould has been encouraged by the addition of preformed organic groups as medium supplements: phenoxymethyl penicillin or penicillin V is probably the best known of these, but allylmercaptomethyl penicillin or penicillin O is medically prominent, because it causes less allergic response than benzyl penicillin.

The penicillins themselves are most unstable and relatively insoluble. They are therefore converted into various salts and esters by the substitution of sodium, calcium, potassium, procaine, alkyl groups, etc., in place of the hydroxyl group or hydrogen of the carboxyl group.

Stability

Penicillin is the least stable of the commonly used antibiotics.

1. Moisture. Penicillin is hygroscopic, although the potassium salts is less so than the sodium or calcium salt, and deterioration by hydrolysis is rapid. When dry, the salts will keep for 2 to 3 years, but when in unbuffered aqueous solutions the salts lose about 16 per cent activity at 4°C and 78 per cent at 24°C, in 7 days. It follows that solutions, once made up, should be used immediately or kept in a refrigerator and used within 7, but preferably 4 days.

2. pH. Acids and alkalis cause rapid deterioration in the potency of penicillin solutions. Following the determination that penicillin is most stable when solutions are kept within a range of pH 6 to 6·5, phosphate and citrate buffers are frequently used to stabilize solutions. Sodium citrate will maintain 90 per cent of sodium benzylpenicillin solution potency for 3 days at 24°C (slightly above room temperature), and for 2 weeks at 4°C. With procaine penicillin suspensions, citrate buffering enables loss to be reduced to no more than 5 per cent in 2 years at room temperature. Strong solutions tend to deteriorate more rapidly than weak solutions.

3. Temperature. As already indicated in the preceding two paragraphs, deterioration rate increases with temperature, and even dry powders will not withstand autoclaving or heating at 100°C for any length of time. All solutions for injection must therefore be prepared by aseptic methods.

Table XIV. The five main penicillins

Pharmaceutical name		Chemical systematic name	Side chain R
British	American		
I	F	Pent-2-enylpenicillin	$CH_3 . CH_2 . CH : CH . CH_2$—
	Dihydro F (flavicidin)	n-Amyl penicillin	N—C_5H_{11}—
II	G	Benzylpenicillin	$C_6H_5 . CH_2$—
III	X	*p*-Hydroxy-benzyl-penicillin	$HO . C_6H_4 . CH_2$—
IV	K	n-Heptylpenicillin	N—C_7H_{15}—

4. Oxidizing. All oxidizing agents rapidly destroy penicillin. Reducing agents are similarly destructive and should not be used with penicillin.

5. Enzymes. The stability of aqueous solutions referred to above applies only to sterile solutions. If solutions of penicillin are contaminated by many of the airborne and waterborne penicillin resistant organisms such as *E. coli* and its relatives, the enzyme penicillinase produced by these organisms rapidly destroys the penicillin. Though most resistant organisms produce this enzyme some do not produce it in detectable amounts and the quantity of enzyme produced is no criterion of the degree of organism resistance. Enzyme action and pH are mainly responsible for the destruction of orally administered penicillin.

6. Miscellaneous. Penicillin is markedly affected by a number of heavy metals, amongst which copper, mercury, iron and zinc, are the most important, particularly as zinc compounds are used as 'fillers' in the production of rubber bungs and piping. These metals attack the thiazolidine ring. Alcoholic groups, and thiol-containing compounds such as cysteine are also antagonistic. The significance of destruction by alcoholic groups is that the keeping of syringes in alcohol is contra-indicated unless they are thoroughly rinsed before use.

Pharmacology

The chemotherapeutic actions of all the salts and esters of penicillin are fundamentally the same, any difference between the penicillins being concerned with absorption and excretion rather than antibacterial activity.

Absorption and Administration Routes

The *inorganic salts* of penicillin—the sodium salt is the one most frequently used—are absorbed very rapidly from parenteral injection sites when aqueous solutions are used. Adequate blood levels are achieved within 30 min from intramuscular sites and in a slightly longer time following subcutaneous injection: there is very little reason for administering penicillin by the intravenous route, especially as thrombophlebitis may result from repeated injections of concentrated solutions. Most penicillin, as with other parenterally administered antibiotics, is administered by the intramuscular route.

Delayed absorption is frequently required to maintain adequate antibiotic concentrations in the blood and tissues, especially as excretion is rapid and the slowing of excretion unpracticable. There are two methods of delaying absorption:

1. If inorganic salts are suspended in oil vehicles, the water soluble salts will diffuse out slowly and continuously over a given period. Official injections of sodium or calcium benzylpenicillin suspended in arachis oil with beeswax were widely used for a short time.

2. If complex, sparingly soluble organic salts such **Procaine Benzylpenicillin,** B.Vet.C. or **Benzathine Penicillin,** B.P., U.S.P., are given as aqueous suspensions with buffering and suspending agents, absorption is also slow and blood levels may be maintained for 24 to 48 hr with one injection.

If these two principles are combined so that 'insoluble' penicillins are suspended in oil with aluminium monostearate, blood levels may be maintained for several days with one injection.

With all these 'long-acting' penicillins, the dosage must be large enough to provide a depot throughout the required period.

The *insoluble pencillins* are, because of their low solubility, only very slowly absorbed. Procaine benzylpenicillin is the most popular and in aqueous suspension will establish therapeutic blood levels in 2 to 3 hr and maintain them for 24 hr, when given in correct doses. Should high levels be required quickly, a 'loading 'dose of a water soluble inorganic salt may be given simultaneously. With benzathine penicillin, adequate blood levels may be maintained for 7 days by one single injection because of the very slow rate of absorption: acute and severe cases, if treated with this salt, should be 'primed' with either a dose of a soluble inorganic salt, a dose of the procaine salt or a combination of both, otherwise therapeutic blood levels may not be reached for many hours.

The *oral route* is not very satisfactory for the administration of the inorganic or procaine penicillins as these compounds are largely destroyed by gastric acidity and bacterial action. If large doses of buffered salts are given the absorption rate is sufficient to achieve therapeutic blood levels. The advent of phenoxymethylpenicillin (penicillin V), which is remarkably stable when given orally, has made the oral route more practicable and it has found a place in small animal practice, most probably as a means of maintaining blood levels after establishing them by parenteral injections.

Blood Levels

The blood level required for all normal uses is 0·03 to 0·05 units per ml blood (0·02 to 0·03 μg per ml). These levels are bacteriostatic: 0·5 units per ml is bactericidal. These levels in themselves are not important except that they are an indication of the levels in the tissues where the bacteria are multiplying, unless, of course, septicaemia (multiplication of bacteria in the blood stream) is present. Once the tissue levels are adequate they may be retained long after the blood levels have dropped below 0·03 units per ml. On this supposition intermittent massive doses were recommended, between which there would be times when no penicillin would be detectable in the blood. This technique is not generally used.

Barriers

1. Blood–brain barrier. Penicillin does not diffuse through into the cerebrospinal fluid unless the meninges are inflamed, and even then massive intramuscular or even intravenous doses are required to achieve any degree of therapy. It is advisable to give an initial intrathecal loading dose with isotonic solutions of penicillin, but great care must be taken as penicillin can cause severe collapse and death when administered by this route. Solutions containing 20 000 to 50 000 units per ml are isotonic.

2. Placental barrier. Penicillin does not readily pass from the maternal circulation to that of the foetus but, as with the blood–brain barrier, if the doses are large enough therapeutic levels can be achieved in the foetal circulation.

3. Intestinal barrier. Penicillin is slowly and irregularly absorbed from the intestines and doses from three to ten times larger than parenteral doses are normally required. Phenoxymethyl penicillin is superior to other penicillins when given by this route.

4. Serous membrane barriers. Penicillin diffuses poorly across serous membranes unless they are inflamed, and even then diffusion is not good. This means that pleurisy and peritonitis are best treated by injections directly into the serous cavities. The same applies to sinovial membranes and, where action within a joint is required, intra-articular injections are indicated.

5. The milk barrier. Diffusion of penicillin into the milk in therapeutic concentrations is achieved only by massive doses given intravenously. It is neither economical nor practical to treat mastitis by this route, though that does not preclude parenteral antibiotic therapy in mastitis where high temperatures and bacterial invasion of the body are present.

Excretion

Following an intramuscular or intravenous dose of a water soluble penicillin, detectable concentrations appear in the urine in a few minutes, and between 50 and 80 per cent can be recovered from the urine in an active form within 6 hr. Part of the remainder is metabolized, while small amounts are lost in the bile and saliva. It has been estimated in renal clearance tests that approximately 20 per cent of the excreted penicillin is lost by glomerular filtration and 80 per cent is excreted by the tubules.

Toxicity

There is no toxicity problem with penicillin in veterinary medicine. Occasionally allergic skin reactions may occur in small animals following previous treatment, and there are a few reports of similar reactions in thin skinned horses. Veterinary surgeons who are themselves sensitive to penicillin must use it with care. Where accidents have happened and penicillin sensitive humans are affected, antihistaminic creams applied locally will counteract skin reactions but severe reactions should be referred to a hospital or doctor. Animals showing allergy should be treated with antihistaminics locally and systematically. Intrathecal injections used in attempts to alleviate neural symptoms of distemper in the dog showed that penicillin was dangerous when used in this way. Experimentally it has been shown that 10 000 units should be the maximum intrathecal dose.

Some laboratory animals, but particularly guinea-pigs, show a specific sensitivity to penicillin.

Action

Penicillin has powerful bacteriostatic action with minimum blood levels of 0·03 units per ml and is bactericidal at 0·5 units per ml when used against susceptible bacteria. Its action is unaffected by blood elements or the products of tissue breakdown, including pus, nor does it affect leucocyte metabolism. It is, however, advisable to remove dead tissue and pus, as these non-vascular substances restrict the diffusion of the penicillin.

Penicillin is active against Gram-positive organisms and with a few exceptions inactive against Gram-negative organisms. The antibiotic is most effective when bacteria are multiplying and is thought to interfere at two stages of metabolism. The antibiotic interferes with the assimilation and concentration of glutamic acid within the bacterial cell and it has been suggested that penicillin also interferes with the nucleotide balance within the cell. More recently it was suggested that this interference in nucleotide metabolism results in an inability of the bacterial cells to produce sufficient cell wall for their needs, so that they outgrow their walls and become lysed, as the protoplasm flows through the gaps under the influence of osmotic pressures.

If bacteriostatic concentrations are present in the blood and tissues the affected bacteria become very susceptible to attack by the body defences—leucocytes, antibodies, etc., and remain so for several hours after the disappearance of effective blood levels. This time lag for staphylococci is about 4 hr in vitro, but it varies according to the duration of exposure to the antibiotic, the concentration of the antibiotic, and the type of organism.

With bactericidal concentrations, the natural body defences are called upon to a lesser degree, but are still required to remove dead bacteria and to combat the toxic products of bacteriolysis and the toxins which may be produced by the live bacteria. It is very important to remember that *no antibiotic will neutralize bacterial toxins*; with such diseases as those caused by *Cl. perfringens* (formerly *Cl. welchii*)—pulpy kidney, lamb dysentery, etc.—and *Cl. tetani* (causing tetanus) it is essential to give doses of antitoxin sera, even though antibiotics may be given to destroy the bacteria.

Antibacterial Spectrum

Penicillin is a narrow spectrum antibiotic and should normally be considered as effective against streptococci only. It is still the best antibiotic for the streptococci which affect animals. Other Gram-positive bacteria are affected but insensitive strains occur and the general response is less satisfactory. A few strains of Gram-negative cocci which affect man are susceptible but in general Gram-negative bacteria are insensitive to penicillin. Some rickettsial response is shown.

In vitro studies show that the following organisms are sensitive to normal bacteriostatic concentrations of penicillin:

Organism	Disease caused
Streptococci Staphylococci *Corynebacterium pyogenes*	Mastitis Systemic and local infections in all species
Corynebacterium renale	Pyelonephritis in cattle
Erysipelothrix rhusiopathiae	Swine erysipelas
Actinomyces bovis	Lumpy Jaw in cattle
Clostridia	Tetanus, pulpy kidney, lamb dysentery, etc.
Leptospira canicola	Leptospiral nephritis
Vibrio foetus	Vibrionic abortion
Bacillus anthracis	Anthrax.

In vivo, the picture is very different, for pathogens may either produce lethal substances which are not susceptible to antibiotic activity or they may produce tissue reactions which prevent access of the antibiotic to the organism. For example, *Clostridium welchii* type D, which causes 'pulpy kidney' in sheep, multiplies mainly in the intestines, producing a toxin which, when absorbed, causes the typical lesions and death: this toxin is not affected by an antibiotic so that once the toxin is produced the destruction of the organism is of no significance: orally administered penicillin also is mostly inactivated before it can have much effect. Similarly, in staphylococcal and *C. pyogenes* mastitis large amounts of pus, oedema or fibrous tissue, are produced, within masses of which the organisms multiply: any intramammary or systemic penicillin injections are unlikely to reach the organisms quickly enough or in sufficient quantity to materially affect the course of the condition.

Insensitive organisms include:

Pseudomonas species. Non-pathogenic under normal conditions, these organisms occasionally become pathogenic or exist as secondary complicating organism.

Pasteurella and Haemophilus species. These are responsible for pneumonic or septicaemic infections.

Coliform and Salmonella species. Both groups of organisms cause gastroenteritis, septicaemias, and *E. coli* causes up to 10 per cent of mastitis cases.

Brucella species. Bovine abortion, kennel cough in dogs, Malta fever, tularaemia and other diseases of man and animals.

Fungi and most viruses are also unaffected.

Penicillin Resistant Strains

Penicillin is remarkable in that bacterial resistance is very slow to develop. There are very few proven instances of organisms which are not basically resistant, such as penicillinase producers, developing resistance to penicillin. Staphylococci, which are active penicillinase producers, can form up to 70 per cent of infections recovered from cases of mastitis. When this occurs penicillin is of little value in therapy as it is unlikely to give more than a temporary clinical cure.

Clinical Applications

By far the most important animal disease for which penicillin therapy is the prime indication is *streptococcal bovine mastitis*. Other specific diseases for which penicillin is indicated are: anthrax in all species, *erysipelas* infection of sheep, pigs and birds, *strangles* and *joint-ill* of horses, some *clostridial* diseases of cattle and sheep. Penicillin has been used with variable results in *actinomycosis* and *pyelonephritis* of cattle. Local treatment of burns, scalds, wounds, eye and ear conditions, is very effective. Though penicillin usage is not always advisable, elevated temperatures due to unknown causes, and secondary invaders complicating virus diseases frequently respond favourably.

Bovine mastitis

This disease can be caused by a variety of organisms, including streptococci, staphylococci, *C. pyogenes*, *E. coli*, *Pseudomonas aeruginosa*, acid fast bacilli, various yeast like organisms and occasionally PPLO like organisms.

Penicillin was originally used to treat all cases of mastitis, but it is now recognized that its value lies mainly in the treatment of mastitis due to streptococci, particularly *Streptococcus agalactiae*.

To treat mastitis it is usual to prepare penicillin in the form of an oil suspension, which is introduced into the gland via the teat sphincter. Penicillin is used in the sodium or procaine form and doses of up to 300 000 units are use for each affected quarter. It is usual to give three injections at intervals of 24 to 48 hr.

Intramammary injections are introduced after the quarter is milked out. The teat must be cleaned with surgical spirit prior to the introduction of the antibiotic.

Anthrax

This disease which affects cattle, sheep, goats, pigs and dogs, can be controlled by the use of penicillin. The general principles of therapy with penicillin is that a large initial dose must be given. Three doses of 100 000 units of soluble penicillin given at intervals of 12 hr was reported as being effective in forty-two out of forty-four cattle.

***Erysipelothrix* Infections**

This condition is recognized in pigs, sheep and poultry, and the organism *Erysipelothrix rhusiopathiae* is particularly sensitive to penicillin.

Pigs. The mild and the acute septicaemic condition respond well to treatment. It is recommended that penicillin be combined with hyper immune serum, so as to provide an immunity to the disease. A dosage of 1000 units of penicillin dissolved in 0·66 ml antiserum per kg body weight is recommended. The dose should be repeated 24 hr later.

Sheep. It has been suggested that penicillin be used for the treatment of arthritis of erysipelothrix origin in sheep, particularly for the treatment of valuable rams. The dosage routine is similar to that recommended for pigs.

Poultry. The recommended dose is 50 000 units of benzyl penicillin with 150 000 units procaine penicillin injected intramuscularly.

Strangles

This disease, which is caused by *Streptococcus equi*, often responds to penicillin therapy, when the treatment is given at an early stage. Because of the clinical variations found in this disease from upper respiratory tract inflammation to massive abscess formation in the lymph glands, response to therapy is very variable.

A dosage of up to 100 000 units penicillin per kg body weight should be given daily for 5 days.

Clostridial Infections

In these infections penicillin must be administered early in the course of the disease, and where possible injected near the site of bacterial infection.

Tetanus

Where the site of infection is known and unhealed, as it frequently is with docking and castration wounds of sheep and pigs, local dressings with penicillin cream will destroy the local infection: systemic injection will do the same with delayed tetanus where the site of entry has healed over. The usefulness of penicillin apart from curbing the multiplication of secondary invaders, ceases when the organism has been destroyed and the real cure must come from the administration of antitoxins to neutralize the toxins which attack and progress by the nerve trunks. This can be supported by nerve sedatives—magnesium sulphate, chloral hydrate, barbiturates, chlorpromazine, etc., by the use of hexamine to aid diffusion of antitoxin across the blood–brain barrier; by use of protein hydrolysates for feeding small animals; by use of intestinal stimulants in the recovery stages, particularly in ruminants, but most of all by careful nursing to maintain muscular and joint movement and to prevent both hypostatic congestion and excoriation of all bony prominences ('bed-sores'). Penicillin wound dressings are an effective prophylxis.

Blackleg

Clostridium chauvoei, and occasionally *Cl. oedematiens* and *Cl. septique*, cause a muscular and connective tissue 'gas-gangrene' in cattle and sheep. These organisms, which multiply in the affected tissues, do not produce such powerful exotoxins as other clostridia and if maximum doses (5000 units per lb body weight) of penicillin are injected early enough a good percentage of recoveries can be achieved. Treatment must be undertaken before muscle damage is great enough to necessitate slaughter for humane reasons.

As a prophylactic against gas-gangrene and blackleg, penicillin has been used in dog bites and wounds of sheep, and it has been recommended that sheep not vaccinated against blackleg should receive one or two injections of 0·5 mega-units following dystocia.

Botulism

Although *Clostridium botulinum* has been shown to be highly sensitive to penicillin in vitro, this fact is of little clinical significance, for botulism is an exointoxication, the toxin being produced by the organism multiplying saprophytically in dead vertebrates and invertebrates. Ingestion of the toxin alone in very small quantities can be lethal to birds, horses, dogs, etc.

Pyelonephritis

This disease of cattle is typified by destruction of kidney tissue, partly by bacterial attack and partly by pressure of the tremendous amounts of pus produced. The organism responsible is *Corynebacterium renale*, which is sensitive in vitro to penicillin, but the character of the clinical condition is such that results in all except early cases are not encouraging. Where clinical cures have been achieved, bacteriological examination has repeatedly demonstrated the organisms in the urine. Treatment should consist of a minimum daily dose of 1 mega-unit of procaine penicillin, but preferably more, continued for at least a week. The most that can be achieved in all but the very earliest cases is to maintain a cow until the completion of a pregnancy or lactation or to enable fattening and slaughter.

Actinomycosis

'Lumpy jaw' is caused by an organism, *Actinomyces bovis*, which is susceptible in vitro to penicillin. The lesions consist of purulent and granulating masses surrounded by bone produced at the periphery of areas in which bone is being resorbed: the whole jaw bone becomes an enlarged mass of bone honeycombed with 'abscess' cavities filled with pus. Some of these pus filled cavities communicate with the exterior, and sometimes the mouth, through sinuses from which the pus exudes. The pathology of the disease may thus be considered as abscesses in which the capsule is bone and not fibrous tissue, and unless these abscesses can be drained and irrigated with penicillin little improvement can be expected. This agrees with clinical experience, and any benefit is only derived from penicillin injected into the centres of suppuration though this is difficult to achieve, and temporary regression is the best that can be obtained with penicillin.

Actinobacillosis

This condition is caused by a Gram-negative organism *Actinobacillus lignieresi*, which affects soft tissues only, is not markedly sensitive to penicillin, though 'wooden tongue' may benefit from its use due to elimination of secondary invaders; sodium iodide or sulphonamides intravenously are most effective.

Streptococcal Infections in Puppies

This condition, often referred to as 'BHS infection', responds well to injections of the bitch during pregnancy—procaine penicillin on three occasions, at mating, mid-pregnancy and just before parturition. Injection of the puppies at birth with 30 000 to 50 000 units soluble penicillin or procaine penicillin will also prove successful.

Non-Specific Bacterial Infections

Penicillin has been widely used as a routine treatment for fevers, the cause of which is unknown, especially in dogs and cats. In septicaemias such as navel-ill in calves and foals, which are often caused by streptococci (*Streptococcus zoöepidemicus* in foals), penicillin is often very effective, but as the condition in foals is more frequently caused by *Bacterium equirulus* (*Shigella equirulus*) and *Escherichia coli*. it is advisable to use a combination of penicillin, 10 000 units per kg and streptomycin 11 megaunits per kg, or ampicillin or amoxycillin at a dose of 6 mg per kg.

Leptospirosis

Disease caused by *Leptospira canicola*, *L. pomona*, *L. mitis*, *L. sejroe*, *L. icterohaemorrhagiae*, amonst many, are being more frequently met in all species. *L. canicola* causes nephritis in man, pigs and dogs; *L. pomona*, a generalized septicaemia in horse, sheep and goats, abortion in cattle and sows; *L. icterohaemorrhagiae* causes Weil's disease in man, and leptospiral jaundice in dogs. These examples are only a few of those reported, but they show how widespread is leptospirosis and how complex the aetiology. When penicillin was first available it was very widely used in the treatment of leptospirosis in dogs, and later in other species. It soon became apparent that leptospiral jaundice did not respond as well as leptospiral nephritis, but this is partially due to the difficulty of diagnosing *L. icterohaemorrhagiae* infections until they are far advanced and liver damage extensive.

There is no doubt, however, that though penicillin in large doses will produce appreciable clinical response, ampicillin, amoxycillin, streptomycin or tetracycline broad spectrum antibiotics are superior. Penicillin does not eliminate *L. canicola* from the kidney and urine and its use results in subclinical carrier states: other antibiotics eliminate the organism completely.

Skin Applications

Penicillin has been combined with sulphanilamide, sulphadimidine and sulphathiazole for wound, burn and scald dressing. The first two are superior to sulphathiazole for these two sulphonamides diffuse more easily into the tissue around the wound and are therefore more effective.

The proportions are usually 5000 units of sodium or potassium penicillin per gram of sterile powder. Lotions and ointments have been applied to the eye with good effect.

Dosage

Parenteral: 4000 to 10 000 units per kg body weight. The dosage should be given every 4 hr if a soluble salt in water is used, every 12 hr if a soluble salt in oil with stearate or wax, every 24 hr if procaine penicillin is used in an oil suspension. Adult stock should receive the high dosage rate only if the condition is acute. Young stock and small animals should be given the higher dose on all occasions.

Oral: small animals: three to five times the parenteral dose, or 100 000 to 300 000 units every 4 hr, if mineral or procaine salts are used. Phenoxymethylpenicillin is given at the rate of 8 mg per kg body weight three times daily. For pigs and poultry 16 mg per kg body weight twice daily is more convenient.

Official Preparations

Penicillin is available in official form as a cream, eye ointment and eye-drops, ointment, tablets, parenteral injections, and intramammary injections.

Cream of Penicillin, B.Vet.C. (Crem. Penicil.)

This is the cream of the B.P. and, unless otherwise requested, contains 1000 units of benzylpenicillin with sodium citrate as a buffering agent and chlorocresol as a preservative. The base consists of 7 per cent 2/2 emulsifying wax, 5 per cent w/w hard paraffin, and 41 per cent w/w liquid paraffin, with water up to 100 per cent. The proportions of paraffins may be varied according to climate and the need for a hard or soft cream. It should be kept in a refrigerator and used within 7 days.

Eye Ointment of Penicillin, B.Vet.C. (Oculent. Penicil.)

This also is the B.P. preparation and normally contains 2000 units per g of a salt of benzylpenicillin. The base is a sterile mixture of 5 per cent liquid paraffin and 95 per cent soft paraffin. The preparation will keep in cool storage conditions, little deterioration occurs when kept for 12 to 24 months.

Eye-Drops of Penicillin, B.Vet.C. (Gutt. Penicil.)

A soluble penicillin salt is dissolved at the rate of 2500 units per ml in solution for Eye-Drops, B.Vet.C., with 0·5 per cent w/w of sodium citrate as a buffer. The solution contains fungistatic proportions of methyl hydroxybenzoate and propyl hydroxybenzoate in freshly boiled and cooled distilled water. The drops should be stored in a cool place and used within 7 days. If kept in a refrigerator, the solution must be allowed to reach room temperature (about 20°C) at least, before use.

Ointment of Penicillin, B.Vet.C. (Ung. Penicil.)

This also is a B.P. specification containing 1000 units per g of benzylpenicillin in 5 parts liquid paraffin and 95 parts soft paraffin. It keeps well in cool storage conditions.

Tablets of Penicillin, B.Vet.C. (Tab. Penicil.)

These are the tablets of the B.P. containing 200 000 units of sodium or potassium benzylpenicillin. They contain up to 5 per cent by weight sodium citrate, taking the penicillin as reference, and are usually coated to obtain some protection from the gastric juices.

Injection of Penicillin, B.Vet.C. (Inj. Penicil.)

The preparation conforms with the B.P. specification, consisting of soluble salts of amorphous penicillin or crystalline benzylpenicillin, with or without a buffering agent, for fresh solution with Water for Injection. Vials containing the sterile powder are available commercially with 0·1, 0·2, 0·5, 1·0, 10·0 and 200 mega-units. Solutions are best used immediately they are made, but storage at 4°C or less maintains potency reasonably well for 4 days with unbuffered and 7 days with buffered solutions.

Injection of Procaine Benzylpenicillin, B.Vet.C. (Inj. Procain. Benzylpenicil.)

This complies with the B.P. specification, consisting of the procaine salt of benzylpenicillin with a dispersing agent and with or without a buffering agent. It is made into a suspension by adding Water for Injection, B.Vet.C. and shaking. Vials containing the sterile powder are available commercially with 0·3, 0·9 and 3 mega-units. The suspension will keep for 7 days when refrigerated.

Injection of Procaine Benzylpenicillin with Benzylpenicillin, B.Vet.C. (Inj. Procain. Benzylpenicil. c. Benzylpenicil.)

Sometimes referred to as Fortified Injection of Procaine Penicillin G, this conforms to the B.P. specification and contains the previous two injections in proportions of 1:3. The addition of Water for Injection gives a solution and suspension. It is available in vials containing 0·4, 1·2 and 4 mega-units. Once prepared the combination should be used in 4 days even if refrigerated, because of the soluble penicillin content.

Stabilized aqueous suspensions of penicillin containing 0·3 mega-units of procaine benzylpenicillin per ml with buffering and dispersing agents are available commercially. They are stable at room temperature with a shelf life of about 1 year, and are supplied in 10 and 30 mega-unit sizes.

Intramammary Injection of Penicillin, B.Vet.C.

This is a suspension in a bland vegetable or mineral oil, with or without aluminium stearate, of a soluble salt of benzylpenicillin. Normally the penicillin content is 100 000 units.

Intramammary Injection of Procaine Benzylpenicillin, B.Vet.C.

This preparation is similar to the preceding one but contains procaine benzylpenicillin instead of the 'soluble' salts. The official preparation contains 100 000 units, but tubes containing 300 000 units per dose are marketed.

Intramammary Injections of Procaine Benzylpenicillin and Dihydrostreptomycin, B.Vet.C.

The injection is similar to the preceding injections but contains 100 000 units procaine benzylpenicillin and 0·1 g (100 000 units) of dihydrostreptomycin base as the sulphate.

Intramammary Injection of Dapsone and Penicillin, B.Vet.C.

This injection contains 33·3 per cent powdered dapsone with 50 000 units of a 'soluble' benzylpenicillin salt per dose, in a base of liquid paraffin and soft paraffin.

All these preparations should be prepared in a sterile manner, heating the base at 150°C for 1 hr and adding the antibiotic aseptically. Prepared and packed in this manner these products are very stable and are given shelf lives of up to 3 years.

II. THE SEMI-SYNTHETIC PENICILLINS

History

The introduction of penicillin, which was a landmark in the history of medicine, stimulated the search for other biologically active drugs which might be recovered from moulds and fungal growth. Many hundreds have been found, but penicillin particularly, because of its high antibacterial activity and its lack of toxicity, has remained the standard for the ideal antibiotic.

The isolation in a pure state of the 'nucleus' of the penicillin molecule in 1957 led to the rapid development of new penicillins with desirable and specific properties. These could be deliberately prepared by the synthetic addition of an appropriate side chain acid group to the nucleus. The initial work had been aimed at the production and modification of *p*-aminobenzyl penicillin since the side-chain of this substance contains a free amino group which lent itself readily to substitution. The production of *p*-aminobenzyl penicillin in culture media was stimulated by the addition to the culture fluid of the side chain acid, *p*-aminophenylacetic acid or its amide. In the absence of the side chain precursor, the chemical assay of the culture fluid gave consistently higher results than the biological assay. This appeared to indicate that a substance had been formed which behaved chemically like penicillin, but which was biologically much less active.

Further investigation showed that the substance produced by the mould in the absence of a precursor acid was destroyed by penicillinase, and also that benzyl penicillin could be formed by acylation of the culture fluid with phenyl acetyl chloride. It seemed probable, therefore, that the nucleus of the penicillin molecule, without the side chain, was present in the culture medium. This substance 6-aminopenicillanic acid was finally isolated in the crystalline state by purification methods making use of column chromatography.

Once 6-aminopenicillanic acid became available in significant quantities, the way was open for the preparation of a range of 'new' penicillins. This work was pursued with the aim of strengthening the activity of penicillin G.

Penicillin itself has several therapeutic weaknesses:

(1) It has low activity against Gram-negative bacteria.
(2) It is destroyed by acid.
(3) It is destroyed by the β-lactamase enzymes (penicillinases) which are produced by some bacteria.

Research, therefore, was concentrated on the development of new penicillins which would overcome these disadvantages.

The semi-synthetic penicillins developed since 1960 can be considered to lie in three different groups:

1. Narrow spectrum antibiotics which have similar activity to that of penicillin G.

2. Penicillinase stable penicillins.

3. Broad spectrum penicillins, which have a wide coverage including in some cases activity against *Pseudomonas*.

Table XV gives the antibacterial pattern shown by members of each group. Table XVI gives an indication of the antistaphylococcal activity, serum binding and oral absorption of a group of penicillinase stable penicillins.

The broad spectrum semi-synthetic penicillins fall into two further categories: those which have a wide spectrum of activity against Gram-positive and Gram-negative bacteria with a low degree of binding to serum proteins and a good pattern of absorption when given by mouth; and those which have activity against *Pseudomonas aeruginosa* with, in addition, activity against other Gram-negative bacteria and also a good

Table XV. Antibacterial spectra of semisynthetic penicillins*

Organism	Minimum inhibitory concentration† (μg/ml)						
	Narrow-spectrum penicillins			Penicillinase-stable penicillins		Broad-spectrum penicillins	
	Penicillin G	Penicillin V	Phenethicillin	Methicillin	Cloxacillin	Ampicillin	Carbenicillin
Staphylococcus aureus	0·02	0·05	0·05	1·25	0·1	0·05	1·25
Staphylococcus aureus (penicillinase-producing)	R	R	250	2·5	0·25	R	25
Streptococcus pyogenes	0·01	0·02	0·05	0·25	0·1	0·02	0·5
Streptococcus pneumoniae	0·02	0·02	0·05	0·25	0·25	0·05	1·25
Streptococcus faecalis	2·5	5·0	5·0	25	25	1·25	25
Clostridium welchii	0·05	0·1	0·1	1	0·5	0·05	0·5
Neisseria meningitidis	0·02	0·05	0·1	0·25	0·25	0·02	0·05
Haemophilus influenzae	0·5	5	5	2·5	12·5	0·25	0·25
Escherichia coli‡	50	125	R	R	R	5	5
Salmonella typhi	5	125	R	R	R	0·5	5
Shigella sonnei‡	50	250	R	R	R	2·5	12·5
Proteus mirabilis‡	5	50	250	250	R	1·25	2·5
Indole-positive *Proteus*‡	R	R	R	R	R	250	5
Klebsiella aerogenes	250	R	R	R	R	250	250
Enterobacter species‡	R	R	R	R	R	250	12·5
Pseudomonas aeruginosa‡	R	R	R	R	R	R	50

* Beecham Pharmaceuticals, unpublished data.
† Sensitivity of majority of strains. R = MIC > 250 μg/ml.
‡ Resistant strains are encountered.

Table XVI. Antistaphylococcal activity, serum binding and oral absorption of penicillinase-stable penicillins

Compound	Date of first report	Minimum inhibitory concentration, *S. aureus* (μg/ml)		Extent of binding in serum (% bound)	Oral absorption in man
		Penicillin-sensitive	Penicillin-resistant		
Methicillin	1960	1·25	2·5	49	None
Ancillin	1961	0·20	0·80	91	Moderate
Nafcillin	1961	0·20	0·40	87	Poor
Oxacillin	1961	0·15	0·40	93	Moderate
Cloxacillin	1962	0·2	0·35	94	Good
Quinacillin	1964	0·5	1·25	40	None
Dicloxacillin	1964	0·15	0·30	97	Very good
Flucloxacillin	1970	0·15	0·30	95	Very good

degree of stability in the presence of staphylococcal penicillinase.

In the first category are ampicillin, compounds which hydrolyse in the body to give ampicillin (hetacillin, metampicillin, pivampicillin) and analogues of ampicillin (epicillin and amoxycillin). In the second category, i.e. those active against *Pseudomonas*, are included carbenicillin, carindacillin, ticarcillin, sulfocillin, suncillin and eureidopenicillins.

It will be seen that the semi-synthetic penicillins provide a range of activity which covers the major bacteria of significance recovered in the veterinary field. Up to the present a number of semi-synthetic penicillins have been developed in a specific way for animal therapy and these include cloxacillin, oxacillin, nafcillin, ampicillin, amoxycillin, pivampicillin and carbenicillin.

Penicillinase Stable Penicillins

Because the β-lactam ring of the penicillin molecule is easily broken, the antibacterial activity of penicillin was frequently destroyed by Gram-positive and Gram-negative bacteria which produced penicillinase enzymes.

In practice, the most important of these penicillin-resistant bacteria are the penicillinase producing strains of *Staphylococcus aureus*. These organisms which existed in nature before the discovery of penicillin, inevitably tend to dominate a bacterial environment in which penicillin is used for therapy.

When penicillin was first introduced, the majority of pathogenic staphylococci in hospital and in dairy units were sensitive to the antibiotic. However, with the regular use of penicillin in hospitals and in the treatment of mastitis in dairy units, the number of penicillin-resistant staphylococci recovered from infections gradually increased. By 1961 in the United Kingdom about 80 per cent of staphylococcal infections in hospitals were caused by penicillin-resistant strains, and about 73 per cent of similar infections in dairy herds.

The development of this high incidence of resistance to penicillin is considered to be due to selection of the naturally resistant strains by continual use of penicillin.

Cloxacillin (Orbenin)

A number of the new penicillins have proved to be of specific value in the veterinary field.

A big advance in the therapy of penicillin-resistant staphylococcal infections took place with the preparation and investigation of the 3,5-disubstituted 4-isoxazolyl penicillins. Many penicillins of this class have been examined, and two, cloxacillin and oxacillin, are discussed. Both oxacillin and cloxacillin are active against penicillin G resistant staphylococci in vitro, and have the additional advantage of being stable in the presence of gastric acid, so that they can be administered orally as well as by injection.

Absorption and excretion. Both compounds are used as the sodium salts for oral absorption and cloxacillin differs chemically from oxacillin only in the presence of a chlorine atom.

Comparative studies on the absorption and excretion of oxacillin and cloxacillin were carried out in man. The following mean results were obtained:

Type of penicillin	Mean serum concentrations (μg/ml) ½ hr	1 hr	2 hr	4 hr
Cloxacillin	7·2	8·9	4·5	<2·0
Oxacillin	5·4	4·9	1·5	0·3

Cross-over studies were also carried out in fifteen fasting subjects with 500 mg doses of oxacillin and cloxacillin, and the antibacterial activity of the serum was determined against twenty strains of *Staph. aureus.*

Type of penicillin	Antibacterial activity at: ½ hr	1 hr	2 hr	3 hr	4 hr
Cloxacillin	21·3	36·3	15·5	5·8	<2
Oxacillin	17·6	10·7	<2	<2	<2

The figures refer to the mean of the reciprocal of the highest dilution of the serum which still inhibited growth.

It was concluded that sodium cloxacillin is well absorbed when given by mouth, and is active for 4 to 6 hr. The serum concentrations, urinary excretion, and serum antibacterial activity are superior to those of sodium oxacillin. Peak serum levels are obtained about 1 hr after an oral dose, and about ½ hr after an intramuscular injection. Doubling the dose doubles the serum concentration.

Oxacillin and cloxacillin are both metabolized to about 10 per cent in the body. The metabolite of cloxacillin has been isolated and found to have antibacterial activity very similar to that of the parent compound.

Toxicology. The acute toxic dose (LD_{50}) in mice has been found to be more than 5 g per kg by mouth, and 1·2 g per kg by intravenous injection, these values being higher than those for penicillins G and V. Prolonged tests at high dosage were also carried out in rats and dogs, and no haematological, biochemical or histological abnormalities were noted. However, subconjunctival injection of a 50 per cent solution in rabbits produced an opacity of the cornea which persisted for at least 14 days. Cloxacillin is thus contra-indicated for use in the eye.

Large intramuscular doses of cloxacillin were given to pregnant rabbits, and it was found that all the animals gave birth to normal litters.

Antibacterial activity. Both cloxacillin and oxacillin are stable to staphylococcal penicillinase, and active against both penicillin-resistant and penicillin-sensitive staphylococci.

The minimum inhibitory concentrations of cloxacillin against a large number of staphylococcal isolates was determined. A total of 382 cultures tested comprised 152 penicillin-sensitive staphylococci and 230 penicillin-resistant staphylococci. Taking the penicillin-sensitive strains, only one required a concentration of cloxacillin as high as 1·25 μg per ml to inhibit growth, and approximately 95 per cent of the strains showed MIC values in the range 0·05 to 0·25 μg per ml. Of the total of 230 penicillin-resistant strains, only two required a concentration as high as 1·25 μg per ml for inhibition, and 97 per cent of the strains showed MIC values in the range 0·125 to 0·5 μg per ml. The largest single group of strains were those with an MIC value of 0·25 μg per ml. Of the total of 230 penicillin-resistant strains, only two showed an MIC as high as 1·25 μg per ml with cloxacillin, whereas with oxacillin the corresponding number of strains was thirty-six (15 per cent).

Similar results were obtained with a smaller number of staphylococcal strains, and also cloxacillin was found to have a slightly greater activity than oxacillin.

Against the major Streptococci of mastitis, i.e. *Streptococcus agalactiae*, *Streptococcus dysgalactiae* and *Streptococcus uberis*, the minimum inhibitory concentrations of cloxacillin are normally within the range 0·05 to 2 μg per ml. Neither compound is active against *Strep. faecalis* or against Gram-negative bacilli.

Cloxacillin is highly bactericidal to penicillin-resistant staphylococci at concentrations only slightly higher than the MIC values. At a concentration of 2 μg per ml of cloxacillin, a kill of approximately 99·9 per cent was obtained in 7 hr in broth cultures of penicillin-resistant staphylococci.

It has been shown that cloxacillin is extremely stable to penicillinase, and also that cloxacillin is slightly more stable than oxacillin. Both oxacillin and cloxacillin are powerful inducers of penicillinase-formation.

A number of other compounds with activity against penicillinase producing organisms have been investigated, and these include narcillin, dicloxacillin and flucloxacillin.

Cloxacillin Sodium

Cloxacillin sodium is a white or almost white crystalline powder with a bitter taste. It has the formula $C_{19}H_{17}ClN_3NaO_5S,H_2O$. It is soluble at 20°C in 2·5 parts of water and is considerably more soluble in alcohol and chloroform.

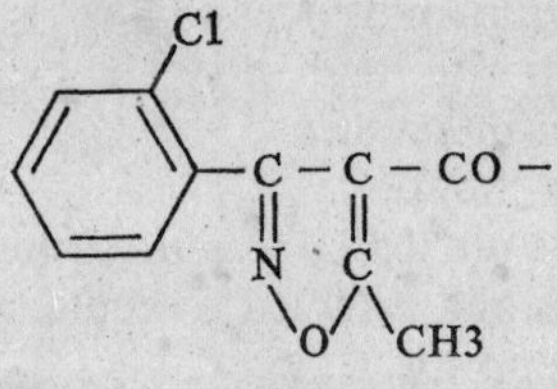

Cloxacillin

Actions and Uses

Cloxacillin has a range of activity similar to that described under penicillin (page 328). Cloxacillin is less active than penicillin against streptococci, but readily achieves adequate blood and milk levels in the body to deal with these organisms. Cloxacillin is active against both penicillin sensitive and resistant staphylococci.

Cloxacillin, when given parenterally in aqueous solution is absorbed rapidly in animals and achieves significant blood levels within ½ hr of injection. It is distributed throughout the body, although only small quantities enter the cerebrospinal fluid. Cloxacillin is particularly indicated for skin and soft tissue infections, wounds and burns, and septicaemic due to Gram-positive infections.

Cloxacillin, because of its activity against penicillin-resistant staphylococci, has become the drug of choice for the treatment of bovine mastitis. It has been studied in many countries, but particularly by the National Institute for Research in Dairying in the United Kingdom as the major antibiotic for use in a control programme for mastitis.

It is used in two ways, as sodium cloxacillin for the treatment of mastitis during lactation, and as the benzathine salt for treatment and prevention of infection during the dry period. Both salts are prepared in an oil base for administration via the teat canal. The sodium salt is given three times at 24-hr intervals and the benzathine salt is placed in all four quarters when the cow is dried off at the end of lactation. This form, developed for the dry cow, lasts in the quarter for 3 to 4 weeks.

It has been demonstrated that treatment during the dry period against staphylococcal infection is three times more effective than treatment during lactation.

Despite continual use for the therapy of mastitis over a 12-year period, staphylococcal resistant strains have not been encountered in the field. The NIRD studies have shown that therapy must be associated with teat dipping and other management processes to ensure successful control of mastitis.

Dosage

Parenteral:
- Small animals: 4–10 mg/kg body weight.
- Large animals: 1–2 mg/kg body weight.

Intramammary:
- Sodium Cloxacillin: 200 mg per quarter.
- Benzathine Cloxacillin: 500 mg per quarter.

Broad Spectrum Penicillins

Ampicillin (Penbritin)

Ampicillin is probably one of the most interesting and valuable of the semi-synthetic penicillins, because whilst it retains the typical Gram-positive activity of the penicillins, it possesses greatly enhanced activity against Gram-negative bacteria.

This penicillin has an antibacterial activity comparable to that of chloramphenicol and the tetracyclines, so it can be described as a true broad spectrum antibiotic.

Physical Properties

Ampicillin in the free acid form is a white crystalline powder, which is sparingly soluble in water. It is stable in acid solution and can thus be given by mouth.

The sodium salt has a much greater aqueous solubility, so that this form of ampicillin is used for parenteral preparations.

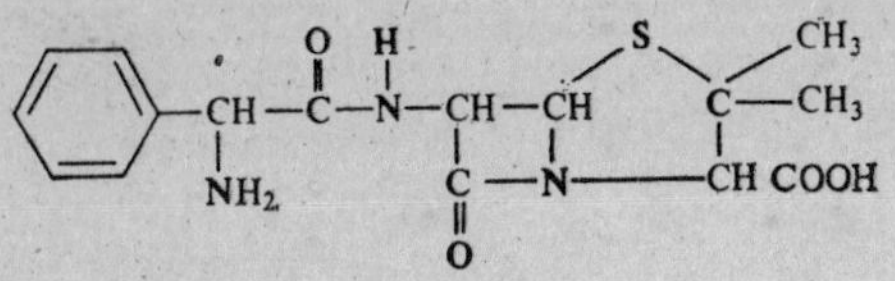

Ampicillin

Absorption, Distribution and Excretion

Ampicillin is well absorbed in most animals after oral administration, and peak serum concentrations are obtained about 2 hr after dosing. If the dose is doubled, the peak concentration will also be doubled.

Animal studies have demonstrated that ampicillin is evenly distributed throughout the body tissues, and is concentrated in the liver and kidneys. Only small quantities enter the non-infected cerebro-spinal fluid.

The drug is excreted in the unchanged form in the urine.

Investigations in rats and in cattle showed that ampicillin is absorbed in the upper part of the small intestine, and is excreted within 3 hr in high concentrations in the bile. The concentration in the bile may be forty times higher than the level in the plasma.

This means that ampicillin may appear twice in the intestines, following an oral dosage. This factor is often of value in the treatment of enteric bacteria, which may be both in the gut and gall bladder.

Toxicology

As with all penicillins, very high doses may be given over a prolonged period without producing any toxic reactions.

Antibacterial Activity

Ampicillin is bactericidal in action and is a true broad spectrum antibiotic, both in vivo and in vitro and is active against a large number of Gram-positive and Gram-negative organisms. These include species of streptococci, staphylococci, *Streptococcus faecalis*, *Cornyebacterium*, *Clostridium*, *Fusiformis* spp., *E. coli*, *Klebsiella*, *Shigella*, *Salmonella*, *Proteus*, *Brucella* and *Pasteurella*.

Resistance

The emergence of resistant strains is stepwise, as in all penicillins. It is destroyed by penicillinase producing staphylococci and some Gram-negative strains.

Actions and Uses

Clinical trials have demonstrated that ampicillin is of considerable value in the treatment of the following clinical entities in farm and small animals:

Cattle
- Calf scours and pneumonia.
- Calf enteritis and septicaemia due to salmonella infections.
- Foul of the foot.
- Mastitis due to *E. coli* infections.
- Metritis and retained placenta.
- Pyelonephritis.

Pigs
- Enteritis and pneumonia associated with *E. coli* infections.
- Erysipelas.
- Metritis.

Sheep
- Contagious foot rot.
- Mastitis.
- Metritis.
- Pneumonia.
- Foot abscess.

Horses
- Enteritis and septicaemia in the newborn and young foal.
- Metritis.
- Respiratory infections, particularly in association with equine influenza.
- Strangles.

Poultry
- Enteritis due to *E. coli* and clostridial infections.

Dogs
- Secondary infection associated with dermatitis.
- Enteritis.
- Ear infections.
- Leptospirosis.
- Pharyngitis and tonsilitis.
- Post operative wounds.
- Respiratory infections.
- Urogenital infections.

Cats
- Secondary infections associated with respiratory viruses.
- Enteritis.
- Ear infections.

Dosage

Oral: 4–10 mg/kg
Parenteral: 2–7 mg/kg

In young animals and in acute infections the upper dose level is recommended. Treatment is normally given at 24-hr intervals, but in very acute infections the full recommended dose may be given twice daily, as there are no problems of toxicity. As with all antibiotics, the earlier therapy is carried out after recognition of a disease entity, the more effective will be the clinical response.

Preparations

Ampicillin is available in powder, tablet, cream, intramammary and parenteral injection form.

Ampiclox

Because *E. coli* infections appear to have become more common in some dairy herds a formulation of ampiclox has been evolved for mastitis therapy.

The effective combination has been found to be ampicillin 75 mg, and cloxacillin 200 mg. The two drugs are combined in a vegetable oil base, and are given for the therapy of lactation infections at intervals of 24 hr.

Amoxycillin (Clamoxyl)

This drug is a close analogue of ampicillin and has the formula (D(−)-α-amino-*p*-hydroxybenzylpenicillintrihydrate) and the structure of the side chain is

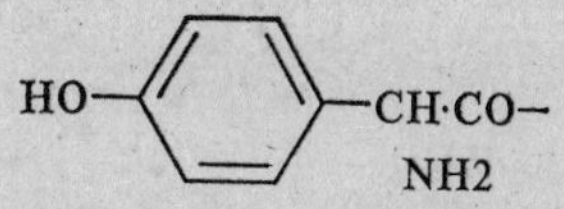

Like ampicillin, amoxycillin is acid-stable and shows an antibacterial spectrum and a level of activity similar to that of ampicillin. Amoxycillin differs from ampicillin in that it is better absorbed when given by mouth in animals; as a result the antibiotic levels reached in the blood and urine are considerably higher than those obtained with ampicillin. The presence of food does not affect the amount of amoxycillin absorbed.

In experimental infections in mice, Acred et al. (1971) demonstrated that the activity of amoxycillin is much greater than that of ampicillin. It has also been shown in vivo that amoxycillin has a more rapid and complete bactericidal effect than ampicillin against certain Gram-negative bacilli. This was demonstrated when the blood levels of the two compounds were similar.

Actions and Uses

Clinical trials in dogs and cats and farm animals have shown a high level of activity against the range of infections mentioned under ampicillin.

Penicillins Active against *Pseudomonas*

Carbenicillin (Pyopen) (disodium α-carboxybenzylpenicillin) was the first penicillin to be described which showed significant activity against *Pseudomonas aeruginosa.*

The structure of the side chain is:

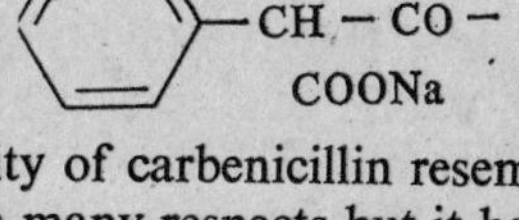

The activity of carbenicillin resembles that of ampicillin in many respects but it has important differences. It is less active against pyogenic cocci and against *Streptococcus faecalis*, but it has a greater stability against staphylococcal penicillinase. Carbenicillin has good activity against strains of *Proteus* and against *Enterobacter*. It is active against *Pseudomonas aeruginosa* at levels of 25–50 μg/ml.

Pharmacology

It has to be administered by injection and gives a peak level at 1 hr. It is rapidly excreted in the urine as with all penicillins.

Actions and Uses

Because the level of activity of carbenicillin against *Pseudomonas* is relatively low, when compared with its activity against other bacteria, it is wise to give the antibiotic by the intravenous route. Knudsen et al. (1967) reported that in man levels of 140 μg/ml were reached in the blood after intravenous administration of 1 g of carbenicillin; this level fell to approximately 50 μg/ml after 1 hr.

Because *Pseudomonas* is not a major cause of disease in animals, carbenicillin and other anti-*Pseudomonas* agents have not been studied in detail in animals.

III. CEPHALOSPORINS

This group of antibiotics is formed by a species of *Cephalosporium* cultivated from a sewage outfall off Sardinia. The culture produces three quite distinct antibiotics, cephalosporin N, a penicillin, cephalosporin P and cephalosporin C.

Two derivatives of these materials are in therapeutic use in the human field: Cephalothin and Cephaloridine.

Cephalothin is 7-(thiophine-2-acetamido)-cephalosporic acid, and is used as the sodium salt.

Cephaloridine is 7-[(2-thienyl)acetamido]-3-(1-pyridylmethyl)-3-cephem-4-carboxylic acid betaine.

Pharmacology

Cephalothin is not absorbed satisfactorily when given orally, so it is administered by intramuscular injection. Like the penicillins it is excreted mainly by the kidneys. Cephaloridine is very similar in behaviour to Cephalothin, but it is slightly more active, is more stable in the body, and is not affected by protein.

Antibacterial Activity

Both Cephalosporins are antibacterial against both Gram-positive and Gram-negative organisms and they have a range of activity similar to Ampicillin. Although they are not readily affected by penicillinase there is a cross resistance with some methicillin-resistant staphylococci.

Clinical Applications

Cephaloridine can be used in the treatment of many infections of domestic animals and is particularly valuable in the therapy of diseases of the respiratory and urinary tracts.

Specific conditions which will respond to Cephaloridine therapy include pneumonia, metritis, nephritis, cystitis, foul in the foot, septicaemia and otitis.

Two cephalosporins—cephalonium and cephoxazole—have been introduced for intramammany therapy and are now under trial.

Dosage

Parenteral treatment: 10 to 20 mg per kg body weight once or twice daily.

Bibliography

ACRED, P., HUNTER, P. A., MIZEN, L. & ROLINSON, G. N. (1971) *Antimicrob. Agents Chemother.*, 416.

BRITISH CATTLE VETERINARY ASSOCIATION (1971) *Control of Bovine Mastitis*. Woking, Surrey: Unwin Bros.

GARROD, L. P., LAMBERT, H. P. & O'GRADY, F. (1973) *Antibiotics and Chemotherapy*. Edinburgh and London: Churchill Livingstone.

LYNN, B. (1965) The semi-synthetic penicillins. *Antibiotica Chemother.*, **13**, 125–226.

KNUDSEN, E. T., ROLINSON, G. N. & SUTHERLAND, R. (1967) Carbenicillin: A new semisynthetic penicillin active against *Pseudomonas pyocyanea*. *Br. med. J.*, **iii**, 75.

PERLMAN, D. (ed.) (1968) *Topics in Pharmaceutical Science*, vol. I, *Penicillins and Cephalosporins*. American Academy of Pharmaceutical Science Symposium. New York: Interscience.

ROLINSON, G. N. & SUTHERLAND, R. (1973) Semisynthetic penicillins. *Adv. Pharmac. Chemother.*, **11**.

WILLIAMS, J. D. & GEDDES, A. M. (1975) *Chemotherapy*, vol. 5, *Penicillins and Cephalosporins*. London: Plenum.

29

Polysaccharide Type Antibiotics, Including the Macrolides

I. STREPTOMYCIN

Streptomycin is a narrow spectrum antibiotic which is official in all the pharmacopoeias. Its salts must contain not less than 700 international units per mg expressed as the dried sulphate. It is now more frequently defined in terms of grams (g), with 1 g of the base equivalent to 1 mega-unit; 1·25 g of the sulphate are approximately equivalent to 1 g of the base.

History

Streptomycin was announced in January 1944 by Schatz, Bugie and Waksman, following a systematic examination of species of *Actinomyces* (sinceren amed *Streptomyces*) organisms for antibiotic production. In their laboratories at Rutgers University, New Jersey, U.S.A., they examined a considerable number of substances. Streptothricin, isolated from *Streptomyces lavendulae* was the first compound to show promise but in common with many other candidates it had problems of toxicity. In September 1943, cultures of a mould, *Streptomyces griseus*, were isolated from an area of heavily manured soil near farm buildings at the University Agricultuıal Station. It soon became apparent that their search, which had concentrated on finding an agent complementary in spectrum to penicillin and therefore active against Gram-negative organisms, was ended. It was later shown that, more important than its Gram-negative activity, streptomycin was highly effective against *Mycobacterium tuberculosis*, and the Nobel Prize 1952 was awarded to Professor Waksman in honour of the achievement.

Following extensive research into the chemical structure of streptomycin, the compound *dihydrostreptomycin* was produced in 1946. It proved on trial to be as effective as streptomycin, though the original claim that dihydrostreptomycin was less toxic than the parent compound did not prove correct, for it was found that the compound is neurotoxic, but that the site of neural damage is different from that of streptomycin.

Manufacture

Streptomycin is now the accepted generic name for several related antibiotics produced by the deep-vat culture method using actinophage resistant strains of *Streptomyces griseus.* (The 'phages' are virus-like organisms which attack and destroy the bacterial cultures.) The strains must be sporulating ones, for non-sporulating strains produce little or no streptomycin. The selection of the most suitable strains for commercial manufacturing processes is of great importance. Some appear to yield little or no antibiotic product, others produce streptomycin while some give rise to other compounds such as grisein or the antifungal candicidin. Once prepared, the base can be turned into the sulphate, hydrochloride or calcium chloride complex, or, by hydrogenation, into dihydrostreptomycin.

Chemistry

Streptomycin is a white crystalline substance with possibly a slight odour and a saline taste. It is very soluble in water but only slightly so in alcohol, ether and chloroform and insoluble in fixed oils, and paraffins.

The structure of streptomycin is complex and consists of the hydroxylated base, streptidine, which is linked glycosidally with a disaccharide type of compound known as streptobiosamine;

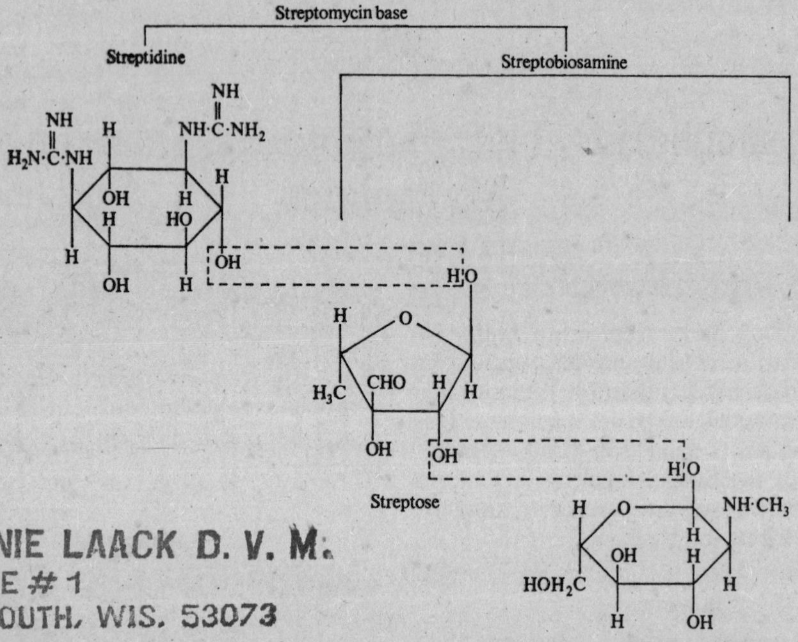

(i) *N*-methyl-1-glucosamine

this compound is in turn formed from a sugar, streptose, and *N*-methyl-1-glucosamine (i).

If streptomycin (ii) is subjected to acid hydrolysis, streptidine and streptobiosamine are produced, but alkaline hydrolysis produces maltol.

Dihydrostreptomycin base differs from that of streptomycin only in the addition of two hydrogen atoms in the streptose structure to change the aldehyde group ·CHO at carbon 3 in to the alcohol group $\cdot CH_2OH$ (iii). From these bases various salts can be formed. Streptomycin sulphate, hydrochloride, and calcium chloride complex, are available as well as dihydrostreptomycin hydrochloride and sulphate.

During biosynthesis, a second antibiotic is produced called mannosidostreptomycin or streptomycin B. It has a lower potency and gives streptidine, streptobiosamine, and mannose on acid hydrolysis, and maltol on alkaline hydrolysis.

Stability

Streptomycin is relatively stable when compared with penicillin, and pure dry powders will retain full potency at room temperature for at least two years.

1. *Moisture.* Streptomycin, but not dihydrostreptomycin, is hygroscopic and deliquesces in moist air. Solutions darken on exposure to light but this darkening does not appreciably affect activity and solutions at pH 6 to 8 will retain

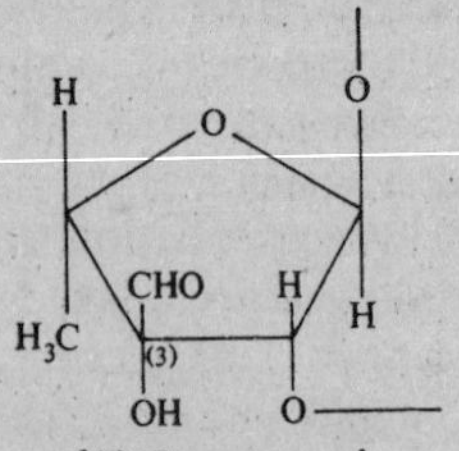

(ii) Streptomycin

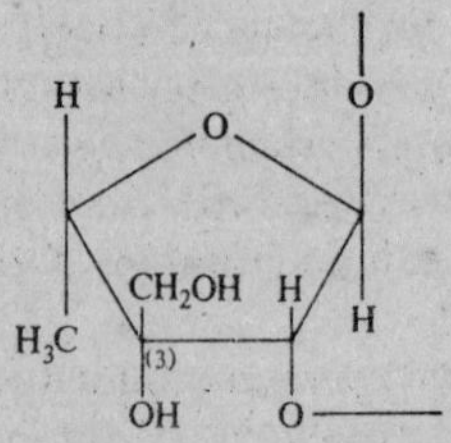

(iii) Dihydrostreptomycin

potency for up to one month at room temperature. Above 28°C potency is rapidly lost. In practice, unbuffered solutions should be used within one week of preparation and should be kept in a refrigerator away from light.

2. *pH.* As with penicillin, strong acids and alkalis cause rapid deterioration. Any pH lower than pH 3 or higher than pH 8 hastens hydrolysis and irreversible destruction. When buffered with sodium citrate, aqueous solutions will keep for at least 1 year without appreciable loss of potency.

3. *Temperature.* Streptomycin is not as susceptible as penicillin to deterioration on raising the temperature, but nevertheless, high temperatures are deleterious and as much as 30 per cent of a solution's potency is lost in 2 weeks at a pH of 6·6 and a temperature of 50°C.

4. *Oxidizing agents.* Most oxidizing and reducing agents affect streptomycin potency.

5. *Enzymes.* So far no enzyme comparable with penicillinase has been discovered, but some resistant strains appear and even strains which need streptomycin and are dependent on it for growth. Resistant strains of bacteria can develop in vitro after only a few passages in the presence of the antibiotic and can appear in a few days in vivo. Resistance is thought to be due to the selection of a resistant population originating from the few resistant individual organisms normally present and generally not to the destruction of the antibiotic. Breakdown of the antibiotic molecule by enzymes produced by the organisms does not appear to take place although some inactivation is attributed to chemical products isolated from *Ps. pyocyana.* The inactivity seems to be against selected bacterial species, the antibiotic retaining its action against others. The lack of an antagonistic enzyme under normal conditions is partially responsible for the intestinal activity of streptomycin, which is greatly superior to that of penicillin.

6. *Miscellaneous.* Streptomycin is inactivated by cysteine and similar sulphydryl-containing compounds but dihydrostreptomycin, which does not contain an aldehyde group, is unaffected. Alcohol does not affect the streptomycin series. Most salts such as sodium and potassium chlorides, sulphates, tartrates and ammonium acetate, reduce activity of the series.

Note: The calcium chloride complex is chemically incompatible with the sulphates of streptomycin and dihydrostreptomycin and with sulphacetamide sodium. These incompatibilities are of practical importance. Any precipitation of calcium sulphate in intrathecal therapy or of calcium sulphacetamide in eye treatment would be extremely dangerous.

Pharmacology

Apart from the difference in the sites where toxicity is apparent, the activity of the streptomycin and dihydrostreptomycin salts is almost identical.

Absorption and Administration Routes

Streptomycin is unusual in that almost no passage across the intestinal mucosa takes place and thus systemic therapy can only be achieved by parenteral injection. Oral forms of streptomycin, frequently combined with sulphonamide drugs and other gut active compounds are used in the treatment of enteritis; the action of the antibiotic is limited to the lumen of the alimentary tract. The absorption of streptomycin is slower than penicillin. Intravenous injections provide maximum blood levels almost immediately, but intramuscular injections take about 60 to 90 min to achieve maximum levels; subcutaneous injections take far longer and this route is therefore not often used. Dihydrostreptomycin should not be administered intravenously because of the greater degree of shock produced.

So far, no adequate means of slowing down the streptomycin absorption rate has been discovered. This is not so important as it is with penicillin for streptomycin is slowly excreted and detectable blood levels normally persists for about eight to twelve hours following intramuscular administration.

Blood Levels

Minimum therapeutic blood levels will vary according to species and strain of organism, but in general 5 μg per ml of serum is satisfactory. As with penicillin, it is the tissue levels which are important and these are maintained for some time after those in the blood have dropped below minimum therapeutic levels. These tissue levels are achieved rapidly, the streptomycin becoming dispersed in the tissue fluids and not entering the cells. The recommended interval between dosage is therefore 12 hr, although 4- to 6-hourly dosing may be necessary to maintain blood levels constantly above the minimum therapeutic level. Blood levels are related but not proportional to the size of dosage.

Barriers

(1) *Blood–brain barrier*. Streptomycin diffuses very slowly into the cerebrospinal fluid unless meningitis is present, when detectable levels can be achieved, but intrathecal injections are necessary to ensure adequate therapeutic fluid levels. Dihydrostreptomycin is unsuitable for intrathecal injection and should not be used. There is, in any case, rarely any indication for intrathecal injection in animals.

(2) *Placental barrier*. Streptomycin passes the placental barrier fairly easily; foetal blood and amniotic fluid levels achieve about 50 per cent of that in the maternal circulation. There appears to be no neurotoxic effect on the foetus.

(3) *Intestinal barrier*. Streptomycin is neither absorbed nor excreted in appreciable amounts through the intestinal mucosa. The extent in the cat and dog is approximately 10 per cent of an oral dose but no appreciable blood levels are achieved. This means that susceptible intestinal diseases can be subjected to maximum antibiotic concentrations.

(4) *Serous membrane barrier*. The only serous membrane crossed by streptomycin in concentrations sufficient to achieve therapeutic levels is the peritoneum. Streptomycin passes readily into normal and ascitic peritoneal fluid though at lower levels than those in the blood. It will not easily pass into the pleural cavity, synovial cavity and pericardial sac, unless inflammatory changes are present, and even then direct injections are advisable. Adequate concentrations in the aqueous and vitreous humours of the eye are achieved following intramuscular injection, but streptomycin will not penetrate the intact conjunctiva. Ulceration of the cornea provides an entrance through which the antibiotic readily penetrates. Aerosol treatment of lung infections has been used for humans and little or no streptomycin is absorbed through the lungs when this method of treatment is used.

(5) *The milk barrier*. Streptomycin is excreted in the milk in detectable and effective quantities. It is therefore of some assistance in severe mastitis of staphylococcal origin, to inject streptomycin parenterally as well as locally. This residue in the milk should be borne in mind when parenteral treatment is being used for a systemic condition. It is essential to ensure that milk is withheld for a period sufficient to avoid contaminating the bulk supply with antibiotic traces.

Excretion

Up to 80 per cent of a single intramuscular or intravenous dose is excreted via the kidney. The excretion rate is maximal during the first 4 hr after a single dose and the majority is excreted within 12 hr. The excretion is by glomerular filtration, and tubular excretion or resorption is almost absent. Some streptomycin is excreted via the liver, and bile concentrations may reach up to 25 per cent of the blood levels.

Toxicity

The toxicity of streptomycin has been widely reported and is of three types: (1) a local sensitization with dermatitis, (2) a chronic, and (3) an acute systemic effect. The *local toxicity* is only of significance to veterinary surgeons who frequently handle powder or solutions and may themselves become sensitized or allergic. Cross-sensitization does not readily occur and dihydrostreptomycin may be used when streptomycin sensitivity has been acquired.

Systemic toxicity is an important factor in both human and veterinary medicine and has been mainly responsible for the limited use of streptomycin in the veterinary field. *Acute toxicity* is characterized by anaphylactoid response with nausea, vomiting, and rapid collapse with loss of consciousness. In many cases, animals will recover after several hours if kept warm and if hypostatic congestion is guarded against: antihistaminics injected parenterally will assist in recovery. This type of shock has been seen in dogs and cats. Dihydrostreptomycin is said to be less dangerous in this way if intravenous injection is avoided. Acute toxicity due to streptomycin has been ascribed to the presence of guanidine grouping in the molecule and also to the formation of an insoluble complex with cephalin.

Chronic toxicity is of two types, one mainly caused by streptomycin and the other by dihydrostreptomycin, though each antibiotic can cause damage of the type ascribed to the other. Streptomycin mainly attacks the cochlear division of the eighth cranial nerve, causing vertigo and, if dosage is continued, an irreversible loss of ability to balance. If dosing is decreased or stopped before symptoms intensify, recovery can take place. Dihydrostreptomycin attacks the auditory division of the eighth cranial nerve but, unlike the cochlear degeneration, no premonitory symptoms are shown and permanent deafness may result if meticulous tests for loss of hearing are not carried out. It seems therefore that dihydrostreptomycin, originally used because it was thought to be non-toxic when compared with streptomycin, is the more dangerous of the two. Both types of damage can occur in animals and large doses should not be administered for longer than three or four days. In human medicine dihydrostreptomycin is used mainly for those patients where the bacteria are sensitive to streptomycin and are suffering from an acute disease not amenable to treatment with another antibiotic. While being retained in the *British Veterinary Codex*, dihydrostreptomycin has been omitted from the *British Pharmacopoeia* in its current edition.

Action

Wherever streptomycin is mentioned in the following text this should be taken as including all the salts of streptomycin and dihydrostreptomycin unless otherwise stated.

Streptomycin may be bacteriostatic or bactericidal in vitro depending in such factors as concentration, pH, culture medium and on the particular organism involved. In vivo, in adequate concentration, the action is usually bactericidal. Unlike penicillin, which is likely to attain maximum bactericidal activity at quite low concentrations, streptomycin shows a progressive rise in antibacterial effect as blood or tissue levels increase. The bactericidal nature of the antibiotic renders it very suitable for use in conjunction with penicillin which is also bactericidal. Many preparations exist containing this mixture. It is active in the presence of blood and tissue breakdown products, including pus, but these fluids slightly reduce the antibacterial effect. Its antibacterial action is increased with increased alkalinity. Streptomycin is most effective as a bactericide against rapidly multiplying organisms but it differs from penicillin in some cases in its inability to destroy resting bacteria. In bacteriostatic concentrations the onus of destroying and removing the bacteria lies with the body's natural defences.

Antibacterial Spectrum

Streptomycin is considered to be a narrow spectrum antibiotic though it has wider powers than penicillin. In the main, it is active against Gram-negative organisms but some Gram-positive bacteria and the *Leptospira* are also affected. In vitro, the following organisms show sensitivity:

Organism	Disease caused
Actinomyces bovis	Lumpy jaw
E. coli and *Salmonella*	Enteritis and septicaemias
Brucella species	Abortion in cattle in pigs Bronchitis in dogs
Corynebacterium	Mastitis, abscesses and septicaemias
Erysipelothrix	Swine erysipelas, synovitis in lambs
Leptospira	Nephritis and jaundice
Mycobacterium	Tuberculosis and Johne's disease
Pasteurella	Pneumonias and septicaemias
Proteus and *Pseudomonas* species	Occasional pathogens
Staphylococci	Mastitis, local sepsis
Vibrio foetus	Abortion in cattle

In vivo susceptibility unfortunately does not follow the in vitro susceptibility, but some guidance may be taken from it. The following organisms and diseases have been reported to respond to streptomycin therapy:

Disease	Causal organism
Actinomycosis (lumpy jaw)	*Actinomyces bovis*
Transit fever	*Pasteurella* species
Calf and pig scours	*E. coli* and *Salmonella*
Mastitis	*E. coli* and staphylococci
Vibrionic abortion in cattle	*Vibrio foetus*
Leptospirosis (hepatic and nephritic)	*L. icterohaemorrhagiae*, *L. canicola*, *L. pomona*, etc.

Erysipelothrix infections in turkeys, various infections secondary to virus diseases, and systemic infections due to pasteurellas, coliforms and salmonellas have also responded.

Some organisms show low sensitivity. They include *B. anthracis* (Anthrax), *Pfeiferella mallei* (Glanders), streptococci, rickettsiae, viruses and fungi.

Bacterial Resistance

Streptomycin resistant organisms tend to dominate when sublethal concentrations are used. This may develop either by the multiplication of resistant mutants or by the destruction of sensitive strains which leaves the field open to the rapid development of resistant strains. This development of resistant strains is common to most antibiotics, but streptomycin is peculiar in that resistance can develop within a week of starting treatment. Dosage should therefore be large enough to ensure rapid destruction of the organisms and administration should not be continued for longer than three or four days if response is inadequate. Resistance to streptomycin also means that an organism will be resistant to dihydrostreptomycin. Cross resistance with neomycin is shown though the reverse is not always true.

Diseases and Streptomycin Therapy

Streptomycin has not been so widely used as other antibiotics because both of its toxicity and of the rapid development of bacterial resistance. There are, however, some diseases for which streptomycin is primarily indicated. These are: *E. coli* infections in all species causing mastitis, metritis, enteritis, cystitis and septicaemias; leptospirosis; *Corynebacterium equi* infections of the horse; and vibriosis.

Bovine Mastitis

Streptomycin alone has been primarily recommended for the treatment of mastitis infections due to Gram-negative organisms such as *E. coli.* It is more commonly used as an intramammary preparation in a variety of combinations of dihydrostreptomycin and penicillin for treating streptococcal and staphylococcal mastitis. The combination is particularly valuable when mixed infections are present. Even so the increased significance of resistant staphylococci in bovine mastitis presents a serious problem to the clinician.

Intramammary infusions of streptomycin have also been successfully used in the treatment of *E. coli* mastitis in a goat.

For combined streptomycin–penicillin treatment, oily intramammary formulations are available commercially. Where no response is shown after the first two doses, treatment with streptomycin should be discontinued and alternative treatment adopted. If bacteriological examination of the organisms has been possible, sensitivity tests may give a guide to the most suitable material to use.

In the absence of information on sensitivity, the broad spectrum antibiotics or the semi-synthetic penicillin cloxacillin may be used, since they are likely to be effective against resistant staphylococci, which are a frequent cause of poor response to treatment. In herds where resistant staphylococci are already known to be well established these antibiotics should be used in initial treatment, before the inflammatory thickening of the mammary duct system has restricted the access of any intramammary preparation to infected foci in the gland.

Specific Systemic and Organic Diseases

Some diseases can be treated with streptomycin in preference to other antibiotics.

Vibriosis

This condition, which is responsible for a considerable amount of abortion during the first three months of pregnancy, is passed from cow to bull and vice versa, by contact and by infected semen. Streptomycin with or without penicillin has been used successfully for the treatment of cows and bulls. In females uterine and vaginal infections due to *Vibrio foetus* have been cured by Plastridge and Easterbrooks and their colleagues, using intra-uterine injections of 1 g streptomycin in about 15 ml sterile distilled water via a glass insemination catheter. This has been achieved irrespective of the stage of the oestrus cycle at the time of injection. Successful treatments of bulls have been reported by local or parenteral antibiotic therapy.

Semen from infected bulls can be rendered non-infective by diluting it 1:10 with a warm diluent containing 1 mg streptomycin base per ml and refrigerating for 6 hr after cooling. It has been reported that optimum concentrations of antibiotic, besides destroying the *vibrio* also increased the number of motile sperms, probably by inactivating contaminating bacteria which are able to affect motility adversely. Vibrionic abortion in sheep has also been treated with streptomycin amongst other antibiotics, but with little success. Vibrionic dysentery in pigs responds moderately but not completely.

Vibrionic dysentery or 'winter scours' in cattle due to *Vibrio jejuni*, previously treated with 60 to 90 ml (2 to 3 fl. oz.) of a 5 per cent copper sulphate solution, would probably respond to large oral doses of streptomycin.

Pasteurellosis

Pasteurella infections of animals can take the form of septicaemias which are rapidly fatal, but chronic and mild infections also occur. *Cattle* suffer from a rapidly fatal septicaemia known as 'shipping fever', 'transit fever' or 'haemorrhagic septicaemia', and calf pneumonia due to *Pasteurella* organisms also exists. Both respond to streptomycin, the adults being given twice daily injections at the rate of 0·25 g per 45 kg or daily injections of 1 g per 50 kg for 2 or 3 days: calves are given twice daily injections of 0·5 to 1 g per kg body weight.

Horses and *rabbits* have recovered from experimental infections with *Pasteurella* organisms when dosed with dihydrostreptomycin at the rate of 4 to 5 mg per kg and 10 mg per kg, respectively every 4 hr.

Leptospirosis

Leptospirosis infections occur in all species and clinical recovery results from penicillin therapy in the case of *Leptospira canicola* and *L. pomona*, but not *L. icterohaemorrhagiae*. It is indicated that a minimum of 10 mg per kg body weight every 12 hr eliminates *L. pomona* from the urine of *calves* and *cattle*. Penicillin does not clear the organisms from the urine. Streptomycin in doses of 0·25 to 0·5 g has also proved effective in stopping the excretion of *Leptospira* in the urine of pigs weighing about 40 kg. In the *dog*, though penicillin gives clinical improvement, streptomycin is superior in both *L. canicola* and *L. icterohaemorrhagiae* infections, completely eliminating the former from the urine: 40 mg per kg body weight in divided doses daily by intramuscular injection are recommended. Treatment should be continued for 4 to 6 days.

Actinomycosis and Actinobacillosis

A few reports of improvement following the use of streptomycin against actinomycosis and actinobacillosis are available.

Coliform–*Salmonella* Group Infections

In the *foal*, *B. equirulis* (*Shigella equiruli*) causes pyosepticaemia with a purulent nephritis and joint ill. Streptomycin is the drug of choice for this infection. It is also effective against septicaemia and enteritis in the foal due to *E. coli* if used in large doses (5 g initially followed by 1 g four times daily for 4 days).

In the *calf*, streptomycin is reasonably active against coliform enteritis but the tetracyclines, neomycin, ampicillin and nitrofurans are among the many preparations which are used to overcome the rapid development of resistance.

Likewise in *pigs*, streptomycin is of value in necrotic enteritis, and *E. coli* scours using 0·25 to 1 g by mouth twice daily or 0·5 to 2 g per pig in drinking water for 2 days.

Corynebacterium equi Infections

This organism is a frequent cause of pneumonia of foals in their first few months of life, and good clinical response is reported to parenteral streptomycin and oral chloramphenicol.

Non-Specific Systemic Diseases

Streptomycin has been used with good effect in a variety of non-specific conditions. Perhaps the most important is cystitis in dogs, which responds even more rapidly if the urine is first made alkaline (if it is not already so) with potassium citrate or acetate. Gastroenteritis and secondary invaders concurrent with the virus infections, especially the distempers of dogs and cats, also respond well. The secondary organisms in otitis of dogs and cats are widely treated with combined penicillin and streptomycin formulations—often intramammary injections.

Miscellaneous conditions treated with streptomycin include a number of diseases of poultry, though the broad-spectrum antibiotics have largely replaced streptomycin. These poultry diseases include contagious coryza and chronic respiratory disease (CRD) of chickens and turkeys, *Erysipelothrix* infections of turkeys and 'quail disease'—an ulcerative enteritis of quails. Tularaemia in dogs due to *Brucella tularensis* has also responded. Cellulitis in cats, caused by bites, responds very well both to local and parenteral administration.

Local Use

Streptomycin is not frequently used for local applications because of the possibilities of skin sensitization.

Dosage

Some suggested doses for particular diseases and species have been mentioned in the therapeutic section, but they can be varied according to the disease, larger doses being given when the organisms are known to be less sensitive or tissue reaction great enough to hinder access of the drug to the organism. To prevent the development of resistant strains, the dose used should be adequate to destroy the sensitive organisms. Fowls require a much larger dose when estimated at the rate of mg per kg body weight and tolerance increases as baby chicks grow older.

The usual dose levels of streptomycin base quoted in the *British Veterinary Codex* are 10 mg per kg body weight by injection; not less than 20 mg per kg body weight orally in young animals and dogs and for intramammary injection in cattle, not less than 100 mg into each quarter.

When streptomycin sulphate is being used 1·25 g of the sulphate are approximately equivalent to 1 g of the base.

A guide to the parenteral use of the drug in various species is given below. The doses are in terms of streptomycin base. With the exception of the chicken injections should be by the intramuscular route.

Horse and foal:
  10 mg/kg body weight
Cow and calf:
  10 mg/kg body weight
Sheep:
  10 mg/kg body weight
Pig:
  10 mg/kg body weight
Dog and cat:
  10–20 mg/kg body weight. Repeat daily (all treatments)
Fowl:
  0·1–0·2 g per bird as a single dose
Chickens:
  2·5–5 mg twice daily

Official Preparations

The official preparations listed in the *British Veterinary Codex* (1965) have been reduced to the Injection of Streptomycin Sulphate, the Injection of Dihydrostreptomycin and the Intramammary Injection of Procaine Penicillin and Dihydrostreptomycin. The Injection of Streptomycin Sulphate is a stable solution, prepared ready for use and with a storage life of 18 months at a temperature not exceeding 20°C. The Injection of Dihydrostreptomycin requires reconstitution and then has a shelf life of 7 days at room temperature or one month when kept below 4°C.

Many of the earlier official preparations have been deleted from the *British Veterinary Codex* but a great variety of both simple and complex formulations exist containing mixtures of streptomycin derivatives with other antibacterial drugs, compounds for symptomatic therapy and adjuvants.

II. NEOMYCIN (FRAMYCETIN)

This antibiotic has a medium spectrum, affecting both Gram-positive and Gram-negative bacteria, strains of *M. tuberculosis*, *Treponema* and *Entamoeba histolytica*. The compound is official in the British and United States Pharmacopoeias as Neomycin Sulphate.

History

Neomycin was discovered by Waksman and Lechevalier in 1949 during their searches for an antibiotic which was effective against streptomycin resistant Gram-negative and acid fast organisms. It was prepared from an actinomycete later identified as *Streptomyces fradiae* found in a soil sample from the farmyard of the New Jersey State Agricultural College (Rutgers University) where *S. griseus*, and thence streptomycin, was discovered.

Manufacture

Neomycin is manufactured by the normal biosynthetic methods for submerged deep-vat fermentation using *Streptomyces fradiae* cultures. Extraction and purification methods are similar to those adopted for streptomycin.

Chemistry

Neomycin is used as the sulphate, a white or slightly yellow crystalline powder which is almost odourless. The powder is hygroscopic and is soluble in water, almost insoluble in alcohol, and insoluble in other organic solvents. The official *United States Pharmacopoeia* preparation of neomycin sulphate must contain at least 60 per cent neomycin base, and 1 mg pure sulphate is equivalent to not less than 650 units.

Neomycin was originally thought to be a mixture of four compounds named neomycins A, B and C, and a further component, *fradicin*. However, later investigations have thrown some doubt on the existence of all these components and the sulphate available commercially appears to be one substance only (*U.S.D.* 25th edn, 1955, p. 884). Currently it is accepted that two neomycins exist—neomycins B and C both conforming to the formula $C_{23}H_{46}O_{12}$.

The partial structure of:

$$C_{12}H_{25}N_4O_5—O—C_5H_7O(OH)_2—O—C_6H_7O(OH)_2(NH_2)_2$$

has been suggested but the complete structure of neomycin is unknown; it appears, with some reservations, to be related to streptomycin, and like streptomycin is a basic compound forming salts, of which the sulphate is the one commonly used.

Stability

Neomycin is one of the most stable of the antibiotics.

1. Moisture. In the dry state neomycin powder is stable for at least 2 years at room temperature. Solutions are also very stable and will retain their potency for at least a year at room temperature; they are also stable for several months at 37°C. Solutions darken on keeping and therefore, in spite of their thermostability they should be refrigerated to maintain appearance.

2. pH. Neomycin is stable over a wide range of pH ranging from pH 2 up to and beyond neutral and has even been said to be most active in alkaline solution (*Antibiotics*, 2nd edn, p. 8, 1952, Pharmaceutical Press). In this property it differs from all other antibiotics.

3. Temperature. Although relatively heat stable, neomycin will not withstand autoclaving or prolonged high temperatures. Solutions are thermostable in all climatic temperatures but will gradually darken in colour.

4. Oxidizing agents. Neomycin is apparently unstable in the presence of these agents, for it has been reported that antioxidants will retard the colour change in solutions buffered to a range of pH 6 to 7.

5. Enzymes. So far no enzyme has been reported which has a specific action against neomycin.

6. Miscellaneous. All compounds containing sulphonic acids or similar compounds will precipitate and inactivate neomycin.

Pharmacology

Absorption and Administration Routes

Neomycin is absorbed to the extent of about 3 per cent of an oral dose and systemic infections cannot be treated by orally administered neomycin. Parenteral injections may be given by all the normal routes but the intramuscular route is the one recommended. In man the parenteral use of neomycin is restricted to those cases where the infective organism is highly virulent, not susceptible in vitro to other antibiotics, but sensitive to neomycin, and only where there is no impairment of renal filtration: this is necessitated by the nephrotoxicity and neurotoxicity of the compound. There is no absorption from the skin so that topical application and oral dosing are the main routes of administration.

Blood Levels

Following intramuscular injection, maximum blood levels are achieved within 1 hr and appreciable amounts are present for from 4 to 8 hr, so that the daily dose is usually split into four doses given at intervals of 6 hr.

Barriers

Neomycin penetrates the blood–brain, the pleural and the peritoneal barriers following intramuscular dosing; it is only slightly absorbed from the intestines and not at all from the skin.

Excretion

Parenterally administered neomycin is fairly rapidly excreted by the kidneys. Orally administered neomycin is in the main excreted in the faeces unchanged.

Toxicity

Neomycin shows a similar and additive toxicity to that caused by streptomycin. It consists of loss of hearing due to irreversible damage to the auditory division of the eighth cranial nerve, and also damage to the kidney causing albuminuria, haematuria, and the formation of renal casts. There is practically no external irritant action even when applied to the eye, and skin sensitivity is practically unknown. A slight laxative action follows oral administration but is not usually detrimental. Neomycin therapy by all routes may be followed in man by an increase in the yeast and fungi population of the skin, mucous membranes, and alimentary tract: so far this has been of no significance in animals.

Action

Neomycin in vitro is bactericidal. The mode of action is unknown but penicillin, bactracin and polymyxin have been successfully combined with neomycin.

Antibacterial Spectrum

In vitro and in vivo activity in experimental animals indicates that neomycin is a broad-spectrum antibiotic with a wide activity against Gram-positive and Gram-negative organisms, and against a number of actinomycetes. In vivo, the spectrum is limited by toxicity considerations. Neomycin has been reported as active against *E. coli*, *Aerobacter*, *Proteus* and the paracolon organisms; it is also effective against *Staphylococcus albus* (*Micrococcus albus*) and *S. aureus* (*M. aureus*), but somewhat less active against streptococci. *Pseudomonas aeruginosa* (*Ps. pyocyanea*) is even more resistant though neomycin is more active against this organism than are the other antibiotics, with the exception of polymyxin and carbenicillin: almost no effects on *Pseudomonas* infections of the urinary tract have been reported. Activity against *B. anthracis* is suggested, and both *Treponema* species and *Entamoeba histolytica* are sensitive. Activity against *Mycobacterium tuberculosis* is more of academic interest in the veterinary field, but the activity of neomycin against strains of organisms resistant to other bacteria, and particularly streptomycin, is of considerable Public Health importance.

Diseases and Neomycin Therapy

Coliform enteritis in all species, urinary infections due to coliforms, and skin, eye and ear infections all respond favourably to neomycin treatment, especially where other antibiotics fail.

There has been little clinical use of neomycin *parenterally* in animals, mainly because of the reported toxicity to the auditory nerve and the kidney. However, in man, dosage of 1 g per day for up to 5 days has not caused more than transient signs of toxicity, if at all, yet bacteriological response has been good following intramuscular injection. Results have been particularly good in treating urinary infections which were resistant to other antibiotics. These reports should encourage the parenteral use of neomycin where response to other antibiotics is not apparent or not anticipated. Large doses are not required for urinary infection treatment.

Orally, neomycin was used experimentally in dogs, usually with complete elimination of *E. coli* from the faeces within 48 hr, though resistant strains appeared if dosing was continued (0·4 g per kg daily in four equal doses) for 5 to 6 days.

Similarly, oral dosage of neomycin in calves and piglets gives good results in the treatment of scours.

In man, oral dosing is used in the following manner to achieve pre-operative disinfection of the colon: the patient is fed on a low residue diet and a purgative given; this is followed by 1 g neomycin sulphate every hour for four doses, and then by 1 g every 4 hr for 1 to 3 days before surgery, but not longer than 3 days. The laxative effect of the neomycin produces slightly increased intestinal activity (*N.N.R.*, 1955, p. 153 summarizing numerous reports in American literature). This provides a minimum administration of 9 g and a maximum of 21 g neomycin sulphate.

For *intramammary* application neomycin is combined with penicillin in some formulations to counteract penicillin resistance and to give a broader spectrum. A dose of 500 mg of neomycin sulphate is used in dry cow formulations.

Topical applications usually contain 5 mg neomycin sulphate per ml or per g, in either aqueous solution or ointment form. The solutions are useful as wet dressings for wounds and eyes, whilst the ointment is useful for a wide variety of superficial conditions, including otitis, corneal ulcers, conjunctivitis and keratitis, furunculosis, abscesses, and dermatitis. It has been used mainly in small animals but similar conditions in farm animals also respond. Combined bacitracin and neomycin preparations are available giving almost complete coverage over the range of Gram-positive and Gram-negative bacteria. Occasionally fungicidal compounds are included in the applications.

Oral administration:

Horse and cattle: 4–7·5 g
Foals and calves: 2–3 g
Lambs and young pigs: 0·75–1 g
Dog: 0·2–0·5 g

These are all daily doses and should be divided into two to four equal doses given at regular intervals. Doses are not standardized and this list is only for general guidance.

Official Preparations

The *United States Pharmacopoeia* contains the official powder, an ointment with not less than 3·5 mg base (equivalent to approximately 5 mg neomycin sulphate per g) and tablets which usually contain 500 mg neomycin sulphate. British Official preparations are confined to the powdered sulphate but ointments and tablets are also available commercially.

Bacterial Resistance

Bacteria become resistant to neomycin slowly.

III. ERYTHROMYCIN

This antibiotic, with an activity against most Gram-positive bacteria, some Gram-negative bacteria, many rickettsiae, as well as activity against some protozoa, is a macrolide. It must contain not less than 900 units per mg and is known commercially as Ilotycin.

History

Erythromycin was first announced in 1952 when it was obtained biosynthetically from the growth of cultures of *Streptomyces erythreus* isolated from soil samples taken in the Philippines. *Manufacture* is by normal deep culture fermentation and extraction according to established principles.

Chemistry

The antibiotic is a white or yellowish white crystalline powder which is odourless and has a bitter taste. Its *solubility* is 1:1000 in water but it is freely soluble in alcohol and ether. Various *salts* and esters are available, the solubilities of which differ so that some are suitable for intravenous or intramuscular injection whilst others may only be used orally.

The *constitution* of erthyromycin is similar to that of the other macrolide antibiotics. Its lactone ring contains fourteen atoms and the two sugars attached to the ring are *desosamine* and *cladinose*: the empirical formula is $C_{37}H_{67}O_{13}N$.

Stability and Pharmacology

The *stability* of erythromycin compares favourably with other antibiotics. As a dry powder it is quite stable at room temperature and refrigerated or frozen solutions do not lose their activity even on prolonged storage (at least 8 weeks), although at room temperature or higher quite rapid deterioration occurs.

Absorption and Administration

Erythromycin is *absorbed* rapidly when administered orally and, as with intravenous or intramuscular dosage, diffusion occurs into most tissues and fluids. The *blood–brain barrier* is not readily passed when the meninges are normal. Diffusion occurs across the placenta and into the pleural and peritoneal cavities. The base is partially destroyed by the gastric acidity and oral forms must either be acid resistant, as with the stearate, or tablets must have an acid resistant coating. *Excretion* is mainly via the kidney, and excretion concentrations are higher than for chlortetracycline at all times, and higher than those of oxytetracycline in the first few hours following administration only. Very high concentrations can be achieved in the urine. Some excretion via the bile occurs and concentrations far higher than those of plasma are achieved.

Toxicity

The toxicity is low and similar to carbomycin in that the nausea and gastro-intestinal disturbances in carnivores are rare.

Action

The *action* of erythromycin is similar to that of carbomycin and it has been recommended in human medicine that its use be restricted to the control of Gram-positive organisms which resist penicillin (mainly staphylococci). In vitro tests indicate that 'stepwise' development of resistance occurs in a manner similar to penicillin, though in vivo resistance develops rapidly. Cross-resistance occurs with carbomycin but not with other antibiotics.

Clinical Applications

Erythromycin has an activity similar to that of penicillin with in addition specific activity against PPLO like organisms.

It can, therefore, be used against penicillin resistant staphylococci, and the other clinical entities described in Chapter 28. Because of its activity against PPLO it has been prepared in a water soluble form for water medication of poultry.

Because of its Gram-positive activity, and the prevalent resistance of staphylococci to penicillin G, erythromycin is prepared for use as *intramammary* formulations for use in both lactating and dry cows.

Dosage

Parenteral therapy: erythromycin lactobionate:

Intravenous injections
- Cattle and horses: 4–8 mg/kg every 12 hr
- Pigs: 4–8 mg/kg every 12 hr
- Dogs and cats: 4–8 mg/kg every 12 hr

Water medication: erythromycin thiocyanate:

Poultry and turkeys: control of C.R.D. Poultry and turkeys: control of sinusitis Poultry and turkeys: control of synovitis	1 part in 10 000 parts of drinking water for 3 days

IV. NOVOBIOCIN

Novobiocin is a medium-spectrum antibiotic with marked bacteriostatic and bactericidal action against Gram-positive organisms. In this it is therefore similar in action to erythromycin. It is known commercially under the name of Biotexin.

Manufacture

The antibiotic, announced in November 1955, is manufactured biosynthetically from at least two actinomycetes, *Streptomyces niveus* and *S. spheroides*.

Chemistry

It is a dibasic acid but is marketed as a light

yellow amorphous monosodium salt or as a purer half-neutralized calcium salt. Though it is more soluble in alkaline solution it is more stable when in acid solutions.

Absorption

Novobiocin reacts on administration in a similar manner to the tetracyclines. It is thus rapidly absorbed and therapeutic blood levels can be achieved following oral, intravenous, intramuscular and intraperitoneal administration. Serum levels are maintained for at least 8 hr following a single oral dose but no appreciable penetration of the blood–brain barrier occurs. In vitro studies have shown high degrees of activity against staphylococci (micrococci). Sensitive Gram-negative organisms include species of *Proteus*, *Pasteurella*, *Salmonella* and *Klebsiella* genera. There is, however, a tendency to develop resistance during therapy as with erythromycin.

Toxicity

Toxicity in man is relatively mild with the exception of a very small percentage of skin reactions and pruritis.

Clinically in man and animals, various streptococcal and staphylococcal conditions have responded favourably: anthrax, mastitis and cellulitis have also shown good responses.

Clinical Applications

This antibiotic has not found a general application in the veterinary field, and is used mainly in combination with penicillin for the treatment of bovine mastitis. It is prepared for intramammary injection at a dosage of 250 mg Sodium Novobiocin per infected quarter.

V. OLEANDOMYCIN

This antibiotic was isolated in 1954 from a strain of *Streptomyces antibioticus*. Its antibacterial activity is very similar to that of penicillin and it belongs to the macrolide group. In animal medicine it has been used to complement streptomycin, in the treatment of streptococcal and staphylococcal mastitis.

VI. SPIRAMYCIN

Spiramycin was isolated in 1954 from a strain of *Streptomyces ambofaciens*. Its antibacterial activity is similar to that of erythromycin and oleandomycin.

It is used for mastitis treatment, for the control of staphylococcal and streptococcal infections. It is less active against Gram-positive organisms than erythromycin.

Like erythromycin and tylosin it is active against the PPLO organism known as *Mycoplasma gallisepticum*. It is used as a soluble powder for the mass treatment of poultry and turkeys.

VII. TYLOSIN (TYLAN)

History

Tylosin is an antibiotic substance with a macrolide structure isolated from a strain of *Streptomycetes fradiae* found in soil from Thailand.

Chemistry

In the dry state, tylosin forms colourless platelets. The melting point range is 128° to 132°C. Aqueous solutions are stable at pH 5·5 to 7·5 at 25°C for a period as long as 3 months. In solution below pH 4·5, tylosin is converted by acid degradation to desmycosin, which is very similar to tylosin. Under more severe conditions, further degradation to neutral sugars occurs.

Solubility

Tylosin base is only slightly soluble in water (5 mg per ml at 25°C). It is soluble in organic solvents such as methanol, ethanol, acetone, chloroform and ether, etc. It readily forms salts with mineral and organic acids which are highly soluble in water, e.g. tylosin tartrate, 600 mg per ml.

Standard

The international unit for tylosin is defined as the activity contained in 0·001 mg of the International Reference Preparation of tylosin.

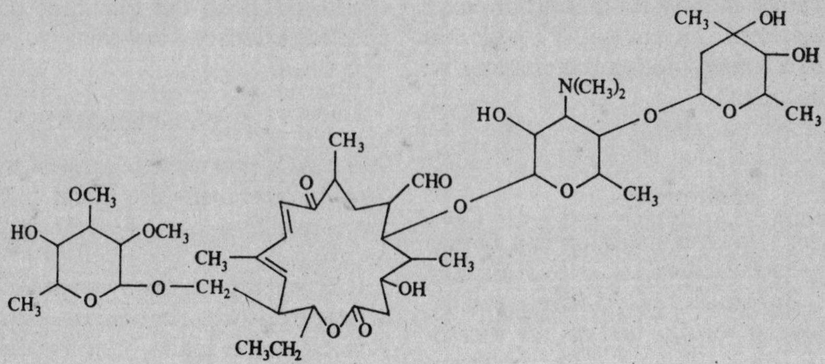

(iv) Structure of tylosin

Antibacterial Spectrum

The antibacterial spectrum of tylosin and its salts is mainly Gram positive. Tylosin, in vitro is very active against pleuropneumonia-like organisms (PPLO) isolated from chickens, turkeys, pigs, cattle, goats and against several species of spirochaetes including leptospira. In animal tests, it is effective against three species of spirochaetes, erysipelas of turkeys, *Haemophilus pertussis*, *Moraxella bovis*, some Gram-negative bacteria and certain vibrios.

Tylosin has a particular action against *Mycoplasma gallisepticum* S6, which is implicated in the chronic respiratory disease (CRD) complex in chickens and infectious sinusitis in turkeys.

Bacterial Resistance

Cross resistance between tylosin and penicillin or tetracyclines has not been demonstrated in vitro, but in tests with *Staphylococcus aureus*, partial cross resistance with erythromycin was found. Pathogenic *Salmonella typhimurium* exposed to tylosin did not develop cross resistance to chlortetracycline or furazolidone in chicken studies.

Pharmacology

Toxicity

Acute, subacute and chronic toxicity studies have shown that tylosin has a wide margin of safety in all species tested. In those animals which have gone to autopsy, no changes have been found attributable to tylosin. Several groups of pigs have been observed with a reaction following treatment with tylosin. This reaction was mild and consisted of an oedema of the rectal mucosa, partial anal protrusion ('rosebudding'), erythema and pruritus which disappeared in a few days. Discontinuation of treatment was rarely necessary and an uneventful recovery followed.

No special precautions are necessary for the protection of the operator when handling tylosin except that, as in the case of most antibiotics, direct contact with the skin should be avoided.

Clinical Applications

Tylosin, as the tartrate salt, may be administered in the drinking water to chickens, turkeys and pigs; it may also be injected subcutaneously in chickens. Tylosin, as the phosphate salt, can be administered orally in a premix form to pigs in their food. Intramuscular injections of tylosin base can be given to cattle, calves, pigs, dogs and cats.

Tylosin tartrate is particularly effective in the treatment of chronic respiratory disease in chickens or for the prevention of CRD at the time of vaccination or other stress. In turkeys, tylosin is an aid in the treatment of infectious sinusitis (IS) and prevention of the respiratory form of IS. In pigs, tylosin can be administered either as the tartrate via the drinking water or as the phosphate via the food (in a premix form) for the treatment or prevention of the enteritis complex of swine when caused by *Vibrio coli* or other susceptible organisms. In vibrionic scouring it is advisable to continue prophylactic treatment with tylosin phosphate once the acute symptoms have been controlled using the tartrate salt. For cattle and calves, intramuscular injections of tylosin are indicated in certain types of pneumonia, foot rot and metritis. In pigs, tylosin is used intramuscularly in cases of erysipelas, pneumonia, dysentery due to *Vibrio coli* and other sensitive organisms. In cats and dogs, intramuscular tylosin is indicated in upper respiratory tract infections, cellulitis, otitis externa, metritis, leptospirosis, secondary bacterial infections associated with virus diseases, and as an aid in combating infections associated with surgical procedures or injuries.

Tylosin should not be used in poultry producing eggs for human consumption. Chickens must not be slaughtered for food within 3 days after injection of tylosin or 24 hr after oral treatment. Turkeys must not be slaughtered for food within 5 days after treatment.

Milk that has been taken from cows during treatment, and for 96 hr after the last treatment must not be used for food. Pigs should not be slaughtered within 21 days of administration of tylosin.

Tylosin is marketed for agricultural use only.

Dosage

Cattle:
  4–10 mg/kg body weight daily, intramuscularly

Pigs, dogs and cats:
  2–10 mg/kg body weight daily, intramuscularly

Chickens and turkeys:
  0·5 g per litre (2·5 g per Imperial gallon) in the drinking water for the appropriate time period.

Tylosin may also be injected subcutaneously into chickens at the rate of 1 ml per kg of birds body weight, maximum dose 2·5 ml. For the treatment of infectious sinusitis in turkeys, tylosin may be injected directly into the sinuses. Inject each sinus with 0·25 or 0·5 ml depending on the severity of the condition. Sinuses may be re-treated in 10 days if swelling persists.

Tylosin may also be administered orally at the rate of 0·2 g per litre (1 g per gallon) of drinking water for the treatment of vibrionic scours in pigs.

Tylosin is also available as a premix with a potency varying from 20 to 80 g tylosin per kg (10 to 40 g per lb.). This is added to feeding stuffs for pigs only.

Preparations

Powder of tylosin tartrate
Injection of tylosin tartrate buffered
Injection of tylosin base
Tablets of tylosin base
Premix of tylosin phosphate gelatinized on carrier
Tylosin tablets
Tylosin for injection

Tylosin for Injection

This is a sterile solution of tylosin base in a mixture of propylene glycol and Water for Injection and contains 4 per cent v/v benzyl alcohol as bactericide. The solution is sterilized by filtration and distributed aseptically into sterilized vials.

Two preparations are available containing activity equivalent to 50 and 200 mg tylosin base per ml.

Tylosin Tartrate Powder

Tylosin tartrate is a white–buff coloured powder. The usual pack sizes contain activity equivalent to 25, 50 and 100 g base.

Tylosin Tartrate Buffered for Injection

This is a sterile preparation of Tylosin Tartrate Buffered filled aseptically into suitable containers. The usual pack sizes contain the equivalent of 3·125, 6·25 and 12·5 g of base activity.

Tylosin Tablets

Tylosin Tablets are prepared from Tylosin Base and Maize Starch B.P. with a small amount of lubricating agent (magnesium sulphate). They contain the equivalent of 200 mg of Tylosin Base activity.

Bibliography

CORCORAN, J. W. & HAHN, F. H. (1974) *Antibiotics: Mechanisms of Action*. Berlin: Springer.

30

Polypeptide Antibiotics

I. POLYMYXIN B

This is the least toxic of five antibiotics known as the polymyxins and designated alphabetically A to E. It is a narrow-spectrum antibiotic with activity against Gram-negative organisms. The antibiotic contains not less than 6500 units in each mg.

In 1947 three groups of workers, two American and one British simultaneously announced the discovery of a new type of antibiotic, later known as the polymyxin group. The compounds discovered by the two American groups proved to be identical and by agreement the substance was called Polymyxin D whilst Polymyxin A was the name given to the compound, originally named Aerosporin, discovered by the British workers. Three more antibiotics were subsequently discovered, among them Polymyxin B (Aerosporin is now used as a trade name for Polymyxin B).

Manufacture

Polymyxins are produced by the usual deep vat culture method using *Bacillus polymyxa* (*aerosporus*). Many different strains of the organisms are used and fermentation conditions must be adapted to the requirements of individual strains. Several medium supplements such as yeast extract are used.

Chemistry

Polymyxin B sulphate is a white or creamy white powder with little or no odour. It is very soluble in water and normal saline but less so in methyl and ethyl alcohol and is insoluble in ether, chloroform, acetone, and hydrocarbons.

The structure of the polymyxins is not completely known but they are considered to be similar to bacitracin in that they are polypeptides, and complex polypeptide structures have been provisionally assigned to them. They are basic in character and form acid salts.

Stability

Polymyxin B sulphate is a relatively stable compound.

1. Moisture. In the dry state this antibiotic will remain stable for long periods. Solutions are stable for at least 6 months when refrigerated and can be heated without loss providing the pH is optimum.

2. pH. Polymyxin is stable in near neutral solutions (pH 5 to 7), but strong acids, and alkalis destroy the potency very rapidly.

3. Temperature. The antibiotic, though classed as thermostable, will not withstand autoclaving and solutions must be sterilized by filtration.

4. Oxidizing agents. Though no information on this point is given in official monographs it may be assumed that such compounds will have a deleterious effect (cf. polypeptide bacitracin).

5. Enzymes. No anti-polymyxin enzyme has so far been reported.

6. Miscellaneous. All bivalent metallic ions block the action of polymyxin by interfering with its absorption by the bacteria, so that magnesium, cobaltous, ferrous and manganous ions, for example, should not be used when formulating compounds containing polymyxin This blocking effect can be partly reversed by citrate ions.

Pharmacology

Absorption and Administration Routes

Polymyxin B sulphate is another antibiotic which is normally not absorbed from the intestines. Parenteral injections must therefore be given to obtain systemic activity. Absorption from either subcutaneous or intramuscular sites is at approximately the same rate as penicillin. A commercially available salt, colistin sodium methane sulphonate, is the drug of choice for parenteral administration. Topical application and oral administration are more commonly used routes.

Blood Levels

As absorption is similar to penicillin, blood levels are achieved fairly rapidly following parenteral administration and are maintained for 6 to 8 hr. The antibiotic circulates mainly in the plasma and is not absorbed by the cells. Four-hourly dosage results in cumulation and the development of high blood levels, as high as 26 units per ml being observed in man. The daily dose of polymyxin in man is given in four equal doses at 6-hourly intervals.

Barriers

Polymyxin B does not penetrate the blood–brain barrier and meningitis must therefore be treated by intrathecal injection (30 000 to 50 000 units once or twice daily in a concentration of 30 000 units per ml normal saline is usual in human medicine). It is not absorbed from the intestines or skin.

Excretion

Very little polymyxin is excreted unchanged either by the kidney, bile or faeces: only 0·3 per cent of a 10 000 unit intramuscular dose was recovered in an active form from a dog's urine over a 24-hr period. This makes it necessary to give relatively large doses parenterally to obtain active levels in the urine. Almost the whole of any oral dose is destroyed in the alimentary canal and only small amounts are recovered from the faeces in an active form.

Toxicity

The toxic effects of the polymyxins are of two types.

1. *Neurotoxic* effects including loss of the sense of balance and of skin sensation in the head and extremities with muscular weakness have been reported in man.

2. *Nephrotoxic* effects with albuminuria, renal casts, and in severe cases haematuria and pyuria are also seen and the non-protein nitrogen content (urea, etc.) of the blood increases. As little as 2·5 mg per kg given intravenously as a single dose will cause acute renal damage in the dog.

All symptoms are reversible and regress when dosing is stopped, providing the doses have not been excessive. Currently manufactured polymyxin B sulphate is far less toxic than that originally produced. When applied topically, concentrations of less than 1 per cent are non-irritant. So far neither acquired bacterial resistance nor skin sensitivity have been seen clinically.

Actions

Polymyxin B sulphate is a bactericidal antibiotic, the mode of action being as yet unknown, but may affect the permeability of the cell. It is synergistic with oxtetracycline, only one-half to one-sixteenth of the normal dose of each being required when used in combination. In vitro synergism has also been shown with chloramphenicol and, to a lesser degree, with neomycin against seventy-two strains of *Salmonella*. Topical action is often improved by combining polymyxin with neomycin or bacitracin. The combination of oxytetracycline with polymyxin B has been widely used in the treatment of mastitis.

Antibacterial Spectrum

Polymyxin is a narrow-spectrum antibiotic with an activity entirely limited to the Gram-negative bacteria. *E. coli*, *Aerobacter aerogenes* (*B. aerogenes*) and related species, *Haemophilus* and *Klebsiella* species, and *Ps. aeruginosa* (*Ps. pyocyanea*) are all susceptible in vitro to concentrations of 0·05 to 2 μg per ml (0·3 to 12 units). Activity against *Proteus* species was comparable with that against *E. coli*, but inferior to that of neomycin. It is active against *Pseudomonas* species.

Diseases and Polymyxin Therapy

This antibiotic has not been used extensively on its own in Great Britain, but in conjunction with oxytetracycline considerable amounts have been used both for intramammary injection and for topical application. In both applications the use of polymyxin has been particularly directed against species of *Pseudomonas*, mainly *Ps. aeruginosa* (*Ps. pyocyanea*).

Bovine mastitis. Polymyxin B sulphate is effective in many cases of mastitis due to *Ps. aeruginosa* (*pyocyanea*), particularly when combined with oxytetracycline, but the combination is not always successful in such cases. The two antibiotics have also been used effectively in a few cases of mastitis due to *Klebsiella* species.

Semen additive. Polymyxin has been used in bulls semen which contained *Pseudomonas* organisms; 50 μg per ml diluent was sufficient to inhibit the growth of the organisms. The addition of polymyxin along with penicillin and streptomycin was sufficient to eliminate all contaminating bacteria. It has also been reported that preputial irrigation with polymyxin eliminated *Pseudomonas* organisms from thirteen bulls, though seven were reinfected later.

Septicaemia. Five litters of puppies with a septicaemia attributed to *Ps. aerogenes* (*pyocyanea*) contracted by suckling were successfully treated by polymyxin.

Topical applications have been especially useful in *otitis externa* in dogs, intramammary formulations containing oxytetracycline and polymyxin B being a popular form of application.

Dosage

The usual dose is 40 000 units per kg body weight orally, or 10 000 units per kg body weight by injection. Give in divided doses.

II. BACITRACIN

This antibiotic has an antibacterial spectrum which follows that of penicillin very closely; many penicillin resistant strains are susceptible to its action. The compound is official in both the British and United States Pharmacopoeias and is one of the few antibiotics of clinical value derived from bacteria.

History

The discovery of bacitracin was announced in 1945 by Johnson and his colleagues, who noticed the bacteriostatic activity of a Gram-positive sporulating organisms derived from tissue removed during treatment of a compound tibial fracture. The organism was of the *Bacillus* group and the patient a Margaret Tracy: hence the name of the antibiotic 'bacitracin'.

Manufacture

Originally prepared from surface cultures of *Bacillus subtilis* (some authorities say the organism is a strain of *B. licheniformis*) the manufacture now follows the general pattern and the deep vat culture method is used. The yield is dependent on the ratio of carbon to nitrogen in the nutrient medium and is greatest at a ratio of 15:1, soya bean meal, starch, calcium lactate and calcium carbonate being used as sources for essential elements (*U.S.D.*, 25th edn, 1955, p. 131). Purification proceeds through various stages including extraction by butyl alcohol, clearing with charcoal, acidification and precipitation with salicylic acid.

Chemistry

Bacitracin is a white to pale buff powder with sometimes a slight odour and a very bitter taste: the insoluble salts such as zinc bacitracin are almost without taste. The powder is hygroscopic and therefore soluble in water; it is partially soluble in ethyl and methyl alcohol and glacial acetic acid, leaving an insoluble residue. It is insoluble in acetone, chloroform, and ether. The purest form, which is used parenterally, must contain a minimum of 50 units per mg antibacterial activity.

There are four substances present in the crude liquors obtained by fermentation. Three of these, bacitracins A, B and C, are active but the fourth, bacitracin F, is almost, if not entirely, inactive as an antibacterial agent.

Structurally, these substances are described as polypeptides which can be represented by:

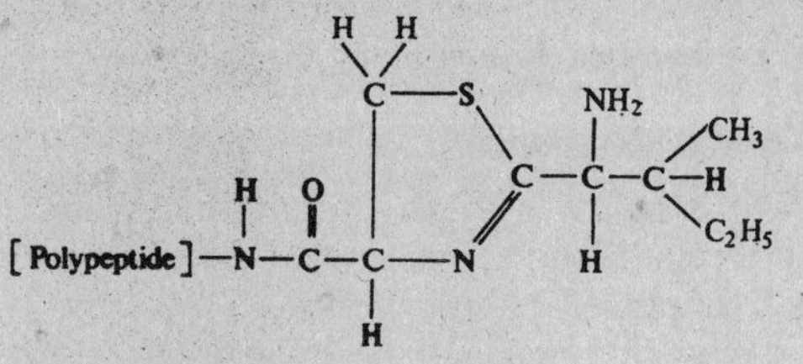

Among the amino acids forming the peptide are phenylalanine, leuceine, glutamic acid, lysine, histidine, cysteine, etc.

It will be noticed that the thiazolidine ring present in penicillin is also present in bacitracin, and as penicillin may be considered as a condensed dipeptide these two antibiotics have been classed, together with polymyxin and the cephalosphorins (close relatives of penicillin), as the peptide antibiotics.

Stability

Bacitracin is slightly more stable than penicillin but is still less stable than most of the newer antibiotics.

1. Moisture. The dry powder is stable at room temperature when dry and will maintain its potency for 15 months at 37°C. Solutions deteriorate rapidly at room temperature but retain 90 per cent potency when kept refrigerated for 2 to 3 months.

2. pH. The optimum pH of solutions is between pH 5 and 7.

3. Temperature. As already inferred the dry powder is stable at normal working temperatures but should the temperature be raised above 30°C deterioration occurs and is rapid at 56°C. All solutions once made should be used immediately or refrigerated.

4. Oxidizing agents. All oxidizing agents destroy bacitracin.

5. Miscellaneous. Heavy metals and their salts will precipitate and inactivate bacitracin, though it is suggested that in some cases precipitation is not accompanied by inactivation; zinc is one of these. Some organic acids will inactivate the antibiotic; tannic, benzoic and salicylic acid are three such acids. High concentrations of sodium chloride are also incompatible with bacitracin solutions owing to precipitation.

Pharmacology

Absorption and Administration Routes

Bacitracin is similarto streptomycin in that almost no absorption takes place following oral dosing or inhalation, and consequently systemic infections can only be treated by parenteral injection. Parenteral injections may be given by any of the normal routes but the intramuscular route is the commonest for veterinary purposes. Following intramuscular administration, absorption is slightly slower than with penicillin; however, excretion is by glomerular filtration and is therefore slower than penicillin so that blood levels are higher and persist longer than with penicillin. Some toxicity occurs, partly due to impurities in the antibiotic produced by the initial manufacturing methods, but also due to the polypeptide nature of the antibiotic and until recently the main use in man has been for topical application. This is also true in animals.

Blood levels

Maximum blood levels may take up to 2 hr to be achieved and adequate levels may be maintained for 6 to 8 hr depending on dosage and administration route.

Barriers

Following intramuscular injection, bacitracin will penetrate into most tissues of the body, but therapeutic amounts do not accumulate in either the pericardial or cerebrospinal fluid. Any infection which develops in a body cavity, though bacitracin passes freely into pleural and ascitic fluid, should be treated by injections directly into the cavity. Little absorption occurs from the intestines or lungs, or from topical administration.

Excretion

Almost all of any absorbed bacitracin is excreted by the kidneys at the rate of glomerular filtration. Most of any oral doses is destroyed in the intestines.

Toxicity

Bacitracin, in the early days of production, was strongly nephrotoxic but since purification has been improved these effects have been reduced sufficiently to allow normal parenteral dosing, providing there is no renal impairment already present. It is advisable to check for albuminuria from time to time when parenteral treatment is being used. Bacitracin is virtually non-toxic when administered orally or topically.

Action

The main antibacterial activity of this antibiotic is due to the bacitracin A component. Bacitracin is a bactericidal antibiotic, the speed of bacteridical action being directly proportional to its concentration, and therefore to its dose. Its activity is unaffected by blood or tissue breakdown products. The mode of action appears to be by interference with cell wall formation.

Antibacterial Spectrum

Bacitracin is a narrow-spectrum antibiotic and is active against a range of organisms similar to that affected by penicillin. In some cases clinical response is not as good as with penicillin, but frequently bacitracin is active against penicillin resistant organisms. The drug is thus effective against most Gram-positive cocci, aerobic or anaerobic, Gram-positive rods and some spirochaetes. It has also been found of value against intestinal amoebiasis in man. Most, but not all, Gram-negative bacteria are unaffected by bacitracin at normal dosage levels.

Bacterial Resistance

The sensitivity of an organism to bacitracin is not related to its sensitivity to other antibiotics such as penicillin and the broad-spectrum antibiotics such as penicillin and the broad-spectrum antibiotics, so that it can be used where these fail, and vice versa. So far, no bacterial resistance to bacitracin of any clinical significance has been reported although it is known that bacterial resistance to bacitracin does exist in human pathogens.

Diseases and Bacitracin Therapy

So far, bacitracin has not been widely used for veterinary purposes in Great Britain but is becoming increasingly popular for local application and as a feed additive.

Feed additive. As a result of the recommendations of the Swann Committee on Antibiotics, (1968) Zinc Bacitracin is permitted for general use as a feed additive for addition to the feed of pigs and poultry in particular for growth promotion purposes.

It has been selected for this purpose as it is little used in medical practice, and it does not show evidence of bacterial resistance or cross-resistance with other therapeutic antibiotics.

Preparations

It is prepared in a number of formulations for feed additive purposes. Two such concentrates contain 55 and 88 g of zinc bacitracin per kg.

31

Tetracyclines

I. CHLORTETRACYCLINE (AUREOMYCIN)

Chlortetracycline was the first of the tetracycline group and the second broad-spectrum antibiotic to be discovered. It is official in the U.S.P. as Chlortetracycline Hydrochloride and is now in the *British Veterinary Codex* under the same name.

History

Further work into the antibiotic producing potentialities of the Streptomycetes following the discovery of streptomycin resulted in the announcement by Duggar in 1948 of a new species of actinomycete *Streptomyces aureofaciens* and a new antibiotic, given the name Aureomycin. The fungus was so named because of the intense yellow colour of cultures; the commercial name of Aureomycin refers directly to the golden-yellow of the antibiotic.

Manufacture

Chlortetracycline is manufactured by deep-culture fermentation as with most other antibiotics, using a strain of mould obtained by mutation following ultraviolet irradiation.

There are several methods of extracting and purifying the substance but the principles are much the same as for oxytetracycline extraction. Use is made of the amphoteric character of the base and the different solubilities of base and salts in various solvents; use is also made of the selective adsorption of base and salts by exchange resins. The final extraction and crystallization is carried out aseptically so that a sterile crystalline salt or base is obtained.

Chemistry

Chlortetracycline base is a golden-yellow crystalline amphoteric compound sparingly soluble in water and with a bitter taste. The hydrochloride is similar in appearance but with a strong bitter taste; it is fairly soluble in water (1:75) but only slightly soluble in alcohol and almost insoluble in other organic solvents.

The structure of the compound is based on the tetracycline nucleus, the hydrogen atom attached to carbon 7 in the nucleus being replaced by a chlorine atom. This substitution is said to be responsible for the instability of this compound when compared with tetracycline and oxytetracycline. This base, in common with tetracycline and oxytetracycline, will form salts

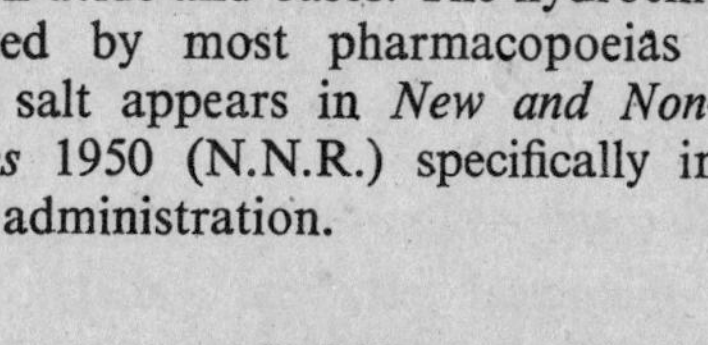

Chlortetracycline

with both acids and bases. The hydrochloride is recognized by most pharmacopoeias and a calcium salt appears in *New and Non-official Remedies* 1950 (N.N.R.) specifically intended for oral administration.

Stability

Had chlortetracycline been more stable (and less irritant on injection) its therapeutic importance would have been considerably greater.

1. Moisture. Chlortetracycline hydrochloride is stable in dry air, and in moisture free containers protected from light is stable for at least 7 months at room temperature. Solutions in water having a pH of 2·5 are relatively stable but cannot be administered. When buffered to neutrality or alkalinity for administration, solutions deteriorate very rapidly losing about 12 per cent of their potency in 30 min at 25°C,

20 per cent in 1 hr and 40 per cent in 2 hr. Solutions must therefore be freshly prepared, for though refrigeration of a solution will maintain potency for slightly longer periods it is of no clinical significance.

Chlortetracycline borate, used for eye infections as a solution, is stable as a dry salt when kept in air, but even this is very unstable in solution.

2. pH. Acid solutions are comparatively stable and at pH 4 will keep in a refrigerator for 2 weeks. At pH 8 the intravenous injection potency is rapidly lost. When buffered with sodium borate (chlortetracycline borate eye drops) solutions have approximately pH 7·5; these solutions are inert after 24 hr but will keep for several days when refrigerated.

3. Temperature. Solutions are susceptible to temperature changes and whereas dilute solutions at pH 2·5 will remain stable for several hours at room temperature, at 37°C they will lose up to 15 per cent potency in 5 hr and 50 per cent in 24 hr.

Oxidizing agents, enzymes, etc. Chlortetracycline is similar to other members of the group in its stability under conditions other than those mentioned above.

Pharmacology

In many respects chlortetracycline has a pharmacology similar to oxytetracycline but differs materially in one or two instances.

Absorption and Administration Routes

Only biological tests are available for estimating blood, tissue and body fluid levels; as chlortetracycline deteriorates rapidly in solution no absolute figures but only relative values can be obtained by these methods.

Following oral administration of single doses, maximum blood levels are achieved in about the same time as those for oxytetracycline—2 to 8 hr—and detectable levels may remain for as long as 24 hr. Increase in dosage is not followed by commensurate increase in blood levels, but levels are maintained for a longer period. Intramuscular injection is painful and is not recommended. Intravenous administration is permissible but care must be taken not to allow perivascular leakage, for irritation locally and thrombophlebitis may result. Small intravenous doses—as low as 25 mg in man—are sufficient to provide satisfactory blood levels which can then be maintained by oral dosage. Maximum levels may be achieved in as short a time as 5 min and detectable levels persist for up to 12 hr after a single dose. Chlortetracycline has also been administered by intraperitoneal injection. Chlortetracycline is suitable for topical applications and is extremely effective when used against susceptible infections.

Blood Levels

In man, blood levels of 2 to 4 μg are considered to be 'therapeutically effective' (*Extra Pharm.*, 23rd edn, 1952 p. 104) but it has been shown (British Patent 654215, 660511) that 0·03 μg is effective against a heavy experimental inoculation of pneumococci. Blood levels are maintained for approximately the same time as similar doses, by the same route, of oxytetracycline.

Barriers

Chlortetracycline rapidly diffuses into most tissues and fluids and is thought to diffuse into the intracellular water of tissues. The remarkable effect on rickettsiae is probably due to this diffusion, in this case into the intracellular fluid of endothelial cells.

1. Blood–brain barrier. Confusion exists concerning the diffusion into the cerebrospinal fluid. It appears that some diffusion take place, but that the concentrations increase in meningitis. This follows the pattern of oxytetracycline.

2. Placental barrier. Diffusion of chlortetracycline across the placental barrier takes place readily and, as with oxytetracycline, appreciable quantities can be detected in the foetal circulation.

3. Intestinal barrier. Absorption from the intestines is rapid but incomplete and therapeutic concentrations in the faeces are easily obtained. The detectable levels may not be quite so high as those given by similar doses of oxytetracycline but that is due in part at any rate, to the instability of solutions of chlortetracycline and the consequent inaccuracy of tests.

4. Serous membranes. The peritoneum and

pleura offer little resistance and detectable quantities appear in peritoneal and pleural fluid within a relatively short time of administration. Intraperitoneal injection has been used for administration, though it has been stated that chlortetracycline causes irritation when given by this route as well as by intramuscular or subcutaneous injection. There is no information on diffusion into the humours of the eye nor across the sinovial membrane.

5. Milk barrier. Chlortetracycline diffuses into the milk in concentrations equal to or higher than those in the blood and combined intravenous and local treatment has been recommended in acute mastitis cases.

Excretion

As with the other tetracyclines, the main excretory routes are the urinary system, the biliary system, and the intestines. Higher concentrations of chlortetracycline than are given by equal doses of oxytetracycline are often found in the bile (also lungs and spleen) but urinary and faecal concentrations are usually lower.

Urinary excretion is from 9 to 15 per cent of a single oral dose during the 24 hr subsequent to dosing, but up to 25 per cent of the total dose may be recovered following repeated dosing (compared with 50 per cent for oxytetracycline). These discrepancies have not been accounted for, oxytetracycline is therefore preferred for urinary infections.

The higher biliary excretion makes chlortetracycline preferred before oxytetracycline for liver infections of which *Leptospira icterohaemorrhagiae* may be quoted as an example.

Details are quoted in *Antibiotics* (2nd ed., 1952, p. 184, Pharmaceutical Press) showing that the excretion of chlortetracycline by the udder persists for 48 hr and the observation is made that the udder probably provides one of the main means of excretion.

Toxicity

Chlortetracycline is slightly more irritant than the other tetracyclines when administered orally, and pigs, cats and dogs may show signs of nausea, may vomit, and may suffer from diarrhoea. Administration may be either in milk or followed by milk, which seems to lessen the incidence of these symptoms. Ruminants have been shown to suffer gastro-intestinal upsets following oral administration, but as with oxytetracycline this is not of any permanent clinical significance, though repeated small doses are less troublesome in this respect than single large doses.

The reference made to fungal growth during the use of oxytetracycline also applies to chlortetracycline, and a fungistatic compound such as the antibiotic nystatin has been used to overcome this trouble in man.

Tetracyclines are deposited in growing teeth and bones, and they should not be used in young animals.

Action

Chlortetracycline is normally bacteriostatic but in higher concentrations and used against the more sensitive bacteria it may also be bactericidal. The method of its bacteriostatic action is an interference with RNA and bacterial protein synthesis by effecting protein induction at the ribosomes by messenger RNA. Though a decrease in antibacterial effect due to antagonism by blood and tissue fluids has been reported, this if of little clinical significance; indeed, the activity in vivo is said to be greater than would be anticipated following in vitro tests.

There has been reported a shortening of blood coagulation time following the administration of chlortetracycline to some animals.

Antibacterial Spectrum

A broad-spectrum antibiotic, chlortetracycline is active against a wide variety of organisms including Gram-positive and Gram-negative bacteria; rickettsial activity is particularly good and it has some antiviral activity. Some protozoa are also affected. It has negligible effect on *M. tuberculosis*, yeasts and moulds. *Proteus* and *Pseudomonas* species are mostly insensitive, the occasional strains are susceptible to higher concentrations than are normally used.

As with other broad-spectrum antibiotics the danger of yeast and mould proliferation can be troublesome.

Bacterial Resistance

Chlortetracycline insensitivity is being met increasingly in clinical practice and resistance follows the pattern of oxytetracycline resistance (*q.v.*). It should be remembered that cross resistance exists within the tetracycline group. Where resistance to one of these broad-spectrum antibiotics exists it is always wiser to use a narrow-spectrum compound to which the organism has a specific sensitivity rather than to rely on larger doses of a related antibiotic.

Diseases and Chlortetracycline Therapy

There is little difference between the disease coverage of chlortetracycline and oxytetracycline, but whereas the latter antibiotic is more effective in the urine and faeces because of its greater concentration, the former antibiotic creates higher concentrations in the lung and spleen and is therefore more effective in pneumonias and septicaemias. Differences exist in degree of effect in some cases, and in administration methods.

Specific diseases which respond to treatment with chlortetracycline include pasteurellosis, strangles, leptospirosis, calf diphtheria, vibriosis in cattle, and infectious feline influenza. Non-specific diseases include mastitis, metritis, gastroenteritis including calf and pig scours, and secondary infections of virus diseases. As with oxytetracycline most of the common bacterial diseases of poultry respond to chlortetracycline treatment.

Mastitis

Following intramammary infusion of 400 mg chlortetracycline, adequate antibacterial levels of 4 μg per ml are maintained for 48 hr, even with regular 12-hourly milkings.

Chlortetracycline is of most value as a treatment of staphylococcal mastitis.

Streptococcal mastitis. Penicillins are preferred for all routine purposes. In acute cases, favourable results have been reported from intravenous injection of 5 g chlortetracycline in conjunction with local udder infusions.

Staphylococcal mastitis has responded well to chlortetracycline intramammary injection. With the increase in mastitis due to penicillin resistant strains, chlortetracycline and other members of the group are valuable. Coliform infections will sometimes respond to chlortetracycline but *Pseudomonas* infections are generally resistant.

Klebsiella mastitis shows good clinical response to chlortetracycline; streptomycin is suggested as a secondary treatment where this failed. The action of chlortetracycline in mastitis thus parallels that of oxytetracycline quite closely.

Specific Systemic and Organic Diseases

Most of these diseases respond in a similar way to oxytetracycline but with some small variations in degree of response.

Strangles

In general, penicillin is preferred due to the greater sensitivity of *Strep. equi* to that antibiotic (0·03 μg* average as against 19·2 μg for chlortetracyclines and 18·3 μg for oxytetracyclines) but nevertheless clinical response has been reported.

Anthrax

Although chlortetracycline has been shown to be effective in vitro against *B. anthracis* no clinical reports are available. However, in view of the good clinical reports on oxytetracycline treatment and the highly successful experimental use of tetracycline in sheep, it is probably that chlortetracycline will be equally effective.

Pasteurellosis

Chlortetracycline is recommended for the treatment of transit fever in cattle and horses and good results have been reported from intravenous injection.

Fowl cholera responded experimentally and chlortetracycline, as well as oxytetracycline, has been used successfully in the field, though furazolidone is preferable to any antibiotic for this and for the *Salmonella* group of diseases in poultry.

Leptospirosis

Bovine leptospirosis has been successfully treated with chlortetracycline and the organism entirely eliminated from the urine. The effect was superior to that of either penicillin or streptomycin.

In *pigs*, various reports indicate that a daily oral intake of about 1·5 g per animal (15 mg per

* Concentration in μg per ml giving 50 per cent inhibition following over night incubation.

kg body weight) will eliminate *L. pomona* infection from breeding sows.

In *dogs*, both *L. canicola* and *L. icterohaemorrhagiae* infections respond to chlortetracycline.

Doses of chlortetracycline in this disease need to be high and should be given early to prevent structural changes in the liver and kidneys.

Coliform–*Salmonella* Group Infections

Many reports and wide clinical usage have indicated the usefulness of chlortetracycline in the treatment of white scours in calves, and enteritis in pigs and dogs, either by intravenous injections or by oral administration; prophylaxis in calf scours is achieved in the same way as with oxytetracycline, dosages for this purpose and for treatment being 11 mg per kg body weight. *E. coli* resistance is, however, becoming much more common.

In *calves*, infections with *Salmonella dublin* have responded very satisfactorily to oral dosage with chlortetracycline.

In *foals*, septicaemic infections with *B. equirulis* (*Shigella equirulis*) have responded well to intravenous injections of chlortetracycline. Prophylaxis is achieved by a single dose at the daily curative rate of 11 mg per kg body weight. However, as streptomycin is effective and the drug of choice the broad-spectrum antibiotic should only be used in streptomycin resistant cases.

In *poultry*, experimental evidence indicates that *Salmonella* infections will respond to chlortetracycline, up to 600 mg per kg being satisfactory against *S. pullorum* if used early. Nevertheless, as with other poultry diseases, furazolidone in mashes will give very good and economical control of salmonellosis, and chlortetracycline, along with other antibiotics, is used mainly for other purposes.

Listeriosis

This disease is typified mainly by meningitis which causes circling in cattle and sheep. The organism implicated is *Listeria* or *Erysipelothrix monocytogenes* which is susceptible in vitro to chlortetracycline but not other broad or narrow spectrum antibiotics including oxytetracycline. Over 60 per cent recoveries in cattle have been reported but no effect is exerted on the disease in sheep: intravenous injections at the daily rate of 11 mg per kg should be continued for 4 days.

Pfeifferella Infections

Chlortetracycline has a marked in vitro effect against *Pf. mallei* which causes glanders, a severe and fatal disease of horses and man. In vitro and in vivo tests on guinea-pigs showed that chlortetracycline was the most active antibiotic in vitro and, with oxytetracycline and dihydrostreptomycin, was able to protect the experimental animals from the disease. Clinical reports are not available.

Poultry

The effects in pasteurellosis (fowl cholera) and salmonellosis have already been mentioned (*q.v.*).

Infectious sinusitis in turkeys has responded to chlortetracycline amongst other antibiotics, also *Chronic Respiratory Disease* (CRD) of chickens, but for the former disease, at least, it is recommended that different antibiotics should be used in succession as the response to each slows down.

Non-Specific Infections

Chlortetracycline has proved of value in the treatment of non-specific enteritis, especially that associated with virus diseases. In virus diseases generally this antibiotic is quite effective in eliminating or reducing secondary bacterial infections. This is of particular value in the canine distemper—hepatitis complex and in the complications of cat distemper or infectious panleucopenia.

Metritis in cattle responds well to chlortetracycline. The soluble 0·5 g tablet has proved valuable in the treatment of retained placentae. The insertion of a 0·5-g tablet every 48 hr from parturition until the placenta sloughs naturally is recommended; this is followed by a decreased incidence of metritis and no impairment of fertility. This is in accord with the effects achieved by oxytetracycline.

Other non-specific conditions such as pneumonia, tonsillitis, bronchitis (including kennel cough), and cystitis, have frequently responded satisfactorily in all species.

Topical Applications

Chlortetracycline has been used successfully in eye conditions as an ointment or a solution of chlortetracycline borate; the latter is used owing to the acidity and irritant effect of the hydrochloride. It is of value in most ocular lesions, penetrating only when the conjunctiva is damaged: it causes less retardation of healing than streptomycin.

As a topical ointment, sometimes in conjunction with systemic therapy, chlortetracycline is useful in *otitis externa*, burns and wounds, abscesses, and other forms of pyogenic infection.

Dosage

Orally: For small animals the generally recommended dose is 20–50 mg per kg body weight daily. Large animals (calves and foals) 10–20 mg per kg body weight.

Preferably given in two equally divided doses at 12-hr intervals and in order to ensure adequate bacterial control of disease it is wise to continue administration for 2 or 3 days after all detectable symptoms have subsided or temperature has been normal for 48 hr.

Intramammary: Cattle: One 6-ml syringe per infected quarter repeated every 24 hours as necessary.

Intra-uterine: Cattle: One to two 0·5-g tablets. Ewes and sows: ½ to 1 tablet.

Parenteral: Not recommended.

Official Preparations

The *British Veterinary Codex* (2nd edn, 1965) recognizes:

Veterinary Chlortetracycline Powder containing not less than 85 per cent of the potency on the label. When the strength of the powder is not specified a powder containing 55 g CTC per kg (25 g per lb.) shall be supplied with a potency of not less than 900 units per mg.

Chlortetracycline Capsules—50 and 250 mg of chlortetracycline.

Veterinary Chlortetracycline Tablets: containing 0·5 g of chlortetracycline are available and comply with the Standard under tablets.

Commercially available in Great Britain are:

Capsules 50 mg and 250 mg.

Soluble tablets containing 0·5 g chlortetracycline hydrochloride.

Soluble powders with chlortetracycline hydrochloride 25 g per lb. and tinted green and with chlortetracycline bisulphate 102·4 g per lb.

Soluble powder 20 per cent chlortetracycline hydrochloride plus 9·1 per cent benzethanium chloride. Specially prepared for addition to drinking water in poultry houses. Add 150 g formula to 15 gallons drinking water.

An intramammary suspension in disposable 6-ml plastic syringes containing 426 mg, etc., hydrochloride in a non-irritantbase.

An ophthalmic ointment in ⅛-oz. tubes containing 10 mg chlortetracycline hydrochloride per g (1 per cent) in a lanolin–petroleum base.

A formulation of chlortetracycline hydrochloride in sesame oil in 6 ml oral doser, each 1 ml containing 50 mg chlortetracycline.

A chlortetracycline hydrochloride powder for topical application in 15 g puffer pack containing 20 mg, etc., per g (2 per cent) and Benzocaine 1 per cent.

5 oz. Aerosol Sprays for topical application containing 1420 mg (1 per cent) chlortetracycline hydrochloride or a 2 per cent Spray coloured with 710 mg (0·5 per cent) Gentian Violet.

II. OXYTETRACYCLINE (TERRAMYCIN)

This antibiotic is one of the most important and widely used antibiotics in veterinary medicine. It has broad spectrum antibacterial action both in vitro and in vivo, and is official as the base and as the hydrochloride both in Great Britain and America.

History

Oxytetracycline was discovered as the result of the coordinated research of a team of eleven workers employed by an American commercial organization. In the course of their investigation they examined approximately 100 000 soil samples submitted to them from all parts of the world, finding seventy-five antibiotic producing moulds. Amongst this number was a previously unidentified species of mould *Actinomyces rimosus*, from which a broad spectrum antibiotic was obtained and named Terramycin. This discovery was announced in 1950.

Manufacture

Oxytetracycline is manufactured by the normal deep culture fermentation methods; approximately 2 kg (about 4½ lb) is isolated from each 1000 gallons of fermentation broth, or about 50 kg (1 cwt) from one 100 000 litre (25 000 gallon) fermentation. Purification is a complex sequence of solution and extraction of base and salts from different solvents following kieselguhr filtration. A crystalline base or the hydrochloride is finally obtained, the former containing not less than 96 per cent of pure base in the official B.P. form, whilst the hydrochloride of the B.Vet.C. must also contain not less than 96 per cent of the salt.

Chemistry

Amphoteric oxytetracycline (oxytetracycline dihydrate) base and its hydrochloride are yellow, odourless crystalline powders with a bitter taste. The base is only slightly soluble in water but the hydrochloride is readily soluble, though some turbidity may result from the release of oxytetracycline base: buffering with sodium glycinate overcomes this. Both are readily soluble in dilute hydrochloric acid, and fairly soluble in methyl and ethyl alcohol, but insoluble in chloroform and ether.

The structure of oxytetracycline (ii) contains tetracycline (i) as a nucleus which is also the basis of chlortetracycline (iii). From these

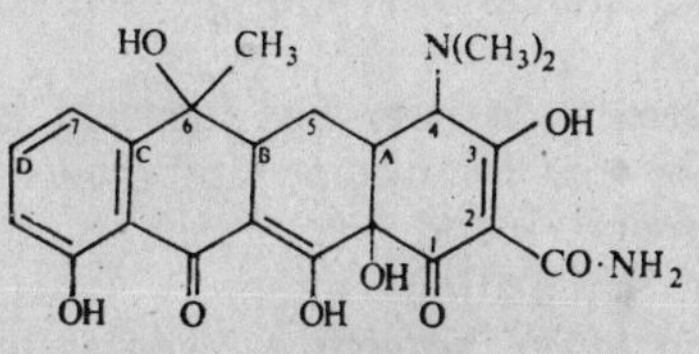

(i) Tetracycline

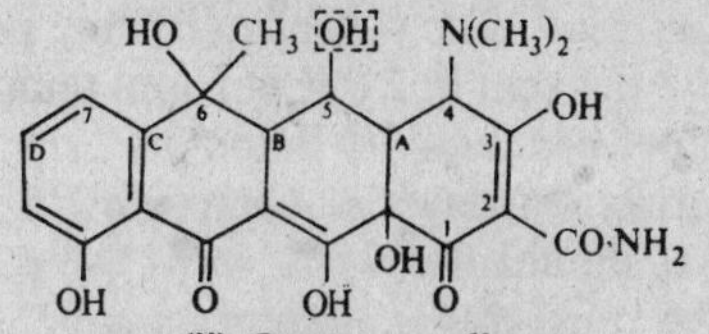

(ii) Oxytetracycline

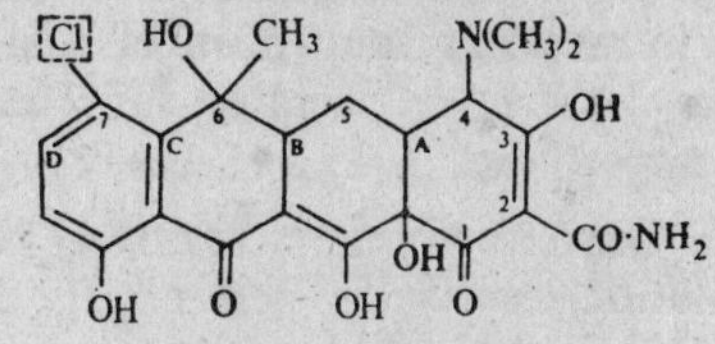

(iii) Chlortetracycline

diagrams it can be seen that oxytetracycline consists of the tetracycline nucleus with a *hydroxyl group* instead of a hydrogen atom at carbon 5 in benzene ring B, whilst chlortetracycline contains a *chlorine* atom in place of the hydrogen atom attached to carbon 7 in ring D.

All three compounds are used as hydrochlorides, which may be represented graphically as follows:

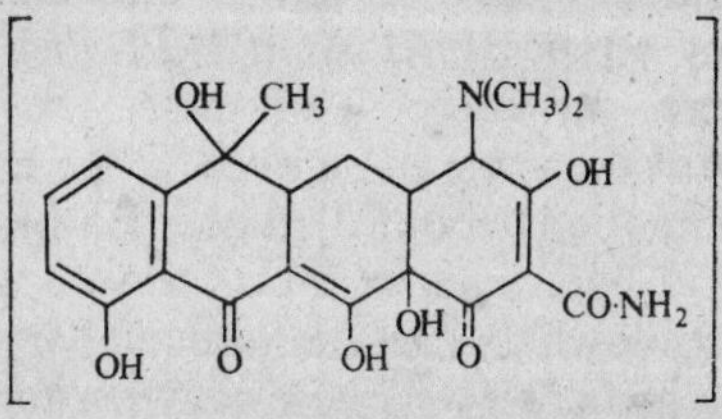

Tetracycline hydrochloride and similarly oxytetracycline and chlortetracycline hydrochlorides

As with the other two members of the series, oxytetracycline is amphoteric and will form salts with either acids or bases.

Stability

Oxytetracycline is very stable when compared with chlortetracycline and penicillin.

1. Moisture. In the dry state this antibiotic is stable for at least 2 years at room temperature. When in solution, however, the situation is not quite so satisfactory for, though unbuffered 1 per cent solutions at a pH of 2·5 will maintain potency for at least 30 days at 25°C, the official U.S.P. parenteral injection, buffered with sodium glycinate and having a pH of approximately 9, loses 2 per cent of its potency (100 mg per ml) in 2 hr and 8 per cent in 24 hr at room temperature. Up to 8 per cent activity is lost in 3 days during refrigeration: 4 days is therefore as long as solutions should be kept and even so they should be refrigerated. Potency is unaffected by the progressive darkening which occurs. Commercial intravenous and intramuscular injections may be kept refrigerated for 2 days and 4 days, respectively.

2. pH. Oxytetracycline is relatively stable in acid solution especially below pH 2·5, but neutral and particularly alkaline solutions deteriorate rapidly. Buffering oxytetracycline with sodium glycinate (to give an alkaline pH) is aimed not at maintaining potency but in maintaining solution, for hydrolysis and deposition of the base occurs in acid solutions of the hydrochloride; this necessary alkalinity is detrimental to stability.

3. Temperature. Solutions of oxytetracycline are no exception to the rule that their stability, though related closely to pH, is also markedly affected by temperature. At pH 2·5 and room temperature unbuffered solutions are quite stable but at the same pH and 37°C the half-life of the antibiotic is about 5½ days. (Half life is a term coined by atomic scientists to indicate the period during which half the radioactivity of an element is lost. This was necessary because rate of activity loss decreased with loss of activity, so that half the activity was lost in 5½ days; complete loss may not follow for several weeks, months or even years. The same principle can be applied to antibiotic potency.)

The *dry* powder will withstand heating for 4 days at 100°C without loss of potency (*Antibiotics*, Pharm. Press, 2nd edn, p. 67) and only 5 per cent potency loss occurs after 4 months at 50°C.

4. Oxidizing agents, enzymes, etc. The tetracyclines appear to be stable in most respects other than moisture, pH and temperature.

Pharmacology

The pharmacology of the base and salts of oxytetracycline is fundamentally the same. The relative insolubility of the base makes it unsuitable for parenteral injection; but its less bitter taste makes it more suitable for oral administration.

Absorption and Administration Routes

Oxytetracycline hydrochloride may be administered by any of the normal routes. It is readily absorbed from the intestines by all mammals but intestinal absorption in poultry is restricted. Effective blood levels are achieved in 2 to 4 hr. By the intravenous route, maximum levels are achieved within a few minutes but intramuscular dosage delays peak blood levels for about 2 hr and levels are not so high though maintained for a longer time. Oily injections intended for subcutaneous injection in poultry should not be used as parenteral injections in mammals. Oxytetracycline is very effective and suitable for topical application.

Blood Levels

In human medicine, 0·5 to 1 μg per ml serum has been accepted as adequate for most purposes in the treatment of infected adults. It has been assumed that this level is adequate in animals also, and this is borne out by clinical experience. These levels are easily maintained by single daily intramuscular doses in cattle, dogs and horses; other species follow the same pattern.

Barriers

Oxytetracycline is most useful in that it readily disperses throughout the body, attaining therapeutic levels in most tissues and fluids within a short time.

1. Blood–brain barrier. Oxytetracycline was originally thought to pass the blood–brain barrier quite easily but, though some will pass the barrier normally, therapeutic levels will only appear in the cerebrospinal fluid when the meninges are inflamed or damaged. Intrathecal injections should therefore be used to initiate treatment.

2. Placental barrier. The placenta is readily passed by oxytetracycline and concentrations in the foetal blood may reach those of the maternal circulation, although it is usually somewhat lower, sometimes even as low as 25 per cent.

3. Intestinal barrier. Absorption from the intestines is reasonably fast but not complete, and in man faecal concentrations as high as 2·5 mg per g have been assayed in an active form. These quantities are several times greater than would be expected from similar doses of chlortetracycline or chloramphenicol.

4. Serous membrane barriers. Therapeutic levels in the pleural and peritoneal fluids can be achieved readily, even from oral dosage, though initial levels should be established by parenteral administration either by intraperitoneal or intravenous injections or both if severe and acute inflammations are present.

5. Milk barrier. Oxytetracycline readily diffuse into the milk and therapeutic milk levels can be achieved by single intravenous doses, but intramuscular dosage is not satisfactory and parenteral administration is not a substitute for local intramammary treatment, especially as milk levels drop sharply after the first milking.

Excretion

The main excretory routes of oxytetracycline are the urinary system and the intestinal route via the biliary system.

The antibiotic appears in the urine within an hour of administration (usually 30 min) and reaches a maximum in about 4 hr in man. About 15 per cent is excreted in the first 24 hr but as much as 50 per cent may ultimately be recovered in an active form. High concentrations—up to 400 μg per ml—may be reached and will persist for at least 24 hr following single intravenous or intramuscular doses. Concentration of oxytetracycline also occurs in the bile, and levels up to eight times those of concurrent serum levels may be achieved. Along with chlortetracycline, oxytetracycline is therefore valuable in treating liver infections. The antibiotic eliminated in the bile serves to swell that already in the intestines and some may be reabsorbed to begin the cycle again.

Toxicity

Acute toxicity is not common with oxytetracycline. Oral therapy in ruminants may cause indigestion during the first few days of treatment, but animals soon become acclimatized and symptoms, including diarrhoea, should slacken within a few days. In small animals, nausea, vomiting and diarrhoea, with lethargy, are the main signs, but toxicity normally results only from large doses and these symptoms are not common.

Tetracyclines are deposited in growing teeth and bones and they should not be used in young animals.

Action

Oxytetracycline is bacteriostatic at low therapeutic levels but in higher concentrations may be bactericidal. The effect, as with all antibiotics, must also depend on the sensitivity of the organism. The method of action is as for all tetracyclines and includes protein synthesis prevention and chelation. The action is not appreciably affected by blood and tissue breakdown products, but in vitro, tests have shown that most antibiotics are less effective against mastitis organisms in milk media. Oxytetracycline is one of these antibiotics; penicillin is far less affected than the others.

Oxytetracycline is synergistic with polymyxin, and most formulations for topical application make use of this phenomenon.

Antibacterial Spectrum

Oxytetracycline is a broad-spectrum antibiotic, both in vitro and in vivo, and is active against a large number of Gram-positive and Gram-negative organisms. Species of streptococci, staphylodocci (micrococci), anthracoids, *Pasteurella*, *Brucella*, *Corynebacteria*, *Erysipelothrix*, coliforms and salmonellae, are sensitive. Some rickettsiae and the larger viruses are also affected and activity against other organisms such as *Pseudomonas*, *Proteus* and *Klebsiella* species is useful but less than other antibiotics such as polymyxin. Though it is effective against some protozoa, oxytetracycline has no activity against yeasts, moulds, and other fungi; this fact is of considerable importance in human medicine, for proliferation of yeasts and fungi has frequently followed broad-spectrum antibiotic therapy and has necessitated further treatment to overcome these secondary invaders. Nystatin (another antibiotic) has prevented this happening when it has been combined with broad-spectrum antibiotics. Such a situation has not so far been reported in animals but clinicians should bear it in mind if broad-spectrum antibiotics are used for prolonged periods.

Bacterial Resistance

It was originally hoped that resistant strains would be slow to appear but that hope has not been fulfilled and resistant strains are found in all domestic animals.

Organisms found resistant to any antibiotic containing the tetracycline nucleus will prove to be resistant to the rest of the group though individual variations of sensitivity do occur. This was clearly demonstrated in 1955. Tetracycline itself may occasionally prove effective where the chloro- and oxy- derivatives have failed, but in principle, where one tetracycline has failed, further treatment should be with non-tetracycline antibiotics or sulphonamides (*B.P.C.*, 1954, Addendum. 1957).

Diseases and Oxytetracycline Therapy

The tetracyclines have similar antibacterial spectra and are used for practically the same purposes. There are, however, differences in degree of action and also in methods of administration.

The diseases which have responded to oxytetracycline are very numerous: some of the more important ones are mastitis due to *E. coli.*, streptococci and staphylococci, transit fever, white scour in calves and enteritis in pigs, strangles, actinomycosis and actinobacillosis, infectious keratitis in cattle, otitis, and secondary invaders in the virus diseases of small animals. Most of the common bacterial diseases of poultry also respond.

Clinical Applications

Mastitis

It is considered that penicillin is the most effective and economical antibiotic for the treatment of streptococcal mastitis, so that it is generally recommended that oxytetracycline be used for mastitis associated with staphylococcal or *E. coli* infections.

Intramammary injections are given at 24-hr intervals, using a dose of 400 to 450 mg per quarter.

Where mastitis is associated with a bacteraemia, local treatment is supplemented with parenteral therapy.

Coliform–*Salmonella* Group Infections

White scour in calves was shown to respond to two doses of up to 0·5 g orally given with an interval of 24 hr. Prophylactically, 0·25 to 0·5 g given on each of the 1st and 2nd days of life will give protection against sensitive strains.

Enteritis in pigs has also responded favourably to oral doses of 50 to 100 mg for suckling pigs given three times at intervals of 12 hr, but this necessitates handling and takes a great deal of time as does the intramuscular injection of 50 mg.

Heavy losses in turkeys due to *Salmonella* infection showed a marked decrease following medication of the food for 5 days at the rate of 250 g per ton, and the drinking water for 3 days at the rate of approximately 710 g (2½ oz.) in 45·5 litres (10 gallons). Food consumption/ returned to normal in 4 to 5 days.

Poultry Disease

Blue comb of turkeys has been successfully treated by medication of the feed (100 ppm in the mash) and drinking water (5 ppm). This gave results within 48 hr, when mortality ceased.

CRD (Chronic Respiratory Disease) in chickens, which is associated with a pleuropneumonia-like organism (PPLO) and similar to, if not the same as, aerosaccitis of turkeys responds well to oxytetracycline therapy. In the former disease 2 ml or a 2·5 per cent oil suspension of oxytetracycline was adequate when given subcutaneously and eliminated symptoms within 7 days. In experimental aerosaccitis, however, 1 part of oxytetracycline in 630 parts of mash was required prophylactically, though less might be required in the field. Infectious sinusitis in turkeys, also related and possibly caused by the same organism as CRD, responds to injection of the mineral oil suspension directly into the infected sinus. It does not respond to oral medication. Contagious catarrh or coryza is also susceptible to subcutaneous doses of oil suspensions, 2 ml being adequate for pullets but 4 ml is required in cockerels.

Erysipelothrix infection in turkeys, shown experimentally to be susceptible to oxytetracycline and benzathine penicillin, along with penicillin and streptomycin combinations, responded without interference with the development of immunity. Oxytetracycline at the rate of 100 mg per kg as an oil suspension, or benzathine penicillin at the rate of 160 000 units per kg in single doses, were both highly effective experimentally.

Anthrax

Over 200 cases of this septicaemic disease which followed vaccination in cattle and horses have been treated. The cases were treated in three groups; penicillin, oxytetracycline and a combination of the two were used in the respective groups. Those treated solely with oxytetracycline appeared to respond more rapidly and more frequently.

Pasteurellosis

Transit fever, and haemorrhagic septicaemia of calves respond very well to oxytetracycline. Ninety-six per cent recoveries from about 4500 calves treated with one or sometimes two treatments were reported.

Experimental fowl cholera in young chickens due to *Past. aviseptica* inoculation has responded to oxytetracycline given intramuscularly. Good results were obtained by single intramuscular injections of an oily suspension at the rate of 25 mg per kg body weight given 6 hr after inoculations. It was more effective in older birds, giving good results on injection 24 hr subsequent to culture inoculation. It was also pointed out that other antibiotics and sulphonamides were less effective and furazolidone ineffective. Sulphonamides were effective in controlling the disease when administered prophylactically, beginning before inoculation; sulphamerazine and sulphadimidine were used at the rates of 0·2 per cent in the mash or 0·1 per cent in drinking water.

Leptospirosis

All tetracycline derivatives are excreted in the urine and bile in high concentrations. It would therefore be expected that *L. icterohaemorrhagiae* and *L. canicola* infections would be highly susceptible to tetracycline treatment. This was confirmed but early treatment is necessary in view of the structural damage which these organisms cause. In cattle, high dosage with intravenous administration is indicated. *Leptospira pomona* carriers in *pigs* have been cleared by oxytetracycline in the food at the rate of 500 g per tonne. In *small animals* the tetracyclines, and in particular oxytetracycline because of the higher levels achieved when compared with chlortetracycline (*U.S.D.*, 25th edn, 1955, p. 953), are superior to other antibiotics but should be given as early after infection as possible and in maximum doses.

Fusiformis Infections

Foot rot in sheep responds to many treatments providing adequate preparation of the feet is effected before the therapeutic is applied. Chloramphenicol aerosol sprays have been widely used, but oxytetracycline is also effective when applied topically in the form of an ointment or aerosol spray.

Foul-in-the-foot in cattle has also responded following intravenous administration of oxytetracycline.

Actinomycosis and Actinobacillosis

Oxytetracycline has been used in both these conditions with favourable responses. Good results in actinomycosis followed the intravenous administration of 2 g daily for 4 days when combined with penicillin and streptomycin. One case of actinobacillosis which had failed to respond to other treatment, made a gradual recovery after three doses of 2 g administered intravenously at intervals of 24 hr. In both types, oxytetracycline should only be used after normal therapy has failed—for economic reasons and because of the possibility of creating tetracycline resistance, though as yet there is no evidence of this arising.

Anaplasmosis

Oxytetracycline given in 2·5-g doses intravenously to ten infected cattle as soon as a temperature rise occurred was responsible for complete recovery in each case. Work in 1957 showed that tetracycline in daily intramuscular doses at the rate of 10 mg per kg body weight, when continued for 10 days, eliminated infection from thirty carrier cattle.

Non-specific Conditions

Oxytetracycline is very useful in all species as an eliminator of the secondary invaders which are usually found in virus diseases. This antibiotic has therefore been widely used either orally or parenterally in the treatment of equine influenza, virus pneumonia of pigs, infectious panleucopenia of cats (cat distemper), and the distemper complex of dogs (including *Brucella bronchisepticus* infections).

Pneumonia or bronchitis in all species and tonsillitis in dogs usually respond. Cystitis in small animals, and metritis in cattle, mares, dogs, and cats, all respond to oxytetracycline therapy following local/or parenteral and oral administration.

Oxytetracycline has been used successfully to restrict infections following abdominal surgery. In the antibiotic sterilization of bulls semen it was found that 20 μg per ml of diluted semen was adequate and gave good results.

Topical Application

The two main conditions for which oxytetracycline is effective externally is in the treatment of infectious keratoconjunctivitis (New Forest disease) in cattle and sheep, and external otitis in the dog.

Oxytetracycline is also useful in many pyogenic skin infections, and in infected wounds or burns.

Foot-rot in cattle and sheep is mentioned earlier in the chapter.

Dosage

Official Preparations

The *British Pharmacopoeia* and the *British Veterinary Codex* both contain monographs on oxytetracycline dihydrate and oxytetracycline hydrochloride. The *British Pharmacopoeia* specifies in addition a tablet containing oxytetracycline dihydrate.

The *United States National Formulary* includes an oral suspension containing oxytetracycline dihydrate, and capsules, an injection and an ophthalmic solution containing oxytetracycline hydrochloride.

Pessets and Tablets

Pessets 500 mg
Tablets 200 and 100 mg

Pessets	*Large animals*
1	per 50 kg body weight Repeat twice daily.
1–2	For intrauterine use.
Tablets	200 mg *Large animals*
2	per 50 kg body weight Repeat twice daily

Terramycin Q Injectable Solution

Terramycin Q is a sterile solution containing the equivalent of 50 mg oxytetracycline per ml as the magnesium complex.

Dosage levels as for injectable solution.

Injectable solution: Given by subcutaneous, intramuscular or intravenous routes. In horse, intramuscular and intravenous routes only.

Horse and foal: 2–5 mg/kg body weight
Cattle and calf: 2–5 mg/kg body weight
Sheep and lamb: 2–7 mg/kg body weight
Young and adult pig: 2–9 mg/kg body weight
Dog and cat: 7–11 mg/kg body weight

III. TETRACYCLINE HYDROCHLORIDE

This is a broad-spectrum antibiotic with similar but not identical activity to that of oxytetracycline and chlortetracycline, which are its derivatives. It is official in the British and United States Pharmacopoeias and the *British Pharmaceutical Codex* 1954 *Addendum* 1957. Tetracycline Hydrochloride must contain not less than 900 units per mg.

History

This antibiotic was discovered independently and in two different ways by three groups of research workers in 1953. Two of the groups arrived at it by the study of degradation products and derivatives of oxy- and chlortetracycline, whilst the third group produced the compound biosynthetically. It is known commercially either as Achromycin or Tetracyn, and is used as the hydrochloride. (*Note:* Achromycin should not be confused with puromycin, an antibiotic produced biosynthetically by *S. albo-niger* and originally named Achromycin.)

Manufacture

Tetracycline is produced commercially by catalytic reduction of the oxy- or chlor-tetracycline derivative. The hydrochloride is obtained by acidifying a solution of the base in n-butyl-alcohol.

Chemistry

Both the amphoteric base and the hydrochloride are yellow, odourless, crystalline powders. The base is slightly soluble (1:2500) and the hydrochloride fairly soluble (1:10) in water. Both base and hydrochloride are almost insoluble in organic solvents but dissolve readily in dilute acids or alkalis. The hydrochloride is slightly hygroscopic and solutions become turbid due to hydrolysis and precipitation of the base.

The graphic structure has already been given (see 'Oxytetracycline').

Stability and Pharmacology

With a solubility slightly less than oxytetracycline (*q.v.*) tetracycline is more stable in solution; it is superior to chlortetracycline in both these aspects.

Tetracycline may be given by all the normal routes but absorption from the gastro-intestinal tract is better than from an intramuscular injection. Blood levels are slightly higher and are maintained longer than those following equal doses of the other two members of the group. Distribution is similar though higher levels are achieved in the cerebrospinal fluid when the meninges are abnormal; urine and bile excretion is high. The antibiotic passes the placental barrier and achieves milk levels of one-quarter to two-thirds that of the blood in humans. Blood and tissue fluids do not inactivate tetracycline.

Resistance to tetracycline occurs in the same way as with the other tetracyclines but, though cross-resistance does occur, tetracycline will occasionally give a good result where the other two have failed. However, where resistance has occurred to one of the tetracyclines, an entirely unrelated antibiotic should be used in preference to another of the tetracycline group. According to experiments in vitro tetracycline is synergistic with streptomycin but usually antagonistic to penicillin when tested against staphylococci. There was no interaction with chloramphenicol or erythromycin.

Actions and Uses

The effects of tetracycline are similar to those of oxytetracycline and chlortetracycline, but variations do occur and tetracycline may prove on the whole to be slightly more active than either of the other two. Toxic side actions are similar but occur less frequently. The number of reports on the clinical use of tetracycline are increasing but much clinical investigation still remains.

Pasteurellosis: transit fever in cattle, sheep and pigs responded to two intramuscular injections at the rate of 2 mg per lb (4·4 mg per kg) body weight given with a 24-hr interval. Eleven out of twelve cows, twenty-one of twenty-two pigs, and eleven of fifteen sheep recovered.

Vibriosis in sheep was treated with single 1-, 0·5- and 0·25-g doses intramuscularly but, as with oxytetracycline and dihydrostreptomycin, it did not influence abortion, even though blood levels were maintained for a period three times longer than with oxytetracycline and the vibrios were susceptible in vitro.

Anaplasmosis carriers were very successfully treated with tetracycline at the daily rate of 11 mg per kg body weight by both intravenous and intramuscular injection: intramuscular treatment daily for 10 days was recommended.

Feline influenza virus did not attack chick embryos even when administration of tetracycline was delayed until 12 to 24 hr before the embryo would normally be expected to die following experimental infection. However, once the virus was removed from the antibiotic's influence it again multiplied freely, indicating a virustatic action. This emphasizes the need for supportive treatment in clinical cases to assist the animals own system to destroy the virus.

Goat pneumonia virus also responds to tetracycline, and also contagious pleuropneumonia of goats.

Dosage

The daily intramuscular dose of tetracycline recommended commercially consists of:

Horse, cow, sheep and pig: 2·2–4·4 mg per kg
Dog and cat: 4·4–11 mg per kg

Doses should be diluted to a strength of 50 mg per ml and doses of more than 0·5 g total should be divided and injected into several different sites. All doses should be given by deep intramuscular injection.

Official Preparations

The *United States Pharmacopoeia* includes tetracycline base, the hydrochloride, capsules, tablets and an oral suspension. The *British Pharmacopoeia* and the 1957 Addendum of the B.P.C. 1954 recognize the hydrochloride only, but the B.P. also recognizes an intravenous and an intramuscular injection, tablets and capsules.

These forms are available commercially; an intravenous injection buffered with ascorbic acid, an intramuscular injection, a 3 per cent topical ointment, an eye ointment and eye drops (as a powder for solution when required), and several flavoured oral preparations are all suitable for veterinary use though not specifically designed for veterinary purposes.

32

Antibiotics of Miscellaneous Structure

CHLORAMPHENICOL (CHLOROMYCETIN, ANIMYCETIN)

This antibiotic is notable because it was the first to be synthesized and manufactured synthetically. It is official in the United States and British Pharmacopoeias and in the *British Veterinary Codex*. Samples must contain not less than 90 per cent of the pure substance.

History

The discovery of chloramphenicol was reported in 1947 by Ehrlich and co-workers following intensive study of the organisms isolated from about 6000 soil samples by Burkholder, a botanist at Yale University. A new species of actinomycete, named *A. venezuelae* after the origin of the soil sample from which it was cultured, was found to produce a highly active antibiotic to which was given the trade name of Chloromycetin by the commercial concern in whose laboratories the work was done. In 1949 the chemical structure and the chemical synthesis were announced along with manufacturing methods, making chloramphenicol the first antibiotic to be manufactured synthetically on a commercial scale.

Manufacture

The biosynthesis from *Streptomyces venezuelae* used initially was similar in principle to that used for the manufacture of other antibiotics. Later the compound was manufactured by chemical synthesis involving ten major reactions achieved in about thirty stages, and commencing with a complex compound *para*-nitrobromoacetophenone. Other methods have been published.

Chemistry

Chloramphenicol, whether manufactured synthetically or biosynthetically, is a white, but sometimes slightly grey or yellow, fine crystalline powder. It has a bitter taste and is only slightly soluble in water (1:400) but soluble in most of the organic solvents.

The *structure* of chloramphenicol is relatively simple, the compound being a derivative of nitrobenzene and of dichloroacetic acid. It is unique, in chemicals of natural origin, in having this aromatic nitro-grouping in its molecule. Defined as D-(—)-threo-2-dichloroacetamido-1-*p*-nitrophenyl-1:3-propanediol, chloramphenicol appears graphically as:

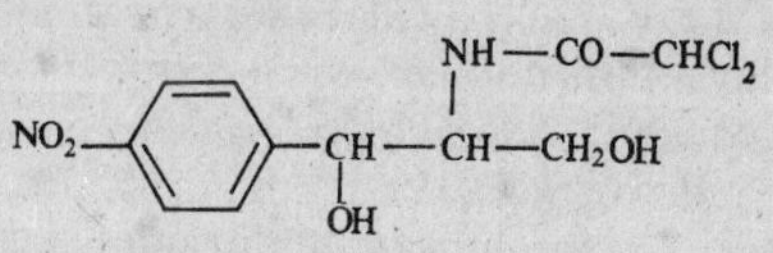

There are four stereoisomers of chloramphenicol but only the D-threo form has antibacterial activity, so that in the formation of derivatives such as the palmitate and cinnamate esters the stereochemical configuration of the propanediol side chain must be retained. The *para*-nitrophenyl group may, however, be changed to the *ortho*- or *meta*-nitro compounds or even be replaced by halogens without eliminating the antibacterial effect, though there is some reduction. It is thus the propanediol D-threostereoisomer which is vitally concerned with the antibacterial activity. The aromatic nitro-group is important in that it represents the possibly toxic portion of the molecule whilst having only a slight effect on the antibacterial properties. It has also been suggested that the —CO . NH group may be implicated in the depression of bone marrow haemopoiesis.

Stability

Chloramphenicol is more stable than either oxy- or chlortetracycline, and the dry powder is stable in air but should be kept in tightly closed light proof containers.

1. Moisture. Aqueous solutions in distilled water will keep for at least 1 month at normal room temperatures. An aqueous solution is unaffected after boiling for 5 hr, providing pH 9 is not exceeded.

2. pH. Solutions are relatively stable between pH 2 and pH 9, but strongly alkaline solutions cause rapid decomposition.

3. Temperature. The powder and solutions are heat stable and solutions will withstand boiling for at least 5 hr.

4. Oxidizing agents. The secondary hydroxyl group is easily oxidized so that the addition of any oxidizing agent is contra-indicated. The nitro-group can be reduced to an amino-group so that reducing agents are likewise contra-indicated.

5. Enzymes. Besides oxidation of the hydroxyl group and reduction of the nitro-group, a third weakness exists in the chloramphenicol molecule which lays it open to enzyme activity. This is the dichloroacetyl side chain which is readily hydrolysed (*U.S.D.*, 2nd edn, 1955, p. 281). However, no specific anti-chloramphenicol enzymes comparable with penicillinase have been reported.

6. Miscellaneous. Monohydrogen phosphate ions, acetic acid, and citrate buffers all assist destruction of the antibiotic by catalytic action.

Pharmacology

Most of the pharmacological information available refers to man and dogs.

Absorption and Administration Routes

Following *oral* administration, chloramphenicol is rapidly and almost completely absorbed, giving very high maximum blood levels about 2 hr after a single dose. Blood levels are proportional to the dose and far exceed those attained by the tetracyclines with similar dosage. Following *intravenous* and deep *intramuscular* injections therapeutic blood levels are achieved rapidly. The absorption per rectum is also sufficient to provide therapeutic concentrations in the system.

Blood Levels

Maximum blood levels take from 2 to 4 hr to develop. Therapeutic blood levels are given as 5 to 10 μg because of the adsorption by serum proteins of a proportion of the circulating antibiotic. It is estimated that as much as 45 per cent of the chloramphenicol in the circulation is bound to protein. Blood levels are proportional to the size of the dose and the period during which the level remains over a given concentration also varies according to dose size; at any given dose the concentration is always maintained longer than the concentration given by similar doses of the tetracyclines: this applies in man to single or multiple dose of 1 g. In dogs, this maintenance of levels may be related to the fact that inactivated chloramphenicol, when excreted into the intestines is hydrolysed back into an active form and can then be reabsorbed.

Barriers

Though the diffusion of chlortetracycline is both rapid and widespread, chloramphenicol is distributed quite as rapidly and diffuses throughout the tissues more widely.

1. Blood–brain barrier. Chloramphenicol diffuses readily into the spinal fluid reaching levels between one-quarter and one-half of the blood levels.

2. The placental barrier. The antibiotic passes across the placenta and achieves foetal blood levels of up to three-quarters of those in the maternal circulation.

3. Intestinal barrier. So efficient is absorption of chloramphenicol from the intestines that for all normal purposes in small animals oral dosage is sufficient. In large animals oral dosage frequently suffices but initial levels are often established by intramuscular injections. Faecal concentrations are appreciable but below those of oxytetracycline in man and presumably in carnivores; no information is available on the faecal content of chloramphenicol in the herbivores, although concentrations expected would be well above those which depress ruminal fermentation in cattle.

4. Serous membranes. Appreciable quantities pass into the pleural cavity and into the peritoneal cavity. Chloramphenicol will diffuse into the peritoneal cavity. Chloramphenicol will diffuse into the aqueous and vitreous humours of the eye (*B.Vet.C.*, 1953) following oral administration. It will penetrate the intact cornea and does not delay corneal regeneration.

5. The milk barrier. Chloramphenicol is excreted in the milk (*B.Vet.C.*, 1953) but mastitis therapy is surer by local infusion, supported by oral or parenteral administration in severe acute cases.

Excretion

The major part of any one dose is excreted in the urine, detectable quantities appearing within 30 min of administration. Part is excreted in a hydrolysed or conjugated form and only 10 per cent is active: concentrations are below those of oxytetracycline and, except in high dosage, lower than those of chlortetracycline. Bile concentrations are also quite high but faecal levels are below those of oxytetracycline and in general about the same as chlortetracycline.

Toxicity

So far no toxic effects other than those normally associated with broad-spectrum antibiotics have been recorded. Occasional nausea and vomiting may occur in dogs and cats but this is far less common than with the tetracyclines. In large animals, some interference with gastro-intestinal flora occurs but is of no more significance than with other similar antibiotics.

In man, blood dyscrasias due to suppression of bone marrow function following prolonged administration, limit the use of this antibiotic.

Action

Chloramphenicol is highly bacteriostatic and can be bactericidal. Its mode of action is an interference with vital protein synthesis at the ribosomes.

Antibacterial Spectrum

The most important aspect of the antibiotic effect of chloramphenicol is its action against all rickettsiae: in addition, it is active against both Gram-positive and Gram-negative bacteria and against some viruses: spirochaetes including leptospirae are sensitive only to relatively high concentrations.

In vitro studies have shown that the blood levels of 5 to 10 μg per ml are adequate for controlling:

B. anthracis
E. coli–Salmonella group
Brucella species
Corynebacterid species
Erysipelothrix species
Pasteurella species
Proteus vulgaris
Ps. aerogenes (*pyocyanea*)
Staphylococci and Streptococci

Clinical response does not follow entirely according to this list.

Bacterial Resistance

Resistance to chloramphenicol developes in a 'stepwise fashion', the maximum increase in one step recorded up to 1951 being eightfold compared with as much as one-thousandfold with streptomycin. In clinical investigations the increments of resistance have occurred at irregular and unpredictable intervals during treatment. In general the number of strains of susceptible species which do not respond is very small. Of eighty-eight strains of organisms isolated from diseased animals and birds, and representing fifteen different species, not one strain showed resistance. Resistance to the tetracyclines does not mean that there will also be resistance to chloramphenicol.

Because of the value of chloramphenicol in the treatment of *Salmonella typhi* infection in human beings, it has been recommended in the U.K. that this drug be used only for very specific problems in the animal field. Its use is now discouraged for oral and intramammary treatment.

Clinical Applications

Chloramphenicol alone, and in combination with dapsone, is used for mastitis treatment. It has the same spectrum as the tetracyclines, and can be used for the treatment of cases of staphylococcal and *E. coli.* mastitis.

Chloramphenicol has been shown to be of value in transit fever, calf scours, *Salmonella* infections (for which it is probably the most useful antibiotic), various secondary invaders in virus diseases, kennel cough, infectious bovine keratitis. It has also been used in conjunction with many other drugs in intramammary injections.

Pasteurellosis

Spectacular results with chloramphenicol in the treatment of transit fever (haemorrhagic septicaemia, shipping fever) were reported in 1952, for 10 g intravenously brought the temperature of a 363 kg (800 lb.) bull down 6°F in 18 hr; a further 10 g returned the animal to normal 24 hr later. As little as 2 g orally in a single dose brought about recovery of a heifer. The majority of the animals recovered in 24 hr.

There is a large variation in the sensitivity of *Pasteurella* strains to drugs. In addition to this strain sensitivity, large differences between in vitro and in vivo susceptibility have been reported, so that variable results might be expected.

Fusiformis Infections

Calf diptheria has been successfully treated by single daily intramuscular injections given at the rate of 4·4 mg per kg body weight for 2 to 3 days combined with local cleansing and dressing with chloramphenicol. Previously the best, but very slow, recovery had been achieved following prolonged treatment with triple sulphonamide compounds combined with high dose penicillin injections.

Foot-rot in sheep has provided chloramphenicol with one of its greatest successes in veterinary medicine.

Chloramphenicol may be applied after paring the feet, either by painting the lesions using a stiff brush, or by using an aerosol spray. Animals should be kept on concrete or wooden floors until the tincture has dried and then turned onto fresh pasture on which no sheep have grazed for at least 2 weeks, but preferably longer. All feet should be reinspected and, if necessary, retreated at 14-day intervals—one, two, or with very severe and chronic lesions, three treatments may be required.

The efficacy of all treatment depends on efficient paring of the hoof, dry conditions. segregation of 'clean' and affected sheep, and the use of 'clean' ground after treatment: these details are aimed at applying the treatment directly to infected tissue, allowing it to heal efficiently and preventing reinfection.

This form of treatment may be discouraged because of the danger of the development of chloramphenicol resistance.

Virus Pneumonia

This condition was reported to respond clinically to chloramphenicol injected twice daily. The main effect is probably obtained against the secondary organisms.

Coliform–*Salmonella* Infections

The clinical response to the treatment of white scour is variable.

Scouring in *pigs* was also shown to resist the tetracyclines but, in many instances, to respond to streptomycin or chloramphenicol. These cases were considered to be *E. coli* infections.

In *salmonellosis*, response is also varied, but chloramphenicol is a useful agent against Salmonella infections. However, furazolidone is preferred in poultry, but clinical reports do indicate that chloramphenicol is relatively successful in mammals; 26·4 to 35·2 mg per kg body weight intravenously in calves affected with *S. dublin* gave good results, and good clinical, but less successful bacteriological, cures were reported in 1953, when twenty of twenty-five cows responded to one intravenous injection and a further three to repeat injections.

Miscellaneous Diseases

Kennel cough and tonsillitis in dogs, frequently due to *Brucella* (*Haemophilus*) *bronchiseptica*, whether concurrent with distemper or not, frequently, and in some cases with spectacular effects, responds to chloramphenicol when administered as capsules or as a palmitate suspension.

Metritis in dogs and in cattle has responded to local treatment and, in some cases in bitches, to oral therapy at the daily rate of 33 mg per kg body weight in divided doses.

Cystitis and other urinary infections respond to chloramphenicol oral treatment, but leptospirosis is in the main resistant, although some cases will respond.

Local application. *Infectious bovine keratitis* due to a Rickettsia responds very well and is considered by some to give the best results. It is effective when applied as an ointment in cattle and sheep if applied once or twice daily for 7 to 10 days. Animals under treatment should be housed and in darkness if possible.

Other eye conditions, wounds, pyogenic infections, and external otitis, have all responded to local therapy, either with a 1 per cent eye ointment or an 0·1 to 0·2 per cent saline or borate buffer aqueous solution.

Dosage

Oral administration:

Calves and foals: 0·5 g two or three times daily

Piglets and lambs: 0·25 g two or three times daily

Dogs: Up to 165 mg per kg body weight daily in three equal and equally spaced doses

Cats: 0·25 g once or twice daily

Capsules or a palmitate suspension may be used.

Intramuscular administration:

Chloramphenicol sodium succinate is used as a 40 per cent solution.

For intravenous injection it is usual to use the succinate as a 10 per cent solution. Dose:

Large animals: 2–4 mg per kg

Sheep, goats, pigs, foals and calves: 4–10 mg per kg

In all severe cases, the higher dosage rates may be used until the acute symptoms have subsided, when the dosage may revert to the lower rates.

LINCOMYCIN (LINCOCIN)

History

Lincomycin was discovered as a result of a search carried out on soil samples received from the U.S.A. and a number of overseas countries. One sample from the area of Lincoln, Nebraska, yielded a species of *Streptomyces* which had new antibiotic activity. Details of the new material were announced by Mason et al. (1962).

Manufacture

Lincomycin is produced by *S. lincolnensis* var. *lincolnensis* n.sp. grown in a fermentation medium consisting of starch, molasses, corn steep liquor, and Wilson's No. 159 animal peptone liquor to which 4 g of $CaCO_3$ per litre is added prior to sterilization.

Chemistry

Crystalline lincomycin hydrochloride is obtained by adding acetone to a concentrated solution of purified lincomycin in aqueous hydrochloric acid, and is recrystallized from water by the addition of acetone.

Crystalline hydrochloride is very soluble in water and soluble in acetone, ethanol and methanole. It has the following structural formula:

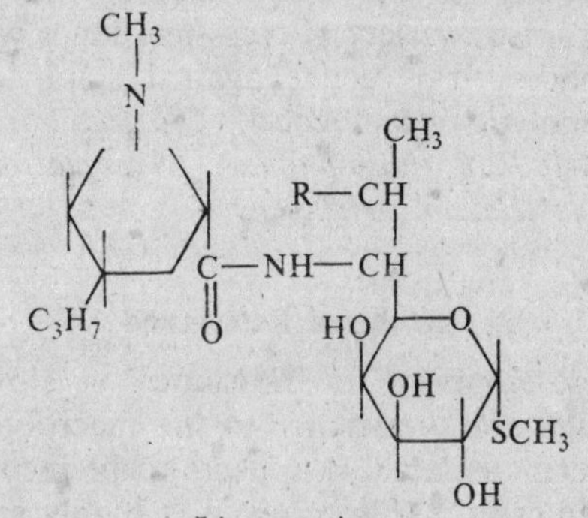

Lincomycin

It is chemically unlike any other major antibiotic.

Pharmacology

Lincomycin is readily absorbed when given by mouth and is also completely absorbed from intramuscular sites. Good levels for activity against Gram-positive organisms are maintained for 6–8 hr. The drug is widely distributed in the body and does not appear to be concentrated in any particular organ. Tissue level studies indicate that bile is an important route of excretion.

In the dog after a single oral dose faecal excretion amounted to 77 per cent of the dose and urinary excretion to 14 per cent. When a single intramuscular dose is given, faecal excretion equals 38 per cent of the dose and urinary excretion 49 per cent. Lincomycin has also been shown to be excreted in the milk of lactating cows and goats.

Toxicity

Lincomycin appears to be a safe compound, and most authors have reported no side effect, apart from diarrhoea. The acute LD_{50} intraperitoneally in mice is 100 mg/kg and orally in rats 15 645 mg/kg. It has been tolerated in dogs at doses of up to 300 mg/kg per day for periods of up to a year.

Antibacterial Spectrum

Lincomycin closely resembles erythromycin in its activity against Gram-positive organisms. It also resembles the macrolides in showing a characteristic variety of cross-resistance with the macrolides. It is active against staphylococci, β-haemolytic streptococci, *Clostridium tetani*, *Erysipelothrix rhusiopathiae*, *Mycoplasma* spp. and *Clostridium perfringens*.

Bacterial Resistance

The development of resistance is slow and although cross-resistance to the macrolides has been demonstrated, it appears to be incomplete and indicates that lincomycin is highly selective in its action.

Clinical Application

Lincomycin is indicated for infections caused by Gram-positive organisms particularly streptococci and staphylococci.

It has been demonstrated to be effective in dogs and cats against respiratory infections and infections of the skin and adjoining tissues. In pigs it has been used for the treatment of infectious arthritis due to staphylococci, streptococci, *Erysipelothrix* and *Mycoplasma* spp.

Dosage

Oral therapy

Dogs and cats 20 mg/kg body weight every 12 hr

Intramuscular and subcutaneous

Dogs and cats 20 mg/kg body weight once a day

Intravenous

20 mg/kg body weight once or twice a day

Intramuscular

Pigs 4–10 mg/kg once daily for 3 to 7 days

Preparations

Tablets. 100 mg tablets of lincomycin hydrochloride monohydrate.

Sterile solution. Each ml contains lincomycin hydrochloride monohydrate equivalent to 100 mg lincomycin base.

Bibliography

MASON, D. J., DIETZ, A. & DE BOER, G. (1962) *Antimicrob. Agents Chemother.*, 554–9.

33

Antibacterials

I. THE SULPHONAMIDES

Introduction

The sulphonamides are a group of complex synthetic organic chemical compounds with chemotherapeutic activity. They have a common chemical nucleus which is essential for antibacterial activity. This nucleus, graphically written as in (i),

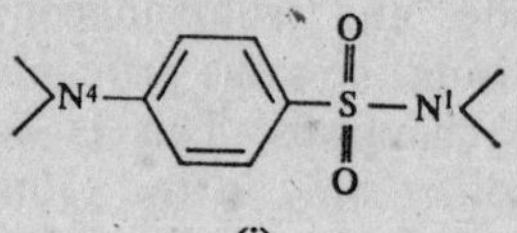

(i)

is very closely related to *para*-aminobenzoic acid, an essential member of the vitamin B complex which can be written as in (ii).

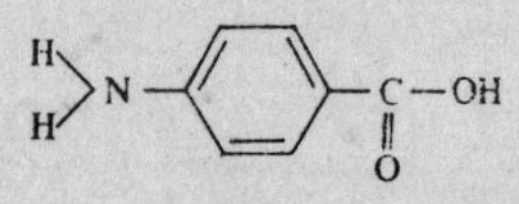

(ii)

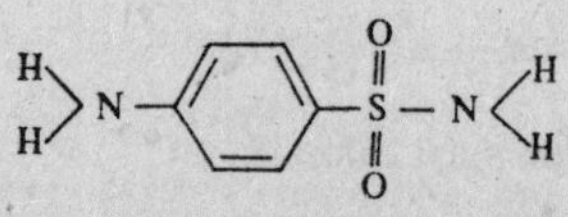

(iii)

This similarity can be emphasized by reference to the simplest member of the sulphonamide group, sulphanilamide or *para*-aminobenzene sulphonamide (iii). Emphasis is required, for it is this similarity which is considered to lead to substitution and interference in cell metabolism by the sulphonamides.

The group can be divided into two main sections according to the site of therapeutic action. These divisions may be termed: (1) the systemic sulphonamides, which are absorbed reasonably well from the intestines, and (2) the so-called 'gut active' sulphonamides which are, in the main, poorly absorbed from the intestines and therefore exert their effect only on the intestinal flora and possibly those organisms situated in the superficial cells of the mucosa.

The more important members of these divisions are set out in Table XV.

History

Sulphanilamide was first synthesized in 1908 by Gelmo, but it was not until 1935, when Domagk, of I. G. Farben Industrie, discovered the ability of certain azo-dyes related to sulphanilamide to protect mice against challenges with streptococci, that attention became concentrated on these products of the dye industry. Prontosil and Neoprontosil, the latter a water soluble red dye which stained everything it contacted, were the first of these azo-compounds to be used therapeutically

Table XV. The members of the two divisions of sulphonamides

Systemic sulphonamides
Sulphanilamide
*Sulphacetamide
Sulphapyridine
Sulphadimidine (Sulphamethazine)
Sulphathiazole
Sulphamerazine
Sulphadiazine
Sulphafurazole (Sulfisoxazole)
‡Sulphaquinoxaline
'Gut active' sulphonamides
†Sulphaguanidine
Phthalylsulphathiazole
Succinyl sulphathiazole
Phthalylsulphacetamide

* Sulphacetamide is similar to sulphanilamide in activity but in Great Britain is rarely used for its systemic effects.

† Sulphaguanidine is absorbed to the extent of 50 per cent or more of any single dose and systemic activity can also be anticipated.

‡ Used only in poultry.

and with good effect. It was stated during the same year in Paris that the activity of the prontosils and other azo-dyes was due to their release of sulphanilamide in the body. This work in Paris was followed in 1936 by work in England which showed that sulphanilamide was less toxic than Prontosil and yet equal in activity.

Since 1936 a great amount of investigation has produced many sylphonamides but only a few have been of therapeutic value, of which a number were British discoveries. Amongst these British discoveries sulphapyridine (M. & B. 693), sulphathiazole (M. & B. 760), and sulphadimidine (Sulphamezathine) are outstanding. In recent years the sulphonamides have been largely displaced therapeutically by the antibiotics. Nevertheless, interest in the sulphonamides continues, especially as synergism and a delay in the development of antibiotic resistance has been shown to occur with some sulphonamide–antibiotic combinations; the concurrent use of sulphadiazine and chlortetracycline or chloramphenicol in the treatment of *Haemophilus meningitis* in man is an example. Sulphonamide research is now concentrated more on finding means of reducing the toxicity of the group or on finding new sulphonamides with reduced toxicity; sulphafurazole (sulfisoxazole) with its low incidence of crystalluria is an example. Another angle of approach is typified by the production of sulphamethoxypyridazine, which has produced significant blood levels in man over 4 days after a single 4 g oral dose. Prolonged blood levels have since been reported in the pig, dog and cat, but strangely enough not in ruminants. Sulphamethylphenazole is another recently introduced long acting sulphonamide.

Manufacture and Chemistry

The sulphonamides are all manufactured synthetically. Those in common use and without any prefix to the 'sulpha-' part of the name, such as sulphanilamide and sulphadiazine, may be synthesized using the same principles and many of the same chemicals. This is because these compounds differ only in the groupings which are substituted for a hydrogen atom of the $—SO_2 . NH_2$ group of the basic compound, sulphanilamide.

Synthesis may be represented by the following scheme:

Stage 1:

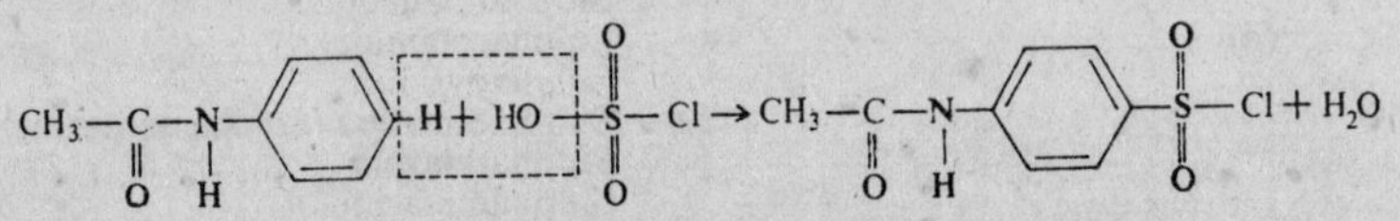

Acetanilide+Chlorsulphonic acid→*N*-Acetylsulphanilyl chloride

Stage 2:

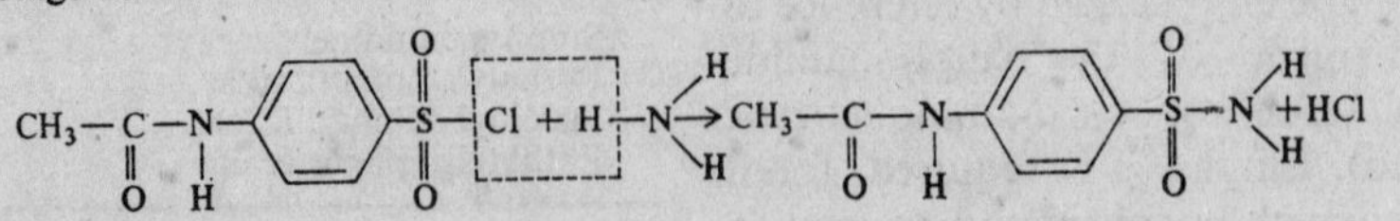

N-Acetylsulphanilyl chloride+Ammonia→*N*-Acetylsulphanilamide

Stage 3:

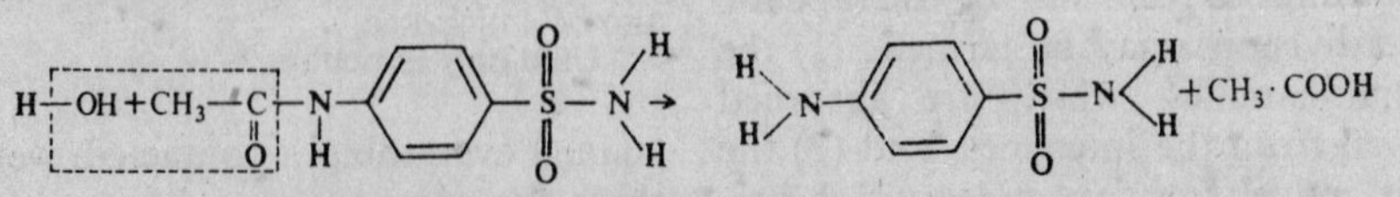

N-Acetylsulphanilamide —hydrolysis→ Sulphanilamide+Acetic acid

The other non-prefixed sulphonamides, with a few exceptions, may be prepared by using various cyclic amines instead of ammonia in stage 2. Some of these amines are set out in Table XVI. The graphic formulae of these sulphonamides can be deduced by attaching the amine minus one hydrogen atom (dotted) to the nucleus derived from *N*-acetylsulphanilyl chloride in stages 2 and 3, e.g.

Phthalylsulphathiazole and the two corresponding salts of sulphacetamide can also be prepared by this use of anhydrides and alcohol.

There are exceptions to these principles, and other methods of manufacture are used, e.g. sulphacetamide is prepared from *N*-acetylsulphanilamide by acetylation of the sulphonamide group using acetylchloride: alkaline hydrolysis later provides a *p*-amino-group at C_4 in the benzene ring.

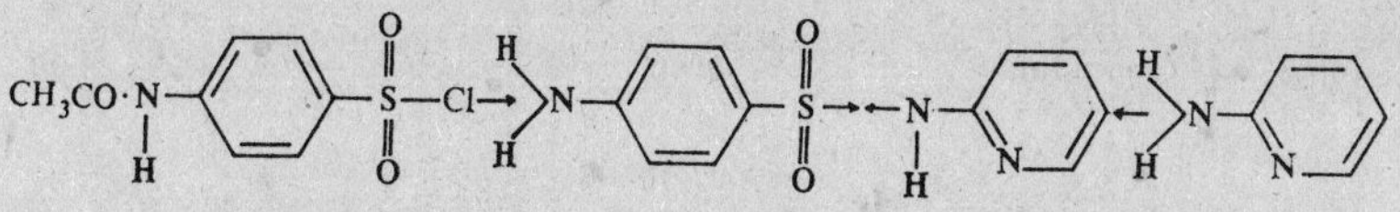

N-Acetylsulphanilyl chloride Sulpha-pyridine 2-Amino-pyridine

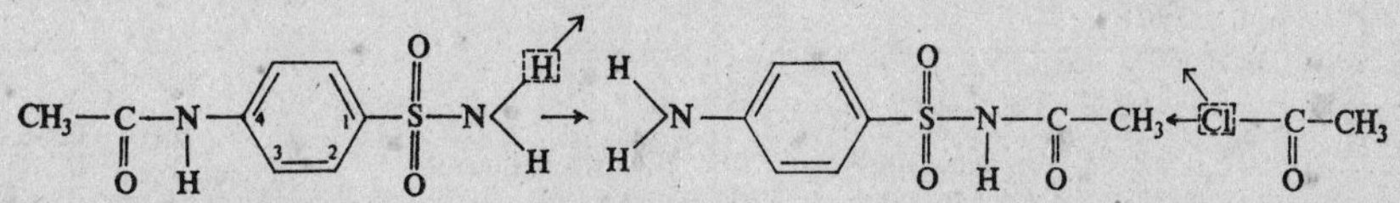

N-Acetylsulphanilamide Sulphacetamide Acetyl chloride

With sulphonamides possessing a prefix, such as succinylsulphathiazole, further reactions are necessary, involving the amino-grouping attached in the *para*-position at carbon 4 of the benzene ring.

This reaction, in the case of succinylsulphathiazole, can be achieved by the use of succinic anhydride and anhydrous alcohol. The process may be represented as follows:

Sodium Salts

The sulphonamides are insoluble as a group and therefore the soluble *monosodium* salts may be used where it is necessary to administer them by injection; these salts, following injection, split off the sodium ions releasing the parent sulphonamide. Their structure consists of a sodium ion substituted for the remaining hydrogen ion attached to N_1 (the nitrogen of the

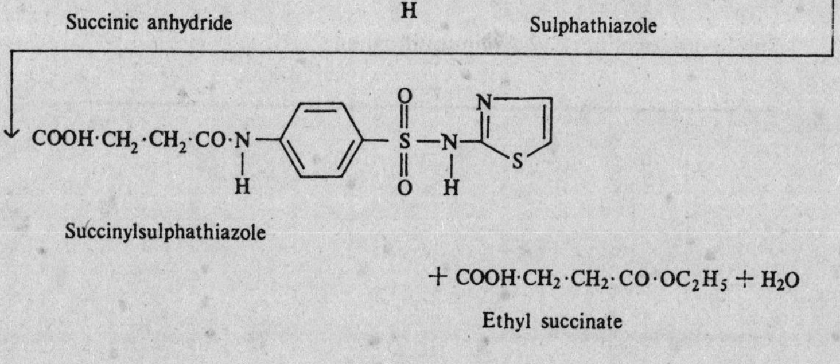

Table XVI. Sulphonamides prepared by using cyclic amines

Sulphonamide	Amine
Sulphanilamide	Ammonia 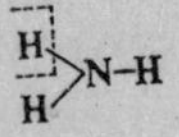
Sulphaguanidine	Guanidine 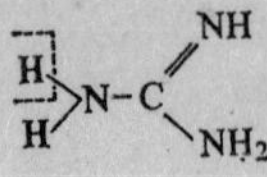
Sulphapyridine	2-Aminopyridine 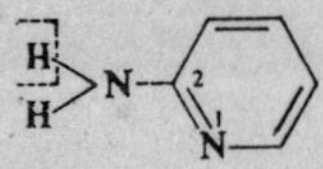
Sulphathiazole	2-Aminothiazole 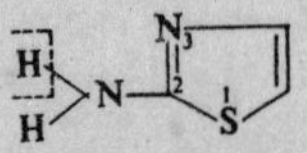
Sulphadiazine	2-Aminopyrimidine 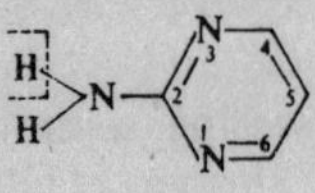
Sulphamerazine	2-Amino-4-methyl-pyrimidine 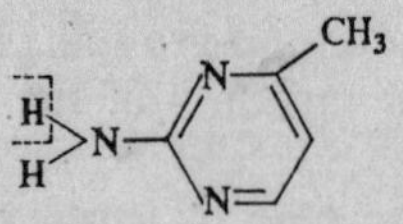
Sulphadimidine	2-Amino-4,6-dimethyl-pyrimidine 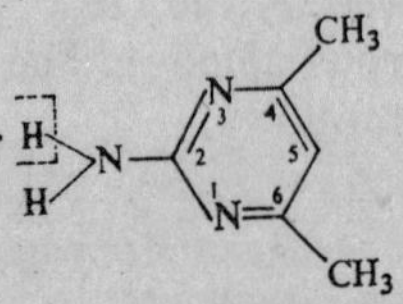
Sulphafurazole	5-Amino-3,4-dimethyl, iso-oxazole 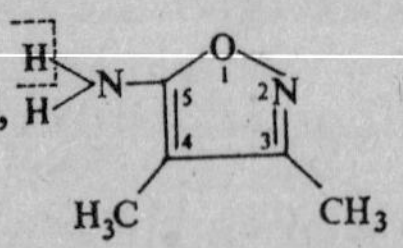
Sulphaquinoxaline	2-Aminoquinoxaline 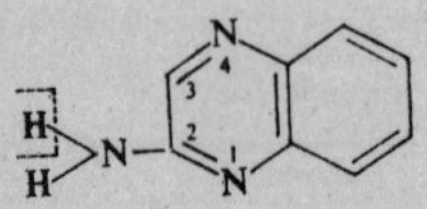

—SO_2NH_2 group of the basic sulphanilamide): e.g.

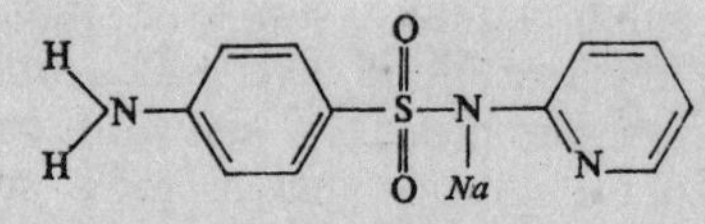

Sulphapyridine sodium

Complex *disodium salts* have also been developed with the commercial names of Soluseptasine, Solupyridine and Soluthiazole, of which the first is included in the B.P.C. as Sulphasolucin. These compounds have a pH of about 8 and may be injected by most of the parenteral routes, whereas the monosodium salts are in general too alkaline (pH 9 to 11) to be given by any route other than the intravenous one. Their structure includes a disodium disulphonated organic group attached at N_4 of the parent compound with the displacement of a hydrogen atom:

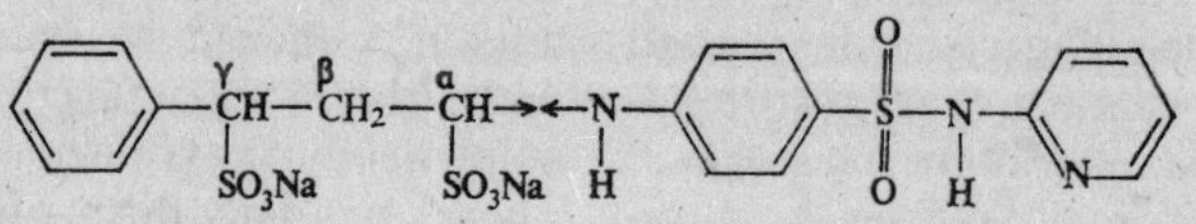

Disodium-disulphonate-phenyl-propyl →← amino-benzene-sulphonamido-pyridine, normally written: disodium 2-(*p*-(γphenyl-propyl-amino)-benzene-sulphonamido) pyridine $\alpha{:}\gamma$-disulphonate.

Chemical and Physical Properties

The sulphonamides as a group are white or off white crystalline powders which darken on exposure to light and should therefore be stored in tins or coloured glass containers. Sulphaquinoxaline differs in being yellow in colour, but the structure of others can be modified to produce a colour without affecting their therapeutic properties. This property, due to the azo dye relationship of the sulphonamides, has been used commercially to provide coloured sulphonamides to unvarying standards without the need for extra time and handling required for colouring by more normal methods. As a group they are mainly odourless but most have a bitter taste.

The sodium salts are similar and differ only in their solubilities.

All these compounds are stable in all respects other than reaction to light and may be sterilized by heat either as powders or solutions.

Solubility

The therapeutic effectiveness of the sulphonamides is dependent to a great extent on their solubility in water, for their concentration in the tissues is governed first by their rates of absorption from injection sites or intestines, and secondly by their concentration in the water present in the various body tissues. Absorption rates are roughly proportional to water solubilities as set out in Table XVII.

The sodium salts are all soluble in water, the monosodium salts achieving solubilities greater than 1:2 at body temperature.

Table XVII. Absorption rates and water solubilities

Sulphonamide	mg per 100 ml	
	In water at 37°C	In blood at 37°C
Sulphacetamide	1587	1970
Sulphanilamide	1460	1970
Sulphafurazole	530 (25°C)	—
Sulphaguanidine	100 (25°C)	—
Sulphathiazole	96	330
Sulphadimidine	75	—
Sulphapyridine	52	60–75
Sulphamerazine	31·8	—
Sulphadiazine	12·7	161
Succinylsulphathiazole	21·0 (20°C)	—

Solubilities in acid urine are similar to those for water, but in alkaline urine the solubilities may be twice as great, and in the case of sulphadiazine may be 100 times greater.

Pharmacology

The members of the sulphonamide group follow a pharmacological pattern common to all members except those which are not absorbed from the intestines, particularly succinylsulphathiazole, phthalylsulphathiazole and the corresponding sulphacetamide compounds. Differences are only of degree.

Absorption and Administration Routes

The *parent sulphonamides* are administered by oral and topical routes only. Following oral administration, absorption rates are roughly proportional to solubilities with perhaps the exception of sulphapyridine, which is slowly and irregularly absorbed in most species, and sulphadimidine which is second only to sulphanilamide and sulphacetamide in its rate of absorption. Maximum blood levels for carnivores, following single oral doses, are usually achieved in a minimum of 1 hr for the rapidly absorbed and a maximum of 8 hr, with an average of about 4 hr for those absorbed more slowly. Absorption rates vary between the species but are related in the main to the bulk of material within the intestines. Thus dogs and cats absorb the sulphonamides rapidly and almost completely, whereas cattle take much longer and the pig and horse are placed intermediately. Birds absorb sulphonamides better than other domestic species.

The *monosodium* salts may be given by intravenous injection but, with the exception of sulphadimidine sodium, it is not advisable to give them by any other parenteral route. Following intravenous injection, or deep intramuscular injection in the case of the sulphadimidine salt, maximum blood levels are rapidly achieved and these salts may therefore be used to institute therapeutic blood levels rapidly, maintenance of these levels being achieved by subsequent oral dosage. These salts may also be used to medicate drinking water but should be suitably buffered to prevent deposition.

The *disodium salts* may be used instead of the monosodium salts, but in addition may be given subcutaneously or intramuscularly. Approximately 5 ml contain the equivalent of 1 g of the parent sulphonamide although sulphasolucin contains less than the other two.

With these solutions of sodium salts the intraperitoneal route is sometimes used.

Blood Levels

As with the antibiotics, blood and tissue levels are of great importance. Blood levels, however, cannot be used for comparing the activity of different sulphonamides, nor is it possible to define an effective blood level which will apply to all sulphonamides, for, as with chloramphenicol (*q.v.*), proportions of most sulphonamides combine loosely with blood proteins. This limits the availability of the compounds to diffuse into the tissues and exert an antibacterial effect. Acetylation, which occurs in the liver and turns the sulphonamide into an inactive form, further reduces the significance of blood levels. Both these factors vary from compound to compound and from species to species. However, levels of 8 to 15 mg per 100 ml blood in man, and 5 mg per 100 ml in animals are taken as being therapeutically effective, though levels of some sulphonamides are still effective when below these standards.

Once adequate blood levels have been established, oral dosing every 8 hr in simple stomached animals and every 12 hr in adult ruminants is sufficient to maintain them, although with the long-acting compounds (such as sulphamethoxypyridazine and sulphamethylphenazole) daily dosing is sufficient to maintain adequate blood levels. Large doses are usually given to establish therapeutic blood and tissue levels initially, but subsequent doses are smaller. Increasing the dose does not increase the blood levels proportionately.

Barriers

The sulphonamides as a group diffuse very widely into the tissues, penetrating into all fluids.

Blood–brain barrier. The penetration of the blood–brain barrier is largely governed by the proportion of sulphonamide which is circulating in the blood combined with protein or in an acetylated form. This can be well illustrated by the pyrimidine derivatives: sulphadimidine when compared with sulphadiazine and sulphamerazine is acetylated more easily and in greater quantity, it is also bound to protein to a greater degree; thus, although sulphadimidine diffuses into the cerebrospinal fluid in appreciable quantities, the levels of active sulphadiazine (two-thirds of blood level) and sulphamerazine (one-half of blood level) are superior and they are preferred for the treatment of meningeal and brain infections. Of the other sulphonamides, sulphanilamide penetrates better than sulphadiazine, sulphafurazole slightly less; sulphathiazole penetrates poorly when compared with all others, including sulphapyridine. Sulphamethylphenazole has a low protein-binding capacity and diffuses into all tissue fluids in relatively high concentrations.

Other barriers. With the exception of the blood–brain barrier, all other natural barriers are passed and effective concentrations appear in all tissues, body cavities, and secretions, including urine, bile and milk; the placenta is also crossed.

Pathological barriers exist in abscess capsules and purulent foci, where vascularity is either minimal or absent.

Excretion

The sulphonamides are excreted mainly via the urine. They are excreted either in an unchanged form, or as a conjugated excretory product which has considerably reduced antibacterial activity. The conjugation takes place in the liver where N_4 acetyl derivatives are formed; these compounds are those formed in stage 2, and from which the sulphonamides were originally obtained in stage 3 of manufacture (see p. 365). The proportion of changed to unchanged sulphonamide varies with the different sulphonamides and also with the different species. In the dog and poultry the amount of circulating and excreted acetyl derivatives is extremely low: it is suggested that processes of acetylation and deacetylation occur at an equal rate so that no appreciable blood level of acetyl derivatives can be accumulated. In other species, deacetylation is the slower process and blood levels of the acetylated sulphonamides build up. The degree of acetylation is important, for with all but the pyrimidine derivatives (sulphadiazine, sulphamerazine and sulphadimidine) and sulphamethylphenazole the acetylated sulphonamides are far less soluble than their parent compounds. Also of importance is the fact that all acetyl derivatives are more soluble in alkaline urine than in neutral or acid urine: it is a common practice in human medicine to administer bicarbonates, citrates and acetates as alkalizers. A third factor of equal excretory importance is the degree of 'renal clearance', which is the amount of a sulphonamide excreted by the glomerulus and not resorbed by the tubules. These three factors, in conjunction with the excretion rate, determine whether or not urine solubilities are exceeded, and consequently whether a particular sulphonamide is safe to use. For example: sulphathiazole is rapidly excreted, up to 90 per cent passing into the urine in 24 hr, has a renal clearance of 87 per cent and an acetylation rate of 20 per cent; the acetyl derivative has a solubility in urine of about 7 mg per 100 ml at pH 5·5 (up to 40 mg per 100 ml at pH 7·5): sulphadiazine is excreted more slowly, probably because of a higher protein adsorption, only 50 per cent passing into the urine in 24 hr has a renal clearance of 31 per cent and an acetylation rate of 25 per cent, the acetyl derivative being soluble to the extent of 20 mg per 100 ml in urine at pH 5·5 (512 mg per 100 ml at pH 7·5). It will be seen that, of these two sulphonamides, sulphathiazole is far less safe to use, for at a given time more sulphonamide, with about the same acetylation rate and an acetyl derivative solubility far below that of sulphadiazine, is present in the kidney. The urine saturation level is therefore more likely to be exceeded and crystallization to occur. This renal crystallization is referred to as 'crystalluria'.

The figures quoted above refer to human sulphonamide excretion; the situation in veterinary medicine is not so serious for acetylation is not so great in animals and in herbivores the urine is normally alkaline (pH 8+). For cats and dogs, urinary alkalization is advisable.

Some sulphonamide is excreted in the bile and appreciable amounts may be found in the milk. Sulphanilamide provides a higher milk level than most sulphonamides, and though it has been said that oral sulphonamide treatment of mastitis has not been satisfactory, this sulphonamide was the first chemotherapeutic of any real value against streptococcal mastitis: until the advent of penicillin, sulphanilamide was very widely and successfully used for this purpose in Great Britain.

Faecal excretion is relatively high, as much as

10 per cent of a single dose of sulphanilamide, and more of other sulphonamides, remains unabsorbed in the intestines; if dosing is excessive or protracted, there is a tendency to cause disturbances in the intestinal flora, and there is also an irritant effect exerted by the compound itself. Continued oral use in cattle causes enteritis with scouring.

Toxicity

The acute toxicity of the sulphonamides reported so frequently in man is of far less significance in animals, possibly because symptoms are mainly subjective—nausea, temporary blindness, giddiness, etc., few of which may be noted by an owner. Skin rashes, photosensitization, haemolytic anaemia and agranulocytosis have all been reported and are due to sensitivity following previous courses of treatment.

Chronic toxicity is of more importance and assumes two forms, one associated with depression of the intestinal flora and interference with vitamin B metabolism, and the other associated with crystalluria.

Continued use of sulphonamides in ruminants results in diarrhoea, dehydration, and debility, with a loss of appetite and milk yield. In normal animals recovery follows cessation of dosage; this recovery is slower in sick animals. These toxic effects are due mainly to the direct irritation of the intestines and to the digestive disturbances caused by the suppression of ruminal flora activity and vitamin B synthesis.

In other species, depression, loss of appetite, nausea and vomiting in those species capable of this reaction, are the usual signs of sulphonamide toxicity.

Crystalluria has been reported in animals but is of less frequent occurrence than in man because acetylation, which produces less soluble, inactive compounds for excretory purposes, is not so great as in man. Nevertheless, crystalluria, with deposition of sulphonamide crystals in the tubules, pelvices and ureters, does occur and results in urinary obstruction and ulceration of the mucosa in contact with the crystals. It produces a decrease in the quantity of urine, whilst any urine passed contains crystalline and tissue casts, albumin and blood.

The causes of crystalluria are numerous, but the condition normally results because dosage has been excessive, either in quantity, frequency or duration; or because the intake of water and its excretion have been restricted or the pH of the urine has become acid. In most cases it is a combination of two or more of these factors.

Treatment consists in all species of stopping dosage, increasing fluid intake, and alkalizing the urine: the last of these is assisted by the fact that sulphonamides in urinary filtrates restrict bicarbonate resorption by the tubules. Where anuria (cessation of urination) occurs increased fluid is contra-indicated. Ureteric irrigation is used in human cases of this type.

All types of toxicity, but particularly crystalluria, are encouraged by conditions which result in dehydration with tissue fluid concentration and a decrease in urine production (oliguria); sulphonamide therapy is therefore contra-indicated in such cases.

Avoidance of crystalluria is achieved by paying strict attention to dosage rates and frequency of administration, and also by stopping administration about 24 hr after the temperature has become normal. Free access to unlimited water supplies is essential. No course of treatment should continue for longer than 7 days. The choice of sulphonamide is of the greatest importance, for those such as sulphapyridine with very low solubility should not be used where more soluble and probably more effective sulphonamides are available.

Where there is a danger of crystalluria and sulphonamides must be used, *triple sulphonamides* should be given. These are combinations of sulphonamides, frequently of the three pyrimidines which possess considerable excretory advantages over equal weights of single sulphonamides. [With combinations of sulphonamides, therapeutic activity is equivalent to the sum of the activity of the individual members, but during excretion the members each act as if they were in solution alone. In this way therapeutic potency is retained without running the risk of exceeding the urinary solubility of any member of the combination.]

Poultry. Toxicity in poultry is indicated by a decrease in egg yield and a thinning of the shell. These symptoms take about 2 weeks to disappear.

Topically, sulphonamides have been shown to retard healing of uncontaminated wounds so that this group of chemotherapeutics should not be used in routine surgery.

Action

The sulphonamides in very low concentrations stimulate cell growth, but in low therapeutic concentrations they are bacteriostatic and in high therapeutic concentration may be bactericidal. They are normally used as bacteriostats and consequently the patient's own resistance and body defences must be encouraged concurrently with sulphonamide therapy.

The sulphonamides act against bacteria by substituting and replacing essential vitamins upon which these susceptible bacteria are actually dependent but which are less essential to their animal hosts. The most important of these compounds with regard to the sulphonamides is *para*-aminobenzoic acid. Interference also occurs in the synthesis of folic acid (pteroylglutamic acid) from *p*-aminobenzoic acid, and the heterocyclic pyridine and thiazole derivatives disrupt nicotinamide metabolism and the working of coenzymes I and II. Synthesis of various amino acids including methionine, a precursor of choline, may also be interrupted. Interference in the metabolism of the vitamin B complex within the bacteria is therefore fairly comprehensive.

As activity is one of competitive substitution, the tissue concentration of a sulphonamide must be kept high enough to prevent access to the *p*-aminobenzoic acid to the bacteria. Consequently, with a standard dose, the fewer the bacteria present the more sulphonamide will be available to approach each individual organism and therefore be absorbed by the bacterium in preference to the *p*-aminobenzoic acid because of its higher concentration. This may be expressed more scientifically 'the activity of a sulphonamide varies inversely with the concentration of the bacteria'. As the sulphonamides interfere with cell metabolism the maximum effect must be exerted when the bacteria are multiplying rapidly; in any infection the bacteria will therefore be most susceptible during the logarithmic phase of the disease, soon after infection. It follows that, for maximum effect, the sulphonamides must be given early in the course of the infection and in maximal doses, and also that the greater and more severe the infection the higher the dosage must be, commensurate with toxicity considerations.

The difference in activity of the various sulphonamides is now thought to be more of degree rather than of selective effect, but nevertheless there is some selectivity which is typified by the activity of sulphathiazole against staphylococci (micrococci). The difference in degree of action has been ascribed to the degree of dissociation of each individual sulphonamide which in turn will be related to the pH of the medium.

Antagonists

The competitive action of the sulphonamides was proved by supplying bacteria with the metabolic products which the organisms are unable to synthesize in the presence of sulphonamides: this eliminated the antibacterial action of the sulphonamides. It is therefore obvious that any of these metabolites will antagonize sulphonamides:

(a) *p*-Aminobenzoic acid and related compounds, particularly the local anaesthetics with a *p*-aminobenzoic acid nucleus such as procaine, butacaine and benzocaine, but also compounds associated with these such as procaine penicillin.

(b) Other members of the vitamin B complex such as nicotinamide, folic acid (pteroylglutamic acid), choline, etc., and precursors of these, including various amino acids, of which glutamic acid and methionine are related respectively to folic acid and choline.

A second group of antagonists are the proteins which combine loosely with the sulphonamides and at least temporarily eliminate their antibacterial activity. Gelatin, albumin, peptone and serum protein all antagonize the sulphonamides; the activity of sulphonamides circulating in blood and tissue fluids is therefore constantly antagonized by the medium in which it is circulating. Associated with this latter group are the products of cell and tissue death, especially pus, which acts also as a non-vascular mechanical barrier.

A number of other compounds including coenzymes, glucose and mercuric chloride are all reported to have antagonistic effects against sulphonamides.

Protagonists

It has been shown that sulphonamides are more active at the higher body temperatures which occur in fevers. It has been shown that urea increases the actions of the sulphonamides by increasing solubility and decreasing their combination with protein.

'Gut Active' Sulphonamides

The members of this group of sulphonamides consisting of sulphaguanidine, phthalylsulphathiazole, phthalylsulphacetamide and succinylsulphathiazole, are generally of low solubility and are poorly absorbed; sulphaguanidine is an exception in that about 50 per cent of any single dose may be absorbed and will therefore not be available for intestinal activity. It has been found, however, that a degree of absorption is useful because of the local effect in the intestinal mucosa during and after absorption. This is important because the bacterial content of the mucosa is frequently very high and purely intra-intestinal activity would be ineffective against these organisms.

The action of the N_4 substituted products is mainly due to their release of the parent sulphonamide after the large intestine, where absorption is minimal, has been reached. Sulphathiazole and sulphacetamide are therefore the active parts of the molecules.

Any part of these sulphonamides which is absorbed reacts in much the same way as do the systemic sulphonamides. Sulphaguanidine, following absorption, has a pharmacological pattern similar to that of sulphanilamide.

The derivatives of sulphathiazole are slightly absorbed—up to about 3 per cent of a single dose, which is insufficient to have either therapeutic or toxic effects. This has, in fact, been considered disadvantageous in that insufficient penetration into the mucosal cells has occurred and bacteria present in these cells have escaped. Phthalylsulphacetamide, on the other hand, is slightly more soluble and is absorbed to a greater extent; penetration of all the tissue of the intestinal wall occurs but no systemic level is achieved nor is toxicity common, owing to the rapid excretion of the compound.

A soluble systemic sulphonamide is often combined with one of these 'insoluble' sulphonamides so that bacterial invasion of the blood and the tissues can be countered at the same time as the intestinal infection is attacked.

In other respects the 'insoluble' sulphonamides act in the same way as the systemic members.

SULPHONAMIDE THERAPY

Although the superior action of antibiotics has reduced the importance of sulphonamides in veterinary medicine, some sulphonamides, of which sulphadimidine is the main one, still retain their popularity because of convenience in dosing and greater economy.

Antibacterial Range

The sulphonamides are, as a group, active mainly against Gram-positive organisms but there is also some activity against certain Gram-negative organisms, a few rickettsiae, and some protozoa. The least active are sulphanilamide and sulphaguanidine, whilst, of the established and official sulphonamides, sulphathiazole is acknowledged as the most active, with the pyrimidine compounds only slightly less so. This, however, is not the order when usefulness is considered, for the low solubility of sulphathiazole deposes it in favour of the sulphapyrimidine group, sulphafurazole and some of the newer more complex sulphonamides. Determination of the relative activity of sulphonamides in vitro is difficult because of many conflicting reports, few of which can be compared because of a lack of uniformity of method. The following activities are generally accepted with sulphanilamide and sulphaguanidine as the least active and sulphathiazole as the most active of the systemic sulphonamides.

1. *Sulphanilamide*, *Sulphaguanidine*, *Sulphacetamide*

   Active against streptococci and *E. coli*, the last being slightly more active than the other two.

2. *Sulphapyridine*, *Sulphadimidine*

   Active in higher degree against streptococci and *E. coli*, but with additional activity against staphylococci, salmonellae and pasteurellae. Sulphadimidine is also active against some rickettsiae and protozoa.

3. *Sulphamerazine, Sulphafurazole*
   These are approximately equal in activity and slightly more effective than group 2, but sulphamerazine has particular activity against *E. coli* and pasteurellae whilst sulphafurazole is almost equal in activity with sulphathiazole against staphylococci.
4. *Sulphadiazine*
   This is almost equal in activity with sulphathiazole but is less active against streptococci and staphylococci and more active against *E. coli.*
5. *Sulphathiazole*
   This is the most active sulphonamide in vitro, possessing the greatest activity against streptococci, staphylococci and pasteurellae.

The more active sulphonamides also have some action against pseudomonas, proteus and clostridial species. This activity is particularly shown by the pyrimidine derivatives.

In man, *sulphathiazole* is used only for staphylococcal infections because its low solubility makes it dangerous. *Sulphadiazine* is preferred as the most active sulphonamide for routine use. In animals, however, sulphadiazine is not absorbed evenly and reliably in all species, and consequently the less active but more reliable *sulphamerazine* becomes first preference. Sulphamerazine does not persist as long as sulphadimidine and requires more frequent dosage: because of this sulphadimidine is widely used in veterinary practice even though its antibacterial activity is slightly lower. The insolubility and toxicity of sulphapyridine and the low antibacterial activity of sulphacetamide and sulphanilamide make their use for routine purposes inadvisable.

This may be summarized as follows:

(1) Routine use: sulphamerazine or sulphadimidine.

(2) Staphylococcal and pasteurellae infections: sulphathiazole or sulphafurazole.

(3) *E. coli*, proteus and pseudomonas infections: sulphadiazine, sulphamerazine.

(4) Urinary infections: sulphafurazole.

The in vivo spectrum of activity of the newer sulphonamides resembles that of the more active of the standard sulphonamides, their advantages lying more in either the activity over prolonged periods and/or their degree of activity which allows the use of smaller dosage rates. These newer compounds therefore have advantages of convenience and reduced toxicity.

The '*gut active*' *sulphonamides* are not clearly differentiated, but sulphaguanidine is of minor importance compared with the sulphathiazole and sulphacetamide substitution products. In human therapy succinylsulphathiazole is the most effective of the 'insoluble' sulphonamides for pre-operative elimination of intestinal bacteria except where diarrhoea is present, when phthalylsulphathiazole is preferred. Succinylsulphathiazole is slightly more toxic and produces semiliquid faeces. In practice, where diarrhoea is nearly always present when enteritis is treated with sulphonamides, phthalylsulphathiazole should give better results. Neither of these compounds allows a compensatory increase of other non-susceptible organisms, which frequently occurs when the tetracyclines are used but a more rapid response can be achieved by combining phthalylsulphathiazole with neomycin.

Phthalylsulphacetamide is said to be very effective in intestinal infections, but insufficient veterinary information is available to compare it with the sulphathiazole derivatives, though the results of human therapy are encouraging.

Sulphaguanidine is much less active than the other three compounds, possibly because of the large proportion which is absorbed.

The antibacterial range of these enteric sulphonamides extends mainly to coli-aerogenes and *Salmonella* species, but also to clostridial organisms as well as coccidia.

Sulphaquinoxaline. This sulphonamide is relatively toxic to mammals and has a narrow margin of safety. Its use is therefore restricted almost entirely to the treatment of poultry. The drug is absorbed quite rapidly when the crop is empty, but somewhat variably otherwise. Once absorbed, excretion is slow and detectable blood levels have been found up to 4½ days after a single dose. Excretion is via the kidneys and faeces. *Toxic* symptoms are a decline in egg yield in laying birds, inappetance, dullness, anaemia and death. Post-mortem lesions include subcutaneous haemorrhages and local necrosis. The causes are thought to be due to antagonism of vitamin K and a drug allergy.

Bacterial Resistance

Strains of bacteria resistant to the sulphonamides are commonly recovered and their incidence depends on the extent of antibacterial therapy applied in a particular field.

Resistance to one sulphonamide usually means that the organism will be resistant to all members of the sulphonamide group. Where this is recognized, chemotherapeutics not related to the sulphonamides must be used, for resistance to the sulphonamides does not necessarily indicate resistance to other chemotherapeutics agents.

Therapeutic Principles in the Use of the Sulphonamides

There are several features of the sulphonamides which play an important part in the general routine of sulphonamide therapy:

1. The effect of a sulphonamide decreases as the number of organisms increases, therefore the indicated sulphonamide should be *given as early in the course of the disease as possible.* For the same reason, *the more severe the infection the larger should be the dose.*

2. Maximum blood levels are achieved in 1 to 2 hr with the most soluble sulphonamides. Where the infection is acute and severe, this time lag between dosing and effect may be fatal, so *in these critical cases treatment should be initiated by parenteral injection* of a monosodium or disodium salt.

3. Because of the low solubility, relatively slow absorption rate, it is usual to *give a large dose initially* at a standard rate according to body weight in order to establish adequate blood levels; these levels, once established, can be *maintained by regularly repeated doses of a smaller size.*

4. The excretion products of the sulphonamides are not very soluble in acid urine and crystallization may occur. *Plenty of water must therefore be available to an animal during sulphonamide therapy* and *urinary alkalizers may be given* concurrently with the sulphonamides.

5. *Any signs of a decrease in urine production, difficulty or pain on urination, or cloudiness and blood in the urine, should be the signal to stop all sulphonamide dosing,* to increase the fluid intake, even to the extent of forcibly administering fluid orally or parenterally, and to institute the administration of urinary alkalizers.

6. Blood levels after the first 2 or 3 days may increase if relatively high level dosing continues. To *decrease the possibility of systemic toxicity, dose sizes should be cut by a half after this initial period.* This also lowers the excretion rate and lessens the danger of crystalluria.

7. Because bacterial resistance can occur, and also because vitamin synthesis by the normal intestinal flora is depressed (this is of special importance in herbivores), *no course of treatment should last longer than seven or eight days.* If no response is shown within the first 2 or 3 days sulphonamide therapy should cease, for no action can then be expected.

8. Because sulphonamides are bacteriostatic in action *dosing should continue for about 48 hr after the disappearance of clinical symptoms.* If this is not observed, a resurgence of the infection may occur, resulting in a relapse. Doses may be further reduced in size or frequency during this period.

The Sulphonamides and Disease Therapy

In most bacterial diseases antibiotics have relegated the sulphonamides to second choice.

Clinical Applications

Specific Systemic and Organic Diseases

Wherever possible antibiotics are now used because of toxicity considerations.

Strangles

As with the antibiotics, variable results are obtained but sulphonamides administered in the early febrile stages will reduce the temperature. Once abscesses begin to form sulphonamides serve only to slow down abscess maturation; they are then contraindicated. With stomatitis and pharyngitis, local applications in electuary form are effective and the electuary should be smeared liberally on the horses teeth three to four times daily. A suitable electuary might contain 30 per cent by weight of sulphanilamide in glycerin, stiffened if necessary with starch. This is somewhat expensive but quite effective.

Joint-Ill

A large proportion of joint-ill cases in foals, calves, pigs and lambs are due to sulphonamide sensitive streptococci and many such cases will respond to sulphanilamide. It is, however, usual to use the more active sulphonamides. Sulphamerazine, sulphapyridine, sulphathiazole, sulphadimidine, as well as the newer compounds such as sulphamethoxypyridazine and sulphamethylphenazole have all been used with good effect.

Pasteurellosis

Haemorrhagic septicaemia of cattle, horses, sheep and pigs all respond to sulphonamides of the heterocyclic type but sulphanilamide is relatively ineffective. As little as 30 g (1 oz.) of sulphamerazine on two occasions, with an interval of 12 to 24 hr, has given a good cure rate. Sulphathiazole is also efficient and sulphadimidine has been used.

Contagious Coryza

This disease, due to an organism *Haemophilus gallinarum* responds to sulphadimidine in drinking water at the rate of 0·2 per cent or in mash at the rate of 0·4 per cent. Sulphathiazole either as the sodium salt, but more commonly as the complex neutral disodium salt, at the rate of 0·1 per cent in the drinking water for 3 to 5 days, is successful in the majority of cases.

Broad-spectrum antibiotics have superseded sulphonamides to a large extent.

Fusiformis Infections

Foul-in-the-foot (foot rot) in adult cattle responds extremely well to sulphonamides and sulphadimidine, as an injection, is the most important.

In *foul-in-the-foot* an intravenous injection of sodium sulphapyridine on one or two occasions with an interval of 4 or 5 days is very effective. The dose is estimated at the rate of 14 g per 100 kg dissolved in sufficient water to make a 12 per cent solution, but it is more usual to give 60 g in 500 ml as a single dose irrespective of weight. Extravascular leakage is dangerous and has caused very severe sloughing. Sulphadimidine sodium causes less damage and is routinely administered by intramuscular injections. It is used as a 33⅓ per cent solution administered at the rate of not less than 37 ml per 100 kg body weight. Sulphamerazine sodium and solutions of the more recent sulphonamides are equally effective.

Actinobacillosis

Wooden tongue will respond in a majority of cases to the use of either sulphamerazine or sulphadimidine.

Coliform–*Salmonella* Infections

White scour in calves and enteritis in pigs, dogs and poultry, have all been treated successfully with the sulphonamides. Sulphaguanidine was first used and was reasonably effective, but later the less toxic and more effective phthalylsulphathiazole, phthalylsulphacetamide and succinylsulphathiazole displaced it.

In calves, both the succinyl- and phthalyl-derivatives of sulphathiazole are effective.

Salmonella infections of poultry have been effectively controlled by sulphamerazine, sulphadimidine, and sulphaquinoxaline, but though birds will survive, many remain carriers and mortalities remain low only during treatment. The sulphonamides are contra-indicated in breeding stock as it has been found that treated birds react to the crystal violet spot test and that the culling of carrier birds is consequently impossible. Furazolidone has become the accepted treatment for *Salmonella* infections in poultry for it has been shown to be capable of clearing B.W.D. reactors.

Non-Specific Systemic Infections

Pneumonia

Perhaps the most spectacular effect of the sulphonamides was the great reduction they effected in the numbers of fatal pneumonia cases in both man and animals. In the *horse*, the majority of pneumonias are due to streptococci which respond extremely well to sulphanilamide. In *cattle* and *pigs*, sulphapyridine and sulphadimidine have been widely and successfully used. In *small animals*, sulphadimidine and sulphathiazole are very effective both for primary infections and infections secondary to viruses.

Metritis

In *cattle*, metritis has been treated successfully with compressed tablets of sulphadimidine or mixed sulphonamides, but here also broad-spectrum antibiotic tablets have proved more efficient in bacterial metritis. The same refers to the *mare*. Parenteral administration of sulphonamides has been found useful in cattle but more especially in *sows*. In the latter species, sulphapyridine and sulphadimidine are equally effective.

Urinary Tract Infections

The urinary tract infections in animals are frequently caused by coliforms and it has been found that in cystitis or metritis the more soluble sulphafurazole and sulphamerazine are of most value. Antibiotics of the tetracycline group are more commonly used owing to their greater activity and lower toxicity, but it occasionally happens that resistance to these antibiotics develops and the sulphonamides can then be used.

Topical application. The use of sulphonamides for wound dressings and during abdominal surgery, as well as intraperitoneal and intrathoracic injections, are no longer recommended in the United States. This is due to tests carried out during the Second World War which indicated that non-infected wounds healed more slowly when dressed with sulphonamides.

Nevertheless, sulphanilamide continues to be widely used as a wound dressing, for which it is more suitable than most of the other sulphonamides owing to its greater tissue solubility.

Besides their successful use as wound dressings the sulphonamides are effective as eye ointments and topical ointments, sulphathiazole with its high activity and wide scope is probably the best of the group. The high solubility and neutral reaction of sodium sulphacetamide has made this compound particularly suitable for eye lotions. A 10 per cent w/v solution is used for superficial effects and a 35 per cent w/v solution for penetration: both are buffered with about 2 per cent w/v of boric acid and should be applied about three times daily. The sulphonamides have been used effectively in infectious keratitis (New Forest disease) in cattle at one extreme and in distemper conjunctivitis in dogs at the other.

It should be remembered that pus actively antagonizes sulphonamides and abscesses should be drained and wounds or inflamed membranes cleared of pus before applying the sulphonamide. It is also important to avoid, if possible, the use of all local anaesthetics of the procaine series, including procaine itself, and benzocaine: where large quantities of sulphonamides of the more active type—sulphathiazole—are used the *p*-aminobenzoic acid released by the anaesthetic is of minor importance.

Dosage

These are given in tabulated form (see Table XVIII, p. 395).

Systemic Sulphonamides

With very young animals it is advisable to administer sulphonamides at two-thirds the usual rates, divide the daily doses into smaller quantities, and give them more frequently. Monosodium salts are given at the same rate as their parent compound; dosages of solutions should be multiplied by their dilution, e.g. sodium sulphadimidine 33⅓ per cent solution will be administered at the rate of 0·6 ml per kg which is three times the dose rate of the dry, pure salt.

'Gut Active' Sulphonamides

(1) Sulphaguanidine
    0·1–0·3 g per kg daily in divided doses
(2) Phthalylsulphathiazole
Succinylsulphathiazole
    0·1–0·15 g per kg in two doses daily
(3) Phthalylsulphacetamide
    Calves: 0·35–0·57 g per kg
    Dogs: 0·15–0·3 g per kg
(4) *Sulphaquinoxaline* (*Poultry Only*)
  Curative:
    In food 0·05 per cent
    In water 0·04 per cent
  Prophylactic:
    In food 0·0125 per cent

Official Preparations

The following *sulphonamides* occur in the B.Vet.C. in the powder form and as a tablet:

Table XVIII. Sulphonamide doses in tablet form

Sulphonamide	Initial dose according to body weight* (g per kg)	Maintenance dose*
Sulphadiazine	0·2	Half initial dose twice daily
Sulphamerazine	0·2	Half initial dose once daily
Sulphadimidine	0·2	Half initial dose once daily
	0·2	Dog, one-third initial dose twice daily
Sulphathiazole	0·2	Half initial dose every 6 hr
Sulphafurazole	0·2	Half initial dose every 12 hr
Sulphanilamide	0·2	One-third initial dose three times daily or half initial dose twice daily
	Small species 0·15	
Sulphacetamide	0·15–0·2	Initial dose repeated daily
Sulphapyridine	0·15	Half initial dose twice daily

* For all species unless stated.

Phthalylsulphathiazole*, Sulphadimidine*, Sulphafurazole, Sulphaguanidine, Sulphamethoxypyridazine, Sulphanilamide, Sulphapyridine*, Sulphaquinoxaline, Sulphathiazole

Tablets are normally of 0·5-g size but those compounds marked * are also available in 5 g sizes.

Sulphacetamide and Sulphacetamide Sodium are also official: the former is now rarely used and the latter only for the preparation of the 10 and 30 per cent official Eye-drops and the Eye Ointment (6 per cent).

The following *sodium sulphonamide salts* are official in the B.Vet.C. as powders only:

Sulphadimidine Sulphapyridine Sulphathiazole

The only *disodium sulphonamide salt*, which is official is Sulphasolucin, derived from sulphanilamide and recognized by the *British Pharmaceutical Codex*.

Sulphathiazole, sulphaguanidine, sulphanilamide and sulphaquinoxaline are not recognized in the *United States Pharmacopia* but appear in the *National Formulary*. There are also variations in the recognition of various sodium salts and pharmaceutical compounds.

In the United States sulphadimidine is known as sulfamethazine (not to be confused with the British commercial name of Sulphamezathine) and sulphafurazole is known as sulfisoxazole.

Commercially the sulphonamides are available in most pharmaceutical forms. Thus, solutions of sulphadimidine, sulphathiazole, and sulphaquinoxaline are available for addition to drinking water for the treatment of poultry, rabbits etc. Compressed tablets for intrauterine insertion are made, eye ointments, topical ointments, sterile dusting powders, injections of sodium and neutral disodium salts, and creams are all available. These may or may not be combined with antibiotics, dyes and other antibacterial substances or potentiators such as urea.

Perhaps the most important commercially available compounds are the triple sulphonamides, the dosage and usage of which depend largely on the individual sulphonamides contained. Combinations may be sulphadiazine, sulphamerazine and sulphadimidine; sulphadiazine, sulphathiazole and sulphamerazine are contained in Trisulphonamide Tablets of the *British National Formulary;* sulphadiazine, sulphamerazine, and sulphapyridine, was one of the earliest of the commercial combinations. The proportions are usually equal, but the B.N.F. combination for example, contains less sulphamerazine than the amounts of the other two, which are equal: the variations take into account slower excretion rates and other factors.

Coccidiosis

The use of the sulphonamides in the control of coccidiosis is discussed in Chapter 38.

Long Acting Sulphonamides

A number of long acting sulphamides were developed during the 1960s, and they included the following:

Sulphamethoxypyridazine, Sulphoremethoxine, Sulphasomizole, Sulphamethoxazole, Sulphasymazine.

These compounds although frequently absorbed readily, are very slowly excreted. The rate of excretion is associated with the degree of protein binding. Their activity against bacteria is frequently different, for instance some have marked activity against staphylococci, but reduced activity against *E. coli*.

The compounds are being used in the veterinary field and clinical investigation will demonstrate their relative value.

Combination of Sulphonamides and Trimethoprim (Trivetrin and Tribrissin)

The sulphonamides have maintained an active place in the armamentarium of drugs which are useful in the control of bacterial infections and it has always been considered possible that their effectiveness could be improved by their use with inhibitors of bacterial dihydrofolate reductases.

This concept was demonstrated in vitro by Hitchings (1961). It was known that the inhibition of folate reductases was a general property of 2,4–diaminopyrimidines and related structures. For antibacterial use it was hoped to develop an inhibitor with a broad spectrum of antibacterial activity.

As a result of studies carried out from 1953–57, trimethoprim, such an inhibitor, was found to be effective and free from side effects and to have a reasonable antibacterial spectrum. It was demonstrated that trimethoprim and sulphonamides potentiated each other with the result that together their antibacterial effects were greatly enhanced. The enhancing effect meant that in addition to an economy in the amount of drug used, the therapeutic index was increased and there was a widening in the range of usefulness of the drugs.

The action of the combination proved to be bactericidal, whereas the components separately were bacteriostatic in action.

Two combinations are in use in the veterinary field trimethoprim/sulphadiazine (Tribrissen) and trimethoprim/sulfadoxine (Trivetrin).

Chemistry

Trivetrin is chemically:

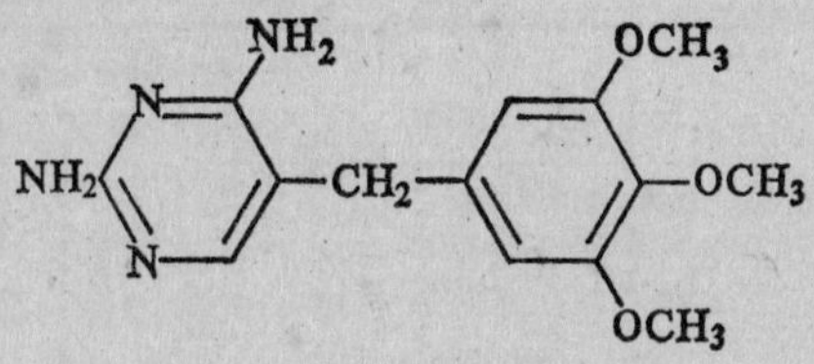

Trimethoprim

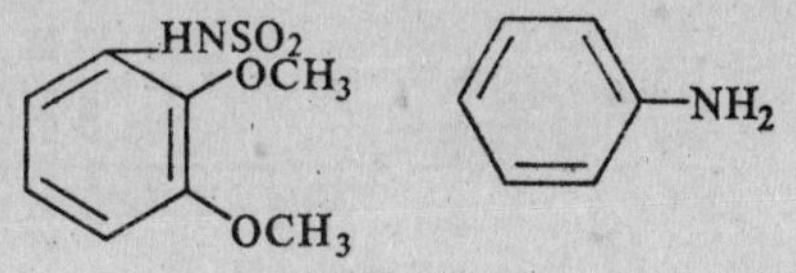

Sulfadoxine

Tribrissen shows the following chemical combination:

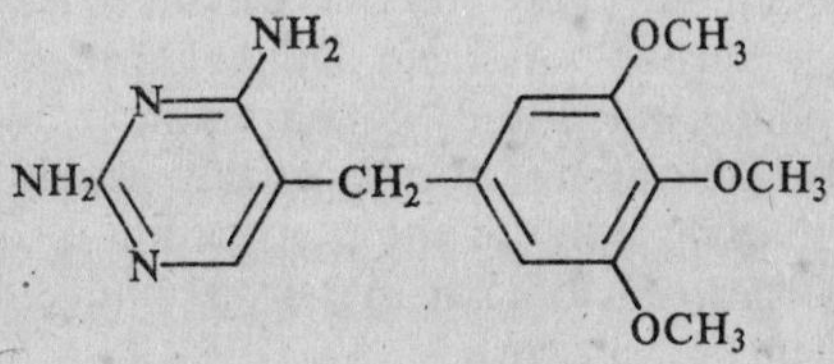

Trimethoprim

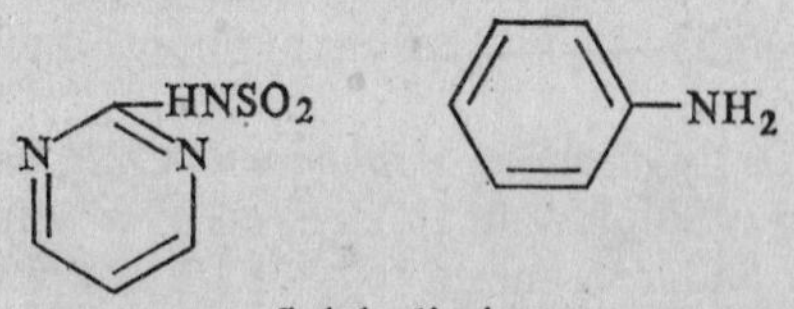

Sulphadiazine

Trimethoprim is combined with either sulfadoxine or sulphadiazine in varying concentrations for veterinary use.

Pharmacology

Trivetrin is normally given by the parenteral route and it is rapidly absorbed from the inoculation site. Therapeutic plasma levels are obtained within 1 hr and peak levels within 4 hr. Both active principles, trimethoprim and sulfadoxine, are distributed throughout the body. In rats trimethoprim appears to concentrate in the kidney, liver, lung and skin. Sulphadoxine is distributed in a similar manner to the other sulphonamides.

Tribrissen is administered particularly by mouth to calves, foals, sheep, pigs, cats and dogs.

Following administration there is a significant antibacterial concentration in the gut. Plasma levels are achieved 2–6 hr after administration.

Trimethoprim is excreted largely in the urine and faeces. Sulfadoxine and sulphadiazine are excreted in the urine, a proportion in acetylated and conjugated forms.

Toxicity

Dogs and Cats. Several studies in dogs have shown that up to ten times the recommended daily dose of the combination is well tolerated. In cats three times the recommended dose can be given with safety.

Farm Animals. Up to ten times the recommended daily dose can be given safely.

Demonstration of Potentiation by Trimethoprim and the Sulphonamides

Potentiation in vitro can be demonstrated by the disc diffusion method. Discs impregnated with trimethoprim and a sulphonamide singly and in combination are applied with plates seeded with the test organisms. Enhanced susceptibility is seen when the zone of inhibition around the combined disc is larger than the zones round either of the single component discs. In general, the greater the difference in zone sizes the greater the degree of potentiation.

Enhanced activity of the combination has been demonstrated by in vivo experiments in induced infections in mice and rats.

Antibacterial Range

The combination is bactericidally active against both Gram-positive and Gram-negative organisms, including *Actinomyces* spp., *B. anthracis*, *Bordetella* spp., *Brucella* spp., Clostridia, *Corynebacterium* spp., *E. coli*, *Haemophilus* spp., *Klebsiella* spp., *Pasteurella* spp., *Proteus* spp., *Salmonella* spp., staphylococci, streptococci and *Vibrio* spp. When discs are used for sensitivity testing of the combination, a disc containing 25 μg trimethoprim and a sulphonamide should be used. They are present in a ratio of 1 : 20.

Clinical Applications

Trials have been reported using both Trivetrin and Tribrissen and the major diseases of dogs and cats and farm animals treated have included infections of the respiratory tract, urinary tract, alimentary tract and genital tract and other bacterial infections.

In general principle the powder and tablet forms, i.e. Tribrissen formulations, are recommended for the treatment of alimentary tract infections. In severe systemic infections it is recommended that a Tribrissen or Trivetrin injectable formulation should be used. Provided the organisms treated are sensitive to the combination, the clinical cure rates obtained are similar to those recorded for the major antibiotics.

Table XIX. Example of the effect of combination on the in vitro MICs of a number of bacteria

Organism	MICs μg/ml			
	Alone		In combination	
	Trimethoprim	Sulphadiazine	Trimethoprim	Sulphadiazine
E. coli	0·63	>250	0·31	0·4
	10	>200	0·31	6·25
Staph. pyogenes	5	>200	1·25	1·56
	5	>200	0·16	6·25
Strep. agalactiae	5	>200	0·04	25
	12·5	>250	0·2	2
Strep. dysgalactiae	3·12	25	0·31	3·12
	3·12	>250	0·4	125
Strep. uberis	0·31	0·2	0·0025	0·2
	1·25	3·13	0·0025	1·56

Dosage

Trivetrin injection (40 mg trimethoprim and 200 mg sulfadoxine per ml). The dosage rate is 1 ml per 15 kg body weight daily. In severe infections the dose may be raised to 1 ml per 10 kg body weight daily. The dose should be repeated daily until the symptoms resolve.

Tribrissen injection (80 mg trimethoprim and 400 mg sulphadiazine per ml). The dosage rate is 1 ml per 30 kg body weight daily or, for severe infections, 1–5 ml per 30 kg body weight daily.

Tribrissen bolus (0·2 g trimethoprim and 1 g sulphadiazine per bolus). The dosage rate is:

Oral 30 mg/kg body weight daily
Intrauterine
Cows and mares: 2·4–4·8 g
Sows and ewes: 1·2–2·4 g

Treatment orally should be repeated daily until 2 days after symptoms have subsided.

Tribrissen Dispersible Powder (0·2 g trimethoprim and 1 g sulphadiazine per 10 g dispersible powder). The dosage rate is 30 mg/kg body weight daily.

Tribrissen oral suspension (80 mg trimethoprim and 400 mg sulphadiazine per ml). The dosage is 15 mg/kg body weight daily. Treatment should continue for 5 days.

Tribrissen tablets (80 mg trimethoprim and 400 mg sulphadiazine per tablet). The dosage rate is:

Dogs: 15 mg/kg body weight twice daily.
Cats: 120 mg daily.

Side Effects and Precautions

1. Occasional temporary swellings at injection sites.
2. Salivation in most cats, if the uncoated tablet is used.
3. Milk from treated cows should not be used for 48 hr after the last treatment with Trivetrin and 72 hr after the last treatment with Tribrissen oral formulation and 7 days after the last treatment with Tribrissen injection.
4. Animals should not be slaughtered until 3 days after the last treatment with Trivetrin, 5 days after the last treatment with oral formulations of Tribrissen and 28 days after the last treatment with Tribrissen injection.

II. NITROFURANS

The nitrofurans are a group of synthetic antibacterial compounds, which are based on the furan ring to which a 5-nitro group has been added. They are bactericidal in action, acting as inhibitors of the carbohydrate metabolism of the bacterial cell.

They have a broad spectrum of activity, being of value against Gram-positive and Gram-negative bacterial infections. They have also proved to be of value in the control of some coccidial infections. Unlike many chemical antibacterials, they have a low tissue toxicity.

Toxic manifestations include peripheral neuritis, gastro-intestinal irritation, depressed spermatogenesis and weight gain and sensitization. These drugs are not used systemically.

Nitrofurazone

Chemically nitrofurazone is 5-nitrofurfuraldehyde semicarbazone:

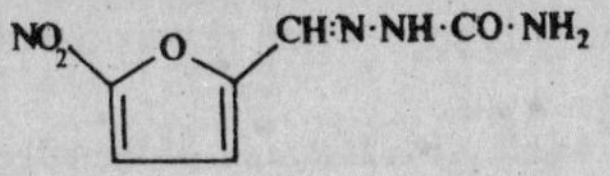

Stability

It is slightly soluble in water, and soluble at 20°C in 600 parts of alcohol.

Antibacterial Spectrum

It is active against a range of both Gram-positive and Gram-negative organisms, and in addition has anti-protozoal activity. Because of this latter action it is used in the control of coccidosis (see Chapter 38).

In the bacterial field it is active against *Salmonella* species but is less effective than furazolidone.

Clinical Applications

***Salmonella cholerasuis* infection.** Nitrofurazone at a concentration in the feed of 0·05 per cent is given to pig for 7 days to control necrotic enteritis. Treatment must be given to all the pigs in the affected group.

Wounds, burns and ulcers. Nitrofurazone is used at a concentration of 0·2 per cent in a cream for the treatment of bacterial infections.

Furazolidone (Furoxone, Neftin)

Chemically Furazolidone is 3-(5-nitrofurfurylideneaminooxazolidin-2-one.

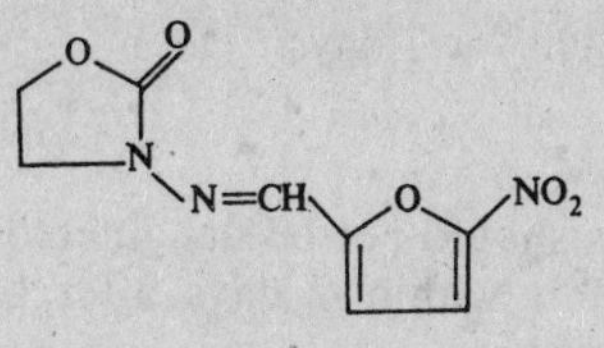

Solubility

Furazolidone is very slightly soluble in water and in alcohol, slightly soluble in chloroform.

Antibacterial Spectrum

Furazolidone has a wide range of anti-bacterial activity which includes: *Clostridia*, *Salmonella Shigella*, Staphylococci and Streptococci, *E. coli*. It is also active against *Eimeria* and *Histomonas*.

Clinical Applications

Furazolidone is usually given by mouth, and although it is poorly absorbed it is effective against many enteric infections.

Salmonellosis. Furazolidone is used for the treatment of *Salmonella* infections in most domestic animals. For poultry it is given in the feed or a concentration of 0·04 per cent for 10 days. In large animals it is given orally at the rate of 10 to 12 mg per kg body weight for 5 to 7 days.

Continual treatment in both poultry and large animals is liable to give rise to resistant strains of *Salmonella*.

Preparations

Furazolidone premix.
Furazolidone suspension.

IV. ARSENICALS

It was discovered that a number of arsenicals when fed to pigs, as feed additives, prevented the continual scouring which is frequently associated with intensive housing methods. The exact mode of action of the compounds is not yet understood but it is assumed that it is associated with anti-bacterial activity.

Arsanilic Acid (AS101)

Chemically arsanilic acid is 4-aminophenylarsonic acid.

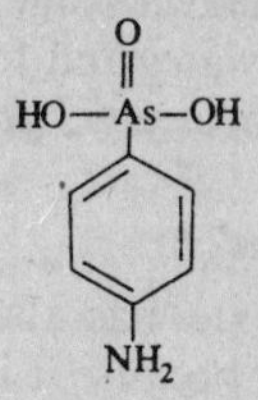

Solubility

Arsanilic acid is soluble at 20°C in 200 parts of water and 40 parts of alcohol. It is a white almost odourless granular powder.

Antibacterial activity

It shows low antibacterial activity in vitro, but is effective for the prophylaxis and treatment of many outbreaks in pigs associated with *E. coli* and *V. coli* infections.

Clinical applications

Neonatal and pre and post weaning. *Scours in pigs*. Arsanilic acid is administered orally in the feed, and is given at a concentration of 250 ppm (parts per million) for up to 3 weeks. The treatment does not eliminate the causal organisms, and does not lead to acquired resistance.

Concentrations of 100 ppm in the feed are given to promote growth.

Coliform septicaemia in poultry. Arsanilic acid is given in the feed at a level of 250 ppm for 5 to 8 days.

Toxicity. The toxic dose for pigs is four times the recommended dose, and in poultry there is an eight-fold safety margin.

Precautions. The treated feed should be withdrawn not less than 10 days before slaughter for human consumption. The floor litter from poultry houses must not be fed to livestock, and the operators must wash their hands before meals and after work.

Preparation

Arsanilic Acid Premix.

Sodium Arsanilate (Pro-gen V)

Solubility

Chemically, sodium arsanilate is the sodium salt of 4-aminophenylarsonic acid: it is soluble at 20°C in 3 parts water and 150 parts of alcohol, and is a white hydroscopic granular powder.

Clinical Indications

Sodium arsanilate has the same activity as arsanilic acid, but because it is more soluble it is suitable for addition to the drinking water.

Dosage

Prophylaxis

Pigs: 10 mg per kg body weight daily for up to 3 weeks

Treatment

20 mg per kg body weight daily for a maximum of 7 days.

Prophylaxis

Poultry: 160 ppm for 7 days

Treatment

320 ppm for 7 days.

Precautions

Animals should not be slaughtered for human consumption within 7 days after cessation of treatment.

Preparation

Sodium Arsanilate Solution.

Bibliography

HITCHINGS, &. &. (1961) A biochemical approach to chemotherapy. *Trans. N.Y. Acad. Sci.*, **23**, 700–8.

JOBARD, A. & SCHEID, J. P. (1971) *Bull. soc. vet. Pract. F.* 27.

REHM, W. F. & WHITE, G. (1970) *Vet. Rec.*, **87**, 39.

34

Vaccines and Antisera

INTRODUCTION

The resistance of an immune individual to the attack of a pathogenic organism or a toxic product of a pathogenic organism, whether it be a helminth, a protozoon, a bacterium, virus or rickettsia or any product of these, is due to a very complex reaction of the cells and fluids of the body. The immunity is due to factors which may be described under two headings: (1) specific immune bodies or antibodies, and (2) non-specific factors, two of which properdin and immunoconglutinin, were referred to by Coombs in his address on 'Immunology in Biology and Medicine' to the 1957 British Veterinary Association Congress.

Both specific and non-specific factors are now receiving a great deal of intensive examination and great strides are being made, but clinically it is only the specific antibody immunity which is used extensively, although substances which stimulate and develop non-specific immunity are not unknown but so far have been used empirically.

Immunity may be *active* or *passive*. Where the animal body is stimulated to produce its own antibody, an active immunity is created which may, with occasional natural contact or artificial boost, persist for a long time as with canine distemper or swine fever, or for as short a time as a year with tetanus or swine erysipelas: where the antibody is merely transferred in the form of antisera from an immune to a non-immune individual this leads to a passive immunity which may persist for up to 3 or 4 weeks, but more usually for 10 to 14 days.

Both active and passive immunities may be conferred naturally and artificially. Any animal which recovers from an attack of a naturally contracted disease will have developed a *natural active immunity: natural passive immunity* is developed usually by the passage of antibodies in the colostrum during the first few days of suckling the young; a good example is the passive immunity against lamb dysentery gained by lambs during their first few hours of life. *Artificial active immunity* is developed when modified antigen (a non-infective pathogenic organism or its products) is injected artificially to encourage the formation of antibodies: *artificial passive immunity* is created when serum from hyperimmunized animals is injected artificially into the non-immune animal.

All these are examples of *acquired immunity*. Where immunity exists in a species, breed or individual from birth and is unrelated to any acquired immunity the term *inherited immunity* is used. Examples of inherited immunity are the relative immunity of man, horses, dogs and cats, to foot-and-mouth disease and rinderpest (cattle plague).

ARTIFICIAL IMMUNITY

The creation of artificial immunity of either an active or passive type is primarily a means of preventing a disease–prophylaxis, but where a disease is already contracted the use of hyperimmune sera may limit the severity and, or, duration of the disease by providing antibodies to neutralize the pathogenic agents. So far, these principles have been applied mainly in the field of microbiology, but in 1957 Jarrett, McIntyre, Jennings and Mulligan of the Glasgow Veterinary School reported on the application of similar principles to helminthology when they announced the development of a vaccine for husk in calves consisting of attenuated living *Dictyocaulus viviparus* infective larvae. Other anthelmintic vaccines have since been investigated but have so far not proved satisfactory in the field.

Passive Artificial Immunity

This type of immunity is created by the parenteral injection of hyperimmune sera which may be used either prophylactically or therapeutically.

A *hyperimmune serum* is a serum which contains a high concentration of specific antibodies and is obtained by repeatedly injecting the donor animal with either increasing doses of a standard antigen (pathogenic organism), or standard doses of an antigen the virulence of which is increased with each succeeding dose, or by a combination of both these methods. These doses of antigen stimulate first the development of a normal immunity, and later a hyperimmunity in which the circulating antibodies are considerably more numerous. Blood is drawn from these animals and subjected to the various processes necessary to produce a sterile, stable, hyperimmune serum. If the donor animal is of the same species as the animal treated, the serum is known as a *homologous antiserum*, whereas if the serum is from a different species it is known as a *heterologous antiserum*. An example of a homologous antiserum is that used in the control of canine distemper virus, whereas lamb dysentery antiserum is normally a heterologous antiserum prepared from horses for use in lambs. Antibodies from homologous antisera persist in the blood longer than those of heterologous antisera.

Serum antibodies are associated with the globulin fraction of the blood and more particularly with the γ-globulin fraction. It is possible to selectively precipitate these fractions and from the precipitated globulin prepare an immune body solution which gives an even stronger protection than a normal or native hyperimmune serum. These are known as *concentrated antisera*. Methods other than precipitation may be used in this concentration process including the use of proteolytic enzymes by which the so-called *refined antisera* are produced.

Antisera

Antisera may be of three types:

1. *Antitoxic sera* in which the immunity of the donor has been developed by the injection of a modified or unmodified toxin. A *toxin* is a soluble poisonous substance produced by a pathogenic organism and remains in the culture medium after the organism has been filtered out. If a toxin is modified in some way to reduce its poisonous potential it is known as a *toxoid;* toxoids are usually prepared from culture media solutions by formalinization or precipitation with alum, producing a formol toxoid and an alum-precipitated toxoid respectively. Antisera prepared against toxins or toxoids are known usually as *antitoxins;* tetanus antitoxin and botulinus antitoxins are well known examples.

2. *Antibacterial sera* in which the antigen has been a suspension of bacterial bodies. These antibacterial sera may contain antibodies which induce specific reactions by the bacteria when the two are brought into contact. There are thus *agglutinins* which cause agglutination or clumping together of the bacteria, *precipitins* which precipitate the bacteria present in a suspension to which an antiserum is added, *bacteriolysins* which lyse or liquefy the bacterial bodies, and *opsonins* which are distinguished from those in normal serum by being heat stable and have the attribute of inducing the phagocytosis of bacteria by leucocytes—they are sometimes referred to as *bacteriotropins* to distinguish them from the thermolabile opsonins which occur in normal serum.

3. *Antiviral sera* which are similar in method of preparation to the other types but the antigen used is a vaccine prepared from a specific virus.

Hyperimmune native sera are a light to dark amber in colour but solutions of globulins or their derivatives, including concentrated antisera may vary from colourless through a greenish-yellow to a dark amber. Solutions may become turbid with age. In powder form these sera are white or light buff in colour, but when reconstituted their solutions are the colour of the parent serum. All sera should be almost odourless except when preservatives are incorporated.

All liquid sera should be *stored* in light-proof containers at temperatures above freezing but below 4°C if maximum shelf-life is to be obtained and expiry dates stamped on labels are usually estimated on this basis. With officially recognized sera these expiry dates must be determined so that the antiserum contains the stated potency during the whole period between the date of manufacture and the date of expiry.

Toxic effects may arise if sera are used in individuals which are sensitive to 'species factors' present in a particular serum. This sensitivity may be natural or acquired but usually it is propagated by a previous injection of a similar serum. The sensitivity is of an allergic nature and occurs more commonly with heterologous than with homologous antisera. Reactions may be peracute and result in death within a few minutes of serum administration, usually when given by the intravenous route: this type of allergic reaction is referred to as anaphylaxis: it rarely occurs following serum administration to domestic animals. A second type which is less acute, occurs not uncommonly and is typified by increased irritability of the animal, with oedema usually associated with the head, ears and neck. The animal's head may become twice its normal size and the ears thickened considerably: other areas may be affected. The greatest danger is that of asphyxia due either to oedema of the larynx or pressure on the larynx and pharynx by oedema of local tissue. This type of reaction develops within a few hours of administration. A third type, rarely recognized in animals, is *serum sickness*, a condition met more commonly in the human field and typified by urticaria (nettle rash), pains in the joints and a fever. It may occur from a few days to a few weeks after injection of the serum.

The mechanics of the reactions are incompletely known. One theory which is commonly held is that foreign protein stimulates the formation of specific antibodies which are closely associated with tissue cells. A further dose of the antigen results in an antibody–antigen reaction which involves tissue cells due to their close association with the specific antibodies. Damage to the cells results in histamine release with increased permeability of cell membranes including those of capillary vessel endothelioid cells, which in turn results in a great increase in extracellular fluids and hence severe oedema, and in peracute cases at any rate, a drop in blood pressure.

All cases other than the peracute ones may be treated with circulatory stimulants and antihistaminics such as adrenaline, mepyramine maleate and promethazine.

Active Artificial Immunity

Passive immunity is satisfactory where a temporary and immediate protection against a disease is required or where antibodies may aid recovery from an established infection, but where a stronger and permanent, or at least prolonged, immunity is required this may only be achieved by the creation of an active immunity.

Active immunity to toxins, bacteria and viruses can be created by the injections of toxoids, bacterial vaccines and viral vaccines respectively. Immunity normally takes a minimum of 10 to 14 days to develop but in some cases it may be several weeks before full immunity is achieved.

Toxoids

A toxoid is a toxin which has been treated in such a manner that its poisonous properties have been destroyed, or almost so, without materially affecting its ability to stimulate the formation of antibodies in a non-immune individual.

The basic toxoid is prepared by growing the toxin-producing organism on or in a suitable culture medium, filtering off the organisms and treating the filtrate containing the toxins with formaldehyde. This produces a *formol toxoid* or *anatoxin*. Various purification processes are now applied to decrease the non-toxoid protein content and improve the antibody stimulating properties of the toxoid: the most modern toxoids are formol toxoids treated first with proteolytic enzymes, particularly trypsin, then precipitated with aluminium sulphate (alum precipitated toxoid) and lastly adsorbed onto aluminium hydroxide.

Formol toxoids may be distinguished from alum precipitated toxoids quite easily, for the former are clear, colourless or pale yellow solutions, while the latter are white or off-white 'fluffy' suspension which settle, giving a white cloud-like deposit and a clear supernatant; the toxoids resuspend very easily.

Aggressin

This term is used less commonly now that whole culture and bacterial vaccines are replacing the natural culture aggressins in clinical practice. Some bacteria, especially members of the genus *Clostridium* produce substances during their growth which in themselves are non-toxic but when injected along with their respective bacteria increase the invasiveness and pathogenicity of those bacteria. These substances occur in bacterial culture or diseased tissue filtrates, and it is these filtrates which increase the 'aggressiveness' of specific bacteria that are called *aggressins*.

Bacterial Vaccines

Compared as a group with toxoids, bacterial vaccines are less successful in their effects just as antibacterial sera are less effective than antitoxin sera. Nevertheless, many antibacterial vaccines are invaluable, especially when whole culture vaccines are prepared so that exotoxins are also included and a combined toxoid and antibacterial vaccine prepared.

Bacterial vaccines may consist of:

1. *Suspensions of killed bacteria.* These may be formalinized suspensions equivalent to the formol toxoids and called *anacultures*, or alternatively they may be cultures or suspensions of bacteria inactivated by heat. In both cases the antigenicity of the cultures should not be affected. The term *bacterin* is sometimes used to designate a member of this group of vaccines.

2. *Suspensions of living attenuated bacteria.* The bacteria used in these preparations are in most cases bacteria which have been grown for many generations in conditions which will allow growth but will decrease pathogenicity. Most of these preparations are true vaccines in that they stimulate the production of antibodies which remain long after the vaccine bacteria have disappeared from the body, but in one or two cases such as the BCG anti-tuberculosis vaccine, immunity to infection persists only so long as the bacteria comprising the vaccine remain in the system. This is an example of the phenomenon known as *premunition* which has also been referred to in connection with immunity to protozoan diseases (p. 477).

If a vaccine is prepared using strains of bacteria isolated from the individual which is to be immunized by the injection of this vaccine, it is referred to as an *autogenous vaccine*. Some species of bacteria exist as a number of different strains, each of which varies antigenically from the others. Consequently a vaccine prepared from one or two strains only cannot expect to provide protection against all strains of that particular species of bacterium, whereas a vaccine prepared from strains causing disease in an individual may be expected to stimulate the production of antibodies specific to those strains. To overcome this disadvantage it is now usual to prepare *polyvalent vaccines* (and antisera), in the preparation of which are used as many different strains of the specific organism as can be found.

Viral Vaccines

The principles that apply to bacterial vaccines also apply to viral vaccines, with a few differences. Thus, there exist vaccines which are composed of suspensions of killed virus: there are also vaccines which consist of suspensions of live viruses which have been reduced in virulence by growth under conditions that reduce pathogenicity but hardly, if at all, affect antigenicity. There are, however, vaccines which consist of suspensions of virus particles which are fully virulent in one species of animal, but which are intended for vaccination of an entirely different species for which the virus has relatively little pathogenicity: this situation arises because the viruses which affect the two species are similar antigenically but have a rigid and specific host affinity. This cross immunity phenomenon is most marked in the viruses causing the various animal pock diseases.

The method of preparing virus vaccines varies according to the virus involved and the tissues for which each has an affinity. The primary factor in the preparation of all viral vaccines is the need of all viruses for living cells, without which they cannot grow. The living cells are provided either by living experimental animals, by embryonated eggs or by tissue cultures. When attenuated viruses are used, they may be modified by growing them in an unusual host, e.g. egg adapted canine distemper virus vaccines, or by growing them in tissues (living animals or cultures) and then subjecting them to adverse growing conditions. An example of the second method is swine fever vaccine in which the blood of artificially infected pigs obtained at the height of the disease is incubated with ethylene glycol and crystal violet.

SPECIFIC VACCINES AND ANTISERA

Artificial passive immunity is induced by injecting an antiserum: this passive immunity can be prophylactic or therapeutic in its effect. In the case of active immunity however, the methods of achieving it vary according to the disease and the individual preference of the physician. In general, live organisms give a stronger immunity than dead organisms, and whole cultures provide a wider protection than either bacterial body vaccines or toxoids alone. When live vaccines are used, great care is usually required to ensure that the live organisms neither precipitate a clinical infection in the injected animal nor contaminate the area or room. In addition, it must be remembered that the inoculated animal will excrete living organisms for several days subsequent to inoculation and will therefore be infective to susceptible individuals. These remarks apply particularly to canine distemper virus inoculations.

Because there are not as yet any agreed standards for the major vaccines used throughout the world, it has been decided not to include in this edition details of animal vaccines.

Vaccine standards and changes in the methods of production are always under review so that the reader is referred to the *British Veterinary Codex* and the *British Pharmacopoeia* and their supplements for up-to-date details.

Bibliography

British Medical Bulletin (1969) Vaccination against infectious diseases. **25**, no. 2.

British Pharmacopoeia (*Veterinary*) (1977) London: Pharmaceutical Press (in preparation).

British Veterinary Codex (1965) London: Pharmaceutical Press.

HERBERT, W. J. (1970) *Veterinary Immunology*. Oxford: Blackwell Scientific.

HULSE, E. C. (1968) The standardization and control of veterinary biological products. *Vet. Ann.*, **9**, 1.

35

Fungicides

Fungicides are drugs which destroy parasitic fungi: *fungistatic drugs* are those which prevent growth and multiplication of fungi. Collectively these drugs are often referred to as antimycotic drugs, the majority being fungistatic and not fungicidal.

That fungi cause external diseases is well known; it is less widely recognized that fungi also cause internal diseases—*Aspergillus fumigatus* is one such organism, and is responsible for 'brooder pneumonia' in chickens and 'mycotic abortion' in cattle. Though successful treatment of internal fungoid diseases has been claimed it is the cutaneous fungoid diseases which are most successfully treated: ringworm is the most common and widely known of these fungus skin diseases of animals.

The *ringworms* of animals and birds are caused by species of fungus belonging to two genera—*Trichophyton*, large spored fungi, and *Microsporum*, small spored fungi.

Trichophyton verrucosum var. *discoides* causes the majority of outbreaks in cattle, but is transmissible to man and occurs occasionally in horses, dogs and goats. *T. mentagrophytes* parasitizes dogs and cats, and *T. equinum*, horses. *T. gallinae* is responsible for favus in poultry. *Microsporum canis* causes the majority of cat and dog ringworm, whilst *M. equinum* is responsible for occasional outbreaks in horses.

All these ringworms are transmissible to man, and *Microsporum* species are especially contagious. Cattle ringworm, though transmitted, does not parasitize people as readily as it does cattle. Lesions in man occur both in hirsute and hair free areas, contrary to the belief held for many years that *Tinea capitis* was not caused by ringworm of animal origin. In recent years numbers of reports have appeared drawing attention to the high incidence of human ringworm caused by fungi from animals. The Public Health aspect of the disease is of considerable importance, whereas the economic importance to the farmer is yet not clearly defined in Great Britain though some economic loss might be inferred from Canadian work.

Treatment

The whole field of veterinary mycology is under intense investigation and the increased knowledge of the conditions of infection, establishment and development of lesions and their response to changes in environment, should lead to more rational methods of treatment.

It is probable that skin humidity and temperature have a great bearing on the establishment and proliferation of dermatophytes. Any agent, therefore, which can affect these two factors can be expected to modify the growth of ringworm. This hypothesis is supported by the improvement in cattle ringworm when a mild iodism is created by the injection of sodium iodide intravenously. Other factors, however, are involved and the creation of iodism is not the complete answer. In many cases a temporary improvement is shown no matter what type of dressing is used, but sooner or later a state of equilibrium is reached and only a change to a new type of dressing will induce further progress. Several changes may be required before a cure is effected. This reaction is shown particularly when the older mercurial or iodine dressings are used, but some of the more recent compounds such as **Dichlorophen,** B.Vet.C., are less likely to produce such reactions. It is also noticeable that a dressing which gives admirable results in one individual will not be so successful in a second animal affected with the same type of fungus. This may be due to a variety of causes; differences in skin metabolism, skin secretions, humidity, temperature and tissue antibodies may each play a part. The occurrence of spontaneous recovery also increases the difficulties of making a true assessment of the value of antimycotic drugs in the field.

Treatment of ringworm should therefore not be commenced with preconceived ideas. If the response to a dressing is unsatisfactory, a different type of fungicide should be resorted to whilst remembering that the ineffective dressing might be extremely effective in other cases and therefore should not be discarded permanently.

Many cases of ringworm due to species of trichophyton frequently exhibit lesions covered by a thick hard crust, and most types of ringworm cause an increase in skin scales and serous exudates. These barriers must be overcome if the antimycotic drugs are to be used efficiently. There are several ways of achieving this:

1. Fungicides capable of penetrating the barriers can be used: some mercurials are reasonably efficient in this way and fatty acid derivatives are also useful.

2. Adjuvant drugs which soften and aid exfoliation of keratin (keratolytics), such as **Salicylic Acid,** B.Vet.C., can be used in conjunction with a fungicide.

3. Vehicles which aid penetration are helpful: animal fats and alcohol are good examples.

4. Oral and parenteral administration may replace topical application.

5. In all cases where topical application is used, the removal of the barrier of scabs and the scrubbing and cleaning of the area with detergents (unless these are antagonistic to the fungicide) are essential if the best results are to be achieved. The dangers of contagion frighten attendants and these procedures are often neglected.

In farm animals, particularly in the case of cattle, treatment of ringworm is the exception rather than the rule, so that effective therapy is only likely to follow if the farmer can be given a complete economic rationale for a course of treatment. He must also be made aware that the environment, in the form of wooden structures or dirty damp farm buildings, can be the area where ringworm infection is maintained year by year.

ANTIMYCOTIC DRUGS

A great number of drugs have been used for the treatment of ringworm, probably because most are only fungistatic and not very active so that drug resistance on the part of the parasite is easily achieved. Modern derivatives of some of these, as for example, **Monosulfiram,** B.Vet.C., and other new type compounds however, are strongly fungistatic or even fungicidal, and most are sufficiently powerful to kill the organism before resistance can develop.

1. Old type antimycotic drugs comprise:
   - Mercury preparations, organic and inorganic
   - Iodine preparations
   - Copper sulphate
   - Sulphur preparations, organic and inorganic
   - Benzoic and salicyclic acids
   - Rosaniline dyes
2. New type ringworm therapeutics include:
   - Dichlorophen
   - Fatty acid derivatives
   - Salicylanilide and its derivatives
   - Antibiotics

Old Type

Mercury Preparations

Mercury preparations of an inorganic nature have been used for many years as antimycotic dressings.

All inorganic mercury preparations are fungistatic but, though active against both large and small spore ringworms, their toxicity restricts their use mainly to cattle ringworm.

Yellow Mercuric Oxide, B.Vet.C. has been used for ringworm occurring around the eyes where more active but highly irritant dressings are not advisable. It is usually applied as a 1 per cent ointment: 'Golden Eye Ointment' (**Eye Ointment of Mercuric Oxide,** B.Vet.C.) is the preparation, used as a daily application. Though once widely used for the treatment of corneal and eyelid ulcers, this eye ointment, apart from it's antimycotic use, is retained only for the dressing of the eyelids in cases of purulent conjunctivitis where there is a tendency for the upper and lower lids to stick together. Such a condition may occur in canine distemper, but even in this disease sulphonamide and antibiotic ointments will prevent the inflammation developing to a purulent stage. Elsewhere on the body **Mercuric Iodide,** B.Vet.C. as a 5 per cent ointment and **Ammoniated Mercury,** B.Vet.C. as a 2·5 per cent ointment (White Precipitate Ointment) are useful anti-fungal preparations. **Ointment of Mercury,** B.Vet.C. (Trooper's or Blue Ointment) containing 30 per cent of metallic mercury has been used with good effects but the danger of absorption and the advent of more efficient drugs has made it obsolete.

Of the organic mercury preparations, **Phenylmercuric Acetate,** B.Vet.C., the chloride, the **Nitrate,** B.Vet.C., and the more complex phenylmercuric derivatives have distinct antimycotic action when applied in ointments and alcoholic solutions of strengths from 0·05 to 0·1 per cent. The compounds are applied daily or with intervals of up to 1 week between dressings.

Iodine Preparations

Though used for many years, they still retain a place in ringworm therapy, particularly as an alternative to dressings which have ceased to be effective. Iodine is also useful in that it can be applied in alcoholic solutions with a brush and penetrates sufficiently to make prior preparation of mild lesions unnecessary so that contact with the lesions is avoided. **Strong Solution of Iodine,** B.Vet.C., brushed on lesions at intervals of 5 to 7 days, and **Weak Solution of Iodine,** B.Vet.C., used at shorter intervals, are reasonably effective in both *Trichophyton* and *Microsporum* infections but, as with mercury preparations, continued use may result in severe irritation and dermatitis. **Strong Ointment of Iodine,** B.Vet.C. is also used.

Perhaps the most useful form of iodine treatment of ringworm is that of sodium iodide injected intravenously in aqueous solution into cattle. For calves of 6 months 30 g, and for adults of 2 years and upwards 65 g, with doses proportionate to age between these quantities, should be injected once only. This treatment is particularly effective in generalized ringworm.

Copper Sulphate, B.Vet.C.

This is widely used by farmers as a home remedy for ringworm, when it is used as an ointment or paste in strengths of up to 5 per cent, or as aqueous solutions of 1 to 2 per cent applied daily.

The compound is quite strongly fungicidal, partly by virtue of its astringent and caustic nature, and partly by the specific effect of the copper ion. This effect is utilized in some commercial fungicides which combine copper and fatty acids—copper undecylenate is particularly effective.

Sulphur Preparations

These in the form of ointments with or without other compounds such as resorcinol, have been used against ringworm in all species, but with poor results, and it was not until complex organic sulphur compounds were devised that any real effect was achieved. **Monosulfiram,** B.Vet.C., is one of the most active fungicidal organic sulphur compounds. For cattle ringworm it is applied with a brush as a 25 per cent alcoholic solution following the removal of scabs and thorough cleansing of the lesions. Dressings should be repeated every 2 or 3 days. The recovery rate is slow. For small animal ringworm, solutions of 2·5 per cent are safer though slower in effect.

Benzoic Acid, B.Vet.C.

Apart from forming balsams when combined with resins (cf. Balsam of Tolu) and acting as an expectorant in cough mixtures, Benzoic Acid, B.Vet.C., also has bacteriostatic and fungistatic properties. These last actions are exerted in acid solutions when the pH is lower than pH 5. It is therefore used in industry as a preservative of fruit and similar foods. It has been used for very many years as a ringworm dressing with salicyclic acid in the form of Whitfield's ointment or **Compound Ointment of Benzoic Acid,** B.Vet.C. This ointment contains 6 per cent benzoic acid and 3 per cent salicylic acid. It is most effective against *Trichophyton* cattle ringworm in man, but it has been pointed out that the benzoic acid played a very small part in the action.

Salicyclic Acid, B.Vet.C.

Salicylic Acid is a white crystalline powder, almost odourless, with a bitter-sweet taste. It is almost insoluble in water but soluble in alcohol and alkali citrate solutions.

Actions and Uses

Salicylic acid is an antipyretic but is rarely used as such because of its irritant nature; sodium salicylate and acetylsalicylic acid (aspirin) are preferred. Salicylic acid is also antiseptic, with an activity approximating to phenol, but has relatively no toxicity owing to its poorer penetration. The major use of salicylic acid is as a *keratolytic* when it softens the keratin layer of the skin and aids exfoliation without causing other than slight and beneficial irritation to underlying tissues. During the casting of the keratin layers ringworm mycelia are also cast and the activity of the acid as a fungistatic is mainly due to this effect.

Salicylic acid is more efficient when applied in water-miscible creams and it is now incorporated with more active antimycotic agents for its keratolytic powers.

Salicylanilide

Derivatives of salicylic acid are amongst the most active of fungistatics and fungicides. *Salicylanilide* and various *chlorosalicylanilides* are amongst the newest antimycotic discoveries and have been widely used in human mycosis. Salicylanilide is a white or pinkish-white crystalline substance which is odourless, slightly soluble in water but freely soluble in most organic solvents.

Actions and Uses

Salicylanilide is effective against *Microsporum* species and some *Trichophyton* species. Its fungistatic action in vitro is approximately eight times that of undecylenic acid and is also very high in vivo (ringworm agents frequently exhibit little fungistatic activity in vitro and yet are effective in vivo). The compound is applied as an ointment in strengths of up to 5 per cent; higher concentrations may be irritant. Daily applications are usual. To overcome the irritant effect of the salicylanilide it is frequently combined with mono- and di-chlorosalicylanilides and undecylenic acid. The chlorosalicylanilides are from 5 to 150 times more active in vitro than salicylanilide, depending on the compound: this gives them a minimum of forty times greater activity than undecylenic acid.

Crystal Violet, B.Vet.C.

Crystal violet, a member of the *rosaniline group* of dyes, is an effective fungistat against *Microsporon* and *Trichophyton species*; and is well known as dark greenish purple metallic looking, coarse crystals which are soluble to 1:200 in water, fully soluble in alcohol and chloroform, but not in ether. It is hexamethylpararosaniline hydrochloride, whereas methyl violet is mostly the pentamethyl derivative with some hexa- and tetra-methyl derivatives. Gentian violet is a mixture of methyl and crystal violets.

Actions and Uses

The methylrosaniline compounds are actively bacteriostatic and bacteriocidal to Gram-positive organisms; they are also anthelmintics and are used in the human field against oxyurids and *Strongyloides* species. They are efficient fungistatic agents, mainly used for ringworm in small animals.

As an antibacterial agent, crystal violet is used on wounds and burns in conjunction with brilliant green and either acriflavine or proflavine, as a jelly or aqueous solution.

Compound Paint of Crystal Violet, B.Vet.C.

An aqueous solution: 3 per cent tragacanth is used to form a gel base in which crystal violet is incorporated at a strength of 1 per cent.

The compound is also used in 0·5 per cent aqueous solutions as an ear dressing for otitis externa in the dog, mainly because of its action against *Pseudomonas pyocyanea*.

Crystal violet is used in preparing attenuated bacterial and virus vaccines of which swine fever–crystal violet vaccine is an example.

For fungistatic purposes it is painted on to ringworm lesions as a 1 per cent alcoholic solution. Daily dressings control the lesions within a few days and recovery usually follows in 2 or 3 weeks.

The dye is very persistent, and in common with all dyes used externally, has the disadvantage of staining the dresser's hands and any object contacted. The use of 95 per cent alcohol as a vehicle makes application easier by evaporating quickly and leaving the dye adherent to the skin.

New Types

New fungicides and fungistats are appearing in numbers, and only a few of those in common use are mentioned below.

Salicylanilide and its derivatives have already been mentioned in relation to the parent compound salicylic acid.

Dichlorophen, B.Vet.C. (G4, Antiphen)

This compound is an effective taenicide and as such is referred to in the chapter on anthelmintics which affect cestodes. Besides this important action, dichlorophen is also one of the most useful fungicides available. At a concentration of 2 per cent it is used in ointment form and in alcoholic solutions. One of its main advantages is the low number of dressings required. Though daily dressings or applications every third day give the best and most rapid results, bovine ringworm has been cleared up following two dressings only.

Dichlorophen is also bactericidal (for *Staphylococcus aureus*) with a phenol coefficient of 100 at 37°C. This is useful when lesions are infected. The compound is non-irritant and in many cases even relieves irritation within a few hours of the first application. *Toxicity* is very low and the compound is effective against both *Trichophyton* and *Microsporon* species. With these attributes dichlorophen is therefore suitable for the treatment of all types of ringworm in all domestic animals.

Fatty Acids and Their Salts

Two saturated acids, propionic and caprylic, and one unsaturated fatty acid, undecylenic acid, are used in antimycotic dressings. Their activity in vitro is slight, whereas in vivo activity is in some cases quite strong. Both the acids and their salts are used, either together or alone. The compounds are frequently used in conjunction with other drugs, partly because of their antimycotic action, and partly because they aid penetration. Undecylenic acid, for example, is a component of human sweat, and it would be expected to disperse and penetrate readily when applied to human skin. The related fatty acids act similarly. They are usually applied in ointment or powder form.

Undecenoic Acid

Undecenoic acid, also known as undecylenic acid; is an oily yellow liquid with a characteristic fatty odour. The acid is almost insoluble in water but soluble or miscible with organic solvents and oils.

Actions and Uses

Undecylenic acid is fungistatic mainly to *Microsporum* species and can be used in concentrations up to 10 per cent: more tends to be irritant, and it is therefore commonly combined with its zinc or copper salts in ratios 2 to 5 per cent with 15 to 20 per cent of the salt. It may be used as a cream in a water miscible base or as a powder with talc.

Zinc Undecenoate, B.P.C.

A white greasy powder with a characteristic odour and is insoluble in water or alcohol but miscible with oil. It is used in antimycotic ointments and powders in strengths varying from 10 to 25 per cent for its antipruritic and fungistatic effect. It is active mainly against *Microsporum* species.

Sodium Propionate, B.P.C.

A white crystalline solid with a pungent acetic acid odour. It is deliquescent and soluble in water, but slightly soluble in alcohol.

Actions and Uses

This fatty acid salt is used in the baking industry to control mould growth, but it is an efficient fungistatic and fungicidal agent against both *Trichophyton* and *Microsporum* species as well as possessing bactericidal properties. It is used in ointment form at strengths of up to 15 per cent, and as aqueous solutions of up to 10 per cent strength. The compound is most efficient when in an acid pH and it is therefore combined with propionic and other fatty acids and their salts.

Sodium propionate is also used extensively in up to 240 g (8 oz.) oral daily doses for acetonaemia. Field reports show the compound to be a useful addition to the number of drugs already used for this condition, particularly if used in conjunction with other glucogenic drugs.

Caprylic Acid

Used mainly in the form of its sodium and zinc salts—sodium and zinc caprylates or caproates. The salts are also known as octoates. The sodium salt is a water soluble yellow granule but the zinc compound is a white insoluble powder. Both are used as fungicides and are reported to be the most active of the fatty acid antimycotic agents so far produced. They are effective against all species of ringworm and are applied as ointments containing up to 15 per cent either of one or a mixture of the two.

These compounds are used mainly in small animal treatment.

Antifungal Antibiotics

The majority of antibiotics are produced by fungi, so it is not surprising that the majority of fungi are resistant to such compounds. Several active fungicides have been recovered from soil surveys.

Nystatin (Mycostatin 20)

This antibiotic was discovered by soil survey members of the New York State Department of Health, and given the name *Nystatin* for this reason. It is formed by *Streptomyces noursei.*

The pure substance has the form of white crystal needles, with the chemical formula $C46H_{77}NO_{19}$. The crystals are only slightly soluble in water and it is stable in the form of a dry powder.

Activity. It is active against *Candida albicans.*

Clinical application. The activity of nystatin is mainly a local one, and it can be used to control *Candida* infections in the mouth and alimentary canal. It is also active in the ears and on the skin. The average dose in human beings is 500 000 units. It can be used for the treatment of yeast infections in the udder at a rate of 100 000 units per quarter.

Amphotericin B

This antibiotic was derived from *Streptomyces venuzuelae.* In its pure form it is separated as deep yellow crystals. The dry powder is stable.

Activity. It is active against *Candida albicans.*

Clinical application. For local use as with nystatin.

Griseofulvin (Grisolvin)

Griseofulvin is (+)-7-chloro-4,6-dimethoxy-coumaran-3-one-2-spiro-1′-(2′-methoxy-6′-methylcyclohex-2′-en-4′-one), an artificial substance produced by the growth of certain strains of *Penicillium griseofulvium.*

Solubility. Insoluble in water. Soluble at 20°C in 300 parts of dehydrated alcohol.

Activity. It is effective against *Trichophyton* species, *Microsporon* species—the causal agents of ringworm in cattle horses, pigs and cats. The antibiotic is given orally and is absorbed from the gastro-intestinal tract into the blood stream. It becomes incorporated in the keratin, and the cells of skin and hair become resistant to the fungi.

Dosage

A daily dose of 15 to 20 mg per kg body weight is recommended. Treatment should be continued for 10 to 14 days in large animals. Small animals may require treatment for 6 to 12 weeks.

Preparation. Griseofulvin tablets contain 125 mg of active principle. A less purified form known as Griseofulvin mycelium is available for large animals.

Bibliography

WINNER, H. I. (1973) *Current Antibiotic Therapy.* Edinburgh and London: Churchill Livingstone.

36

Insecticides

An insecticide, is a compound which will destroy insect parasites. In veterinary medicine, however, the nâme is applied only to those substances which are active against parasites living on the skin of animals. These parasitic organisms are metazoan animal parasites, known as ectoparasites.

The ectoparasites of domestic animals include
1. The *Arachnida*, or relatives of the spider, which are:
- (a) The *ticks*, which affect mainly cattle and sheep, but some species also parasitize horses, dogs and poultry.
- (b) The *mites*.
    - (i) *Family Sarcoptidae* includes the mange mites which cause chorioptic, notoedric, otodectric, psorergetic, psoroptic and sarcoptic mange, in the different species, and also the depluming itch and scaly leg mites.
    - (ii) *Family Demodicidae* includes the parasite causing demodectic mange.
    - (iii) *Family Gamasidae* includes the northern and red mites of poultry.
    - (iv) *Family Trombidiidae* includes the harvest mite, the larva of which causes so much trouble in dogs and cats throughout the summer.

2. The *Insecta* or flies and their relatives, comprising three orders, Diptera, the flies; Anopleura the lice; and Siphonaptera, the fleas.

A. Order Diptera.
- (a) Suborder *Athericera*. This group contains several families of free living flies, some of which feed on the blood of the host, whereas others are only parasitic during the larval stages of their life story. Amongst the former are the *Glossina* species, or tsetse flies, whilst among the latter are species of *Lucilia*, the sheep maggot or blow fly, species of *Hypoderma*, the warble fly, *Gasterophilus*, the horse botfly, and *Oestrus ovis*, the sheep nostril fly.
- (b) Suborder Pupipera. In this group belong the *keds* or *Family Hippoboscidae*. The sheep ked, *Melophagus ovinus*, is wingless, but the horse ked, *Hippobosca equina*, and the deer ked, *Lipoptena cervi*, are both winged.

B. Order Anopleura. These are the *lice*, wingless creatures classified according to whether they
- (a) bite the host: suborder *Mallophaga*, which includes the lice of the dog, sheep, horse, cow, goat, cat and also poultry

or
- (b) suck the blood of the host: suborder *Siphunculata*. These lice affect the pig, horse, cow, sheep, goat, dog and man.

C. Order Siphonaptera. These are the *fleas*, which suck the blood of the host, injecting irritant substances which slow up blood clotting. They are wingless insects and can attack domestic and farm animals. The ‘chigger flea’ affects pigs as well as man and the dog.

ECTOPARASITES

These parasites can be broadly divided into three main groups: (1) those which migrate in their larval stages through the tissues of the host, (2) those which burrow into and live in the superficial skin layers, and (3) those which live on the skin, though biting through into the deeper tissues.

(1) Parasites of the first group include horse bots, cattle warbles and sheep nostril fly larvae. These parasites are now treated during the larval migrating stage, by the application of organophosphorus compounds topically or in some cases by oral dosing.

(2) The second group consists of parasites which burrow into the skin or penetrate deep into the hair follicles. They are the mange mites, and surface applications are at present the best form of treatment. Insecticides can be used in vehicles which contain detergents and keratin softening agents such as the alkali hydroxides and sodium carbonate, or oleaginous lubricating agents, all of which will aid penetration of the active agent. As in the case of the treatment of the other ectoparasites, feeding or injecting the host with therapeutic agents has been used with moderate success, but toxic effects in some cases have restricted the wide use of these routes.

(3) The third group of parasites which live mainly on the skin surface, lice, poultry mites, fleas, keds, ticks and the larvae (maggots) of *Lucilia* species, are all controlled effectively with surface sprays or dips.

THE IDEAL INSECTICIDE

An insecticide should: (1) destroy all parasites at every stage of their life history; (2) be rapid in action; (3) be non-toxic, or of low toxicity, to the host; (4) be compatible with the various bases and vehicles required for application; (5) be economically practicable; and (6) provide a minimum residue in or on the host and be rapidly destroyed in the environment.

RESISTANCE TO INSECTICIDES

Just as antibiotic resistance has become a great problem in bacterial disease therapy so has resistance to insecticides become of major importance in the control of ectoparasites. Tick resistance is common both in larvae and adults and in strains where it develops, the resistance is transmitted genetically, sometimes by dominant, sometimes by recessive, and sometimes by partially recessive genes. Nevertheless, resistance of blowflies to dieldrin in Australia can be bred out in six generations which have been reared in the absence of the compound. Investigations into parasite resistance are proceeding in many countries and it would appear that insecticides can be grouped according to whether resistance is due to a single gene or more than one. Thus one type of genetic resistance is common to benzene hexachloride, aldrin and dieldrin. Resistance to dicophane and certain organophosphorus compounds however may be related to two or more genes. It is also possible that in some cases parasites may possess genes capable of conferring monotype and multitype resistance. Certain mosquitoes possess BHC, dieldrin, and dicophane resistance, and some blowflies show BHC, dieldrin and malathion resistance. Work is now proceeding into the investigation of compounds which will inhibit the enzymes which destroy the insecticide in the insect body, and which will therefore maintain the susceptibility of the parasite. *N*,*N*-di-n-butyl-*p*-chlorobenzene sulphonamide is such a compound and protects dicophane from destruction when combined with it.

In the W.H.O. monograph on insecticide resistance by Brown and Pal (1971) it is stated that 'resistance in the field has become a measurable quantity. Greater understanding of the nature of resistance has also been achieved, and it has become generally realised that resistance results from a change in the genotype composition of an insect population induced by a selection pressure exerted by insecticides. When resistance develops to a given insecticide, spraying with the insecticide is no longer continued, but a change is made to a different compound.' Because new insecticides are not always available, an insecticide with a resistance pattern may still have to be used and in such a case it must be used with discretion.

OLD TYPE INSECTICIDES

Many of these are still in effective daily use but have been grouped together, mainly because of their non-specific character and, in general, their lower activity.

Arsenic, phenol and cresol, have been widely used as sheep and cattle dips, but they are being supplanted by insecticides which are safer and more effective.

Arsenic has been used at a strength of 0·1 per cent sodium arsenite for repeated rapid dipping, and 0·3 per cent for less frequent dipping. In Great Britain the yellow dips of arsenic and sulphur were common until they were replaced by the more modern insecticides. A typical formula for the preparation of one of these is:

℞ Arsenic trioxide Potassium carbonate Sublimed sulphur	*āā* 20 lb.
Soft soap	4 lb.

Stirred and boiled in 10 gallons of water and diluted to 100 gallons, this gave an arsenite concentration of 0·2 per cent. The main disadvantage of these dips was the danger of poisoning from skin absorption, particularly on hot days.

Sulphur

In its various forms sulphur has been used both internally and externally throughout the centuries. Its irritant laxative action when administered orally is now rarely resorted to, though some will still remember the 'brimstone and treacle' weekly dose. The laxative effect is due to the production of irritant sulphides by reaction of the sublimed sulphur or brimstone with the intestinal acids and alkalis. Externally, sulphur was the best anti-mange dressing available for veterinary use until the advent of the chlorinated benzene derivatives. It is still used extensively for routine baths for dogs. The various compounds may be applied as ointments or oily suspensions at a concentration of 1:8:as powders diluted with talc, kaolin, etc.: as dispersible bath powders combined with soaps and wetting detergents as aqueous solutions of 1:8 up to 1:60, combined with lime, arsenic, phenol, etc., for sheep dips. For area disinfestation and disinfection, fumigating candles can be used.

Sublimed Sulphur, B.Vet.C. (Sulphur sublim.)

This lemon yellow slightly gritty powder, which is tasteless and odourless when freshly prepared, is commonly known as 'flowers of sulphur'. When kept for a short period a slight odour and acid taste develops due to oxidation. It is insoluble in water and alcohol.

Precipitated Sulphur, B.Vet.C.

Precipitated Sulphur, B.Vet.C. (Sulphur Praecip. or Milk of Sulphur) is a pale yellow smooth powder without odour or taste and is used as a substitute for sublimed sulphur on all occasions, but particularly where there is a danger of sulphur dermatitis.

Monosulfiram, B.Vet.C.

Known commercially as Tetmosol, this is probably the most outstanding of these organic sulphur compounds. It is a yellow, or yellowish brown, crystalline powder which is insoluble in water but soluble in alcohol. It has a faint odour and a bitter taste.

Actions and Uses

Monosulfiram is active against fleas, lice, ticks and mites, and is also a potent fungicide. In all cases, it is advisable to scrub the skin clean of all scales and scabs first, applying the dressing while the skin is still moist but not wet.

For use against ectoparasites a dilution of 1:10 of the 25 per cent alcoholic solution is brushed well into the skin. In farm animals if the condition under treatment is mange, the dressing should be repeated every 2 or 3 days until the condition is cured, but in cats and dogs three dressings should be sufficient.

Surface living ectoparasites can easily be controlled by washing with a 5 per cent soap at suitable intervals—2 to 3 weeks.

Mesulphen, B.Vet.C.

Commercially known as Sudermox, is a yellow oily preparation which in emulsion form is very useful as an antipruritic, a skin softener, and an insecticide. The compound should be massaged into the affected area or applied with a brush, making sure that the dressing contacts the skin and is not merely applied to the coat. Dressings should be repeated every 2 to 3 days until the condition is cured. Animals should not be dressed completely on any one occasion and extra care should be taken when applying the compound to cats. The preparation is neither so cheap, so easy to apply, nor so persistent as the newer chlorinated benzenes.

Derris, B.Vet.C. and Lonchocarpus, B.Vet.C. (Rotenone)

Derris is the rhizome and root of leguminous plants originally obtained from the East Indies,

Malaya, Burma and Siam area, but other sources were exploited during the war. From these rhizomes are made *Prepared Derris*, B.Vet.C., which is a powder containing about 5 per cent of rotenone, the main active ingredient and insecticide.

Actions and Uses

Rotenone is an effective insecticide which acts more rapidly than benzene hexachloride, but less rapidly than pyrethrum. It is not persistent and must therefore be used frequently: an interval of 7 to 10 days is optimal. The substance acts on the parasite both by surface absorption and following ingestion, and is therefore termed a 'contact and stomach poison'.

Prepared derris may be applied as a wash or dip with 5 to 20 per cent soap or detergent to assist in wetting the powder. In this form it is active against all the 'superficial' parasites, ticks, keds, fleas and lice, and has been used against these parasites in all species, but is now mainly used against sheep keds and lice, and fleas and lice in dogs. In this form also it was used against warbles in cattle and was specified under the Warble Fly (Dressing of Cattle) Order 1948, when it should contain 9.5 g of derris resins or equivalent, and 25 g of soap, per litre of dressing. Applications were made four times between 15 March to 30 June directly into the warbles in the skin. The wash was scrubbed on, sprayed on, or the animals were dipped. This Order was revoked in June 1964 in favour of vigorous advisory campaigns for the use of derris and organophosphorus compounds along with other control methods for warble fly control.

Derris is also used in dusting powder form for small animals, when it is diluted to 1 per cent rotenone, with talc and similar diluents. It is applied in this manner infrequently in large animals because of the danger from 'burning' where the dressing concentrates in drainage channels through the coat when the animals are exposed to rain.

Rotenone itself is a white crystalline powder now used mainly as a 2 per cent solution in cod-liver oil, or similar bland oil, as an alternative dressing in demodectic mange of dogs.

Lonchocarpus, B.Vet.C.

Lonchocarpus was used as an alternative to derris in World War II and owes its activity to a 3 per cent rotenone content. It is therefore used in exactly the same way as derris. None of these preparations are effective in mange.

Toxicity

Derris and lonchocarpus were originally used by natives as a fish poison and their effect is one of nervous system paralysis preceded by convulsions. Death results from respiratory paralysis. River pollution by waste dips should be avoided, for fish are extremely susceptible to rotenone poisoning. It is unlikely that toxicity will be seen in mammals under normal conditions.

Pyrethrum Flower, B.Vet.C. (Pyrethrins)

When powdered, these dried flower heads of a chrysanthemum produce a compound known for many years as *Dalmation insect powder*. The active principle, mainly in the form of *pyrethrins* should be present in a concentration of at least 1 per cent. An extract, **Pyrethrum Extract,** B.Vet.C., contains approximately 25 per cent pyrethrins. Two pyrethrins are known, of which pyrethrin I is the more active in aqueous suspension, but in other forms the two are equal in effect. Two other compounds, Cinerin I and Cinerin II and have a similar action.

Actions and Uses

Pyrethrins are contact poisons which have a rapid paralytic action preceded by muscular excitation and convulsions, giving a 'knock-down' effect, from which some insects may recover. However, for this 'knock-down' effect they are frequently combined with the slower acting but more lethal chlorinated benzene derivatives. Pyrethrum powder was once extensively used on its own as a treatment for fleas and lice. More recently, synergistic agents, such as piperonyl butoxide have been introduced. These intensify and prolong the effect of the pyrethrins and form a useful insecticidal combination. Concentrations of 0·2 per cent pyrethrins in kerosene are widely used as sprays on farms for fly control.

Benzyl Benzoate, B.Vet.C.

This compound exists as an oily liquid or a crystalline colourless solid, with a typical aromatic odour and burning taste; it is insoluble in water but soluble in alcohol and other organic solvents.

Actions and Uses

Benzyl benzoate is effective against most external parasites in domestic animals, and has been used against sarcoptic, otodectic, and demodectic mange, in the form of a 25 per cent emulsion, and, as an alcoholic solution, against fleas, lice, and harvest mites on dogs. It is also active against ringworm fungi.

When the whole of an animal is affected, small areas only should be dressed at a time, or a diluted bath should be used. For bathing, the alcoholic solution should be diluted by 50 per cent water and rubbed into the animals coat after a thorough wetting. After a 10 to 15 min interval the greater part of the dressing should be washed off and the animal rubbed dry.

When applied as a 25 per cent emulsion in the treatment of sarcoptic mange the animal must be clipped and the dressing massaged well into the skin. This is time consuming and unnecessary with the newer dressings; moreover, it has no residual effect so that weekly dressings are required.

Toxicity

Benzyl benzoate is a nervous system depressant, and toxic effects include vertigo, incoordination, coma, and death from respiratory and cardiac depression. Where vertigo is noticed, immediate washing followed by dosing with caffeine and other stimulants are measures which are normally effective. Though widely used as an ear dressing in cats against *Otodectes cynotis*, its use in this respect is not very safe and it is not recommended for cats.

SYNTHETIC INSECTICIDES

The first of the modern insecticides were **Dicophane,** B.Vet.C., commonly referred to as DDT, and **Gamma Benzene Hexachloride,** B.Vet.C., both of which were developed during World War II, and are chlorinated hydrocarbon compounds. Since then, many new chlorinated hydrocarbons have been produced, some related directly or indirectly to these first two. Amongst these, methoxychlor, chlordane, toxaphene, aldrin and dieldrin were outstanding. Extensive surveys of the tissue storage of some of these compounds particularly dieldrin and related substances, has shown that the compounds are cumulative. There is reason to believe that residues may not increase after a certain level has been reached and no chronic toxicity effects have so far been demonstrated unequivocally. Many governments have banned the use on food animals of the more persistent chlorinated compounds and restrict the use of others pending the production of positive evidence in either direction.

Parallel in development there has appeared a series of organophosphorus compounds which have superseded many of the chlorinated hydrocarbons.

Both groups of insecticides are toxic but the phosphorus compounds which act by inhibiting cholinesterase enzyme breakdown of acetylcholine can be more dangerous. Toxic symptoms are group specific, but post-mortem pictures vary widely within each group. Excitement, convulsions and high body temperatures typify the symptoms of poisoning with the chlorinated hydrocarbons, whereas dyspnoea, profuse salivation and stiffness are the outstanding symptoms shown in poisoning due to the phosphorus compounds.

Toxic symptoms usually only occur when the insecticide has been wrongly mixed or has been used on animals in a very poor physical condition.

Carbamate Insecticides

A third class of insecticidal compounds has been introduced in the veterinary field in the form of the carbamates. These carbamates, whilst they are relatively non-toxic, are insecticidal by virtue of their anti-cholinesterase activity and are therefore pharmacologically similar to the organophosphorus compounds.

Under the commercial name of Sevin (a carbamate), carbaryl has been used against fowl mite in the form of dusting powders: a tablespoonful of a 50 per cent dust applied along the middle of the back immediately after milking is safe and protects cattle against horn fly for 15 days. It is used as a 0·125 per cent spray against cattle ticks and a 0·5 per cent spray will kill lice and protect cattle for about 20 weeks: such sprays and intramuscular injections at the rate of 15 mg per kg body weight are not effective against warble fly larvae. Tolerance in ticks has already been reported.

Although toxicity does not appear to be as great as that of the organophosphorus compounds the inactive cholinesterase complex formed with carbaryl cannot be reactivated by aldoximes (e.g. 2-PAM) and atropine is therefore the obligatory antidote. The same care taken with organophosphorus compounds should be excercised during the use of carbaryl.

ORGANOCHLORINE INSECTICIDES

Although the organochlorine insecticides are not used now in many countries, a description of their activity has been maintained in this edition, so as to record the historical development of the modern insecticides.

Dicophane, B.Vet.C.

Dichloro-diphenyl-trichloroethane or DDT is a white or off-white crystalline powder which is usually odourless. It is impure and specified to contain not less than 75 per cent of the active agent. Insoluble in water for normal purposes, it is soluble to the extent of 2 per cent in alcohol and 10 per cent in fixed oils. Other organic solvents readily dissolve the compound.

Actions and Uses

Dicophane is a neurotoxic contact poison. It may also act as a stomach poison following ingestion. The precise method of action is not known but it affects both the motor and sensory nervous system in a manner common to the chlorinated hydrocarbons as a group. The first signs of toxicity, which appear immediately on contact with an aerosol spray, is a frantic attempt to rub off the substance followed by general excitability, incoordination, paralysis and death. The 'knock-down' action on flies of pure aerosol sprays takes several minutes to take effect which compares unfavourably with the immediate action of those containing pyrethrins.

The toxic effect on parasites is shown by a considerable increase in irritability of the host as the parasites increase their pre-paralytic activity. This period may last up to 30 min or more depending on the type of preparation and concentration used. Thus, if powders are used the period will be longer than if oily dressings, which increase absorption, are used. The toxic effect of dicophane on *small animals* has made the compound unpopular for dogs and cats, but it is occasionally used against fleas and lice as a bath powder containing 0·2 per cent, or a dusting powder containing 5 per cent, though only for healthy adult dogs. Cats, puppies and birds, are susceptible to poisoning and, if bathed with dicophane, animals should be thoroughly rinsed after a few minutes. Dusting powders should be brushed out of the coat after a period of 15 to 30 min.

In *large animals*, DDT has been of immense value in controlling lice, keds and ticks, on sheep, cattle, and horses, and has also assisted in clearing areas of disease bearing insects such as mosquitoes, house flies, etc. It is used in dusting powders containing 5 per cent but more especially in sprays and dips in concentrations varying from 0·2 to 1 per cent depending on the parasite and species to be treated.

Toxicity

Absorption of dicophane through intact skin is not of great significance unless oil solutions are used. In small animals particularly, the licking of dicophane from dressed areas may cause trouble, but the greater danger results from dressing young animals with dusting powders in confined spaces. In large animals, toxic dangers are slight, particularly as they are the least susceptible species, but sufficient may be excreted in milk to cause toxicity to children, and great care should be taken when dressing milking cows. Symptoms are those expected of neurotoxic agents, hyper-excitability, convulsions, paralysis and death. Post-mortem signs are similar to those of chloroform poisoning with fatty changes of the liver. Pentobarbitone is recommended to control the nervous symptoms and subanaesthetic doses should be combined with cessation of the dicophane treatment and removal of dressings by washing or brushing. Calcium borogluconate intravenously in doses of 2 to 3 g is an effective antidote in dogs.

Benzene Hexachloride, B.Vet.C.

The compound is a synthetic preparation containing approximately 12 per cent of the insecticidal gamma isomer. It is a light buff powder with a disagreeable and persistent odour; it is not soluble in water but partially soluble in the organic solvents. This impure compound is used commercially for the preparation of dusting powders and dips.

Gamma Benzene Hexachloride, B.Vet.C. (Lindane, Gammexane)

This is the pure insecticidal gamma isomer and is a white crystalline solid with a slight but characteristic odour. It is insoluble in water but soluble in most of the organic solvents.

Actions and Uses

The action of gamma benzene hexachloride against insects is reported to be through interference with inositol metabolism. The substance is essentially a contact poison and the effect on the parasite is similar to dicophane but paralysis occurs at a slightly faster rate. A 'knock-down' agent, usually pyrethrin, is used in most BHC fly sprays. As with dicophane, animals will show increased irritation following dressing, but some of this is due to irritant isomers if the impure benzene hexachloride is used.

In *small animals*, benzene hexachloride preparations have been used extensively, but, although the pure gamma isomer is four times less toxic than dicophane, there is still some danger of toxicity. Dogs may be bathed in concentrations of up to 0·2 per cent using dispersible commercial bath powders, but cats and puppies are best dressed with dusting powders containing from 0·2 to 0·5 per cent gamma isomer which, if the whole body is dressed, should be brushed out of the coat after about 30 min. This will prevent the licking and swallowing, or the inhalation, of toxic quantities. Dusting powder of Gamma Benzene Hexachloride and Pyrethrum, B.Vet.C. contains 0·5 per cent of the isomer and 0·4 per cent pyrethrins. BHC is also used in ear dressings in concentrations of 0·1 per cent. Higher concentrations may cause toxic reactions in cats. For localized areas, creams containing up to 0·2 per cent gamma isomer are frequently used.

The compound is effective against all parasites of small animals, especially sarcoptic mange.

Fleas and lice are easily eradicated with BHC particularly if bedding and kennels are sprayed or washed with suspensions or solutions of up to 0·25 per cent strength.

In *large animals*, benzene hexachloride has been of great value. Its lower chronic toxicity and its earlier discovery made it more popular than dicophane. It is the insecticide of choice for the treatment of mange in all species.

In *sheep*, dips and sprays at a concentration of 0·06 per cent have proved extremely effective against scab, and enabled single dipping to be adopted in Great Britain. Dips based on benzene hexachloride were reintroduced in Great Britain in 1975 because sheep scab had reappeared after the mite had apparently been eliminated in 1949–50. No other insecticide has proved to be as effective in the control of sheep scab. Against *Ixodes ricinus*, the common British sheep tick, benzene hexachloride has also proved to be the best agent available. Keds and lice have been almost eradicated in some areas.

In *cattle*, BHC is as useful as it is in sheep but resistance has developed in a few strains of *Boophilus* ticks.

Sheep dips containing BHC have remained stable as long as 6 years, but they should incorporate copper sulphate at the rate of 0·03 per cent as a bacteriostat. *Erysipelothrix* growth in dips has been responsible for post-dipping lameness, particularly in lambs.

Poultry houses have been kept free of flies by the use, every 2 to 3 days, of an electric fumigator dispersing 0·22 mg gamma isomer per ft^3.

Residual effects are not maintained quite so long as with dicophane owing to the more volatile nature of BHC. For normal purposes 3 to 6 weeks is the average period of protection but occasionally protection may persist longer. Weather conditions, method of application, and type of parasite, all modify the residual period.

Toxicity

The chronic toxicity of Benzene Hexachloride, B.Vet.C. is greater than that of the pure gamma isomer, but acute toxicity is not so severe. This stems from the relatively rapid excretion rate of the gamma isomer, and the tissue retention of the beta isomer. In general, the toxicity is slightly greater than that of dicophane, but the symptoms and treatment of both are the same.

Dieldrin ($C_{12}H_8Cl_6O$)

This complex chlorinated hydrocarbon is a white crystalline odourless compound, insoluble in water but soluble in most organic solvents. It is stable in alkaline solution unlike most other similar compounds with the exception of its near relative **Aldrin** ($C_{12}H_8Cl_6$).

Actions and Uses

Dieldrin is a stomach and contact poison. Its action is slow, taking as long as 24 hr to kill adult sheep blow flies, and from 24 to 36 hr to kill newly hatched larvae. Residual efficiency was very great, for all larvae that hatched over a period of 24 weeks were killed. This was an improvement on DDT which, though preventing oviposition persisted for only 8 to 19 weeks. In practice, dips containing 0·05 per cent dieldrin protect against bodystrike for 3 to 4 months and against crutch strike for 8 to 10 weeks. Another property not possessed by other members of the chlorinated hydrocarbon group is its ability to diffuse down the wool fibres as they grow. This meant that the compound could be applied as a surface low pressure spray, and yet maintain efficient larvicidal action.

Dieldrin has also been used experimentally as a systemic insecticide.

Dieldrin promised to be a most valuable insecticide when its full potential was established, but the rapid build-up of tissue residues has resulted in governmental bans in most countries of the world including Great Britain.

Before its withdrawal from New Zealand and Australia resistant strains of *L. cuprina* (sheep blowfly) had developed—some strains were as much as 400 times more resistant than standard susceptible strains.

Toxicity

Relative to dicophane, this insecticide is about three times as toxic, with an oral LD of 85 compared with dicophane's 250. However, when these figures are related to a minimum dipping concentration of 0·2 per cent for dicophane and 0·05 per cent for dieldrin the true toxicity relationship becomes apparent. Symptoms and treatment, are similar to those of other chlorinated hydrocarbons.

Toxaphene (Coopertox)

Toxaphene, also known as chlorinated camphene, with a probable formula of $C_{10}H_{10}Cl_{18}$ is an amber, wax-like solid, with an odour of pine. It is insoluble in water but soluble in organic solvents. Strong alkalis cause it to decompose.

Actions and Uses

Toxaphene is both a stomach and contact poison, but requires a knock down agent if used as a fly spray. Its principal use is for cattle tick sprays or dips, used in concentrations of 0·25 to 0·55 per cent. Its value is mainly because of its residual effect, which is superior to that of any other similar dressing for cattle. It has become an important factor in stock control in some areas of Africa where the tick-borne East Coast Fever has hitherto been a great problem. Toxaphene provides sufficient residual protection to allow cattle to be moved through the infective zones in safety, as all the ticks are destroyed before the *Theileria* they carry can be injected into the blood stream of the cattle during feeding. It has been reported that it gives the best tick control of all substances tried in both wet and dry seasons in Northern Nigeria: 0·37 per cent sprays were used. Toxaphene is also satisfactory for the control of keds.

Toxicity

In general, its acute toxicity is up to four times greater than dicophane but its chronic toxicity is less. Symptoms are indicative of medullary stimulation and consist of dyspnoea, cyanosis, muscular tremors, and convulsions. Chronic toxicity is the same as for all other chlorinated hydrocarbons. Treatment follows the same principles as for other members of the group.

Chlordane

Chlordane with a basic formula of $C_{10}H_6Cl_8$, is a contact, stomach and fumigant, insecticide falling between dicophane and BHC in most of its attributes. It is widely used as an alternative dip against sheep and cattle ticks but is not commonly used for other purposes.

ORGANOPHOSPHORUS INSECTICIDES

Because of the withdrawal of a number of the chlorinated hydrocarbon insecticides (particularly aldrin and dieldrin) from use against the ectoparasites of livestock, an enormous amount of research has been carried out on the organophosphorus insecticides, with the aim of finding materials which would provide good protection against arthropods in particular, without at the same time producing undesirable side effects or tissue residues. On pp. 435–8, a description is given of organophosphorus compounds used as anthelmintics.

The organophosphorus compounds are now used against warble fly, *Oestrus ovis*, and *Gasterophilus* larvae, sheep blow fly, adults and larvae cattle ticks and keds and lice in sheep. More than 50 000 organophosphorus compounds have been investigated for insecticidal activity, and there is now a range of products which have passed field studies and are available commercially. They have many properties in common: (a) they have low mammalian toxicity, (b) few have very marked persistency, (c) they are stable for short periods.

General Properties of the Organophosphorus Compounds

Physical characters

Organophosphorus insecticides may be yellow to deep brown, oily liquids, or yellow to white crystalline powders. Most have characteristic garlic-like odours.

Actions

The organophosphorus insecticides are both contact and stomach poisons, acting by combining with, and neutralizing, cholinesterase. Owing to their lipid solubility they are rapidly *absorbed* by all normal routes. The compounds are excreted mainly as metabolic breakdown products via the urine. In the tissues the organophosphorus compounds are subject to enzyme attack which results in hydrolysis and oxidation. In mammals this enzyme attack causes detoxication of the compounds, but fortunately such enzymes do not occur in insect tissues in quantities sufficient to inactivate the insecticides. Certain organophosphorus compounds inactivate the hydrolytic enzymes selectively, and can therefore be used to intensify the activity of those compounds which are more susceptible to hydrolysis. Thus malathion and EPN (*O*-ethyl-*O*-(4-nitrophenyl)phenylphosphonothionate) are together fifty times more effective (and toxic) than either used separately. This is a true example of synergism. Oxidative enzyme reactions are also most important in that they may be responsible for converting inactive compounds into active anticholinesterase compounds; these are compounds which react with cholinesterase in competition with the normal substrate. They form very stable combinations which either hydrolyse only slowly or not at all, so that normally produced acetylcholine continues to exert its stimulant effects in an uncontrolled manner.

Toxicity

All organophosphorus compounds are toxic to man and animals by virtue of their cholinesterase blocking activity. This action can be counteracted by the use of atropine and similar compounds, or by the reactivation of the enzyme by the use of aldoximes such as 2-PAM. Most organophosphorus compounds are rapidly excreted and tissue storage is minimal. In general it is usual to protect animals from stress for about 2 hr after oral dosing. Delayed toxicity does occur occasionally during the first two days after dosing, and it is, of course essential that dosing with insecticides and anthelmintics both of which may have a cholinesterase inhibiting action is not carried out simultaneously.

Consumer risk. An interval of 7 days between treatment and slaughter is usually adequate to ensure consumer safety. There are exceptions to this, and the manufacturer's directions should be closely followed.

Risk in handling. When applying preparations rubber gloves and other protective clothing should be worn, and inhalation of spray and skin contamination avoided. Should contamination occur the insecticide should be washed off immediately and medical assistance called if contamination is gross.

Wild life and fish hazard. Wild life and particularly fish are susceptible to organophosphorus poisoning; waste dip or empty containers should be carefully disposed of so as to avoid stream, pond and pasture contamination.

Field Use of Organophosphorus Compounds

The organophosphorus compounds are used for animal treatment in various ways:

(1) *External application.* Dipping, spraying, jetting, etc.

(2) *Systemic application.*

(a) Pour on method. When this method of treatment is used, the action is systemic. The insecticide in solution or suspension is poured on to the back of the animal and is absorbed through the skin and then disseminated to other parts of the body.

(b) Feed additive. Addition to the feed for several days.

(c) Intramuscular injection.

(d) By stomach tube.

External Application

Dipping and Spraying

Compounds used for local skin and wool action and applied by dip or spray (sheep blowfly, and ked and lice control):

Insecticide	Per cent concentration	Control from (weeks)
Bayer 9037	0·01	8
Bronophos	0·02	8
Carbophenothion	0·042	8 (as a dip)
Chlorphenvinphos	0·02	12 (as a spray)
Chlorphenvinphos	0·04	26 (as a dip)
Diazinon	0·025	6–8 (as a dip or spray)
Phosalone	0·2	8 (as a dip)

Blow fly control as a separate treatment is carried out in different ways in different countries. In Australia jetting along the back and in the crutch area is the methods of choice, and is more effective than spraying. When jetting the spray fluid is directed under pressure (50 to 100 lb./in^2) parallel to the wool staples, and reaches the skin level along the back and runs off down the sides.

In New Zealand and the United Kingdom, it is usual to dip sheep for blowfly control shortly after shearing, so that a complete coverage of the treated animal is obtained. Treatment at this time is also expected to control lice and ked infestations.

The following compounds are used for the control of the major tick infestations in the U.S.A., Australia and Africa and are applied either by dipping or spraying. The intervals used between treatments will depend on the type of tick infestation being treated. One-host tick (*Boophilus microphus*) infestations should be treated at fortnightly intervals, whereas two- and three-host tick infestations may require treatment every 5 days. In association with spraying or dipping it is usual to spell the pastures so that an opportunity is given for the larval stages to die out on the ground.

Insecticide	Per cent concentration
Coumaphos	0·02
Chlorphenvinphos	0·2
Diazinon	0·5
Ethion	0·1

Systemic Application

The systemic action of the organophosphorus compounds is used particularly for the control of warble fly in cattle, *Oestrus ovis* in sheep and *Gasterophilus* larvae in the horse. A wide range of compounds is now used for the control of these infestations by the systemic route.

Pour on Treatment

Insecticide	Dose rate (mg/kg)
Coumaphos	10
Fenchlorphos	150
Fenthion	7·5
Ruelene	25
Trichlorphon	50

Feed Additives

Compounds are used for the control of oestrus ovis infestations, and given as feed additives:

Insecticide	Dose rate; single dose (mg/kg)
Famophos	100
Fenchlorphos	100
Ruelene	125
Trichlorphon	100

Administration by Stomach Tube

For the control of *Gasterophilus* spp. in horses, the compounds are administered as a single dose by stomach tube:

Insecticide	By stomach tube (mg/kg)
Dichlorvos	50
Ruelene	100
Trichlorphon	75

OVERDOSAGE WITH ORGANOPHOSPHORUS COMPOUNDS

Occasionally animals will receive an overdose of an organophosphorus compound, or be in a poor state at the time of dipping and thus susceptible to toxicity. When this occurs affected animals should receive *atropine sulphate* at a dose of 1 mg per kg body weight. Atropine is a specific antidote to overdosage with a cholinesterase inhibitor.

Bibliography

O'BRIEN, R. D. (1967) *Insecticides: Action and Metabolism.* New York: Academic Press.

British Pharmacopoeia (*Veterinary*) (1977) London: Pharmaceutical Press (in preparation).

British Veterinary Codex (1965) London: Pharmaceutical Press.

BROWN, A. W. & PAL, R. (1971) *Monograph on Insecticides.* Geneva: WHO.

37

Anthelmintics

INTRODUCTION

Because anthelmintics are classified according to the families of parasites sufficient parasitology will be mentioned to provide a background for discussion of the drugs and their uses.

Drugs which are grouped under the heading of anthelmintics are intended to act against parasites which inhabit the alimentary tract, its associated structures and such organs as the liver, lungs, and the blood circulation. They are thus distinguished from those drugs which act against external parasites and are called insecticides. Those drugs which act against the parasites other than helminths which inhabit the tissues and blood of animals have not yet been grouped under any one heading.

There are, of course, parasites with complicated life histories which at various stages exist in two or even three areas. Examples are certain tapeworms which live in the tissues of one host as cystic immature stages, and in the adult worm stage inhabit the alimentary tract of those animals of which the first host is the natural prey. The horse bot fly lays eggs on the hairs of the host's coat; these hatch into larvae which penetrate into and migrate through the tissues, reach the stomach and are finally passed in the faeces from which, in due course, are released the adult flies.

Each parasite will be mentioned, wherever possible, according to its habitat when destruction or control is attempted, e.g. tapeworms as alimentary parasites, trypanosomes and histomonas as tissue parasites, and warble larvae as external parasites. The first will therefore be mentioned in this chapter whereas the other two will be dealt with elsewhere.

Most parasites of the intestinal tract and structures such as the bile ducts are classified under three main headings.

1. *Cestodes or tapeworms* are segmented flat worms with a complex life history involving an intermediate host which may be either warm or cold blooded. *Taenia solium* in the immature cystic form lives in the body tissues of the pig, whilst the adult worm lives in the intestine of man. In the group with cold-blooded intermediate hosts is *Dipylidium caninum* which lives in the alimentary canal of dogs and cats in the adult form, but in dog and cat fleas in the immature cyst stage. The clinical significance of these worms is related more to aesthetic considerations than to pathological potential.

2. *Trematodes or flukes* are non-segmented, mainly flat worms with a complex life history involving snails. The adults of the important species normally live in the bile ducts of ruminants. They are responsible for considerable losses, to the livestock industry mainly because of slow and poor development of affected animals, and to the meat industry by reason of damaged livers and poor carcases.

3. *Nematodes or roundworms* are elongated, tubular worms which inhabit the stomach and intestines of both domestic and wild animals and birds. A small group, the 'lungworms', are found in the lung tissues and bronchi. Their life cycle involves free living forms and frequently immature forms which migrate through the host's tissues before reaching maturity in the intestines or lungs.

This group includes species which are of the greatest clinical significance.

THE PARASITE/HOST RELATIONSHIP

The whole story of the relationship between worm parasite and host is not yet known, but is even now seen to be as complex as that between bacteria and viruses and their hosts. Indeed the picture is very similar.

The establishment of significant infection or infestation in an animal or bird is determined by a series of inter-related factors.

1. *Parasite dose*. Just as the establishment of a clinical infection by bacteria or viruses depends to some extent on the number of organisms taken in by the host within a given period, so does the number of infective eggs or larvae taken in by the host determine whether an infestation becomes established and of clinical importance, although in some cases large doses induce the development of strong resistance by the host.

2. *Parasite significance*. Some families, species, and even orders, show different levels of danger and affinity for the different species of animals. Some orders or parasites are of less significance than others. Tapeworms, for example, are comparatively insignificant clinically when compared with the strongyles or ascarids. In general those parasites inhabiting the stomach and early part of the small intestines are the most pathogenic.

3. *Host resistance*. Adults are far more resistant than young stock. Considerable research is taking place into the host resistance aspect of parasitism. Resistance to parasitization can be considered as due to two related phenomena.

In the first place resistance is associated, with an antigen–antibody reaction in which the antibody may be inherited, or acquired by reaction to non-lethal doses of eggs or larvae.

In the second place there would appear to be an individual resistance proportional to general health and nutrition. It is rare to find animals on a good nutritional plane succumbing to parasitic infestations except possibly in the case of lung worm infestations due to *Dictyocaulus*. It would seem that a healthy well-fed animal exerts some influence on the parasites infesting it and controls to a certain extent the number of immature parasites which develop from the infective stages taken in. Also the reproductive rate of the mature parasites is affected so that the pasture and general surroundings are not so heavily contaminated with infective eggs or larvae.

It must be realized that, though this chapter concerns the use of drugs to free animals from intestinal parasites, this is only one aspect of the treatment of intestinal parasitism. In this way therefore, anthelmintics are no exception to the rule that drugs are not the only, or even the most important, weapon in the battle against disease. Adequate diet, housing, cleanliness, exercise, pasture management, etc., are some of the factors involved in the prevention and control of parasitism, as well as other forms of infection.

Finally it should be said that the mere presence of parasites within the alimentary canal is not an indication that they are causing trouble. Usually it is only when several parasites are noticed in the faeces on several occasions, or when considerable numbers of eggs are counted that concern may be felt. (It has been stated that 3000 adult *Haemonchus contortus* females have been found in the abomasum, and that 9000 eggs per g have been counted in the faeces of apparently normal adult sheep.) It is, however, always wise to administer anthelmintics to remove parasites known to be present even if the animal is showing no signs of their effect. Where the animal is obviously ill other causes of illness should always be ruled out before anthelmintics are used.

CHARACTERS OF AN ANTHELMINTIC

An ideal anthelmintic should:

1. Have the widest possible therapeutic index. (This is the ratio of effective dose to toxic dose, e.g. therapeutic index of 1:2 is very close or narrow, as one part will destroy the parasite whereas two parts will destroy the host. A therapeutic index narrower than 1:4 is not considered safe, and a compound with such an index is not safe to use as a therapeutic agent.) Many anthelmintics have narrow ratios but the newer ones for the control of both gastro-intestinal worms and liver fluke have proved to be remarkably safe in use.

2. Have a wide spectrum of activity.

3. Be active against both the mature and the immature stages of the worm.

4. Not require any interruption in the normal life of the animal, nor cause any check in its development.

5. Be acceptable to the animal so that administration causes no concern.

6. Be economically acceptable as part of a total control programme.

THE COMMON HELMINTHS OF DOMESTIC ANIMALS AND AGENTS USED AGAINST THEM

The helminths are grouped in their three categories by anatomical form:

1. The flat segmented tapeworms or cestodes,
2. The flat, non-segmented flukes or trematodes.
3. The roundworms or nematodes which are circular in cross-section. This group can be further divided into two sub-groups, the strongyloid and the ascaroid worms.

The Horse

Three species of the *tapeworm*: *Anoplocephala* species are found in the small intestine. *Flukes* do not normally affect horses, but members of the third group, the nematodes, are the most important helminths of horses.

Strongyloids can cause severe symptoms; anaemia, dysentery and death may follow severe infestations. Four genera are involved

1. *Strongylus*
2. *Triodontophorus*
3. *Trichonema*
4. *Trichostronglyus*

The first three ('redworms'), occur in the large intestine and the fourth, which is one of the few helminths capable of passing from horse to cow, to sheep and in the reverse order, in the small intestine. A fifth genus is *Dictyocaulus*, a lungworm genus not of much significance in the horse.

Treatment of these strongyloids during the last two to three decades has been by the administration of phenothiazine, but more recently it has been shown that piperazine will remove 100 per cent *Trichonema* adults, 50 to 60 per cent of *Triodontophorus* species, and a similar percentage of *Strongylus vulgaris* adults, though other *Strongylus* species are less affected. These figures approximate to those given by 30-g doses of phenothiazine. Thiabendazole, because of its lack of toxicity and its high efficiency against the large strongyles, *Triodontophorus* species and oxyurids, has become a major equine anthelmintic. Chenopodium, sodium glycarsamate and carbon tetrachloride are now obsolete.

Ascaroids in the horse are of less importance. Three genera are involved:

1. *Ascaris*
2. *Oxyuris* (seatworms, whipworms or pinworms)
3. *Strongyloides*

Piperazine compounds are extremely effective against immature and mature *Ascaris* species. Benzimidazole, pyrantel, dichlorvos and thiabendazole are now the drugs of choice for the treatment of the major helminths of the horse.

Ruminants

The parasitic genera affecting cattle, goat and sheep are almost identical, and treatments, with few exceptions, vary in dosage only.

Tapeworms, One genus (*Moniezia*) occurs in the small intestines of ruminants, this rarely causes trouble, and then only temporarily. Yomesan (Bayer 2353) has proved to be effective against *Moniezia* at a high dose rate. The bunamidine salt is also now available.

Flukes. Two species, *Fasciola hepatica* and *Dicrocoelium dendriticum* are responsible for considerable extensive and severe liver disease in cattle and sheep. *Fasciola* is the one mainly implicated. A range of anthelmintics active against both larvae and adult stages of *Fasciola* are now available in sheep and cattle.

Roundworms. *Stongyloids:* as in the horse it is this group which causes the greatest trouble, and considerable losses occur annually from unthriftiness and death of stock. Species from eight strongyloid genera may be found in the stomach, small, and large intestines. The stomach worms consist of species of *Haemonchus*, *Ostertagia* and *Trichostrongylus*, all of which can, and do, cause severe losses. The small intestine worms are species of *Nematodirus*, *Bunostomum* (hookworms) and *Cooperia*, of which the first two are important pathogens. Some species of the genus *Trichostrongylus* also occur in the small intestine and can be dangerous pathogens. Two trichostrongyloids, *Chabertia* and *Oseophagostomum* (nodular worms), occur in the large intestine. Both are of pathological significance.

The anthelmintic hitherto of widest application was phenothiazine, to which *Bunostomum* and *Nematodirus* have a low sensitivity. Thiabendazole and other new anthelmintics have replaced phenothiazine as the anthelmintic of choice for the treatment of strongyles in cattle and sheep and established themselves as the greatest advances in anthelmintic chemotherapy since the discovery of phenothiazine. *Bunostomum* species are sensitive to thiabendazole, and the compound has the considerable advantage over phenothiazine in that it eliminates *Nematodirus* infestations in addition to being more efficient against mature and immature forms of the majority of other gastrointestinal parasites of ruminants; above all it is virtually non-toxic. Tetramisole, another of the newer anthelmintics, is active against enteric helminths and can in addition be given by injection. So far no anthelmintic vaccine against gastro-intestinal nematodes has yet been introduced commercially following investigations stimulated by the development of the husk vaccine.

The lungworms consist of three strongyloid genera, *Dictyocaulus*, *Protostrongylus* and *Muellerius*. The first, which is the most pathogenic, is the only one of the three which occurs widely in Great Britain, causing 'husk' in cattle. Intensive research into husk prophylaxis and treatment produced the world's first commercial vaccine against a worm parasite. In addition two chemicals, cyanacethydrazide and diethylcarbamazine have been developed to treat infestations which have become patent. Deaths still occur following treatment, often due to the destruction of parasites and blockage of the lung airways, and it is still felt that more work is needed to establish optimum doses and to determine relative efficacy against mature and immature *Dictyocaulus*. Tetramisole, levamisole and thiabendazole also exert a clinically beneficial action on *Dictyocaulus* infestations.

The Pig

No *tapeworm* is normally found in the intestines of the pig in Great Britain. *Diphyllobothrium latum* is occasionally found in the small intestine but man is the definitive (usual) host for this tapeworm.

Flukes do not normally occur as a major infestation in pigs.

Roundworms of both strongyloid and ascaroid genera occur in pigs. *Strongyloids* are of three main genera:

1. *Hyostrongylus* species are found in the stomach of pigs and can be quite troublesome.
2. *Oesophagostomum* (nodular worm) species cause lesions in the colon which are responsible for appreciable losses of sausage skins.
3. *Metastrongylus* species are the lungworms of pigs.

Piperazine compounds are effective treatments for *Oesophagostomum* infestations.

The *ascaroid* species are the most important helminths of pigs. *Ascaris* species and *Strongyloides* species are the two generally involved. They normally inhabit the small intestine. Piperazine is 100 per cent effective against adults and intestinal larvae of *Ascaris lumbricoides*.

Dichlorvos, levamisole, mebendazole and fenbendazole have all been demonstrated to be active against the major helminths in pigs.

The Dog and Cat

Tapeworms are very common in the dog but are less frequently seen in the cat. Four genera are implicated:

Taenia species, *Echinococcus*, *Dipylidium* and *Diphyllobothrium*. *Taenia* and *Dipylidium* are the only two occurring normally in Great Britain. All are inhabitants of the small intestine. They are rarely more than a nuisance to the animals but are important from the Public Health aspect as household pets may act as a reservoir for the infestation of the food of animals and man.

Dichlorophen is a safe cestode anthelmintic but the disintegration of the worms caused by the drug has been reputed to cause allergic reactions, and moreover owners, and veterinary surgeons too, are more satisfied if they are able to count the number of worm heads voided after treatment. Bunamidine and niclosamide (Yomesan, Mansonil) are now used in dogs. Ill effects from tapeworms are not often seen and treatment is rarely required in dogs under 1 year of age.

Flukes are of no significance in the dog and cat in Great Britain. **Roundworms** are more pathogenic than tapeworms but it is mainly puppies that require treatment. *Strongyloides* species occurring in the small intestine are *Ancylostoma* and *Uncinaria* species (hookworms), whilst *Aleurostrongylus* and *Oslerus* species occur in the respiratory tract. The best treatment for *Uncinaria* would appear to be piperazine but this compound is on an average only 40 per cent effective against *Ancylostoma*. Thenium, dichlorvos and diethylcarbamazine are all now used for the control of *Strongyloides* species.

Against the lungworms an effective treatment has yet to be discovered.

Whipworms should be treated with phthalofyne, and butyl chloride may also be used.

Ascaroids occurring in the dog and cat consist of species of *Toxocara* and *Toxascaris*.

Piperazine compounds again are 100 per cent effective.

A *filaroid* species, *Dirofilaria immitis*, the adult stages of which live in the heart and major blood vessels of the dog, is treated with thiacetarsamide sodium (Caparsolate) and diethylcarbamazine citrate (Dirocide). Treatment is both preventive and therapeutic.

Poultry

The *tapeworms* of poultry live in the small intestines and belong to four genera. *Davainea* and *Raillietina* species are considered of greater importance than members of the other two species, *Amoebotaenia* and *Hymenolepis*.

The standard treatment was carbon tetrachloride but more recently it has been found that di-n-butyl tin dilaurte and tin oxide are very efficient, though the latter was more toxic than the dilaurate. Investigation into the action of proteolytic enzymes has indicated that further research may provide a new range of safe and efficient anthelmintics.

Flukes have been reported in ducks but no significance is attached to them in poultry.

Roundworms comprise the *Strongyloides*, *Syngamus* (gapeworm) and *Amidostomum* (gizzard worm of geese), neither of which are easily treated. Reports indicate that thiabenzadole, tetramisole and mebendazole are very useful in the treatment of *Syngamus* infestations.

Ascaroid genera affecting poultry are those of *Heterakis* (caecal worm) and *Ascaridia*, found in the small intestine.

A genus of the order *Trichinelloidea* is also found in poultry in this country. This is the genus *Capillaria*, species of which may be found in the oesophagus, crop and small intestine of poultry. Pathogenicity is reported to be serious. Experiments on pigeons showed that faecal counts of 4000 eggs per gram were invariably associated with symptoms, and that piperazine was over 90 per cent effective against *Capillaria* species.

Since poultry were intensively reared, helminths have ceased to be of real economic importance.

THE USE OF ANTHELMINTICS

It will be apparent from examination of these lists of parasites and drugs that certain drugs have an action specifically against parasites within an order or class. Partly by chance, and partly by industry and observation, modern research has provided us with highly efficient drugs specific to both important orders of the class Nematoda as well as the class Trematoda.

It will be seen that thiabendazole is highly efficient and specifically active against strongyloid species, piperazine is very efficient and active against ascaroids as well as being extremely effective against some strongyloid species, whilst hexachloroethane and other drugs is very effective against the trematodes.

It is common practice now to mix two drugs, e.g. a flukicide and a nematicide, so that the farmer can dose for both types of worms at the same time. This is the equivalent of giving a broad spectrum effect with an antibiotic.

COMPARISON OF THE EFFICACY OF DIFFERENT ANTHELMINTICS

It has been suggested that in a textbook of this type, it would be useful to compare the relative merits of different therapeutic agents, as indicated by field trials and other control procedures. In the case of anthelmintics, particularly with those new drugs that have been introduced during the last few years, it has not been considered that such information would add to that already given in the text, as when efficiency is indicated for a drug, it means an activity in the region of 95–100 per cent. The effective use of anthelmintics is related so closely to management systems that a difference of efficiency of between 95 and 100 per cent has little practical relevance, especially when all the key management preventive techniques are practised. In the case of liver fluke control, for instance, a drug which is relatively inefficient against larval stages, such as carbon tetrachloride, can be shown to be effective in controlled trials where eradication of the snail intermediate host is carried out in addition to fluke therapy.

It has to be appreciated that all anthelmintics are effective only for the short time that they are in the body and that a grazing animal can become re-infested within a few days of treatment.

Efficiency against Gastrointestinal Worms

A paper by Anderson and Dobson (1975) discusses the importance of anthelmintic efficiency when related to strategic control schemes for helminthiasis in sheep. This paper from Australia illustrates how the use of anthelmintics which are effective against gastrointestinal worms has to be related to management systems, climatic conditions, and the susceptibility of the animals to infestation. It is because of these variations that the veterinarian has to use his skill in defining which particular routine will suit a farming system.

Efficiency against Liver Fluke

In Holland, where a national control programme for liver fluke infestations is carried out, it has been found that when controlling liver fluke infestation in cattle, it is possible to use an anthelmintic which is mainly active against the adult fluke. Such limited activity would not always be considered adequate in the control of fluke infestations in sheep in the hills of Wales or Scotland. Armour and Bruce (1973) describe the strategic use of rafoxanide in the control of liver fluke infestations in sheep.

In cattle infestations, liver fluke disease manifests itself as a mainly subclinical infestation because acute attacks by larval stage are uncommon, whereas in sheep infestations the larval stages of the fluke can under suitable climatic conditions (heavy rains and very moist herbage) cause an acute fatal disease.

It is because of these differences that fluke control programmes in cattle and sheep may require different emphasis and may have to be initiated at different times in the year.

Above all it has to be appreciated that climate, animal population and the state of the grazing all play an important part in the effective use of even a very active anthelmintic.

The aim of modern therapy is to prevent infestation by controlling larval infestations. This is achieved by strategic dosing during known periods of larval activity.

CLASSIFIED ANTHELMINTICS

The drugs are grouped according to the class or order of helminths against which they are active. Where overlapping of activity occurs drugs will be related to the class against which they are most commonly used.

It is common with the available anthelmintics to have a therapeutic drug which will be active in a number of classes. Where this occurs the activity will be mentioned under the particular drug.

Cestoda

Control of cestodes cannot be achieved by drug administration only, and therefore measures aimed at preventing contact between primary and intermediate hosts must also be used. Thus dogs and cats should not be fed on raw meat, while fleas and other insect intermediate hosts should be destroyed and suppressed wherever possible.

Organic Drugs of Synthetic Origin

Hexachlorophene and dichlorophen have both been used in recent years for the treatment of tapeworm infestations. Their use against other species of helminths gives poor results.

Bunamidine Salt

Bunamidine hydrochloride (*N*,*N*-dibutyl-4-hexyloxy-1-naphthamidine hydrochloride (Scolaban)) is now the major salt of bunamidine used for the control of tapeworm. It has replaced the hydroxynaphthoate. It is a white crystalline powder soluble at 20°C in 200 parts water and in 2 parts of alcohol. The hydrochloride is irritant especially to the eyes and has to be handled with care. The drug is used for the treatment of tapeworm infestations in dogs and cats. A single treatment is effective against *Taenia* species, *Dipylidium caninum*, and adult forms of *Echinococcus granulosus*. A second dose should be given in *Echinococcus* infestations after an interval of 6 weeks to eliminate those worms which were immature at the first treatment. Dogs and cats should be dosed on an empty stomach. Vomiting may follow treatment, but this does not effect the efficiency of the drug.

Dosage

Dogs: 2–5 kg 100 mg 5–10 kg 200 mg
Cats: 100 mg

Dichlorophen, B.Vet.C.

This is known as Di-phenthane 70 and is available in various proprietary compounds.

It is a chlorinated derivative of phenylmethane and consists of a white powder with a phenolic odour and a combined phenolic saline taste. It is almost insoluble in water, but very soluble in alcohol and ether.

Dichlorophen is an effective bacteriostat, a fungicide and an anticestode agent.

As a *fungicide*, dichlorophen is widely used as a preservative. In the Second World War canvas and webbing equipment was impregnated with this compound to good effect. Insolubility in water ensures persistence. It has a very wide veterinary application in the treatment of mycosis.

As a *tapeworm treatment* dichlorophen is used in the dog and cat against *Taenia* and *Dipylidium*, but its effect on *Echinococcus* is variable.

Dosage

Sheep:
0·5 g per 2·5 kg body weight
(adults 5–15 g, lambs 2–4 g)

Dogs:
0·3 g per kg body weight

Cats:
0·1 to 0·2 g per kg body weight

The dosages are somewhat bulky and detract from the usefulness of the drug.

The drug can be administered in tablet or suspension and is best given after overnight fasting. No purgative is needed.

It should be impressed on the client that the tapeworms are digested and passed in an unrecognizable form often 6 to 8 hr after dosing. This is considered by many practitioners to be undesirable as tangible results are more impressive and more easily assessed. Possible allergic reaction to the products of worm disintegration is also feared.

Toxicity

Toxicity is low but the vomition and diarrhoea associated with most anthelmintics occur occasionally.

Niclosamide (Yomesan, Mansonil)

Niclosamide is 5-chloro-*N*-(2-chloro-4-nitrophenyl)-2-hydroxy-benzamide. It is a cream coloured tasteless powder. It is slightly soluble in water, but is soluble at 20°C in 150 parts of alcohol. It is prepared in tablet form for dogs and cats. Each tablet contains 500 mg niclosamide. When used at the recommended dosage it has a wide margin of safety.

Actions and Uses

It is a taenicide which eliminates scolex and body. The head may not be observed, but it is eliminated with the rest of the worm and is often destroyed by the disintegrating action of the drug on the scolex and neck. It is effective against tapeworms in dogs and cats, and against *Moniezia* spp. in cattle, sheep and goats. It is administered as a drench to cattle and sheep, and as tablets to dogs and cats.

Dosage

Cattle: 50 mg per kg body weight
Sheep and Goats: 100 mg per kg body weight
Dogs and Cats: administer orally at the rate of 1 tablet (500 mg) per 3 kg/body weight preceded by an overnight fast.

Dye Compounds

The triamine dyes, particularly methyl violet, are reported as being more effective than the diamine compounds such as brilliant green when used for *Dipylidium* infestations of the dog, but the acridine dyes are more efficient and practical. Mepacrine Hydrochloride, B.Vet.C. in doses of 20 to 40 mg per kg is effective against *Didylidium* and *Taenia* species, but even at the rate of 50 mg per kg is useless against *Echinococcus*. Aminacrine Hydrochloride, B.Vet.C. has been reported to be safe and effective in cats at the rate of 200 mg per kg. Further investigations into the effect of the acridine dyes are in progress.

Inorganic Drugs

Lead Arsenate, B.Vet.C.

An insoluble white powder which has been introduced in recent years almost specifically for the treatment of *Moniezia* infestations of sheep and cattle, but it has been proved ineffective against the fringed tapeworm *Thysanosoma actinoides* which is very common in sheep in various parts of the world (Australia, America, etc.).

Dosage

Dosage in calves, sheep and goats is 0·5 to 1 g and in adult cattle 1 to 2 g. It is not safe to use in poultry but in these doses toxicity is not likely in ruminants. The dose is usually administered in a capsule and no fasting is required; it may be followed by an oleaginous purgative.

Tin Compounds

Tin compounds are being widely investigated as anthelmintics in all species. Stannous oxide and di-n-butyl tin dilaurate are in use, the dilaurate in doses of 1 ml, or at the rate of 0·5 ml per kg body weight, is reported as being highly efficient against tapeworms in poultry, particularly against *Davainea* species.

Tin compounds have also been used in man and dogs, with good effects reported against *Taenia saginata*, *T. solium* and *Dipylidium caninum*. A dose of 0·5 g stannous oxide daily for 8 days was given to human patients.

Trematoda

It cannot be emphasized too strongly that the use of drugs is not in itself the complete answer to fascioliasis. A very strong economic case has been made for the application of a combined programme of regular anthelmintic treatment in association with a detailed scheme for snail control. In many cases the problem responds more completely to methods aimed at destroying the intermediate snail hosts, both by copper sulphate pasture dressings and by improving land drainage. This should be combined with periodic herd and flock dosing to gain the best results.

Organic Drugs of Synthetic Origin

Carbon Tetrachloride, B.Vet.C.

This drug is a heavy, clear and colourless volatile liquid with an odour similar to chloroform. It has a burning taste, and though non-inflammable will produce phosgene gas when ignited. The liquid is practically insoluble in water but miscible with organic solvents and oils.

Actions and Uses

The main use of carbon tetrachloride is in the treatment of *Fasciola hepatica* infestations of sheep. It is of no value against immature flukes in normal doses and is not effective against *Dicrocoelium dendriticum.*

Carbon tetrachloride is absorbed slowly but the presence of animal or vegetable fats and oils will increase the absorption rate. Once absorbed and circulating, the drug is excreted at various points—the lungs, the kidneys, but mainly through the liver. The biliary system returns the active drug metabolites to the intestines, and in so doing the mature and nearly mature adult flukes in the bile ducts are destroyed. The immature flukes in the liver tissue are unaffected and consequently monthly dosing is required where heavy infestations occur.

Dosage

Sheep: 1–3 ml

The drug can be administered as a drench and mixed in 300 or 600 ml of liquid paraffin, but capsules are often a more suitable form of administration. Care should be exercised that capsules should not be broken in the mouth, for the tetrachloride has similar anaesthetic properties to chloroform. No fasting or purgatives are necessary but dosing is usually carried out before morning feeds, and can be followed 2 to 3 hr later by a saline purgative. It is advisable, before flock or herd dosing with carbon tetrachloride and similar chlorinated hydrocarbons, to withhold concentrate foods for a week or so prior to dosing, and to try the effect on a few of the worst affected individuals in the group. Flock susceptibility is responsible for frequent losses in sheep where dosing with carbon tetrachloride is not adequately controlled. Since about 1960 the parenteral administration of carbon tetrachloride alone or in mineral oil solution has been widely practised. Solutions in liquid paraffin or in purified soft paraffin, usually with a local anaesthetic are preferred. In Great Britain mixtures of carbon tetrachloride with soft paraffin containing butyl alcohol as local anaesthetic have been widely used in sheep. Doses in sheep should provide a maximum of 4 ml carbon tetrachloride and in cattle a maximum of 30 ml at the rate of 1 ml per 9 kg body weight. Lameness following injection is common and local necrosis persisting for at least 6 weeks has been reported. Experimental evidence indicates that fibrous encapsulated lesions may develop in muscle tissue. Pigs, horses, and cats are not suitable species for dosing with this drug.

Toxicity

Toxicity is typical of the group of the carbon and chlorine compounds and is due to the considerable damage which can be caused to the liver. Death is usually due to liver failure following extensive fatty changes, but earlier symptoms are hyperaesthesia and convulsions, followed by coma. Hypocalcaemia and low liver glycogen content are predisposing causes and treatment of toxicity should consist of intravenous calcium borogluconate and dextrose injections, whilst the detoxicating activities of the liver should be reduced by cutting the protein intake.

Hexachloroethane, B.Vet.C. (C_2Cl_6) (HCE, Avlothane)

This white or off-white crystalline powder has an odour faintly resembling that of camphor. It is almost tasteless, insoluble in water, slightly soluble in alcohol but soluble in most other organic solvents. Up to 5 per cent wetting or dispersing agent is usually included to aid administration.

Actions and Uses

Absorption of hexachloroethane is fairly rapid and its metabolism follows a similar course to carbon tetrachloride except that the majority is excreted via the liver. Although it was claimed that hexachloroethane attacked immature flukes, more recent trials show that immature flukes, and particularly those in the liver tissue, are unaffected. All *Fasciola* species are affected. Hexachloroethane is also effective against *Haemonchus* and *Trichostrongylus axei* in the stomach of ruminants but not against intestinal parasites.

Dosage and Administration

Cattle:
15–100 g at the rate of 10 g per 50 kg body weight
Sheep and Goats:
8–15 g

Administration is by drench or stomach tube and the precautions adopted to prevent toxic symptoms with carbon tetrachloride should be observed. Dosing should continue at monthly intervals where repeated infestation may occur.

Toxicity

This is relatively low but herd and flock susceptibility does occur. Treatment is the same as for all chlorinated hydrocarbons.

Tetrachlorodifluoroethane

This compound is widely known as Freon 112. It has been investigated as a treatment for liver fluke in both cattle and sheep at dose rates of 100 to 300 mg per kg body weight in sheep and up to 660 mg per kg in cattle. The compound is given in liquid paraffin as a drench. Adult flukes were effectively destroyed but immature flukes were inconsistently affected. Some commercial samples showed a complete lack of effect. This was shown to be due to the presence in the sample of a high proportion of inactive isomers. Deaths in cattle have been reported and it is probable that if the isomer problem is solved the product will still be unsuitable for cattle.

MODERN EFFECTIVE FASCIOLICIDES

The discovery of diamphenethide, with its high activity against immature flukes, is a considerable advance in the therapy of fluke infestations, particularly in sheep. It has made possible therapy early in the autumn before the flukes have become established in the liver. If early treatment is carried out, the hill sheep has a better chance of surviving pregnancy and poor winter grazing conditions.

The range of fasciolicides now available for therapy provides the drug armoury necessary for the establishment of effective fluke control programmes, even under the most difficult conditions.

Diamphenethide (Coriban)

Di-(2-(4-acetaminophenoxy)ethyl)ether (VII) is a smooth white thixotropic opaque liquid which sets to a soft gel.

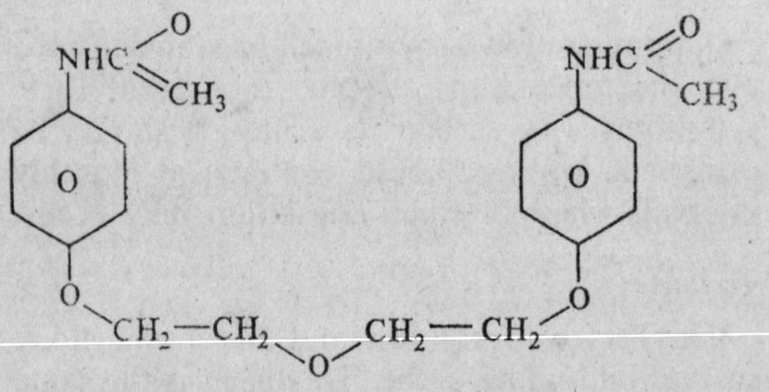

Diamphenethide

Actions and Uses

The drug is active against mature and immature *Fasciola hepatica*. It is given orally as a drench.

Field trials have demonstrated that a dose of 100 mg per kg was almost 100 per cent effective against *F. hepatica* between 1 and 6 weeks and 80 mg per kg was 100 per cent effective against 3–50-day-old *F. gigantica*.

Dosage

Sheep: 14 ml for 23–31 kg body weight
21 ml for 32–40 kg body weight
28 ml for 41–61 kg body weight
43 ml for 62+ kg body weight.

This dosage can be used for both prevention and cure of fascioliasis.

Withholding Period

Do not slaughter within 7 days of therapy.

Nitroxynil (Trodax)

Nitroxynil is 4-hydroxy-3-iodo-5-nitrobenzonitrile. It is a light yellow solid, slightly soluble in water used as a 20 per cent solution.

Actions and Uses

The drug is active against *F. hepatica*, mature and immature flukes not less than 4 to 6 weeks old. It is also effective against *F. gigantica*.

It can be given either orally or by subcutaneous injection. It is claimed to be more effective when given by the subcutaneous route.

Dosage

Sheep and cattle: 1 ml of 20 per cent solution for 20 kg body weight (10 mg/kg).

Withholding Period

Do not slaughter within 30 days of therapy.

Oxyclozanide (Zanil)

Oxyclozanide is (3,3′,5,5′6-pentachloro-2,2′-hydroxy benzanilide zanil). Oxyclozanide is used for the control of *Fasciola hepatica* in sheep and cattle. It is administered by mouth in a suspension containing 3·4 per cent w/v.

Dosage

Sheep: 10–20 mg per kg body weight
Cattle: 10 mg per kg up to a maximum of 3·4 g

Tribromosalan (Hilomid)

This product consists of a mixture of equal parts of 3,5-dibromo salicylic acid-4′-bromanilid and 9-bromosalicylic-4′-bromanilid. It is a white tasteless powder, which is given as a drench. It is active against both adult and immature stages of *Fasciola hepatica*.

Dosage

Sheep: 30 mg per kg body weight—12 weeks old fluke; 60 mg per kg body weight—immature fluke.

Rafoxanide (Flukanide, Ranide)

Rafoxanide is a salicylanide compound 3-5 diiodo-3′-chloro-4′-(*p*-chlorophenoxy) salicylanide.

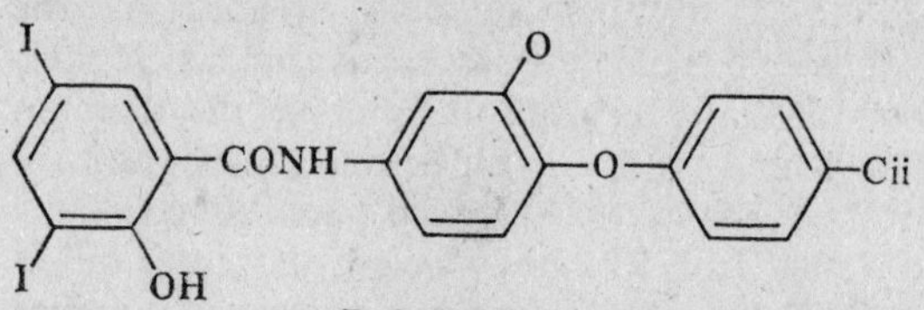

Rafoxanide

The pure drug is an off-white crystalline powder with a high melting point, 173–7°C.

Pharmacology

Rafoxanide is well absorbed by cattle and sheep with peak plasma levels occurring between 24 and 48 hr after dosing. There is a good correlation, following a single dose, between drug blood levels, efficacy and toxicity. Rafoxanide is not metabolized by cattle and sheep to any detectable degree. The half life varies from 5 to 10 days in sheep.

Actions and Uses

Rafoxanide is active against 99 per cent of adult and immature *Fasciola hepatica*, more than 99 per cent adult and up to 91 per cent immature of *Fasciola gigantica.*

Dosage

15 mg/kg removed 90 per cent of 4-week-old flukes.

Dosage recommended after a range of field trials:

Sheep and Cattle: 7·5–10 mg/kg.

Toxicity and Tissue Residues

Rafoxanide has been demonstrated to be safe and effective at the recommended dose.

Nematoda

The control of this class of parasite follows the pattern of other classes. Mere administration of drugs is valueless without husbandry methods aimed at reducing food and water contamination, whether in the animal accommodation, or on pasture. In some cases, as for example in strongyle infestations of the horse, veterinarians are relying more and more on small repeated doses of a drug to depress egg laying by the female worms and at the same time reduce hatchability: by this means pasture infestation can be kept at a minimum. Correct grazing husbandry related to adequate but not overstocking, mixed species grazing and periodic harrowing to break up and expose to light and air the faecal droppings of the animals will assist in maintaining a safe level of pasture contamination.

Research has demonstrated that larval infestations established in the gut can form a major source of subclinical infection and emphasis is now placed on the development of anthelmintics which have a high activity against the larval stages.

It is important too to establish a control programme of therapy which is appropriate for the particular system of management and type of pasture. Control programmes can therefore differ from country to country, and even within a country in continents such as the U.S.A. and Australia. The main aim of such control programmes is to dose animals at a time which will prevent larval infestations. It has been shown particularly in the case of *Ostertagia* infestation that the damage done by the larvae when establishing themselves in the gut wall is very considerable.

A. ANTHELMINTICS AFFECTING STRONGYLES

Bephenium Embonate B.Vet.C. and Bephenium Hydroxynaphthoate B.P.C. (Frantin)

These two compounds are salts of a complex quaternary ammonium compound with respective empirical formulae of $C_{57}H_{58}N_2O_8$, H_2O and $C_{28}H_{29}NO_{14}$ Both are almost odourless, yellow crystalline powders which have a bitter taste and are insoluble in water but slightly soluble in alcohol. For veterinary use they are marketed as dispersible powders for the preparation of cattle and sheep drenches. The embonate was the first to be used but the hydroxynaphthoate 2 parts salt contain 1 part bephenium base) is now the preparation which is most commonly used.

Actions and Uses

The first reports of the veterinary use of bephenium compounds appeared in 1958 and 1959 and showed that the embonate was effective at 250 mg per kg body weight dosage rate against *N. filicollis* and *N. battus* in sheep. Activity against other nematodes of sheep—*Haemonchus*, *Ostertagia*, *Oesophagostomum* and, to a lesser extent, *Trichostrongylus* and *Cooperia* species—were also demonstrated. The hydroxynaphthoate compound, whilst being equally efficient against *Nematodirus* species, is more active than the embonate against other species of nematodes.

Dosage and Administration

Cattle, Sheep: 250 mg per kg body weight.

For the dosing of sheep it is recommended commercially that the contents of a 250 g bottle should be mixed in 2·27 litres water and 45 ml given as a drench or by gun to each lamb, or 75 ml to each adult sheep, irrespective of weight. Prophylactically, during the nematodiriasis season, dosing should be repeated every 3 weeks. Therapeutically, the same dosage scale should be used, with a repeat dose, if necessary, 3 weeks later.

Because of their relatively low activity against helminths other than *Nematodirus*, the bephenium drugs have been largely replaced by thiabendazole, levamisole etc. They are still used, however, in areas where *Nematodirus* infestations can provide a special problem.

Toxicity

It has been claimed, and reported, that the bephenium compounds are non-toxic at many times the effective dose, but occasional toxic symptoms at twice the recommended dose in sheep can occur. Salivation and vomiting occur fairly frequently in dogs but these symptoms do not appear to affect the efficiency of the product —capsules are superior to tablets in this respect.

(*N*:*N*-Dimethyl-*N*-2-phenoxyethyl-*N*-2′-thenyl-ammonium)*p*-chlorobenzene sulphonate (Ancaris)

The search for compounds active against hookworms has produced a number of entirely new types of compounds. This is one of a series now collectively referred to as thenium compounds. It was shown that the compound was highly effective against *A. caninum* and *U. stenocephala* when given on two occasions (morning and evening) in doses at the rate of 20 to 25 mg per kg body weight. As the compound is only partially active against *T. canis* commercial preparations combine the thenium compound with a piperazine compound in proportions of 1:2. Weaned puppies and adults of less than 2·5 kg body weight should be dosed at half the adult rate and unweaned puppies should be given doses at half the adult rate if over 2·5 kg or a quarter of the adult rate if under 2·5 kg body weight. Animals should be fasted for about 3 hr before and after each of the two doses.

Methyridine (Promintic)

The compound has a chemical name of 2(methoxy-ethyl)pyridine. It is a colourless liquid with a characteristic odour; it causes a burning sensation when placed on mucous membranes and if splashed into eyes should be washed out with copious amounts of water. As methyridine is readily absorbed through intact skin it must be washed off immediately. The fluid has a strong solvent action on fat and rubber and cannot therefore be stored in rubber capped containers.

Actions and Uses

Methyridine, with an action against gastro-intestinal roundworms has been developed as an anthelmintic mainly for cattle and sheep for administration either orally, by subcutaneous injection or by intraperitoneal injection. When used for oral administration methyridine should be diluted with 70 per cent of water, whilst for injection methyridine should be diluted with 10 per cent of water. Following administration the drug is rapidly absorbed either from the intestines, the peritoneal cavity or the injection site; a high proportion is excreted, or re-excreted, into the intestines exerting its anthelmintic action during the process. Excretion is complete in about 24 hr.

The drug acts by causing muscular paralysis due to neuromuscular block which is irreversible. This action makes the compound extremely effective against the intestinal worms of *cattle* and *sheep*, irrespective of the route of administration. Adult *Trichostrongylus*, *Nematodirus*, *Oesophagostomum*, *Bunostomum*, *Chabertia* and *Trichuris* species are highly susceptible and can be cleared within 16 hr. However, the activity against stomach worms (*Haemonchus* and *Ostertagia*) is irregular, owing to the variation in abomasal pH. Reports also indicate that methyridine is very effective against immature forms (99 per cent effective against 7-day-old forms but only 92 to 93 per cent effective against 14-day-old immature forms) of *Haemonchus* and *Trichostrongylus*.

The drug will eliminate *Capillaria* infestations from *chickens* following either individual oral dosing or drinking water medication. *Heterakis* and immature ascarids were not affected.

All reports indicate that methyridine is a highly efficient anthelmintic and that striking clinical improvement follows immediately on its use. But for the advent of thiabendazole the drug could have been of considerable significance in spite of the very narrow therapeutic ratio and the local irritant effect. As it is the greatest value of the drug is probably its activity against *Trichuris* species, against which thiabendazole is only moderately active.

Dosage

The standard dosage rate for all species is 200 mg per kg body weight. It is very important that this rate is not exceeded. *Chickens* should be given drinking water medicated at the rate of 0·2 to 0·4 per cent methyridine.

Administration

For *cattle* and *sheep* the oral subcutaneous and intraperitoneal routes may be used although local reactions occur at the injection site in a large proportion of those animals injected subcutaneously, and it has been indicated that the intraperitoneal route is not advisable in sheep. It would therefore seem advisable to use the intraperitoneal route for cattle and the oral route for sheep. *Chickens* should be given medicated drinking water for a period of 8 hr, but no ill effects have followed its use for 24 hr.

Toxicity

Methyridine possesses both a local and a general toxicity. The *local* toxicity is shown by mild to severe local tissue reaction and necrosis at the site of subcutaneous injections in a very high proportion of cases. So numerous were the reports of local reaction that the intraperitoneal route was developed for routine administration of the drug to cattle. *General* toxicity results from overdosage and may also follow normal dosage if the drug is given 'within a short time' of the administration of diethylcarbamazine. Twice the normal dosage rate has caused deaths, the toxic effects being irreversible—no antidote is known.

Organophosphorus Compounds

During recent years organophosphorus compounds, originally developed as systemic insecticides, have been recommended and used as anthelmintics, mainly in ruminants. The marketing of a large number of closely related complex chemical compounds each of which has been given a number of different registered names and identified by chemical formulae expressed in a variety of ways, has created a very confused situation. Their therapeutic vagaries have made the position even more complicated. Nevertheless, their wide range of activity and high efficiency under some circumstances has resulted in their widespread investigation and clinical evaluation. Certain basic group characteristics have been identified; they can be summarized as follows; '. . . the dose rate must be adjusted closely to body weights to ensure adequate anthelmintic activity and to minimize toxic effects; toxicity is unpredictable even when the prescribed dose is used; safety margin is small and the restricted range (due to inherent selectivity as well as the necessity to maintain a 'safe' dose level) requires a precise differential diagnosis before their selection for use in a flock. These compounds are not generally useful against *Oesophagostomum* spp., *Chabertia ovina* or *Nematodirus* spp., and against some other species activity depends in part on their being swallowed into the abomasum'. The position with regard to cattle is very similar.

Now that certain governments are banning the use of some hydrocarbon insecticides (aldrin, dieldrin and their related compounds) following grave international concern over the possibility of toxicity resulting from tissue storage of the chlorinated hydrocarbon insecticides, greater use of the organophosphorus insecticides could considerably increase the incidence of toxicity following worming. Greater care should therefore be exercised to ensure that organophosphorus insecticides and anthelmintics are not used simultaneously.

A wide range of organophosphorus anthelmintics are available and have been recommended. They are listed below under the names by which they are most commonly known.

Coroxon—*o*,*o*-diethyl, *o*-(3-chloro, 4-methyl, 7-coumarinyl)phosphorate.

Coumaphos—*o*,*o*-diethyl, *o*-(3 chloro, 4-methyl 7-coumarinyl)phosphorothioate.

Haloxon—*o*,*o*-di(2-chloroethyl), *o*-(3-chloro, 4-methyl, 7-coumarinyl)phosphate.

Neguvon—*o*,*o*-dimethyl,2,2,2-trichloro, 1-hydroxyethylphosphonate.

Ruelene—4-tert-butyl, 2-chlorophenylmethyl, methylphosphoramidate.

Dichlorvos—2,2 dichlorovinyl dimethyl phosph.

Very few organophosphorus anthelmintics have stood the test of field experience and because of their low therapeutic ratio they have been replaced by the newer anthelmintics described in the text.

Coumaphos

This compound is variously known as Asuntol, Bayer 21/199, Co-Ral and Muskatox. Its chemical formula may be written either as above or with the word 'umbelliferyl' substituted for the 'coumarinyl' group. It is available commercially in the form of a wettable powder for oral administration.

Actions and Uses

At dosages which are almost certain to result in occasional toxic symptoms Asuntol has a wide, though unreliable, range of activity, in common with all organophosphorus compounds, but when the dosage is reduced to its optimum level the species of worms for which it can be recommended are reduced to *Trichostrongylus*, *Haemonchus*, and *Cooperia* species. Other species are irregularly affected, e.g. *Oesophagostomum* and *Nematodirus* species. Coumaphos, whilst being more active, is also more toxic than Neguvon, an earlier product of the same laboratories.

Dosage and Administration

Cattle: 15 mg per kg body weight
Sheep: 8 mg per kg bodyweight

The direct entrance of the compound into the abomasum plays an important part in the activity of the compound and the oesophageal groove should therefore be closed by premedication with sodium bicarbonate and copper sulphate in the respective species.

Toxicity

Coumaphos causes a marked decrease in the blood cholinesterase levels and appears to be more toxic than Neguvon. In *cattle* severe symptoms and one death occurred in forty-six cattle dosed at 30 mg per kg body weight, but only very mild symptoms in 100 head dosed at 20 mg per kg have been reported, these results were compared with those obtained using Neguvon, which caused severe distress but no deaths at 180 mg per kg body weight. In *sheep* some toxic symptoms may be expected in a proportion of cases even when recommended dose levels are used.

Treatment

Atropine is the only antidote widely available although others are being developed. Sheep should be given up to 6 mg per kg body weight of atropine sulphate in divided doses, commencing with 2 mg per kg in aqueous solution: one-third of this should be given intravenously in a 10-ml volume and the rest subcutaneously in the same volume. Similar treatment should be given every 4 to 6 hr, according to response, until the maximum is reached.

Coroxon and Haloxon (Loxon)

Coroxon and *Haloxon*—which are very closely related to Coumaphos—have been produced following the search for compounds with greater activity but reduced toxicity. Coroxon, in combination with phenothiazine, was reported to be a safe, synergistic combination and extremely active in sheep. A mixture providing 2 mg of Coroxon and 200 mg phenothiazine per kg body weight was reported highly active against *Haemonchus*, *Trichostrongylus* and *Ostertagia* species and *Strongyloides papillosus*.

Haloxon was reported in 1962 following work in cattle, sheep, carnivorous species and rodents. It showed marked activity against *Haemonchus*, *Ostertagia*, *Trichostrongylus*, *Cooperia*, *Strongyloides*, *Nematodirus* and *Bunostomum* species in sheep and cattle at a recommended dose of 35 to 50 mg per kg body weight. Activity against the immature stages of some of these species was noted. Investigation into the toxicity of this compound in laboratory animals, dogs, cats, monkeys, poultry and sheep revealed neurotoxic effects in poultry and sheep at high dosage rates (above 530 mg per kg body weight). A therapeutic index of 3 to 4 for old ewes and 5 to 7 for young lambs was established by these same experiments.

Haloxon has now been widely used for treatment of cattle and sheep, and is a safe anthelmintic for general use.

Neguvon

The compound is known as Bayer L13/59, Dipterex and Dylox. It is known chemically as *o,o*-dimethyl,2,2,2-trichloro,1-hydroxyethylphosphonate or, alternatively, *o,o*-dimethyl, 1-hydroxyethyl,2,2,2-trichlorophosphonate. The use of the compound in cattle and sheep has been widely reported. The manufacturers recommend the product for the treatment of Bots in horses, of *Haemonchus*, *Oesophagostomum*, *Bunostomum*, *Cooperia*, *Ostertagia* and *Trichostrongylus* species in cattle, of *Haemonchus contortus* and *Oestrus ovis* (Nasal grub) in sheep, and of *Ascaris* in pigs.

In *sheep* the compound is effective at the comparatively safe dosage rate of 55 mg per kg against *Haemonchus* species, but relatively ineffective against other species. Increase of range and degree of activity can be obtained by pharyngeal swabbing with 10 per cent copper sulphate solution, but this will also increase toxicity. Occasional individuals will show no, or only poor, response following standard dosage.

Toxicity

Though less toxic than coumaphos, Neguvon can produce typical organophosphorus toxicity symptoms approximately 1 hr after dosing and it has been recommended that cattle should be kept under observation during the immediate post-dosing period so that atropine or 2-PAM (2-pyridine aldoxime methiodide) at the rate of 50 mg per kg body weight suspended in physiological saline, may be given immediately untoward symptoms appear.

Ruelene

Chemically the compound is 4-tert-butyl-2-chlorophenylmethyl,methylphosphoramidate. Known variously also as Dowco 132, Dow M1261, Montrel and Kempack (a 35 per cent oil-based solution), this compound was originally developed as a systemic insecticide to replace the more toxic Etrolene (Ronnel, Trolene, E.T.-57). However, significant anthelmintic effect had already been noted following the oral use of Etrolene in cattle against warble fly larvae (*Hypoderma* species) and subsequent investigation confirmed the clinical observations. These findings led to examination of Ruelene for anthelmintic activity. Numerous reports have indicated that Ruelene possesses satisfactory anthelmintic activity. In calves older than two weeks, cattle over 6 months and adult sheep may be treated with doses up to safe maxima of 25, 50 and 150 mg per kg body weight, respectively. At these dosage levels high efficiency may be expected against *Haemonchus* and *Cooperia* species, whilst efficiency against *Ostertagia*, *Trichostrongylus*, *Bunostomum*, *Nematodirus* and other species is irregular and often unsatisfactory at relatively safe dosage levels.

External or pour-on applications have been used successfully, and high efficiency in cattle against *Cooperia* species when applied at the rate of 100 mg per kg body weight has been reported, but unsatisfactory action on other species. In sheep, it was reported that pour-on treatment was erratic and unsatisfactory.

Above the dose levels recommended symptoms of *toxicity* can be expected, often after a delay of 18 hr, but usually of only 2 or 3 hr. Cattle or sheep which are housed or hand fed are more likely to suffer from toxic effects. However, this drug is generally considered less toxic than most organophosphorus anthelmintics.

Phenothiazine, B.Vet.C.

Although this anthelmintic has now been largely replaced by more active anthelmintics, it is still used as a cheaper form of therapy. It has been considered right, therefore, to retain this description of its activity.

This most important anthelmintic is a green amorphous powder which darkens on exposure to light. It is insoluble in, and is not wetted by, water, but is reasonably soluble in alcohol and acetone. As the compound is not 'wetted' by water it is used mainly in the form of a dispersible powder formed by the addition of approximately 5 per cent of a detergent. When dispersed in water a uniform suspension is formed which settles slowly but is easily redispersed by stirring.

It was first reported in 1940 and later confirmed by various workers that the particle size of any sample markedly influenced the activity of phenothiazine. Since that time further work has served to further confirm the finding. Maximum efficiency follows if the compound has a particle size of 70 per cent not less than 5 μm, 90 per cent not less than 10 μm, and the rest a maximum size of 30 μm diameter. It is now accepted that the very fine particles are required for efficiency against the smaller species, whilst the larger particles are more effective against the larger species, found in the large intestines. It is probable that any increase in activity will also be accompanied by an increase in toxicity.

Actions and Uses

The pharmacology of phenothiazine is still debated, but the accepted theory agrees with most of the known facts. There appear to be variations between the species, and the greater toxicity amongst some species is probably related to such differences. In general, there remains within the intestines about 50 per cent of any dose and it is this unabsorbed proportion which is responsible for the anthelmintic effect. It has been shown that phenothiazine is ingested by the parasite and, by using a labelled sulphur (^{35}S) isotope, absorption through the cuticle of some parasites has been proved. It also appears that *Ascaridia galli* at any rate shows a selective absorption of the drug, for the phenothiazine was absorbed more rapidly and in greater quantity by the parasite than by the surrounding mucosa. Approximately five times more drug is taken in by the parasite than is absorbed by an equivalent weight of mucosa. This drug becomes fixed in the parasite tissues. The precise way in which the lethal effect is exerted is not known, but it is not related to the quantity absorbed, for some parasites unaffected by the drug will take into their tissues an amount lethal to susceptible worms, and yet show no signs of toxicity.

Mature parasites only are destroyed, which necessitates repeated dosing to destroy the parasites that were immature at the time of the previous dosing.

What happens to the drug during the time it remains in the intestines is debated, some workers saying that there is a change into phenothiazone, others that phenothiazine sulphoxide is formed, and still others who say that there is little change. Following absorption, quantities of the oxidation product phenothiazine sulphoxide occur in the portal circulation and further oxidation occurs in the liver, with the production of leucophenothiazine and other substances which are excreted mainly in the urine and bile. The two compounds leucophenothiazine and leucothionol are colourless when excreted but are oxidized in the atmosphere to the brown-red dyes phenothiazone and thionol. These are fast dyes and indelibly stain wool. It is therefore important, as these compounds appear in the urine within 30 min of oral dosing, to disperse flocks immediately after administration, otherwise staining and ruining of many fleeces can occur. Staining will also occur if phenothiazine suspensions are split on the fleece during administration. Discoloration of milk may also occur, but aesthetic considerations are the only ones of any consequence. It is interesting that milk containing these phenothiazine excretory products is reported to keep fresh far longer than normal milk; this is no doubt due to the bacteriostatic effect of the dye compounds. Clients should always be warned, if using the drug for the first time, that excretions and secretions will be coloured.

Phenothiazine is used against strongyle parasites in ruminants, horses and related species, poultry and in some laboratory animals such as rabbits. There are no indications for its use in the dog and cat, and the dog in particular is very susceptible to poisoning. The drug is not used in human medicine because of a similar susceptibility. There is rarely, and since piperazine has proved effective against nodular worms (*Oesophagostomum*) it might almost be said never, any indication for its use in pigs, which as a species are susceptible to toxic effects.

In the *horse* phenothiazine is effective against the redworms—*Strongylus*, *Triodontophorus* and *Trichonema* species. Against *Trichostrongylus axei* phenothiazine is not consistent, nor is it satisfactory against oxyurids and must be combined with chenopodium and other drugs.

In *ruminants* the genera of stronglyes most affected are *Haemonchus*, *Ostertagia* and *Oesophagostomum*; the effect on trichostrongyles is variable probably because of the variation in particle size in the different batches of commercial phenothiazine, but they respond to the higher doses as do *Cooperia* species. *Nematodirus* species do not respond satisfactorily to phenothiazine treatment. Some *Bunostomum* species respond and some workers report activity against *Chabertia* species, but no effect is exerted on tapeworms or flukes. Phenothiazine is unusual in that it affects both immature and mature forms of *Haemonchus contortus*.

Dosage and Administration

Single dose treatment:
- Horse: 10–30 g at 3 g per 50 kg
- Cattle: 20–60 g at 10 g per 50 kg
- Sheep and goat: 5–40 g
- Poultry: 0·25–1 g

Repeated doses:
- Horse: 2–5 g
- Foals: 1–2 g
- Cattle: 0·5–5 g at 0·5 g per 50 kg
- Sheep: 0·25–0·5 g

Young foals, calves under 3 months, lambs under 1 month, and heavily pregnant females should not be dosed. Doses for sick or debilitated animals should be calculated with care. No purgation is required.

Phenothiazine may be administered in the food, as a drench or by stomach tube, in tablet form or as salt licks.

Horses. In horses, it is preferable to administer the compound in food wherever possible, but, though described as a tasteless and odourless powder, some horses will refuse medicated mashes even after starvation. In these cases electuaries or administration by stomach tube are the methods most commonly used, although rigid gelatin capsules are often a very useful means of giving the powder.

As blood horses are susceptible to phenothiazine poisoning practitioners are very chary of administering full doses and prefer to give half doses on each of two successive days. Debilitated animals, even when this course is adopted, may show signs of toxicity, for complete excretion does not take place before the second dose is given so that a cumulative effect is exerted.

Arising from this, prophylactic doses of up to 300 mg (5 gr.) in the food daily have become very popular, and reports of such doses administered continuously for 5 to 10 years indicate that no toxic effect occur in the host and no phenothiazine resistance is developed by the parasites. This method of treatment is valuable in that: (a) egg production by adult females in the intestines is reduced, (b) hatchability of eggs passed in the faeces is considerably reduced, and (c) that, of the infective larvae which are taken in by the horses, relatively few reach maturity. It also means that pastures carry a relatively minor parasite load and that the danger to foals running with adults is considerably reduced.

Sheep. In sheep, drenching guns with a continuous feed from canisters on the backs of the operators and with oesophageal nozzles, are used for flock dosing with phenothiazine dispersible powders. This means that doses are injected directly into the oesophagus and staining of the fleece during administration is minimal. Another popular way of giving phenothiazine to sheep is in the form of tablets—an elongated tablet weighing about 5 g is the usual one—which are administered by balling gun, forceps or by hand. Sheep unfortunately soon learn to hold the tablet and cough them out when released back into the flock. This means that tablets are discovered when the flock is dispersed and the offenders impossible to identify.

Repeated daily administration of 0·5-g doses has been shown to be as effective as in horses but individual dosing is impracticable, and in Great Britain the salt lick requirements of sheep are insufficient to provide adequate dosage levels of phenothiazine. In America the use of 1 part phenothiazine to 9 or 10 parts mineral or plain salt has been given up to 11-year trials and proved satisfactory. Where salt licks have proved unsuccessful, mineral and phenothiazine food supplements in the ratio of 1 part phenothiazine to 18 or 20 parts mineral supplement have proved of use.

Where intermittent dosing operates, administration should take place at approximately monthly intervals with emphasis on late winter and spring dosing, when a seasonal increase of infective larvae on pasture occurs. Where low level dosing is used it is advisable to give therapeutic doses once in spring and once in the autumn in order to eradicate or lessen the adult parasite burden.

Cattle. In cattle, the story follows much the same pattern as in sheep, but on the whole larger doses—up to 25 mg per 0·5 kg body weight—are required against parasite species other than *Haemonchus* and *Oesophagostomum*. The usual method of dosing cattle with phenothiazine is in the form of a drench. Salt licks containing the drug have yet to be proved of value and low level dosing is not yet generally adopted.

Poultry. The low toxicity of phenothiazine for poultry makes it a useful anthelmintic for birds. Individual dosing with 0·5 g pills or tablets for fowls, and 1 g for turkeys, is the safest and most accurate method of administration but medication of dry or wet mash feed is the most practical way of treating birds. The food is medicated once a month.

Toxicity

Phenothiazine is toxic in two ways. The first is by causing a haemolytic anaemia which affects all species except sheep and goats, but is of most importance in horses and, to a lesser extent, cattle. The cause of the haemolysis is not known but it has been suggested that in horses the phenothiazine and its derivatives are no more than activators of an enzyme lysolecithin which is normally found in horse blood. This enzyme has been shown to lyse red blood cells in vitro. Signs and symptoms are those typical of a haemolytic anaemia—dullness, lack of appetite (anorexia), weakness, enlargement of the heart and vital organs such as liver and kidneys, frequent and weak pulse, some oedema and jaundice. Severe enteritis also occurs. All tissues deteriorate rapidly following post mortem.

Treatment is symptomatic only, with extensive use of whole blood transfusions where practicable. No antidote is known.

The second type of toxicity results in the occurrence of photosensitization. It occurs most frequently in calves but lambs have also shown lesions on exposed lightly pigmented skin areas such as the muzzle, ears and eyes, and particularly the cornea. The cause of the photosensitization is reputed to be due to the saturation of the liver conversion processes and the passage of phenothiazine sulphoxide into the general circulation. When exposed to bright sunlight within 24 to 48 hr of dosing, irradiation of the affected and exposed tissues results in a process similar to severe sunburn and all non-pigmented skin areas become inflamed and frequently slough: corneal opacity and ulceration is a manifestation of the same phenomenon.

Treatment, where exposure has occurred, consists of antihistamine applications with local and general sedatives, in conjunction with protection from further exposure to direct sunlight.

Thiabendazole (Omnizole, Equizole)

Chemically the compound is 2-(4′-thiazolyl)-benzimidazole with an empirical formula of $C_{10}H_7N_3S$. It is a stable, white crystalline compound without appreciable odour or taste. It is available commercially with the name Thibenzole as both a wettable powder and as a suspension for oral administration to ruminants, and as a pallatable, easily suspendable powder for horses. It is now prepared in many different forms, particularly pellets, so that it can be readily added to the feed.

Actions and Uses

The use of isotope labelled thiabendazole has shown that, following oral administration to lambs at the rate of 50 mg per kg body weight, 75 per cent of the dose was recovered from the urine and 14 per cent from the faeces in 96 hr. The substance is distributed throughout most of the body tissues; it is almost completely converted into the 5-hydroxyl derivative before being excreted without further changes or after conversion into the glucuronide or sulphate ester. None of these metabolites are dyes and therefore do not stain the wool or hair, as do those of phenothiazine.

Since the first announcement of the discovery of a new anthelmintic reports from all over the world have confirmed the outstanding character of the substance when used against the strongyles of cattle, sheep, goats, pigs, horses and birds. The discovery of thiabendazole proved to be the biggest advance in strongyle therapy since the advent of phenothiazine. Activity against *Hyostrongylus* species, *Oesophagostomum* species, migrating ascarids and trichinosis in pigs and against most of the intestinal nematodes of dogs is reported.

In the *horse*, the compound gives over 90 per cent removal of mature oxyurids, *Strongylus* and *Triodontophorus* species, and a high percentage of immature forms. In one investigation mature forms of seventeen species of small strongyles were completely removed as were all mature oxyurids, *S. vulgaris*, *S. equinus* and high percentages of *S. edentatus*, *Parascaris equorum* and all immature forms.

In *cattle* and *sheep* almost 100 per cent activity is exerted against the common gastro-intestinal strongyles, but the action on *Trichuris* species is not great. These results have since been confirmed by many workers in many different countries. No wool staining, milk taint or other unwanted side actions have been reported. To control *Dictyocaulus filaria* a dose of 75 mg per kg body weight is necessary. The top dose 100 mg per kg body weight is used for the control of *Nematodirus* infestations.

In poultry and pheasants a dose of 0·1 per cent the diet is recommended for the control of gapes.

Dosage and Administration

Horse: 50–100 mg per kg body weight
Cattle: 50–100 mg per kg body weight
100 mg per kg body weight may be used in severe infestations 30 g is the normal maximum dose
Sheep: 50–100 mg per kg body weight
Pigs: 50 mg per kg body weight

The oral route must be used in all species, using the suspension where forced administration is necessary and the powder or pellet where the drug is to be incorporated in the feed.

Toxicity

Thiabendazole, in addition to its remarkable activity against strongyles, has virtually no toxicity. All research workers have repeatedly stressed this aspect, doses up to twenty times the standard dose having been given to ruminants without serious trouble.

Morantel Tartrate (Banminth)

Morantel tartrate is a *trans* isomer of 2[2 (3-methyl-2-thienyl)vinyl]-2-methyl-1,4,5,6-tetrahydropyrimidine tartrate.

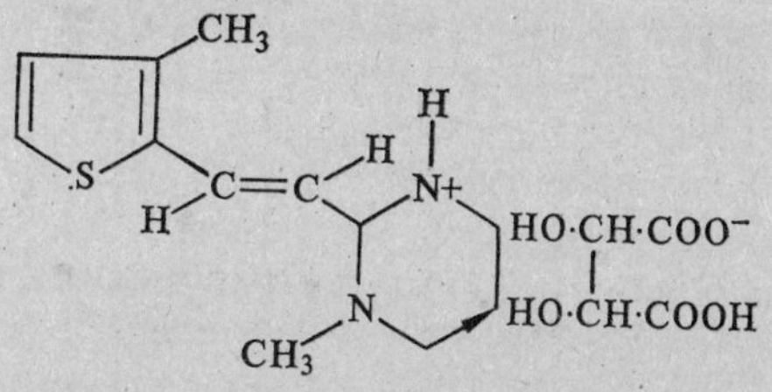

Morantel tartrate

It is a pale yellow crystalline solid of uniform appearance. It is readily soluble in water and insoluble in ethyl acetate and benzene.

Actions and Uses

It is a broad spectrum anthelmintic with activity against many adult and immature gastrointestinal worms in sheep and cattle. These are *Haemonchus*, *Ostertagia*, *Trichostrongylus*,

Cooperia, Nematodirus, Chabertia and *Oesophagostomum.*

Dosage

This form is prepared for use in combination with diethylcarbamazine to obtain activity against lung worm. In this case the dosage, in the proportions 2 per cent morantel tartrate to 5·9 per cent diethylcarbamazine, is:

Sheep: 7 ml for 7–17 kg body weight
14 ml for 18–37 kg body weight
21 ml for 38–50 kg body weight
28 ml for 50+ kg body weight
Cattle: 28 ml per 50 kg body weight.

The drug is available in the form of a suspension.

Toxicity

Morantel tartrate is a very safe anthelmintic and over seven times the therapeutic dose did not produce toxic symptoms.

Drug Compatibility

It can be given in the presence of drugs for the control of flukes.

Pyrantel Tartrate (Strongid)

Pyrantel tartrate is *trans*-1-methyl-2(2-(2′-thienyl)vinyl)-1,4,5,6-tetrahydropyrimidine tartrate.

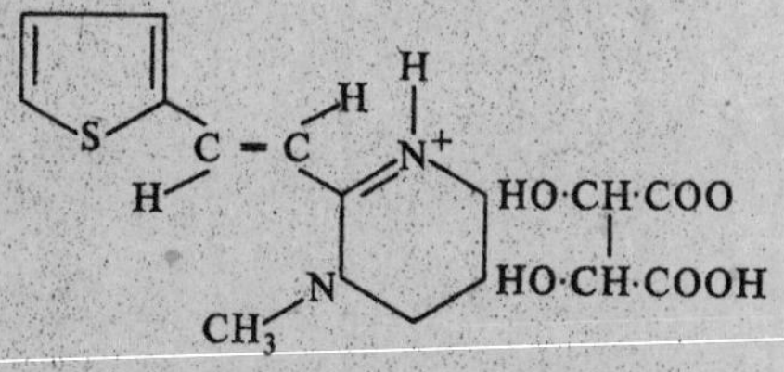

Pyrantel tartrate

It is an off-white, free flowing granular material containing 50 per cent w/w pyrantel tartrate in an inert carrier.

Actions and Uses

It is a broad spectrum anthelmintic and is indicated for use in the horse for the control and treatment of infestations due to *Strongylus vulgaris*, small strongyles, *Oxyuris equi* and *Parascaris equorum*. It should be given in warm water by stomach tube.

Dosage

Horse: 11 mg pyrantel tartrate/kg body weight.

Toxicity

Pyrantel tartrate is a very safe anthelmintic and has a therapeutic ratio of greater than ×20.

Tetramisole Hydrochloride (Nilverm)

Tetramisole hydrochloride, is the hydrochloride of 2,3,5,6-tetrahydro-6-phenylimidazo-[2,1-*b*]-thiazole.

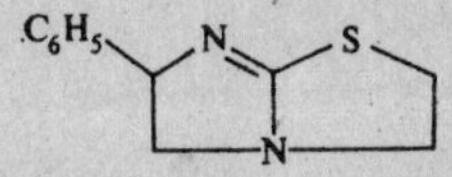

The chemical is soluble at 20°C in 5 parts of water, and in 10 parts of methyl alcohol and is a white crystalline powder.

Action and Uses

Tetramisole is active against the main gastrointestinal worms of cattle, sheep and goats. It is also active against *Dictyocaulus* the lung worm. At normal doses the drug is active against both adult and larvae stages.

Pigs: Tetramisole is active against *Ascaris, Hydrostrongylus, Oesophagostomum* and *Metastrongylus.*

Cats: The drug has been reported to be active against *Aelurostrongylus abstrusus*, the lungworm of the cat.

Dosage

Cattle, sheep, goats and pigs: 15 mg per kg body weight.

Cats: 5 mg per kg body weight daily for 6 days.

Toxicity

Overdosage in cattle and sheep give rise to transient signs of head shaking, salivation, and licking of the lips.

Parbendazole (Helmatac)

Parbendazole is methyl 5-butylbenzimidazol-2-ylcarbamate.

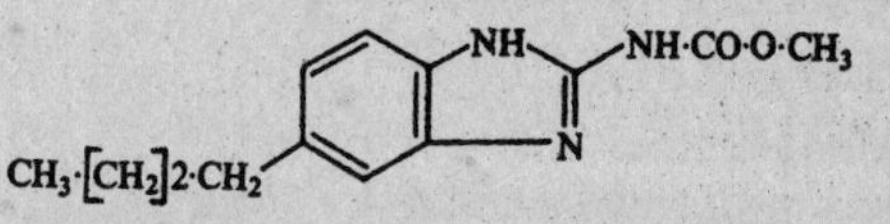

It is insoluble in water, but soluble at 20°C in 900 parts of alcohol. It is a white crystalline powder.

Actions and Uses

Parbendazole is active against *Bunostomum, Chabertia, Cooperia, Haemonchus, Nematodirus, Oesophagostomum, Ostertagia, Strongyloides, Trichostrongylus* and *Trichuris.*

It is used in the treatment of gastro-intestinal worm infestations in cattle and sheep and is active against adult and many larval species.

Dosage

Cattle: 30 mg per kg body weight
Sheep: 20–30 mg per kg body weight

Toxicity

Pregnant cattle should not be treated with Parbendazole.

Levamisole Hydrochloride (Ripercol, Nilverm)

Levamisole is (1-2:3:5:6-tetrahydro-6-phenyl-imidazo(2,1-6)thiazole hydrochloride and is a broad spectrum anthelmintic.

Pharmacology

Levamisole is rapidly absorbed and carried to all parts of the body. Peak blood level occurs within 1 hr of injection and tissue levels persist for up to 5 days. The drug is excreted mainly through the faeces and urine.

Mode of Action

Most nematodes are expelled within 24 hr; low levels of the drug act as ganglion stimulants and give rise to muscular paralysis, whereas higher levels interfere with carbohydrate metabolism.

Actions and Uses

Levamisole is a broad spectrum anthelmintic and is active against adult and larval stages of *Haemonchus, Ostertagia, Trichostrongylus, Cooperia, Nematodirus, Bunostomum, Oesophagostomum, Metastrongylus, Ascaris, Hyostrongylus* and *Trichuris* and against the lung worm *Dictyocaulus*. It has been examined against both artificial and natural infections and has the advantage of activity at a relatively low level of drug. Because of its activity against the larval stages, it has widened the whole field of the strategic dosing of cattle and sheep, to prevent larval infestation.

Dosage

As with all modern anthelmintics prepared for the treatment of farm animals, Levamisole is offered in a variety of formulae to fit in with practical methods of dosing. It is available as an oblet, pellet, soluble drench, wettable powder for addition to drinking water and injectable solution.

Oral dose:
Cattle: 7·5 mg/kg body weight
Sheep: 7·5 mg/kg body weight
Pigs: 7·5 mg/kg body weight
Poultry: 18–36 mg/kg body weight
Subcutaneous injection:
18·2 per cent solution: 2 ml per 50 kg/body weight.

Precautions

Cattle treated orally should not be slaughtered within 48 hr of treatment. The anthelmintic should not be given to dairy animals of breeding age. Animals treated by injection should not be slaughtered within 7 days of treatment.

Mebendazole (Telmin)

Mebendazole (R17 635) is methyl-5(6)-benzoyl-2-benzimidazole carbamate.

C'16$H_{13}N_3O_3$

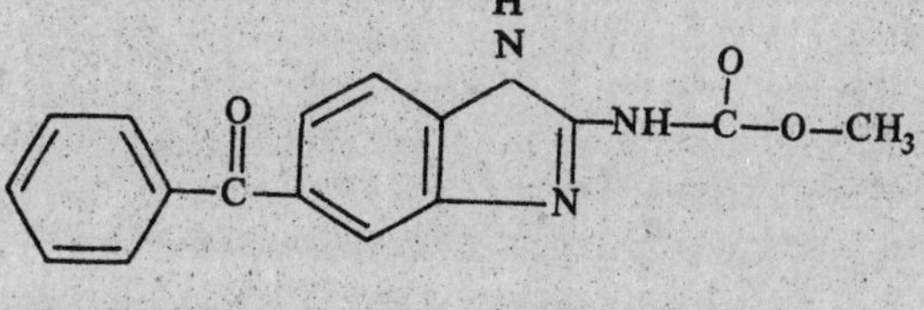

Mebendazole

It is a grey-white to pale yellow powder that is practically insoluble in water and most solvents.

Pharmacology

Detailed studies of the action of the drug have shown that it inhibits irreversibly the uptake of glucose by the worms. This leads to the starvation of the worms and the death of the parasite.

Actions and Uses

It is a broad spectrum anthelmintic, particularly useful in the control of helminths of the horse. It is active against *Parascaris equorum*, *Ascaris suum*, *Strongyloides ransomi*, *Strongylus equinus*, *Strongylus* spp., *Triodontophorus*, *Trichonema*, *Trichuris suis*, *Oesophagostomum* spp. and *Oxyuris equi*, showing 90–100 per cent activity. It is also reported as being active against *Syngamus trachea* in poultry.

Dosage

Horses: 5–10 mg/kg body weight
Pigs: 5 mg/kg body weight

It can be given as a liquid preparation either orally by syringe or added to the food; the medicated dose for pigs is 31 ppm in the feed.

Toxicity

The therapeutic index is higher than 40. It is poorly absorbed; more than 90 per cent remains in the stomach and intestines and is subsequently excreted unchanged in the faeces. The absorbed portion is broken down in the liver and excreted in the urine.

B. ANTHELMINTICS AFFECTING ASCARIDS

Piperazine Hydrate, B.P.C. (Oxacine)

This drug has been known for many years as a treatment (of doubtful value) for gout. In 1954 the first veterinary publications began to appear indicating that a drug had been identified which would be a considerable advance over all others for the treatment of ascariasis. These early indications have been fully substantiated.

The hexahydrate is physically unstable, being very deliquescent, and consequently the drug is used in the form of the more stable salts: the adipate (from adipic acid), the citrate, the phosphate, sulphate and hydrochloride are all satisfactory. A further compound with carbodithioic acid is also available commercially. All are white crystalline powders, and, except for the adipate, are soluble in water, have a saline taste but no odour.

Actions and Uses

All the piperazine compounds have similar effects, with slight variations. Their action is one of narcosis of the parasite which can them be removed from the intestines by normal peristalsis, possibly augmented by a laxative or purgative, which should preferably be of a saline nature. Should the parasite not be removed and the piperazine be entirely eliminated the parasite will recover from the narcosis and return to normal activity. This can be demonstrated by washing a paralysed worm in normal saline and observing the subsequent recovery which takes place in an hour or two.

The parasites affected are mainly ascarids and nodular worms of all species of domestic animals; they respond with almost 100 per cent efficiency. There is a variable effect on hookworms and strongyles, but no effect on whipworms (e.g. *Trichuris* species), or tapeworms.

In the *horse* piperazine adipate removes almost 100 per cent of mature and immature *Parascaris equorum* and adult *Trichonema* species, over 80 per cent of adult *Oxyuris* species, and up to 60 per cent of *Triodontophorus* species and of *Strongylus vulgaris*. There is no effect on other *Strongylus* species, *Habronema* species or tapeworms. Against the strongyles it is indicated that these results were similar to those achieved by maximum doses of phenothiazine. An equivalent effect has been reported with the normal dose of the adipate, citrate, phosphate and carbodithioic acid compounds when tested in horses, whilst the carbodithioic acid compound was equally effective against the small strongyles and the ascarids at about half the normal dose.

Similarly, reports have shown that piperazine is equally effective in exotic equine species such as zebras.

In *cattle* and *sheep* piperazine has been used but is only indicated for ascarids and nodular worms—*Ascaris vitulorum* and *Oesophagostomum* species; the former does not occur in Great Britain.

In *pigs* piperazine compounds are extremely efficient against *Ascaris lumbricoides* and *Oesophagostomum* but the action on *Hyostrongylus* is not so good.

In *dogs* and *cats* the action is 100 per cent efficient against *Toxocara* and *Toxascaris* species, up to 75 per cent against hookworms and the drug has practically no effect on whipworms (*Trichuris* species) or tapeworms. The heartworm *Dirofilaria immitis* responds very well. Its effect on zoo canine and feline species is equally good at doses proportional to their weight. In *poultry Ascardia galli* responds extremely well but *Heterakis gallinae* is not quite so susceptible.

This drug is the drug of choice for the treatment of ascariasis in all species.

Dosage and Administration

Horse and Cow:
: 0·2 g per kg body weight. Maximum 80 g (foals 30 g, yearlings 60 g)

Sheep and Goat:
: 4–8 g per 10 kg

Pig:
: 1 g per 10 kg live weight in each of three successive feeds or 3 g per score if dosed individually

Dog and Cat:
: 0·11 g per kg body weight (puppies under, 2·5 kg, cats and kittens, 0·25 g maximum)

Poultry:
: 0·3 g approximately, for each adult bird (about 32 mg per kg body weight) given in each of two successive feeds or in drinking water for 2 days

No fasting is required but the drug should preferably be administered to hungry animals. A purgative is not essential but laxatives assist in expelling the paralysed worms which otherwise will appear in the faeces after about 24 hr and continue to be excreted for several days. This continued excretion is particularly seen where doses are divided and administered over a period.

The drug is administered either in the food or drinking water. Most *horses* will accept the drug in bran mashes but, if necessary, aqueous solutions may be administered by stomach tube or the drug may be given as an electuary. In *pigs* the usual method for mass dosing is to divide the calculated dose into equal parts and administer them in two successive, either wet or dry, feeds. For individuals, single doses may be given. Medications of the drinking water for several days has also given good results. *Dogs* and *cats* are usually dosed with tablets. *Poultry* are normally dosed by medication of the drinking water on each of two successive days; no other drinking water should be available.

In all cases separate worming pens should be used, for the eggs within voided parasites are viable and reinfection can occur if regular pens are contaminated.

Toxicity

The toxicity of piperazine compounds is normally very low. Occasional neurotoxic symptoms and emesis occur in small animals but these are usually of a transient nature. The drug can be administered to pregnant animals and also to those animals suffering from gastroenteritis which makes the anthelmintic almost unique.

Tetrachloroethylene, B.Vet.C.

This is a liquid similar in most respects to carbon tetrachloride.

Actions and Uses

Tetrachloroethylene is most active against the ascarids of horses, pigs, dogs and cats, but its main value lies in its effect on hook-worms. An appreciable effect on horse bots is also reported. It is effective against many strongyles of sheep and cattle, particularly *Haemonchus* species, but there are numbers of reports of severe toxicity in cattle and as it has few advantages over phenothiazine its use is not recommended for these animals. It is used extensively in sheep in various parts of the world.

Dosage and Administration

Horse and Cow:
: 15–30 ml at the rate of 0·1 mg/kg

Sheep:
: Up to 9 months (2·5–5 ml). Adults up to 7·5 ml

Pig:
: 2·5–10 ml

Dog:
: 1–5 ml

The administration details are similar to those for carbon tetrachloride. Capsules are used and mineral or vegetable oil mixtures are also used, although fats and oils are not recommended as they increase the absorption rate, which is normally low. A saline purgative an hour or so after administration is useful in dogs, but it is said that omission of a purgative is followed by better results, even though absorption is greater.

Toxicity

Similar to that of carbon tetrachloride but to a lesser degree. Tetrachloroethylene should not be used in animals debilitated for any reason, nor should it be given to pregnant animals. Cases of toxicity in the dog are relatively common, mainly from overdosage. Should a capsule break in the mouth during administration the danger of anaesthesia is far less than with carbon tetrachloride.

Treatment consists of the lowering of protein and fat diets and the increase of carbohydrate foods. Choline chloride orally or injected can also be used, and calcium, injected as borogluconate, should be given in addition to a high calcium level in the diet before and after dosing.

Toluene ($C_6H_5:CH_3$)

A light, clear, almost colourless and highly inflammable liquid with a benzene-like odour.

Actions and Uses

Toluene, also known as toluol, is from 87 to 100 per cent effective against ascarids and hookworms, and 75 per cent effective against whipworms in the dog and cat. Several reports have indicated a good effect on ascarids, including immature forms, and on *Gasterophilus intestinalis* (horse bot fly larvae) in the horse. It is safer than carbon disulphide and is therefore to be preferred. The drug has yet to be fully evaluated.

Dosage and Administration

Horse: 0·4 ml per kg body weight
Dog and Cat: 0·2 ml per kg body weight

In the *dog* and *cat* toluene is best given in capsules, but solution in fixed oils is permissible. It should be given fasting to obtain the best results. It can be combined with dichlorophen to give comprehensive treatment.

In the *horse* the drug should be given by stomach tube and fasting is not essential. In the United States it is administered to foals as young as 12 weeks and repeated at 6 months, and then periodically during the yearling period.

Toxicity

Toluene is relatively non-toxic when used in the recommended doses.

Dichlorvos (Atgard, Task Equigard)

2,2-Dichlorovinyl dimethyl phosphate is a volatile organophosphorus compound which is readily destroyed by oxidizing agents and by hydrolysis. For use as an anthelmintic it has to be prepared in a resin pellet in such a way that it remains protected until it reaches the small bowel, where on release it is active against a range of helminths.

Pharmacology

During treatment, because of the effect of organophosphorus compounds on the enzyme acetylcholinesterase, a drop in circulating cholinesterase can be detected. Normal values are regained within 5 to 10 days.

Single or weekly repeated doses at recommended levels have not produced abnormal changes in the major blood parameters.

Actions and Uses

Dichlorvos is active against *Toxocara*, *Toxascaris*, *Ancylostoma caninum*, *Ancylostoma tubaeforme*, *Uncinaria stenocephala*, *Hyostrongylus rubidus*, *Oesophagostomum* spp., *Ascaris suis*, *Trichuris suis*, *Parascaris equorum*, small strongyles, *Oxyuris equi* and *Strongylus vulgaris*. It is therefore a valuable anthelmintic for the treatment of dogs and cats, pigs and horses.

Dosage

Dogs and Cats: 10 mg/kg body weight
Pigs: 7·5 mg/kg body weight
Horses: 8 mg/kg body weight

Precautions

Do not use when animals are being treated with an organophosphorus insecticide.

Drugs which Affect Whipworms (Trichinelloids—genus *Trichuris*)

Phthalofyne (Whipcide)

This compound is known commercially as Whipcide, and chemically as 3-methyl-1-pentyn-3-yl sodium phthalate. It has been described as the drug of choice for the treatment of whipworm (*Trichuris vulpis*) infestation, which, whilst it does not occur in Great Britain is nevertheless an important parasite in many other parts of the world. Investigations during the last decade have shown the preparation to be highly efficient and safe to use. Administration after a light meal is advisable as the food reduces the tendency to vomit. Doses of up to 300 mg per kg intravenously gave very good results though causing transient side effects. Vomiting, ataxia and depression for up to 6 hr occurred in a high proportion of cases. However, investigators are agreed that the drug is well tolerated even by pregnant bitches. A strong offensive faecal odour is created, but apparently this can be controlled by simultaneous dosing with chlorophyll.

ANTHELMINTICS AFFECTING LUNG WORM

The lungworm of cattle, sheep, and pigs are responsible for considerable economic losses annually, Prior to the development of a lungworm vaccine, and effective chemotherapeutic agents, affected animals were treated by means of intratracheal injections and various forms of inhalants. Turpentine and phenol were frequently used as the active principles of injectable formulations. Two effective drugs were introduced in 1957 and 1958 respectively, cyanacethydrazide and diethylcarbamazine citrate. Other drugs such as tetramisole, levamizole and carbendazole have activity against lung worms in addition to their action against gastrointestinal worms.

Cyanacetohydrazide (Helmox, Dictycide)

Cyanacetohydrazide has the formula

$$CN.CH_2.CO.NH.NH_2.$$

This substance is a white, crystalline powder which is soluble in water but deteriorates fairly rapidly in solution—commercial preparations, although stabilized, retain full activity for only 2 to 3 weeks following reconstitution by the addition of hot sterile water. A yellow discoloration sometimes develops on keeping but such solutions may still be used.

Actions and Uses

It is available for oral or injectible use for the treatment of lungworm disease (parasitic bronchitis) in cattle, sheep and pigs. It has been shown that the compound is active against adult *D. viviparus* in cattle, *D. filaria* and *Protostrongylus rufescens* in sheep, and *Metastrongylus* species in pigs. Removal of the adult worms is variable but clinical improvement is often appreciable. Action on *Muellerius* species and *Neostrongylus linearis* in sheep is negligible, although results against other lungworm species in sheep appear to be better than those achieved in cattle.

It would seem that, whilst cyanacetohydrazide can be useful in established infections and in reducing environment contamination, it is of little value during the time when the larvae are migrating and when most of the tissue damage is occurring. The narrow therapeutic index makes the compound even less useful as a routine treatment for lungworm infestations.

Dosage and Administration

As the therapeutic index is narrow animal weights and dosages must be estimated as accurately as possible.

All species:

subcutaneously: 15 mg per kg body weight
orally: 17·5 mg per kg body weight

These doses should be repeated on each of three sucessive days; such courses may be given monthly during periods of great risk. Maximum doses by any route of 5 g for cattle of 300 kg or more and 0·4 g for sheep, goats and pigs of 25 kg and over should be strictly observed.

Toxicity

Transient local irritation and pain at the site of injection have been reported. Injection of three times the normal dose is fatal and twice the recommended dose can cause symptoms varying from loss of appetite to convulsions. At the recommended dosage levels the incidence of untoward effects is negligible.

Toxic symptoms are similar to those of strychnine poisoning and similar symptomatic treatment is indicated—chloral hydrate or barbiturates. Pyridoxine (vitamin B_6) is the physiological antagonist.

Diethylcarbamazine Acid Citrate, B.Vet.C. (Banocide, Caricide)

This complex methylpiperazine derivative, with the empirical formula $C_{16}H_{29}N_3O_8$, is a white crystalline, odourless powder which is freely soluble in water. Originally introduced as a treatment for filariasis and ascariasis in dogs and cats and now used as such in the U.S.A. and other countries where filariasis is an important infestation, the compound is now mainly of veterinary significance in the U.K. for its action in the treatment of parasitic bronchitis in cattle and sheep, when it is usually given as a neutral solution for either injection or oral administration. Solutions containing the equivalent of 40 per cent w/v of the acid citrate are available commercially.

Actions and Uses

In *cattle*, diethylcarbamazine is used for the treatment of *Dictyocaulus viviparus* infestations. Doses of 22 mg per kg body weight on 3 successive days by intramuscular injection are used routinely, but a single dose at the rate of 44 mg per kg was reported to give better respiratory relief. The drug is considered of value in early cases but relatively ineffective in mature, established cases. This assessment was supported by the results of research carried out by government workers in Northern Ireland. Response in *sheep* affected with *D. filaria*, seems to be much the same as that shown by calves with *D. viviparus* in that immature forms are susceptible whereas adult worms are more resistant.

In *dogs* and *cats*, diethyl carbamazine has been superseded in the U.K. by piperazine as a treatment for ascariasis, but the compound can be used effectively in dogs affected with heartworm (*Dirofilaria immitis*). Oral therapy over a prolonged period is advisable.

Dosage and Administration

Cattle and Sheep:

22 mg per kg body weight daily on three successive days

Dogs:

22–44 mg per kg body weight three times daily by mouth for 3 to 5 weeks

Administration to cattle and sheep may be by intramuscular injection, or the second and third doses may be given orally. Dosing can be repeated monthly during high risk periods.

Toxicity

The compound is relatively non-toxic. The LD_{50} was calculated as 485·3 mg per kg body weight in cotton rats and it has been indicated that doses of 440 mg per kg were non-toxic to calves, giving a therapeutic index of 20.

Bibliography

ANDERSON, N. & DOBSON, R. J. (1975) Anthelmintic efficiency related to strategic control schemes for helminthiasis in sheep. *Aust. vet. J.*, **51**, 67.

ARMOUR, J. & BRUCE, R. G. (1973) The prophylaxis of ovine fascioliasis by the strategic use of rafoxanide. *Vet. Rec.*, **92**, 83.

ARMOUR, J. & CORBA, J. (1972) The anthelmintic efficiency of diamphenithide against *Fasciola hepatica* in sheep. *Vet. Rec.*, **91**, 211.

BAKER, N. F. & FISK, R. (1972a) Levamisole as an anthelmintic in calves. *Am. J. vet. Res.*, **33**, 1121.

BAKER, N. F. & FISK, R. (1972b) Administration of the anthelmintic levamisole in drinking water for cattle. *Am. J. vet. Res.*, **33**, 1399.

BAKER, N. F. & WALTERS, G. T. (1971) Anthelmintic efficacy of carbendazole in cattle. *Am. J. vet. Res.*, **32**, 29.

BENNETT, D. G., BICKFORD, A. A. & LUND, J. (1974) Safety evaluation of mebendazole in horses. *Am. J. vet. Res.*, **35**, 1003.

CAMPBELL, N. J. & HOTSON, I. K. (1971) The anthelmintic efficiency of dioxanide and rafoxanide against *Fasciola hepatica* and *Haemonchus contortus* in sheep. *Aust. vet. J.*, **47**, 5.

CAMPBELL, W. C. & BUTLER, R. W. (1973) Efficacy of cambendazole against tapeworm and roundworm infections in lambs. *Aust. vet. J.*, **49**, 517.

DICKERSON, G., HOWEENIST, M. & KINGSBURY, P. A. (1971) A chemotherapeutic agent for all stages of liver fluke disease in sheep. *Br. vet. J.*, **127**, xl.

DRUDGE, J. H. & LYONS, E. T. (1972) Critical tests of a resin-pellet formulation of dichlorvos against internal parasites of the horse. *Am. J. vet. Res.*, **33**, 1365.

ECKERT, J. & BISIG, D. (1971) Prufung der Wirksamkeit von Cambendazol auf unreise Studien von Trichostrongyliden des Rindes. *Berl. Münch. tierarztl. Wschr.*, **84**, 392.

GIBSON, T. E. (1972, 1973, 1974) *The Veterinary Annual.* Bristol: John Wright.

HARFENIST, M. (1973) *Pesticide Sci.*, , 871.

HASS, D. K. (1973) Critical evaluation of dichlorvos tablets in puppies. *Vet. Med. small anim. Clin.*, **68**, 900.

HASS, D. K., ALBERT, J. R., PILLOW, B. G. & BROWN, L. J. (1973) Dichlorvos gel formulation as an equine anthelmintic. *Am. J. vet. Res.*, **34**, 40.

HASS, D. K. & YOUNG, R. (1973) Anthelmintic efficacy of dichlorvos therapy in pregnant swine. *Am. J. vet. Res.*, **34**, 195.

KATES, K. C. & COLGLAZIER, M. L. (1973) Experimental development of a carbendazole-resistant strain of *Haemonchus contortus* in sheep. *J. Parasit.*, **59**, 169.

LAPAGE, G. (1968) *Veterinary Parasitology*. Edinburgh: Oliver & Boyd.

LYONS, E. T., DRUDGE, J. H., LABONE, D. E. & TOLLIVER, S. C. (1972) Field and controlled test evaluations of levamisole against natural infections of gastrointestinal nematodes and ring worms in calves. *Am. J. vet. Res.*, **33**, 65–71.

McKENNA, P. B. (1974) The anthelmintic efficacy of thiabendazole and levamisole against inhibited *Haemonchus contortus* larvae in sheep. *N.Z. vet. J.*, **22**, 163.

PRESIDENTE, P. J. A. & KNAPP, S. E. (1972) Anthelmintic efficacy of rafoxanide against immature *Fasciola hepatica* in calves. *Am. J. vet. Res.*, **33**, 1603.

REINECKE, R. K. & LE ROUX, D. G. (1972) Anthelmintic activity of mebendazole in equines. *J. S. Afr. vet. med. Ass.*, **43**, 287.

ROWLANDS, D. A. T. (1974) Diamphenithide prophylaxis of ovine fascioliasis. *Vet. Rec.*, **95**, 547.

SAUPE, E. & NITZ, K. J. (1972) Die Anthelmintischen Wirkung von Mebendazol. Ein Feldversuch an Pferden. *Berl. Münch. tierarztl. Wschr.*, **85**, 21.

SNIJDERS, A. J., HORAK, I. G. & LOUW, J. P. (1973) Trials with rafoxanide. 6. The effect of repeated and single treatments with rafoxanide against *Haemonchus contortus* and *Oestrus ovis* in sheep. *J. S. Afr. vet. med. Ass.*, **44**, 251.

SNIJDERS, A. J., LOUW, J. P. & SERRANO, F. M. H. (1971) Trials with rafoxanide. 1. *Fasciola gigantica* in cattle in Angola. *J. S. Afr. vet. med. Ass.*, **42**, 249.

SOULSBY, E. J. (1968) *Helminths, Arthropods and Protozoa of Domesticated Animals*. London: Baillière Tindall & Cassell.

STOYE, M., ENIGK, K. & BÜRGER, H.-J. (1971) Die Wirkung von Cambendizol bei experimentalen *Ostertagia* und *Cooperia* Infektionen des Rindes. *Tierärztl. Umsch.*, **26**, 108.

TURTON, J. A. (1974) Controlled trials to determine the efficacy of levamisole against *Ostertagia circumcincta* and *Trichostrongylus colubriformis* in lambs. *Res. vet. Sci.*, **16**, 152.

WALKER, D. & KNIGHT, D. (1972) The anthelmintic activity of mebendazole: a field trial in horses. *Vet. Rec.*, **90**, 58.

WARDLE, R. A. & MCCLEOD, J. A. (1968) *The Zoology of Tapeworms*. New York: Hafner.

38

Coccidiostats

Coccidiosis, although an ever present disease in any intensive poultry rearing unit, is now controlled by the addition of low levels of coccidiostats to the daily ration. Since the sulphonamides were first used as specific coccidiostats, many new and efficient drugs have been developed during the last 20 years.

Most of the new drugs are investigated first for their value in the control of coccidiosis in broilers, and later developed for the control of coccidial infections in adult poultry stock and other animals.

COCCIDIOSIS

This is a protozoan parasitic disease in which the protozoa may enter and destroy cells in the liver (rabbit) and intestines (cattle, sheep, goats, pigs, dogs, cats, rabbits and fowls); and one species attacks the caecum in the fowl. There are two genera, *Isospora* which affect the small intestine of dogs, cats and man, and *Eimeria*, species of which cause lesions in all the sites and species of animals mentioned above.

The parasite has a life history which involves two types of multiplication, an asexual method or schizogony stage which is in itself cyclical, and a sexual type or sporogony stage. During the schizogonous stage multiplication can occur with fantastic speed and whole stretches of mucosa, in the cells of which the *schizonts* or asexual dividing structures grow, may slough as the mucosal cells are burst open by the swelling schizonts which in turn themselves burst, releasing *merozoites* to attack other cells. This leaves a raw bleeding surface which provides an easy route of access to the body for pathogens already in the intestines. The sporogonous stage is of less importance to the host, for fewer cells are damaged. During this phase a female cell or *macrogamete* is fertilized by a *microgamete* and forms a capsulated *zygote*, which is then passed out in the faeces (or urine, in renal coccidiosis of geese and swans) and is known as an *oocyst*. The oocyst is therefore the means of transmission of infection to other individuals, and the damage caused during their formation in the first host is relatively small when compared with damage caused by the sporozoites they release and the schizonts formed from them in the new host.

The majority of therapeutics suppress the production of the second generation schizonts in the dangerous asexual stage of reproduction, but some influence is exerted on gametocyte formation.

The main sign of coccidiosis in all species is diarrhoea, which may be intermittent in the mild chronic cases; the faeces may vary in colour from yellow to bright red with blood clots and shreds of tissue, depending on the severity and acuteness of the attack. The appetite is usually maintained, but as the digestive processes are affected by the damage caused in the mucosa, animals become debilitated, and emaciated in the more severe cases; anaemia and death follow. All the signs associated with this type of condition in the various species will occur; for example, raised tail, caked hindquarters, dropped-in flanks, staring coats with loss of colour, and a drooping, miserable appearance in badly affected calves; or caked vent feathers, huddled and ruffle-feathered stance with drooping wings in poultry.

Coccidiosis Control

As the methods of poultry husbandry have become more intensive, the problems of coccidiosis control have become more complex, partly because research workers continue to discover new species of pathogenic intestinal coccidia. The early work in the study of coccidiosis was concentrated on the control of *Eimeria tenella*, the causal organism of caecal coccidiosis. This species constituted the major cause of coccidiosis, and it was also a very convenient parasite for experimental studies.

Most of the fundamental studies were therefore carried out using this organism, and the basic principles of therapeutic control were designed to control caecal infections. Because chemotherapy has to develop in parallel with intensive methods of animal husbandry, it has often proved practical to translate directly laboratory findings to field application.

The sulphonamides were the basic drugs first used on coccidiosis, and it was with them that it became possible to prove that interference with the life cycle of the parasite by means of a chemotherapeutic agent could lead to disease control. The sulphonamides were unique in the early work in that it was shown that they could be used both to treat diseased birds, and also to prevent the development of acute infections.

With the development of large poultry units, it soon became clear that the treatment of recognized infections was unlikely to be effective, as the losses would be too high. The aim, therefore, in the development of new coccidiostats is to evolve drugs which can be added to the mash at low levels, and which will control both caecal and intestinal coccidiosis. Suitable drugs must have a low toxicity, so that they can be given to the birds for prolonged periods. It was also considered important that a drug should not interfere with the stages of the parasite which were responsible for the stimulation of immunity.

This aspect is not considered so important now in the case of broiler production, as these birds are kept on a drug until a day or two before slaughter, so that their state of immunity is unlikely to be challenged.

Drug absorption. It is only in the case of the sulphonamides that full information is available on the relationship between drug absorption and coccidiostat efficiency. It does appear, however, that it is desirable for a coccidiostat to be absorbed into the gut if it is to have a maximum effect.

Drug resistance. Resistance to drugs can occur in coccidiosis control as in other protozoal and bacterial infections.

The factors which appear to favour the development of resistance are the continued exposure of large populations to the drug, and the use of the drug to suppress growth of the organism rather than eliminate it. Both these conditions inevitably occur under intensive conditions of poultry management. This means that new coccidiostats tend to have a limited life.

COCCIDIOSTATS

In view of the basic work carried out with the sulphonamides, this group is considered first.

Sulphonamides

Sulphanilamide and *sulphapyridine* were the first of the group to be shown to have activity against coccidia, and they were found to be active against several intestinal species, but not the two main ones *E. tenella* and *E. necatrix*.

Sulphaguanidine was the first to prove of practical value. When it was given continuously at a concentration of 0·5 per cent in the food it controlled infections due to *E. praecox*, *E. mitis*, *E. maxima*, and *E. hagani*. Higher concentrations 1 to 1·5 per cent were necessary to control *E. tenella* and *E. necatrix*, subsequently other members of the group were found to be active against coccidia. Work in 1945 showed that *sulphadimidine* could be added to the drinking water to treat established infections.

Finally two sulphonamides proved to be particularly effective in the control of coccidial infections, sulphadimidine (sulphamethazine) and *sulphaquinoxaline*.

Treatment in poultry. Sulphadimidine and sulphaquinoxaline are the only sulphonamides considered to be therapeutically and economically effective in the treatment of coccidiosis in chickens and turkey poults. Sulphaquinoxaline is slightly more palatable and is preferred for turkeys because of this.

In caecal coccidiosis the drugs were given in the food or drinking water using one of two methods. Medication was continuous for 5 days or intermittently on a 3-day on/2-day off basis. The intermittent method was accepted as the best method for administering sulphonamides to poultry, and it consisted of two medication periods, each of three days duration, divided by a 2-day period when normal food and drinking water was provided. This intermittent method of dosing served two purposes; the first was relatively incidental in that cumulation and toxicity due to the sulphonamide was avoided; the second was considered to be of extreme importance. It is fairly certain that when deaths or faecal changes are first noticed in a flock there will be a large proportion of its members unaffected by the coccidia; on a continuous method of administration it was considered that the coccidia would not become established in these birds and no immunity could develop. Thus when treatment was stopped these birds could again become susceptible to infection and further outbreaks would occur.

The aim using the sulphonamides was to allow first stage schizogony to progress normally; this stimulated the production of antibodies by the host and the disease was then terminated at the level of second stage schizogony by the use of drugs. The second stage selectivity of the sulphonamides would have been lost if dosage was too high or if high concentrations occurred through cumulation, and first stage schizogony was also suppressed. With this suppression the stimulation of antibody formation would cease and with the cessation of treatment the bird, still susceptible, would once more be at the mercy of any further infection.

Sulphadimidine

Sulphadimidine is administered most commonly in the drinking water at the rate of 60 ml of a 16·66 per cent solution of the sodium salt in 4·5 litres (1 gallon) of water. This provides a concentration of 0·2 per cent.

Sulphaquinoxaline (Embazin, Sulfacox)

This may be given in drinking water at the rate of 60 ml of a 10·23 per cent solution of the sodium salt in 13·5 litres (3 gallons) which provides a concentration of 0·0424 per cent; or in mash at the rate of 113·6 g of a premix in each hundredweight (51 kg) of mash, which gives a concentration of approximately 0·05 per cent sulphaquinoxaline. Sulphaquinoxaline is also used prophylactically in the food of chickens and turkeys at the rate of 28·4 g premix to each hundredweight of food for the first 8 weeks of life (chickens); the concentration provided is 0·0125 per cent. With the recognition of resistant strains of coccidia it is now usual to combine sulphaquinoxaline with other drugs such as amprolium to cover a wide range of coccidia.

Duodenal coccidiosis is treated in a similar manner using sulphaquinoxaline. Preventive medication is suggested as being the best method of controlling the disease (U.S.D. 25th edn, 1955, p. 1026) but a warning is given that medication should cease four days before killing to allow time for sulphonamide elimination.

In all cases of curative therapy, treatment should be initiated immediately signs of disease are shown if the best results are to be obtained.

Sulphonamide resistance by some strains of coccidia has been reported and such resistance has been attributed to the widespread use of subinhibitory levels in medicated food stuffs.

In addition the sulphonamides, however, had one disadvantage which was of considerable importance where long term dosage had to be carried out, they could produce toxic symptoms. Therefore, a search was made for alternatives.

PYRIMIDINE DERIVATIVES

This group of anti-coccidial drugs acts by interfering with folic acid metabolism in the parasite but as sulphonamide activity is on the same metabolic pathway, the two groups are therefore synergistic; for this reason they are frequently used together. There are three main compounds in the group, **Amprolium Hydrochloride,** B.Vet.C., **Diaveridine,** B.Vet.C. and **Pyrimethamine.** The first was originally used alone but like the other two compounds it is now common to combine it with a sulphonamide usually sulphaquinoxaline.

Amprolium Hydrochloride, B.Vet.C. (Amprolmix)

This water soluble, white, odourless powder is chemically 1-(4-amino-2-propylpyrimidin-5-ylmethyl)-2-methylpyridinium chloride hydrochloride. A hydroxide is also available. The powder is soluble in 2 parts of water, and in 170 parts of alcohol.

Actions and Uses

Amprolium, is very active as a preventive agent against *E. tenella* and *E. necatrix* and can be used as a therapeutic agent, although its activity against *E. brunetti* is low. The compound exerts its action by preventing the parasite utilizing aneurine (vitamin B_1). Amprolium toxicity is indicated by aneurine deficiency symptoms but this should not be seen under normal conditions.

Dosage

For prophylaxis in broiler stock amprolium is fed continuously at the rate of 0·0125 per cent. Where natural immunity is required stock should be fed levels of 0·004 to 0·008 per cent. Vitamin K in the diet at the rate of about 2 mg per kg (2 g per tonne) decreases mortality. Vitamin K_1 is superior to menaphthone (Menadione) for this purpose. In combination with sulphaquinoxaline, amprolium can be used to provide adequate control of all five major species of *Eimeria*, the concentrations in feed being 0·006 per cent of each drug. In drinking water concentrations of 0·018 per cent sulphaquinoxaline, and 0·024 per cent amprolium should be used. In cattle amprolium in doses between 0·125 and 0·5 g per head have been used for the control of *E. zurnii* infections in calves.

Diaveridine, B.Vet.C. (Darvisul)

The compound is a white, odourless and tasteless powder which is only slightly soluble in water. Chemically it is 2,4-diamino-5-(3,4-dimethoxybenzyl)pyrimidine.

Actions and Uses

Diaveridine has coccidiostatic activity by virtue of its interference with folic acid metabolism. Synergistic combinations with sulphadimidine and sulphaquinoxaline are available commercially. It was demonstrated that concentrations of 0·001 per cent diaveridine and 0·008 sulphaquinoxaline gave protection against *E. tenella*, *E. necatrix*, *E. acervulina* and *E. maxima*. The combination does not interfere early in the parasite life-cycle and strong immunity is developed. The combination with sulphaquinoxaline is relatively non-toxic and may be fed at the recommended levels to young stock and laying hens for long periods without danger.

Pyrimethamine

This compound, is 2,4-diamino-5-(*p*-chlorophenyl)-6-ethylpyrimidine. It interferes with folic and folinic acid metabolism and is therefore mainly used as a synergist to sulphonamides. At a concentration of 0·005 per cent with sulphadimidine (0·05 per cent) or sulphaquinoxaline (0·0125 per cent) for 3 to 7 days complete protection against *E. tenella* experimental challenge can be obtained. However, it has been shown that the combination with sulphaquinoxaline is less effective for this species than Nicarbazin and Zoalene.

OTHER COCCIDIOSTATS

Nicarbazin (Nicoxin)

This pale yellow powder is only slightly soluble in water and is virtually odourless. Chemically it is a complex of *pp'*-dinitrocarbanilide and 2-hydroxy-4,6-dimethylpyrimidine. Continuous feeding of the drugs at the rate of 0·0125 per cent in the food—or 0·01 per cent where challenge is likely to be low—provides protection against *E. tenella*, *E. necatrix* and *E. acervulina*. Immunity development is usually unaffected. Changes in egg quality, hatchability and shell pigmentation can occur, so the dosing of laying replacement stock should cease when birds approach the point of lay.

Nitrophenide

Nitrophenide is 3,3'-dinitrodiphenyl disulphide. The compound is incorporated in feed at the rate of 0·0125 per cent although higher concentrations have been used quite safely. *E. tenella* is adequately controlled at this level. Higher levels up to 0·04 per cent, near to the safety limit, are required to control *E. necatrix*. At these levels toxic effects could occur if mixing is poorly carried out. Nitrophenide with its narrow therapeutic index, is now rarely used for coccidosis control, as there are more efficient drugs available with safer therapeutic indices.

Zoalene

This compound has two chemical names—2-methyl-3,5-dinitrobenzamide and 3,5-dinitro-*ortho*-toluamide.

The compound is used at the rate of 0·0125

per cent in food and 0·015 per cent in the drinking water. At these concentrations *E. tenella*, *E. necatrix*, *E. acervulina* and *E. maxima* infections in chickens are controlled in broilers.

The drug has also proved of value in the control of *E. adenoeides*, *E. gallopavonis*, and *E. meleagridis* infections in turkey poults. Resistance to zoalene can develop, and cross- resistance with nitrofurazone has been recorded.

Nitrofurans

The nitrofurans group of antibacterials have been discussed in Chapter 32, but these drugs also have activity against coccidia and two of them, nitrofurazone and furazolidone, are used for this purpose.

Nitrofurazone

Nitrofurazone is mainly of value in the prevention of coccidiosis in poultry, and its main activity has been demonstrated to be against *E. tenella* infections. A level of 0·005 per cent is recommended in the food for continuous medication. Levels of 0·01 per cent and 0·02 per cent in the drinking water and feed respectively have been recommended for curative treatment. Mixtures of nitrofurazone and furazolidone are also recommended, and these contain 0·0063 per cent of nitrofurans.

Furazolidone

Furazolidone although active on its own against *E. tenella* and *E. necatrix* is not usually prescribed for the treatment or prevention of coccidiosis. A preparation containing 25 per cent nitrofurazone and 3·6 per cent furazolidone is used for the control of caecal and intestinal coccidiosis. The mixture is fed continuously at a concentration 0·006 per cent in the ration for prevention of infection, and at a level of 0·025 per cent for treatment.

Methyl Benzoquate

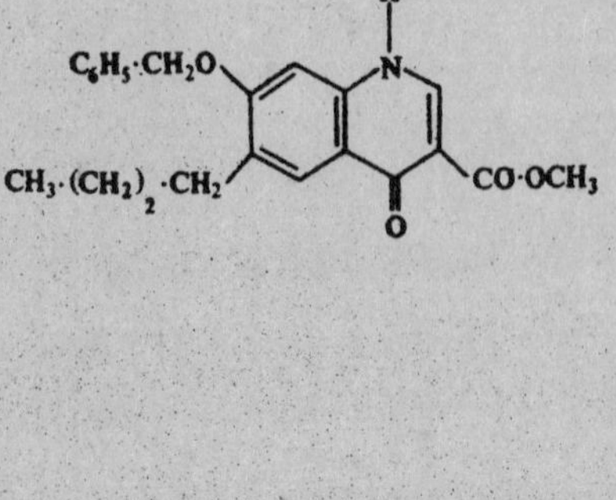

This is a coccidiostat of the quinolate group which is active against species of *Eimeria* which affect chickens. This includes, *E. tenella*, *E. necatrix*, *E. brunetti*, *E. maxima*, *E. acervulina* and *E. mivati*.

Methyl benzoquate is used for the prevention of coccidiosis in broiler chickens, and is administered continuously from day old to slaughter. It is incorporated as a premix containing 4 per cent of methyl benzoquate in the feed to produce a final concentration of 0·002 per cent. At this concentration there is no residue hazard.

Clopidol (Coyden 25)

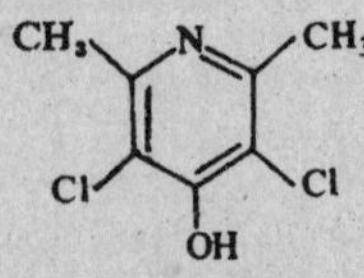

Clopidol is also a coccidiostat of the quinolate group. It is active against *E. tenella*, *E. necatrix*,

Table XX. Coccidiostats now in common use in broilers

Name of drug	Dosage
Sulphaquinoxaline	0·0125% in the food
Nitrofurazone	0·005–0·01% in the food
Mixture of nitrofurazone and furazolidone	0·0063% nitrofurans in the food (0·0055% nitrofurazone+0·0008%) furazolidone
Nicarbazin	0·0125% in the food
Zoalene	0·0125% in the food
Darvisul	Mixture of diaveridine and sulphaquinoxaline 0·01% in the food
Amprolium	0·0125% in the food
Amprolmix	0·0125% amprolium in the food; 0·0008% ethoprobate in the food
Pancoxin	0·008% amprolium; 0·006% sulphaquinoxaline; 0·0005% ethopobate in the food
Clopidol	0·0125% in the food
Decoquinate	0·004% in the food
Methyl Benzoquate	0·002% in the food
Monensin	0·010–0·012% in the food
Robenidine	33 ppm in the food

Table XXI. Coccidiostats used for the treatment of coccidiosis in adult stock*

Name of drug	Dosage
Mixture of nitrofurazone and furazolidone	0·0126% 7 day course in drinking water
Nitrofurazone	0·01% in the drinking water
Mixture sulphadimidine 6·4% and dimethyldipyridyl 1·35%	0·0195% sulphadamidine and 0·004% dimethyldipyridyl for 6 to 10 days
Mixture diaveridine 3·2% and sulphaquinoxaline 2·8%	0·005% diaveridine and 0·004% sulphaquinoxaline in an interrupted schedule 3 : 2 : 3
Soluble formulation of amprolium and ethopobate	0·024% amprolium and 0·0016% ethopobate

* Data from: Joyner, L. P.

E. brunetti, E. maxima, E. acervulina, E. mivati, E. mitis and *E. praecox.*

It is used for the prevention of coccidiosis in broilers and replacement birds up to 18 weeks in age. It is administered from day old to slaughter, and is administered in the feed at a concentration of 0·0125 per cent. It should not be given to birds producing eggs for human consumption.

Decoquinate

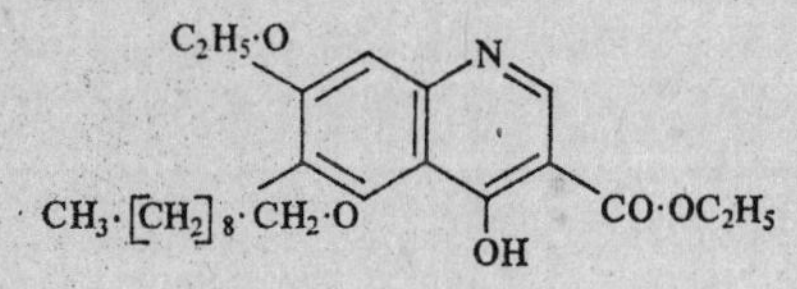

Decoquinate is a quinolate group coccidiostat which is active against *E. tenella, E. necatrix, E. brunetti, E. maxima, E. acervulina, E. mivati, E. mitis* and *E. hagani.*

Decoquinate is used for the prevention of coccidiosis in broiler chickens. It is administered from day old to slaughter by adding an 8 per cent premix in the feed to produce a final concentration of 0·004 per cent.

Tables XX and XXI indicate the changes that are taking place in the chemotherapy of coccidiosis. Drug mixtures are often proving the only effective answer to the wide range of coccidial species encountered in intensive poultry houses. These mixtures are aimed at enhancing the activity of the sulphonamide, and at widening the activity to include a range of species of coccidia.

Monensin Sodium (Elancoban)

This is a new coccidiostat which is obtained by fermentation from the mould *Streptomyces cinnamonensis.* It is active against *E. necatrix, E. tenella, E. acervulina, E. brunetti, E. maxima* and *E. mivati.*

Chemistry

It is a straw-coloured coarse powder and is prepared in the form of a premix for addition to animal feeding stuffs. Each kilogram of premix contains monensin sodium equivalent to 200 g monensin activity. The carrier is soya bean meal.

Pharmacology

Monensin acts at the cell level in the gut at an early stage in the life cycle (the first asexual cycle) by preventing the development of first generation schizonts. Its action is primarily coccidiocidal and it is believed to act by interfering with intracellular ion transport.

Results of Trials

Clarke et al. (1974) demonstrated in European trials that at a strength of 0·012 per cent in the feed, monensin gave significant improvements in body weight, feed conversion and control of coccidial lesions in the presence of naturally occurring infections in broilers.

Dosage

Laboratory and field trials have demonstrated that monensin should be present in the feed at levels between 0·010 and 0·012 per cent; 0·01 per cent is recommended for standard additions to the feed and 0·012 per cent where a severe coccidial challenge is recognized.

Precautions and Side Effects

Monensin is suitable for addition to broiler feeds and should not be included in the diet of laying birds. It must be withdrawn from the feed 72 hr before slaughter.

It is advised that operators preparing feeds should avoid contact with the premix and should wear dust masks.

Robenidine (Cycostat)

Chemistry

Robenidine is 1,3-*bis*(*p*-chlorobenzylidene-amino) guanidine hydrochloride.

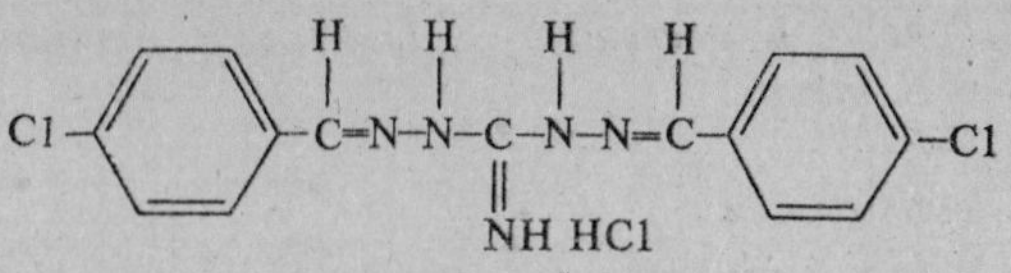

Robenidine

It is an ivory-coloured solid which darkens on exposure to light. It is insoluble in water, slightly soluble in methyl and ethyl alcohol and soluble in chloroform. It is prepared as a premix which contains 6·6 per cent robenidine in a cereal meal carrier.

Pharmacology

Robenidine has a dual mode of action; it is both coccidiostatic and coccidiocidal. It is active against *E. necatrix*, *E. tenella*, *E. acervulina*, *E. brunetti*, *E. maxima* and *E. mivati*. Experiments have shown that, unlike monensin, it does not affect the early stages of the coccidia, but has coccidiostatic activity in the cell against the late first generation schizonts. It then has coccidiocidal activity on the second generation schizonts.

It is suggested that this dual mode of action has an advantage in coccidiosis control because there is a lessened risk of relapse if the birds should go off their feed at some stage and so obtain inadequate levels of drug for a short period.

Result of Trials

Kennett et al. (1974) showed that at a dose of 30 g/ton of diet robenidine prevented clinical disease in the presence of eight laboratory species of *Eimeria*.

Dosage

A dose of 0·5 kg of Cycostat (6·6 per cent robenidine) per ton of complete feed supplies an active level of 0·0033 per cent.

Precautions and Side Effects

Robenidine is suitable for addition to broiler feeds and should not be included in the diet of laying birds. The recommended dose should not be exceeded. Treated feed must be withdrawn 5 days before slaughter.

Bibliography

CLARKE, M. L., DRAZ, M., GUILLOTEAU, B., HUDD, P. L. & STOKER, J. W. (1974) Efficacy studies with robenidine, a new type of anticoccidial, in the diet. *Avian Path.*, **3**, 25–32.

KENNETH, R. L., KANTOR, S. & GALLO, A. (1974) European field evaluation of monensin, a new anticoccidial agent. *Poultry Sci.*, **53**, 978–86.

39

Antiseptics and Disinfectants

The difference between antiseptics and disinfectants is not clearly defined. In practice *antiseptics* are agents used externally on man and animals to kill or prevent multiplication of infective agents—bacteria, viruses, protozoa, etc. They are therefore bactericidal or bacteriostatic. *Disinfectants* are looked upon as agents in the main too toxic to apply to the body but useful in destroying pathogens contaminating inanimate objects—buildings, feeding and cooking utensils, harness, transport vehicles, surgical instruments, the atmosphere, etc. Many undiluted antiseptics will act as disinfectants, while many disinfectants when diluted can be used as antiseptics.

Antiseptics and disinfectants may be classified as follows:

(1) Oxidizing agents:
- (a) Peroxides
- (b) Potassium permanganate
- (c) Halogens

(2) Reducing agents:
- (a) Sulphur dioxide
- (b) Formaldehyde

(3) Metallic compounds:
(4) Acids and alkalis
(5) Alcohol
(6) Phenols, cresols and xylenols.
(7) Miscellaneous organic compounds
(8) Dyes
(9) Detergents

The methods of assessment used to determine bactericidal or bacteriostatic action are not very satisfactory. The activity of an antiseptic is at present determined in relation to the effect of an equal volume of a standard phenol solution, and is expressed as a 'phenol coefficient'. This coefficient may be expressed as a Rideal–Walker coefficient or a Chick–Martin coefficient, depending on the test used to assess antibacterial effect. Both are in vitro tests, the former determined by a straight mixture of bacterial suspension and antiseptic dilution, whilst the latter test involves the mixing of the test antiseptic with a suspension of yeast cells or similar organic material before the bacterial suspension is mixed with the antiseptic. This supposedly simulates the organic matter which will contact the antiseptic in vivo, during its use as an antiseptic; but for a true evaluation, additional information is necessary, including data relating to skin reactions, systemic toxicity, reactions to different infective organisms, and suitability for specific requirements. It follows that some antiseptics may fulfil requirements in respect of some of these conditions and yet be useless otherwise and vice versa. For example, if anionic detergents—ordinary soap is one—are used on the skin and followed by a cationic antiseptic such as **Benzalkonium Chloride Solution,** B.Vet.C., the cationic antiseptic in aqueous solution has a greatly reduced activity. If, however, the skin is swabbed with alcohol subsequent to washing with soap, or if the cationic antiseptic is dissolved in a mixture of 50 per cent ethyl alcohol, 10 per cent acetone and water (so called 'tincture' of benzalkonium chloride), the activity of the antiseptic is improved considerably.

OXIDIZING AGENTS

The oxidizing agents can be further subdivided into those which release gaseous oxygen, *the peroxides*, and those which cause oxidation without the release of oxygen gas: the halogens, and potassium permanganate. These and the reducing agents are rapid in action.

Solution of Hydrogen Peroxide, B.Vet.C.

An aqueous solution of hydrogen peroxide is colourless, and odourless, but has a characteristic taste; it bubbles in contact with the oral mucous membrane. Hydrogen peroxide is relatively stable when slightly acidified but decomposes in alkaline media and in contact with organic matter and metals. The official solution is a 20-volume one: solutions of 10 volumes and 100 volumes are also available; the former is official in the *United States Pharmacopoeia* but is less stable than the 20-volume solution, whilst the latter is dangerous if heated or if organic matter is introduced into the container—it will explode. The concentration expressed in 'volumes' relates to the volume of oxygen released by a single volume of the solution.

Actions and Uses

Hydrogen peroxide has a two-fold action. The *first* is an antiseptic and deodorant action by virtue of the strong oxidizing effect of the nascent oxygen produced in contact with organic tissue. The oxygen is produced, in part at least, by the action of the enzyme catalase. The *second* action is a mechanical cleansing action resulting from the rapid release of oxygen.

These actions persist only as long as oxygen is being produced: thus, as the end of bubbling is the end of the action, it is possible to keep a close check on its activity.

A 50-volume solution required about 5 min to kill anthrax spores and 1 min to kill pyogenic cocci. Diluted four times, this solution would not destroy even cocci with a 30-min exposure. Activity in the presence of organic matter is transient and considerably reduced, mainly because of the speed of oxygen release. The peroxides are thus only of value for their mechanical effect and some mild antiseptic action. They are used in purulent pocketing wounds.

Sodium Perborate

A white crystalline powder, will decompose in solution to give sodium metaborate and hydrogen peroxide, which in turn will gradually evolve oxygen unless heated when oxygen comes off more rapidly. Magnesium, sodium and strontium peroxides will evolve oxygen but are rarely used in veterinary medicine.

Potassium Permanganate, B.Vet.C.

Potassium permanganate consists of dark, odourless, purple crystals with a metallic lustre which are soluble in water giving pink to deep purple solutions depending on concentrations. Solutions, particularly in contact with organic matter, have a typical and distinct odour.

Actions and Uses

Solutions of potassium permanganate have strong oxidizing properties without the release of gaseous oxygen and are antiseptic; they are also astringent. Solutions of 1:1000 ($\frac{1}{2}$ gr. per fl. oz., approximately) are used as wound and mouth lotions for antiseptic and deodorant purposes, whereas 5 per cent solutions are astringent and have been used to reduce excessive granulation: footbaths of this strength have been used to harden racing greyhounds' feet. The antiseptic action is rapidly lost in contact with tissue and solutions turn brown; brown solutions are inactive. The staining of tissues is one of its main disadvantages though the stains can be removed by oxalic acid and sodium metabisulphite solutions.

Two actions which are made use of, though the advisability of such usages is debated, are that of antidotal effects against most alkaloidal poisons which have been swallowed—morphine, strychnine and acontine are examples—and also that of antidotal effect against snake bite. For the former use aqueous solutions of 1:5000 (0·2 g per litre or 1 gr. per $\frac{1}{2}$ pt.) are recommended, whilst snake bite must be treated by making a cruciate incision over the bite into which the permanganate crystals are rubbed. As the action in both cases is one of destruction of the toxic substance by oxidation, the permanganate must be administered immediately, certainly within 5 min of poisoning, if the treatment is to be of any value.

Where crystals are mixed with *Solution of Formaldehyde*, B.Vet.C., in the approximate proportions of 1:2, formaldehyde gas is rapidly evolved. This process is made use of in fumigation.

The Halogens

The antiseptic halogens are chlorine and iodine, both of which are still used to some degree though giving place in some situations to chloroxylenol and quaternary ammonium compounds.

Chlorine

This halogen is a strong oxidizing agent besides being extremely toxic. It possesses antiseptic and deodorizing properties, even in the very dilute concentrations required to avoid toxicity. It is not used in the form of a gas but mainly as unstable hypochlorites which release the chlorine slowly when exposed to the atmosphere or organic material. With organic material it forms chloramines by combining with amino groups of proteins. These chloro-amines are themselves slowly and mildly antiseptic due to their slow release of chlorine.

Chlorinated Lime, B.Vet.C.

Known also as bleaching powder, is a dull white powder with a characteristic odour; it is only partially soluble in water and alcohol. It contains not less than 30 per cent w/w of available chlorine, but decomposes slowly taking up carbon dioxide and releasing chlorine; calcium oxide (quicklime) acts as a stabilizer. The compound is a powerful disinfectant and deodorizes so long as free chlorine is available. It is used on drains, buildings, etc., but should not be used in dairies because of the taint quickly acquired by milk. Thirty-eight grams per litre of water is the usual farm strength, but frequently a handful of the powder is scattered around drains and on the gratings. Care should be taken to avoid contamination of clothing, for the compound bleaches most dyes and rots cloth. Metals are also corroded.

Solution of Chlorinated Lime and Boric Acid, B.Vet.C.

Chlorinated Lime and Boric Acid Solution (Edinburgh Universities' solution—Eusol) is prepared from *Eupad*, a stable powder consisting of equal parts of chlorinated lime and boric acid. Three per cent aqueous suspensions shaken and filtered after standing a give solution which is effective for up to 2 weeks.

Surgical Solution of Chlorinated Soda, B.Vet.C.

This is better known as sodium hypochlorite or Dakin's solution, contains approximately 0·5 per cent w/v of available chlorine. It is not very stable and is commercially available in a concentrated form which keeps much better.

Both eusol and the sodium hypochlorite solution are used as lotions for wounds and burn dressings. All chlorine releasing agents are rapidly inactivated by organic matter.

Sodium hypochlorite is recommended for use as a teat dip for mastitis control at a concentration of 4 per cent active chlorine. In this form it is used to prevent cross-infection during the milking process and its use forms part of a major programme for the control of mastitis.

Iodine, B.Vet.C.

Solutions of iodine have been used for many years as antiseptic wound dressings. Iodine itself is an oxidizing agent though less active than chlorine. It is antiseptic in dilute solution but irritant and has a tendency to retard healing.

For antiseptic and mild irritant action, the **Weak Solution of Iodine,** B.Vet.C.—the common tincture of iodine which contains 2·5 per cent iodine in an alcoholic solution of an equal amount of potassium iodide—and the **Aqueous Solution of Iodine,** B.Vet.C. (Lugol's) iodine are used, the latter mainly for intra-uterine and vaginal irrigation after dilution to strengths of up to 0·2 per cent.

The bluish-black 'metalescent' crystals of iodine colour all solutions brown and also stain the skin: solutions of ammonia or sodium thiosulphate will remove these stains.

Combinations of iodine with non-ionic surface-acting agents, referred to as *iodophores* are available commercially for surface antisepsis and disinfection. They are claimed to provide 80 per cent available iodine in a relatively non-toxic, non-irritant and non-staining form. Increased antibacterial, antiviral and antifungal properties are claimed and the preparations are recommended commercially for foot-rot in sheep, ringworm in cattle, the treatment of wounds and for preoperative skin preparation. As disinfectants, iodophores have been used to cleanse and disinfect cages and kennels, surgical equipment and theatres. For direct skin treatment commercial solutions are used undiluted, but for routine preoperative and general antiseptic purposes 1 per cent dilutions are suggested. Solutions are active as long as the iodine amber colour remains. They are used in mastitis-control programmes for udder washing before milking and teat-dipping after milking.

Iodoform, B.Vet.C.

Analogous to chloroform with a formula of CHI_3, iodoform is an intense lemon-yellow crystalline powder with a penetrating and persistent odour. It is almost insoluble in water, slightly soluble in alcohol, but reasonably soluble in ether and chloroform, and soluble in oils.

Actions and Uses

Iodoform is a persistent antiseptic and deodorant. The antiseptic action according to Sollmann, who quotes a number of authors, 'is probably by liberating iodine'. This does not occur in vitro but only on contact with organic tissue or secretions. The iodine is evolved so slowly that only mild antiseptic effects result, but there is stimulation of tissue cells with drying of wound exudates and a mild local analgesia. In addition a mild phagocytosis results.

It is used in dilutions of 1:4 or 1:8 with zinc oxide, talc, kaolin, etc., as a dusting powder in seborrhoeic conditions of horses.

REDUCING AGENTS

Two agents, sulphur dioxide and formaldehyde are conveniently discussed under this subdivision of antiseptics. They are both primarily fumigant antiseptics, though they both have other applications. **Sulphur dioxide** is liberated as a gas when sulphur is ignited. It has been, and is still occasionally, used as a fumigant disinfectant for animal houses, to destroy bacteria and viruses, parasites and vermin. For every 3m^3 (100 ft^3) of air 0.5 kg (3 lb.) of sulphur should be burned. For full effect, the atmosphere should be moist and all ventilation should be sealed. It is common to spray the walls, ceilings and floors with water before lighting the sulphur. The dioxide dissolves in the water to form sulphurous acid and it is the reducing action of this acid which is bactericidal. Specially made sulphur candles can be obtained or sulphur may be mixed with $\frac{1}{40}$ of its weight of charcoal, which helps it to burn. Buildings should be sealed for at least 12 hr, but preferably 24 hr, subsequent to fumigation.

As sodium metabisulphite, sulphur dioxide is used as a preservative for foods, as a preservative antioxidant for adrenaline or procaine and adrenaline injections of the B.P. and B.Vet.C. amongst others, in strengths of 0·1 to 0·2 per cent.

Sulphur dioxide has a number of disadvantages which detract from its usefulness as a fumigant. It attacks and corrodes metal, rots fabrics, and bleaches dyes.

Solution of Formaldehyde, B.Vet.C.

Everyone is familiar with the colourless liquid with an irritant vapour and a pungent odour which is universally used as a preservative for dissection specimens. On standing, polymerization occurs and a white deposit of paraformaldehyde forms in spite of the alcohol which is left in the solution to delay the process. The solution contains approximately 40 per cent formaldehyde gas.

Actions and Uses

Externally, formalin is strongly astringent and antiseptic. If applied undiluted to the skin, protein precipitation occurs and the skin becomes hardened: it has been used in a similar manner to potassium permanganate for hardening greyhounds feet but should be applied diluted 1:10 to give a 4 per cent formaldehyde solution, which is the normal preservative strength. The main external action is as a fumigant disinfectant or as an aqueous disinfectant spray.

For fumigation of buildings it is usual to mix potassium permanganate and solution of formaldehyde in a ratio of 3:5, using 45 to 90 g ($1\frac{1}{2}$–3 oz.) of potassium permanganate for each 3m^3. Rooms to be fumigated should be closed for at least 10 hr and the best procedure in small animal surgeries and waiting rooms is to fumigate each day after evening surgery so that the rooms may be closed overnight. Care should be taken that splashing does not occur, for gas evolves very rapidly and hot solution may be splashed over a fairly wide area—12 to 14 in. mortars make ideal receptacles. The permanganate, preferably in 'fumigating block' form, should be placed in the generating vessel and the formalin poured on just before vacating and closing the rooms. Spraying a water mist in the rooms prior to fumigation assists the action of formalin. Neutralization of the gas is quickly achieved by ammonia.

Vermin are not killed by formaldehyde fumigation, but fabrics and dyes are unaffected so that for disinfectant purposes it is preferred before sulphur dioxide. It is an accepted method for disinfecting imported straw, clothing, cattle cake, grain, etc., under the Foot and Mouth Disease Order. Hay and straw stacks, clamps of roots, etc., which cannot be fumigated, may be sprayed with a 2 per cent formaldehyde solution (the official solution diluted 1 part in 20 parts). Stacks should be 'pulled' after spraying and the 'pullings' burnt.

Formaldehyde is also used in the preparation of formalinized vaccines from bacteria and viruses.

Disinfection by fumigation is giving place to aerosol space disinfection by the continuous or semicontinuous use of electrically heated and driven aerosol generators. Various agents are used including propylene and triethylene glycols.

METALLIC COMPOUNDS

Compounds of mercury, silver and copper are the main metallic antiseptics and disinfectants.

Mercury Compounds

Though mercuric chloride, iodide and cyanide are all antiseptics and disinfectants, the considerable danger attaching particularly to their antiseptic use has made them unpopular. They are far too irritant for normal usage and organic mercurials were devised to overcome this attribute. The most useful of these are phenylmercuric compounds of which the nitrate and the acetate are the most active.

Phenylmercuric Nitrate, B.Vet.C. and Phenylmercuric Acetate, B.Vet.C. (1953)

Both these compounds are white or almost white crystalline powders which are not very soluble in water.

Actions and Uses

The actions of these two compounds is mainly bacteriostatic. They are less active in the presence of serum and other proteins and do not kill spores. Their antibacterial action is attributed to interference with essential enzyme activity by combining with sulphydryl groups, and it is the phenylmercuric ion which is mainly responsible. The compounds are used in concentrations of 0·03 to 0·06 per cent for swabbing the skin, preferably in a solution buffered with either an organic or inorganic acid. Buffering is required as the phenylmercuric compounds are more stable in acid solutions and any breakdown of the organic ion to produce mercuric and mercurous ions increases irritation. Alcoholic solutions are frequently used because the salts are more soluble, but the added antibacterial effect of the alcohol is also useful.

Buffered solutions are relatively non-irritant, non-toxic, do not stain, and are non-corrosive to to all common metals except aluminium, apart from the effect the acid buffer may have. They are widely used in the veterinary field.

Thiomersal, B.Vet.C.

Also known as Merthiolate, a cream crystalline powder very soluble in water and reasonably soluble in alcohol, is another organic mercurial antiseptic. Its actions and uses are similar to those of the phenylmercuric compounds in a strength of 0·1 per cent, but the compound is reputedly not reduced in activity by serum proteins and is therefore widely used as a bacteriostat in antisera. The *Mercocresols*, are a group of antiseptics used in the U.S.A. consisting of a combination of phenylmercuric and cresol compounds and having the combined effect of a cresol and a phenylmercuric salt but the low toxicity of the phenylmercuric component. A mixture of 0·19 per cent of each component is applied undiluted in alcoholic solutions to superficial wounds and preoperative skin sites, but dilutions of 1:5 to 1:20 are recommended for mucous membranes.

ACIDS AND ALKALIS

The strong mineral acids can be used as disinfectants, but their corrosive actions limit their usefulness. *Boric acid* is a weak acid which was formerly used as an antiseptic for irrigating the conjunctival sac. It possesses only a limited antibacterial action.

Several alkalis are in common use as disinfectants, e.g. sodium hydroxide, sodium carbonate and quicklime. Most bacteria are inactivated above pH 9, but a particular property of the alkalis is their anti-viral acrivity. A 4 per cent solution of sodium carbonate (washing soda) is employed in the cleansing and disinfection procedures required after an outbreak of foot-and-mouth disease.

ALCOHOL

Seventy per cent w/v ethyl alcohol is well-known for its ritual use on a pre-injection skin swab. The point that all antiseptics require a finite time for maximum effect is well made when it is realized that for 70 per cent ethanol the time required runs into minutes. Nevertheless, its solvent properties are of value in that much surface contamination is removed.

PHENOLS AND CRESOLS

The use of the pure phenols and cresols and simple formulations of them are obsolescent as antiseptics but they are retained for disinfection. They are occasionally used internally as anti-zymotics. Newer derivatives of the phenols and cresols, such as the various chlorinated cresols and the phenylmercuric compounds are now more commonly used for antiseptic purposes.

Phenol, B.Vet.C.

Pure phenol usually exists as white crystals which become pink on keeping and take in water due to their deliquescence. It has a distinct characteristic odour and is soluble in 12 parts water, soluble in alcohol and also in other organic solvents, including glycerin and vegetable oils. The compound is extremely caustic and dangerous so that any skin contaminations should be immediately dressed with glycerin and then washed off.

Actions and Uses

Externally, phenol is caustic, antiseptic, disinfectant and parasiticidal. For antiseptic purposes phenol is now no longer used, due to the corrugation and numbing of the skin by the 0·2 to 1 per cent solutions formerly used.

As a disinfectant, phenol is 'approved' by the Ministry of Agriculture for use in premises subject to a disinfection order under the Diseases of Animals Acts—as a disinfectant wash and incorporated at a strength of 0·5 per cent in lime wash for walls. Phenol is still used as a bacteriostat in various injections, and in vaccines and sera. The *British Pharmacopoeia* cites 0·5 per cent phenol as a suitable bacteriostat and preservative for multidose injection packs.

Cresol, B.Vet.C.

This compound is an impure liquid mixture of cresols and phenols obtained from coal tar distillation. The straw coloured liquid which darkens with age and exposure is less soluble in water than phenol but soluble in the organic solvents.

Actions and Uses

The actions of cresol are similar to those of phenol, but toxicity is lower. Cresol is very effective against acid-fast bacteria, only slightly so against viruses, and has no effect on spores. It is, however, very useful in that organic matter has relatively little effect on its bactericidal powers. In practice, **Solution of Cresol with Soap,** B.Vet.C., better known as 'lysol', is preferred because of its greater solubility in water. A concentration of 0·5 or 1 per cent has been used to wash out the uterus of cows following retained placentae: a similar concentration has been used for 'wet' surgery.

Many of the 'Black Fluid' and 'White Fluid' disinfectants used on farms and on veterinary premises are cresol compounds; they are quite effective, many of them having phenol coefficients of R.W. 18 or more, and dilution factors of 30 to 50. They should not be used in dairies, for the persistent phenolic odour quickly taints milk and milk products.

Toxicity

Phenols and cresols are especially toxic to cats and dogs: washing of cat cages with phenols or cresols, and even carbolic soap used for bathing dogs have resulted in death. Symptoms are those of lethargy, coma or convulsions followed by coma, with death supervening as a result of respiratory or cardiac failure. Treatment consists of bathing with liberal amounts of warm water where skin absorption is suspected. Where the compound has been licked off paws or skin, gastric lavage with water or limewater is recommended though not very easy to carry out. Intravenous and subcutaneous saline injection and respiratory stimulants are useful. External contamination with concentrated phenols or cresols should be swabbed with glycerin.

Perhaps the compounds which have done more to out-date phenol and cresol as antiseptics have been the *chloroxylenols*. Two are in common use throughout the world—parachlorometaxylenol (PCMX) and dichlorometaxylenol (DCMX). These antiseptics as a group are greatly reduced in efficiency by contact with organic matter and have low Chick–Martin values.

Chloroxylenol, B.Vet.C.

This is xylenol with a chlorine atom in the *para*-position. The *ortho*- and *meta*-compounds are less active. Chloroxylenol is a white or creamy white crystalline substance with a slight phenolic odour. It is insoluble in water (1:3000) but soluble in organic solvents and soap solutions—**Solution of Chloroxylenol,** B.Vet.C. is a 5 per cent chloroxylenol solution with 10 per cent terpineol and a castor oil soap in alcohol and water.

Actions and Uses

This chlorinated dimethylphenol is a highly effective bactericide against some Gram-positive bacteria. Streptococci are most susceptible, staphylococci less so, but many Gram-negative organisms such as *Pseudomonas* and *Proteus* species are hardly affected. The Rideal Walker phenol coefficient when properly formulated is 62 against *B. typhosum* (*Salm. typhi*), but solutions of 5 per cent, such as the official one, have phenol coefficients of about 3, that is they are three times more effective than a standard solution of phenol. The main advantages of these solutions are their low irritant properties and very low toxicity. In the majority of people an undiluted 5 per cent solution causes no irritation apart from that following strong detergent activity due to the alcoholic soap-solution base. A few individuals are hypersensitive to the chloroxylenols and cannot use them.

Solution of chloroxylenol can be used both as an antiseptic and disinfectant but its expense and the fact that the action is greatly reduced by organic matter make Solution of Cresol with Soap preferable for disinfectant purposes.

As an antiseptic, chloroxylenol solutions are very widely used in Great Britain. For pre-operative skin sterilization a dilution of 1:4 is recommended, but for wound cleansing and irrigation of the uterus and vagina 1 to 2 per cent solutions are better (approximately 10 to 20 ml per litre—2 teaspoonfuls to 1 tablespoonful in 1 pt. of water).

Dichlorometaxylenol, B.P.C.

This is a more recent discovery, with approximately four times the activity of the single chlorine compound (R.W. phenol coefficient *Salm. typhi* 250, *Staph. aureus* 180 and *Strep. pyogenes* 160), but in addition there is similar activity against many other organisms including *Esch. coli* and *Ps. pyocyanea.* Fungicidal activity is also appreciable. Toxicity is very low (50·7 g per kg body weight have been given to mice before producing ill effects), and the lower concentrations required because of greater activity reduce it even further.

Dichlorometaxylenol is formulated in a similar way to the monochlor compound, but using approximately 25 per cent of the concentrations. It is used in 2 per cent concentrations in surgical soaps and compares favourably with hexachlorophane. Toilet soaps contain 1 per cent for deodorant and antiseptic properties.

Hexachlorophene, B.Vet.C.

Similar in bactericidal, physical, and chemical, properties to the chloroxylenols but is a diphenolic compound and has only a slight odour. The R.W. coefficient is approximately 125. The major use of hexachlorophane is in surgical soaps, when it is used in strengths of 0·5–2 per cent, at which concentrations it is almost non-irritant.

MISCELLANEOUS ORGANIC COMPOUNDS

The two important compounds, Chlorhexidine Hydrochloride, B.Vet.C. and Dequalinium

Chloride, B.P. are both complex organic compounds of relatively recent discovery now widely used.

Chlorhexidine Hydrochloride, B.Vet.C.

This compound is a white or off-white crystalline powder, odourless but has a bitter taste. It is soluble to approximately 0·25 per cent in water and alcohol and 2 per cent in propylene glycol.

Actions and Uses

Chlorhexidine is a highly active antiseptic against many Gram-positive and Gram-negative organisms; the inhibitory concentration in vitro against *Esch. coli* is 1:100 000; *S. dublin* and *Vibrio cholera*, 1:300 000; *Staph. aureus* 1:1 000 000 and *Strep. pyogenes* 1:2 000 000. In vivo activity is also very high. The activity is well maintained in the presence of tissue fluids and pus: it is active in the presence of other antiseptics such as quaternary ammonium compounds but is incompatible with anionic compounds including soap. Where prolonged action is required the relatively insoluble hydrochloride is used either as 1 per cent dusting powders or creams. The diacetate and the gluconate are used if solutions are required: for pre-operative skin preparation 0·5 per cent alcoholic or 1 per cent aqueous solutions may be used, but for general antiseptic washing purposes, e.g. dairy use, concentrations of 1:5000 are adequate; eye lotions should contain 1 part in 1000. In metritis 1 g pessaries may be inserted into the uterus direct or dispersed in about 500 ml water and injected by catheter. The *toxicity* is low and local irritation is uncommon.

Dequalinium Chloride, B.P.

Both the chloride and the acetate are white or creamy-white powders which are odourless but have a bitter taste. The chloride is relatively insoluble (slight in cold water, 1:30 in hot water, 1:200 in propylene glycol) but the acetate is very soluble (1:2 water, 1:12 alcohol).

Actions and Uses

Both compounds are highly active against a wide range of Gram-positive and Gram-negative organisms but anti-fungal activity is also possessed against some species of *Trichophyton* and *Microsporum*. Tissue fluids do not diminish the activity but as with chlorhexidine the dequalinium salts are incompatible with soaps and other anionic detergents. The compounds have been used as 0·4 per cent creams and in powder form for infected wounds, burns and similar skin lesions.

DYE COMPOUNDS

Dyes used for medicinal purposes have been classified under six headings:

1. The *azo-dyes*, related to the sulphonamides and typified by *scarlet red* which is used for stimulating epithelial growth in clean, dry, but static, wounds and ulcerations.
2. The *rosaniline dyes*, which are coal tar derivatives, and are typified by *crystal violet*.
3. The *acridine dyes*, also coal tar derivatives, are exemplified by *acriflavine*.
4. The *fluoroscein dyes* which have a restricted application.
5. The *phenolphthalein dyes* which include the *phenolsulphonphthalein* used in kidney filtration tests.
6. A few dyes which do not easily fall into any of these categories but are of some interest. Methylene blue is one of these.

Of these six groups of dyes, only two are of any value as antiseptics: the rosaniline dyes, brilliant green and crystal violet, and the acridine dyes, acriflavine, euflavine and proflavine.

The Rosaniline Dyes

These dyes are also referred to as triphenylmethane dyes, **Crystal Violet,** B.Vet.C. has been mentioned in the chapter on fungicidal agents, where its activity against Gram-positive organisms is described. The dye is rarely used solely for its antibacterial effect but it is frequently incorporated in skin applications as a colouring substance with bacteriostatic effect and occasionally to mark the area and ensure compliance with instructions. When used on intact skin as a lotion or paint it is best applied as a spirit dressing especially on small animals; this ensures rapid drying with localization of the dye and less risk of staining the owners clothing and furnishing fabrics. Aqueous solutions should be applied to wounds, abraded skin and burns, for spirit dressings are painful. Concentrations of 1 to 2·5 per cent are used in alcoholic solutions and 0·5 per cent in aqueous solutions, frequently with equal quantities of brilliant green (**Paint of Brilliant Green and Crystal Violet,** B.Vet.C.).

Brilliant Green, B.Vet.C.

This is a water and spirit soluble golden crystalline substance. Water and alkaline solutions are green but turn reddish yellow when acidified with hydrochloric acid.

Actions and Uses

Brilliant green, though active against Gram-positive organisms in common with the other basic dyes of the rosaniline series, is mainly of value for its selective bacteriostatic action on *Esch. coli. Externally*, it is of value in 0·05 to 1 per cent aqueous solutions for infected wounds and burns, and 0·5 per cent alcoholic solutions, with or without equal quantities of crystal violet, can be used for pre-operative skin sterilization.

Internally, brilliant green was used extensively for gastroenteritis in calves and dogs, but since the advent of broad spectrum antibiotics its popularity has decreased. For *white scours* of calves the dye is most frequently used with bismuth salicylate as the unofficial compound, mixture of bismuth salicylate with brilliant green; less frequently it is given as **Mixture of Brilliant Green and Kaolin,** B.Vet.C. (1953). The former contains bismuth salicylate, kaolin and 1:5000 brilliant green, and the latter 0·2 per cent of brilliant green with approximately 14 per cent of kaolin in water. The recognition of infective drug resistance could lead to the 'rediscovery' of agents such as brilliant green.

The dye stains the faeces and is not popular with dog owners because of the damage done to floor coverings. In calf scours, acute severe cases do not respond rapidly enough and antibiotic and sulphonamide therapy is preferred, especially if combined with koalin and administered concurrently with treatment to combat dehydration.

Acridine Dyes

These dyes were of considerable importance as bacteriostatic antiseptics but, as with the other non-specific bacteriostats, their importance has waned with the introduction of sulphonamides and antibiotics. They are, however, still used, frequently in conjunction with antibiotics and sulphonamides, because of their lack of toxicity and non-irritant character, and also because of continued activity in the presence of serum, though blood and pus do cause some decrease in activity. Phagocytosis is very little affected. It follows, that in wounds characterized by copious serous exudates but little bleeding or pus formation, these dyes will be very useful. The drugs are mainly active against pyogenic organisms, particularly streptococci and staphylococci, whilst they possess little activity against Gram-negative organisms, and *Ps. pyocyanea* and *Proteus* species are highly resistant. Antibacterial action is thought to be due to interference with the amino acid sequence of DNA.

The important members of the group are Euflavine (neutral acriflavine), Acriflavine, B.Vet.C., Aminocrine Hydrochloride, B.Vet.C., and Proflavine Hemisulphate, B.Vet.C.

Acriflavine, B.Vet.C.

This is an orange-red crystalline powder, odourless, but with an acid taste. It is soluble in water, alcohol, and glycerin, only slightly soluble in saline, but insoluble in organic solvents other than those mentioned. Dilute aqueous solutions of acriflavine are yellow with a green fluorescence. The neutral compound is a mixture of chlorides precipitated by the addition of sodium chloride to a neutral solution of acriflavine. Acriflavine itself is a mixture of the hydrochlorides of two acridine derivatives and is acid in solution.

Actions and Uses

The neutral compound is often preferred for wound dressing because it is less irritant than the hydrochlorides but otherwise its actions are the same.

Externally, acriflavine is a slow acting bacteriostat against pyogenic Gram-positive organisms; it is applied to the skin as Acriflavine Emulsion, B.P.C. containing the dye at a 1:1000 (0·1 per cent) dilution, as a 0·1 per cent aqueous or 0·1 per cent saline solution, or as a jelly containing the same amount of dye. The last of these has been used extensively as a burn treatment. Impregnated bandages and gauze are also used, but it has been pointed out that the dye has an affinity for staining cotton and its activity is thereby reduced.

Internally, dilutions of 0·01 per cent acriflavine were used extensively as udder infusions for the treatment of mastitis before the advent of penicillin and sulphanilamide, and were considered to be the best treatment then available.

Intra-uterine injections containing acriflavine in water and glycerin at dilutions of 1:5000 to 1:10 000 were frequently used for 'the whites' or leucorrhoea, and pessaries containing 0·1 per cent used.

Proflavine Hemisulphate, B.Vet.C.

The characters of proflavine are similar to acriflavine, though the crystals are somewhat darker in colour and are less soluble.

Proflavine is less potent than acriflavine but is quicker in action and is less toxic. It is rarely used in veterinary medicine for anything other than skin and wound antisepsis in the form of dusting powders and solutions in concentrations of up to 0·2 per cent. Combination with various sulphonamides are popular.

Aminacrine Hydrochloride, B.Vet.C.

A yellow powder which is slightly more soluble than proflavine. Its effect, toxicity and irritant properties, lie between those of acriflavine and proflavine.

Incompatibility

All the acridine dyes are antagonized by hypochlorites and these two types of antiseptics should not be used together.

The Fluorescein Dyes

Of this group, only **Fluorescein Sodium,** B.Vet.C. is used to any extent. When one drop of a 2 per cent solution, which is reddish brown, is placed in the eye at the outer canthus dilution with tears occurs and the eye becomes green and fluorescent. Any corneal ulcerations show up as light green fluorescent patches which are easily seen without a lens; foreign bodies show up as dark areas within the fluorescent zone. The dye is normally drained off through the lacrimonasal duct and fluorescent staining will appear at the nostril within a minute of application to the eye unless the duct is blocked. This diagnostic aid is most frequently used in small animals but is equally efficient in other species. More recently, up to 30 ml of a 10 per cent autoclaved solution have been injected intravenously in horses suspected of suffering from *periodic ophthalmia* or iridocyclitis; if the disease is present, fluorescence is noticeable in the aqueous humour within 10 min. The same phenomenon is shown in dogs with intra-ocular inflammation if up to 4 g (60 gr.) are given orally in aqueous solution (it dissolves in an equal volume of water, but more dilute solutions are preferable); up to 20 min may elapse before fluorescence. As with the acridine dyes, photosensitization occurs and animals should be protected from direct sunlight for up to 2 days after dosing.

The Azo-Dyes

Scarlet red or Sudan IV is the only azo-dye used in veterinary medicine, if the coloured sulphonamides are excluded. It has been used in clean, granulating wounds to stimulate epithelial growth, and it is recommended that a 4 per cent ointment is used, alternating every 48-hr with a bland emollient ointment. Stronger concentrations may cause necrosis and, if applied over too long a period, irritation may occur. It is of little value in indolent ulcers.

DETERGENTS

This group of compounds is part of a large group collectively referred to as *surface-active agents*, which all have the property, in some degree, of reducing surface and interfacial tensions—the properties of emulsifying agents. Other properties of surface active-agents are those of 'wetting', spreading, penetrating, foaming and, in some cases anti-foaming, and cleansing activity. One agent may possess one or more of these attributes and perform these several functions at the same time. Thus *detergents* are classified as emulsifying and cleansing agents which frequently possess antibacterial properties, though the term has become a common household name for a chemical cleansing agent. Some detergents have stronger antibacterial properties than others, but antibacterial effect is not related directly to surface-activity.

Surface-active agents are of two main types, (1) those which ionize; ionic; and (2) those which do not, non-ionic. The detergents belong to the ionic group and are of two types: (1) those which are active in mildly acid solution by virtue of active anions, *anionic detergents*, and (2) those which are active in alkaline solution and in which the detergent property is vested in the cation, *cationic detergents*.

The antibacterial activity of detergents is greatly reduced by the presence of organic matter such as blood, serum and pus, and ionic detergents have little or no effect on viruses, spores, or fungi.

Anionic Detergents

Included in this group are the alkaline earth salts of organic acids with high molecular weights, sulphate alcohols and sulphonated oils. Examples of these used in the veterinary field are the soaps (sodium or potassium oleates, etc.), calcium and ammonium mandelates, bile salts and sodium lauryl sulphate.

Only the soaps and sodium lauryl sulphate are of any significance as antibacterial and cleansing agents, and their antibacterial activity, which is not great, extends only to Gram-positive bacteria, so that for most purposes stronger and more comprehensive antibacterial agents must be incorporated with them.

Soaps have been used as detergents for many years. Their cleansing properties are widely known and are due mainly to the emulsification of grease and the loosening of keratin, dirt and similar debris. Their antibacterial properties are low and any benefit arising from pre-operative skin cleansing arises from removal of dirt and debris to which bacteria are adherent. Soaps are therefore combined with stronger antibacterial agents such as alcohol, hexachlorophane and phenylmercuric nitrate. There are two official soaps.

Hard Soap, B.Vet.C.

This is sometimes referred to as castile soap, is an off-white, almost odourless, solid substance obtained by the reaction of a vegetable oil with sodium hydroxide. It consists mainly of sodium oleate, is moderately soluble in cold but freely soluble in hot water, and is almost completely soluble in alcohol. When dry, it is easily powdered and can then be incorporated in bath powders. Curd soap, which is similar, consists of sodium soap made from animal fats.

Actions and Uses

Hard soap is a surface tension reducing agent and can be used as an emulsifying and dispersing agent. One gram (15 gr.) will emulsify 30 ml (1 fl. oz.) of a volatile oil such as turpentine, or 60 ml (2 fl. oz.) of a fixed oil, and the soap is used for this in the preparation of soap liniments. Its cleansing properties are well known and evidence is available to demonstrate some degree of antibacterial effect. It is used, amongst other things, as a lubricant enema or rectal suppository in the treatment of faecal impaction. Pharmaceutical hard soap is used as an excipient for pills and horse balls, and as a detergent basis for insecticidal shampoo powders.

Soft Soap, B.Vet.C.

A brown to yellowish-white petroleum-jelly-like substance, the lighter colour being a purer product. It is frequently coloured green with such dyes as chlorophyll. The compound is a potassium soap of a vegetable oil and is soluble in water and alcohol. It may contain free caustic and is usually considered unsuitable for skin cleansing because of this.

Actions and Uses

Soft soap can be used as an emulcent in much the same way as the hard soap, and 16 g (240 gr. will emulsify 30 ml (1 fl. oz.) of a volatile oil and 60 ml (2 fl. oz.) of a fixed oil. Many liniments are made using these proportions. Alcohol and ether soaps contain soft soap—**Soap Spirit**, B.Vet.C. contains 65 per cent of soft soap and is used in the preparation of liquid shampoos. As with hard soap, the potassium compounds have some antiseptic action which is enhanced by the inclusion of spirit and other antibacterial compounds.

Sodium Lauryl Sulphate, B.Vet.C.

That is a white or yellowish fine powder or crystals with a faint odour. It is irritant to mucous membranes. The compound is soluble in water giving turbid solutions but is only partially soluble in alcohol.

Actions and Uses

Sodium lauryl sulphate is a surface-active compound with useful wetting and detergent properties. It is used as a wetting and dispersing agent in cattle drenches containing phenothiazine, hexachloroethane, or light vegetable powders, and in insecticidal dips and sprays containing gamma benzene hexachloride and dicophane. It is also used extensively in shampoos and dips, sometimes combined with soaps. The compound is active in reasonably acid or alkaline media and also in hard water, as the calcium and magnesium salts of the alkyl sulphate are also soluble. The inclusion in soap powders retards the formation of insoluble calcium and magnesium soaps and therefore maintains activity.

As an emulsifying agent, sodium lauryl sulphate is used in many ointment bases including **Emulsifying Ointment,** B.Vet.C. and **Emulsifying Wax,** B.Vet.C.

The antibacterial properties of the compound are typical of anionic detergents. No effect is exerted on Gram-negative organisms.

All anionic detergents are antagonized by cationic compounds and vice versa.

Cationic Detergents

These compounds are sometimes referred to as 'reversed soaps'. They consists therefore of neutral salts of bases with high molecular weights, and include the quaternary ammonium compounds and the rosaniline and acridine dyes. In contrast to the anionic compounds, these cationic detergents are more active in alkaline media.

As a group, these agents are active against both Gram-positive and Gram-negative bacteria, but higher concentrations are required against the latter group.

The dyes, which have been referred to previously, possess useful antibacterial effects, but the most important cationic detergents and antiseptics are the quaternary ammonium compounds.

The Quaternary Ammonium Compounds

These compounds can be regarded as ammonium (NH_4) compounds in which the four hydrogen atoms have been replaced by organic groups. If one of these organic groups is a long chain fatty or paraffin-like chain a typical cationic detergent is produced which has strong antiseptic properties, but if two groups are long chains a cationic detergent is produced with very strong antiseptic powers and greater stability over the pH range. The two types are more effective in alkaline solution and activity is reduced in the presence of pus, blood and serum, due to their adsorption and chemical reaction with a large variety of substances, including proteins. They are, for example, completely adsorbed by charcoal. The double chain compounds are claimed to be less reactive with organic matter.

Quaternary ammonium compounds have no effect on viruses and though some effect against spores is claimed for them this should be discounted in practice. When used as disinfectants for instruments and utensils these should first be heat sterilized to destroy spores, after which, provided all organic matter has been scrubbed off, the quaternary ammonium antiseptics will maintain sterility. It is advisable with many of these antiseptics to include antirust compounds such as sodium nitrite in the solutions. Some antifungal action is exerted.

The quaternary ammonium detergents form a film on the skin which is strongly antiseptic externally but unreactive internally so that pathogenic bacteria may exist under the films. This indicates that sterilization of operation sites by this group of compounds alone is almost impossible. Anionic detergents, if used first and thoroughly rinsed off, will assist, but possibly the best method to overcome this phenomenon is to include alcohol in the quaternary ammonium solutions.

Cetrimide, B.Vet.C.

Cetrimide is a compound of the quaternary ammonium group of detergents which is very widely used. It is a fluffy white or creamy-white powder which is soluble in water and alcohol. It is a strong foaming agent.

Actions and Uses

These are typical of the group. Solutions of 1 per cent are used for skin cleansing and wound or burn dressings; 0·1 per cent concentrations are used for solutions in which udder cloths, milkers' hands and utensils are rinsed and 0·5 per cent creams are applied to the teats and hands as prophylactic procedures in mastitis control. Instruments are kept in 1 per cent aqueous solutions containing 0·2 per cent sodium nitrite to prevent rusting.

As with many of these cationic compounds toxicity and irritant properties are relatively insignificant in the concentrations and quantities normally used.

Benzalkonium Chloride Solution, B.Vet.C.

Another typical cationic detergent with strong antibacterial properties. For skin sterilization, a 0·1 per cent alcoholic solution should be used, but for mucous membranes and deep wounds, 0·01 to 0·05 per cent concentrations are better. Instruments require 0·02 per cent for storage, with 0·5 per cent sodium nitrite as an antirust agent.

Note: It is important to remember that soaps are antagonistic to all cationic detergents. Where soap is used to clean the skin prior to operations it should be well rinsed off and the area preferably swabbed with alcohol before applying any cationic antiseptic.

Ampholytic Surfactant Biocides

This group of compounds, usually called the ampholytes, are based on dodecyl-di(aminoethyl)-glycine. Like the soaps and quaternary ammonium compounds (quats), the ampholytes dissociate into ions in water. The active ion of a soap is the anion and its main property is as a cleansing agent. The quats are powerfully antibacterial and in their case the cation is the active group. The ampholytes yield anions, cations and zwitterions and possess both biocidal and detergent properties and are exemplified by the Tego series.

The ideal properties of a disinfectant are that:

(a) It should be rapidly effective.
(b) It should have a broad spectrum of activity including: bacteria, viruses and fungi.
(c) It should not allow the emergence of resistant populations.
(d) It should not be inactivated by protein.
(e) It should be minimally toxic.
(f) It should be non-staining and non-corrosive.
(g) It should not harm the skin by direct action or sensitization.
(h) It should be odourless and deodorizing.
(i) It should possess residual action after rinsing.
(j) It should have detergent properties.
(k) It should be simple and economical in use.

Of all the compounds mentioned in this chapter, the ampholytes come closest to possessing all these properties and it is not surprising that they are now in widespread use in laboratory animal colonies, in sterile manufacture areas in the pharmaceutical industry, in foodstuffs manufacture and in the brewing industry.

The Antibiotics and Sulphonamides

Certain antibiotics are too toxic for systemic use and are therefore reserved for topical application, e.g. Bacitracin and Polymyxin. Although those used systemically are presented as topical forms, their use should be avoided in order to minimize problems of resistance and sensitization. Some sulphonamides are also used topically, e.g. Sulphacetamide. These compounds are fully discussed in Chapter 33.

Bibliography

British National Formulary (1966) British Medical Association: Pharmaceutical Society of Great Britain.

40

Miscellaneous Chemotherapeutics

In this chapter a number of chemically unrelated groups of compounds appear. They are mainly anti-protozoal chemotherapeutics but many of the compounds also have antibacterial actions. The compounds are grouped under the following headings: the sulphones; the diamidines; the amino-phenanthridium compounds: the quinapyramine compounds and other trypanosomidal compounds; miscellaneous antipiroplasm compounds; anti-histomoniasis compounds, including the nitrofuran group of compounds.

THE SULPHONES

The sulphone chemical grouping has been known for many years, the ethyl sulphones (including sulphonal, a hypnotic) having been introduced as long ago as 1888 by Baumann and Kast, but it was the discovery that 4:4-diaminodiphenylsulphone was very effective in protecting mice from streptococcal infections that modern research methods were applied to the sulphones as a group.

In the veterinary field it was found that the compound now officially known as Dapsone in the *British Veterinary Codex*, was well tolerated by cattle. This stimulated further work by Francis and his colleagues who have shown in numerous publications the value of dapsone in the treatment of bovine mastitis and coccidiosis. Concurrent with this work various sulphone derivatives have been developed which have been used in the treatment of human diseases due to acid fast organisms: their use in tuberculosis has lapsed in favour of antibiotic therapy but they are used extensively in the treatment of leprosy.

Dapsone, B.Vet.C.

This compound may be considered as the 'parent sulphone' and bears the same relationship to the sulphone group that sulphanilamide does to the sulphonamide group. Chemically it consists of two amino-benzene groups joined by a sulphone (—SO_2—)bridge:

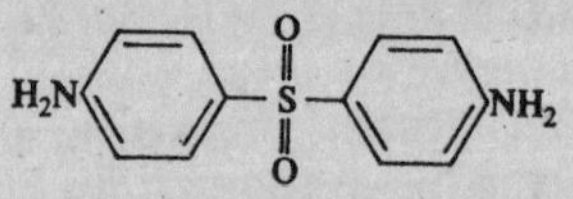

Dapsone

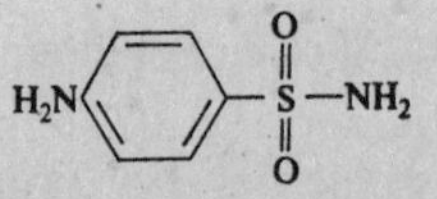

Sulphanilamide

The close relationship with the sulphonamides is immediately apparent. The compound exists as a white odourless crystalline powder which is almost insoluble in cold water, soluble in alcohol, acetone and in dilute hydrochloric acid solutions.

Pharmacology

The *absorption* rates following oral administration are similar to those of sulphanilamide, sulphamerazine, and sulphadimidine, whilst blood levels are maintained for long periods: a single oral dose at the rate of 0·2 g per kg body weight provided blood levels of over 1 mg per 100 ml in a calf for 5 days. *Milk levels* following oral dosing are disappointing, being only a third of the blood levels, whilst those of sulphanilamide are slightly higher. Nevertheless, dapsone is about three to four times more active than sulphanilamide and the levels following intramammary injection persist for several days.

Excretion via the kidney is slower for dapsone than any of the other sulphones. The *toxicity* of dapsone restricts its use to bovines, in which species it is no more toxic than sulphapyridine or sulphathiazole. Dapsone is very toxic in sheep.

Action and Uses

The *action* of the sulphones is considered to be similar in method to that of the sulphonamides. It is bacteriostatic and coccidiostatic. The drug is mainly active against Gram-positive organisms, particularly streptococci, against which it is about three times more potent than sulphanilamide, whilst when compared with the heterocyclic sulphonamides it appears to be one to two times less active. However, dapsone is much less active than these same heterocyclic compounds against the Gram-negative organisms.

There are only two diseases in which dapsone is used: they are streptococcal mastitis and coccidiosis of cattle. For the latter use the high dosage makes it a very expensive treatment: the oral *dosage* consists initially of 0·1 g per kg body weight, followed by half this dose daily for up to 5 days.

In *mastitis*, dapsone may be used alone or in conjunction with penicillin as an oily suspension. The *dosage* is 8 to 10 g, though in newly calved cows 4 g is usually adequate, in the form of a 33·3 per cent mineral oil suspension injected into infected quarters on two or three occasions at 48-hr intervals. When combined with 50 000 or 100 000 units of penicillin one dose is often sufficient. It is essential to leave dapsone preparations in the udder for as long as possible without milking (24 to 36 hr has been recommended) if the best results are to be obtained, mainly because of the low milk solubility and the prolonged action. Dapsone has for this reason been recommended in 12-g doses for dry cows in which the dapsone compound is injected and left until calving. Milk clots and slight irritation have been reported following the introduction of dapsone into the udder, but these effects are of a temporary nature. The intramammary injection has also been used effectively in rickettsial keratoconjunctivitis (pink-eye) of cattle and sheep.

The *official* preparations consist of Dapsone, B.Vet.C., Intramammary Injection of Dapsone, B.Vet.C. which is a 33·3 per cent suspension in a mixed paraffin base, and a similar preparation containing in addition 50 000 units of penicillin.

Commercially there are available dapsone powder, an intramammary injection containing 4 or 10 g of dapsone with 100 000 units of penicillin, and also one containing 0·15 g chloromycetin and 0·3 g dapsone per tube. These antibiotic–dapsone preparations make use of the synergistic action which is shown by such combinations. Use of chloromycetin for mastitis has now been largely discontinued.

THE DIAMIDINES

These are a group of chemotherapeutic compounds found to have a trypanocidal activity related directly to the guanyl (that is, the amidine) organic group, $NH{:}C(NH_2)—$. The presence of two of these groups separated by a paraffin hydrocarbon chain was found to give a high trypanocidal effect which increases as the length of a paraffin chain is increased and reaches maximum at undecane diamidine (i).

$$\underset{\displaystyle NH_2}{NH{=}\underset{|}{C}}—C_{11}H_{22}—\underset{\displaystyle NH_2}{\underset{|}{C}{=}NH}$$

(i)

Investigations were extended from these alkyl diamidines to the aromatic or benzene ring diamidines, and it was found that if two benzamidine groups

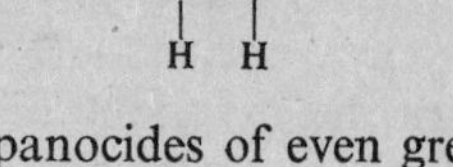

were joined, either by: (a) an oxygen atom, (b) a dioxyalkyl chain $—O—C_nH_{2n}—O—$, or by (c) an ethylene linkage

$$—\underset{H}{\underset{|}{C}}{=}\underset{H}{\underset{|}{C}}—$$

a series of trypanocides of even greater activity was produced. It was also found that these compounds possessed activity against other protozoa such as the babesias, and in addition antibacterial activity similar in character to that of the sulphonamides, but of a lesser degree; the action is thought to be one of interference with oxidation processes within the bacteria. More recently a further group has been developed (d) where the linkage is

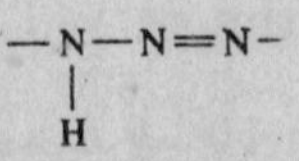

This group of compounds has all the activities of the ethylene linkage series but to a greater degree.

When administered internally the diamidines cause an arterial dilatation and a lowering of blood pressure; there is also an increase in tone and movement of intestinal muscles. Kidney and liver damage are the main toxic effects: these effects result in vomiting, diarrhoea, jaundice, etc.

The main examples of aromatic diamidines conforming to the group formula (ii), where '*x*' may be any one of the four linkages a, b, c or d,

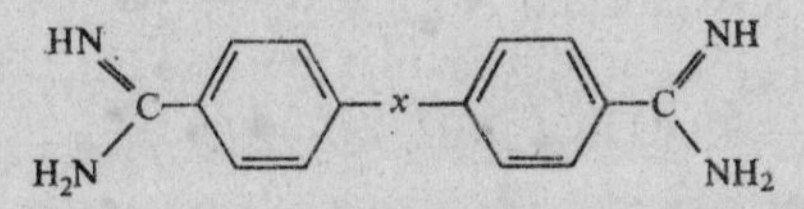

(ii)

mentioned above, are Phenamidine where x = a, Propamidine where x = b, '*n*' being 3, Stilbamidine where x = c, and diminiazene where x = d. The formulae for these four are therefore as in (iii), (iv), (v) and (vi).

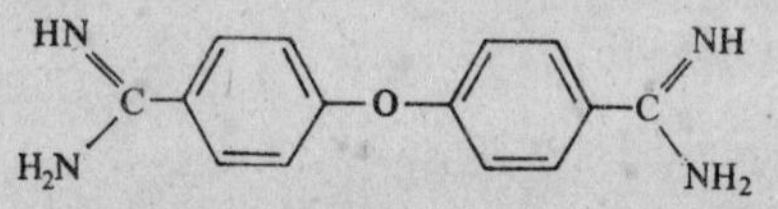

(iii) Phenamidine

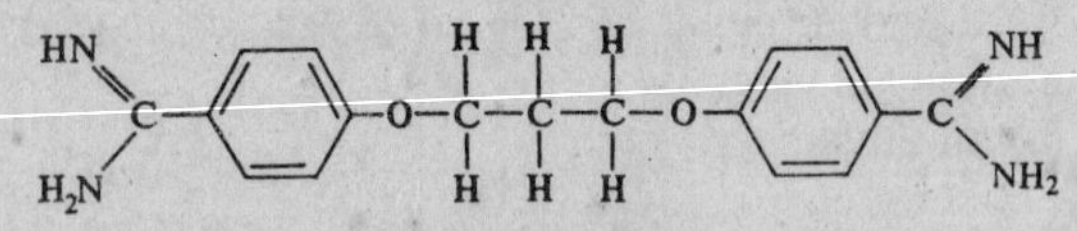

(iv) Propamidine

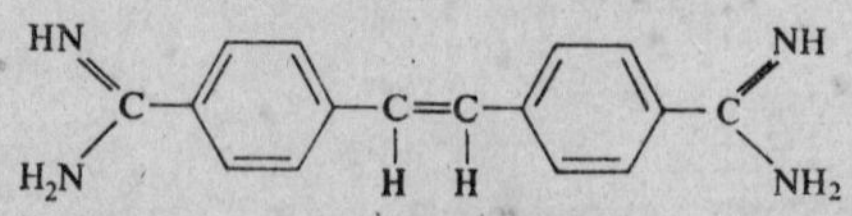

(v) Stilbamidine

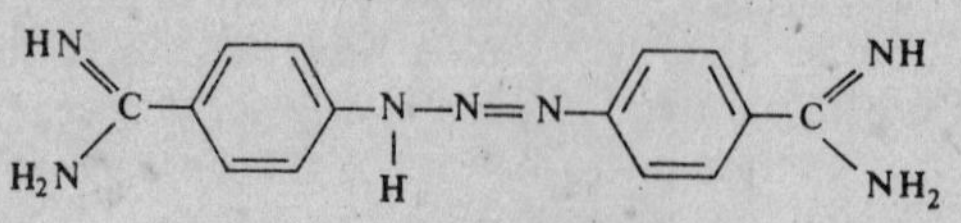

(vi) Diminazene

A further compound similar to Propamidine in formula but where '*n*' is 5 has also been widely used as a trypanocidal drug: the propane of propamidine becomes pentane, giving the name pentamidine.

The diamidines have been used either as free bases or as one of three salts, the dihydrochloride the dimethylsulphonate or the isethionate (dihydroxyethane sulphonate). In the veterinary field the isethionates are used most frequently. Substitution products have also been used, the dibromopropamidine isethionate probably being the best known of these.

Diminazene Aceturate, B.Vet.C.

This complex compound is an odourless yellow powder which is relatively soluble in water (1:14) but only slightly so in organic solvents.

Introduced commercially under the name of Berenil this compound was reported to have anti-trypanocidal activity. This activity has since been confirmed. Other actions common to the group, babesicidal and bactericidal actions (mainly *Brucella* and *Streptococcus* species) are also exerted by the compound. It was suggested that diminazene might be of value where resistance to ethidium, homidium and quinapyramine has developed.

As a trypanocide, diminiazene is effective against *T. vivax* and *T. congolense* but is less effective against *T. brucei*, necessitating an increased dosage rate against this species.

Actions and Uses

Dosage is in the form of a compound injection which consists of 8·75 per cent phenazone and 7 per cent diminazene azeturate and may be administered either by the intramuscular or subcutaneous routes within 5 days of preparing the solution or, if stored in a refrigerator, 14 days.

For *Babesia* infections single doses at the rate of 3·5 mg per kg body weight (or 0·05 ml per kg of the compound injection) should be adequate in horses, cattle, sheep and dogs. Disappearance of symptoms may be expected within 24 hr.

For *Trypanosoma* infections the same dosage as for *Babesia* may be given for *T. vivax* and *T.*

brucei infections, but the dosage rate should be doubled for *T. brucei* infections. The dosage rate may be increased to 8 mg per kg in resistant cases.

Local reactions may occur at the sites of injection: they may be severe in horses. Apart from this, serious side effects are not to be expected at normal dosage rates.

Phenamidine Isethionate, B.Vet.C.

This compound is a white odourless compound with a bitter taste. It is soluble in water but almost insoluble in organic solvents.

Although the compound has been used in the treatment of demodectic mange in dogs, this practice is largely discontinued because of the toxic effects and the variability of response. Its use is therefore confined to the treatment of *Babesia* infections in cattle, horses, and dogs. It is administered subcutaneously to horses and cattle in doses of 0·03 ml of a 40 per cent solution per kg body weight (1·5 ml per cwt), and to dogs at the rate of 0·3 ml of a 5 per cent solution per kg body weight (0·136 ml per lb). Action in cattle is rapid and premunity is established: single doses are usually sufficient, but doses may be repeated after 5 to 6 days in canine *Babesia* infections. *Stilbamidine* has been used in human trypanosomiasis but the less toxic *pentamidine* is now preferred even though its activity in some cases is less. Neither is now used for veterinary purposes to any appreciable extent, though in *Anaplasma marginale* and *Theileria dispar* infections in cattle, when pentadimine, at the rate of 4 mg per kg given intramuscularly, was combined with an equal amount of euflavine given intravenously: good results were claimed. It has a strong action against *T. brucei* and *T. evansi*, neither of which responds to the phenanthridium compounds, though they are reasonably sensitive to quinapyramine.

Propamidine Isethionate, B.Vet.C.

This white, hygroscopic, odourless, bitter tasting powder is soluble in water, reasonably so in alcohol and glycerin, but insoluble in other organic solvents. The halogen derivative *Dibromopropamidine Isethionate*, B.Vet.C., is similar in its characters but its solubility in water is greater, and in alcohol less, than that of its parent compound.

These compounds were, but are no longer used for *Babesia* infections in dogs, their indications being for topical and intramammary use only. They are applied as wound dressings in the form of water miscible creams or gels containing 0·15 per cent, and are active against Gram-positive and Gram-negative organisms.

Other Trypanocidal Drugs

The phenanthridium amine compounds, Quinapyramine (Antrycide) and suramin can be conveniently mentioned here.

The amino-phenanthridium compounds. The main preparations in use are the dimidium, homidium and pyrthidium bromides.

Dimidium Bromide, B.Vet.C. (1953)

This compound has been omitted from the 1965 B.Vet.C. in favour of Homidium salts (*q.v.*).

Chemically, dimidium bromide consists of 3,8-diamino-5-methyl-6-phenylphenanthridinium bromide. Previous chemical representation gave it as 2,7-diamino-10-methyl-9-phenylphenanthridinium bromide. It is a dark red crystalline bitter tasting, odourless substance which is soluble in water to the extent of 1:250; it is soluble in alcohol at a ratio of 1:70. Because of this low solubility in cold water the compound is used as a freshly prepared supersaturated solution at a concentration of 1 to 2 per cent. This may be given by the subcutaneous, intramuscular, or intravenous route: large reactions occur following subcutaneous injection, but reaction is relatively slight following intramuscular injection: the latter route appears to be preferred by most workers. The compound causes a slight fever during the first few days after dosage and delayed photosensitization occasionally occurs, though recovery usually follows improved grazing. This situation in Sudan does not apply to Nigeria where large scale toxicity has occurred. This latter report is difficult to reconcile with reports of a few cases of photosensitization in the Gold Coast, where better grazing was also followed by the absence of any outbreaks. Liver damage, to which the photosensitization is marked during the first nine weeks subsequent to doses of 1·25 mg per kg body weight (N. Rhodesia, Report 1952).

The chemotherapeutic action of the phenanthridium compounds including dimidium bromide is believed to be due to interference with cell division. The drug is markedly active against *T. congolense*, but less active against *T. vivax*, and inactive against *T. brucei* and *T. evansi*. The first three cause *nagana* or *tsetse fly disease* in almost all species of domestic animals in tropical Africa, whilst the last named causes *surra* in most species,but occurs in the non-tsetse areas—North Africa, India, the Americas and the Philippines. The diseases are transmitted by biting and blood sucking flies.

Dimidium bromide is mainly a curative drug but protection has been conferred for an average of 138 days. The *British Veterinary Codex* 1953 mentions 28 days as the protective period.

Drug resistance develops but the organisms may still be sensitive to homidium bromide or quinapyramine (Report: Kenya, 1953). The reverse does not hold good, for an organism resistant to quinapyramine may also be resistant to dimidium bromide.

Dosage

Dosage of dimidium bromide varies from 1 to 1·5 mg per kg body weight (51 to 76·5 mg per cwt) in *horses* and *cattle*. In *dogs* the lower rate of 1 mg per kg is usual.

Homidium Bromide, B.Vet.C.

Originally known as *ethidium* or R.D. 1572, homidium is 3,8-diamino-5-ethyl-6-phenyl-phenanthridinium. It is available as the bromide and chloride (Novidium) which differ only in the increased water solubility of the chloride. Homidium has a greater trypanocidal action than dimidium against *T. congolense* and *T. vivax*; unlike dimidium it has some activity against *T. brucei* but, like dimidium, is inactive against *T. evansi*. Some prophylactic action is obtained. Toxicity is less than dimidium and photosensitization does not occur (Animal Report, Kenya 1953), but severe local reactions can result from subcutaneous injection so that the deep intramuscular route should be used wherever possible, when local swellings and lameness are relatively slight and transient. Both compounds have been used extensively for more than ten years and resistant strains of *T. congolense*, at least, have been reported.

Dosage

Dosage of a 1 per cent aqueous solution is at the rate of 0·25 to 1 mg per kg body weight (12·75 to 51 mg per cwt). Higher rates do not usually cause more than transient liver damage and 10 mg per kg are used, providing freedom from infection for up to 5 weeks: doses greater than 2·5 mg (127·5 mg per cwt) may be expected to cause severe local reactions at injection sites. Following injection the parasites disappear from the blood within 24 hr possibly accompanied by a febrile reaction. Whilst the prophylactic period is normally about a month periods of up to 4 months have been reported.

Pyrithidium Bromide, B.Vet.C.

Known commercially as Prothidium and R.D. 2801, and with a proposed international non-proprietary name of Pyritidium Bromide this preparation is 3-amino-8-(2-amino-6-methyl-pyrimidin-4-yl-amino)-6-*p*-amino phenyl phenanthridine-5,1′-dimethobromide. It is administered by the subcutaneous or intramuscular routes as a prophylactic against *T. congolense*, and *T. vivax* and *T. brucei*, when protection may be as long as 6 months, or, in areas of great risk, as short as 2 months. Local swelling at injection sites can be expected.

Dosage

Dosage should be at the rate of 2 mg per kg body weight as a 1 or 2 per cent solution. Heavier dose rates should only be used where test dosing has indicated safety under local conditions. Otherwise deaths from liver damage may be experienced.

The Quinapyramine Compounds

There are two compounds, the sulphate which is rapidly absorbed and mainly trypanocidal with low prophylactic value, and the chloride which is more slowly absorbed and has a strong prophylactic effect. Combined as the 'Antrycide Pro-salt' or 'Pro-salt R.F.' the two compounds are used extensively for trypanosome therapy and prophylaxis.

Quinapyramine Chloride, B.Vet.C. and Sulphate, B.Vet.C.

These salts of quinapyramine, known commercially as Antrycide, are white to pale yellow crystalline powders which are odourless but have a bitter taste. The chloride is soluble to the extent of 2 per cent in boiling water but is only slightly soluble in cold water. On the other hand the sulphate is water soluble (1:2). Both are almost insoluble in organic solvents. Because of their different solubilities these two compounds have complementary effects against trypanosomes when combined in a single injection. The soluble compound has short but rapidly acting therapeutic effect, whereas the long acting slowly absorbed chloride is prophylactic. The sulphate provides initial protection and eliminates subclinical infections, whilst the chloride continues the protection after the effect of the sulphate has disappeared.

Quinapyramine is active against *T. congolense*, *T. evansi* and *T. vivax* in cattle, horse species, camels and dogs. *T. equiperdum* is also susceptible as well as *T. equinum*. *T. brucei* is less affected; action against *T. simiae* is reported. The action is attributed to the inhibition of growth and cell division.

Toxicity is not so important as with the phenanthridium compounds except in young stock, where overdosage may cause trembling, sweating and salivation, followed, in severe cases, by an increase in respiration and heart-rate, collapse, and death. In normal dosage, symptoms, when they occur, are only of a mild and transitory nature.

Dosage

Dosage need not be assessed meticulously except in young stock, and animals weighing less than 3 cwt (150 kg).

Quinapyramine Sulphate:

Animals weighing: up to 150 kg (3 cwt) individually assessed at the rate of 4·4 mg per kg body weight (2·27 mg per lb.).

150–200 kg (3–4 cwt) body weight: 1 g
200–350 kg (4–7 cwt) body weight: 1·5 g
Over 350 kg (7 cwt) body weight: 2 g

These doses are given subcutaneously as a 10 per cent solution, though in horses, where sloughing may be severe, it is advisable to inject intramuscularly a 5 per cent solution, or to divide the dose into two or three injections at different sites.

Sulphate/chloride prophylactic mixture

This is available in powder form consisting of three parts sulphate and 2 parts chloride. The B.Vet.C. (1965) solution contains 10 per cent of the sulphate and 6·67 per cent of the chloride, and is administered by subcutaneous injection at the rate of 0·025 ml per kg body weight, each animal's weight being estimated individually. Indurations and reactions may persist for several months following subcutaneous injection.

Protection against dourine transmission has been conferred by injecting stallions with the prophylactic mixture on one or two occasions in the breeding season, an interval of 90 days between injections being satisfactory. Transmission during service can also be prevented by injecting mares at least 18 days before the day of service.

Protection against *T. congolense* infection in cattle has been conferred for an average of 230 days (almost twice that of dimidium bromide) In areas where risk is high the protection period is much lower and is even at higher dosage rates and is taken to be between 2 and 3 months for practical purposes. Doses repeated every 2 months can prevent deaths from *T. congolense* or *T. vivax* for at least 18 months (Annual Report N. Rhodesia 1952). Suramin will considerably increase the protective power of quinapyramine. as well as that of ethidium.

An interesting report indicated that a solid immunity was achieved during 28 months of protective treatment with quinapyramine. This immunity was solid to either artificial or natural infection.

Resistant strains develop in animals which become infected as the tissue levels fall but parasites may not be detected in the blood, though they should be apparent in gland smears. Chronically infected cases of this type deteriorate in condition and treatment with other agents such as suramin often fails. However, it has been shown that quinapyramine resistance is not permanent, and that resistance to 5 mg per kg doses may be lost after about 110 days (a *T. congolense* strain resistant to 8 mg per kg and then passaged through rats was used in the experiments reported). This indicates that quinapyramine resistance may be of less importance than previously thought.

Compounds related to quinapyramine, including one commercially known as Tozocide, are being developed. Experimentally they are less toxic and more active than the parent compound. It is probable that these also will benefit by combination with suramin.

Suramin, B.Vet.C.

This is a complex aromatic organic compound with an empirical formula of

$$C_{51}H_{34}O_{23}N_6S_6Na_6.$$

This sodium salt is a white or pinkish white hygroscopic powder with a solubility of 1 in less than 1 of water, though it is almost insoluble in organic solvents.

It has marked curative, but less useful prophylactic properties against *T. evansi* (the cause of surra) in horses, camels, cattle and dogs, against *T. brucei* (nagana) in horses, dogs and, sometimes cattle, and against *T. equinum* which causes Mal de Caderas in equidae. It has also been used against *T. equiperdum*, the cause of dourine in horses. The compound has no activity against *T. congolense*, *T. vivax* or *T. simiae*.

The future of suramin would seem to be as a synergistic potentiator of the phenanthridium and quinaldine derivatives, for when suramin is combined with such compounds the period of protection is approximately twice that of the compound alone. The reason for this is unknown, but may be associated with the protein binding capacity of the suramin.

Toxicity is a real danger, for the therapeutic index is very narrow: horses and donkeys are very susceptible but camels are quite resistant. Symptoms are those associated with liver, kidney, spleen and adrenal gland damage.

Dosage

Therapeutic dosage

Horse: 7–10 mg per kg body weight (0·35–0·5 g per cwt)

Camel: 8–12 mg per kg body weight (0·4–0·6 g per cwt)

Cattle: 12 mg per kg body weight (0·6 g per cwt)

The dose in horses may be repeated up to three times at weekly intervals.

Prophylactic doses for the three species vary from 1 to 2 g repeated at 10-day intervals.

These doses are given as a 10 per cent solution.

Other drugs which have been used for their trypanocidal action are: **Antimony and Potassium Tartrate**, B.Vet.C. (Tartar Emetic) in intravenous doses of 3 to 5 mg per kg for horses and cattle (0·15 to 0·225 g per cwt), and 1 to 3 mg per kg for dogs and cats (0·45 to 1·36 mg per lb.); *Stibophen*, U.S.P., a sodium salt of a complex antimonial organic compound; and earlier still, several dyes such as **Trypan Red** and **Trypan Blue.** These compounds are neither so safe nor so effective as the more modern drugs.

Davey recommended the following drugs for the treatment of trypanosomiasis in the different species.

Curative therapy

T. congolense and *T. vivax*
- Cattle: Ethidium bromide or quinapyramine methyl sulphate
- Horse: *T. congolense:* quinapyramine methyl sulphate 3 mg/kg
  T. vivax: quinapyramine methyl sulphate 5 mg/kg
- Dog: *T. congolense:* quinapyramine methyl sulphate 3 mg/kg

T. simiae
- Pig: Quinapyramine methyl sulphate. (Up to 5 mg/kg, repeated)

T. evansi, *T. equinum* and *T. equiperdum*
- Horse: Quinapyramine methyl sulphate—up to 5 mg/kg

T. evansi
- Cattle: Quinapyramine methyl sulphate—3 mg/kg (India)
- Camel: Either suramin or quinapyramine methyl sulphate

Prophylaxis

The prophylactic quinapyramine salt mixture is the only drug recommended. It should be effective against *T. congolense*, *T. vivax*, *T. evansi*, *T. equiperdum* and possibly *T. equinum*.

Antipiroplasm Drugs

The use of phenamidine and diminazine has already been mentioned. Other compounds *bovis* or *B. gibsoni*. The drug is relatively non-toxic but stains tissues and secretions, including milk, a blue-green colour which may persist for several weeks.

Dosage by intravenous injection only:
Horse and Cow: 1–4 g (15–60 gr.)
Sheep: 0·5–1 g ($7\frac{1}{2}$–15 gr.)
Dog: 20–150 mg ($\frac{1}{3}$–$2\frac{1}{2}$ gr.)

Solutions should be freshly prepared and given at body temperature in concentrations of from 1 to 2 per cent. Shock may result if injections are given too rapidly. Subcutaneous injections cause sloughing.

Other compounds used included arsenicals such as neoarsphenamine, and antimony preparations have also been used.

Amicarbalide, diminazene, phenamidine and quinuronium preparations are now used in preference to these older compounds.

ANTIHISTOMONIASIS DRUGS

This group of compounds has developed considerably during the last decade, prompted by the needs of the rapidly expanding poultry rearing industry with its concentration on intensive rearing methods. Some of the more important compounds have been classified into four groups: the nitrofurans; the nitrothiazole derivatives, the pyrimidine derivatives and a miscellaneous group of 'nitro' compounds.

NITROTHIAZOLE DERIVATIVES

There are three compounds of this group now available commercially, Acinitrazole, B.Vet.C., Aminonitrothiazole, B.Vet.C., and nithiazide.

Acinitrazole, B.Vet.C.

This is 2-acetamido-5-nitrothiazole (vii). It is a buff or brownish-yellow powder with only a slight odour and taste: it is only slightly soluble in water. Acinitrazole is known commercially as Entramin A, Enheptin A, Trichorad and Tritheon. The therapeutic and prophylactic effect against blackhead in turkeys (histomoniasis) is similar to aminonitrothiazole but acinitrazole is preferred because of its low toxicity even when fed for long periods.

Dosage

Dosage for therapeutic purposes is at the rate of 0·08 per cent in the feed for fourteen days: propylactic use should follow therapy. For prophylaxis half the therapeutic concentration should be used and all classes of birds should be medicated, for treated birds do not develop immunity. Individual turkeys may be dosed with capsules at the rate of one capsule (0·1 g) for birds up to 3 kg, two capsules for those from 3 to 5 kg and three capsules for those over 5 kg.

Aminonitrothiazole, B.Vet.C.

The chemical formula is 2-amino-5-nitrothiazole (viii) and the compound is a greenish, or orange-yellow, bitter tasting, odourless powder only slightly soluble in water.

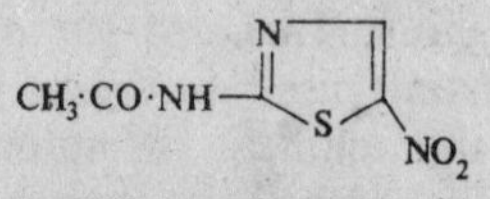

(vii) Acinitrazole

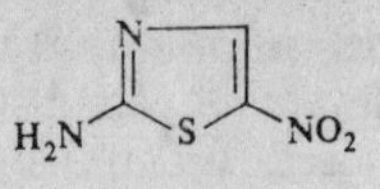

(viii) Aminonitrothiazole

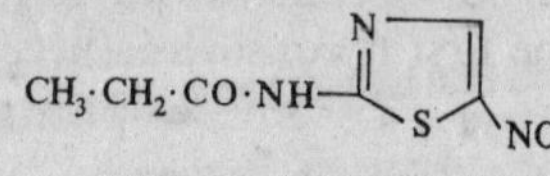

(ix) Nithiazide

The compound is used both therapeutically and prophylactically against blackhead (histomoniasis) in turkeys. This is a disease in which the liver is badly damaged by the attack of a protozoan parasite *Histomonas meleagridis:* infestations are contracted by ingestion of the eggs or adult worm of *Heterakis gallinae*, which itself parasitizes fowls and acts as an intermediate host for the protozoan. Until the advent of aminonitrothiazole there was no drug capable of controlling the infection once contracted, though sodium acetarsol is a useful prophylactic when given before infection occurs and administered continuously. In 1949 McGregor introduced aminonitrothiazole and demonstrated its superiority over all other compounds available at that time. It is used extensively wherever turkeys are reared. The compound must be used continuously to give maximum protection. It acts on both parasite and worm intermediate host.

Treatment is achieved by feeding mash medicated at the rate of 0·1 per cent for 14 days when *prophylaxis* should commence by reducing the level to 0·05 per cent or by giving water medicated at the rate of 0·03 per cent with a soluble salt. Medication should continue for at least 20 weeks. Individual treatment may be achieved by giving 100 mg capsules to birds *after feeding*. Unlimited water supply should be available at all times to offset the toxic effects which are especially dangerous in young stock and are mainly associated with the kidneys.

Nithiazide

Nithiazide known commercially as Hepzide, is 1-ethyl-3-(5-nitro-2-thiazolyl)-urea (ix). It is used for the prevention (0·025 per cent) and treatment (0·04 per cent) of histomoniasis and hexamitiasis in poultry by feed medication. Soluble powders are also available commercially for drinking water medication. When nithiazide is used for treatment the high level medicated feed should be given for a seven day period after which low level prophylactic feeding should be continued for a further 2 weeks or so. Commercially higher levels are recommended for prophylaxis in turkeys (0·04 per cent): nithiazide is available as a 22·4 per cent premix of which 40 oz. and 64 oz per ton feed are recommended for fowls and turkeys, respectively.

Other histomonastats include dimetridazole.

Dimetridazole (Emtryl)

This compound, known as Emtryl or 8595 R.P., has the chemical structure 1,2-dimethyl-5-nitroimidazole (x). The compound was shown in 1961 to have high activity—up to three times greater than acinitrazole—as a histomonocide.

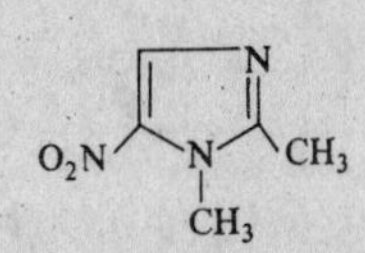

(x) Dimetridazole

Numerous publications appearing since that date have confirmed both the therapeutic and prophylactic value of the compound against *H. meleagridis* infections. Concentrations of 0·01 to 0·1 per cent are satisfactory food levels for prophylaxis in turkeys and 0·0125 to 0·05 per cent for chickens, but the highest levels may result in a reduction in egg size. The high levels are satisfactory for therapeutic purposes over periods of about 5 days after which the lower levels should be continued for prophylactic effect. It has also been shown that 0·025 per cent in the drinking water will in 5 days control established infections. Tissue residues are not detectable 48 hr after drug withdrawal.

With the identification of *Treponema hyodysenteriae* as a key causal agent of swine dysentery, a search was made for drugs which would be active against this organism. It was found that dimetridazole had a specific action on the spirochaete. The drug is now added to the feed or water of pigs, as a growth promoter and a specific in the treatment of swine dysentery.

Emtryl Premix: 22·5 per cent dimetridazole.

Dosage: Prevention of swine dysentery 1 kg per ton of feed
Treatment 2 kg per ton of feed

Bibliography

ALEXANDER, T. J. L. (1969) The clinical signs, diagnosis and control of swine dysentery. *Vet. Rec.*, **85**, 59.

PART IV

THE ESSENTIALS OF PHARMACY IN VETERINARY PRACTICE

41

Prescription Writing

It is essential that there should be a standard language common to both prescribers and dispensers. Formerly, when it was considered inadvisable to disclose the ingredients of medicines, all prescriptions were written in Pharmaceutical Latin—a relic of the days when Latin was the written language of the Arts and Sciences. Much ignorance and charlatanism was disguised by this means, but, as it is now more usual to explain the treatment to clients, there is no need for secrecy and today English is more widely used in prescription writing. Latin is retained for certain parts of prescriptions, mainly for ease and speed in writing but partly because it is picturesque. Appendix I (p. 505) gives a list of commonly used latinizations.

The increase in the use of English in prescription writing has occurred since the change of the Imperial standard system of weights and measures in 1878 from the Apothecaries' to the Avoirdupois system. More recently a change over to the continental Metric system has been undertaken with a view to international standardization. The tenth edition of the *British Pharmacopoeia* uses the metric system exclusively. Therefore, as three different systems of weights and measures may be met, a knowledge of the relationship of these systems is necessary to facilitate conversions of weights and measures from one system to another. The key to these conversions are the equivalents:

1 kilogram (kg) = 15 432 grains (gr.) = 2·2046 pounds (lb.)
and 1 litre (l.) = 1·7598 pints (pt.)
or 28·412 millilitres (ml) = 1 fluid ounce (fl. oz.)

A conversion table is added as Appendix II.

WEIGHTS AND MEASURES

The Metric System

The Metric system is the Continental system of weights and measures which, after some delay, has been adopted as the international standard system. Weights and measures are expressed in multiples of ten, the standards being the *kilogram* and the *litre* for mass and fluids, respectively.

It is essential in the Metric system that the decimal point should be strongly indicated, the absence of numerals before or after the point being shown by a nought. (The B.P. also recommends that in prescriptions a capital 'G' should be used to indicate 'grams' so that confusion with 'gr.' for 'grains' should not arise.) The 'ml—millilitre—0·001 litre' has superseded the 'cubic centimetre—c.c.' in fluid measures. Numerals should always precede the weight or measure abbreviation:

45·0 g or 45·0 ml
2·0 kg or 2·0 l.
0·4 g or 0·4 ml

As all drugs now have their dosages expressed in metric weights and measures, this should be regarded as the system of choice.

The Apothecaries' System (Imperial)

The survival of the earliest of the three systems, the Apothecaries', persists mainly because the majority of pharmaceutical balances were supplied with apothecaries' weights. This type of weight, in various sizes from ¼ gr. to 240 gr., is provided because the Avoirdupois system recognizes no denomination between 1 gr. and 1 oz.

It is important to indicate clearly, by the way in which the prescription is written, which system of weights and measures is intended. In the Apothecaries' system the weight denomination is usually expressed as a symbol and is followed by the number in Roman numerals. These numbers should have a line drawn over them and above this, to serve as a double check, a dot should be placed for each unit numeral. If the last numeral is a unit then it should be prolonged downwards to indicate that it is the last figure, viz. eight Apothecaries' ounces—$\overline{\text{viij}}$. Fluid measures are similarly written except that the symbols are preceded by the letters 'fl.' viz. seven fluid ounces—fl.-$\overline{\text{vij}}$.

Apart from the half grain which is written, $\overline{\text{ss}}$ fractional quantities are written normally—gr. $\frac{1}{8}$.

It should be noted that although the unit weight of 1 gr. is identical in all Imperial systems (including the jewellers' Troy system) the Apothecaries' ounce and pound differ considerably from the corresponding Avoirdupois weights.

Both the Apothecaries' and Avoirdupois systems should be regarded as obsolete.

The Avoirdupois System (Imperial)

The Avoirdupois system was the official system of weights and measures in the United Kingdom and was tied to the standard pound. This pound is equivalent to 7000 gr. or 16 oz. It will be seen that each ounce is therefore equivalent to 437·5 gr. and is considerably smaller than the Apothecaries' ounce. This is why it is so essential to indicate the system of weights in use when writing prescriptions. Avoirdupois weight numerals should always be written in arabic figures and these should be followed by the correct abbreviations:

One-fiftieth of a grain (avdp.) = $\frac{1}{50}$ gr.
Sixty grains = 60 gr.
Two ounces = 2 oz.
Two pounds = 2 lb.

It is inadvisable to use abbreviations other than these.

Fluid measures are identical in both the Apothecaries' and Avoirdupois systems, but the rules relating to the positioning and types of symbols and figures used in prescription writing follow those for the weights of the corresponding system.

Note: American Standards are not the same as the corresponding Imperial Standards and consequently the weights and measures are not identical.

Approximate Equivalents

These are convenient for rapid conversion of quantities from the Imperial to the Metric system and vice-versa. They chould not be used where toxic drugs are involved.

Household Equivalents

These measures are only very approximate, but are useful to indicate *fluid* quantities to a client, and dosages on labels are usually expressed in these household measures (i.e. 1 teaspoonful=5 ml.

On no account should these equivalents be applied as estimates of *mass*—volumes of equal weights, solids naturally differ considerably according to their individual densities.

Calculation of Percentage Solutions

The calculation of the quantities required to make percentage mixtures of either solids, liquids or solutions, is easy if the metric system is in use, for 1 g or 1 ml of solid or liquid A with 99 g or 99 ml of solid or liquid B will make a 1 per cent mixtures of solids or liquids. If either solid or liquid is measured by weight, the percentage mixture is indicated as w/w (weight in weight): if liquids alone are used and measured by volume the percentage mixture is represented by v/v (volume in volume): if a solid, measured by weight, is dissolved in a liquid, measured by volume, the percentage mixture is indicated as w/v (weight in volume).

If the Avoirdupois system is in use, the situation is more involved. Weight in weight mixtures are straightforward in that, if 4·375 gr. of a solid C are made up to 437·5 gr. with solid D, a 1 per cent w/w mixture of solid C in solid D is given. A v/v mixture is also straightforward in that 4·8 m in 480 m is a 1 per cent v/v mixture. On the other hand, as the fluid ounce contains 480 m., a 1 per cent w/v mixture of solid C in liquid E requires 4·375 gr. of solid C to be dissolved in a small quantity of liquid E of which, after the solid is dissolved, more is used to make up the volume to 1 fl. oz. (480 m). This means that 1 gr. is dissolved in approximately 110 m. (i.e. 480÷4·375). If the liquid is water, these same proportions hold good for w/w percentage solutions, for 1 fl. oz. of water weighs 437·5 gr.

PRESCRIPTION WRITING

With the increase in variety and expense of pharmaceutical preparations 'treatment by prescription' is becoming much more common in veterinary practice.

Although the deciphering of badly written and poorly expressed prescriptions is a skilled art very ably practised by dispensing chemists, the *Poison Regulations* and *Misuse of Drugs Act* rightly lay down strictly defined requirements and recommendations. Veterinary prescribers should conform to these standards so that interpretation may be rapid, easy and correct, resulting in minimum delays for the client in obtaining the medicines. Writing should always be neat and legible.

Printed prescription forms are permissible and save a great deal of time. These will usually have as a heading the business name and address of the prescriber, and possibly the '℞' sign. At the bottom, the words 'For Animal Treatment Only' may be printed and other refinements may be added as the individual wishes.

Prescription Essentials

1. The **Name** and **Address** of the **Prescriber** are required for reference in cases of doubt or mistakes.

2. The **Date** must be included for the same purpose, and a prescription reference number may also be included.

3. The **Superscription.** This is the symbol ℞ (The Roman symbol for Jupiter and originally intended to invoke his blessings and aid in the action of the drugs), which is now the symbolic order to the dispenser to 'take' the drugs indicated and is held to mean *recipe—take thou.*

4. The **Inscription.** This consists of the names and quantities of the drugs written in full English or, more quickly, in Latin abbreviations. Where an official preparation is to be dispensed this should be indicated by the letters (B.Vet.C.), (B.P.) or (B.P.C.), respectively. Indications of percentages and individual drug quantities are not then necessary.

Where proprietary preparations are prescribed it is usual to state the proprietary name and the manufacturer.

Choosing Drugs

The selection should be based on a rational diagnosis of the disease, correlated with a knowledge of the applied pharmacology of drugs.

It is usual to build an inscription on four main pillars:

1. **the basis** or main active agent of the prescription,
2. **the adjuvant** which has a complementary effect,
3. **the corrective** which counters any undesirable tendencies of the first two agents, and
4. **the vehicle** or excipient, usually a bulky agent but often therapeutically inert. It may, however, possess qualities which make a corrective unnecessary and often has a flavour if used internally. The flavour may disguise objectionable tastes and should be acceptable to the species to which the patient belongs, e.g. glucose, honey, treacle for dogs, cats and pigs; liquorice and aniseed for ruminants and horses.

These four guiding principles are based on a hypothesis attributed to Asclepiades a Grecian physician who settled in Rome during the second century B.C. This hypothesis suggests that a medicine should 'Cure Quickly, Safely, and Pleasantly'. These words correspond directly to the four divisions just mentioned.

An example is that of a gastro-intestinal sedative useful in the treatment of simple disturbances of the digestive system in dogs.

℞

Bism. Carb.	5 gr.
Sod. Bicarb.	3 gr.
Mag. Trisil.	10 gr.
Dextros.	20 gr.
Ft. pulv. 1.	

Bismuth carbonate, a relatively slow acting antacid, is assisted by the rapidly acting bicarbonate. Both these, on meeting the acid in the stomach, release carbon dioxide gas which distends that organ. This distension results in the reflex secretion of further acid and other gastric juices. This is the so-called 'alkaline stomachic' action. To counter this secondary acid production, magnesium oxide or trisilicate can be included, both of which act as antacids without secondary stimulation. The trisilicate has other properties such as toxin adsorption and a mucosal protective action by mechanical coating. The glucose, besides adding bulk, has a sweet flavour, a soothing, protective and lubricant—demulcent—action. It is also a readily available source of energy.

Choosing Doses

The dosage of drugs in regular use will soon become familiar, but should the use of an unfamiliar drug be contemplated, reference to a standard work is necessary. Remember that, with few exceptions, low doses can always be increased, but an overdose of a drug may lead to its recovery by post-mortem analysis. Remember also that pharmacists are trained to scrutinize prescriptions carefully and to verify any abnormally high does of a poisonous substance. Should such a dose be included intentionally in a prescription, it may be time-saving to indicate your intentions to the dispenser by following this dose by the letters Q.R. (*Quantum recte*—the correct quantity).

Species dosages. Formerly it was possible to estimate the dosage for the species as follows: Horse ≡ 1, cow ≡ $1\frac{1}{2}$; sheep and goat ≡ $\frac{1}{5}$, pig ≡ $\frac{1}{8}$, dog ≡ $\frac{1}{16}$, cat ≡ $\frac{1}{32}$. These ratios are only approximate and must not be strictly applied to toxic drugs, particularly those to which certain species are susceptible. Again, reference to an authoritative publication is the best course of action.

Vehicle quantities. For **powders** the quantity of vehicle will be determined by the volume of powder capable of being swallowed by the individual patient. Any discrepancy between the volume of the active drugs and the optimum total volume for the species or individual will be made up by the necessary amount of vehicle.

The main species in which powders are administered directly and not in food are the dog and cat. In these species the approximate maximum weights are 120 gr. and 30 gr., respectively, though these are rare extremes, 30 to 40 gr. and 10 to 15 gr. being more normal. In cattle, powders of up to 4 oz. have been recommended.

For ointments the vehicle quantity will depend on the required concentration of the active ingredients.

With **mixtures,** the fluid vehicle quantity calculation is a little more complicated. The size of the container, which should always be completely filled by the medicine prescribed, rigidly controls the total bulk. The amount of vehicle used for each dose put into the bottle will therefore be determined by the number of doses to be dispensed, and by the volume of the insoluble solids or fluid medicines in each dose. The vehicle is used to bring up the volume to a suitable quantity for filling the container. It is usual to include extra doses of the medicine should those required for the treatment not fill the most suitably sized bottle. These extra doses will allow for wastage.

So far, single doses of individual drugs have been considered, all of which, added together, make one dose of a compound medicine. To complete the Inscription the frequency and duration of administration must be determined and the individual drug doses must be multiplied by the number of doses of the compounded medicine needed. This calculation must not be left to the dispenser.

The Subscription

1. The **Subscription** is that part of a prescription which directs the dispenser during the compounding of drugs. Latin abbreviations are usual, although full English is not precluded: e.g. M. et div. in pulv. 12—Mix and divide into twelve powders; Ft. pulv. 1. Mitt. 12—Make one powder. Send twelve.

There are few complications in writing the subscription, for which there exists a restricted number of Latin phrases and words. It has been said earlier that these indicate clearly and concisely the manipulations to be carried out, the type of pharmaceutical preparation to be made and the quantity or number of packs to be sent. The latinizations are contained in Appendix I.

2. The **Signature,** symbolized by the abbreviation Sig. which is included on *all* prescriptions immediately below the subscription, is an anglicization of *signetur*—Let it be labelled. It will, therefore, instruct the dispenser in all the details to appear on the label as directions to the owner, e.g.

Sig. **For Animal Treatment Only**
Powders For Cow: Ear No. H.E. 9254
One powder to be given twice daily in food

Latin abbreviations may again be used but,

where scheduled drugs (those controlled by the *Poison Regulations* and *Misuse of Drugs Act*) are included, no latinization or abbreviation of **'For Animal Treatment Only'** is allowed. Latin translations of many directions are very involved, so that inaccuracies are more likely, and it is therefore recommended that, in all but the simplest of directions, unabbreviated English should be used.

3. **Prescriber's signature and qualifications.** These are needed for reference and to indicate the personal acceptance of responsibility in the prescribing of the drugs, for where 'controlled' drugs are incorporated a preparation may not be sold except on the direction of someone officially entitled to prescribe under the Statutory Regulations previously mentioned—the Poison Rules and the *Misuse of Drugs Act*.

4. Where a prescription does not contain a controlled drug that prescription may be dispensed as frequently as the client wishes, **unless** instructions are included to prevent this. The Latin abbreviation **non-repet.**—let it not be repeated—will tell the chemist to impound the prescription, so preventing further dispensings. Should a number of repeats be advisable then the prescriber may indicate the number of repeats required, and if he so wishes, the interval between dispensing, or even specific dates. For some drugs special regulations are in force regarding repeat dispensings.

The following prescription incorporates the information given so far.

This form is only a recommendation which follows that given in the regulations controlling drugs and need not be strictly adhered to providing the details mentioned are included, but it should not be forgotten that ease in reading and dispensing will be helped by uniformity of style.

Identity of the Patient

The words 'For Animal Treatment Only' are required by law to appear only on prescriptions containing 'scheduled drugs', but it is a wise precaution to include this in all 'signatures', and hence on all labels. No confusion can then occur between human and veterinary medicines should the two groups be kept in the same cupboards, as they may well be where the animal involved is a household pet. Even greater safety follows if the particular animal is identified, e.g.:

Sig. For Animal Treatment Only
Mixture for Poodle

The Quantity of Each Dose

This will appear on the label expressed in units easily measured by the client. Fluid and semi-fluid medicines will have doses expressed in household measures. Occasionally powders will also be expressed in terms of teaspoons or tablespoons, but unless mass therapy is involved, as with piperazine treatment of pigs and poultry, it is wiser to weigh and wrap each powder dose individually. No discrepancy can then occur due

Practice Name and Address		A. Veteran & A. N. Other M's.R.C.V.S. 270/272 High Street Manderton Sussex
Date	10 September 1976	
Script Ref. No.	Ref. No. 1235	
Superscription	℞	
Inscription	Acid Sodium Phosphate 10 g Sodium Chloride 60 g	
Subscription	Fiat pulv. l. Mitte 6.	
Signature	Sig. For Animal Treatment Only Powders for Roan Cow, Ear No. H.125. One powder to be given as a drench in 500 ml of warm water morning and evening	
Veterinary Surgeon and Qualifications		A. Veteran M.R.C.V.S.
Client's Name and Address	L. Gentle Castle Farm Broadbeach Sussex	
Repetition Instructions		Non-repet.

to misinterpretation of instructions; a level teaspoonful can contain widely differing quantities depending on whether the powder is packed closely or not.

Route of Administration

The choice of the route of administration is governed by the factors discussed in Chapter 3. In the context of this section this direction will inform the owner how and where the medicament is to be applied.

Administration Vehicles

Vehicles used in the administration of drugs must be compatible chemically, physically and therapeutically: obviously no client should be advised to mix aqueous solutions into oil or oily solutions into water.

Frequency of Administration

Frequency of dosing will depend on a number of factors, each of which must be considered when making a decision:

1. The character of the disease. If the disease is very acute and severe, then, with due regard to toxicity considerations, frequent massive doses are indicated to save the animal's life. Where subacute conditions are involved less urgent and heroic measures are necessary, and, providing adequate tissue levels are achieved, high and frequent dosage is not necessary.

2. The character of the drug. Some drugs are absorbed and excreted more rapidly than others and the dosage of these drugs must be sufficiently frequent to maintain adequate tissue concentrations: aqueous solutions of penicillin are an example. Toxicity of the drug also has a bearing on dosage frequency, for it is usual to give doses of sulphonamides frequently for the first 2 days, when the disease is at its height; subsequently, for a further 2 or 3 days, the dosage frequency can be decreased as the tissue levels are stabilized and the disease controlled. This decrease in frequency is necessary as excretion of the sulphonamides is slow and toxic levels would soon be built up if a high frequency of dosage were maintained.

3. Animal physiology. This may affect frequency. Ruminants, by virtue of the depôt mass in the rumen, need only twice daily oral dosage but parenteral dosage follows normal practice.

4. The ability of the owners to administer the drug at the required interval. The correlation of the dosage frequency with the owners daily routine is an important factor, particularly in farming, where working hours are often governed by weather and daylight. Interference with farming routine should be as little as possible, otherwise dosing will be neglected and the results of treatment erratic and unpredictable.

Duration of Treatment

This depends entirely on the nature of the disease and **the number of doses** will vary accordingly, but a sufficient number to allow for wastage and flexibility of revisiting should be prescribed.

42

Dispensing

This chapter is intended to deal only with that part of pharmacy in which medicinal substances are mixed, packed and labelled. It is essentially a practical subject, but as more becomes known of the chemical and physical reactions involved in making some preparations, such as emulsions and dispersions, the theoretical aspect of dispensing is becoming a study in itself. These theoretical aspects will be mentioned only superficially.

REQUIREMENTS FOR DISPENSING

The first requirement in dispensing is a knowledge of the weights and measures. A reference to Chapter 41 will serve as a reminder of the three systems of weights and measures relevant to Great Britain.

Because the British Government successfully negotiated the inclusion of the United Kingdom as a full member of the Common Market (European Economic Community) the impetus of the change from Imperial Systems of Weights and Measures usage to those of the Metric system was recently accelerated by the adoption of the metric system by the pharmaceutical industry.

The normal dispensing balance is designed to weigh quantities up to 2 oz. and has a left-hand 'closed' pan to take weights and a right-hand 'open' pan, suspended from the beam side only so that powders may be placed on and removed from the pan without the inconvenience of a second suspending bar. Note that the drugs are weighed directly onto the scale pan: an exception is made with metal pans when mercurial and similarly corrosive drugs are to be weighed, but as metal pans are used only on the oldest type of balance this exception will not apply very frequently. There are two types of balance, class A and class B. Class A balances are accurate to 3 mg ($\frac{1}{20}$ gr.) and are used for weighing small quantities of toxic drugs—that is quantities of, say, 60 to 300 mg (1 to 5 gr.). Class B balances are usually accurate to 12 mg ($\frac{1}{5}$ gr.); greater accuracy than this is not required for drugs other than highly active compounds. Nowadays the dispensing balance is being replaced by the laboratory type.

The maintenance of a pharmaceutical balance is not involved, but the life of weights and balances can be prolonged by observing two rules—that no weighing of powders shall commence until the weight drawer beneath the platform is closed, and that no powder is added to, or removed from, the scale pan whilst the beam is up: by conforming to these rules the spilling of powder is largely avoided and the corrosion which this frequently causes is also avoided. Weights are manipulated by hand, and beyond an occasional wipe no cleaning is recommended.

For amounts larger than 2 oz. various types of commercial balances are used.

Fluids are rarely weighed in dispensing and various sizes of *glass measures* are available, calibrated in apothecary volumes—minims, fluid drachms and fluid ounces. Metric measures and combined Imperial and Metric measures are available. All are intended to be read with marks at eye level and volumes are measured using the bottom of the meniscus: in most measures the main calibrations are repeated on the reverse side to assist in level reading.

Manipulation of solids and semi-solids is achieved by the use of different sizes of *spatulae* graded according to length. These usually have wooden handles though in some of the older types horn was used; the blades are metal and are similar in appearance to painters' palette knives with parallel semi-sharp edges and a rounded end. No metal spatulae should be used with mercurial compounds; bone or non-corrodable synthetic spatulae are designed for this purpose.

Where reduction of solids to a finer state (known as trituration) is required this is done by means of a *pestle and mortar*. The mortar is a lipped bowl made of unglazed porcelain for normal purposes and of glass for the management of poisons: the non-porous surface of the glass avoids the contamination of substances by traces of poison which would be left behind from previous triturations with a more porous surface.

Mixing of powders, the making of emulsions, and various other compounding manipulations are also carried out in a mortar.

A special type of mortar has been used in the past for compounding the mass used in pill making by hand. This mortar is double ended; one end is deeply concave and is used for triturating and mixing the powders, whilst the other end is shallow and is used for incorporating the massing fluid which was most commonly syrup of liquid glucose. Pill making by hand is archaic and these mortars are now rarely seen except in teaching schools and museums.

Sieves for ensuring a standard particle size are now rarely used except in manufacturing. They are now made of stainless steel mesh for grades up to 100, but tinned copper was used previously and may still be seen. For grades finer than 100, silk screens are used. The number or grade of a sieve describes the number of meshes which occur along the sides of an inch square of the sieve.

The *British Veterinary Codex* defines the particle size of powders of various grades:

A *coarse* powder:
All the powder passes through a No. 10 sieve and up to 40 per cent passes through a No. 44

A *medium coarse* powder:
All the powder passes through a No. 22 but only 40 per cent through a No. 60

A *medium fine* powder:
All passes a No. 44 but not more than 40 per cent passes a No. 85 sieve

A *very fine* powder:
All passes a 120 mesh silk sieve

An *ultra-fine* powder:
90 per cent of the particles not greater than 5 μm and none greater than 50 μm, determined by microscopic examination

If a powder is given a number that powder must *all* pass through a sieve of the same number.

All dusting powders should pass a No. 60 sieve, otherwise particles may cause irritation and thereby aggravate any skin condition. This is very important with such powders as sublimed sulphur which have a tendency to be gritty and cause irritation.

PRINCIPLES OF PRESCRIPTION DISPENSING

When dispensing, one should work methodically, accurately, quickly and neatly. Working to a standard method frequently ensures that the remaining three requirements are automatically complied with.

Methods will vary according to individual ideas but the following scheme may act as a guide.

1. Read the prescription and check the doses. Note any incompatibilities and make sure that the prescription complies with the regulations attaching to poisons and other controlled drugs.
2. Write the label.
3. Make sure all dispensing utensils are clean.
4. Prepare the ingredients, compound and fill into the required container(s).
5. Recheck the label against the prescription, label and wrap.
6. Clean dispensing utensils and bench.

DISPENSING POWDERS

Trituration and Mixing

Powders used in the veterinary field may be one or several solids mixed together in a finely divided state. They may be intended for direct administration as a powder, as with dusting powders and many small animal medicines; they may be given mixed in water either as a solution or suspension, e.g. cattle drenches; they may be dissolved in a vehicle and administered as a parenteral injection, or they may be used in the preparation of other forms of pharmaceutical compounds such as electuaries, ointments, tablets, pills and horse balls. The manipulation of powders therefore forms a large part of dispensing.

The first stage in dispensing powders is to choose a clean, dry mortar which is suitable for the volume of powder to be triturated and mixed. Commencing with the powder of smallest volume, each powder is tipped into the mortar in small quantities, each addition being thoroughly triturated and mixed before any more is added. When any powder is particularly coarse it may be triturated separately prior to mixing into the combination. If the powder is adhesive, as with kaolin or Armenian bole, trituration is usually unnecessary and should be avoided where possible: during mixing, however, some trituration is unavoidable and the powder will be compressed on the sides of the mortar; should this happen the caked powder must be scraped off with a spatula and thoroughly mixed and triturated with the succeeding powder. If colouring agents such as the Armenian bole are included in a formula or prescription they are usually the smallest volume of powder included: they should be mixed in after the first of the active ingredients has been triturated in the mortar, otherwise a large amount of the colour will be lost on the sides of the mortar. The usual reason for the inclusion of colour, apart from its disguising effect, is to ensure that the powders as a whole are adequately mixed. Where bulky powders are to be mixed into powders of small volume this should be done slowly, a small quantity of the bulk being added at a time with thorough trituration and mixing between each addition.

Note: If powders with innocuous ingredients are small in volume (0·5 to 1·5 g or 2 to 20 gr.) it is usual to include in the mass quantities of each ingredient sufficient to make one more powder. This allows for losses in trituration and manipulation. It should not be done with toxic drugs.

Crystalline substances should either be very finely ground or dissolved in a volatile fluid and incorporated into the rest of the powder in much the same way as one mixes cement.

If toxic drugs are to be included these should be added and mixed in last to avoid duplication.

If explosive drugs such as potassium nitrate are to be incorporated in compound powders or used alone they should not be triturated. Vigorous trituration of the mixed powders should be avoided.

Division

When triturated and mixed, the bulk may be divided into individual powders in one of two ways. The most accurate method, which should be adopted for all powders containing toxic or highly active drugs, is by weighing each powder individually. This is laborious and time consuming so that where accuracy is not essential division by eye is acceptable. This should be carried out 'scientifically' and not by heaping small quantities onto their respective wrapping papers and adding or subtracting until approximately equal heaps are obtained. If a flat rectangular heap is made of the powder this can be fairly accurately halved with a spatula, the two halves being separated by moving one half laterally; this can usually be done without disturbing the symmetry of the two rectangles. These two main rectangles may now be subdivided at right angles in a similar manner. This can be continued indefinitely with perhaps a little tidying between each division. With practice accuracy to 65 mg (1 gr.) can be achieved.

Once divided, each powder can be transferred to a wrapping paper using one spatula as a fixed buffer and another spatula to lift. By this means the catapulting of powders across the bench is avoided.

Poisons, individually weighed, should be added last, immediately before wrapping, so that again, duplication is avoided: mixing is achieved by repeatedly turning the powders over with a spatula.

Fractional doses. It may happen that doses of a drug are required which are fractions of a grain, as for example 5 mg ($\frac{1}{12}$ gr.) arecoline hydrobromide; this may be incorporated in a capsule with 0·2 g (3 gr.) chlorbutol for the treatment of tapeworms in a dog. There are several methods of doing this but the following is relatively simple.

Weigh out accurately 65 mg (1 gr.) of the hydrobromide and place it in a 3 inch glass mortar. Weigh out 0·7 g (11 gr.) of an 'inert' powder such as lactose and tip it into the mortar. Mix, and weigh out a series of 65 mg (1 gr.) powders; there should be eleven complete powders, the twelfth being short due to manipulation losses. Each of the complete 1 gr. powders will contain 5 mg ($\frac{1}{12}$ gr.) of the hydrobromide.

Wrapping

Powders which are intended for administration by a client must be wrapped individually. Several different procedures can be discovered by reference to any standard pharmaceutical text which details counter or dispensing techniques. Which of the methods is used is immaterial so long as a sift proof double fold is made with the free edges of the wrapping papers. For neatness it is advisable to use only one method, which ensures uniform shapes and sizes.

Powders not prepared for dispensing as such may be processed further and incorporated in: (1) mixtures and lotions, (2) ointments, (3) emulsions, and (4) pills and horse balls.

DISPENSING MIXTURES AND LOTIONS

In veterinary practice the vehicle for fluid medicines is usually water; lotions, however, may have glycerin, spirit, propylene glycol, polyethylene glycol and any one of a host of other organic solvents.

The method of dispensing the internal and externally applied fluids is the same.

After having read the prescription and written the label, etc., all the powders should be weighed out and prepared separately. *Soluble solids* should then be dissolved one by one in a portion of the vehicle in a glass measure, solution being hastened by the use of warm water and vigorous stirring. When dissolved the solution is made up to volume and poured into the bottle. *Slowly soluble solids* such as exciccated magnesium sulphate should be triturated in a warm mortar with some of the warmed vehicle and transferred to the measure when dissolved. *Insoluble solids* if for internal administration may be triturated in a mortar either with 4 ml (60 m.) mucilage of tragacanth per 30 ml of final mixture, or 0·6 g (10 gr.) Compound Powder of Tragacanth, B.Vet.C. per 30 ml of mixture. Some of the vehicle should be added in the latter case until a cream is formed. With a lotion this addition of suspending agent is deleted, a cream being formed with some of the vehicle only. The suspension is diluted with more vehicle and transferred to a measure, and the mortar washed with some of the vehicle, the washings being added to the contents of the measure. The correct volume is made up with more vehicle. *Solid poisons* should be incorporated immediately before adding the vehicle. *Fluid drugs* are added directly into the measure, the measure(s) used for these fluids being washed with some of the vehicle and the washings added to the mixture already in the bulk measure. *Fluid poisons* should be added directly to the bottle just before topping up and corking.

In practice, where speed is more important than great accuracy, it is common to dispense straight into the bottle. In this case the bottle must be dry, and wherever possible the solids should be funnelled into the bottle while it is dry, otherwise clogging of the bottle neck may occur. The *soluble solids* are therefore poured into the bottle after trituration and mixing: some of the vehicle is then added, the cork screwed into the neck and the solids dissolved by vigorous shaking: when this has been achieved the bottle is uncorked and the vehicle added until the bottle is filled to the junction of shoulder and neck. *Insoluble* solids should be funnelled into the dry bottle, a little vehicle added and the bottle allowed to stand until the fluid has 'wetted' all the powder; more vehicle can then be added and the bottle shaken vigorously after corking. When evenly dispersed the volume can be made up. If the powder is not first wetted it may clog a corner of the bottle and never be dispersed correctly. Where *soluble* and *insoluble* powders are to be mixed the soluble powders should be dissolved first and the solution added to the bottle after the insoluble solids have been dispersed. Fluid drugs are added in the same manner as for the normal procedure except that they are dispensed directly into the bottle. *Eye lotions* and *eye drops* should be prepared with a sterile vehicle.

Note: Suspending agents must not be used in the dispensing of any type of veterinary *Lotion.*

When the mixture is in the bottle the cork should be pressed in with a screwing action and given a final tap on the bench—this should not be a hard knock or the bottle may fracture causing, at the least a flooded bench, and at the worst a severe laceration of the hand holding the bottle.

Labelling and wrapping follow. For approved wrapping techniques refer to standard pharmacy texts. To estimate the size of paper required (this refers to any object to be wrapped, not just bottles) the width of the paper should be two-thirds the width of the the base plus the the height of the object, plus two-thirds the width of the top—this provides sufficient paper to overlap at top and bottom and to cover the height of the object.The length of the paper should be one and a half times round the bottle, box, etc. This provides enough to go round the bottle and yet leaves sufficient to fold over and fasten in the free ends of the paper.

DISPENSING OINTMENTS AND CREAMS

Next in frequency of dispensing come the ointments, though, along with the general trend away from personal dispensing, the compounding of ointments is left more and more in the hands of commercial manufacturers.

Ointments with a pure paraffin base are not popular now because their immiscibility with body fluids makes them difficult to apply if the skin is wet, though some drugs make this type of base essential for reasons of stability (e.g. penicillin), slow absorption, or slight penetration as with mercuric iodide blisters. The same difficulty of application also applies when 5 per cent lanolin is included in the base.

The majority of ointments are now made with water miscible bases and differ from creams only in their firmer consistency. These water miscible bases may be of:

1. *A compound type* containing two phases of an emulsion, namely the fat phase, which may be of animal, vegetable or mineral types, or a mixture of these, and an emulsifying phase consisting usually of a detergent.

2. *A compound type* containing all three phases of an emulsion. These emulsions or creams are miscible with water in all proportions, water being the continuous phase of the oil-in-water emulsions (*q.v.*). The consistency of preparations of this kind can be varied by increasing the proportion of water.

3. *A simple type* which is often a water soluble organic base consisting of a mixture of polyethylene glycols. An official example consists of a mixture of equal parts of polyethylene glycols 4000 and 400, the latter being fluid and the former a solid with the consistency and appearance of paraffin wax. The consistency of the base can be varied by changing the proportions of the two glycols.

There are two methods of preparing ointments, referred to as: (1) the fusion method, and (2) the trituration or levigation method. The *fusion method* consists in melting the fatty components of the base on a water bath in decreasing order of melting points to avoid overheating. Drugs, if fat soluble, should be dissolved in the fat base, whereas insoluble solids should be added just before the base begins to solidify, stirring being continued until the ointment is set. Frequently, however, the insoluble solids are incorporated after the base has set by means of trituration in a mortar or by levigation with a spatula. Ointment making by *trituration* and *levigation* differs only in that with the latter method the mixing and smoothing of the medicaments through the base is done on an ointment slab of marble or glass, using a spatula to turn and rub the base and drugs. In both cases a small amount of the base must be added to all the powder and mixed and smoothed in before taking a further portion of the base: by this means an evenly mixed ointment is obtained. Fluids should be mixed in drop by drop.

Ointments with the two phase compound type of base are usually prepared by the fusion method, any insoluble drugs being added by levigation or trituration.

Those with the three phase compound base should always be prepared by a mixed method of fusion and trituration. By this method all the fat phase and fat soluble drugs, including the emulsifying agent should be heated on a water bath and melted together in order of decreasing melting points. The water is heated separately to about 60°C and a mortar warmed with boiling water; the melted fats are then poured into the mortar, the water added and the whole stirred until set. Water soluble components may be dissolved in the water though this may occasionally interfere with correct emulsification; these compounds can therefore be added after emulsification but whilst the ointment is still fluid, or by levigation when cold.

The simple type of ointment base is combined with the medicaments by trituration or levigation.

Eye ointments are prepared by using the official base which has previously been heated for 1 hr at 150°C. All utensils can be sterilized either by autoclaving or by wetting with alcohol and burning off.

Packing and Wrapping

Ointments may be packed into cardboard boxes—chipette boxes made from twisted card stuck together, or 'corruganza' seamless boxes stamped out of sheets of card. These are cheap and convenient but are neither so serviceable nor hygienic as ointment jars or collapsible metal tubes. The collapsible metal tubes may be filled in the same way that the jars are filled, by pouring while the ointment is still fluid, but once set the ointment can still be filled into them: this is achieved either by introducing the ointment in small quantities and tapping down, or better by wrapping the ointment in a tube of greaseproof paper, pushing the open end of the cylinder into the metal tube as far as the tube shoulder, and flattening the last 2cm of the tube with a spatula which is kept in position as the paper is withdrawn leaving the ointment within the tube. The tube end is then sealed by turning the flattened end over twice and corrugating the seal.

Wrapping of an ointment box or jar is achieved in much the same way as for a bottle, the walls of the box or jar being considered as top and bottom for purposes of estimating paper size and of wrapping.

DISPENSING EMULSIONS

An emulsion is a relatively stable mixture of two normally immiscible liquids. If oil and water, the two substances commonly used in veterinary emulsions, are shaken together they will rapidly separate to give an oil layer above a water layer: this is partly due to the difference in surface tensions between the two liquids but is also associated with their different densities. If a third substance is introduced which will coat the globules of one of these substances preventing their coalescence, and which will form a transitional buffer bridging the physical differences between the two substances, then a dispersal of one substance within the other can be made semi-permanent: this is an emulsion.

The three phases or compounds forming an emulsion are known as:

1. The *disperse or internal phase*, which is the substance split into small globules and dispersed throughout the second phase.
2. The *continuous or external phase* in which the globules are dispersed and which, within limits can be used to dilute the emulsion, and
3. The *emulsifying agent or emulgent*.

Two general types of emulsions can be made with oil and water. The first is the *oil-in-water* emulsion in which the disperse phase is oil and the continuous phase is water. This type of emulsion is the one commonly used in veterinary medicine. The second is the *water-in-oil* emulsion in which the previous position is reversed; some ointments containing lanolin (wool fat) are partly water-in-oil emulsions, the lanolin acting as the emulsifying agent. In the main, water-in-oil emulsions are used internally whilst the oil-in-water emulsions may be used either internally or externally.

Whether an emulsion will be water-in-oil or oil-in-water depends mainly on the emulgent used, and to a lesser degree the methods of preparation. Thus an emulgent which is essentially lipophilic (fat loving) will produce an emulsion with oil as the external phase, and one which dissolves more easily in water will produce an oil-in-water emulsion with water as the continuous phase.

Oil-in-Water Emulsifiers

Carbohydrates:
Acacia, tragacanth, pectin, dextrin, malt, agar and chondrus (Irish moss) mucilage
Proteins:
Egg albumin and yolk, casein, gelatin, condensed milk
Soaps:
Monovalent alkali metal soaps (sodium, potassium and ammonium soaps), soaps of some amines particularly triethanolamine

Water-in-Oil Emulsifiers

Saponins:
- Tincture of Quillaia

Soaps:
- Polyvalent alkaline earth soaps (calcium, magnesium)

Alcohols:
- Many of the complex alcohols—alcohols of wool fat, lanolin, cetyl and stearyl alcohols.

In the last few years research has produced several series of new emulsifying agents. Most of these agents are esters of polyhydric alcohols and long chain fatty acids, for it has been found that hydroxyl (—OH) and carboxyl (—CO . OH) groups are hydrophylic and long chain hydrocarbon groups are lipophilic: if the balance of these two different characteristics is varied, substances predominantly either lipophilic or hydrophilic can be formed, and by combining one substance from each of these groups a compound emulsifying agent can be produced.

Another type of emulsifying agent is the sulphated alcohol derived from long chain fatty acids. Sodium lauryl sulphate is an example. Quaternary ammonium detergents are also used for oil-in-water emulsions.

Methods of Making Emulsions

There are several methods of making emulsions, but the principle involved is the same for all methods. This consists of breaking the disperse phase into the smallest globules possible and dispersing them evenly throughout the continuous phase. This may be done manually with a pestle and mortar, or mechanically as with a colloid mill in which a rotating disc exerts considerable shearing action on the globules, or as with a ram-jet type of homogenizer in which a plunger forces the three phases through a small aperture under high pressure.

The method used in practice on very rare occasions will usually be that of the pestle and mortar. This will produce only medium grade emulsions with a globule size of from 5 to 20 μm, whereas a homogenizer will reduce this to less than 2 μm. *Internally administered* emulsions will usually contain acacia if prepared in practice. There are two ways in which they can be prepared, a primary emulsion being made first in both cases. The first is the *dry gum* method in which the oil is placed in a mortar and the dry gum lightly and quickly stirred into the oil, followed quickly by the water, vigorous trituration continuing until the emulsion is complete. The second is the *wet gum* method in which a mucilage is first prepared with water and gum, and the oil triturated into the mucilage a few drops at a time.

The dry gum method is rapid but produces a less elegant and less stable emulsion.

To form these *primary emulsions* acacia, water, and oil must be used in proportions which may vary only slightly. These proportions differ according to the viscosity of the oil, for the lower the oil viscosity the higher must the proportion of gum, and also water, for these last two remain in a permanent ratio of 1:2. In some cases better emulsions may be obtained by increasing the viscosity of the water and gum mixture by adding tragacanth, methyl cellulose or sodium alginate. The proportions of the three phases are:

	Oil	Water	Emulgent (gum)
Fixed and mineral oils	4	2	1
Volatile oils	4	4	2

With the oleoresin type of oil—turpentine—more acacia is needed, equal parts of oil and gum being used.

If mucilage of acacia is used in the wet gum method, allowance must be made for the waters contained in the mucilage, and the quantities become:

	Oil	Water	Emulgent (mucilage)
Fixed and mineral oils	4	$\frac{1}{2}$	2
Volatile oils	4	1	4
Oleoresins	4	2	8

If emulgents other than acacia are used the proportions of the three phases will depend on the emulsifying properties of each emulgent. A guide to these different proportions will be found in the *Pharmaceutical Pocket Book*, published by The Pharmaceutical Society.

It is essential when preparing emulsions that

all utensils be spotlessly clean and uncontaminated, for emulsions will 'crack' (break down) very easily if either dirt or chemical contamination is present, especially if volatile oils are the contaminants.

Once the primary emulsion is made drugs may be added either directly or dissolved in a little of the diluting water. The emulsions can then be made up to volume.

Externally Applied Emulsions

These are liniments and are usually prepared in veterinary medicine using soaps as emulgents. These soaps may be used as such or made in the process of emulsification.

The method normally used is the so called 'bottle method', when the ingredients are put into the bottle which is then shaken until emulsification occurs.

The commonest liniment prepared is 'white liniment' or 'ammonia and turpentine liniment', containing turpentine, ammonium carbonate, ammonia, or ammonium chloride, soap and water. The soap may be hard soap or soft soap but the proportions vary according to which is used, for 1 g (15 gr.) of hard soap, or 15 g (½ oz.) of soft soap, are required to emulsify 30 ml (1 fl. oz.) of a volatile oil or 60 ml (2 fl. oz.) of a fixed oil.

In the liniment mentioned, the oil is poured into a clean cold bottle; the soap is dissolved in water on a warm bath, cooled, poured into the oil, and shaken to mix; the ammonium carbonate in solution is then added and the mixture rocked until the creamy white emulsion forms. Vigorous shaking is unnecessary and, if the emulsion does not form immediately, standing on the bench for a few minutes followed by a gentle rocking will usually achieve the required effect. The volume can then be made up with water.

Emulsions may be prepared by this method on a small scale, or whole winchesters may be prepared at a time.

A typical formula is

℞	Ammonium carbonate	8 g (120 gr.)
	Turpentine oil	30 ml (1 oz.)
	Soft soap (Green)	15 g (½ oz.)
	Water to	240 ml (8 fl. oz.)

In what has been called the *alkaline method* of making liniments, the soap is made first by adding caustic potash to, say, rape oil in the ratio of 30 ml (1 fl. oz.) of oil to 15 ml (½ fl. oz.) potassium hydroxide solution (5 per cent KOH). This is shaken up and the oils added—approximately 60 ml (2 fl. oz.) of a volatile and 120 ml (4 fl. oz.) of a fixed oil. This is shaken well to form an emulsion and the other drugs added in solution; the whole is then made up to volume after a further shake; shake again and the liniment is complete.

When eggs are used—not a common practice for veterinary liniments—15 ml (½ fl. oz.) of egg yolk will emulsify about 60 ml of a volatile oil and 120 ml of a fixed oil (2 and 4 fl. oz. respectively). One egg yolk is about 15 ml in volume. To make the liniment, it is usual to triturate the yolk in the mortar, triturate the oil into it and then add the aqueous solutions of the solid constituents in small quantities, triturating continuously. This type of emulsion is stable over a far wider range of pH than other types of emulsion, and acetic acid is often added with the intentions of neutralizing any excess caustic used in making the soap.

Stability of emulsions depends on a number of factors which vary according to the emulgent used. The effect of pH on emulsions has been investigated, but without a great deal of information being produced. However, egg yolk emulsions, as already stated, will withstand the addition of relatively high proportions of acid: tragacanth emulsions are most stable at pH 2, whilst acacia emulsions are stable over a range of pH 2 to 10. The lack of 'creaming' or separation of a supernatant from concentrated emulsions is related to the size of the globule, and for this reason the mechanically prepared emulsions are the most permanent. Complete 'cracking' may be produced by the addition of electrolytes or alcohols. These two types of compounds should therefore be avoided in emulsion making.

Dispensing Pills

Pills are never made in practice now, and the procedure is only of historical interest. The essentials of the procedure consist of pressing the excipient, or binding substances, into the solid drugs in the shallow end of a pill mortar after the solids have first been prepared by trituration and mixing in the deep end of the mortar. When all the excipient has been thoroughly dispersed throughout the powder the mass is rolled into a sphere by hand and transferred to a cutting machine platform. The sphere is then rolled with the back of the cutter, any tendency to stick being prevented by dusting both platform and roller with an inert powder. When the cylinder is long enough to cover the requisite number of flutings on the cutter, it is transferred to the cutter flutings of the platform, the cutter placed over it and tapped down sharply. This cuts the pill cylinder into cachets (small pillow-like masses) which are returned to the machine platform and rolled together with a wooden, flat cup roller.

Dispensing Horse Balls

This procedure is also more of historical than practical interest. Horse balls may be prepared either by:

1. A *fusion* method in which fats or resins are melted and other solid ingredients added just before the mass solidifies. Whilst still malleable the mass is rolled into cylinders of 1·3 to 2 cm diameter and chopped into lengths of about 5 cm. Any tendency of the mass to stick to the ointment slab or spatulae being avoided by dusting both with an inactive powder:

2. A *trituration* or *levigation* method similar to the preparation of ointments.

Horse balls containing aloes are usually prepared by the first method whilst the once common tonic ball containing ammonium carbonate and powdered nux vomica is prepared by the second method.

When the cylinders have been made they should be wrapped first in grease-proof paper and then in an outer wrapping paper. Only the outer paper is removed before administration.

Horse balls are *wrapped* in a special way. The paper is cut to size allowing twice two-thirds the diameter plus the length of the cylinder and approximately twice round. The paper is then folded over at one end and placed on the bench with the fold underneath and away from the dispenser. The ball is then rolled in the paper beginning at the nearest edge and rolling away; this brings the fold on the outside. The ends are then turned in. This type of wrapping is suitable for both internal and external wrapping.

Bibliography

The Chemists' Veterinary Handbook. London: International Textbook.

The Pharmaceutical Pocketbook. London: Pharmaceutical Press.

43

Legal Obligations

The production, supply and use of medicines are controlled by various Acts of Parliament. Those of particular interest to veterinary surgeons are:

1. The Medicines Act 1968.
2. The Misuse of Drugs Act 1971.
3. The Methylated Spirits Regulations 1952 made under the Customs and Excise Act 1952.
4. The Protection of Animals (Anaesthetics) Acts 1954 and 1964.

THE MEDICINES ACT 1968

This Act, with the regulations made under it, is a comprehensive measure replacing most of the previous legislation on the control of medicines for human use and for veterinary use. The Act is administered by the Health and Agriculture Ministers of the United Kingdom, who are advised by a consultative body established under the Act called the Medicines Commission and various expert committees.

It aims to ensure the safety, quality and efficacy of medicines, principally by establishing a licensing system which applies to the marketing, manufacture, importation and distribution of medicinal products. There are three types of licences: product licences, manufacturer's licences and wholesale dealer's licences. The product licence is held by the person responsible for the importation or the composition of the product to which it relates and controls its marketing.

Enforcement of the Act and regulations made under it is the duty of the Ministers and is exercised through the Medicines Inspectorate which acts as their main field force, or by arrangement with Local Authorities and the Pharmaceutical Society of Great Britain.

The Act establishes rights of entry for persons authorized in writing by an enforcement authority and provides for sampling at all stages of manufacture and distribution of medicinal products.

Limited exemption allows a veterinary surgeon or a veterinary practitioner to supply medicinal products for administration by him or under his direction to an animal or herd which is under his care. He may also prepare remedies for use, or for supply to a client or another practitioner for supply to his client. The exemption does not apply to the preparation of poultry vaccines. Other vaccines may be prepared only if autogenous and plasma or serum only if intended for use within the same herd as the source animal. Similar exemption allows pharmacists to prepare remedies in accordance with a prescription given by a practitioner but as with practitioners the exemption does not apply to the preparation of poultry vaccines or to plasma or serum except as indicated above.

The Act applies to medicated animal feeding stuffs but those prepared in accordance with a practitioner's prescription for a client are exempt.

Except for certain products whose purity or potency cannot be tested by chemical means exports are generally exempt from the Act. Practitioners may not import medicinal products except under licence.

Animal Test Certificates are required before a new drug can be tested in the field. The issue of the certificate is dependent upon the receipt of details of the pharmacokinetics of the drug in the type of animal on which the trial is to be carried out. Information is also required on the type of formulation to be used, its stability etc.

Product licenses are subsequently issued after the drug has been proved under field conditions and a complete description can be given of the drug which will allow the preparation of a data sheet, which will be available for the veterinarian's use.

The product licence gives permission for the product to be sold, supplied, imported or exported. Licences are also necessary for the production and wholesale distribution of medicines.

Manufacturers are required to conform to official standards or approved specifications with regard to the quality of their products and to follow further requirements governing their description, labelling and advertisement. They are obliged to supply the practitioner with a data sheet (which must contain information as specified under the Act) at maximum intervals of 15 months for each product promoted. The licensing requirements will replace the licensing requirements of the Therapeutic Substances Act 1956 and Diseases of Animals Act (Part II), 1950 when these are repealed.

Labelling

Medicinal products prepared or dispensed by a practitioner or supplied to a client must be labelled to show the following particulars:

1. The name of the person having possession or control of the animal or herd and the address of the premises where the animal or herd is kept, or the address of one such premises.
2. The practitioner's name and address.
3. The date dispensed.
4. The words 'For animal treatment only', except that where the container is so small that it is not reasonably practicable to show such words they may be omitted.

There are detailed labelling requirements for medicated animal feeding stuffs and for medicinal products for administration in medicinal tests on animals.

THE MISUSE OF DRUGS ACT 1971

Substances controlled under the Misuse of Drugs Act 1971 (controlled drugs) are listed in four schedules to the Misuse of Drugs Regulations 1973. Those listed in Schedule 1 and those in Schedule 3 are subject to less stringent control than those in Schedule 2, whilst possession and supply of those in Schedule 4, e.g. cannabis and lysergamide, is restricted to persons authorized by the Secretary of State.

Subject to certain exemptions it is unlawful for any member of the public to possess or supply drugs which are fully controlled under the Act. The exemptions include certain authorized persons who handle controlled drugs in the course of their duties while they are acting in the course of those duties; they are practitioners (defined in the Act as doctors, dentists, veterinary surgeons and veterinary practitioners), pharmacists, hospital matrons and, in certain circumstances, ships' masters. Any person may have a controlled drug in his possession for administration for medical, dental or veterinary purposes in accordance with the directions of a practitioner provided a false statement has not been made for the purposes of obtaining the supply or prescription.

Supply for Personal Use

A practitioner obtaining supplies of controlled drugs must order them on a written requsition which:

1. Is signed by him.
2. States his name, address and profession.
3. Specifies the purpose for which the drug is required and the total quantity to be supplied.

12 January 1977 A. N. Other M.R.C.V.S.
12 Cardington Road
Fanfield
Essex

Please supply for use in the above veterinary practise:

or

Please for veterinary use:

Injection of Morphine Sulphate 25·0 ml
(containing 65·0 mg Morphine Sulphate in 2·0 ml)

A. N. Other M.R.C.V.S.

If the drug is not collected personally the person collecting it for the practitioner will require written authorization to do so in addition to the requisition. The supplier must be reasonably satisfied that the documents are genuine.

In emergency, controlled drugs may be supplied to a practitioner without a requisition provided he gives an undertaking to furnish a signed requisition within the twenty-four hours next following in which case the regulations require him to deliver the requisition to the supplier in accordance with the undertaking.

Prescriptions

It is an offence to issue a prescription containing a controlled drug unless, in the case of a prescription issued by a veterinary surgeon or a veterinary practitioner:

1. It is in ink or otherwise indelible.
2. It is signed and dated by the person issuing it.
3. It specifies the address of the person issuing it.
4. It has written thereon the words 'For animal treatment only'.
5. It specifies, in the handwriting of the person issuing it, the name and address of the person to whom the controlled drug prescribed is to be delivered, the dose to be taken, the strength where appropriate, and the total quantity to be supplied in both words and figures.
6. In the case of a prescription for a total quantity intended to be dispensed by instalments, it contains a direction specifying the amount of the instalments and the intervals to be observed when dispensing.

The following prescription will serve as a guide:

A. N. Other M.R.C.V.S.
12 Cardington Road
Fanfield,
Essex.

℞ 12 January 1977

Eye drops of Cocaine, 1% 10ml (ten millilitres).

Sig. For Animal Treatment Only.
Instil a few drops into the eyes every 4 hours.
Repeat on 19 January 1977

A. N. Other M.R.C.V.S.

A. N. Farmer Esq.
Leeds Farm
Fanfield
Essex

A controlled drug may not be supplied on a prescription:

1. Before the date specified in the prescription.
2. Unless the prescription complies with the above requirements.
3. Unless the practitioner's address is an address within the United Kingdom and the supplier is satisfied that the prescription is genuine.
4. Later than 13 weeks after the date specified in the prescription, except that where the prescription contains a direction that specified instalments may be dispensed at stated intervals the person dispensing it must not supply the drug otherwise than in accordance with that direction and the occasion on which the first instalment is dispensed must not be later than thirteen weeks after the date specified in the prescription.

The person dispensing the prescription must retain it in every case.

Records

The practitioner must keep a register in which particulars of controlled drugs obtained or supplied (whether by way of administration or otherwise) must be recorded. Entries must be in chronological sequence and in the form shown below.

The register must not be used for any other purpose and must contain a separate headed section for each drug used. Entries must be made on the day or the day next following the day on which the drug was obtained or supplied, in ink or otherwise so as to be indelible.

No cancellation, obliteration or alteration of any such entry should be made. Corrections should be made only by dated marginal notes or footnotes.

Date	Name of Supplier	Address	Quantity	Drug and form	Signature
13 January 1977	A. Mann Ltd	2 Walsall Road Fanfield Essex	25·0 ml	Morphine Sulphate Injection 65·0 mg per 2 ml	A. N. Other

Date	Client	Address	Animal	Quantity	Drug and form	Signature
20 January 1977	C. Sawyer	12 Ross Way Fanfield	10-year-old fox-terrier Male	5·0 mg tablets	Methadone hydrochloride 6 tablets	A. N. Other

The register must be preserved for a period of two years from the date of the last entry. The register and stocks of drugs are subject to inspection by authorized persons. Controlled drugs may be destroyed only in the presence of and in accordance with any directions given by a person authorised by the Secretary of State.

The practitioner must ensure that, so far as circumstances permit, controlled drugs are kept in a locked receptacle which can be opened only by him or by a person authorized by him.

A practitioner who prescribes or supplies controlled drugs in an irresponsible manner may have his authority withdrawn by the Secretary of State.

The importation or exportation of controlled drugs is prohibited except under a licence issued by the Secretary of State.

THE METHYLATED SPIRIT REGULATION, 1952

These regulations are designed to prevent the public obtaining and drinking industrial methylated spirits or mineralized methylated spirits either before or after redistillation.

A veterinary surgeon or practitioner may purchase up to 4 gallons of mineralized methylated spirits and ½ gallon industrial methylated spirits at any one time; the latter may be obtained only on a signed order. If authorized (on a permit issued by a Customs and Excise Officer) up to 4 gallons of industrial methylated spirits may be obtained from a pharmacist, but should more than 4 gallons be required at one time this must be obtained from a Methylator on authorization from a Customs and Excise Officer.

Industrial methylated spirits may be prescribed but not more than one pint will be dispensed either undiluted, diluted with water, or as an ingredient of some formulation. Any prescription or order will be fulfilled only once and that must occur within 7 days of the date on the prescription or order.

There are no regulations governing the quantity of methylated spirits kept on veterinary premises except where local bye-laws limit the stocks of inflammable material on such premises. There is also no regulation to prevent a veterinarian obtaining ½ gallon of industrial methylated spirits daily from any or every chemist in an area.

THE PROTECTION OF ANIMALS (ANAESTHETICS) ACT, 1954, 1964

The 1964 Act amended that of 1954 in respect of the castration of male animals, the dehorning and disbudding of cattle and the docking of lambs.

The main provisions of the Act are that all surgical operations must be carried out with an anaesthetic, but that the type and method of anaesthesia are left to the decision of the veterinary surgeon and not specified, as previously. The Act states that 'the expression "Animal" shall not include a fowl, another bird, fish or reptile'.

Exemptions are as follows:

1. The making of injections or extractions with a hollow needle.
2. Experiments authorized under the Cruelty of Animals Act, 1876.
3. First Aid emergency treatment.
4. Docking and dewclaw removal in puppies before their eyes are open.
5. Any minor operation performed by a veterinarian normally without anaesthetic because of the speed of the operation or its painlessness.
6. Any minor operation which is not customarily performed only by a veterinary surgeon or practitioner.

The 1964 Act makes it obligatory to use an anaesthetic for castration of:

1. All male dogs, cats and horses.
2. Bulls and rams over 3 months of age and pigs over 2 months of age. If rubber rings are used for castration they must be applied during the 1st week of life.

The Act also makes it obligatory to use an anaesthetic for dehorning of cattle, the disbudding of calves unless carried out by chemical means during the 1st week of life, and the docking of lambs tails unless carried out by means of a rubber ring applied during the first week of life.

PART V

APPENDICES

APPENDIX I

LATIN ABBREVIATIONS

āā: *ana*, of each.
a. c.: *ante cibum*, before food.
ad: up to.
add.: *adde*, add thou.
ad lib.: *ad libitum*, at pleasure.
admov.: *admove*, apply; *admoveatur*, let it be applied.
ad us.: *ad usum*, according to custom.
aeq.: *aequalis*, equal.
agit.: *agita*, shake; or *agitatus*, having been shaken.
altern. d.: *alterno die*, every other day.
ante: before.
applic.: *applicandus*, to be applied.
aq.: *aqua*, water.
aquam ad—: water up to—.
aq. dest.: *aqua destillata*, distilled water.
aq. ferv.: *aqua fervens*, warm or hot water.
aq. frigid.: *aqua frigida*, cold water.
aq. steril.: *aqua sterila*, sterile water.
bln.: *balneum*, bath.
b.i.d.: *bis in die*, twice daily.
c.: *cum*, with.
cap.: *capiat*, let him take.
circ.: *circa*, around; or *circiter*, about.
coch. amp.: *cochleare amplum*, a table-spoonful.
coch. mag.: *cochleare magnum*, a table-spoonful.
coch. med.: *cochleare medium*, a dessert-spoonful.
coch. min.: *cochleare minimum*, a tea-spoonful.
coch. mod.: *cochleare modicum*, a dessert-spoonful.
d. in p. aeq.: *divide in partes aequales*, divide into equal parts.
det.: *detur*, let it be given.
dieb. altern.: *diebus alternis*, every other day.
div.: *divide*, divide.
dos.: *dosis*, a dose.
et: and.
extempl.: *extemplo*, immediately.
ft.: *fiat* (*fiant*), let it (them) be made, e.g. ft. mist.: *fiat mistura*, let a mixture be made.
id.: *idem*, the same.
infric.: *infricetur*, let it be rubbed in; *infricandus*, to be rubbed in.
m.: *misce*, mix.
m. d. u.: *more dictu utendus*, use as directed.
m. et n.: *mane et nocte*, morning and night.
mitt. *mitte*, send.
mor. dict.: *more dicto*, as directed.
non: not.
o. m.: *omni mane*, every morning.
o. n.: *omni nocte*, every night.
p. c.: *post cibum*, after food.
part. aeq.: *partes aequales*, equal parts.
parv.: *parvus*, small.
phial. agit.: *phialam agita*, shake the bottle.
post cib.: *post cibum*, after food.
pulv.: *pulvis*, a powder; or *pulverizatus*, powdered.
Q. R.: *quantum rectum*, the correct quantity.
q. s.: *quantum sufficiat*, *quantitas sufficiens*, *quantum satis*, sufficient.
quat.; *quater*, four times.
rep. or repet.: *repetatur*, let it be repeated.
s.: *sine*, without.
si op. sit: *si opus sit*, if necessary.
sig. *signetur*, let it be marked (labelled).
solv.: *solve*, dissolve.
stat.: *statim*, immediately.
suff.: *sufficiens*, sufficient.
t. d. s.: *ter die sumendus*, to be taken three times a day.
t. i. d.: *ter in die*, three times a day.
trit.: *tritura*, triturate.
ut dict.: *ut dictum*, as directed.
ut supr.: *ut supra*, as above.

APPENDIX II

WEIGHTS AND MEASURES

Metric Measures

WEIGHTS (mass)

1 gamma (γ)	=	1·0 microgram (μg)	
1000 gamma	=	1·0 milligram	(1·0 mg)
1000 milligrams	=	1·0 gram	(1·0 g)
1000 gram	=	1·0 kilogram	(1·0 kg)

MEASURES (fluid)

1 microlitre (λ)	=	1·0 microlitre (μl)
1000 microlitres	=	1·0 millilitre (ml)
1000 millilitres	=	1·0 litre (l.)

Approximate Imperial—Metric Equivalents

WEIGHTS (mass)

1 gr.	=	65·0 mg	100·0 mg	=	1½ gr.
60 gr.	=	4·0 g	1·0 g	=	15 gr.
1 oz. (avdp.)	=	28·0 g	100·0 g	=	3½ oz. (avdp.) 3¼ oz. (apoth.)
1 oz. (apoth.)	=	30·0 g	1·0 kg	=	2¼ lb. (avdp.)
1 lb. (avdp.)	=	453·0 g	50·0 kg	=	110 lb. (avdp.)

MEASURES (fluid)

60 m.	=	3·5 ml	1·0 ml	=	15 m.
1 fl. oz.	=	28·5 ml	100·0 ml	=	3½ fl. oz.
10 fl. oz.	=	284·0 ml	1·0 l.	=	1¾ pts.

APPENDIX III

NOTES ON THE TREATMENT OF POISONING

In veterinary practice, poisoning cases may arise from the accidental ingestion of a variety of poisonous plants or from the ingestion of food or water which has been contaminated by industrial effluent or one of the chemicals used in agriculture, e.g. crop protection agents. In addition, the occasional malicious administration of a 'classical' poison may be encountered. Nowadays, it must also be borne in mind that the increasing potency of drugs often carries an enhanced toxicity or side-effect potential. The toxicity hazards of the drugs themselves have been described already and it is not proposed to list all known poisons and antidotes, but rather to describe approaches to the management of poisoning cases using, for the most part, drugs described in previous chapters. The only compounds to be described in any detail in this appendix are those which are used only as antidotes in specific types of poisoning.

For some toxicants specific antagonists are available. If administered early these can be life-saving, e.g. nalorphine and mephenesin. However, the identity of the poison if often unknown and in these circumstances more general measures have to be taken. These are usually attempts to remove unabsorbed poison from the animal and its environment and the application of measures designed to support life while allowing the animal time to metabolize and/or excrete the absorbed fraction.

General Procedures

1. If it is possible, the poison should be removed from the animal by washing if on the skin and by emesis and/or gastric lavage if in the stomach. Saline purgatives and diuretics may help if the agent has already left the stomach. More sophisticated techniques are exemplified by peritoneal dialysis and the use of an artificial kidney. If necessary, samples of suspect material, vomit, stomach washings, dialysis fluids, blood or urine should be collected for analysis.

2. Non-absorbed poison may be amenable to chemical neutralization, e.g. ferric hydroxide for arsenic, soluble sulphate for lead or to adsorption, e.g. by activated charcoal.

3. When the identity of the toxicant is unknown or when no antidote is available, symptomatic treatment should be instituted, e.g. the control of convulsions with CNS depressants, respiratory depression with analeptics or artificial respiration, pain with analgesics and hypotension with sympathomimetics.

4. Measures to minimize or relieve tissue damage should be instituted, e.g. the use of anti-inflammatory agents locally. The infusion of glucose saline to ensure adequacy of liver glycogen stocks, and the infusion of fluids and electrolytes or plasma volume expanders, form part of a general supportive or shock control programme.

5. Drugs which aid in detoxification procedures may be given, e.g. methionine and cysteine.

Antidotal Mechanisms

When the identity of the poison is known, it may be possible to institute specific antidotal therapy. There are two major mechanisms by which antidotes can exert their beneficial actions.

1. They may modify the fate of the poison within the body.

2. They may interfere at the site of action of the poison.

These generalizations indicate that antidotal activity is simply a rather dramatic example of the usual mechanisms of drug action. As drug action has been discussed in Part I, it is necessary at this point only to exemplify the relevant mechanisms.

1(a). Absorption of the poison may be prevented by changing its physical state, e.g. a soluble poison may be rendered non-available by causing it to be precipitated (conversion of lead acetate to lead sulphate by magnesium sulphate), by causing it to be complexed (the chelation of iron by sodium ferrocyanide) or by causing it to be adsorbed (the adsorption of gases by activated charcoal).

1(b). Distribution of the poison may be limited by the administration of an antidote which, because it resembles the poison, would share the same volume of distribution as the poison. The chloride ion and the bromide ion distribute throughout body water in a ratio which depends on the amount of each which is absorbed. By increasing the chloride intake, bromide is gradually replaced in the ECF by chloride. Poison which has become distributed and may even be at its site of action can be removed by making available a binding site of greater affinity. In this manner the chelates will remove certain metal ions from the body in an inactive form, e.g. BAL will chelate mercury.

1(c). The inactivation rate of absorbed poison could be increased by supplying agents involved in the degradation pathway, e.g. thiosulphate acts as a sulphur donor in the conversion of cyanide to thiocyanate, methionine is a source of methyl groups in transformations resulting in methylation.

1(d). The rate of excretion of the poison may be increased, e.g. administration of calcium will increase the rate of withdrawal of strontium and radium from bones and thus increases their rates of excretion. The use of urinary acidifyers or alkalizers will increase the rate of excretion of poisons which are basic or acidic in character, e.g. salicylates will ionize more freely in a urine alkalinized with sodium bicarbonate and so their resorption will be decreased.

Apart from those agents which simply destroy tissue, i.e. the corrosive poisons, it is realistic to think of poisons as acting at a particular receptor. The methods of drug antagonism as described in Part I, will then be seen to be fully applicable to poisons.

2(a). An antidote may occupy a receptor 'protectively' and so prevent the access of the poison to its site of action. In this way atropine can block the effects of poisons which have parasympathomimetic or anticholinesterase activity.

2(b). An antidote may displace the poison from the receptor on a competitive basis, e.g. the reversal of curare by neostigmine, morphine by nalorphine and coumarin derivative anticoagulants by vitamin K.

2(c). An antidote may act because it is the substrate for which the poison was an analogue. Many coccidiostats are inhibitors of folic acid synthesis, e.g. sulphonamides replace the normal substrate *para*-aminobenzoic acid (PABA) and so block the synthetic pathway. If PABA is supplied it competes with the amount of sulphonamide present in a quantitative manner, i.e. by adding greater quantities of PABA, the effect of the sulphonamide can eventually be prevented. If, however, the end product of the synthetic pathway, folic acid, is made available, the sulphonamide effect is overcome but on a non-competitive basis as the antidote (folic acid) obviates the need for synthesis.

2(d). An approach often used in the treatment of poisoning is to administer drugs whose effect is the opposite to that produced by the poison, e.g. the CNS stimulant action of procaine can be controlled by barbiturates, the bradycardia typical of gamma benzene hexachloride poisoning in dogs can be removed with atropine, while the cardiac arrhythmias of digitalis overdosage can be controlled by β-adrenergic blocking drugs or procainamide.

BAL and Arsenic

Arsenic was widely used in therapy in many forms and for purposes ranging from the control of blackhead in turkeys to the control of the ectoparasites of sheep. This was one reason why arsenic poisoning occurred among the domesticated species. It has now been replaced by more effective and less toxic agents, but the recognition of transmissible resistance to antibiotics in enteric organisms may result in the reappearance of arsenicals for the prevention and control of scours.

In the body, arsenic is found in association with proteins and it is believed that it attaches to the sulphydryl (SH) groups of the sulphur-containing amino acids. It was this affinity which lead to the rational synthesis of BAL (British Anti-Lewisite) or dimercaptopropanol as an antagonist to arsenic-containing war gases, e.g. Lewisite. BAL has two free SH groups and it is assumed that it competes for arsenic with tissue SH groups, especially those of the tissue respiratory enzyme cofactor α-lipoic acid. Each molecule of BAL chelates with one of arsenic. The resulting non-toxic complex is stable and is excreted as such in urine.

A *chelating agent* is a molecule which is capable of binding a metal ion into its structure in such a way that the metal may be pictured as being held between a pair of claws and thus becomes part of a ring-structure. The atoms which bind the metal are called *ligands* and they donate electrons to the metal to form coordinate bonds. The metal ion may be held by more than one chelating molecule or by more than one pair of ligands within the one molecule and is thus locked into a multi-ring structure.

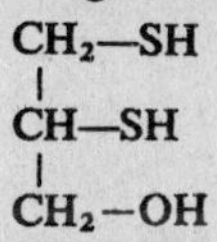

Dimercaprol

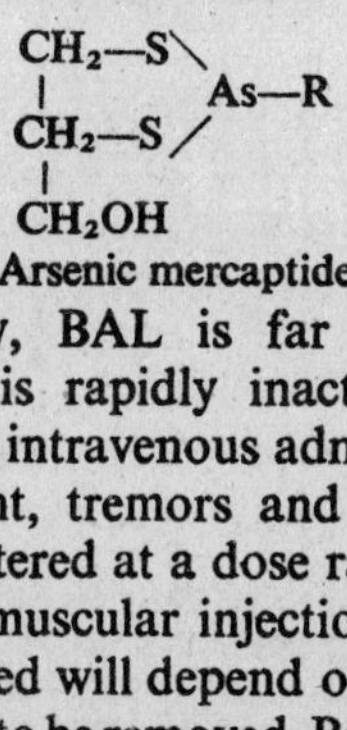

Arsenic mercaptide

Unfortunately, BAL is far from an ideal antidote as it is rapidly inactivated, irritant, unstable and its intravenous administration may cause excitement, tremors and convulsions. It may be administered at a dose rate of 2 to 3 mg per kg by intramuscular injection, but the final total administered will depend on the amount of arsenic which is to be removed. BAL has also been used in cases of mercury and antimony poisoning.

Ca-EDTA and Lead

The occurrence of lead poisoning in domesticated animals should decline in frequency as white pigments other than lead become more widely used in paints. The habitual licking of their surroundings by calves was frequently a cause of poisoning when lead-based paints had been used.

BAL forms a stable 5-membered ring with arsenic while EDTA holds calcium in four 5-membered rings, i.e. it is a tetradentate chelating agent.

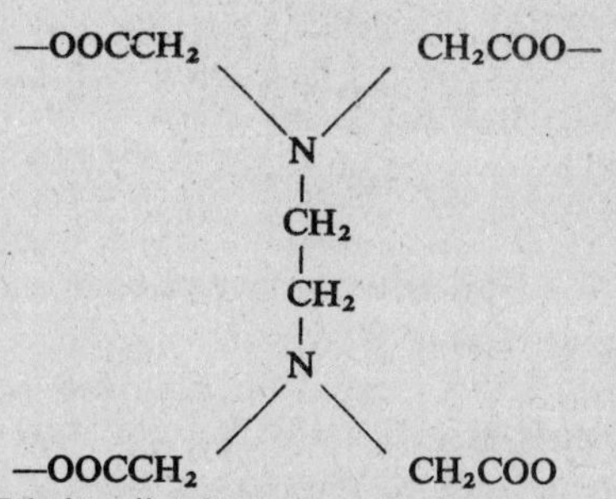

Ethylenediamine tetracetate (EDTA)

EDTA itself cannot be given to animals because it would chelate calcium. This has been utilized, however, in the experimental induction of hypocalcaemia. When there is need for a heavy metal antidote, calcium EDTA is used, thus avoiding the problem of ionic calcium loss. The affinities with which a chelate will bind various metals depend upon the atomic structures of the available metals. Metals with higher binding affinities tend to displace those with lower affinities and it is by this means that lead is able to replace calcium in Ca-EDTA. The resultant lead chelate remains water-soluble and is cleared safely and easily by functioning glomeruli.

Sodium Calciumedetate, B.Vet.C., is described as the calcium chelate of disodium ethylenediamine tetra-acetic acid.

Dosage suggested for the horse is 1 g per 13·6 kg (30 lb.) given as a 2 per cent solution by intravenous drip daily for 3 days, followed by an interval of 3 to 5 days and then a further 3-day course. Following experience with cattle it is probable that intermittent dosing will be just as successful as drip techniques. In calves single doses of from 3 to 6 g according to size, given intravenously, are to be preferred: a second dose may be given 2 or 3 days later if necessary. As the drug does not cross membranes readily the intravenous route is used. Lead accessible from the ECF is therefore that most rapidly removed. This antidote is relatively non-toxic but may cause kidney damage.

Chelates and Radioactive Metals

The rate of removal of radioactive elements can sometimes be hastened by chelates. This manoeuvre is of value when the element has a long half-life because the radiation dose can then be much reduced. Di-sodium calcium edetate will complex uranium and the closely-related DTPA (diethylene triamine penta-acetic acid) chelates plutonium.

Desferrioxamine and Iron

This agent forms a very strong chelate with ferric iron and is capable of removing iron from certain physiological chelates, e.g. ferritin and transferrin, but not from vital sites, e.g. haemoglobin and the cytochromes. While desferrioxamine is administered parenterally, sodium ferrocyanide is a chelator of iron suitable for oral administration.

D-Penicillamine and Copper

This is an orally administrable chelating agent which is a derivative of cysteine. It is able to bind cuprous copper in a five-membered ring. BAL will also chelate copper, but is a more toxic drug than penicillamine.

P-2-AM and Organophosphorus Compounds

The organophosphorus compounds are now widely used in veterinary medicine in the control of a variety of parasites, and are also commonly used as agricultural pesticides. This generous exposure of the organophosphorus compounds means that poisoning cases occur amongst the domesticated animals, and especially in ruminants as this species normally has a rather low cholinesterase reserve.

Organophosphates inactivate cholinesterase, so allowing the accumulation of acetyl-choline at all sites of release in the body. The early use of certain oximes, e.g. P-2-AM, can hydrolyse the recently phosphorylated but not yet stabilized enzyme complex. Such drugs are called cholinesterase reactivators and their use is necessary as the rate of hydrolysis in the body is very slow, i.e. unaided recovery will depend on the rate of synthesis of new cholinesterase. Pralidoxime (*N*-methylpyridinium 2-aldoxime or P-2-AM) was synthesized as an agent which would have affinity for the binding sites of cholinesterase, have the ability to break the enzyme phosphate bond and form a stable, inert complex with the organophosphate. The value of pralidoxime has been demonstrated in many experiments in which treated animals are shown to tolerate greatly increased doses of organophosphate. Atropine usually increases this tolerance. In therapy atropine may be given repeatedly at twice the normal rate together with anticonvulsants.

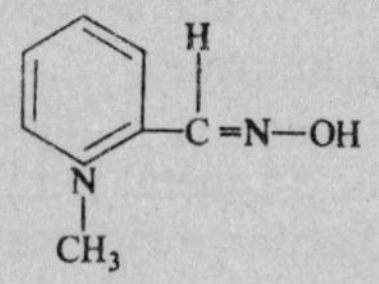

Pralidoxime (P-2-AM)

Thiosulphate, Nitrite and Cyanide

The treatment of cyanide poisoning needs to be undertaken very rapidly if it is to be successful. The approach is two-fold.

Thiosulphate solutions administered intravenously accelerate the detoxication of cyanide by acting as sulphur-donors for the reaction normally catalysed by the enzyme cyanide–thiosulphate transferase. Cyanide is converted in this way to harmless thiocyanate, i.e. CN^- to CNS^-.

Nitrites convert haemoglobin to methaemoglobin, which, in turn, combines with cyanide to form the stable complex, cyanmethaemoglobin. The cyanide ion complexes with the ferric ion of methaemoglobin. The sequential administration of nitrite and thiosulphate increases the LD_{50} of cyanide in a manner which indicates synergistic activity between the antidotes.

In cyanide poisoning, treatment can rarely be attempted, but where it is possible a combined treatment with sodium nitrite and sodium thiosulphate may be tried. In horses and cattle, 10 ml of a 20 per cent solution of sodium nitrite should be injected intravenously followed immediately by up to 50 ml of a 20 per cent solution of thiosulphate: sheep require 10 ml of a 10 per cent nitrite solution and 20 ml of a 10 per cent thiosulphate solution: in the dog, sodium nitrite in doses of 2·2 mg per kg body weight (1 mg per lb.) as a 1 per cent solution should be followed by a 25 per cent solution of sodium thiosulphate at the rate of 1·1 g per kg (0·5 g per lb.). All injections should be given slowly and repeated as necessary.

Bibliography

BOYLAND, E. & GOULDING, R. (1974) *Modern Trends in Toxicology*, 2. London: Butterworths.

CLARKE, E. C. G. & CLARKE, M. L. (1975) *Veterinary Toxicology*, London: Baillière Tindall.

CURRY, A. S. (1972) *Advances in Forensic and Clinical Toxicology*. Cleveland, Ohio: CRC Press.

FORSYTH, A. A. (1968) *British Poisonous Plants*, Bulletin No. 161. London: HMSO.

OEHME, F. W. (ed.) (1975) *Toxicology for the Small Animal Practitioner*. Philadelphia: Saunders.

RADELEFF, R. D. (1970) *Veterinary Toxicology*, 2nd ed. Philadelphia: Lea & Febiger.

General References

AMA Drug Evaluations (1975) Acton, Mass.: AMA Publishing Sciences Group.

Blood, D. C. & Henderson, J. A. (1974) *Veterinary Medicine*, 4th ed. London: Baillière Tindall.

Bowman, W. C., Rand, M. J. & West, G. B. (1969) *Textbook of Pharmacology*. Oxford: Blackwell Scientific.

British National Formulary (1974–6) London British Medical Association.

British Pharmacopoeia (*Veterinary*) (1977) London: Pharmaceutical Press (in preparation).

British Veterinary Codex (1965) London: The Pharmaceutical Press.

Catcott, E. J. & Smithers, J. F. (1970) *Bovine Medicine and Surgery*. Wheaton, Ill.: American Veterinary Publications.

Catcott, E. J. & Smithers, J. F. (ed.) (1972) *Equine Medicine and Surgery*. Wheaton, Ill.: American Veterinary Publications.

Catcott, E. J. & Smithers, J. F. (ed.) (1973) *Progress in Canine Practice*. Wheaton, Ill.: American Veterinary Publications.

Conn, H. F. (1976) *Current Therapy*. Philadelphia: Saunders.

Crossland, J. (ed.) (1970) *Lewis's Pharmacology*, 4th ed. Edinburgh and London: Livingstone.

Dunne, H. W. & Leman, A. D. (1975) *Diseases of Swine*, 4th ed. Ames, Iowa: Iowa State University Press.

Goodman, L. S. & Gilman, A. (ed.) (1975) *The Pharmacological Basis of Therapeutics*, 5th ed. New York: Macmillan.

Grunsell, C. S. G. & Hill, F. W. G. (1975) *The Veterinary Annual*. Bristol: Wright Scientechnica.

Hungerford, J. G. (1970) *Diseases of Poultry*. Sydney: Angus and Robertson.

Jensen, R. (1974) *Diseases of Sheep*. Philadelphia: Lea & Febiger.

Kirk, R. W. (1974) *Current Veterinary Therapy, Small Animal Practice*, 4th ed. Philadelphia: Saunders.

Meyer Jones, L. (1965) *Veterinary Pharmacology and Therapeutics*, 3rd ed. Ames, Iowa: Iowa State University Press.

Siegmund, O. H. (ed.) (1973) *The Merck Veterinary Manual*, 4th ed. Rahway, N.J.: Merck.

Wilkinson, G. T. (1966) *Diseases of the Cat*. Oxford: Pergamon.

International and Proprietary Names of Drugs

Adrenal Hormones

Betamethasone	Betsolan
Cortisone acetate	Cortistab; Adreson; Cortelan
Dexamethasone	Cortisol; Decadron; Dexadreson
Flumethasone	Fluvet
Methylprednisolone	Depo Medrone
Prednisolone	Cordelta; Deltacortil; Predsol
Triamcinolone	Vetalog

Anabolic Steroids

Nandrolone phenylpropionate	Nandrolin

Anaesthetics

Butobarbitone	Butobarbital; Neonal; Soneryl
Cyclopropane	Trimethylene
Halothane	Fluothane
Lignocaine hydrochloride	Lignostab; Xylocaine
Mepyramine maleate	Anthisan
Methohexitone sodium	Brietal Sodium
Methoxyflurane	Penthrane
Pentobarbitone sodium	Arbunox; Nembutal; Sagatal
Phenobarbitone	Gardenal; Luminal
Thialbarbitone sodium	Kemithal Sodium
Thiopentone sodium	Intraval Sodium; Pentothal
Trichloroethylene	Trilene

Analgesics

Pethidine hydrochloride	Demerol; Dolantal; Dolantin
Phenazone	Antipyrin
Phenylbutazone	Butazolidin

Anthelmintics

Bephenium hydroxynaphthoate	Alcopar; Frantin; Nemex
Bunamidine hydrochloride	Scolaban
Coumaphos	Asuntol
Cyanocetohydrazide	Helmox; Dictycide
Diamphenithide	Coriban
Dichlorophen	G.4; Dicestal; Teniathane
Dichlorvos	Atgard; Task; Equigard
Diethylcarbamazine	Banocide; Caricide; Heteragan
Haloxon	Loxon
Hexachloroethane	HCE; Avlothane
Levamisole	Ripercol
Mebendazole	Telmin
Methyridine	Mintic; Promintic
Metriphonate	Dyvon; Trichlorphon; Dipterex; Neguvon
Morantel	Banminth
Naphthalophos	Rametin
Niclosamide	Mansonil; Marneson; Yomesan
Nitroxynil	Trodax
Oxyclozanide	Zanil
Parbendazole	Helmatac
Phenothiazine	PTZ; Phenovis
Piperazine adipate	Coopane; Entacyl
Piperazine citrate	Antepar; Pipricide
Piperazine hydrate	Oxacine
Piperazine phosphate	Antepar; Helmezine
Phthalofyne	Whipcide
Pyrantel tartrate	Strongid
Rafoxanide	Flukanide; Ranide
Ruelene	Etrolene
Tetrachlorodifluoroethane	Freon 112
Tetramisole hydrochloride	Nilverm
Thenium	Ancaris
Thiabendazole	Thibenzole; Equizole; Omnizole
Tribromosalan	Hilomid

Antibacterial Agents (Non-Antibiotics)

Arsanilic acid	AS 101; Maize Pro-Gen
Dapsone	Avlosulfon; Udolac
Diaveridine	Darvisul
Dibromopropamidine isethionate	Brolene; Brulidene
Diminazene aceturate	Berenil
Furazolidone	Furoxone; Neftin; Tricofuram
Nitrofurantoin	Furadantin
Phthalylsulphathiazole	Thalazole
Sodium aminarsonate	Pro-Gen V; Sodium Arsanilate
Sulphacetamide	Sulfacet; Albucid; Vetrisul
Sulphadimidine	Sulphamethazine
Sulphamethoxypyridazine	Midicel
Sulphamethyl-phenazole	Vesulong
Sulphonamide/trimethoprim	Trivetrin; Tribrissen
Sulphapyridine	M & B 693
Sulphapyrimidine	Debenal
Sulphaquinoxaline	Embazin
Sulphathiazole	Thiazamide
Thiomersal	Merthiolate

Antibiotics

Cephalosporin	Cephaloridine; Ceporin; Keflodin
Chloramphenicol	Chloromycetin; Animycetin

Chlortetracycline	Aureomycin
Erythromycin	Erythrocin; Ilotycin
Lincomyein	Lincocin
Neomycin	Neobiotic
Novobiocin	Tiotexin; Albamycin; Biotexin; Cathomycin
Oxytetracycline	Terramycin
Penicillin	Benzapen; Crystapen
New Penicillins	
Cloxacillin	Orbenin
Ampicillin	Penbritin
Amoxycillin	Clamoxyl
Carbenicillin	Pyopen
Polymyxin	Aerosporin
Spiramycin	Rovamycin
Streptomycin	Streptovex; Stryzolin
Tetracycline hydrochloride	Achromycin; Tetracyn
Tylosin	Tylan

Anti-Convulsants

Pentobarbitone sodium	Nembutal; Arbunox
Phenytoin sodium	Alepsin; Epanutin; Eptoin; Phenantoin
Primidone	Mysoline

Antihistamines

Antazoline hydrochloride	Antistin
Chlorpromazine hydrochloride	Largactil
Diphenhydramine hydrochloride	Benadryl
Mepyramine maleate	Anthisan
Promethazine hydrochloride	Phenargan
Trimeprazine tartrate	Vallergan

Anti-Inflammatory Agents

Betamethasone	Betnelan
Cortisone acetate	Andreson; Cortistab
Dexamethasone	Decadron; Dexadreson; Dextelan; Deronil

Antiseptics

Aminacrine hydrochloride	Acramine; Pentaflav; Acramine Yellow
Benzalkonium chloride	Roccal; Zephiran; Marinol
Cetrimide	Cetavlon; Biocetab; Vesagex
Chlorhexidine hydrochloride	Hibitane

Antispasmodics

Pethidine hydrochloride	Dembrol; Dolantol; Dolantin
Phenylbutazone	Butazolidin
Trimeprazine Tartrate	Vallergan

Antithyroid Agents

Methylthiouracil	Muracil
Propylthiouracil	Propacil

Babesicides

Amicarbalide isethionate	Diampron
Diminazine aceturate	Berenil
Quinuronium sulphate	Acapron; Acaprin; Pirevan; Ludobal

Cerebral and Medullary Stimulants

Bemegride sodium	Megimidine Sodium
Caffeine	Theine; Trimethylxanthine
Nalorphine hydrobromide	Lethidrone
Nikethamide	Coramine; Nicamide

Coccidiostats

Amprolium hydrochloride	Amprolmix
Clopidol	Coyden 25
Decoquinate	Deccox
Diaveridine	Darvisul
Dimitolmide	Zoalene
Furazolidone	Furoxone; Neftin; Tricofuram
Mepacrine hydrochloride	Atebrin; Mecryl; Metoquin
Methyl benzoquate	Statyl
Monensin	Elancoban
Nicarbazin	Nicoxin; Nicrazin
Nitrofurazone	Furacin; NFZ; Nefco
Robenidine	Cycostat
Sulphadimidine sodium	Sulphamethazine Sodium; Vesadin
Sulphaquinoxaline	Embazin; Sulfacox

Diuretics

Caffeine	Trimethylxanthine; Theine
Chlorthiazide	Chlortride
Frusemide	Lasix
Hydrochlorthiazide	Direma; Esidrex; Vetidrex
Theobromine sodium salicylate	Diuretin

Ecbolics

Ergometrine maleate	Ergotrate
Oxytocin	Pitocin; Syntocinon
Pituitary (posterior injection)	Infundin; Pitibulin

Fungicides

Chlorbutol	Chloretone
Dibromopropamidine isethionate	Brolene; Brulidene
Dichlorophen	G4; Anthiphen
Griseofulvin	Grisolvin; Fulcinforte
Nystatin	Mycostatin 20
Thiomersal	Merthiolate

Histomonacides

Aminonitrothiazole	Enheptin A; Enheptin T
Dimetridiazole	Emtryl
Nifursol	Salfuride

Insecticides

Benzene hexachloride	Gammexane; Lindane; Lorexane
Benzyl benzoate	Spasmodin; Ascabiol
Carbaryl	Sevin
Carbophenothion	Trithion
Chlorfluvinphos	Birlane
Coumaphos	Asuntol
Crufomate	Ruelene
Derris	Tuba Root
Dichlorvos	Nuvan
Dicophane	DDT
Dimpylate	Diazinon
Dioxathion	Delnav
Fenchlorphos	Trolene
Metriphonate	Dyvon; Trichlorphon; Neguvon
Toxaphene	Coopertox

Metabolic Hormones

Thyroxine sodium	Eltroxin

Narcotics

Diethylthiambutene hydrochloride	Themalon

Pituitary Hormones and Substitutes

Adrenocorticotrophin	Corticotrophin
Chorionic gonadotrophin	Antuitrin; Follutein; Lutormone; Prolan; Apocrine V
Oxytocin	POP; Pitocin; Syntocinon
Serum gonadotrophin	Antostab; FSH; Gonadyl; Serogan; Apocrine

Reproductive Hormones

Hexoestrol	Hexettes
Medroxyprogesterone acetate	Veramix Sponge
Megestrol	Ovarid
Methyl testosterone	Seoval; Orandrone
Oestradiol	Oestradiol Implants
PMS	Veramix Plus
Progesterone	Protormone; Piaponin; Progestin
Stilboestrol	Pabestrol

Sedatives

Chlorbutol	Chloretone

Tranquillizers

Acepromazine maleate	Acetylpromazine
Chlorpromazine hydrochloride	Largactil
Diethylthiambutene hydrochloride	Themalon
Mephenesin	Lissephen; Myanesin
Promazine hydrochloride	Sparine
Promethazine hydrochloride	Phenergan
Trimeprazine	Vallergan

Trypanicides

Diminazene aceturate	Berenil
Homidium bromide	Ethidium
Pyrithidium bromide	Prothidium
Quinapyramine	Antrycide
Suramin	Antrypol; Naganol; Moranyl

Vitamins

Aneurine hydrochloride	Davitamon B; Vitamin B1
Ascorbic acid	Vitacee; Proscorbin
Biotin	Vitamin H; Coenzyme R
Calciferol	Vitavel D; Sterogyl
Folic acid	Pteroylglutamic Acid
Inositol	Bios I
Menaphthone sodium bisulphite	Vitavel K
Nicotinic acid	Davitamon PP; Niacin; Nicovel; Pellagrin
Pyridoxine	Vitamin B6
Tocopheryl acetate	Davitamon E; Dal-Tocol
Vitamin A	Avoleum
Vitamin B_2	Riboflavine; Lactevan; Lactobene
Vitamin B_{12}	Cyanocobalamin; Distivit B12
Vitamin G	Riboflavine

Therapeutic and Pharmacological Index

This index is a guide to the action and uses described in the body of the book and is complementary to the main index and the list of International and Proprietary Names of Drugs. It does not present an exhaustive list of therapeutic substances under each disease, but it is hoped that it will assist in the selection of drugs for specific disease entities. The inclusion of a number of pharmacological headings, in addition to the disease headings, has entailed some overlapping with the main index, but it is hoped that their inclusion will increase the usefulness of the index.

The entries under each heading are arranged in alphabetical order, and the number following an entry indicates the page or pages on which information on the action and uses of the substance may be found. It should be noted that the inclusion of a substance under a disease heading does not necessarily imply that that substance is always the only one to be used in the treatment of the disease. So many diseases are due to a number of factors that any index can act only as an aid to effective diagnosis and treatment.

Additional sources of information on drugs in the United Kingdom are the IVS and the ABPI publication which is a compendium of data sheets.

General Index